SOIL MICROMORPHOLOGY

SOIL MICROMORPHOLOGY

Proceedings of the second international working-meeting on soil micromorphology

Arnhem, The Netherlands, September 22 – 25, 1964

EDITED BY

A. JONGERIUS

Soil Survey Institute, Wageningen, The Netherlands

ELSEVIER PUBLISHING COMPANY

AMSTERDAM LONDON NEW YORK

1964

ELSEVIER PUBLISHING COMPANY
335 JAN VAN GALENSTRAAT, P.O. BOX 211, AMSTERDAM

AMERICAN ELSEVIER PUBLISHING COMPANY, INC.
52 VANDERBILT AVENUE, NEW YORK, N.Y. 10017

ELSEVIER PUBLISHING COMPANY LIMITED
12B, RIPPLESIDE COMMERCIAL ESTATE
RIPPLE ROAD, BARKING, ESSEX

LIBRARY OF CONGRESS CATALOG CARD NUMBER 64-23406

WITH 225 ILLUSTRATIONS, 8 COLOURED PLATES AND 33 TABLES

C.H. EDELMAN
1903–1964

Dr. Cornelis Hendrik Edelman, Professor of Regional Soil Science at the State Agricultural University in Wageningen, The Netherlands, died on May 15th, 1964, at the age of 61 years. With his death, soil science lost one of its greatest leaders.

Professor Edelman was well known all over the world. He played an important role in the International Society of Soil Science, and fulfilled many international missions for FAO and UNESCO, as well as for several governments. His students practise soil science in a great number of countries.

Professor Edelman was the teacher of nearly all Dutch soil scientists and of many foreign colleagues. In 1945, he promoted the establishment of the Dutch Soil Survey Institute and was its first Director.

However, it is not only for the aforesaid that we commemorate Professor Edelman, but also for the fact that he was the promotor of soil morphology investigations in The Netherlands. He felt confident that morphologic research is of greatest importance and even indispensable for the further development of soil science, and can also contribute considerably to the study and solution of practical agricultural problems concerning the soil. Since 1950, the year that the work started, Professor Edelman continuously stimulated and propagated the extension and application of morphologic research. This led amongst others to the establishment of the Micropedology Section at the Soil Survey Institute in 1957, which was recently followed by the foundation of the Department of Soil Morphology at the State Agricultural University, Wageningen. For a few years, he also personally worked on morphologic problems and it was his desire to give a circumstantial exposition of his ideas at the Working-Meeting.

Professor Edelman would undoubtedly have played an important role in the further development of soil morphology in the coming years. With great gratitude and deep emotion, I therefore dedicate this book to his memory.

A. Jongerius

INTRODUCTION

In the early part of the thirties, Prof. Dr. W.L. Kubiena introduced microscopic and stereo-scopic investigations of undisturbed soil, and thus a totally new and, for the further development of soil science, very important research conception was born.

The work of Professor Kubiena has richly borne fruit. His enthusiasm and his very many publications have induced many people to practice soil micromorphology. Particularly after about 1950, the importance of this branch of soil science increased rapidly and hence the need for contact among the various workers. This resulted in the First Working-Meeting on Soil Micromorphology, held in 1958 in Braunschweig–Völkenrode (Germany) and organized by Prof. Dr. H. Frese and Dr. H.–J. Altemüller, both of the Institut für Bodenbearbeitung der FAL.

Every since, soil micromorphology has made progress in many respects. The number of micromorphologists has increased, new fields of investigation have been entered, old ones extended, new methods developed, etc. Therefore, at the request of Professors Kubiena and Frese and Dr. Altemüller, a Second International Working-Meeting on Soil Micromorphology was organized, this time in Arnhem, The Netherlands. Professor Kubiena presided at the meeting.

The papers presented at the second Meeting are collected in this book. As was the case with the contributions to the first Meeting, these papers cover many facets of micromorphologic research.

The first group of papers concerns micromorphologic investigation of decomposition of organic matter and humus formation as well as microscopic biology. Since 1958, these topics have made great strides.

The second group consists of many papers dealing with the "classical" topics of soil formation and soil classification. It is gratifying to see that so many soils from nearly all over the world were discussed. It is obvious that with the aid of micromorphology, insight into the said topics can be extended considerably.

The third group consists of papers dealing with micromorphologic results in the field of paleopedology.

As to methods, preparation techniques and micromorphometrics were accentuated in 1958. Several papers in this book show that especially concerning the latter topic, considerable progress has been made since then. Amongst others, it is now possible to characterize soil structure very exactly with the aid of micromorphometry. Other remarkable developments are the application of phase contrast microscopy to thin section investigations and the progress in colour photography.

Regarding the possibility of the application of micromorphology to that part of soil science which is directed at agricultural practice, gratifying developments have begun, as appears from some of the papers. The same can be said of non-agricultural soil science, viz. soil mechanics.

The participation in the Meeting, both concerning the number of nationalities represented and the number of participants, was greater than in 1958. This is also apparent from the number of papers. The published papers of 1958 amount to 28, and now to 46; in 1958, the authors represented eight nationalities, and now thirteen. This is partly a result of the contributions of colleagues from Australia, Canada and New Zealand, and partly because of the growing number of European countries where micromorphology is practised.

We may repeat that micromorphology and its applications have been strongly developed during recent years; nevertheless we can state that this is just the beginning. Without a doubt, soil micromorphology has a great future.

DR. IR. F.W.G. PIJLS

Director of the Soil Survey Institute

Wageningen, The Netherlands

EDITORIAL

At the conclusion of the editorial work, I wish to thank all the members of the Dutch Soil Survey Institute who were concerned with the preparatory work for this book. Without the enthusiastic assistance of the type room, drawing office and photography section, the editor could not have completed the work in time.

I am much indebted to the following institutions, factories and firms: "Landbouwhogeschoolfonds", Wageningen: Koninklijke Nederlandsche Heidemaatschappij, Arnhem; Grontmij, De Bilt; Robert Blohm, Hamburg; Gevaert, Antwerp; Klaiber, Schwenningen (Neckar); Ernst Leitz, Wetzlar and Nierstrasz, Amsterdam; Microscopes Nachet, Paris; N.V. Cocheret, Arnhem.

Without their financial assistance, it would have been impossible to publish all the presented papers in their unabridged forms.

CONTENTS

SOIL FORMATION AND CLASSIFICATION

PALEOPEDOLOGY

METHODS AND APPLICATIONS

PERORATION

THE ROLE AND MISSION OF MICROMORPHOLOGY AND MICROSCOPIC BIOLOGY IN MODERN SOIL SCIENCE

W.L. KUBIENA

Bundesforschungsanstalt für Forst- und Holzwirtschaft, Reinbek bei Hamburg (Deutschland)

We owe this fortuitous working-meeting to a series of favourable circumstances. One of these arose from the fact that the town from which this meeting is being directed disposes of two independent research departments of micromorphology. Wageningen has become a center of micromorphological research which is able to give much stimulation to the visitor, be it in regard to equipment, new techniques or particular research projects. This is due to the ability of a man who disposes of a particular talent for organization and who has kept all of us vigorously involved in preparing the meeting. He has given valuable new impulses for the collaboration of the different working branches. The research department established by Dr. A. Jongerius is one of the best-equipped at present. However, its particular importance is due to the fact that it belongs to the Dutch Soil Survey Institute, directed by Dr. F.W.G. Pijls. It is therefore devoted to a particular specialization, a purpose guided by continuous practical requirements. It is able to satisfy these requirements, as can be seen by its rapid growth, by the esteem it is held in, and by the popularity which micromorphology has gained in The Netherlands during the last decade. The department of Micromorphology of Ir. S. Slager belongs to the Institute of the former Professor Dr. C.H. Edelman of the State Agricultural University of Wageningen, whose early death has been a very great loss to all of us.
Dr. Edelman as a scientist has seen his goal to be to help man and therefore he always gave preference to that which increased productivity, diminished need and raised prosperity. It has been a great and noble goal. We may ask today what might has caused him, in this connection, to see micromorphology as a tool to attain it. We know that his main interest has been research of soil structure and, in collaboration with S. Slager, the application of particular methods to field soils. Why is it that he and so many other soil scientists working in the most varied branches, have increased more and more their interest to micromorphologic research within the last decade? Is it the mere application of the microscope as a working tool? Is it the mere application of microscopic techniques? Or is it more than that, and if it is, will it bring about new results and new possibilities of research for soil science? Would soil science develop in the same way without soil micromorphology? Would it attain the same standard of knowledge if micromorphology had never appeared? These are the questions which I shall try to treat and answer in the following article.

NATURE OF MICROMORPHOLOGIC RESEARCH

Which is the principle of micromorphology? We should not be too afraid to use the designation morphology. This term, created by J.W. von Goethe, designates the science of the form and of natural formations in general, including the causes and laws of their genesis. But where in soil science (apart from profile morphology) have properties, laws or genesis of form ever been in question? For my students I used to apply a particular analogy in order to make the morphologic principle of research understood: soils in their multiplicity, their variability of dynamics and biology, are comparable to watch works of very different constructions. We are able to investigate a watch in very different ways. We can put it into a mortar and pound it to a very fine powder. The chemical analysis of this powder gives us complete information on the nature and quantity of the different metals used for the construction of the whole, and perhaps also on the nature of the stones which fill out the pivot bearings. Of course this analysis cannot tell us anything about the existence of the numerous small driving wheels and checking wheels, the kind and number of springs, screws, fuses, pivots, chain links, levers, small metal wires, bridge elements, etc. of which the works of a watch are composed. For this purpose a mechanical analysis would be necessary which would sort all these isolated components (and not only according to size), and investigate them either in groups or individually. It is easy to understand that by this analysis it will not be easy to conclude from the nature of these isolated components the function of each of them or from this the function of the whole, or to establish a hypothesis of the connection of the individual elements in the whole. A third mode of analysis of such a watch would investigate its works without destruction, endeavour to examine every part in its place and determine the nature of their connection. This would be necessary to conclude their function from a better basis. There still exists a fourth possibility for analysis, and that is to investigate the works in a state of motion and to examine their elements, their position and their role in the course of action and counteraction.

The principle of the last two modes of analysis represents that of micropedology, i.e. the pinciple of undisturbedness and the principle of functional investigation, in part by direct observation of function. The soil is also an object which is in a state of constant change and whose interior is in constant motion. The fact that the investigation of the soil by these principles at the same time represents microscopic research is necessarily caused by the facts that the individual elements of the soil only become visible under the microscope, and that the deciding events take place in microscopic dimensions.

Is the comparison with watch works a correct one, or does it rather complicate the matter? It is a very simplified comparison, for the soil is much more than watch works; it is a whole world which differs from ours only because it is one of much smaller dimensions. We would not know more of our own world if we could only see it from the distance of the stratosphere. The soil is an organized space of life like our own living space, a "living system", as S.A. Waksman puts it. In spite of this, our picture of the watch works allows us to show the essence of the morphologic principle.

Morphology refers not only to the inanimated world as in the case of geomorphology, but also to the animated and living world. The term had originally been created by Goethe in the sense of morphology and morphogenesis in the development of organisms. Only later did it obtain the wider, present meaning. In soils the organic part, in its final effects, cannot be separated from the inorganic one. How far this interlacing of the inorganic and the organic takes place, and how important it is for practical soil science, could only be fully shown by micromorphology. I shall therefore particularly come back to this subject later on. Micromorphologic research because of its name, can easily be taken for one whose mission is only to describe forms and structures, and whose final goal is completely absorbed in the establishment of a kind of histology of the soil or a doctrine of structure. To be sure, micromorphology deals with soil structure to a great extent, and it owes many valuable results to this fact. However, a glance at the great multiplicity of subjects encompassed by our working-meeting shows that it is by no means completely absorbed in it. It encompasses nearly all branches of soil research, from soil microbiology, soil zoology, humus chemistry and soil weathering to soil mineralogy and soil micromorphometry. The connection between them is easily understood if we think of our comparison with the watch works. We are not contented with the mere investigation of the composition of the works, but also want to know their function and the role of every element in the operation of the whole. For micromorphological research the soil not only represents a composition of constructural elements, but also a composition of organisms, and finally a composition of biological, chemical and physical processes. This manifold but always characteristic framework of the most different components, phenomena and events, which only in its particular combination forms the entity of a soil, this organized multiplicity to investigate, is our mission, each in his own specialization.

THE MORPHOLOGIC PRINCIPLE

The many-sidedness of our work and the great variety in specialization of the members of our congress lead us to understand the concept of the morphologic principle. If the composition of the phenomena to be investigated is given from the beginning by nature, we can never get hold of it by some kind of synthesis but only by continuous analysis. A mineralogist not working according to the morphologic principle will state that a soil contains considerable amounts of goethite besides plenty of amorphous iron hydroxides. A mineralogist applying the morphologic principle will try to determine in which parts of the microscopic fabrics of the soil the crystal aggregates of goethite are precipitating, in which form they are deposited, in which development phases they pass to their final stages, by which macroscopic environmental conditions they are produced, or to which particular circumstances within the microscopic soil fabric they owe their formation. A soil zoologist not working according to the morphologic principle, but wishing to investigate the particular fauna of a special soil variety, will determine those forms which seem to be typical for that soil by a comparison of species lists. A soil zoologist investigating the same soil variety according to the morphologic principle will determine the animal species most typical for it by more or less direct research in the interior of the soil; he will select the most interesting animal forms according to the importance of their biological efficiency; he will follow their life customs, investigate their living spaces, their environment and food requirements, their role in the decomposition of particular plant and animal residues, in the humification of decomposition products and their binding of inorganic components of the soil.

The micromorphologist sees undisturbed entities, or at least undisturbed sections of them, in the soils. The only reality which offers everything to him is the undisturbed and living soil in nature which has been left in its corresponding environment. Several specialists see the necessity to begin their work in the interior of the soil in nature by the application of direct microscopy, and to return again to the place where they started. Frame samples are sections of pedological entities, as are thin sections and microtom sections prepared from them. It is essential for the method of the micromorphologist that he searches for the entity so that he can see it with his eyes and conceive it in its reality. Every insight into a thin section which the microscope reveals to him allows him to recognize the integrity of the fabric. He does not only see all that which is already known to him, but also that which has resisted research up to the present. The unknown and unexplored stimulates him and reminds him constantly of what is missing for an understanding of the whole. The undisturbed framework of building elements and functional efficacies of the soil therefore have a particular organizing faculty for entire soil research. It is like a register in the form of a net pattern of lines, in which the investigator may put all his results, and in which the empty spaces give notice of what remains to be cleared up by himself or his colleagues. His kind of registration, however much it can take place only in the spirit, will always be done at a predetermined point, i.e. it will be produced with full precision and not at any imprecise or hypothetically assumed spot.

Micromorphology needs a great number of particular research techniques. However, it has not only taken over techniques from other natural sciences or adapted them for its purposes from petrography, metal microscopy, plant anatomy, microbiology, microchemistry and microphysics, but it has also created numerous techniques of its own. How much this particular branch of micromorphology is still developing will be noticed in a series of particular contributions to our present meeting. But the micromorphologist does not only need particular techniques for his work, for in most cases he also finds himself in a position to apply other new ways leading to knowledge. The microscopic techniques allow him to perceive an overwhelming number of unknown details. Where should he begin with his investigation? The beginner tends to lay hold of all, to describe all he is able to distinguish. But such descriptions are tiring and will scarcely be read by anybody. It is very important for the micromorphologist to select *the essential* from an abundance of material, and to present it with particular emphasis. Essential criteria are found by continuous comparative investigations of a multiplicity of soils. It is a matter of finding the particular, unequivocally typifying criteria; this kind of research follows the *principle of differential diagnosis*.

However, this *principle of comparative investigations* must be applied for the investigation of any microscopic phenomenon for a number of soils.

It will always be a matter of the entire spectrum, consisting of different stages of development or different variations produced by environmental conditions of differing intensity. Every phenomenon presents itself in weakly pronounced, typical, and degenerated phases. They are all generally recognized easily by their well-marked and strongly differentiated micromorphology. To the micromorphologist the soil collection in most cases is not a set up for exhibition but working equipment, one of the most important details of his research department. Of similar importance is his collection of thin sections to which he can return at any time when new questions arise.

All these new techniques, implements and principles of research must lead to new results. The novelty of the results, the unusuality of the newly found characters, present their main feature; the methods of micromorphology and micropedologic biology do not replace the current macrochemical and macrophysical methods, but at the same time they themselves have become irreplaceable. Of particular interest to us is the value of the results, their role in the progress of soil science, and the possibilities they open for its further development.

SOIL LIFE AND HUMUS RESEARCH

Beginning wit soil life and its effects on humus formation, we can already point to an important achievement: the thin section of humus formations permits like no other means the recognition of the biological efficiency of a soil or habitat in a most clear and simple way. This refers to the decomposition and humification of the organism residues, to the kind and intensity of the combination of humus substances with inorganic components, and to the biological influence on structure formation. In view of the clearness and great efficiency of these possibilities, it is surprising that humus thin sections have not yet found common application in the practical taxation of soils and habitats. Most suitable for taxation purposes is the investigation of super sections which are 10–15 times larger than normal preparations (generally 6 x 8 cm, i.e. the size of the sample frame). In most cases they allow the examination of the entire humus layer including förna, F- and H-horizon in a single preparation (vertical section). If the preparation is produced automatically by grinding machines, it does not take any particular effort to cut an impregnated frame sample at intervals of 5 mm in horizontal plates, and to prepare a separate thin section from each of them. The valuation of the intensity of biological decomposition and humification is firstly performed by the determination of the humus form. Since every humus form corresponds to a particular biology, it is an expression of the mode and intensity of decomposition and humification of the organism residues and fixation of mineral substances at the same time. Micromorphology can claim that only by its principles has it been possible to succesfully continue the development of the doctrine of humus forms, initiated in very fortunate manner by P.E. Müller. The investigation of thin sections, as well as investigations in the field of micropedologic biology, permit a much more detailed analysis and a much more exact distinction and determination of humus forms than former methods. While Müller distinguished three humus forms, at present we know of at least eighteen easily distinguishable and identifiable humus forms and varieties, as given in my last review on the subject: *The Soils of Europe*. The further completion of the taxation of soils and habitats is possible to every advisor by comparative micromorphologic analysis. They allow him to establish particular efficiency and valuation scales for the soils of his working district. By the use of morphometrical methods, a valuation by quantitative methods will also be possible in the future.

Humus research, in my opinion, has received a series of new and decisive impulses from micromorphology. As the most important ones I would like to cite the following:

(1) Micromorphologic analyses have stated that in all biologically active humus forms the soil fauna represents not only the accompanying, but *the decisive agent* of decomposition and humification. Coprogenous and non-coprogenous humus micromorphologically is nearly always easy to distinguish. The knowledge of dropping forms in general allows one to recognize the role of particular animal groups in the decomposition of organism residues. In the coprogenous part of the humus better decomposition and humification dominates by far. Valuable humus forms are inconceivable without the particular participation of the soil fauna in their genesis. The soil fauna is also entirely responsible for the mixing and combining of the organic substance with inorganic parts of the soil.

(2) Micropedology can show up in detail the dominating role of the soil fauna in humus formation. Morphologically specialized zoology is able to study the role of every decisive animal form in the transformation of organism residues. This goes so far that even the succession of different animals and their effects in decomposition can be determined. The research method has been completed particularly by G. Zachariae, who primarily applies the investigation of super sections combined with observation in situ and in vivo.

(3) The method of micropedology allows one to investigate the biology and effectiveness of soil fungi and actinomyces in situ and in vivo. It permits, within microscopic dimensions, the determination of their claims in regard to environment conditions, food requirements, their antagonistic effects and their role in the decomposition of particular organic substances and tissue elements. At the same time, it could be testified that fungi in the soil produce particular forms of growth which are different from those developing in artificial media. Mycological soil studies are essentially facilitated by micropedology, for through the aid of micromanipulation (by hand or apparatus) conidia, spores and fragments of mycelium can, without particular effort, be directly transferred from the soil space to artificial media for the preparation of pure cultures. In addition, by the microscopic analysis of thin sections the development of soil fungi and their effect on the decomposition of tissue elements can be investigated advantageously.

(4) With the aid of debris preparations and microtome sections, micromorphology allows one to examine the decomposition products and humic substances in humus forms by direct investigation, i.e. without the extraction and application of solvents. Their determination or characterization can be achieved by dye reactions, microchemical tests or micro-optical methods. Because of the preserved morphological connection, continuous comparative investigations permit one to not only determine the origin of the decomposition products, but also to follow the genetical sequence of transformations.

(5) From micromorphological investigations, it could be testified that the importance of soil fauna applies not only for their dominating role in humification, but also for their role in structure formation in the same way. Almost all aggregate formation in the soil is produced by the action of soil animals. This influence is not limited to the humus horizons, but also refers to the (B)- and B-horizons. In addition to aggregate formation, a similar influence can be stated in the space formation. The only non-biological aggregate and space types of structure-forming character can almost only be found in the form of fragment structures (produced by a decrease in volume when drying), crumbs in earthy rotlehm (by "rote Vererdung"), and aggregates and endurations produced by lateritic precipitation.

SOIL BIOLOGY AND SOIL TYPE[1]

Soil science has an immense amount of microbiological and pedozoological publications at its disposal. But the material available would not suffice to explain the particular biology which contributes to the genesis of particular soil types. Up to now textbooks of soils have brought either nothing or only very general notes on the characteristics which allow one to distinguish the biology of a podzol, braunerde, pseudogley, tropical braunlehm, solonchak, solonetz, smonitza, chernozem, sierozem, dystrophic, oligotrophic, eutrophic and sapropelitic anmoor from each other. We know that all these types show particularly great contrasts in their soil life, and that their humus formations have been developed to very striking products of these differences, but our biological knowledge in this respect is very limited. Since the essence of the soil consists in its being a biological system, soil science will remain incomplete without that knowledge. Soil science urgently needs the development of a particular working branch for its progress: the *special soil biology*, the biology of the soil types, sub-types and varieties. Such a particular working branch cannot confine itself to only species lists and organism numbers of soil types, because in this way nothing is yet said on the soil life itself, or on the effectiveness of the different organisms involved. But, in view of the tremendous abundance of organism forms and the high numbers of organisms in most soils, how can such research be performed? Would the life of a scientist be sufficient to conceive the biology of even one single soil profile? Could a summarizing synthesis of the biologies of the different organisms of the soil - without considering their mode of combining with each other and without knowing their mutual checking and stimulation - ever lead to a real knowledge of the entire biology of a soil?

It was necessary to develop this pessimistic picture in order to point out the particular role of a micromorphologically oriented biology of the soil, and its indispensability for modern soil science. The picture of a great abundance of phenomena, and of an overwhelming predominance of details, is the same here as in the other branches of micropedology. The mode of overcoming these difficulties, with the application of the same methods,

[1] The expression "soil type" is used in the original sense of W.W. Dokuchaev, and represents an equivalent to "great soil group".

is as possible here as there: micromorphologically oriented soil biology does not start from a species list and organism numbers, no matter how useful they otherwise might be, but from the biological investigation of the most important decisively characterizing phenomena. In what manner this becomes possible will be demonstrated in papers of both microbiological and zoological specialists in the course of this meeting. A special soil biology can only start from the micromorphology of the soil. Tracing the essential biogenic phenomena, it examines the course of their genesis by direct investigations in the interior of the undisturbed soil, and the effectiveness of the organisms decisively involved. Its goal is not a synthesis of results obtained outside the soil, but the analytical investigation of the natural entanglement of the processes. It can confine its investigation to any limit desired, since the smallest contribution of this kind will also represent a section of the real soil life and elucidate a point of its framework. In reality this kind of research has already begun. Its progress, a great deal of which depends on the general progress of soil science as an independent natural science, will be determined only by the number of microbiologists and zoologists working in this particular branch of soil biology. The specialization of their research requires that they also devote their interest to micromorphology and humus research at the same time. In this connection, it can be said that progress in the knowledge of humus forms will also depend greatly on the investigations of such morphologically specialized microbiologists and zoologists.

MICROMORPHOLOGY AND SOIL SYSTEMATICS

Systematics always represent a touchstone for the stage which a science has reached corresponding to its phase of development. In opposition to a classification, which serves particular purposes of practical application and which according to their specialization imposes intentional simplifications, systematics see as their goal the establishment and further evolution of a general system able to connect all branches of specialization, in short, all sciences dealing with soil research. My predilection for soil systematics has particularly fostered my predilection for micromorphology. The latter has given to every scheme of division, be it classification or systematics, the knwoledge of a number of very characteristics soil properties, unknown up to the present. In most cases where doubts arise as to where to order particular soil forms according to conventional properties or profile morphology, micromorphological analysis can bring the decision. In my opinion an up to now unattained perfection of insight into the inner construction of a soil has brought to all soil research the very decisive knowledge that not only every soil type, sub-type and variety, but also every one of their horizons, corresponds to a particular microscopic fabric. The pattern presented by a thin section is characteristic for a soil type in the same way as its profile. However, it does not simply represent an additional means for its characterization, it reveals entirely new aspects of the innermost essence of a soil. I am of the opinion that there exist whole groups of soils whose different forms can be unequivocally be distinguished only by the aid of micromorphology. Complete evidence in this respect exists in the case of all the highly weathered soils of the ever humid and alternately humid tropics. The distinction of the rubefied, pseudogleyed and indurated lateritic soils, and those forms which have been left without all these transformations (soils which belong the braunlehm type of the traditional nomenclature), is very difficult with other methods, but easy with micromorphological analysis. The subtype earthy braunlehm which is characterized by a frequent loss of silica and a remarkably loose structure does not show any traces of laterization. It can only be recognized with complete security micromorphologically. However, in their eutrophic varieties these soils belong to the most fertile formations of the tropics. The distinction of hardened laterite varieties and "Raseneisenstein" formations would also be very imperfect without the thin section technique. Furthermore, by micromorphology it has been possible to distinguish ten rendsina and twelf ranker varieties, ordered according the grade of their development and the particular influence of environmental conditions, in both cases by their kind of humus formation. The micromorphologic distinction of humid and arid rendsinas (xero-rendsinas), by the ratio of clastic and recrystallized calcium carbonate in the upper soil horizons, seems important to me. The association of loess soils, widely distributed in the temperate zones of the northern and southern hemispheres which are intensely used for agriculture, show a great variability in microscopic fabrics. These fabrics, produced by recent and former environment conditions, are so manifold that a complete study of their possible development can only be performed by thin section analysis.

In a way similar to that of systematics and soil classification, micromorphology also facilitates the exact distinction, checking and delimitation of the mapping units of the soil survey. Therefore many survey institutes already dispose of micromorphology departments or laboratories, such as here in The Netherlands, in England, in various other European countries, and also in the United States and New Zealand. The usefulness of thin section analysis for research in the field of soil tilling and soil technology was proved by the fifteen year old application at the "Institut für Bodenbearbeitung" of the Agricultural Research Center of Braunschweig-Völkenrode, under W. Frese and H.J. Altemüller. It leads our thoughts back to this place where we have seen so much, and benefited so greatly by our successful first working-meeting on soil micromorphology.

FABRIC FORMATION AND CLIMATIC ENVIRONMENTAL CONDITIONS OF THE HABITAT

In the same way as the dependence of the soil type and soil profile on environmental conditions had been detected by W.W. Dokuchaev, a similar dependence was determined for micromorphology by continuous investigations spread over almost the entire world. With many soil formations and habitats, surprising details in their differences can be recognized by differences of the soil fabric. This is particular true for soils of the humid tropics where the degree of humectation and aeration, and temporary water logging with and without reduction, can be recognized by comparative fabric analysis. These different degrees become visible by different degrees of fabric transformation in the direction of brown granulation and floculation ("braune

Vererdung"), pseudogleyization, gleyization and beginning laterization. Climatic measurements are impossible in the tropics, even with a limited number of locations, not to mention the great number of habitats required for a practical advisory service in tropical agriculture. The interpretation of comparative micromorphological analysis is perfected by continuous checking of the results of cropping yields, physiological damages, and the nature and intensity of the development of particular plant diseases. This procedure has shown itself to be very useful and is still capable of improvement.

MICROMORPHOMETRY AND STRUCTURE RESEARCH

Structure analysis, according to my nomenclature, means that part of fabric analysis which deals with the faculty of agricultural soils for aggregate and space formation. The peculiarity of aggregate and space formation possesses a soil in different ways and rates. As a rule it represents, besides nutrient contents, the most deciding factor for fertility. More than ever structure research deals with problems of water economy of the soil, and particularly with water conductivity in connection with the alternation of structure in different horizons, and with the interrelation of structure and water capacity. In structure research micro-*morphometric* analyses have increased greatly. From a morphometry with the aid of diagrams, produced by designing over the frosted glass plate of a table projector, a new *photometric* technique was developed by which photograms instead of designed diagrams are produced, as well as quantitative evaluations performed by photometrical analyses. As with micromorphology in general, morphometric analysis did not develop in competition with physical soil analysis. Both complement, but cannot replace, each other. Micromorphometry has enriched soil science by a new research method and a series of new, quantitatively determinable soil properties. Among these are: the aggregate content on a morphometrical basis, the aggregate size relation, the aggregate type relation, the pore content and floculation degree of aggregates, the space content on a morphometrical basis, the space size relation, the space type relation, the inner surface area of a soil, the morphology and morphometry of the micropores and intergranular spaces, and finally the quantitative values of the homogeneity or heterogeneity of the structure of a soil.

As opposed to mechanical structure analysis, morphometry has the advantage that the structure of the soil is completely preserved against destruction by impregnation. This is of particular importance with aggregate and space analysis, since aggregates frequently show extremely complicated, filigreelike shapes. The combination of simple aggregates to aggregate complexes is often established by very fragile bridges which, although extraordinarily water stable, show very little stability against mechanical destruction. In a similar way valuable crumb aggregates, rich in pores with high floculation degrees and having high water stability, show an easy mechanical breakdown which makes the application of mechanical aggregate analysis to these structurally most valuable soils impossible. Between the stability and the form of the aggregates exists a close interrelation. By using comparative physical determinations of the water stability together with morphological analyses, this allows one to obtain detailed evaluations of the aggregate forms. Since a soil can possess very different aggregate forms showing different stabilities, the ratio of aggregate forms can be used for general taxation of the soil structure. On the whole the structure photogram presents such a valuable proof of the healthy inner construction and the biological capacity of a soil for the expert, that its role for the advisory service can be compared with the role of X-ray diagrams of the human lung for public health service. In the photogram all unfavourable phenomena of structure formation also appear with total clarity and they can be recognized at once by their nature. The examination of a detrimental property can be expressed by the possibility of quantitative photometric determinations. A great advantage for the application of photometric technique in experimental stations and advisory laboratories is given by the automatization of the procedure, and the possibility of using it even in cases of extensive sample material with relatively low man-power. This has already begun with the almost completely mechanized preparation of thin sections, (by greatly effective grinding machines) which has recently been improved by the Vestopal technique of H.J. Altemüller, and has ended with the latest photometric techniques of W. Beckmann. Well-organized photogram records, which can be established for a country, district of other administration unit, allow for a continuous comparison, reclassification and standardization of the results. The great predictive value of the photogram also makes it useful for the investigation and checking of mapping units of agricultural soil surveys and the mapping of forest habitats.

MICROMORPHOLOGY AND SOIL FERTILITY

Over more than eight decades soil fertility and soil genesis and classification developed separately in many countries. Only in recent years has a growing tendency for collaboration been observed, which is clearly expressed in the organization of joint meetings. The advisory laboratories formerly established schemes for the quantities of fertilizers to be applied on the basis of nutrient analyses. However, when a careful check on the correctness and productiveness of the prescribed quantities of fertilizers was performed, surprising discrepancies could be found in many cases. It was noted that the fertilization effect of the same amounts of fertilizer could be very different according to the soil and environmental conditions. Different papers grew from experiments on the fertilization effect under different water conditions, pH-values, humus contents, humus forms and also different soil types. In all cases the interdependence could not be doubted. On the other hand, these inter-relations roused the interest of specialists on soil systematics and cartography, and they began to co-operate on these problems. In fact the conditions seem to be the same as for the general behaviour of the soil: not only a single factor influences the fertilizer effect (although some factors used to come to the foreground) but all the factors together. The soil does not consist of one factor, or of a sum of factors, but represents a complicated and interdependant tissue of factors, in which every single factor stipulates and influences the other. Hence every soil acts as a whole, as an entity of factors. To determine the effectiveness of the soil as a unit of factors is practically impossible. However, there are aspects of the whole soil which are suitable to represent this effectiveness in a sufficiently proper way. These are the profile morphology, the humus form, and the micromorphology. Dates on the profile morphology and the humus form are indispensable for even a cursory characterization of a soil; they can be obtained without any particular effort. Almost all the decisive factors are contained in the micromorphology, or correlate with characteristics expressed by it. A close collaboration of soil micromorphology and soil fertility therefore promises fruitful results. In my opinion the inclusion of structure photograms into these investigations appears to be of particular importance, since besides the nutrient contents, many decisive factors for soil fertility can be recognized in them.

MICROMORPHOLOGY AND PALEOPEDOLOGY

The necessity of paleopedology for the progress of soil science does not need to be particularly accentuated today. Agriculturally used soils, which are entirely formed by recent weathering, in many countries are less frequent than those whose development is of a polygenetic nature, whose genesis had begun at least partly under environmental conditions of a former geological period. Occasionally they are soils in situ, but much more frequently they are present in the form of redeposited soil material, erosion sediments of soils. Rarely they are soils which continue their development in the former way. In most cases the recent environmental conditions are so different from those of the former geological periods concerned that a characteristic transformation takes place by their influence. How can all these details be recognized and determined?

Firstly, a detailed knowledge of the presently most important soils and their corresponding environmental conditions is necessary, primarily for those of the tropics and subtropics. Furthermore, soils which only occur in residues must also be recognized and unequivocally determined. The necessity to recognize soil residues in sediments is closely connected with this, as well as those soil remnants which occur in the form of insoluble components in limestones. Finally, the determination of the kind of soil transformation by climatic change is of essential importance. While the transformation of a soil, produced under former environmental conditions, takes place primarily on the soil surface and in the upper horizons, other changes of a different nature occur in fossil soils situated at great depth, entirely removed from the influence of the biosphere. These changes by diagenesis, instead of soil forming processes, must be recognized and clearly distinguished from the latter. All these investigations can only be carried out insufficient or not at all without micromorphological analyses. A successful development of paleopedology, or even only a satisfactory analysis of a polygenetical soil, will therefore depend greatly on the application of micromorphologic techniques. The results are not only important for soil science but make soil science a valuable auxiliary science to geology and particularly to stratigraphy. Layers lacking fossils of organisms can contain abundant residues of soils which can be evaluated stratigraphically and can yield information on the nature of the former environmental conditions, related to the layers. In this respect they are not only able to substitute for the lack of fossils, but can also complete the evidence of fossils by providing conclusions on the particular living conditions of the habitat. Earth and life history can be advantageously supplemented by these pedological data. In the coincidence of the geological, paleontological, paleobiological and paleopedological phenomena exists the possibility for control of the solidarity of life and habitat.

CONCLUSIONS

In the preceding article I did not touch upon all the fields of application of micromorphology, nor have I cited all the important contributions produced by micromorphologists. I only was able to give some examples. Much will become evident by itself and perhaps obtain much more significance than all that I have exposed here with particular emphasis. But the quoted examples will already be sufficient to draw the following conclusions:

Modern soil science will not be able to continue in the future without micromorphology, its technique, and its principles of research. Its results would otherwise remain too much on the surface, and would insufficiently penetrate into the interior. It would inadequately serve for the understanding of the whole, and be overly contented by single achievements and fragmentary successes. It would have to miss important and sometimes decisive details of the soil, and would never catch sight of the real appearance of the soil which, in its essence, begins only in microscopic dimensions. It would never, even using the microscope, obtain the impression of an undisturbed soil with its great multiplicity of microscopic constructions, but would only deal with little piles of debris. It would never pursue soil life directly but could only make indirect contact with it in round-about ways. However, we know that micromorphology must also not come to a standstill. It must try to incorporate the most modern technique to a close co-operation with methods which must proceed by other principles, but which bring about great advantages for the knowledge of particular details of the soil. But we know that we are always guided more in our research than the former soil science by one element: the element of form. Our formula in brief is not "substance", but "substance and form". Substance alone is something partial, and with form alone this is also true. The whole depends only on the combination of both. This fact, expressed briefly, is the basis of the indispensibleness of micromorphology in modern soil science.

But soil science without micromorphology and micropedologic biology would be deprived of another advantage: our kind of work makes the investigation of the soil more attractive, more inspiring and more interesting. My 35 years of experience with micromorphology lead me to say today, as on the very first day, that by means of our kind of research investigating the soil is not work, it is pleasure.

DÜNNSCHNITTUNTERSUCHUNGEN ÜBER DEN ABBAU LIGNIFIZIERTER GEWEBE IM BODEN

ULRICH BABEL

Bundesforschungsanstalt für Forst- und Holzwirtschaft, Abteilung Bodenkunde[1], Reinbek bei Hamburg (Deutschland)

EINLEITUNG

Hinter den morphologischen Umwandlungen in sich zersetzenden Pflanzengeweben stehen chemische Umwandlungen der Pflanzenstoffe. Deshalb sind mikroskopische Untersuchungen geeignet zur Bearbeitung von chemischen Fragen der Humusbildung (siehe dazu Babel, 1964).

MATERIAL UND METHODE

Hier wird über morphologische und mikroskop-chemische Untersuchungen berichtet, die die Umwandlungen von Buchenblattadern in Rohhumus zum Gegenstand hatten. Blattadern bestehen aus morphologisch und chemisch stark differenzierten Teilen; das Mikroskop macht sichtbar, wie diese Differenziertheit verschiedenes Verhalten bei der Zersetzung hervorruft. Die Umwandlungen der verschiedenen Pflanzenstoffe zu braunen, huminstoffartigen Substanzen kann direkt beobachtet und mit Hilfe optischer und mikrochemischer Methoden näher untersucht werden.

In dem untersuchten Rohhumusprofil, das fast nur aus Buchenblättern gebildet wird, haben die Blattadern einen beträchtlichen Anteil am Ausgangsmaterial der Humusbildung (Angaben zum Profil - Hausbruch bei Hamburg - siehe Babel, 1964). Die Adern und ihre verschieden weit abgebauten Reste wurden unter dem Stereomikroskop aus den aufeinanderfolgenden Schichten des Rohhumusprofils ausgelesen. Zur morphologischen Untersuchung wurden Paraffinschnitte hergestellt. An diesen wurden auch verschiedene optische und mikrochemische Eigenschaften ermittelt. Die wichtigsten von diesen waren: Als Lignin-Nachweise: die Phloroglucin-Reaktion, die weissblaue Eigenfluoreszenz des Lignins, das Absorptionsmaximum bei 280 mμ; als Zellulosenachweis: die Doppelbrechung; ausserdem die Färbung mit Rutheniumrot (vor allem als Pektin-Nachweis), Färbung mit Schiffs Reagens (als Nachweis für Aldehyde), Löslichkeit in Natronlauge (analog makrochemischen Humusextraktionen) und in Natriumhypochlorit (zur Kennzeichnung des Abbaugrades).

---

[1] Leiter: Prof. Dr. W. Kubiena.

ERGEBNISSE

Die frischen Blattadern

Dem Prinzip der Arbeit entsprechend waren zunächst die frischen Blattadern zu untersuchen, um die späteren Umwandlungsstadien als Abbauprodukte bekannter Ausgangsmaterialien ansprechen zu können. Die Ader des frischen Buchenblatts besteht aus einem kollateralen Leitbündel, das von einem Sklerenchymring umgeben und oben und unten von einem Kollenchym begleitet ist (vgl. dazu Abb.1 und Tafel A1, die Ader aus der Streuschicht lässt die ursprüngliche Anordnung des Gewebes noch erkennen). Das Phloem ist durch Stege verholzter Zellen, die Xylem und Sklerenchym verbinden, in mehrere Gruppen geteilt und umgibt das Xylem meist etwa zu drei Vierteln. Das Xylem ist gegliedert in einen kleinerzelligen, schwächer verholzten zentralen Teil, in mehr oder weniger deutliche zu den Stegen führende Strahlen und in den Phloemgruppen benachbart liegende Gruppen grosser Gefässe. Im Xylem und noch stärker im Sklerenchym ist die Gliederung der Zellwand in Mittelschicht (das ist Mittellamelle und Primärwand) und Sekundärwand durch die höhere Lichtbrechung der Mittelschicht leicht zu erkennen (Nomenklatur nach Frey-Wyssling, 1959). Die Unterscheidung dieser Wandteile ist wichtig, weil sie sich bei der Zersetzung ganz verschieden verhalten.

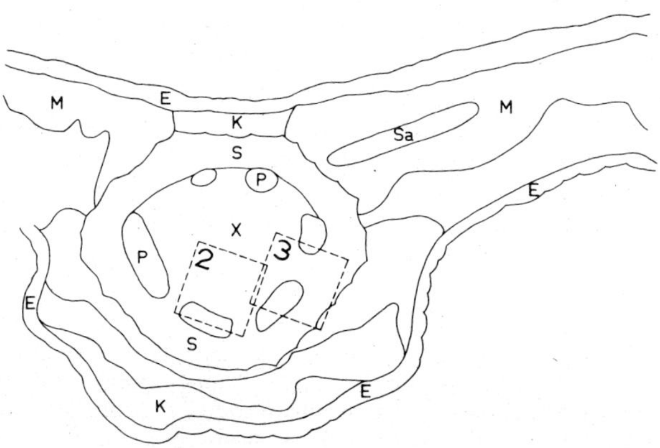

Abb.1. Buchenblattader aus der oberen Streuschicht eines Rohhumus (siehe Tafel A1). K = Kollenchym; S = Sklerenchym; P = Phloem; X = Xylem; E = Epidermis; M = Mesophyll; Sa = Teil einer angeschnittenen Seitenader. 2 und 3 = ungefähre Lage der Ausschnitte von Tafel A2 und A3.

Im Kollenchym sind die Zellwände durch einen hohen Pektin-Gehalt charakterisiert; die Zellulosefibrillen liegen wenig dicht und sind schlecht orientiert. Das Phloem enthält neben Zellulose bekanntlich die zur Gruppe der Hemizellulosen gehörige Callose. Sklerenchym und Xylem haben Zellwände, die fast nur aus Zellulose und Lignin aufgebaut sind; dabei ist der Ligningehalt in der Mittelschicht und besonders in den Zwickeln zwischen den Zellen höher als in der Sekundärwand; auf die so entstandene höhere Dichte in der submikroskopischen Struktur der Mittelschicht geht auch die höhere Lichtbrechung und Streuung zurück. Zellinhaltstoffe spielen in allen diesen Geweben mengenmässig keine Rolle. Die verholzten Gewebe (Sklerenchym und Xylem) bestehen schon in der lebenden Pflanze aus toten Zellen; sie enthalten höchstens in sehr geringen Mengen denaturierte Reste von Zellinhaltstoffen, die im Sklerenchym oft eine Art von Tertiärwand bilden (vgl. Meier, 1955).

Die Abbauvorgänge

I. Schon bei Blättern aus der oberen Streuschicht (L_1, siehe Tafel A1), die ein halbes Jahr am Boden gelegen haben, ist das Phloem weitgehend zersetzt; nur gelbe Massen sind zurückgeblieben an den Rändern der Löcher, die so entstanden sind. In der unteren, 2–3 Jahre alten Streuschicht (L_2) sind auch diese verschwunden.

II. Die Kollenchyme sind ebenfalls schon im L_1 stark zersetzt. Die Wände werden vom Zellumen ausgehend gelb gefärbt, und im gleichen Masse, wie die Färbung zu weiter aussen gelegenen Wandteilen fortschreitet und sich intensiviert, nimmt eine Aufrauhung der inneren Zellwandteile und ihre Umbildung zu körnig-warzigen Belägen zu.

Vor allem wurde die Zersetzung von Sklerenchym und Xylem untersucht. Diese Gewebe sind die hauptsächlichsten Ligninlieferanten des Rohhumusprofils Hausbruch. Da dem Lignin eine bedeutende Rolle als Ausgangsmaterial für die Bildung natürlicher Huminstoffe zugeschrieben wird (Flaig et al., 1963) ist ihre Untersuchung also von besonderem Interesse.

III. Im Sklerenchym von Blattaderresten aus der oberen Streuschicht sind Mittelschicht und Sekundärwand fast immer unverändert (Stadium 1; schematische Darstellung; siehe Tabelle I). Nur eine sehr dünne, innerste Wandschicht, die - wie oben erwähnt - als Tertiärwand bezeichnet werden kann, ist oft gelb geworden. Später - bei Adern aus der unteren Streuschicht und aus der Vermoderungsschicht - färbt sich die Sekundärwand gelb (Stadium 2) und hebt sich gleichzeitig oder bald darauf von der Mittelschicht ab, so dass sie als Schlauch (im Querschnitt als Ring) im Zellumen liegt (Stadium 3), der weiter verquillt und dann im Querschnitt als gelbbrauner Klumpen zu sehen ist (Stadium 4; siehe Tafel A2). In den am stärksten

	1	2	3	4	(5)
Alter	frisch, nach 1/2 Jahr	nach 1 Jahr und später			
Morphologie	 farblos	gelb	braungelb	gelbbraun	
Doppelbrechung	+	+	(+)	(?)	
Fluoreszenz	+	+	(+)	(?)	
280 mμ - Max.	+	(+)	(+)	× S.U.!	
Phloroglucin	+	?	–	–	
NaOCl	nicht gelöst	entfärbt, z.T. gelöst		entfärbt, überwiegend gelöst	

Tabelle I. Abbau der Sekundärwände im Sklerenchym. Zellulosenachweis durch Doppelbrechung; Ligninnachweis durch Fluoreszenz, Absorptionsmaximum bei 280 mμ, Phloroglucin.

zersetzten Blattaderresten aus der unteren Vermoderungsschicht und der oberen Humusstoffschicht ist oft zu sehen, dass dann schliesslich alle Sekundärwandreste verschwunden sind (Stadium 5). Dieses Stadium ist allerdings nicht regelmässig zu beobachten; offenbar werden im Boden die Aderreste normalerweise schon vorher von Tieren aufgenommen. Während dieser ganzen Veränderungen der Sekundärwand behält die Mittelschicht ihre ursprüngliche Form und bleibt farblos. Höchstens ganz zuletzt wird sie stellenweise zerdrückt, und manchmal sind dann an ihrer Oberfläche kleine, unregelmässige Flecken gelber Substanzen zu sehen, von denen aber nicht gesagt werden kann, ob sie Reste der Mittelschicht oder der Sekundärwand sind. Die wichtigsten optischen und mikrochemischen Eigenschaften der Abbaustufen der Sklerenchym-Sekundärwände sind in Tabelle I zusammengestellt.

Dem morphologischen Umwandlungsgrad der gelben Sekundärwandreste entspricht weitgehend ihr Verhalten. Die Ringe (Stadium 3) zeigen noch Doppelbrechung und Fluoreszen und werden von Natriumhypochlorit zwar entfärbt, aber nur teilweise gelöst. Bei Stadium 4 zeigen die Reste keine oder fast keine Doppelbrechung unf Fluoreszenz mehr. Von Natriumhypochlorit werden sie vollständig gelöst. Weitere neue Reaktionen der gelben Teile, die die unveränderten Sekundärwände nicht zeigen, sind Färbbarkeit durch Rutheniumrot und durch Eisen (III)-Salz/Blutlaugensalz; in Natronlauge sind die gelben Stoffe teilweise löslich. Mit Schiffs Reagens sind die unveränderten Wände undeutlich, aber die gelben und braunen Sekundärwandreste beträchtlich färbbar.

In der Mittelschicht bleiben Fluoreszenz und Phloroglucin-Reaktion bis zuletzt unverändert. Die Doppelbrechung nimmt dagegen vom Stadium 4 an etwas ab. Ab Stadium 3 lässt sich die Mittelschicht mit Rutheniumrot und mit Schiffs Reagens schwach anfärben.

IV. Im Xylem sind die Abbauvorgänge und die Reaktionen der Zellwandreste prinzipiell ähnlich (siehe Tafel A3). Aber hier ist schon bei Adern aus der oberen Streuschicht eine recht starke Gelbfärbung zu sehen. Teilweise verfärben sich wie im Sklerenchym zuerst die Sekundärwände; sie werden dabei rauh und manchmal körnig, während eine Ringbildung seltener ist. Teilweise verfärben sich aber zuerst die Mittelschichten, was im Sklerenchym nie vorkommt. Von der unteren Streuschicht an verschwinden die gelbverfärbten Wandteile, was im Falle der Mittelschichten zur Mazeration der Sekundärwände führt. Im Falle der Sekundärwände bleibt ein noch recht gut zusammenhängendes, aber dünnwandiges, weil nur aus Mittelschichten bestehendes Gewebe zurück. Dieses Stadium ist in der Vermoderungsschicht nicht selten zu finden (siehe Tafel A2). Dann sind gewöhnlich die kleinerzelligen, schwächer lignifizierten zentralen Teile des Xylems schon völlig verschwunden, auch ohne dass Spuren von Tierfrass beobachtet werden können.

In den optischen und mikrochemischen Eigenschaften unterscheiden sich die Xylemreste - bei Berücksichtigung der teilweise andersartigen Morphologie des Abbaus - in den meisten Punkten nicht vom Sklerenchym. Jedoch sind schon bei frischen Blattadern die Mittelschichten gut mit Rutheniumrot färbbar.

DISKUSSION

Die Unterschiede in den Zersetzungserscheinungen, die in den Geweben der Buchenblattader beobachtet wurden, entsprechen der morphologischen und chemischen Differenzierung der frischen Ader. Die morphologische Untersuchung zeigt, dass zuerst die pektin und hemizellulosereichen Gewebe - Kollenchym und Phloem - angegriffen werden; es folgen die lignifizierten Wände im Sklerenchym und Xylem. Am schwersten werden die sehr ligninreichen Wandteile der letztgenannten Gewebe - die Mittelschichten und die Zwickel zwischen den Zellen - abgebaut. Es ist jedoch hervorzuheben, dass reine Zellulosewände - aus dem Mesophyll - ähnlich lange erhalten bleiben (vgl. Babel, 1964).

Dass die Sekundärwände im allgemeinen eher als die Mittelschichten morphologische Veränderungen zeigen, wurde auch von Meier (1955) in Versuchen mit Holzvermorschungspilzen - und zwar bei allen Fäuletypen - beobachtet. Die Ausnahme von dieser Regel, die im Xylem zu beobachten war, wo stellenweise zuerst die Mittelschicht abgebaut wird, entspricht einer primären chemischen Verschiedenheit der Mittelschicht des Xylems von der des Sklerenchyms, die durch positive Rutheniumrot-Reaktion nachgewiesen wurde. Offenbar hat die Xylemmittelschicht entweder einen höheren Pektingehalt oder ist das Pektin wegen geringerem Ligningehalt und daher weniger dichter Feinstruktur der Zellwand für das Reagens und für Abbaureaktionen leichter zugänglich.

Der chemische Abbau ist in der Regel ausser durch strukturelle Veränderungen durch Verfärbungen der Zellwände und Zellwandteile zu erkennen. Gleichzeitig verschwinden in den verholzten Zellwänden - wie die optischen und mikrochemischen Reaktionen gezeigt haben - Zellulose und Lignin. Die Neubildungen sind daher offenbar Umwandlungsprodukte aus jenen. Sie sind nach ihrer gelben bis braunen Farbe und ihrer Löslichkeit in Natronlauge als Huminstoffe anzusprechen. Die Löslichkeit in Natriumhypochlorit haben sie gemeinsam mit anderen braunen, auch im chemischen Bau huminstoffartigen Substanzen der Rohhumusdecke, nämlich Blattbräunungsstoffen und Phlobaphenen (Babel, 1961, 1964). In den tieferen Schichten des Rohhumusprofiles verschwindet ein Teil der Neubildungen wieder, wie besonders an den Sekundärwänden des Xylems nachgewiesen wurde. Die für das Problem der Entstehung stabiler Humusstoffe aus Lignin wichtige Frage, ob die Neubildungen nur aufgelöst und verlagert oder aber völlig mineralisiert werden, ist noch ungeklärt. Eine Auflösung und Verlagerung ohne stärkere chemische Veränderungen ist jedoch unwahrscheinlich.

Der Abbau des Lignins

Der chemische Abbau des Lignins geht aus der Zusammenstellung der Reaktionen in Tabelle I hervor. Die Tatsache, dass die Phloroglucin-Reaktion vor den beiden anderen Merkmalen ausfällt, lässt - da diese Reaktion für die Seitenkette der Lignin-Bausteine charakteristisch ist- -darauf schliessen, dass der Abbau des Lignins an der Seitenkette beginnt. Das ist eine Bestätigung makrochemischer Ergebnisse (Literatur s.u.). Auch die positive Reaktion der zersetzten Sekundärwände mit Schiffs Reagens, also der Nachweis, dass Aldehydgruppen gebildet werden, entspricht den Ergebnissen anderer Untersuchungen über den Ligninabbau.

Die Chemie des Ligninabbaus (Brauns und Brauns, 1960; Flaig et al., 1963) ist vorwiegend an pilzinfizierten Hölzern untersucht worden. Das Lignin wird depolymerisiert und gleichzeitig oxydiert. Die so entstehenden Abbauprodukte polymerisieren dann zu braunen, huminstoffartigen Substanzen. Als natürliche monomere Lignin-Abbauprodukte wurden unter anderem nachgewiesen: Coniferylaldehyd, Vanillin, Protocatechusäure, Brenzcatechin. Die Oxydation und Polymerisation von Brenzcatechin zu Huminstoffen unter Verhältnissen, wie sie auch in der Natur möglich sind, wurde ebenfalls nachgewiesen.

Die Tatsache, dass huminstoffähnliche Produkte am Orte des Lignin-Abbaus auftreten, also offenbar topochemisch gebildet werden, spricht für in der angegebenen Literatur geäusserte Annahmen, dass die Polymerisation der Abbauprodukte beginnen kann, ehe das Lignin völlig depolymerisiert und die Seitenkette ganz entfernt ist.

Der Abbau der Zellulose

Der Abbau der Zellulose ist durch die Abnahme und das schliessliche Verschwinden der Doppelbrechung nachgewiesen (Tabelle I), was auf die Zerstörung der Feinstruktur der Zellulosefibrillen schliessen lässt. Wieweit Zelluloseabbauprodukte an der Bildung der braunen huminstoffähnlichen Substanzen in den Lignin-Zellulose-Wänden teilnehmen, ist noch nicht geklärt. (Theoretisch gibt es Möglichkeiten dafür über die Bildung von Furfurol und ähnlichen Verbindungen). Während auch die ligninfreien Wände von Phloem und Kollenchym bei der Zersetzung gelbe Substanzen wenigstens intermediär zu bilden vermögen, wurden jedoch an den reinen Zellulose-Wänden des Mesophylls keine Verfärbungen festgestellt.

In den lignifizierten Sekundärwänden verlaufen, wie Tabelle I zeigt, Lignin- und Zelluloseabbau simultan. (Zu vergleichen sind Doppelbrechung und Fluoreszenz, welch letztere hier als Ligninmerkmal besser geeignet ist als die Phloroglucin-Reaktion). Nach Meier (1955) handelt es sich demnach um den Typ der Simultanfäule. Die Bedeutung des Zellulose/Lignin-Verhältnisses als Rottequotient (Schobinger, 1958) zeigt sich vor allem in der schlechten Zersetzbarkeit der Sklerenchym-Mittelschichten. Es haben jedoch auch dort strukturelle und chemische Änderungen eingesetzt, wie die Änderungen einiger Reaktionen zeigen. Die Reaktion mit Rutheniumrot kann hier vielleicht als Nachweis der Bildung karboxylischer Gruppen (bei der Zellulosezersetzung?) gewertet werden, (entsprechend der Färbbarkeit von Oxyzellulose mit Rutheniumrot); vielleicht ist sie auch nur eine Folge der Strukturlockerung, die durch Abnahme der Intensität der Doppelbrechung nachgewiesen ist.

Ergänzend sei bemerkt, dass die morphologischen Erscheinungen bei der Zersetzung anderer Blätter und Nadeln prinzipiell den hier beschriebenen ähneln.

Die Untersuchungen waren von mikromorphologischen Beobachtungen an Bodendünnschliffen angeregt worden. Es war zu sehen, dass morphologische Studien - mit entsprechenden methodischen Änderungen und Ergänzungen - zu Aussagen führen können, die von humuschemischem

Interesse sind. Da die Arbeiten an den natürlich auftretenden Gefügebestandteilen gemacht worden sind und zu deren botanischer und chemischer Deutung beitragen, können sie nun auch wieder bei mikromorphologischen Untersuchungen eine Hilfe sein.

ANERKENNUNG

Ich danke Herrn Prof. Dr. W. Kubiena, der mir die Anregung zu dieser Arbeit gegeben hat, und der Deutschen Forschungsgemeinschaft, die sie durch Sachbeihilfen unterstützt hat.

SUMMARY

Microscopical investigations were applied to chemical problems of humification, especially to the role of different plant substances in humus formation. The morphological alterations during decomposition of the veins of beech leaves in raw humus were investigated by paraffine sections. By means of histological (microchemical and optical) methods, it was attempted to obtain information about the chemical processes involved.

The chemical differentiation of the tissues results in a marked difference in decomposition phenomena. Very highly lignified parts of walls begin to decompose last; also, pure cellulose walls (of the mesophyll) are preserved in very old layers of the humus cover. In the lignified walls, lignin and cellulose decompose simultaneously (the lignin decomposition seems to begin in the side chain; aldehyde groups are formed). At the same time, newly formed substances appear, which have the colour and solubility of humic substances ("Huminstoffe"). Parts of these disappear in later stages of decomposition, which means that part of the lignin decomposition products either dissolve or mineralize completely.

Attempting a botanical and chemical interpretation of the morphological components, the investigations also made a contribution to the micromorphology of humus covers.

LITERATUR

Babel, U., 1961. Mikroskopische Humusuntersuchungen. Vortrag auf der Tagung der Deutschen Bodenkundlichen Gesellschaft, Wien, unveröff., 8 pp.

Babel, U., 1964. Humuschemische Untersuchungen eines Buchen-Rohhumus mittels mikroskopischer Methoden. Z. Pflanzenernähr., Düngung Bodenk., im Druck

Brauns, F.E. and Brauns, D.A., 1960. Chemistry of lignin. Suppl. Vol. Academic Press, New York–London, 804 pp.

Flaig, W., Salfeld, J. und Haider, K., 1963. Zwischenstufen bei der Bildung von natürlichen Huminsäuren und synthetischen Vergleichssubstanzen. Landwirtsch. Forsch., 16 : 85–96.

Frey-Wyssling, A., 1959. Die pflanzliche Zellwand. Springer, Berlin–Göttingen–Heidelberg, 367 pp.

Meier, H., 1955. Über den Zellwandabbau durch Holzvermorschungspilze. Holz Roh-Werkstoff, 13 : 323–338.

Schobinger, U., 1958. Chemische Untersuchungen über die Umwandlungen von Weizenstroh-Lignin im Laufe der Verrottung. Diss., Eidgenöss. Tech. Hochschule Zürich, Zürich, 80 pp.

FARBENBILDER

Tafel A1. Buchenblattader aus der oberen Streuschicht (L_1) eines Rohhumus (Dünnschnitt, x 160). Schematische Skizze s. Abb.1. Phloem stark zersetzt (Löcher); Kollenchym mittel zersetzt, unten zerrissen; Xylem mit beginnender Zersetzung; Sklerenchym fast unverändert. Im Mesophyll Blattbräunungsstoffe aus Zellinhaltstoffen.

Tafel A2. Xylem (Oben) und Sklerenchym einer Buchenblattader aus der unteren Vermoderungsschicht (F_2) eines Rohhumus (Dünnschnitt, x 1100). Im Xylem sind nur noch die Mittelschichten erhalten, teilweise fehlen auch diese; im Sklerenchym sind die Sekundärwände zu gelben bis braunen Ringen und Klumpen umgeformt (Stadium 3 und 4), die Mittelschichten sind kaum verändert.

Tafel A3. Das Xylem einer Buchenblattader aus dem Übergang von Vermoderungs- zu Humusstoffschicht eines Rohhumus (Dünnschnitt, x 1100). Teilweise ist die Sekundärwand (z.B. grosse Zelle Mitte), teilweise die Mittelschicht (z.B. im rechten oberen Bildteil) zu braunen Stoffen umgewandelt.

THE ROLE OF THE FUNGUS CENOCOCCUM GRANIFORME (SOW.) FERD. ET WINGE IN THE FORMATION OF MOR

F.H. MEYER

Federal Research Organization for Forestry and Forest Products, Reinbek near Hamburg (Germany)

WHAT FACTORS REGULATE THE FORMATION OF MOR?

The mor profile is characterized by the accumulation of decaying plant debris on the mineral soil (Kubiena, 1953). This results, a.o., from the low activity of micro-organisms. Humus respiration, that is, the consumption of oxygen per gramme of organic material, is a measure of the intensity of decomposing organic material by fungi and bacteria. Under *Fagus sylvatica* in the upper fermentation layer of mor, it reaches 3/4 of the mull-value, while in the humus layer it attains only 1/4 of the mull-value (Meyer, 1959).

The low microbial activity is closely related to the many factors which account for the formation of mor, such as:

(a) Unfavourable properties of litter. For example: an ample C/N-ratio (Wittich, 1943); a high lignin content (Grosskopf, 1928); a low number of bases (Wittich, 1952) (Laatsch, 1957), and a high tannin content, which precipitates proteins and thus renders them resistant to attack (Handley, 1954).

(b) Disadvantageous soil properties: poverty of bases and nutrients, sand with low water capacity, close texture and checked ventilation.

(c) Climatic factors: a cool and rainy climate (Hesselman, 1925); low temperatures hamper the microbial activity and high rainfalls favour eluvation of nutrients.

(d) Economic measures play a role in the vegetation influenced by man. Only the upper layers of soil are included in the circle of nutrients in a stand of planted trees with shallow rooting, whereas nutrients illuviated into the deeper layers are lost to the circle of nutrients (Hartmann, 1961). Soil temperature is decreased in a stand with a dense canopy in comparison to an open stand, as is microbial activity.

THE OCCURRENCE OF HYPHAE OF CENOCOCCUM IN MOR

Besides the factors mentioned under (a)–(d) the fungus *Cenococcum graniforme* plays a particular role in the formation of mor. Lihnell (1942) advocates aligning *Cenococcum graniforme* in the ascomycetes, since the structure of the wall of their sclerotia resembles the perithecium wall of the ascomycetous genus *Cephalotheca*. *Cenococcum graniforme* is capable of existing saprophytically, as well as in symbiosis with trees. Among the mycorrhizal fungi, *Cenococcum graniforme* shows the widest host range

(Trappe, 1962); it is a symbiotic partner of nearly all the trees in the forests of the boreal and temperate zones. The hyphae of *Cenococcum graniforme* are characterized by thick walls and a blackish-brown colour, and their diameter generally varies between 4–6 μ. It occurs in all forest soils, but in an abundant quantity in mor. The fungus does not seem to form a mycorrhiza with *Calluna*, so the following statements are only valid for forest humus.

P.E. Müller, the father of humus micromorphology, observed a great deal of blackish-brown mycelium in mor of Danish beech forests as early as 1887. Based on our present knowledge, the mycelium mentioned by Müller (1887) might belong to the fungus *Cenococcum graniforme*. Before the role of this fungus in the formation of mor is discussed, its ecology should be briefly described.

THE RESISTANCE OF CENOCOCCUM AGAINST DROUGHT

Investigations of the ecology of *Cenococcum graniforme* were initiated in the autumn of 1959, when the influence of summer drought on mycorrhizas was studied in the surroundings of Hamburg. During the examination it was found that in soils which had suffered greatly from the drought the quantity of *Cenococcum* mycorrhizas had increased in contrast to those with hyaline hyphae. It could be noticed repeatedly that hyphae of *Cenococcum graniforme* established themselves on the surface of light-coloured mycorrhizas, the cortex tissue of which was entirely shrunken under the influence of drought, whereas mycorrhizas still intact were free of blackish-brown mycelium (Fig.1). *Cenococcum graniforme*, therefore, seems to possess a higher resistance to drought than other mycorrhizal fungi.

Witkamp (1960) reports that his counts of mycelium in soil in November 1959 (soil suspension slide method) gave a percentage value of 91–100% thick and brown mycelia.

In order to study the influence of water supply on the occurrence of *Cenococcum graniforme*, the following experiment was conducted: Young plants of *Fagus sylvatica* were cultivated in a green-house in earthenware pots filled with different humus forms (mull, moder and mor), and watered as follows: series A received water every time that it was necessary, series B received water one half as often as A, series C received water one third as often as series A, and series D received water one fourth as often as series A. The result of this experiment is shown in Table I.

It can be seen from Table I that *Cenococcum graniforme* behaves towards increasing soil drought in a different manner than other mycorrhizal fungi. While the share of the latter is reduced strongly, the percentage of *Cenococcum* mycorrhiza remains either unchanged, or even slightly greater (in moder). The differential sensitiveness of mycorrhizas to water deficiency results in the relative enrichment of *Cenococcum graniforme* during drought. Similar results were obtained by Worley and Hacskaylo (1959).

Table I further indicates that *Cenococcum graniforme* is also present in mull, although not in the same amount as in mor. Here too there is a relative enrichment of *Cenococcum* with increasing drought. Since *Cenococcum graniforme* plays a minor role in mull under natural conditions, the water economy might be more balanced in mull than in mor.

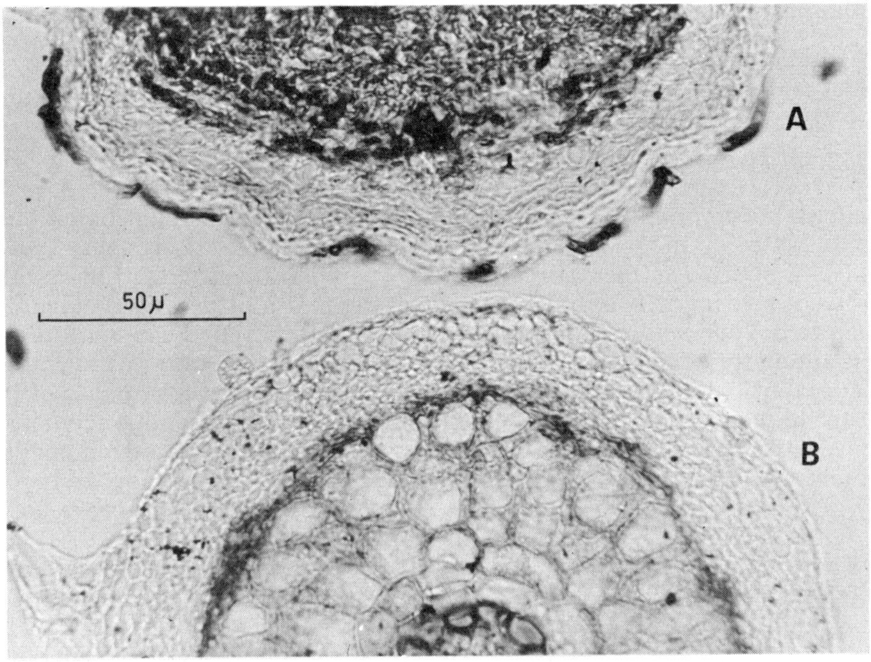

Fig.1.A. Mycorrhiza that had suffered under the drought of the summer 1959. The tissue of the cortex is shrunken and in the fungus sheath *Cenococcum graniforme* has established itself. B. Intact mycorrhiza without colonization of *Cenococcum graniforme*.

TABLE I

Influence of different moisture content of soil on the development of mycorrhizas of two-year-old beech seedlings (cultivated in the humus layer). The numbers indicate the percentage of root tips converted into mycorrhizas.

	Mull		Moder		Mor	
Watering	*Cenococcum graniforme* mycorrhiza	Other mycorrhizas	*Cenococcum graniforme* mycorrhiza	Other mycorrhizas	*Cenococcum graniforme* mycorrhiza	Other mycorrhizas
every time	5	75	15	60	30	40
every 2nd time	5	55	15	35	30	20
every 3rd time	5	30	20	10	30	10
every 4th time	5	30	20	5	30	5

As shown in Table I the water factor has a decisive significance in the distribution of *Cenococcum*. Hence in the following paragraphs the water economy of mor will be compared with that of mull. The occurrence of mycorrhizal fungi is determined less by average values of water content in humus than by extreme values. This is especially true for periods of dryness lasting for only a short time; even a short period of dryness may kill the tender hyphae of many mycorrhizal fungi. An extreme period of drought may change the fungus flora of tree roots to a great degree; *Cenococcum graniforme*, being more resistant to drought, then gets the upper hand and the roots harbouring it may only be very slowly recolonized by the other mycorrhizal fungi.

THE WATER ECONOMY OF MOR

In order to get an idea of which layers of the different humus forms the supply of water is most endangered during the periods of drought, the intensity of rooting in the various layers has been investigated in detail. The intensity of rooting in the different layers gives a hint as to the capability of trees to absorb water from these layers in correspondence with their rooting intensity. The root tips of *Fagus* are converted into mycorrhizas to a great extent, which facilitates the uptake of water; in those parts of the root distal to the root tip the water uptake decreases by increasing suberization of root cortex (Kramer, 1956). Therefore the number of root tips per unit of volume was counted in the different layers of soil. In respect to the capacity of water uptake, the number of root tips ecologically gives a better scale of measurement than the length or the weight of the roots.

The intensity with which *Fagus sylvatica* roots in the diverse horizons of mull, moder and mor soils is shown in Fig.2 (concerning the properties of the sample plots cf. Meyer, 1959, 1962). From Fig.2 it follows that the intensity of rooting in the different humus forms is rather different. In mull, *Fagus sylvatica* develops a maximum of 556 root tips per 100 ml of soil (H2). In mor, on the contrary, 45,600 root tips (upper F3) develop. This extraordinarily great difference in the intensity of rooting results, among others, from the nutrient supply of the soil. According to the findings of Kern et al. (1961), the total length of tender roots is closely related to the nutrient content of the soil: "The higher the available supply of nutrients, the less is the amount of tender roots necessary for its exploitation".

However, not only does the number of root tips show considerable deviations between the different humus forms, but the distribution of roots in the different layers does too. As Fig.2 shows, in the eutrophic brown earth the roots are spread in nearly the same manner in all the layers; in the podsolic brown earth there is a tendency to prefer the fermentation-layer, while in podsol nearly all tender roots occur in the fermentation-layer, and the layers beneath are only weakly or not at all traversed by roots. In the mineral soil of podsol the nutrient supply is so low that the roots first develop in the A-horizon, where nutrients become available to plants in the course of decomposition of litter. This concentration of roots near the surface leads to quick exhaustion of water in A-horizon in drought periods, while in eutrophic brown earth all soil layers are utilized to nearly the same degree.

CAUSES FOR THE ABUNDANCE OF CENOCOCCUM IN MOR

Fig.2 shows, furthermore, that the share of *Cenococcum graniforme* increases with the rooting intensity. The highest number of *Cenococcum graniforme* mycorrhizas occurs in the F3 layer of mor (podsol), which is

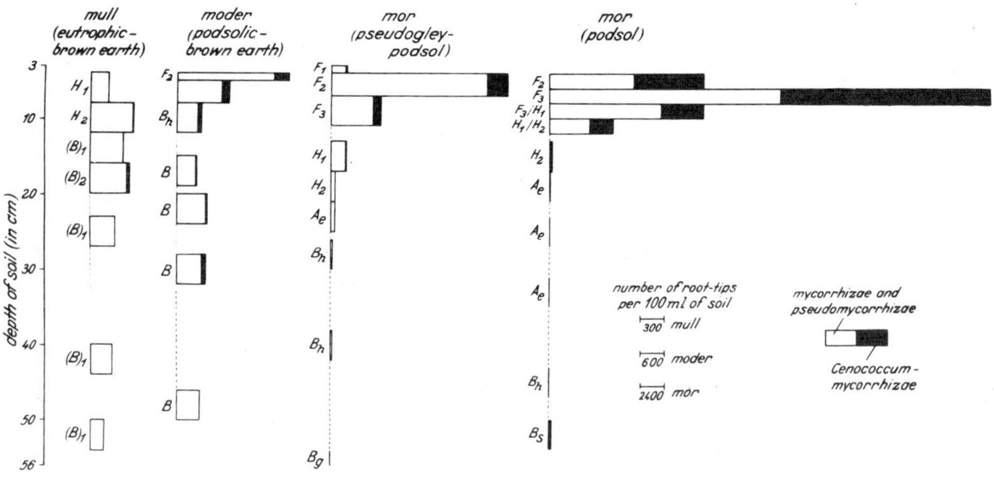

Fig.2. Intensity of rooting of *Fagus sylvatica* in various layers of eutrophic brown earth, podsolic-brown earth, pseudogley-podsol and podsol. Nomenclature of soils and soil horizons according to Kubiena (1953).

most intensely interwoven by roots. This enormous development of *Cenococcum* might particularly be traced back to the unbalanced water economy in the strongly rooted layers. Besides this, the resistance capacity of *Cenococcum graniforme* to certain antibiotics may play a role, for in mor antibiotics are produced by fungi in larger quantities (Brian, 1949). For example, Levisohn and Parry (1960) found that the development of many mycorrhizal fungi is inhibited by antibiotics formed by *Alternaria tenuis;* only *Boletus subtomentosus* and *Cenococcum graniforme* (= *Mycelium radicis nigrostrigosum*) were not influenced by these substances.

Not only do the mycelia of *Cenococcum* occur in great quantities in mor, but the black-coloured sclerotia do too. Sclerotia are to be found in mull and moder as well, but in smaller quantities and as small-sized ones. The sclerotia are predominantly formed in such layers where the conditions of existence become more disadvantageous to *Cenococcum graniforme*, viz. in the zone of transition between the humus-rich upper soil and the humus-poor mineral soil. The thin sections of this zone are characterized by numerous sclerotia. The distribution of the sclerotia in the different layers of podsol is shown in Table II.

From the preceding paragraphs it follows that the abundance of *Cenococcum graniforme* in mor must be traced back to the concentration of roots in the A-horizon in the first place. This fact, characteristic of mor, implies that the water content of mor decreases to a high degree in drought periods, and this facilitates the relatively better growth of *Cenococcum graniforme*.

TABLE II

Distribution of sclerotia of *Cenococcum graniforme* in the podsol profile.

Mor/Pseudogley-Podsol		Mor/Podsol	
Layer	Number of sclerotia per 100 ml of soil	Layer	Number of sclerotia per 100 ml of soil
F1 (3– 4 cm)	0	F2 (4– 6 cm)	0
F2 (4– 7 cm)	1	F3 (6– 8 cm)	8
F3 (7–11 cm)	2	F3/H1 (8–10 cm)	35
H1 (13–17 cm)	14	H1/H2 (10–12 cm)	89
H2 (17–21 cm)	251	H2 (13–17 cm)	161
Ae (21–25 cm)	16	Ae (17–21 cm)	203
Bh (26–30 cm)	9	Ae (23–27 cm)	134
Bh (38–42 cm)	2	Ae (31–35 cm)	51
Bg (54–58 cm)	0	Bh (43–47 cm)	47
		Bs (50–54 cm)	75

CENOCOCCUM IN THIN SECTIONS OF MOR

The observation of thin sections shows that the blackish-brown hyphae of *Cenococcum* contribute to a great deal of the material of mor, especially in the lower fermentation and the upper humus layers, the numerous hyphae of *Cenococcum* are a characteristic feature of thin sections of mor, (cf. Plate A4 and A5). *Cenococcum* does occur in the B-horizon too (Plate A6).

The enrichment of the dark hyphae is not only favoured by periodic drought, but also by the fact that they undergo decomposition more slowly than the tender hyaline hyphae. Martin et al. (1959) investigated the disintegration of 25 microscopic fungi and found that the light-coloured hyphae are destroyed more quickly than the darker ones. The investigation of thin sections also supports the opinion of a slow decomposition of *Cenococcum* hyphae. For we see that after the death of *Cenococcum graniforme* mycorrhizas the decay of root tissue proceeds at a rapid rate, while the mycelia of the fungal sheath are still visible in the form of dark rings; later these rings disintegrate into pieces. A further reference to the difficult decompossibility of the *Cenococcum* hyphae is furnished by the data on humus respiration (Meyer, 1959). In the upper fermentation layer of mor (podsol) the litter of beech is attacked with an intensity of 78% of that of mull, whereas in the humus layer of mor, with its high content of *Cenococcum* hyphae, the rate of decomposition amounts to only 24% of that of mull. Thus, the "permanent humus" of mull is more intensively decomposed than the material of the H-layer of mor.

In the lower part of the humus layer a great many fragments of dark hyphae are conspicuous, the contours of which often appear blurred in thin sections. This may be a hint that a slow decomposition of dark hyphae takes place here or that the hyphae undergo autolysis (cf. Plate A4).

DISCUSSION

The findings of the present investigation appear to contribute an important factor to the problem of mor formation. No doubt poverty of bases in soil and the other previously mentioned factors play a decisive role in mor formation. However, mor formation is not only the result of a slow decay of litter. As seen above, in the course of litter decomposition there is an intensive production of dark hyphae, which are apparently more difficult to decompose than the litter itself. Consequently, the lower part of the A-horizon of mor not only consists of the organic material derived from litter, but new organic material has been added to such a degree that the properties of the humus layer are considerably changed.

In mor the organic litter material is not only enriched by dark hyphae during its decomposition, but an enrichment of other organic materials, neither included in the litter nor derived from it, also takes place (namely that of rootlets). To what degree the great quantities of roots influence the chemical composition of certain layers may be shown by the values of C/N-ratio as given by Meyer (1959). In the mull with poor root contents, the C/N-ratio gradually decreases during litter decay. In the mor it first decreases, and there rises again in the intensively rooted layers (F3 and H1). This increase might be partly traced back to the additional organic material of the roots.

The concentration of roots near the surface of the soil is important for the development of *Cenococcum graniforme*. In podsol this shallow rooting is determined by a deficiency of nutrients. In addition, in pseudogley–podsol the rooting is impeded in subsoil through unfavourable texture and low oxygen content. Though the properties of förna and upper fermentation layer (pH-value, N-content, number of bacteria) in the pseudogley–podsol plot resemble those of moder or partly even those of mull, mor formation takes place, and the humus layers of pseudogley–podsol and podsol exhibit the same feature by the dominance of hyphae of *Cenococcum graniforme*.

The fungus *Cenococcum graniforme* seems to be widely distributed. It was found by various authors in humus forms characterized by periodic drought or nutrient deficiency. Thus Mikola and Laiho (1962) report that *Cenococcum* is a typical mycorrhizal fungus in Finnish spruce forests which especially occurs in the surface layers. Göbl (cited in Schiechtl and Stern, 1963) found in samples taken from *Abies cilicica*-dry forests of Taurus predominantly black mycorrhizae, which resemble those that are formed by *Cenococcum graniforme*. According to Wright (1963) *Cenococcum graniforme* is favoured by the low moisture content of soil in central Oregon.

SUMMARY

The mycelia of *Cenococcum graniforme* play an important role both quantitatively and qualitatively in mor, and exhibit a significant micromorphological feature in the humus layer of mor. The intensive development of *Cenococcum graniforme* in mor is promoted a.o. by periodic drought which is caused by the concentration of roots in the upper layers of podsol. The accumulation of roots in the A-horizon of mor is again a consequence of the deficiency of nutrients or oxygen in the subsoil.

REFERENCES

Brian, P.W., 1949. Studies on the biological activity of griseofulvin. Ann. Botany (London), 13 : 59–78.

Grosskopf, W., 1928. Wie verändern sich stofflich und morphologisch die Fichtennadeln bei der Bildung von Auflagehumus in geschlossenen Fichtenreinbeständen. Tharandt Forstl. Jahrb., 79 : 343–362.

Handley, W.R.C., 1954. Mull and mor formation in relation to forest soils. Forestry Comm. Bull., 23 : 1–115.

Hartmann, F., 1961. Grundsätzliches zum Problem der Waldernährung. Allgem. Forstz. 72 : 1–4.

Hesselman, H., 1925. Studier över barrskogens humustäcke, dess egenskaper och beroende av skogsvarden. Statens Medd. Skogsforskningsanstaltet Stockholm, 22 : 169–552.

Kern, K.G., Moll, W. und Braun, H.J., 1961. Wurzeluntersuchungen in Rein- und Mischbeständen des Hochschwarzwaldes (Vfl. Todtmoos 2/I–IV). Allgem. Forst. Jagdz., 132 : 241–260.

Kramer, P.J., 1956. Roots as absorbing organs. In: W. Ruhland (Editor), Encyclopedia of Plant Physiology. Springer, Berlin–Göttingen–Heidelberg, 3 : 188–214.

Kubiena, W.L., 1953. Bestimmungsbuch und Systematik der Böden Europas. Enke, Stuttgart, 392 pp.

Laatsch, W., 1957. Dynamik der mitteleuropäischen Mineralböden. Steinkopff, Dresden–Leipzig, 280 pp.

Levisohn, I. and Parry, M.E., 1960. Researches in mycorrhiza and soil mycology. Rept. Forest Res., Forestry Comm. London, 1959 : 91–94.

Lihnell, D., 1942. *Cenococcum graniforme* als Mykorrhizabildner von Waldbäumen. Symbolae Botan. Upsalienses, 5 (2) : 1–19.

Martin, J.P., Ervin, J.O. and Skepherd, R.A., 1959. Decomposition and aggregating effect of fungus cell material in soil. Soil Sci. Soc. Am. Proc., 23 : 217–220.

Meyer, F.H., 1959. Untersuchungen über die Aktivität der Mikroorganismen in Mull, Moder und Rohhumus. Arch. Mikrobiol., 33 : 149–169.

Meyer, F.H., 1962. Die Buchen- und Fichtenmykorrhiza in verschiedenen Bodentypen, ihre Beeinflüssung durch Mineraldünger sowie für die Mykorrhizabildung wichtige Faktoren. Mitt. Bundesforschungsanstalt Forst- Holzwirtsch., 54 : 1–73.

Mikola, P. and Laiho, O., 1962. Mycorrhizal relations in the raw humus layer of northern spruce forests. Commun. Inst. Forest. Fenniae, 55, 18 : 1–13.

Müller, P.E., 1887. Studien über die natürlichen Humusformen und deren Einwirkung auf Vegetation und Boden. Springer, Berlin–Göttingen–Heidelberg, 324 pp.

Schiechtl, H.M. and Stern, R., 1963. Studien über die Entwaldung im Kilikischen Ala Dag (mittlerer Taurus in Kleinasien). Ber. Naturw. Med. Ver. Innsbruck, 53 : 173–192.

Trappe, J.M., 1962. Fungus associates of ectotrophic mycorrhizae. Botan. Rev., 28 : 538–606.

Witkamp, M., 1960. Seasonal fluctuations of the fungus flora in mull and mor of an oak forest. Mededel. Inst. Toegepast Biol. Onderz. Natuur, 46 : 1–51.

Wittich, W., 1943. Untersuchungen über den Verlauf der Streuzersetzung auf einem Boden mit Mullzustand, II. Forstarchiv, 19 : 1–18.

Wittich, W., 1952. Der heutige Stand unseres Wissens vom Humus und neue Wege zur Lösung des Rohhumusproblems im Walde. 2. Aufl. Schriftenreihe Forstl. Fak. Univ. Göttingen, 4 : 106 pp.

Worley, J.F. and Hacskaylo, E., 1959. The effect of available soil moisture on the mycorrhizal association of Virginia pine. Forest Sci., 5 : 267–268.

Wright, E., 1963. Ectotrophic mycorrhizae on pine seedlings in Oregon. Ecology, 44 : 173–175.

COLOURED ILLUSTRATIONS

Plate A4. Thin section of humus layer (H2) of pseudogley-podsol. The humus layer contains many hyphae and fragments of hyphae. Some of the fragments appear blurred. In the centre of the figure the remains of a *Cenococcum graniforme* fungal sheath is visible; x 300.

Plate A5. Humus layer of mor. *Cenococcum graniforme*-mycorrhiza, the decay of which is in progress. The root tissue has been decomposed and the fungal sheath has not yet been disintegrated; x 300.

Plate A6. *Cenococcum graniforme*-mycorrhizae in the Bh-horizon of podsol. Crossed nicols.

MICRO-MILIEU STUDIES IN THE A-HORIZON OF A MULL-LIKE RENDSINA

I. SZABO, M. MARTON and G. PARTAI

Institute of Soil Research and Agrochemistry, Hungarian Academy of Sciences, Budapest (Hungary)

INTRODUCTION

In order to gain an insight into the special microbiology of soil types and their varieties, investigations should particularly be developed in two directions. Firstly, direct observation and study is necessary of those micro-organisms which perform conspicuous and characteristic transformations in the pore-space system of soils, and which play a role, along with the fauna, in the formation of characteristic humus forms (Kubiena, 1953, 1962). Secondly, the taxonomical determination of the isolated microbes is indispensable according to modern differential diagnostic criteria, since only the latter renders the real comparison of microbe-associations of different soil types and humus forms possible.

In this paper we intend to provide some data concerning the first type of investigation by presenting the results obtained in the course of micro-milieu studies in the humus-horizon of a mull-like forest-rendsina. Our aim, first of all, has been to establish the most important types of substrate colonization of soil micro-organisms and, in this connection, to clarify some terminological concepts which in our opinion are essential for the exact microscopic-biological description of soils. Moreover, we have endeavoured to gain some information about the factors which cause the extremely heterogeneous spread of microbes in even the smallest micro-localities of soils. Finally, by also basing our work on our earlier investigations (Szabó et al., 1964) of the micro-fauna of this soil, we have tried to answer a question much discussed of late (Volarovich and Tropin, 1963; etc.) as to whether electron-microscopic examinations are able to demonstrate organisms (among them animals of special structure) whose existence is still unknown.

METHODS

1. Observation in the field, under natural environment, by means of a magnifier.

2. Direct examination of the soil by means of an incident light microscope.

3. Simple Rossi-Cholodny slide technique: Examination of stained and unstained slide preparations under ordinary light and under phase contrast illumination.

4. Modified Rossi-Cholodny slides with agar-film coating. The nutritive

composition of the latter: $(NH_4)_2SO_4$-agar without C-source, glucose–asparagin agar, pepton agar, etc.

5. Agar-film soil-microcultures.
6. Soil-microcultures without nutrient-agar-film.
7. Tribe's cellophane membrane method.
8. Electronmicroscopic examination. The soil suspensions were prepared by shaking in distilled water, 0.01 *N* NH_4OH, and phosphorous tungstic acid, then dripping, for different time intervals. In addition, shaking for five hours in 1% phosphorous tungstic acid and preparation of C-replicas.
9. Preparation of soil thin sections by the use of Polikon ES, an unsaturated polyester cast-resin produced in Hungary. Examination under ordinary and polarized light.
10. Comparative micromorphological studies between microbe pure-cultures isolated from soil particles and colonies of these organisms observed under natural soil conditions.

THE EXAMINED SOIL

A mull-like rendsina formed on limestone under a mixed forest crop of *Carpinus betulus, Quercus cerris, Q. pubescens* and *Q. petraea*. Horizons: L-layer, F-layer, A-, A/Ca, Ca/C, C-horizon (designation according to Kubiena, 1953). The humus form of the A-horizon is mull-like rendsina moder: small clay content, the constant presence of microscopic plant-residues of recognizable structure, mineral particles (mainly calcite) mostly closed into excrements.

COLONY-TYPES OF SOIL MICROBES

(Substrate colonization by soil bacteria, actinomycetes and fungi)

Before the microscopic biological description of the soil examined is given, we turn to the characterization of the types of microbe colonies, and their varieties, that we observed. We feel bound to emphasize that the typing presented here is of general validity, and can be applied to the description of growth- and colonization-forms of microbial associations of other soils, although the occurrence of individual types in various soils may be significantly different.

Solitary cells (Fig.6), spores, fragments of hyphal filaments. They are found on the surface or inside of the soil particles. They generally fail to develop because of the unfavourable local conditions. They are usually inactive.

Miniature colonies. Bacterial colonies consisting of 4–20 cells, or not or only little branching actinomycetal filaments 5–10 μ in length. They are evoked by the diminished possibilities for growth on the surface, as well as inside, of the soil particles.

Linear colonies. These colonies consist of a long single cell-line of bacteria, usually among plant-fibres.

Compact colonies (Fig.6). Bacterial colonies or compact skein-like masses of hyphae are usually grown in a heap (sometimes of more than 20

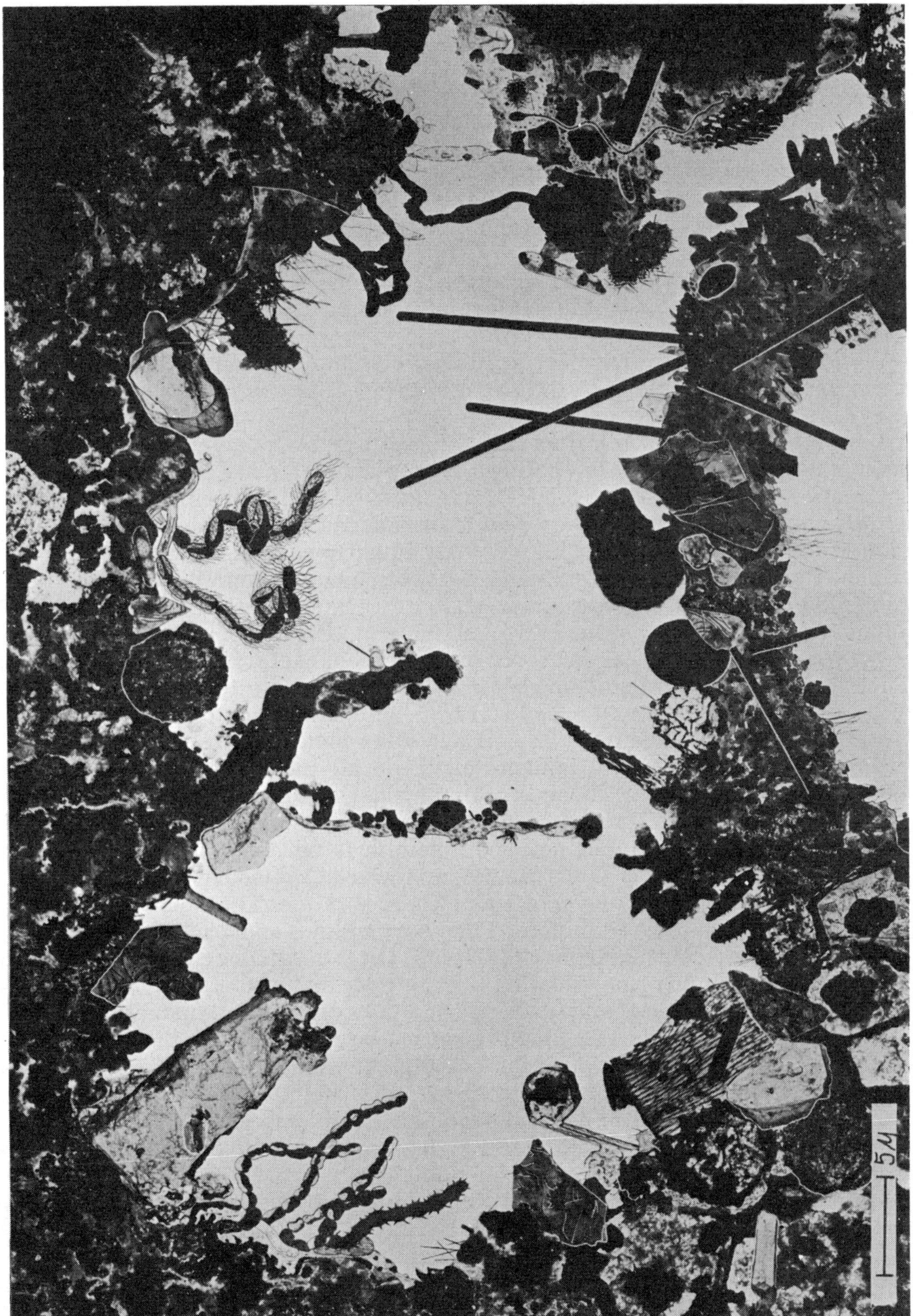

Fig.1. Micro-milieu reconstruction of mull-like rendsina moder. Based exclusively on electronmicroscopic investigations and photographs.

cells) on the surface of particles rich in nutriment.

Plate colonies (Fig.6). The colony-forming bacteria grow in two dimensions, usually on the surface of plant residues or on mineral particles.

Cover-like colonies. The soil particles, rich in nutriment, are completely overgrown.

Bridge-formation (Fig.2). The connection of soil particles by elastic "bridges" which are formed by tightly sticking cells. They are usually pillar-like, arched and have a broad base at the sticking point.

Peg-formation. Bacterial colonies piled up on organic residues especially rich in certain nutrients form "pegs", dripstone-like growths protruding into the pore spaces.

Procession-like growth. Elongated, ribbon-like, two-dimensional bacterial colonies usually on the surface of plant particles. They have an irregular run.

Diffuse colonies. Diffuse, scattered colonies of numerous bacterium cells appearing to belong to the same species. The colony-form, in the first place, is that of moving bacteria.

Meniscus-formation. Bacteria, or possibly actinomycetes, may provide meniscus-like linings in the hollows if they grow in pore angles.

Mining-growth. The hyphal filaments of actinomycetes and fungi grow slowly into the interior of the organic or inorganic components of the soil. The width of these canals equals or only slightly surpasses the diameter of hyphal filaments. The bacteria and Nocardia usually form hollows and replace the consumed organic materials with cells. The mining can take the form of a long-shaped sack, or it may have multipled openings, etc.

Ring-formation. The bacteria grow in a characteristic ring-like fashion. Such growth can only be observed on Rossi-Cholodny slides.

Scouting (searching)-growth. From the compact colonies of actinomycetes and fungi developed on soil particles rich in nutriment, several extremely elongated hyphal filaments protrude far into the soil, searching for new nutriment sources. These scouting hyphal filaments penetrate through those areas of the soil which contain no or only insignificant amounts of nutriment ready to be taken up. Their growth is rendered possible by the nutriment-flow directed from the colony towards the searching hypha-tips. If the germinating actinomycete spore finds nutriment in small amounts it only develops scouting filaments. These organisms approach strange colonies also with scouting hyphal filaments. On the influence of the metabolites of the strange colony, the scouting hyphae may turn away and grow in another direction, or they may grow around the colony, or penetrate into it.

Indifferent penetration. It is observed quite frequently that actinomycetes or fungi grow through the colonies of each other or of bacteria without any detectable morphological sign of stimulation or inhibition.

"Turfing"(Fig.3). Fungi and especially actinomycetes mining inside soil particles, develop a rich turf-like aerial mycelium-cover towards pore space. In certain cases the whole surface of aggregates may be covered. It occurs that the "turf" consists of strips (micro-fields) of different sporophore types.

Network formation (Fig.2). Fungi and actinomycetes form a rich network among soil particles, practically connecting them.

Inhibited colony formation. Microbial colonies may modify the growth

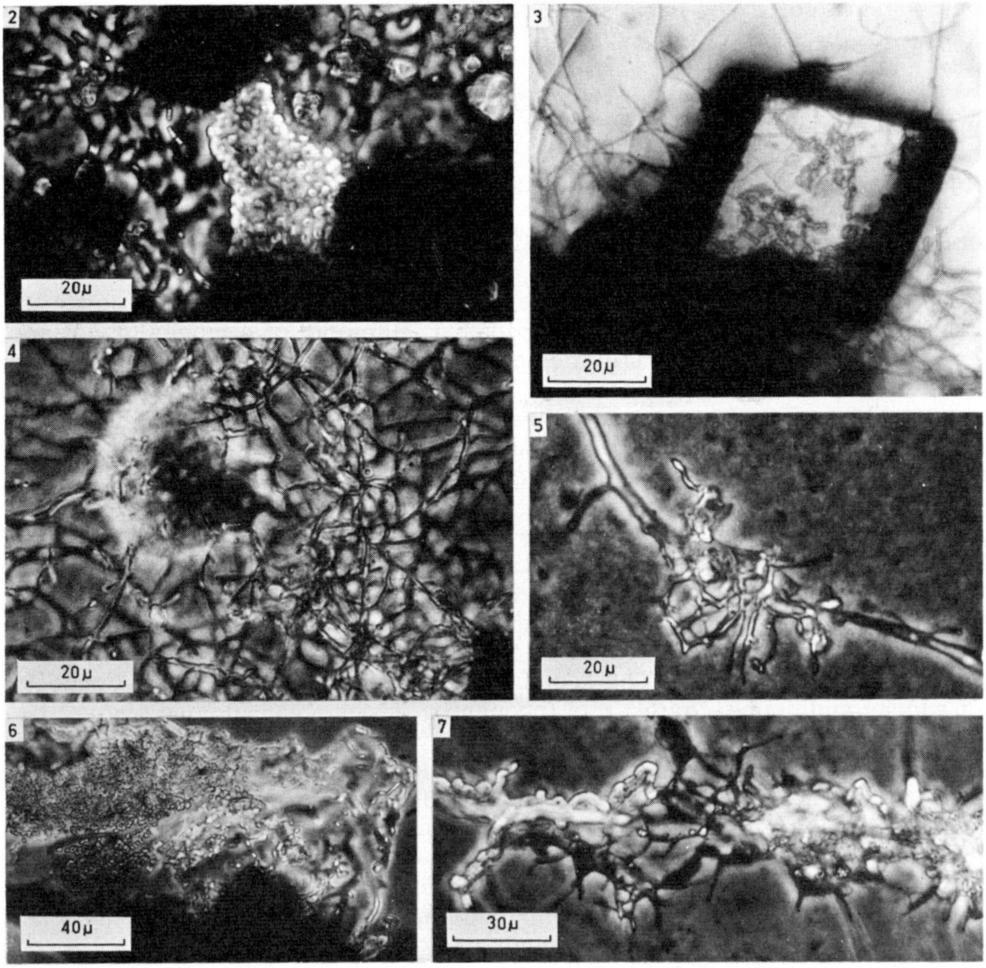

Fig.2–7. The most important colony-types of soil microbes in the humus horizon of a mull-like rendsina.

Fig.2. Bridge-formation and actinomycete-network.

Fig.3. Turfing of actinomycetes.

Fig.4. Devoured (enclosed) colony.

Fig.5. Hanging *Streptomyces*-colony on *Penicillium* scouting hyphal filament.

Fig.6. Solitary cells, compact and plate colonies.

Fig.7. Succession of a colony: *Penicillium* hypha-filament→*Streptomyces* sp.→mucous coccus-like cells.

conditions of adjacent colonies. The sensitive organism may get around the antagonistic colony by procession-like growth, leaving an empty inhibition halo over. In other cases asymmetric colonies appear due to mutual inhibition.

"Devoured" (enclosed) colonies (Fig.4). In the compact colony of a soil microbe one or more, compact or loose, colonies of other microbe species may be found enclosed. Such "devoured" colonies may be formed when the former actually grows around it, and in this case the "devoured" colony is usually destroyed. If the "devoured" colony later begins to grow inside a big colony of e.g., actinomycetes, and lives off it as a parasite, it destroys its "host" from within.

Hanging colonies. (Fig.5). They are usually the result of parasitic growth conditions. On the "scouting hyphal filament" of a fungus or actinomycete another actinomycete or bacterium begins to grow, forming a compact focus. If this compact focus consists of bacteria then seen from above, it resembles a bead strung on a thread. The hanging colonies are mostly actinomycetes occurring on filaments of mould fungi. It should be noted that both the hanging colony and scouting hypha are located at the surface of soil particles. Essentially "true hanging colonies", formed when the hyphal filament connects two soil particles through the air space of the pore and the hanging colony is in the air space itself, are extremely rare. Our opinion in the cases of hanging growth that we deal with is that this form of parasitism occurs when the population of the colony consumes the nutriment of the hypha for a prolonged period instead of rapidly destroying it.

Inactive 'colonies'.The accumulation of vegetative cells, and especially of spores and spore-chains, can frequently be observed in pore angles, fissures of mineral particles, recesses of crystals, etc. These "colonies" come into existence on the influence of fine mechanical effects (micro-erosions) and are not metabiotic microbe associations.

Hyphospheres. The hyphal filaments of actinomycetes and fungi may be accompanied by other micro-organisms, usually actinomycetes and bacteria, which are tightly grown upon them. In the majority of cases there is metabiosis and stimulation which, depending upon the environmental conditions, may turn into parasitism or antagonism.

Rhizospheres. This should be regarded as a special milieu in which micro-organisms may develop all the forms of growth described above. Thus they may connect the root-surface with soil particles by "bridge-formation". They may develop "plate" and "cover-like" colonies on the surface of roots, and may form compact colonies or grow diffusely. The most important property of microbial colonies, regarded as characteristic rhizosphere-inhabitants, appears to be the fact that, at least at one point (the "basis" of the bridge, etc.), they are in direct contact with either the capillary roots or with the cells of root surface.

Floating and drifting colonies. In the presence of high water content floating and drifting colonies appear in the L- and F-layer. These colonies consist of bacteria and may take the shape of chains, packets, bunches, etc. They are usually accompanied by solitary floating and drifting bacteria.

In the aforementioned cases the colonies of soil microbes were grouped according to their external appearance (morphology). However, in the biological microscopic description of soils the morphological features of focus-forming cells and the changes occurring in time should also be taken into account.

Homogeneous and heterogeneous colonies. A colony may be regarded as "homogeneous" if, as shown by direct microscopic examination, the cells represent the same morphotype or differ in external form but have a common origin proved by continuous observation. The heterogeneous colonies may be built up from several species. In distinct heterogeneous colonies the populations of the different species are separated from each other sector-wise within the focus. In mixed heterogeneous colonies the different species are found in a mixed mass.

The already described "devoured colony", as well as the *covered heterogeneous colony* and *basket formation* appear to be special cases of distinct heterogeneous colonies. The colony of a soil microbe, stuck on a soil particle and then covered on the outside by one or more cell-layers of another microbe, is referred to as a covered heterogeneous colony. In case of basket formation, the compact bacterial colony is surrounded by loosely-woven but tightly sticking actinomycete mycelia in the form of a basket. This important colony-type is usually the result of strictly metabiotic processes.

Simple development of colonies. The local growth of cells, the formation of the colony and then the exhaustion of the nutriment source is followed by the appearance of resting reproductive bodies (spores, sclerotia, oidiospores, chlamidospores, etc.) and by lytic processes (autolytic reactions, bacteriolytic processes, sensitivity to phages, etc.).

The succession of colonies (Fig.7). From its formation until its destruction the microbial colony may exhibit the characteristic population change, the succession of different morphotypes. In the simplest case the colony consists of one single species and the observed different morphotypes belong to a coherent line of development. Thus, for instance, the mycelium-filaments of Nocardias may be transformed into a mass of cocci, etc. In a more complicated case, different, although cyclic, developments accompanied by morphological changes not exhibiting species follow each other in the colony. Finally, those cases, in which several species, developing in a special morphocycle, follow each other during the span of live of the focus, can hardly be analysed by means of direct microscopic examination. The difficulties are even enhanced by the appearance of misshapen formations, and by the frequent presence of teratological forms, etc., due to the interaction of species. As a rule, the species emerging in a later period destroy the aged earlier population and even use them as a nutritive source. It may occur that a colony will consist of the population of the active cells of one species, and of numerous spores of another species. According to our observations, in the course of the decomposition of organic materials of the soil, the changing order of microbe associations is not only determined by the order of exhaustion of the qualitatively different organic material-fractions of the substrate being broken down. In many cases the following sequence is characteristic of the decomposition process: Fraction I of the soil-substrate→micro-organism-A; micro-organism-A as substrate→ micro-organism-B; micro-organism-B as substrate→micro-organism C, etc. Fraction II of the soil substrate→micro-organism G; micro-organism G as substrate→micro-organism-H, etc. Our observations suggest that the order of species feeding on the biomass of the micro-organism which split the soil-substrate itself may frequently be the same. According to this concept, micro-organism-A is not identical with -G, but -B is already identical with -H, etc.

THE MICROSCOPIC BIOLOGICAL DESCRIPTION OF THE HUMUS HORIZON

The constituents of the horizon

The excrements (coprogenic aggregates) are usually large (>1 mm) as shown by thin section examination, and inside them the mineral particles and dark-coloured humic matters form a mosaic-like structure. This is mainly due to the intensive earth-worm action (without mull-formation!). A smaller amount of excrements comes from isopodes, glomerides, julides, etc. An even smalle part of the coprogenic aggregates is of homogeneous structure and contains exclusively dark, humificating organic material (excrements of plant-eating mites, etc.). In the main portion of A-horizon the not-fresh excrements are crumbled, eroded and fallen into pieces. The most important source of the organic material-accumulation taking place here is the litter of tanbark oak and hornbeam. This is consumed by *Bibio* sp. (Diptera) larvae living in the L-layer in a large mass, the excrements of which form the main bulk of the F-layer. The latter is driven through the intestinal canal of earthworms whicl increase the amount of A-horizon in this manner. According to light- and electron-microscopic morphological analysis of the material of coprogenic aggregates and the humus horizon, the components can be divided into five groups:

(1) *Amorphous coprogenic humus* exhibits no structure, even at 40,000 fold magnification.

(2) *Microstructures of vegetable origin*: fragments consist of a few cells derived from tissue-portions of very different origins. Inside the intact cells the plasm is usually shrunken and dark, and it is on the way to humification, probably due to its own polyphenoloxidases. In addition, there are pollen grains, especially smooth pollen grains (*Quercus*), fragments of various hair-formations and thus of bifurcate piles (pilus ramosus; *Quercus*), membrane residues of winged fruit (samara; *Acer*), various vegetable stabilizing shells prepared by microbial decomposition, etc.

(3) *Microstructures of animal origin*: These consist of fine debris of tarsi, tergite-, sternite- and pleura-laminas of arthropods, chitin-pile formations, corneal portion of insect eyes, antenna-fragments, wing-scales of Lepidoptera, wing veins and "cells" of Hymenoptera, tail-like springing organs of Collembola, etc. There are also prepared, crumbled, fantastic-shaped masticating apparatuses and abandoned shells of Rotifera, residues of corroded hair- and feather-formations of birds and mammals, crumbled spiral chitin-stabilizers of the trachea-canaliculi of insects, etc.

(4) *Microstructures of microbial origin*: These include great amounts of cell membranes of lysed bacteria, dead actinomycetes, hyphal filaments with smooth, spiny or hairy surfaces, sometimes ciliated cell-membrane portions of dead ciliates with cirruses and ciliary membranes (*membrana adoralis*), shell-like structures of Thecamoebae, broken off flagella of flagellates and bacteria, the origin of which can only be established, even

in the case of electron-microscopic investigations, if they are in connection with the cells or are in the neighbourhood of cells, etc.

(5) *Inorganic microstructures*: There are great amounts of fine mineral particles; among them, mainly in the form of primary rhombohedron, ditrigonal scalenohedron or primary hexagonal prisms, or in a combination of these, corroded or excellently preserved calcite crystals. Moreover, calcite needles, which rapidly disappear from the field of vision of the light microscope on the addition of diluted hydrochloric acid: they are usually 1 μ wide and 20–30 μ long. However, as shown by electronmicroscopic investigations, a whole range of needle-types of different sizes can be found in this soil, the width of which varies from 0.01 μ–1 μ. There are also shells and their fragments of Foraminifera set free from the limestone (parent rock) by disintegration and weathering, etc.

THE MICROFLORA

The components listed above form an almost fantastic mixture-microstructure inside the coprogenic aggregates, on their surface or in the space between them, whose details are mostly unrecognizable. Everywhere in this miraculous "structure-cemetery" micro-organisms search for possible sources of food. In the micro-localities of the humus horizon actinomycetes prevail throughout the year. These actinomycetes mine inside structured and coprogenic organic materials, forming homogeneous and (mostly) heterogeneous "turfs" and networks on their surfaces. The turfing of actinomycetes, according to our investigations, appears to be the most important and characteristic colony-type of microbes in the mull-like rendsina moder. The overwhelming majority of actinomycetes is also proved by the results of the determination of the number of different micro-organisms in the A-horizon carried out by the soil dilution plate method. The true picture, obtained by direct examination and by the simple Rossi-Cholodny slide technique, is substantially modified however, if the slides are covered with nutrient agar film, and especially if this film contains proteins (peptone, tryptone) too (cf. Garrett, 1963). In this case a great number of bacterial colonies and searching filaments of fungi, and especially the former, appear which are not characteristic of the humus horizon at all. Our estimate is that 90% of the surface of the excrements is covered, with different densities, by turfs of actinomycetes. The hyphal filaments of actinomycetes form a network primarily in the micro-fissures inside the excrements, while the space between the aggregates is interwoven in a rare occurrence by hyphae (searching filaments) of fungi (mostly *Penicillium*, *Aspergillus* and *Mucor*). Although numerous actinomycete species have been isolated and determined by us from the humus horizon of this soil (Szabó and Marton, 1964) for several years, they could nevertheless not be recognized by direct microscopic examination at their place of origin. Only in a few cases when the sporophores were extremely characteristic, as was the case with the multi-curved spiral conidiophores of *Streptomyces chartreusis* sitting on long axes, could we establish that a *Streptomyces* species well-known to us formed micro-fields in heterogeneous turfs of actinomycetes grown on the surface of aggregates. Generally, however, only the turfs of *Rectus flexibilis* sporophores and of sterile aerial mycelium-filaments could be observed, and these have very little to tell the taxonomist. By the use of other methods, however, we have been able to establish that the actinomycete populations of excrements of different origin are far from being the same. Moreover, they may be highly characteristic of the animal species. In fact, if the possibly fresh excrement of different animal species were to be distinctly examined and their microflora isolated, pronounced differences in the species-composition could be observed. For example, in the intestinal canal and in the freshly piled up excrements of the already mentioned *Bibio* sp.-larvae, in sharp contrast to the average actinomycetes flora of the soil, the permanent occurrence of only four and five species, respectively, could be detected which were, however, characteristic of these biotopes. The results of these investigations will be published in detail elsewhere.

According to our observations the young "searching hyphal filaments" of fungi go through the actinomycete turfs by "indifferent penetration". Only a few hyphospheres on these filaments appear to be composed mainly of bacteria. On the other hand, actinomycetes appear on the aged hyphal filaments of fungi, firstly only in the form of "hanging colonies" and then covering the whole length of the hyphal filaments. In a few cases we observed that the hanging colonies of actinomycetes grown on mould hyphae were attacked and destroyed by cells resembling micrococci, but finally even the latter fell victim to actinomycetes.

The sporadic occurrence of solitary cells and spores can be detected on almost every micro-object. On crystals, on minute mineral particles, "miniature-colonies" at the most can be observed, and these are presumably oligocarbophylic and oligonitrophylic organisms. On cellulose fibres few linear bacterial colonies, and mostly longitudinal actinomycetal colonies, can be found. The main characteristic of Nocardia and bacteria appears to be the presence of loose or compact colonies, which in many cases reflect the specific morphological conditions of the species. The bacterial "bridge"- and "meniscus"-formations are relatively common. The number of Azotobacters in the humus horizon is high, as shown by the results of soil dilutions carried out on N-free nutrient agar plates, although cells of this type may be found in the soil only in miniature colonies. In the fresh excrements of various arthropods colonies sometimes resembling the cells of *Bac. mycoides* could be seen. Otherwise this species is very frequent in the humus horizon. The number of coccoidal bacteria is low, and they live in miniature-colonies or in loose foci. On the other hand, the frequent occurrence of coccus-shapen actinomycetal spores may be found locally in the micro-fissures of mineral particles (inactive "colonies").

The number of "devoured" colonies is not great and these are, with few exceptions, the colonies of extremely small rod-shaped bacteria surrounded by a characteristic mucous cover and enclosed by actinomycetal mycelia. Sometimes, in a deceiving way, *Nocardia*-growth may take the form of "devoured-colonies" when the disintegration of mycelia into cocci has already begun inside the colony, while it has not yet done so on the periphery (homogeneous, morphocyclic development exhibiting colony).

In general, the number of bacteria is greater on fresh excrements than on the already "mummified" aggregates. Among bacteria the cells of *Caulobacter,* and especially of *Pseudomonas*-type, are very frequent, as shown by electronmicroscopic investigations. It could be established only by the aid of parallel comparative studies on pure cultures that a part of the bacterial-like foci occurring in the humus horizons belongs to a white *Nocardia* species forming few if any branching filaments. It should be sharply distinguished from the colonies of the morphocycle exhibiting *Nocardia paraffinae* living in the same place, although in appreciably smaller numbers.

THE SOIL WATER FAUNA

In the A-horizon, disregarding its 1–2 cm top-region, active Rotifera have never been observed, although 22 species of these organisms can be found in the L- and F-layer (Szabó et al., 1964). The minimum water requirement of these organisms far exceeds the maximum soil water content that we measured in the A-horizon in the course of the year. Essentially the same statements are true for Ciliata; 29 species of Testacea live on this soil in the L- and F-layer, but in the humus horizon they are very rare. Finally, the sporadic presence of the active forms of flagellates and amoebae in the A-horizon has also been observed; 39 flagellate and 23 amoeba species can be found in the different horizons of this soil by means of classical culture methods (among the latters mainly *Amoeba-*, *Nuclearia-*, *Vahlkampfia-*, *Naegleria-*, etc., species).

Finally we should note that mycorrhiza observations were out of the scope of our present investigations. The strong rhizospheral effect of numerous plants (on a 100 m^2 parcel nearly fifty species) was observable in the humus horizon. The morphological description of microbial cells and colonies found on the roots of different plant species will form the subject of another paper. The colonies, mainly of bacteria, living on the root fibrils (pili radicis) or more generally on the surface of roots, may be identified from a colony-morphological point of view with one or more of the types described above.

DISCUSSION

On the basis of our results it can be established that the examinations carried out by means of an electron-microscope, though having revealed a whole range of the most different and most bizarre microstructures, have failed to enlarge the list of microfauna of this soil, which had already been compiled with the aid of light microscopy. The formation of large amounts of in their origin hardly recognizable microstructure masses is due to the co-operation of several passages of soil-materials through the intestinal canal of animals and decomposition processes. These processes only lead to the formation of the mull-like rendsina moder in the soil studied by us, and the lack of clay-minerals prevents the completion of moulding, to wit the formation of true mull (see Kubiena, 1953; Altemüller, 1957; Jongerius and Pons, 1962). The coprogenic aggregates are populated by micro-organisms with characteristic colony-types. In our opinion the typifying of the colonies of soil microbes and the application of these type-designations to the microscopic biological description of the different soils will provide further possibilities and aid for an exact differentiation of soil-types. Fehér (1946) has shown by a great many glass-slide examinations that actinomycetes prevail in the sand soils of Sahara which are extremely dry and poor in nutriment, but their colonies always exhibit one special growth type: solitary, long and thin "searching filaments". In certain chernozem and meadow soils the "bridge-" and "meniscus-" formation of bacteria appears to be the characteristic colony form, etc. At this point we feel bound to recall the indispensible need for microscopic biological observations to always be supplemented by the parallel study of microbe pure cultures. Even should our examination technique develop to a high degree, the direct recognition of bacterium and actinomycete species will still only be possible in few cases. Nevertheless, colony morphological data will be of special value when its comparison with the physiological abilities of colony-forming micro-organisms will become possible.

Similarly, it also seems necessary to perform experimental population-morphological investigations with known species in sterile soil samples. Two different species may influence each others colony-formation in quite different ways depending upon the food conditions prevailing in the micro-localities, the produced metabolites, the time advantages in growth given to one or another organism and other factors. Szabó et al. (1956) have shown that a *Streptomyces* sp.-strain, which is able to produce antibiotics even in a sterile soil sample (the presence of the inhibitor substance is proved by paper chromatography), gives rise to an extremely varied colony-morphological picture when cultured together with a sensitive *Mycobacterium mucosum*-strain in a soil culture. In the immediate vicinity of the thick, compact mycelium mass of actinomycetes the *M.mucosum* failed to develop. At a certain distance it occurred only in very dense, large foci, although still in the neighbourhood of *Streptomyces*, and at even greater distances it also grew sporadically. Finally the *Mycobacterium* formed hyphospheres on solitary searching-filaments. All these phenomena are due to the fact that the cells can tolerate higher antibiotic concentrations in compact populations, while the low concentration produced by searching filaments near the hyphae has a stimulating effect, etc.

SUMMARY

Based upon the results of the microscopic biological examination of the humus-horizon of a mull-like rendsina, the occurrence of soil microbes in the mull-like rendsina moder may take the form of the following colony-morphological types: solitary cells; miniature colonies; linear colonies; compact colonies; plate colonies; cover-like colonies; bridge formation; peg formation; procession-like growth; diffuse colonies; meniscus formation; mining growth; ring formation; scouting (searching)-growth; indifferent penetration; turfing; network formation; inhibited colony formation; devoured (enclosed) colonies; hanging colonies; inactive colonies; hyphosphaera. In addition, the colonies may be of homogeneous composition. The latter form either distinct or mixed heterogeneous colonies. The special cases of heterogeneous colonies are the "covered heterogeneous colony" and "basket formation". By means of microscopic observations simple and combined development of colonies can be distinguished. The most complicated case of colony development and succession occurs when several species, each exhibiting a life-cycle connected with the morphological change of their cell-form, follow each other. The most important and characteristic colony-type of the mull-like rendsina moder seems to be the "turfing" and mining-growth of actinomycetes.

REFERENCES

Altemüller, H.J., 1957. Bodentypen aus Löss im Raume Braunschweig und ihre Veränderungen unter dem Einfluss des Ackerbaues. Diss. Univ. Bonn, Bonn, 250 pp.

Fehér, D., 1946. Der Wüstenboden als Lebensraum. Bericht der französisch-ungarischen Sahara-Expedition im Jahre 1936. Röttig-Romwalter, Sopron, 128 pp.

Garrett, S.D., 1963. Soil fungi and soil fertility. Pergamon, London, 165 pp.

Jongerius, A. und Pons, L.J., 1962. Einige mikromorphologische Bemerkungen über den Vererdungsvorgang im niederländischen Moor. Pflanzenernähr., Düng., Bodenk., 97 : 243–255.

Kubiena, W.L., 1953. Bestimmungsbuch und Systematik der Böden Europas, Enke, Stuttgart, 392 pp.

Kubiena, W.L., 1962. Wesen, Ziele und Anwendungsgebiete der mikromorphologischen Bodenforschung. Pflanzenernähr., Düng., Bodenk., 97 : 193–205.

Szabó, I., Oroszlán, I. und Marton, M., 1956. Die Ursachen einer herdartigen Koloniebildung der Bodenmikroben. Agrokem Talajtan, 5 : 183–192.

Szabó, I. und Marton, M., 1964. Zur Frage der spezifischen Bodenmikrofloren. Ein Versuch zur systematischen Bestimmung der Strahlenpilzflora einer mullartigen (Wald-) Rendsina. Zentr. Bakteriol., Parasitenk., Abt. II, in press.

Szabó, I., Marton, M. und Varga, L., 1964. Untersuchungen über die Hitzeresistenz, Temperatur- und Feuchtigkeitsansprüche der Mikro-organismen eines mullartigen Waldrendsinabodens. Pedobiologia, in press.

Volarovich, M.P. and Troping, V.P., 1963. An electron-microscope study of peat microflora. Mikrobiologiya, 32 : 281–287.

MICROMORPHOLOGIE ET PÉDOZOOLOGIE EXPÉRIMENTALE

Contribution à l'étude sur plaques minces de grandes dimensions d'un sol artificiel structuré par les Lombricides

C. JEANSON

Centre National de la Recherche Scientifique, Paris (France)

INTRODUCTION

L'étude, dans les sols en place, de l'activité de la faune et de sa participation à la genèse et à l'évolution du sol présente de nombreuses difficultés. Celles-ci viennent de la disparition plus ou moins marquée des conditions initiales et de la variation des facteurs du milieu. Les phénomènes biologiques se superposent aux phénomènes physico-chimiques et c'est souvent un effet global qui est saisi dans le milieu naturel. Il n'est pas possible d'évaluer l'action propre de chaque groupe d'animaux terricoles.

Pour approcher ce problème d'une manière plus précise, une méthode expérimentale (Jeanson, 1960, 1963) est appliquée à l'élevage de Lombricides. Le choix de ce groupe est basé sur l'importance de sa biomasse[1], du brossage[2] des horizons qu'il réalise dans la plupart des sols des climats tempérés et de l'intérêt pour la pratique agricole que pourrait présenter la connaissance de la structure qu'il détermine dans un milieu donné.

L'étude micromorphologique des coupes longitudinales dans des élevages de Lombricides permet de comparer le milieu initial au milieu final et d'évaluer l'action d'une espèce sur un milieu artificiel soumis à des facteurs contrôlés.

METHODES

Élevage

Principe

Un milieu artificiel dont l'humidité, la température et l'éclairement sont constants recoit une espèce déterminée de Lombricides pendant un certain temps.

---

[1] 120 individus ou 60 g/m^2 sur une profondeur de 20 cm soit 600 kg/ha dans une chênaie (Bornebusch, 1950).

[2] Les galeries de *L. terrestris* Lin. vont jusqu'à 3 m de profondeur. Les rejets de surface peuvent atteindre 62 tonnes/ha/an pour les sols agricoles ($\sim \frac{1}{2}$ cm) (Evans, 1948).

Dispositif expérimental

Des tubes de verre de 30 cm de hauteur et de 10 cm de diamètre maintiennent le milieu d'élevage sur un plan d'eau constant de 3 cm. L'humidification se fait par ascension capillaire. L'évaporation est limitée en maintenant l'atmosphère saturée en surface par une éponge humide. Les élevages sont placés dans une pièce à température constante (18–20° C) à l'obscurité pour éviter le développement des algues sur les parois.

Milieu initial

Les matériaux utilisés pour la série d'expériences considérée sont: la terre d'un horizon d'accumulation d'un sol de limon, de la luzerne, du carbonate de calcium.

L'horizon-B a les caractéristiques physiques et chimiques suivantes (Bastisse, 1951, p.729).

TABLEAU I

Horizon d'accumulation d'un sol de limon (Versailles)

Analyse physique (‰ de terre sèche)			Analyse chimique (‰ de terre sèche)	
sable grossier	$> 200\ \mu$	8	CaO des carbonates	3,36
sable fin	200–20 μ	559	Bases échangeables CaO	4,9
limon	20– 2 μ	160	MgO	0,35
argile	$< 2\ \mu$	247	K_2O	0,13
calcaire		6	P_2O_5 assimibilable (acide citrique 2%)	0,07
matière organique		7	N_2O_5 du N total	1,89
colloides humaniques		2		
pH		7	Carbone total (Méthode d'Anne)	3,00

La terre peut être soit employée seule comme milieu d'élevage, soit mélangée à 1% de farine de foin de luzerne et à 1% de carbonate de calcium en poudre sur toute la hauteur de la colonne ou sur une moitié seulement selon les essais.

La structure du milieu initial est de deux types: soit grumeleuse avec des agrégats inférieurs à 2 mm, soit compacte sans agrégats. Elle est conditionnée par le mode de remplissage des tubes. Les agrégats terreux ou les matériaux mélangés sont versés d'un mouvement giratoire par petites pelletées pour répartir les particules d'une manière homogène directement dans les tubes pour la structure grumeleuse et dans 10 cm d'eau pour la structure compacte.

Espèces

Deux individus de taille et de poids voisins appartenant aux genres *Lumbricus terrestris* Linné et *Allolobophora icterica* Savigny, espèces fréquentes dans les sols de limon cultivés de la région de Versailles, sont élevés de 1–19 mois selon les essais. En fin d'expérience, ils sont extraits du milieu par le courant électrique pour arrêter leur activité à un moment déterminé et éviter que leur présence ne perturbe les mesures physiques et les dosages chimiques ultérieurs.

Contrôle des élevages

En cours d'expérience, à travers les parois des tubes, on peut suivre les modifications progressives de la structure et de la porosité (Jeanson, en préparation), en fonction du développement du réseau de galeries périphériques.

Étude micromorphologique

Fabrication des plaques minces

Après extraction des lombrics et dessiccation du milieu d'élevage à l'étuve, les blocs de terre sont imprégnés par une résine polyester qui les plastifie sans modification de la structure, puis découpés longitudinalement et polis selon la méthode de Jongerius et Heintzberger (1963). Les plaques minces obtenues ont 15 x 8 cm, 10–20 μ, elles couvrent les 3/4 de la section longitudinale des élevages (20 x 10 cm), la base de ceux-ci étant en contact direct avec le plan d'eau n'est jamais exploré par les lombrics et ne figure pas sur les plaques minces.

Mesure des pores

Dans le milieu final, on distingue des pores de deux catégories de taille: des pores d'un diamètre inférieur à 2 mm dûs aux fermentations microbiennes et des pores supérieurs à 4 mm représentant les sections des galeries ou des loges d'habitation des lombrics.

Les mesures des pores se font sur des négatifs obtenus par projection de l'image donnée au microscope polarisant (nicols croisés). Les clichés grossissant 4,3 x les plaques minces sont recouvertes d'une plaque de plexiglass quadrillée à maille de 5 mm. Le diamètre des pores est mesuré par le réseau et le nombre de pores de chaque catégorie compté sur chaque ligne horizontale. Cette technique de mesure, simplifiée par rapport à celle utilisée par Jongerius (1962), se base sur les principes rappelés par Hennig (1958).

RÉSULTATS

Deux cas choisis dans une série d'expériences sont décrits à titre d'exemple. Les observations sont faites au microscope polarisant et sur négatifs photographiques.

Micromorphologie réalisée par Lumbricus terrestris Lin.

Conditions expérimentales

Deux individus sont élevés pendant un mois dans une colonne d'horizon-B à structure grumeleuse.

Structure du milieu d'élevage

Aspect général. Le négatif (Fig.1) de la plaque mince montre un réseau important de galeries (sections transversales et longitudinales) réparti à tous les niveaux de l'élevage avec une légère concentration à la partie inférieure. La présence de galeries plus nombreuses en profondeur indique une humidité supérieure plus favorable aux animaux. Dans cette zone la structure devient progressivement compacte par suite du tassement.

Distribution dimensionnelle des pores. La répartition des pores est donnée dans le Tableau II. Les pores inférieurs à 2 mm sont dûs à la structure initiale du milieu, les autres à l'action de *L. terrestris* Lin. La classification dimensionnelle du diamètre des pores donnée aux Tableaux II et III est déduite de celle des pores sur Fig.1–3 (5, 10, 20, 40 mm) compte tenu du grandissement 2,2.

Structure caractéristique

Les parois des galeries sont formées d'un manchon à grains fins contrastant nettement avec celui de la terre environnante. Ce manchon est dû au tri

---

Fig.1. Structure réalisée par *Lumbricus terrestris* Lin. (en un mois à partir d'un milieu initial à structure grumeleuse); x 2,2. (1) Section longitudinale d'une galerie. (2) Section transversale. (3) Paroi: manchon à grains fins. (4) Zone de tassement entre deux galeries. (5) Micropores < 2 mm dus à la structure initiale du milieu. (6) Cassure due au démoulage. (7) Arrachements étoiles dus au polissage. (8) Base de l'élevage, structure compacte due au tassement.

Fig.2. Structure réalisée par *Allolobophora icterica* Sav. (en dix-neuf mois à partir d'un milieu initial à structure compacte enrichi en luzerne); x 2,2. (1) Section transversale de galerie, paroi avec manchon à grains fins. (2) Galerie avec une paroi à deux zones concentriques (3 et 4). (3) Pellicule de matières humifiées. (4) Stries concentriques d'oxydes ferriques. (5) Loge d'habitation. (6) Déjections. (7) Micropores dus aux fermentations microbiennes.

Fig.3. Structure réalisée par *Allolobophora icterica* Sav. (en dix-neuf mois à partir d'un milieu initial à structure compacte enrichi en luzerne et en carbonate de calcium dans sa moitié supérieure); x 2,2. (1) Pellicule de matières humifiées. (2) Loge d'habitation avec une paroi à 3 zones (1, 3, 4). (3) Carbonate de calcium. (4) Stries concentriques d'oxydes ferriques. (5) Micropores.

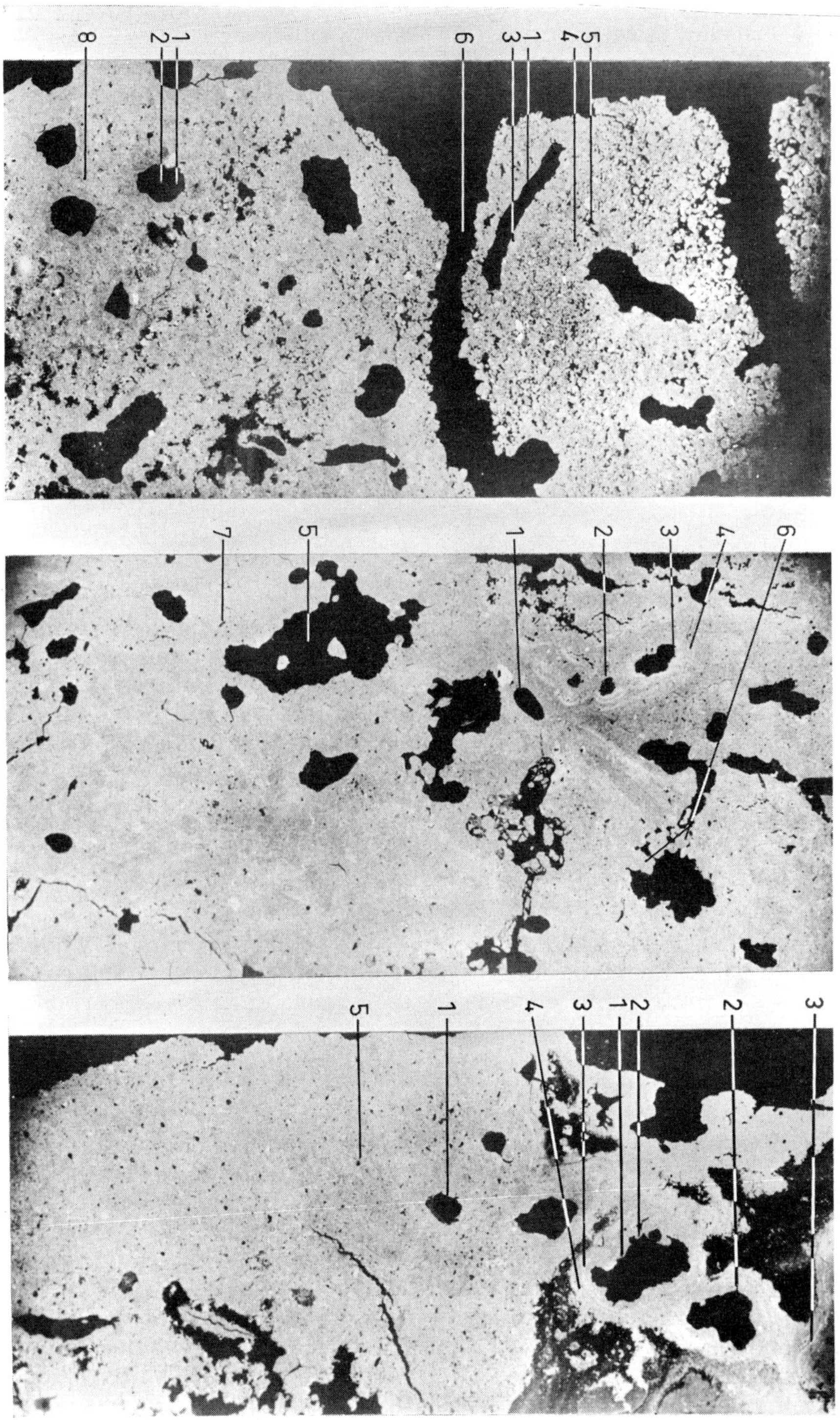

TABLEAU II

Porosité d'un élevage de *L. terrestris* dans un milieu à structure grumeleuse[1]

Taille des pores (mm)	Porosité de chaque classe de pores (%)	
	moitié supérieure	moitié inférieure
<1,1	16,2	9,8
1,1 –2,3	0,9	10,2
2,3 –4,65	6,2	5,8
4,65–9,3	5,6	6,3
>9,3	0	0
Total	28,9	32,1
Porosité total	61,0	

[1] cf. Fig.1.

effectué par le lombric et au tassement résultant de son passage.

Les lombrics tapissent, en effet, leurs galeries de produits issus de leur tube digestif contenant des particules triées. Cette observation est confirmée par les résultats de l'analyse physique de la terre des parois. On note l'absence de déjections dans les galeries, *L. terrestris* Lin. est une espèce rejetant ses terricules de préférence en surface.

Micromorphologie réalisée par Allolobophora icterica Sav.

Conditions expérimentales

Deux vers sont élevés pendant dix-neuf mois dans une colonne d'horizon-B enrichi de 1% de farine de foin de luzerne, soit sur toute sa hauteur (Fig.2), soit dans sa moitié supérieure de 1% de luzerne et de 1% de carbonate de calcium pulvérisé (Fig.3).

Structure des milieux d'élevage

Aspect général. L'activité des lombrics est localisée dans les zones contenant de la luzerne: les galeries (diamètre 4–6 mm) et les loges d'habitation (diamètre 8–10 mm) sont dans la moitié supérieure de l'élevage (Fig.3) ou sur toute la hauteur de l'élevage (Fig.2).

Les déjections sont surtout présentées dans les loges d'habitation. Ceci confirme le comportement d'*Allolobophora*, espèce déposant ses terricules en grandes parties dans le sol. Les pores d'un diamètre inférieur à 2 mm sont dûs aux fermentations microbiennes favorisées par la présence de luzerne, matière organique facilement décomposable. Cette structure est dite en "mie de pain" (Hénin, 1960).

Distribution dimensionnelle des pores. La taille des pores et leur répartition selon la profondeur sont données par le Tableau III.

TABLEAU III

Porosité des élevages d'*A. icterica* dans des milieux à structure compacte

Taille des pores (mm)	Porosité de chaque classe (%)			
	matière organique sur toute sa hauteur[1]		matière organique sur la moitié supérieure[2]	
	moitié supérieure	moitié inférieure	moitié supérieure	moitié inférieure
<1,1	0,88	0,59	0,5	4,2
1,1 –2,3	1,79	2,39	2,2	0,4
2,3 –4,65	2,10	3,70	6,4	1,8
4,65–9,3	2,12	2,16	12,0	0
>9,3	1,30	0	0	0
total	8,19	8,84	21,1	6,4
Porosité totale	17,03		27,5	

[1] cf. Fig.2.
[2] cf. Fig.3.

Structures caractéristiques

Au manchon de fines particules tapissant les parois des galeries et les loges d'habitation s'ajoutent des zones concentriques de matières humiques, de carbonate de calcium et d'oxydes de fer.

Deux zones concentriques. Dans les élevages contenant 1% de luzerne une pellicule noirâtre de un millimètre se forme sur la face interne du manchon à grains fins. Elle est formée de matière humique noire et de terre initiale, sa teneur en carbone est de 35% en valeur relative supérieure à celle de la terre avoisinante. Cette zone est discontinue et partiellement visible sur les clichés. Elle est formée de débris végétaux digérés par le lombric rejetés avec les déjections (Fig.2).

Une zone de couleur rouille de 4 mm pui succède. Elle est formée de stries successives d'oxydes ferriques qui peuvent commencer dans le manchon à grains fins et s'étendre au-delà. En effet, la matière organique en fermentation provoque dans la terre en cours d'élevage une solubilisation du fer et l'apparition de taches bleu-verdâtres dûes au fer ferreux caractéristique du "gley". Il précipite en fer ferrique et s'oxyde au contact de l'air des galeries en stries successives.

Trois zones concentriques. En présence de carbonate de calcium et de luzerne, une zone d'un millimètre s'intercale entre la zone noire et la zone rouille. Elle est formée d'un précipité continu de carbonate de calcium, qui peut avoir deux origines: soit produit par le métabolisme du ver, soit par une précipitation du bicarbonate de calcium (le CO_2 des fermentations solubilisant le CO_3Ca du milieu initial) au contact de l'air (Fig.3).

L'origine de cette formation est en cours.

CONCLUSION

L'étude micromorphologique des plaques minces faites dans un sol artificiel où sont élevés des lombrics permet de préciser les conditions de la formatio[n] et la structure des parois des galeries en relation avec la présence de matière organique. La variété des conditions expérimentales conduit malgré tout à une homogénéité des résultats.

Quelle que soit l'espèce élevée et le milieu, un manchon à grains fins suivi d'une zone de tassement entoure chaque galerie. Par contre, les matériaux du milieu d'élevage conditionne la nature des précipitations chimiques groupées en zones concentriques autour du manchon.

La double origine biologique et physico-chimique des parois de galeries de deux espèces de Lombricides, mise en évidence expérimentalement, permet d'interpréter les formations analogues dans le milieu naturel.

SUMMARY

Micromorphology and experimental soil zoology: a contribution to the investigation by means of mammoth-sized thin sections of artificial soil structurated by earth-worms

This paper deals with the micromorphological investigations of two earthworm culture experiments in a texture-B material of loess loam.

Experiment I. *Lumbricus terrestris* Lin. reared during one month in an anaerobic environment without organic material. The initial soil structure was characterized by aggregates <2 mm.

Experiment II. *Allolobophora icteria* Sav. reared during nineteen months in an anaerobic environment with organic matter. Soil structure was destroyed at the beginning of the experiment as a result of the mixing of components (soil, lucerne and $CaCO_3$).

At the end of the experiments, it appeared that: (1) The distribution of the network of earthworm tunnels in the cultures varied with the position of the organic material. This network determined the macroporosity, whereas the microbiological fermentation caused the microporosity. (2) The tunnels were covered by a continuous layer of fine particles. The layer had a thickness of about one-tenth of the tunnel diameter and was formed by the earthworms from selected particles, which were plastered to the tunnel wall. (3) In the culture experiment containing 1% lucerne and 1% $CaCO_3$, the mentione[d] layer was surrounded by three concentric zones consisting of humic materia[l] $CaCO_3$ and iron-oxides respectively. The humic zone had a biological origin, the other two were caused by physico-chemical factors.

BIBLIOGRAPHIE

Bastisse, E.M., 1951. Dix-huites années d'études lysimétriques appliquées à l'agronomie. Ann. Agron., 2 : 727–781.

Bornebusch, C.H., 1950. Soil fauna and its importance in soil type formation. Trans. Int. Congr. Soil Sci., 4th, Amsterdam, 1950, 1 : 173–184.

Evans, A.C., 1948. Studies on relationships between earthworms and soil fertility. Ann. Appl. Biol., 35 : 1–13.

Henin, S., 1960. Le profil cultural. Soc. Éd. Ingr. Agr., Paris, 230 pp.

Hennig, A., 1958. Kritische Betrachtungen zur Volumen- und Oberflächenmessung in der Mikroskopie. Zeiss-Werkzeitschrift,30 : 78–86.

Jeanson, C., 1960. Étude expérimentale de l'action de Lumbricus herceulus (Savigny) (Oligochète lumbricide) sur la stabilité structurale des terres. Compt. Rend., 250 : 3041–3043.

Jeanson, C., 1961. Sur une méthode d'étude du comportement de la faune du sol et de sa contribution à la pédogénèse. Compt. Rend. 253 : 2571–2573.

Jeanson, C., 1963. Perspective de la pédozoologie expérimentale. In: J.A. Moore (Editor), Proc. Intern. Congr. Zool. 16, Wash., D.C., 2 : 311.

Jeanson, C., en préparation. Essai de pédozoologie expérimentale: Morphologie d'un sol artificiel structuré par les Lombricides.

Jongerius, A., 1962. Optic volumetric measurements on some humus-forms. In: J. v.d. Drift and J. Doeksen (Editors), Soil organisms. North Holland, Amsterdam, pp.137–148.

Jongerius, A. and Heintzberger, G., 1963. The preparation of mammoth-sized thin sections. Soil Surv. Papers, Neth. Soil Surv. Inst., 1 : 37 pp.

WELCHE BEDEUTUNG HABEN ENCHYTRAEEN IM WALDBODEN?

G. ZACHARIAE

Schloss, Reinbek bei Hamburg (Deutschland)

EINLEITUNG

Enchytraeen sind in allen mitteleuropäischen Waldböden vorhanden. Sie gehören zu den wenigen Tierformen, die bei nährstoffarmer Streu oder ungünstigem Klima noch vorkommen. Sie bleiben aber auch in günstigen und reich besiedelten Standorten konkurrenzfähig, weil sie schon in der von Tieren wenig besuchten L-Schicht Nahrung finden und andererseits Material fressen, welches anspruchsvollere Arthropoden und Regenwürmer als Kot ausscheiden. Nur Trockenheit schadet ihnen (Nielsen, 1955a, b). Sie sind noch bei 0°C aktiv, solange das Substrat feucht ist, und können darum unter dem Schnee oder der vereisten Streuoberfläche tätig sein und sich sogar fortpflanzen. Aus gefrorenen Proben der Waldbodendecke kommen nach Auftauen oft Enchytraeen hervor. Bei Frost gehen die meisten Würmer aber in tieferen Schichten, soweit sie nicht absterben; dabei flüchten sie offenbar nicht vor dem Eiszustand, sondern vor der Austrocknung, die besonders zum Winteranfang unter der Frostschicht erfolgt. Wahrscheinlich überdauern sie Frost und sommerliche Trockenzeiten vorwiegend im Schutz der Ei-Kokons und gleichen die Verluste vermöge ihrer kurzen Entwicklungsdauer bald aus. In Gebieten mit ausgeprägtem Wechsel der Jahreszeiten erreichen sie die höchsten Individuenzahlen in Spätsommer und Herbst und dann wieder im Frühjahr.

Unsere Waldböden enthalten zahlreiche Arten mit im einzelnen verschiedener Lebensweise. Je nach dem Mikroklima suchen sie innerhalb des A-Horizontes höhere oder tiefere Schichten auf, wenn auch einige Arten bestimmte Bereiche bevorzugen (O'Connor, 1957; Springett, 1963). Zu diesem Thema können die Arten vernachlässigt werden; es genügt, die sichtbaren Spuren ihrer Tätigkeit zu unterscheiden.

AKTIVITÄTSBEREICHE DER ENCHYTRAEEN IM WALDBODEN

Bei langsamer Streuverarbeitung und entsprechend starker Auflage gibt es drei Bereiche, in denen Enchytraeen Nahrung aufnehmen und charakteristische Exkremente hinterlassen, wie Abb.1 für den Fichtenwald zeigt; für Laubstreu gilt das gleiche. Bei rascher Zersetzung sind diese Vorgänge meistens auch vorhanden, aber räumlich nicht so klar getrennt. Sie werden nach der Reihenfolge im Profil besprochen.

(a) Im Laubwald findet man bei feuchtem Wetter stets Würmer zwischen

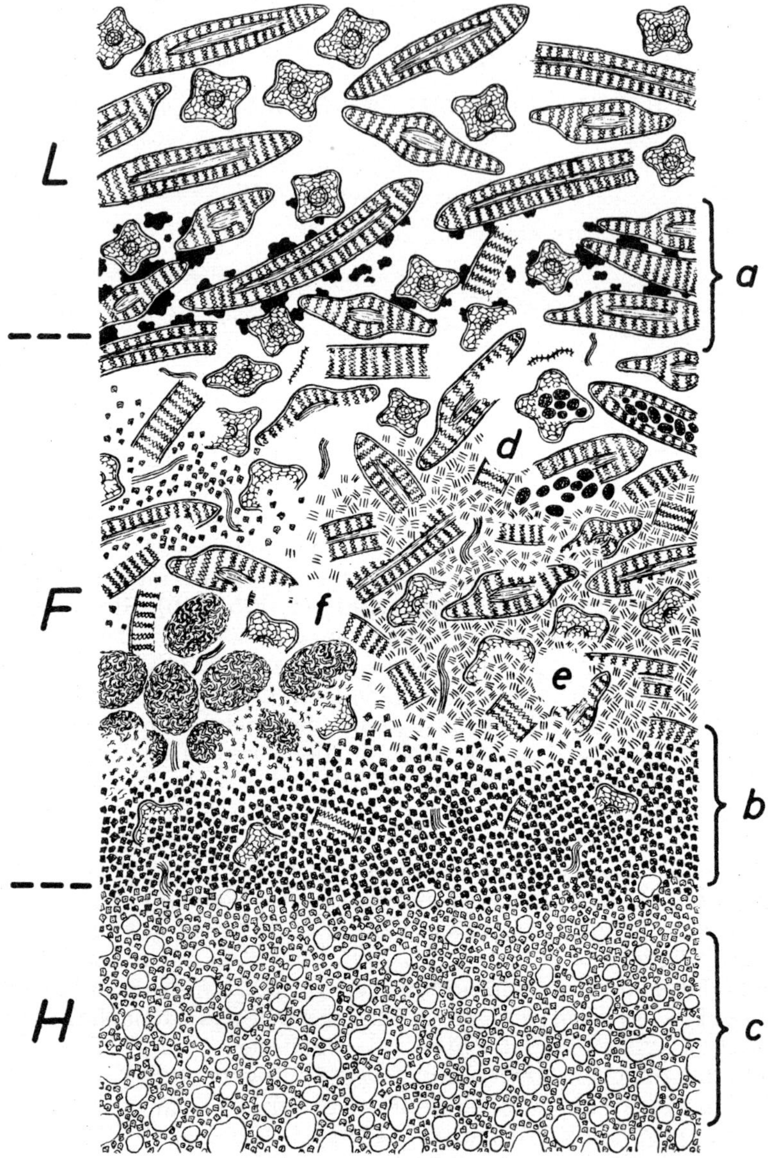

Abb.1. Beispiel eines Humusprofils, in dem alle Bereiche mit Enchytraeenspuren deutlich erkennbar sind. (Die einzelnen Bildteile sind vereinfacht und in ungleichem Massstab gezeichnet). (a) Schwarze, strukturlose Kotmassen nach Aufnahme von Mikrobenschleim und feinen organischen Resten; (b) Körnige, rein organische Losung nach Verzehren von moderndem Arthropodenkot; (c) Körnige Losung aus humoser Feinerde zwischen gröberen Sand körnern; (d) Losung von in Nadeln minierenden Oribatiden; (e) Exkremente von Lycoriidenlarven; (f) Kotballen von Tipulidenlarven. (Nach Beobachtungen und Dünnschliffen im Standort FA Stauffenburg/Harz, Holenberg Abt.52 b; Fichte auf Buntsandstein.)

den gut erhaltenen Blättern der L-Schicht. Sie fressen den Schleimfilm mit Bakterien und Protozoen, organischen un mineralischen Staub, Pollen und die Kotreste von dort lebenden Collembolen, Milben und kleinen Dipterenlarven. Leicht zersetzliche Streu beknagen sie mit Zähnen oder Schableisten im Mund auch selbst und können sie fein skelettieren; bei Rotbuche und Eiche ist dies nur selten und erst in älteren Schichten möglich. Die meist schwarze Kotmasse enthält wie die Nahrung kaum gröbere Bestandteile. Wird sie in einzelnen Krümeln abgesetzt (Abb.2A), so ist sie oft nicht sicher von Collembolenkot (Zachariae, 1962, Abb.4; 1963, Abb.1) zu unterscheiden. Zwischen dichtliegenden Blättern legen die Würmer Röhren aus Kot an, in die sie sich bei Trockenheit und Nachtfrost zurückziehen (Abb.2A). Die Regenwürmer der Gattung *Dendrobaena* bauen gleiche, nur grössere Röhren.

Im Nadelwald leben die Enchytraeen ebenso, können hier aber niemals unzerkleinerte Streu fressen. Die Kotröhren sind zwischen den Nadeln (Abb.2B) nicht so deutlich wie auf Blättern, wo sie in einer Ebene liegen. In stärkeren Auflagen entsteht oft 0,5-1,5 cm unter der Streuoberfläche eine charakteristische Schicht von schwarzen Exkrementen (Abb.1, 2B, 3). Weil die dichtliegenden Nadeln und das oft reichliche Pilzmycel auch Collembolenkot auffangen (Zachariae, 1962, Abb.5; 1963, Abb.3, 4), ist nur nach Beobachtung im Gelände abzuschätzen, welchen Anteil jede Tiergruppe an dieser Kotmasse hat. Meistens ist jener der Enchytraeen grösser.

In Laub- und Nadelwald ist diese Kotform umso deutlicher, je langsamer die Streuverarbeitung abläuft. Bei schnellerer Zersetzung werden diese Spuren bald durch grössere Arthropoden und Regenwürmer zerstört.

(b) In der F-Schicht wird die Streu gewöhnlich von verschiedenen Dipterenlarven oder Diplopoden fein zerkleinert. Deren modernder Kot liefert den Enchytraeen ein weiteres Nahrungssubstrat. Wahrscheinlich resorbieren sie nur leicht verdauliche Abbauprodukte und einen Teil der Mikroflora; jedenfalls lässt ihr Kot immer noch Strukturelemente der Pflanzengewebe erkennen. Die Exkremente sind zwar wie immer dünne, kurze Würstchen, zerbrechen aber sofort und ergeben ein körniges Pulver (Abb.1; Zachariae, 1962, Abb.6). Im Schliff allein ist dieses oft schwer von der Losung kleiner Dipterenlarven und mancher Oribatiden (Tafel A7) und dem zerfallenen Kot grösserer Arthropoden zu unterscheiden; man muss schon im Gelände auf räumliche Anordnung, Frassbild und Farbe achten.

Dieser Enchytraeenkot tritt kaum in Erscheinung, wo Regenwürmer alle Streureste der F-Schicht in ihre groben Exkremente umwandeln. Er fehlt auch, wenn die Leistung der zerkleinernden Arthropoden ausfällt. Wenn aber die Zerkleinerung gesichert ist und Regenwürmer selten sind oder fehlen, wird das feine Material oft restlos von Enchytraeen verarbeitet (Abb.1). Dieser Fall ist unter Fichte häufig, im Laubwald seltener. An Standorten, wo Nährstoffarmut oder ungünstiges Klima die mikrobielle Zersetzung verzögern, kann das Kotpulver der Enchytraeen eine deutliche Schicht bilden (Tafel A7). Es liefert dort das Material für die oft mehrere cm mächtige Auflage über dem Mineralboden. In dieser meistens sehr sauren und von Pilzen beherrschten Schicht sind die Würmer fast nie mehr tätig. Die Teilchen, aus denen sie im Schliff zusammengesetzt erscheint, sind wahrscheinlich erhaltene Gewebereste in der Bissgrösse der zerkleinernden Arthropoden. Es wäre also nicht richtig zu sagen, solche

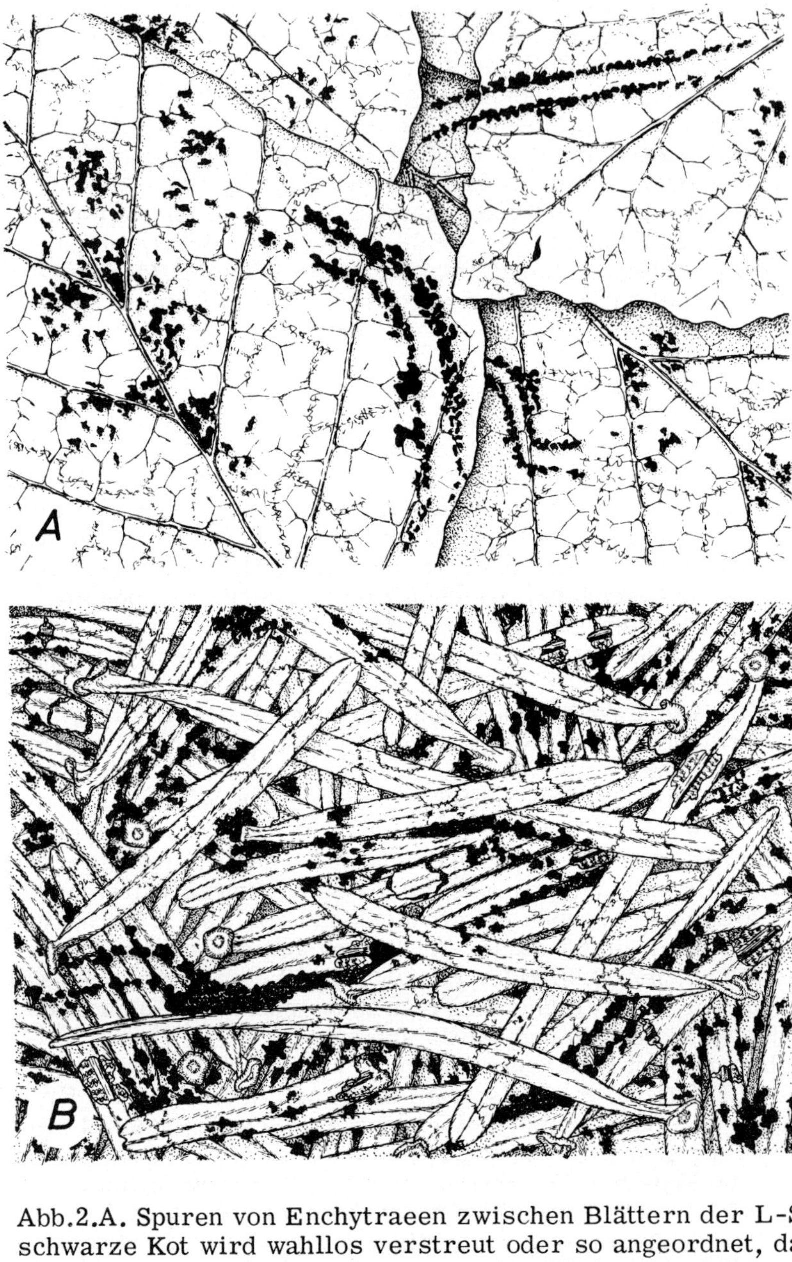

Abb.2.A. Spuren von Enchytraeen zwischen Blättern der L-Schicht. Der schwarze Kot wird wahllos verstreut oder so angeordnet, dass er mit dem oberen und unteren Blatt zusammen eine Wohnröhre bildet; x 2,7. (FA Buchenau/Bayer. Wald, Im Gränk Abt.V/1 a; Buche auf Granit; 19-8-62.)
B. Schwarzer Enchytraeenkot zwischen Nadeln der unteren L-Schicht. Zwischen den dicht, aber unregelmässig liegenden Nadeln sind Wohnröhren nur selten erkennbar. Solche Kotmassen sind in Tannenstreu oft viel reichlicher als unter anderen Coniferen. Der von Blattläusen erzeugte "Honigtau", in geringerem Masse auch das engere C/N-Verhältnis der Streu, fördern

wahrscheinlich die Mikroflora, welche den Würmern zur Nahrung dient; x 2,8. (Fbz. Bonndorf/Schwarzwald, Gdw. Ebnet, Eichholz Abt. II/1; Tanne auf Granit; 11-6-63.)

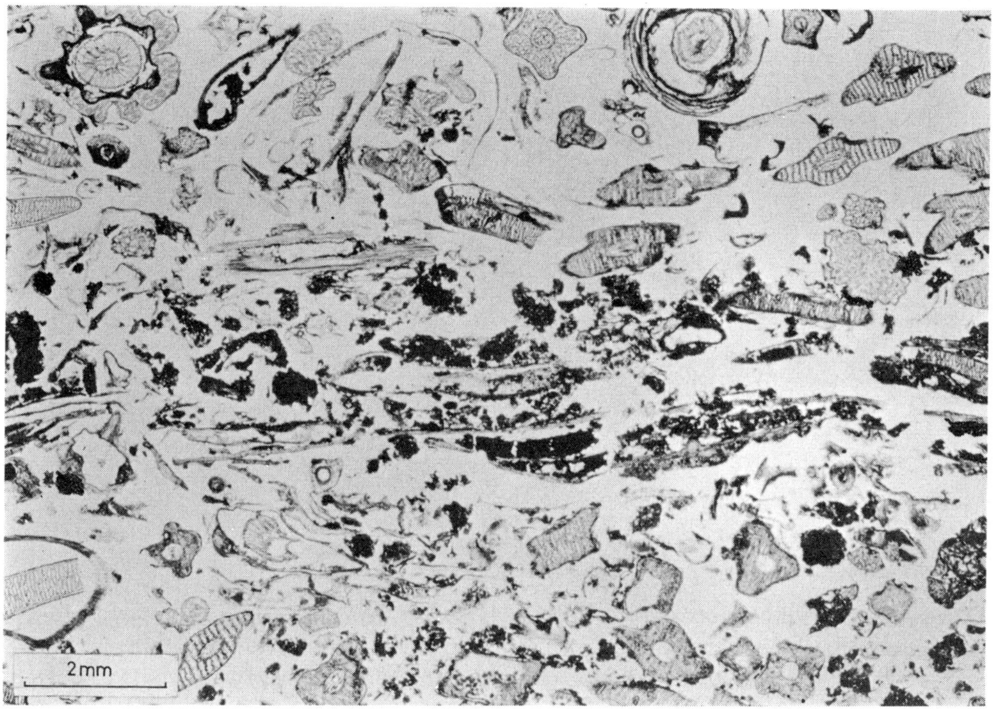

Abb.3. Die Schicht mit strukturlosen Exkrementen in der unteren L-Schicht, erzeugt von Enchytraeen nach Verzehren von Mikrobenschleim, organischem Detritus und Staub im gleichen Bereich. Dünnschliff; x 6,8. (FA Zwiesel-Ost/Bayer. Wald, Goldquelle Abt.IV/1; Fichte auf Gneis; 25-4-60.)

Auflagen bestehen aus Enchytraeenkot; sicher aber gehen sie aus solchem hervor.

Betont sei, dass hier wie auch bei (a) die Menge der Kotspuren keine Schlüsse zulässt über Wohndichte und Aktivität der Würmer. Wenn sich deren Lösung ansammelt, so bedeutet dies nur, dass die weitere Zersetzung durch Mikro-Organismen und Tiere sehr langsam erfolgt.

(c) Schliesslich leben die Enchytraeen auch in den obersten Schichten des Mineralbodens. Zweifellos ernähren sie sich dort von verwesender organischer Substanz und der Mikroflora, die in der humosen Erde enthalten ist. In die Kotbrocken der Regenwürmer fressen sie Gänge, die sie hinter sich mit Lösung füllen (Abb.4). So bereiten sie den Zerfall der oft sehr kompakten Kotmassen in kleinere Krümel vor; er wird durch Quellen und Schrumpfen, Frostspaltung und die wühlende Fauna vollendet. Auch Feinerde, die schon mehrfach den Regenwurmdarm passierte, wird von ihnen noch bearbeitet (Abb.5, Tafel A8). Die Struktur unserer besten Waldböden entsteht, indem aus Regenwurmkot doch Zerfall und ständige Lockerung

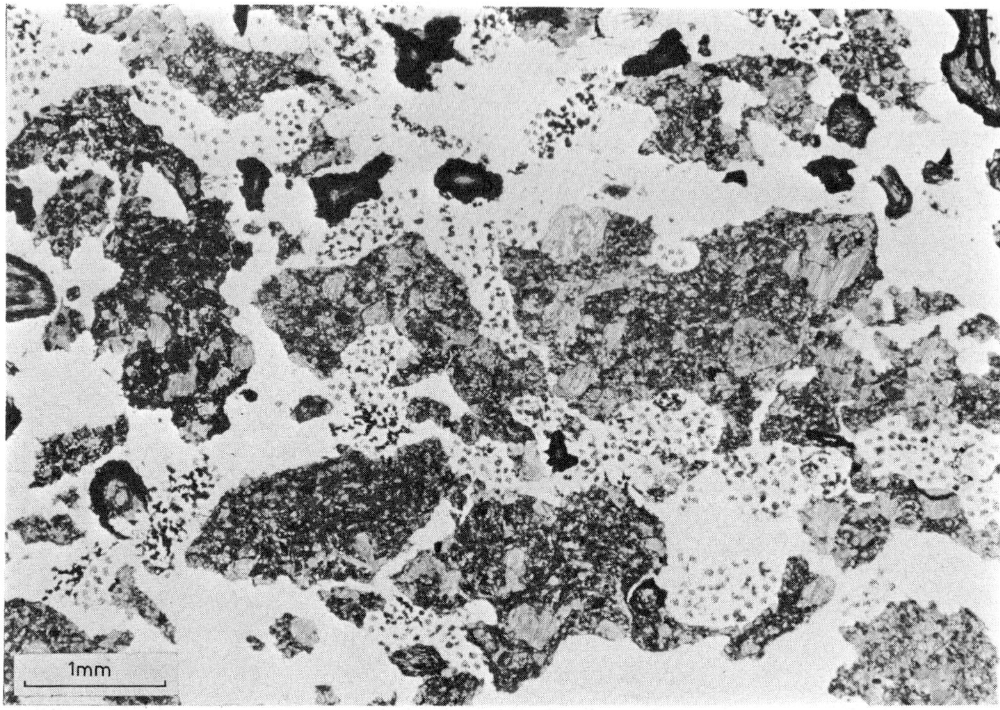

Abb.4. Kotbrocken von *Lumbricus terrestris* in der mineralischen H-Schicht. Die Gänge und Höhlen wurden von Enchytraeen gebohrt und mit ihrer Losung gefüllt Dünnschliff; x 13,5. (Fbz. Bonndorf/Schwarzwald, Gdw. Wellendingen, Steinahalde Abt.I/12; Tanne-Fichte auf Granit; 31-10-62.)

stabile Krümel hervorgehen; dazwischen und an deren Oberfläche liegen feinste Körnchen, die wenigstens zum Teil Enchytraeenlosung sind.

Bei besonders starker Regenwurmfauna werden diese Spuren durch ständige Durcharbeitung immer wieder verwischt, obwohl auch dort Enchytraeen reichlich vertreten sind (Abb.6). Ihre Losung fehlt im Mineralboden aber auch dann, wenn keine organischen Reste eingemischt sind, was diese schwachen Würmer selbst nicht bewirken können.

In Beständen auf stärker sandigem Lehm bewohnen häufig, ausser einigen *Lumbricus rubellus*, keine Regenwürmer den Mineralboden. Dann ist oft die ganze humose Feinerde zwischen den Mineralkörnern zu Enchytraeenkot geformt (Abb.1; Tafel B1). Dieser Zustand ist unter Laub- und Nadelbäumen sehr häufig.

Diese Beobachtungen gelten zunächst für mitteleuropäische Waldböden. Erwähnt sei aber, dass vermutlich auch die Alpine Protorendsina und der Alpine Pechmoder (Kubiena, 1953, pp.48–49, 214–218) überwiegend Enchytraeenlosung enthalten. Früher wurde angenommen, dass der Humus dort aus Collembolenkot besteht; das ist bei Form und Menge der erkennbaren Teilchen ebenso unwahrscheinlich wie im Wald (Zachariae, 1963). Diese Frage wird an anderer Stelle genauer untersucht.

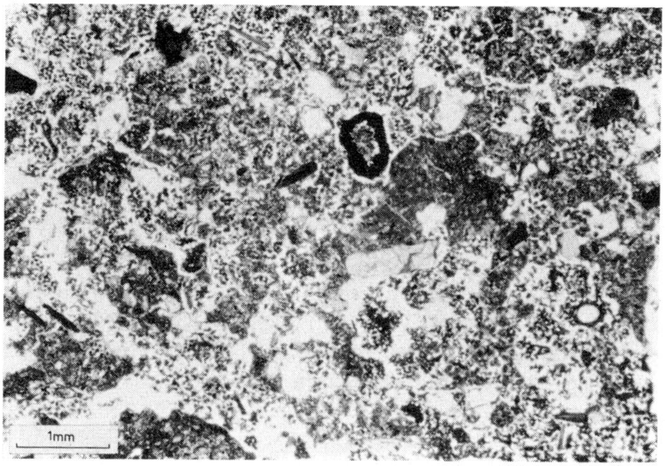

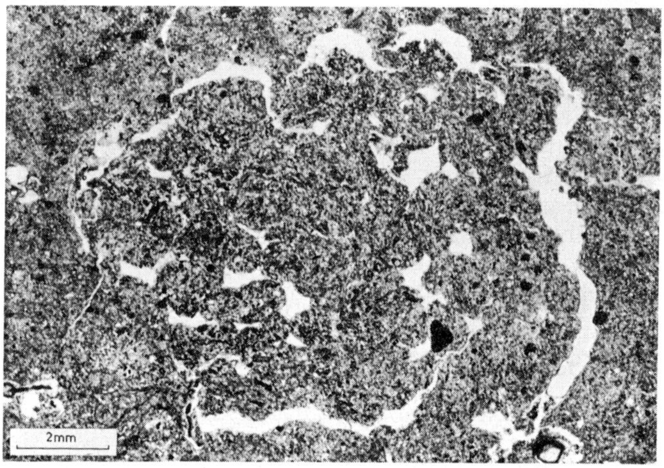

Abb.5. Erde mit verschiedenen Humusgehalt im Übergang von der mineralischen H-Schicht zum B-Horizont. Fast die ganze Erde wurde von Enchytraeen zu Losung verarbeitet. Dabei erfolgte selbst auf kleinstem Raum keine Vermischung des verschiedenen Materials; nur deshalb konnte das in Tafel A8 sichtbare Farbmuster erhalten bleiben. Dünnschliff, Durchlicht. (FA Zwiesel-Ost/Bayer. Wald, Schaufelhütte Abt. IV 18; Fichte auf Gneis; 25-4-60.)

Abb.6. Kotmassen von Regenwürmern *(Allolobophora)* in der mineralischen H-Schicht. Die ungewöhnlich starke Tätigkeit von Regenwürmern in diesem Standort beseitigt die Spuren der Enchytraeen sofort wieder, obwohl auch diese reichlich vorkommen. Die Zergliederung und Lockerung der Erde wird darum nicht wirksam. Dünnschliff; x 6,8. (Fbz. Badenweiler/Schwarzwald, Gdw. Auggen, Steinacker Abt. I/4; Buche auf Löss über Muschelkalk; 5-6-63.)

DIE BEDEUTUNG DER ENCHYTRAEEN IM WALDBODEN

Für mehrere Vorgänge im Waldboden könnten die Enchytraeen Bedeutung haben. Sie werden nachfolgend einzeln besprochen.

Beeinflussen Enchytraeen Verbeitung und Zusammensetzung der Mikroflora?

Für die Infektion der Frischstreu sorgen wohl Wind und Regen. Die Würmer kommen dafür kaum in Frage, weil sie offenbar nur an solchen Pflanzenresten Nahrung suchen, die schon mikrobiell besiedelt sind. Von dem schwarzen Kot (a) geht nie eine stärkere Zersetzung der Blätter oder Nadeln aus. Weil sie - wie unten gezeigt - auch keine Durchmischung bewirken, haben sie für die Verbreitung von Mikroorganismen wohl keine Bedeutung. Ob ihre Frasstätigkeit die Mikroflora verändert, wurde noch nicht untersucht.

Tragen Enchytraeen zur mechanischen Zerkleinerung bei?

Dies trifft nur bei leicht zersetzlichen Streuarten zu und wurde an solchen auch im Laboratorium festgestellt (Schaerffenberg, 1950; Dunger, 1958.) In unseren Wirtschaftswäldern, die vorwiegend aus Nadelhölzern, Rotbuche und Eiche bestehen, sind sie auf flüssige, stark erweichte (a) oder schon zerkleinerte Nahrung (b, c) angewiesen. Es wurde eindeutig beobachtet, dass dort eine Streuzerkleinerung nur von Arthropoden und in bestimmten Fällen von grossen Regenwurmarten durchgeführt wird.

Durchmischen Enchytraeen organische und Mineralsubstanz?

Erfahrungsgemäss schreitet die mikrobielle Zersetzung rasch fort, wo sich Streu und Mineralsubstanz innig berühren; wenn Bodentiere dazu verhelfen, erfüllen sie eine wichtige Funktion. Bei Enchytraeen konnte ein Transport von Mineralerde in die auf dem Boden liegenden L- und F-Schichten (a, b) nirgends beobachtet werden. Aber auch wenn sie humose Erde fressen (c), bleibt das Material am Ort, während der Wurmkörper sich darüber hinwegbewegt. Wo in einem Bodenbereich verschiedenes Material durch Regenwürmer oder andere Ursachen grob vermengt wurde, bleibt das am Dünnschliff sichtbare Farbmuster auch nach der Bearbeitung durch Enchytraeen erhalten (Abb.5; Tafel A8).

Wenn die Enchytraeen in den Kotbrocken von Regenwürmern gebohrt haben (Abb.4), schrumpft ihre Lösung beim Trocknen häufig stärker als die umgebende Masse. Die Gänge folgen meistens den dunkleren Partien. Hieraus und aus polarisationsoptischen Vergleichen kann man schliessen, dass die Enchytraeen auch in der Erde solche Stellen bevorzugen, die reicher an organischer Substanz sind. Man könnte also fast von einer Entmischung sprechen, jedoch ist das in diesen Dimensionen belanglos. Die Vermischung fördern sie jedenfalls nicht.

Tragen Enchytraeen zum chemischen Abbau der Pflanzenreste bei?

In Nadelbeständen findet man häufig nebeneinander Plätze mit einer dicken Schicht von kaum bearbeiteten Nadeln und andere, wo Enchytraeenkot (b) ebenso hoch angesammelt ist, nachdem die Streu vorher von Dipterenlarven zerkleinert wurde (Tafel A7). Dies entspricht der allgemeinen Erfahrung, dass Tierspuren ungleich, als ein langsam wechselndes Mosaik, über den Waldboden verteilt sind. Solche Fälle mit unterschiedlicher Enchytraeenleistung zeigen, dass die Würmer die Zersetzung nicht wesentlich voranbringen; entscheidend ist offenbar die Aktivität der Mikroflora. Aus dem Kotpulver (b) können sogar mächtige Auflagen entstehen, welche die forstliche Praxis dann zum Rohhumus rechnet. Ob die Enchytraeen für deren schlechte Eigenschaften verantwortlich sind, könnte nur in Zusammenarbeit mit Humuschemie und Mikrobiologie geklärt werden.

Fördern Enchytraeen die Speicherung von Nährstoffen?

Der in der L-Schicht abgesetzte Kot (a) ist sehr dauerhaft; es ist aber unbekannt, ob er schon beständige Humusstoffe enthält. Weitere Hinweise fehlen. Selbst bei Regenwürmern ist nach neueren Arbeiten fraglich, ob sie direkt zur Humifizierung der Streureste beitragen (Dunger, 1963). Diese ist wohl mehr eine Leistung von Mikroorganismen; die Tiere nehmen soweit Einfluss darauf, wie sie deren Aktivität fördern. Das gilt wahrscheinlich auch für die Enchytraeen. Weil sie die Mikroflora kaum begünstigen, wird ihr Wert in dieser Hinsicht gering sein.

Die Enchytraeenpopulation selbst kann ein bedeutendes Nährstoffdepot darstellen. Bei gutem Humuszustand ist ihr Anteil an der Bodenfauna zwar immer gering, in Nadelwald mit stärkerer Auflage aber kann sie nach Nef (1957) mit 10 g/m^2 ein Drittel der gesamten Biomasse vertreten und unter den einseitigen Bedingungen vieler Nadelstreuflächen sogar 13–25 g/m^2 erreichen (Nielsen, 1955a; O'Connor, 1957). Der N-Gehalt beträgt 2–3% (Nielsen, 1961; O'Connor, 1963); dort ist also mit einem Stickstoffvorrat je ha von 2–3 kg, in Extremfällen 5–7,5 kg, zu rechnen. Vermutlich geht noch mehr jährlich durch die Population hindurch, indem die Würmer Nährstoffe von der Mikroflora übernehmen und an Räuber (Chilopoden, Carabiden, Staphyliniden, Vögel) weitergeben. In dieser Hinsicht haben sie also gerade für ungünstige Standorte grossen Wert.

Beeinflussen Enchytraeen die Struktur des Bodens?

Bei langsamer Streuzersetzung ist oft das gesamte angereicherte Material zu Enchytraeenkot (b) geformt bzw. daraus hervorgegangen. Ob dessen charakteristische Struktur die schlechten Eigenschaften solcher Auflagen verstärkt oder mildert, ist unbekannt.

Wertvoll ist zweifellos die Zergliederung der kompakten Regenwurmexkremente (c) in humosen, lehmigen Böden. Hier tragen die Enchytraeen entscheidend zur Entstehung einer guten Krümelstruktur bei. Wo die Leistung von Regenwürmern geringer ist, können die Enchytraeen die ganze humose Feinerde zwischen den Sandkörnern prägen (c). Der so entstehende Strukturzustand ist häufig bei Humusformen, die man als Moder bezeichnet;

sein Wert für die Forstwirtschaft lässt sich noch nicht beurteilen. Nach Trappmann (1954) besitzt die Enchytraeenlosung eine erhöhte Stabilität; dies sit wohl wie beim Regenwurmkot eine Nebenwirkung der Mikrobentätigkeit (Scheffer und Ulrich, 1960, pp.204–207).

Wo Regenwürmer fehlen, können die Enchytraeen nicht wie jene Lüftung und Drainage in dichtem Unterboden schaffen. In tieferen Schichten bohren sie anscheinend nur, wenn diese durch andere Tiere erschlossen wurden und der Gehalt an organischen Resten hoch ist.

Für die Ziele der forstlichen Humuswirtschaft haben die Enchytraeen also nur gesicherte Bedeutung bei der Nährstoffspeicherung in der Biomasse und der Ausbildung einer Krümelstruktur in guten Böden. An einer Begünstigung der Mikroflora, der Streuzerkleinerung und der Durchmischung von Bodenmaterial haben sie praktisch keinen Anteil. Ihre Mitwirkung bei chemischem Abbau und Humifizierung der Pflanzenreste ist nicht eindeutig geklärt.

ANERKENNUNG

Diese Beobachtungen sind Teil einer Untersuchung, die mir die Deutsche Forschungsgemeinschaft durch ein Stipendium ermöglicht. Ihr sowie Herrn Prof. Dr. Kubiena, Reinbek, und zahlreichen Beamten privater und staatlicher Forstämter sei für die Unterstützung herzlich gedankt.

SUMMARY

Of what significance are Enchytraeidae in the forest soil?

The Enchytraeidae are unpretentious and live in optimal habitats as well as in less suitable forest soils. They are active and reproduce even at 0°C, and therefore they reproduce both under snow or on a frozen surface. Should unfavourable conditions prevail, population reduction will be balanced by fast development afterwards.

Three zones can be distinguished where they feed and produce excrements in forest soil: (a) Among the preserved litter of the L-layer they feed on the film containing microbes and detritus. The droppings will be scattered or used for their characteristic concealing tubes. (b) They consume decaying Arthropod faeces in the F-layer and convert them into granular droppings which, should decomposition be slow, yield raw humus-like masses. (c) They eat humose earth and thus pierce and cleave the compact excrements of big earthworms. Where earthworms are rare the entire fine material between the sand grains is often moulded by the Enchytraeidae.

The Enchytraeidae accumulate nutrients in the biomass; in favourable soils they contribute to the development of a crumb structure. They are of practically no significance for encouraging microflora, litter comminution, and blending of soil substances. Their assistance in chemical decomposition and humification of plant residues is not satisfactorily evident.

LITERATUR

Dunger, W., 1958. Über die Zersetzung der Laubstreu durch die Boden-Makrofauna im Auenwald. Zool. Jahrb. (Syst.), 86: 139–180.

Dunger, W., 1963. Leistungsspezifität bei Streuzersetzern. In: J. Doeksen and J. van der Drift (Editors), Soil Organisms. North-Holland, Amsterdam, pp. 92–102.

Kubiena, W.L., 1953. Bestimmungsbuch und Systematik der Böden Europas. Enke, Stuttgart, 392 pp.

Nef, L., 1957. État actuel des connaissances sur le rôle des animaux dans la décomposition des litières de forêts. Agricultura (Louvain), 5: 245–316.

Nielsen, C.O., 1955a. Studies on Enchytraeidae. 2. Field studies. Natura Jutland., 4/5: 1–58.

Nielsen, C.O., 1955b. Studies on Enchytraeidae. 5. Factors causing seasonal fluctuations in numbers. Oikos, 6: 153–169.

Nielsen, C.O., 1961. Respiratory metabolism of some populations of Enchytraeid worms and freeliving nematodes. Oikos, 12: 17–35.

O'Connor, F.B., 1957. An ecological study of the Enchytraeid worm population of a coniferous forest soil. Oikos, 8: 161–199.

O'Connor, F.B., 1963. Oxygen consumption and population metabolism of some populations of Enchytraeidae from North Wales. In: J. Doeksen and J. van der Drift (Editors), Soil Organisms. North-Holland, Amsterdam, pp.32–48.

Schaerffenberg, B., 1950. Untersuchungen über die Bedeutung der Enchytraeiden als Humusbildner und Nematodenfeinde. Z. Pflanzenkrankh. Pflanzenschutz, 57: 183–191.

Scheffer, F. und Ulrich, B., 1960. Lehrbuch der Agrikulturchemie und Bodenkunde. 3. Humus und Humusdüngung. Enke, Stuttgart, 1: 266 pp.

Springett, J.A., 1963. The distribution of three species of Enchytraeidae in different soils. In: J. Doeksen and J. van der Drift (Editors), Soil Organisms. North-Holland, Amsterdam, pp.414–419.

Trappmann, M., 1954. Beitrag zur Biologie und Ökologie von *Enchytraeus buchholzi* Vejdovsky 1879 (Annelida, Clitellata). Thesis Tech. Hochschule, Braunschweig.

Zachariae, G., 1962. Zur Methodik bei Geländeuntersuchungen in der Bodenzoologie. Z. Pflanzenernähr. Düng. Bodenk., 97 (142): 224–233.

Zachariae, G., 1963. Was leisten Collembolen für den Waldhumus? In: J. Doeksen and J. van der Drift (Editors), Soil Organisms. North-Holland, Amsterdam, pp.109–124.

FARBENBILDER

Tafel A7. Frische Losung von Lycoriidenlarven (oben), die modernde Nadeln der F-Schicht frassen, Kotballen von minierenden Oribatiden in einem Nadelstück (links) und freiliegend (Mitte) und die Schicht aus älterer Losung von Enchytraeen (unten). Die Würmer haben die von Dipterenlarven zerkleinerten Pflanzenreste weiterverarbeitet. Wenn alle Tierspuren älter und verpilzt sind, lassen sie sich wesentlich schwerer unterscheiden. Dünnschliff x 27. (FA Friedrichsruh/Sachsenwald, Söhren Abt.82; Fichte auf glazialem Geschiebe; 30-3-62.)

Tafel A8. Erde mit verschiedenem Humusgehalt im Übergang von der mineralischen H-Schicht zum B-Horizont. Regenwürmer (*Allolobophora rosea*) haben hier Erde von beiden angrenzenden Bereichen gefressen und beim Kote vermischt. Cf. Abb.5., Dünnschliff, Auflicht; x 27. (FA Zwiesel-Ost/Bayer. Wald, Schaufelhütte Abt.IV/8; Fichte auf Gneis; 25-4-60.)

Tafel B1. Zu Enchytraeenlosung geprägte Feinerde zwischen Sandkörnern in der mineralischen H-Schicht. Die Sandkörner wurden teilweise durch polarisiertes Licht farbig sichtbar gemacht. Dünnschliff; x 27. (FV Panker/Ostholstein, Strezer Berg Abt.12a; Buche auf glazialem Geschiebe; 6-4-62.)

SOIL FAUNA AND SOIL PROFILE IN SOME INLAND-DUNE HABITATS

J. VAN DER DRIFT

Institute for Biological Field Research (Itbon), Arnhem (The Netherlands)

INTRODUCTION

Among the environments where the initial stages of profile development can be studied are the inland-dunes. Notwithstanding the fact that shifting sands on the Veluwe have been checked in the last hundred years by planting Scotch pine stands, there are still several sand dunes which provide an opportunity to study the development of vegetation, fauna and soil profile on these bare sands.

The successive vegetation types on these sands are fairly well known, and those of the Veluwe were recently described extensively by Stoutjesdijk (1959). The inland-dune sand soils of the Veluwe were studied by Schelling (1955), and their suitability for afforestation was evaluated. The fauna in these soils has not yet been studied before. It may be expected that the soil fauna in the shifting sand is absent, or at best very scarce. In the initial vegetation types the soil fauna will also be rather poor. It seems possible to trace the specific effects of the few animal groups present, and to investigate their role in the genesis of the soil profile. For this purpose the soil fauna was analysed in sand, and in four stages of vegetation development, at the Oud Reemster Zand in the National Park "De Hoge Veluwe". Thin sections of the soil profiles of this location were also made.

SITE AND SAMPLING PROCEDURE

The Oud Reemster Zand is one of the most southern inland-dunes of the Veluwe. The eastern part consists of a complex of hills and valleys. Big sand drifts ended here. At present sand is transported only on a limited scale. A picture of the sampling site is given in Fig.1. A sandy valley, partly covered by the Algae *Prasiola crispa* Menegh and *Zygogonium ericetorum* Kütz var. *terrestre* Kirchn, is fringed at the foot of the hills by patches of the moss *Polytrichum piliferum* Schreb. and tussocks of grass species in which *Festuca ovina* L. predominates. The flanks of the hills are partly grown over by a patchy vegetation of grasses, mosses and heather, *Calluna vulgaris* Sal. At the top of the hill some old pines, *Pinus silvestris* L., maintain themselves as remnants of a Scotch pine stand which was cleared many years ago. Within this site the following locations were sampled for soil fauna and soil profile study:

Fig.1. Sampling site Oud Reemster Zand, Hoge Veluwe.

(a) sand : bare sand without distinct algal cover; microscopic investigation reveals small colonies of *Zygogonium ericetorum*.
(b) moss : patches of *Polytrichum piliferum* on the lower flanks of the hill.
(c) grass : patches of *Festuca ovina* at the same place.
(d) heather: about 30 cm high shrubs of *Calluna vulgaris*, with only a very thin litter and "humus" layer under them.
(e) pine : under the wide-spread branches of a single *Pinus silvestris* with a litter and humus layer, each about 3 cm thick.

On 5 dates, November 6th and 14th, January 8th, February 25th and March 17th, soil samples of different sizes were taken in each of the habitats. For micro-arthropods the samples measured 20 cm^2, for enchytraeids 100 cm^2, and for macro-arthropods 500 cm^2. The samples were separated into sub-samples according to organic matter content. In general the depth of sub-sample (a), with the relatively highest content of organic remains, was about 1 cm. Sub-sample (b) extended to about 5 cm of depth, and sub-sample (c) from 5–10 cm. The latter sub-sample could be omitted in sand, moss and grass habitats, as the animal population in that layer seems to be either absent or very scarce. The organic layer under pine was separated into two sub-samples, (a1) and (a2), respectively comprising the needle layer-L and the fermentation layer F. Arthropod samples were desiccated in a Tullgren apparatus, and enchytraeids were extracted in a warm water funnel (O'Connor 1955).

Samples for thin sections were taken on December 10th. At each habitat two samples of 15 x 8 x 5 cm were taken, one in a horizontal direction (depth 0–5 cm), and the other in a vertical direction (depth up to 15 cm). The technique used for the preparation of the sections with a thickness of 15 μ is fully described by Jongerius and Heintzberger (1963).

RESULTS

Fauna analysis

Table I indicates the occurrence of the main groups of macro-arthropods in the five habitats. The few species occurring in the sand were met incidentally here. They may cross the sand plain to find a new territory, but they do not live in this habitat. In the moss vegetation next to some incidental species, there are at least larvae of two Collembola and of two Nematocera species which feed on plant debris, and which occur regularly. Therevidae and Carabidae have mainly predacious habits, and Byrrhidae feed on moss. In the other three habitats the number of groups as well as the number of individuals increases considerably, culminating in the pine litter layer. The especially great abundance of Collembola is notable.

In Table II the total numbers of micro-arthropods per 100 cm^2 are given, and the mean numbers of enchytraeids per 100 cm^2. The mesofauna is lacking completely in the sand. In the moss patches the mites are mainly represented by very small Trombidiformes, and the Collembola by Isotomidae. Enchytraeids were not found in the moss. The Isotomidae is the only group of the mesofauna which may be of importance for the breakdown of plant residues in the *Polytrichum* sods. In the grass vegetation the saprophagous Isotomidae and Oribatei are much more numerous than in the moss and the likewise saprophagous Enchytraeidae occur densely. All three groups will play their role in the litter breakdown in this habitat. In the heather the Oribatei is by far the most numerous group, but the Isotomidae and Enchytraeidae have also increased considerably. In the pine litter the Oribatei and Isotomidae remain at about the same level, while the Enchytraeidae have increased still more. The enchytraeid populations in grass, heather and pine are composed of two species: *Cognettia sphagnetorum* Vej., and *Marionina clavata* Nielsen and Christensen. Though the former prevails in the majority of cases the proportion varies widely. The Isotomidae, with few exceptions, consist of *Isotoma sensibilis* (Tullb.).

The species composition of the Oribatei is given in Table III. The species are arranged according to their occurrence in the successive habitats. No single individual was met in the sand. In the moss only five oribatids belonging to four species were found and in the grass five species, one of which had not less than 40 specimens. Numbers of species and individuals increase in heather. Of thirteen species three had more than a hundred individuals, *Tectocepheus velatus*, *Trhypochthonius cladonicola* and *Carabodes minusculus*. In the pine habitat 23 species occur, again with *Tectocepheus* and *Trhypochthonius* being the most abundant species, followed by the root-habiting *Rhysotritia duplicata*.

TABLE I

The occurrence of macro-arthropods in inland-dune vegetations[1]

	Sand	Moss	Grass	Heather	Scotch pine
Dendrobaena octaedra[2] (Sav.)					scarcely
Geophilidae				scarcely	scarcely
Obisium muscorum Leach				scarcely	regularly
Lycosidae			scarcely	scarcely	
Linyphiidae	incidentally	incidentally	scarcely	regularly	regularly
Macrargus rufus (Wider)				scarcely	scarcely
Collembola[2]					
Neanura muscorum (Templ.)		incidentally	incidentally	incidentally	scarcely
Isotoma sensibilis (Tullb.)	incidentally	scarcely	regularly	regularly	abundantly
Orchesella cincta (L.)		scarcely	scarcely	scarcely	abundantly
Entomobrya nivalis (L.)			scarcely	scarcely	scarcely
Entomobrya 5-lineata Börn.			scarcely	regularly	
Diptera Nematocera larv.[2]					
Chironomidae		scarcely	scarcely	scarcely	scarcely
Mycetophilidae		scarcely	regularly	scarcely	scarcely
Ceratopogonidae			regularly		scarcely
Tipulidae			scarcely	scarcely	
Cecidomyuidae			scarcely	regularly	regularly
Diptera Brachycera larv.					
Dolichopodidae/Empididae			scarcely	regularly	regularly
Rhagionidae			scarcely	scarcely	
Therevidae		scarcely	scarcely	scarcely	scarcely
Coleoptera					
Carabidae	incidentally	scarcely	scarcely	scarcely	scarcely
Staphylinidae			regularly	regularly	regularly
Byrrhidae		scarcely	scarcely		
Cantharidae				scarcely	scarcely
Elateridae			scarcely	scarcely	regularly

[1] Key: - - - - - - = occurring incidentally; ——— = scarcely; ═══ = regularly; ≡≡≡ = abundantly.

[2] Feeding, at least partly, on plant debris.

TABLE II

Micro-arthropods and Enchytraeidae in several inland-dune habitats (Numbers per 100 cm^2)

Habitat	Sand		Moss		Grass		Heather			Scotch pine			
subsample	(a)	(b)	(a)	(b)	(a)	(b)	(a)	(b)	(c)	(a1)	(a2)	(b)	(c)
approximate depth (cm)	0–1	1–5	0–1	1–5	0–1	1–5	0–1	1–5	5–10	+6–+3	+3–0	0–5	5–10
Gamasides			5	1	24	6	61	31	1	107	26	2	
Trombidiformes			29			3	18	1		23	11	1	
Uropodina					2		8			17	1		
Tyroglyphidae		1	6		6	2	2	2		44	15	35	
Oribatei			4	2	40	16	1074	119		1055	623	40	2
Acari div.			3		8		63	6		67	12	1	
Poduridae				1	1		1	5	1	2	9		
Tullbergia				11		3	4	5		3	8	16	3
Isotomidae			42		10	174	167	53	9	122	34	13	
Collembola div.					8		4			2	1		
Acari Total		1	47	3	80	27	1226	160	1	1313	689	79	
Collembola Total			42	12	19	177	176	63	10	130	52	29	
Enchytraeidae					100	50	500	150	10	850	200	30	

TABLE III

Oribatei in several inland-dune habitats (numbers per 100 cm²)

Habitat	Sand		Moss		Grass		Heath			Scotch pine			
subsample	(a)	(b)	(a)	(b)	(a)	(b)	(a)	(b)	(c)	(a1)	(a2)	(b)	(c)
approximate depth (cm)	0–1	1–5	0–1	1–5	0–1	1–5	0–1	1–5	5–10	+6–+3	+3–0	0–5	5–10
Tectocepheus velatus (Mich.)			1		32	8	722	80		612	106	2	
Trhypochthonius cladonicola (Willm.)			1				107	6		67	9	1	
Oppia and *Suctobelba* spec.				2	2	4	16	2		7	15	1	
Nothrus silvestris Nic.			1				5	1		19	8		
Rhysotritia duplicata Grandj.					1	1		7		61	39	28	1
Brachychthonius spec.						2	45	4		16	5		
Carabodes marginatus (Mich.)					3		21	10			1		
Carabodes minusculus (Berl.)							114	4					
Chamobates incisus v.d. Hammen							12	5		30	10		
Camisia exuvialis Grandj.							12			7	2		
Platynothrus peltifer (C.L. Koch)							6			30	7		
Camisia spinifer (C.L. Koch)							3	1		23	1		
Camisia segnis (Hermann)							2			1			
Hypochthonius rufulus C.L. Koch										13	15		1
Trichoribates trimaculatus (C.L. Koch)										6	2	1	
Galumna spec.										7			
Steganacarus striculus (C.L. Koch)											6		
Camisia biverrucata (C.L. Koch)										4			
Carabodes labyrinthicus (Mich.)										1			
Thyrisoma lanceolata (Mich,)										1			
Pelops spec.										1			
Damaeus spec.										1			
Adoristes ovatus (C.L. Koch)											1		
Rhysotritia ardua (C.L. Koch)												1	

Soil profile sections

The thin sections were studied with magnifications up to 200 x. Characteristic parts were photographed (Fig.2–17).

Sand

Intermingled with the mineral grains (varying in diameter from 130–260 μ). Various plant fragments occur, mainly pine needles blown in by the wind. The plant tissues are disrupted by shrinkage and mechanical damage. There is no evidence of animal attack on the tissues, nor animal droppings of any kind (Fig.2). In the top cm small patches of algal growth are found, mainly consisting of *Zygogonium ericetorum*, but sometimes accompanied by the larger *Prasiola crispa*. These Algae are disintegrating locally, but animal attack could not be observed (Fig.3).

Moss

In the top cm many plant fragments, together with droppings of macro-arthropods (diameter up to 1,000 μ), Fig.4, and especially of springtails (diameter 25–1,000 μ), occur. Excrements of oribatids and enchytraeids could not be found, and this is in accordance with the failing to record these animals in this habitat. The macro-arthropod excrements fall to pieces and are washed down to 2 cm of depth (Fig.5) together with the bigger springtail excrements. Their disintegration products and the smaller springtail droppings (25–50 μ) are transported down to 4 cm. They are distributed rather evenly between the sand grains (Fig.6, 7). At the lower limit they accumulate into small horizontal bands. Under these bands no organic particles occur.

Grass

The upper 1–2 cm contain plant fragments and a great many excrements of enchytraeids, springtails and macro-arthropods (saprophagous Nematocera larvae). The bigger excrements, with a diameter down to 100 μ, are transported downward to 2 cm at the most. Underneath, small excrements occur both separately and in groups. The latter may be deposited by enchytraeids (Fig.8, 9). Pores between the sand grains are bigger and more numerous than in the previous type. This may be caused partly by the movements of the enchytraeids. As a result of the more porous structure of the upper 6 cm, the small organic particles are washed down to this depth. Dead roots, even in deeper layers, are mainly eaten by phthiracarid mites. Their pellets apparently remain surrounded by the unattacked rootbark during long periods (Fig.10). They are seldom observed free between the sand grains. Small springtails can also take part in the breakdown of roots (Fig.11).

Heather

The top cm consists mainly of plant remains accompanied by large masses of droppings of enchytraeids, collemboles, Nematocera larvae and mites. Down to about 5 cm excrements or parts of them, 100 μ or larger in diameter, occur. From 5–8 cm of depth only small particles occur. They are partly deposited by enchytraeids, and when present they occur in rather distinct strands, partly washed down by rain water.

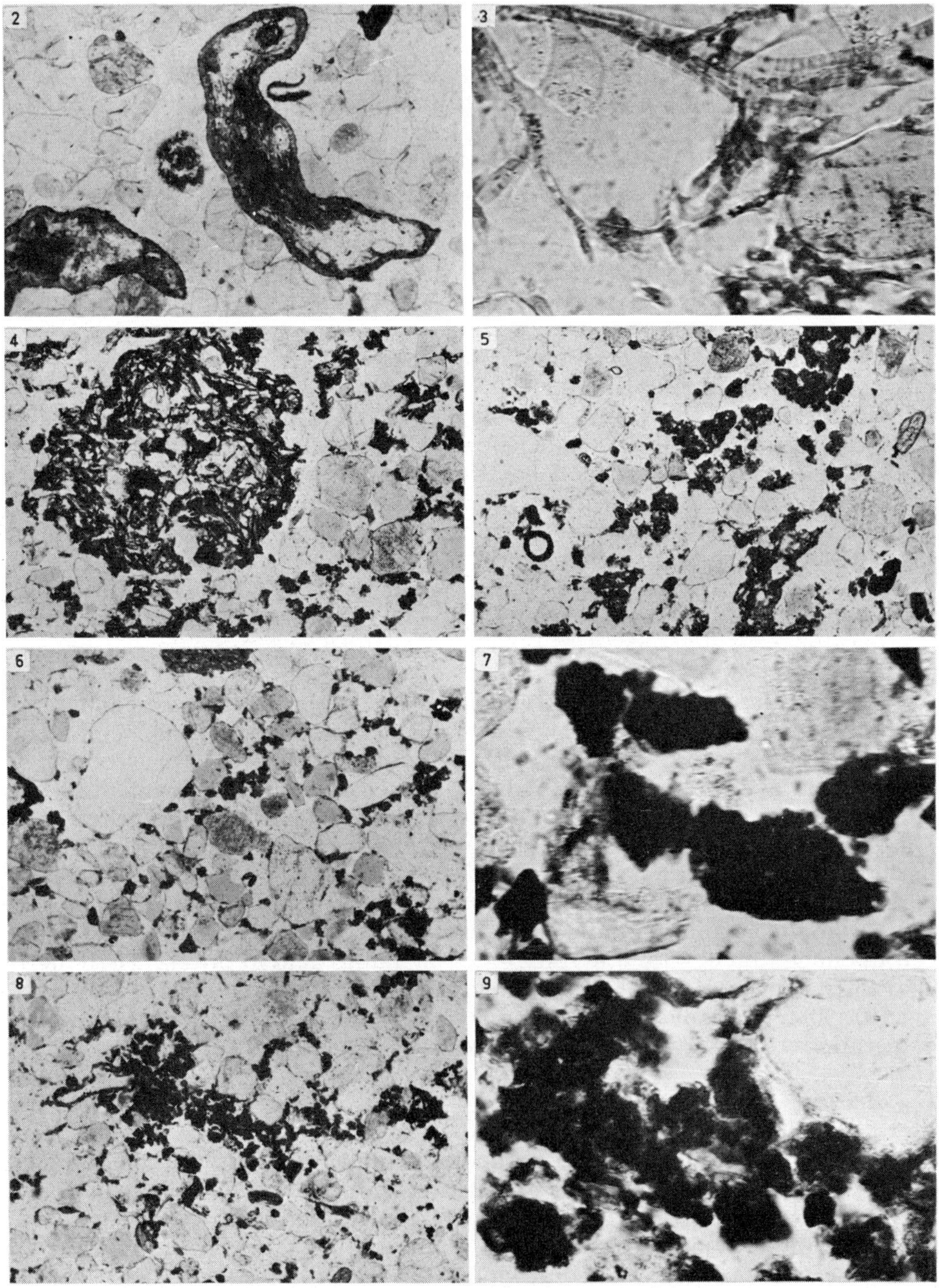
2
3
4
5
6
7
8
9

The deeper in the profile, the smaller the diameter of the organic particles, which may accumulate in small horizontal bands (Fig.12). In this profile as in the others, numerous sclerotia give evidence of the presence of Fungi (Fig.13). The hyphae are visible, with highest magnification, on the sand grains and in the organic particles.

Scotch pine

The litter and humus horizons, each about 2–3 cm thick, contain few sand grains. The grains are blown into this layer or else were taken up by animal activity. The coarse plant fragments in the litter layer are attacked by enchytraeids, mites, collemboles and macro-arthropods, the excrements of which are present in large amounts (Fig.14). Those of enchytraeids measure about 40 μ, while the mite pellets vary from 40–100 μ (Fig.15). The collembole pellets range from 50–200 μ, and those of macro-arthropods from 0.5–1.5 mm. Excrements of the earthworm *Dendrobaena octaëdra* contain small sand grains (Fig.16). The humus layer mainly consists of animal excrements which are attacked by micro-organisms, especially Fungi. The bigger pellets fall to pieces in this horizon. The upper 3–4 cm of the mineral substrate are very rich in organic particles between the grains. They have been partly deposited here as excrements, but the main part is washed down from the A_0-horizon. Accumulations up to 300 μ in diameter, consisting of dark brown amorphous organic material, occur (Fig.17). The lower limit of the occurrence of organic particles is still not reached at a depth of 15 cm.

Fig.2. Transverse section through two pine needles, blown into the sand. No evidence of animal attack; no animal droppings (x 18).

Fig.3. *Zygogonium ericetorum* in the upper cm of the sand. At the bottom right are disintegrating Algae (x 128).

Fig.4. Macro-arthropod excrement, probably a dropping of Nematocera larva in a moss habitat. The porous structure of the pellet indicates initiating disintegration (x 18).

Fig.5. Macro-arthropod excrements washed down several cm into mineral substrate. Moss habitat (x 18).

Fig.6. Evenly distributed pellets of isotomids at 3 cm of depth, washed down from the surface horizon of moss habitat (x 18).

Fig.7. Isotomid excrements, washed down, at greater magnitifation (x 128).

Fig.8. Enchytraeid excrements deposited at 3 cm of depth in the grass habitat (x 18).

Fig.9. Part of the enchytraeid excrements at greater magnification (x 128). Note the difference from the springtail excrements in the moss layer (Fig.7).

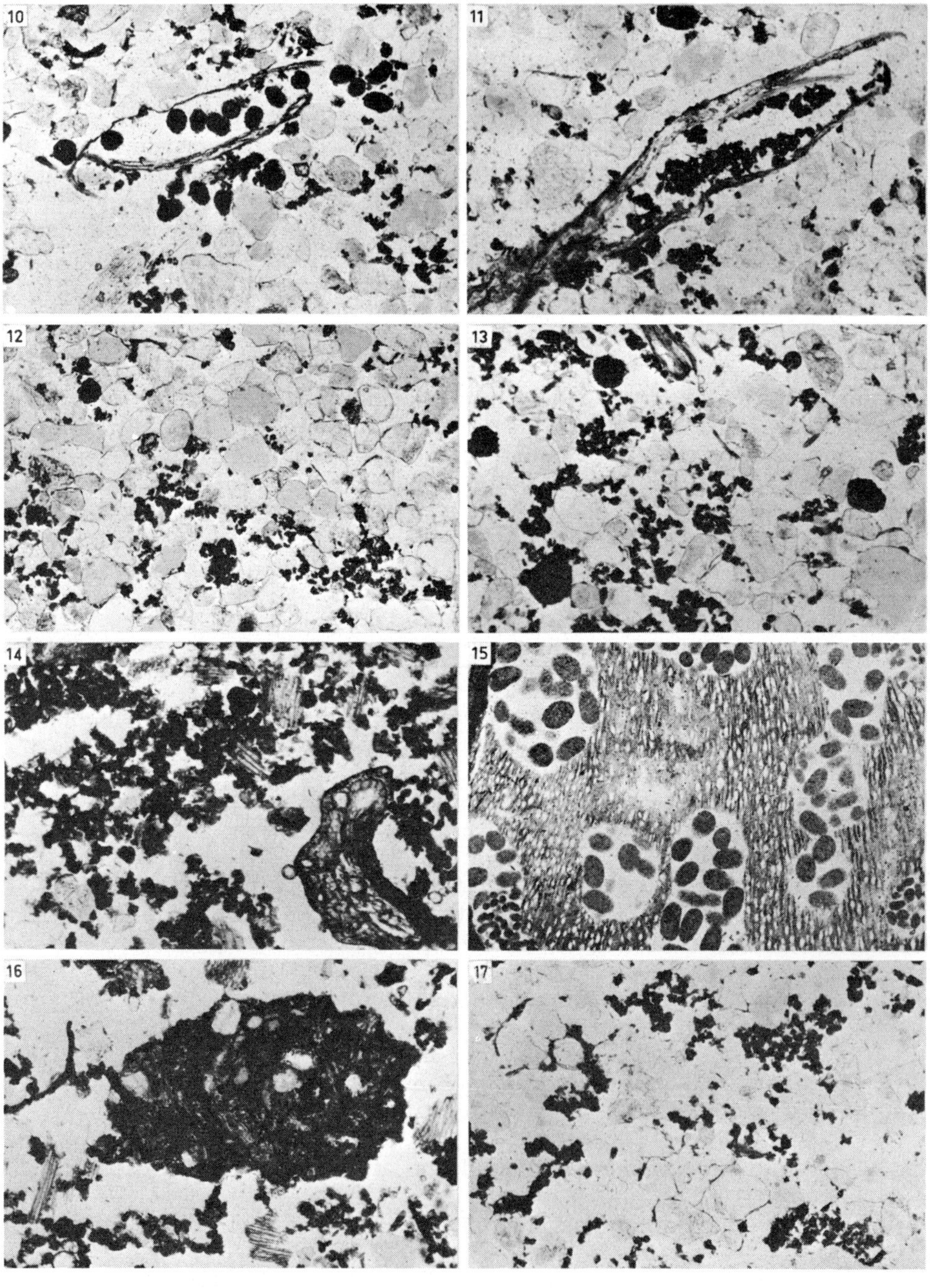
10
11
12
13
14
15
16
17

DISCUSSION

Identification of the droppings of soil animals in the soil profile is a precarious matter. Experiments have shown that the colour, shape and measurements of the pellets are dependent on the type of food and the size of the animal. Species feeding on a wide range of organic debris give particular difficulties in this respect.

By studying very simple habitats, where both vegetation and fauna are composed of a small number of species, it seems possible to connect the droppings in the soil profile with the animal species.

In the inland-dune investigated bare sand evidently produced too small amounts organic matter, and provided too poor meteorological circumstances, for development of even a poor animal community. No animals were encountered in this habitat. The Algae present at the soil surface and the blown-in plant parts apparently were not attacked by any animal.

In the moss habitat the only abundant saprophagous group is that of Collembola, with *Isotoma sensibilis* as the main species. The small pellets must be mainly ascribed to them. As this species was only recorded in this habitat in the upper 1–2 cm, and the excrements occurred down to about 4 cm, it seems likely that these excrements were washed down by percolating water. Their distribution around the sand grains and the occurrence of small horizontal bands at the lower limit also favour this supposition.

In the grass, heather and pine habitats this downward movement of excrements, and of particles from them, also occurs. The depth of this movement increases with the development of vegetation. In the grass vegetation the lower limit of organic matter transported downwards is 6 cm. In

---

Fig.10. Phthiracarid mites leave their pellets in excavated dead roots, the bark of which generally remains unattacked (x 18).

Fig.11. Small springtails (Isotomidae) may take part in the breakdown of dead roots (x 18).

Fig.12. Excrements washed down by rain-water at about 7 cm of depth into a small horizontal band. Heather (x 18).

Fig.13. Four sclerotia at 10 cm of depth, under heather (x 18).

Fig.14. Plant fragments and excrements of enchytraeids (top left), springtails (middle left), and mites (middle right), in the litter layer of a Scotch pine (x 18).

Fig.15. *Rhysotritia duplicata* excavate a piece of wood and leave their pellets (x 18).

Fig.16. Dropping of the earthworm *Dendrobaena octaëdra*, containing several sand grains. Litter layer of Scotch pine (x 18).

Fig.17. Enchytraeid excrements about 2 cm below A_0-layer of Scotch pine. At left, dark brown amorphous, organic accumulations (x 18).

the heather vegetation it increases to about 8 cm of depth, and under pine to more than 10 cm below the original soil surface. However, from the distribution pattern of the organic material it seems evident that part of this material is deposited by animals on the spot. Rather large accumulations, with the longitudinal axis not only being horizontal but also vertical, or in any other direction, suggest that slowly moving animals such as enchytraeids have caused them. Indeed, they occur in the mineral layer, and in experiments with these worms in a small amount of soil between two glass sheets, they proved to concentrate their excrements along their galleries. It seems beyond doubt that enchytraeids contribute to the downward movement of organic debris in the profile. To some degree other real soil inhabitants, like *Tullbergia* and *Isotoma*, may also have some effect in this respect. However their excrements are more scattered, and it is difficult if not impossible to distinguish them from particles carried by water. Phthiracarid mites which penetrate into the deeper profile horizons do so by following dead roots. Their droppings occur in great numbers in the excavated roots, where they seem rather resistent to further decay. They consist mainly of micro plant fragments (up to 25 µ) and hyphae. The most numerous oribatids, such as *Tectocepheus velatus* and *Trhypochthonius cladonicola*, do not leave the slightest trace of their activity. Investigation of the gut contents of *Tectocepheus* revealed that this species mainly eat spores of Fungi and hypha These are the main components of the gut contents of the other Oribatei too, with the exception of the Phthiracaricae. For the breakdown of plant remains they are of minor importance.

ACKNOWLEDGEMENTS

I wish to thank Ir. J.H. van Tuil, director of the National Park, for his permission to perform this study at the "Hoge Veluwe", Dr. A. Jongerius (Soil Survey Institute, Wageningen) for making and photographing the thin sections, and Dr. J.Th. Koster (State Herbarium, Leyden) for identifying the Algae.

SUMMARY

The soil fauna and the soil profile of five habitats in an inland-dune on the Veluwe, The Netherlands, are described. These habitats are: bare sand, a *Polytrichum piliferum* vegetation, a grass vegetation with *Festuca ovina* prevailing, a *Calluna vulgaris* heath, and a stand with some *Pinus silvestris*. In the bare sand no animals are living, and the Algae and blown-in plant remains are not attacked by animals. In the moss vegetation isotomids are the main agents in the breakdown of plant remains, their excrements being carried down to 4 cm of depth by percolating water. In the remaining three habitats the soil fauna increases in numbers and species with the increasing production of organic matter. Concordantly the "humus" containing layer increases in depth. In these habitats Enchytraeidae, Collembola and phthiracarid mites are the main saprophagous agents.

REFERENCES

Jongerius, A. and Heintzberger, G., 1963. The preparation of mammoth-sized thin sections. Soil Surv. Papers, Neth. Soil Surv. Inst., 1 : 37 pp.

O'Connor, F.B., 1955. Extraction of enchytraeid worms from a coniferous forest soil. Nature, 175 : 815–816.

Schelling, J., 1955. Stuifzandgronden (Inland-dune sand soils). Detailed Rept. Forest Res. Sta. T.N.O., 2 (1) : 57 pp.

Stoutjesdijk, Ph., 1959. Heaths and inland-dunes of the Veluwe. Wentia 2 : 1–96.

UNTERSUCHUNGEN AN TORF-DÜNNSCHNITTEN AUS EINEM MOORPROFIL VOM TEUFELSMOOR BEI BREMEN

G. GROSSE–BRAUCKMANN und D. PUFFE

Staatliche Moor-Versuchsstation, Bremen (Deutschland)[1]

EINLEITUNG

Die in einem Moor abgelagerten Torfe enthalten nebeneinander strukturierte pflanzliche Gewebereste und völlig humifiziertes Material sowie mannigfache Übergänge zwischen diesen beiden Extremen. Die Verteilung und räumliche Anordnung dieser Komponenten sowie ihre Mengen und ihr Aussehen wechseln von Torfart zu Torfart ausserordentlich stark; es handelt sich hierbei also um wichtige kennzeichnende Merkmale der Torfe. Zur Erfassung dieses Merkmalskomplexes ist die mikroskopische Untersuchung von in toto geschnittenen Torfproben besonders geeignet. Derartige Untersuchungen entsprechen damit der durch Kubiena in die Bodenkunde eingeführten mikromorphologischen Methode, bei der sonst besonders mit Bodendünnschliffen gearbeitet wird.

Die Fragen, zu deren Lösung die mikroskopische Untersuchung von Torf-Dünnschnitten beitragen kann, betreffen unter anderem den Ablauf der Zersetzungsprozesse bei der Torfbildung; dabei handelt es sich um einen ähnlichen Fragenkomplex wie bei der mikromorphologischen Untersuchung der Humusauflagen von Mineralböden.

In einer vorangegangenen Arbeit (Puffe und Grosse–Brauckmann, 1963), worin auch methodische Einzelheiten über die Herstellung der Mikrotomschnitte angegeben sind, haben wir uns eingehend mit den Ergebnissen mikromorphologischer Untersuchungen an schwach zersetzten Bleichmoostorfen beschäftigt und dabei versucht die Befunde auch im Hinblick auf den Torfbildungsprozess und die dabei ablaufenden Zersetzungsvorgänge auszudeuten. Im folgenden sollen besonders mikromorphologische Befunde an minerotrophen Torfen ("Nicht-Hochmoortorfen": Übergangs- und Niedermoortorfen) vorgeführt und erörtert werden; es werden sich dabei - das sei schon vorweggenommen - zahlreiche Parallelen zu den Hochmoortorfen ergeben.

Bei den untersuchten Torfen handelt es sich um Proben aus verschiedenen Tiefen eines Moorprofils aus dem zum Teufelsmoorgebiet gehörenden "Kollbecksmoor". Dieses liegt etwa 30 km nordnordwestlich von Bremen und befindet sich zur Zeit in industrieller Abtorfung. Es ist auf den in diesem Gebiet weit verbreiteten Talsanden als Versumpfungsmoor aufgewachsen und hat im heutigen ("teilentwässerten") Zustand noch eine Mächtigkeit von 2–4, meist 2,5–3 m. Der Beginn der Moorbildung fällt nach einer vorliegenden pollenanalytischen Untersuchung in die frühe Wärmezeit (Boreal);

---
[1] Direktor: Prof. Dr. W. Baden.

zunächst wurde ein Radizellentorf mit Resten von *Phragmites* und *Menyanthes* sowie einzelnen Holzresten (*Salix*, *Betula*) abgelagert. Im mittleren Profilteil des Moores, dessen Entstehung in den ersten Teil der mittleren Wärmezeit (Atlantikum) fällt, wird der Anteil der Birkenreste in den Radizellentorfen grösser; darin treten auch ausgesprochene Stubbenlagen auf. In den oberen Lagen dieses mittleren Profilteils erscheinen ausser den genannten Pflanzenresten auch solche von der Kiefer und von (minerotraphenten) Sphagnen; dadurch deutet sich bereits der bevorstehende Übergang zum Stadium des ombrotraphenten Hochmoores an, dessen Torfe das oberste Drittel des Profiles einnehmen; der Ablagerungsbeginn dieser Torfe fällt noch in die mittlere Wärmezeit. Das Alter der jüngsten Torfe ist dem Pollendiagramm nicht mit Sicherheit zu entnehmen. Die Hochmoortorfe sind stark zersetzt und entsprechen in ihrem Typus ungefähr dem sogenannten älteren Hochmoortorf ("Schwarztorf"); ein schwach zersetzter ("jüngerer") Hochmoortorf, der sonst in den Hochmooren des Teufelsmoorgebietes regelmässig die obersten Torflagen bildet, ist nicht zur Ausbildung gelangt oder nicht mehr vorhanden.

ERGEBNISSE

Über den Erhaltungszustand von Pflanzenresten im Torf

An den in den Torfen noch strukturiert erhaltenen Pflanzenresten fällt zunächst meist ihr äusserlich erstaunlich guter Erhaltungszustand auf. Manche Reste, wie Moose und Hölzer, sowie Früchte und Samen, erscheinen - verglichen mit dem Zustand der lebenden Pflanzen - kaum verändert; andere, wie bestimmte unterirdische Teile krautiger Pflanzen, scheinen zwar gewisse Veränderungen durchgemacht zu haben, sind aber an kennzeichnenden Merkmalen noch gut anzusprechen (cf. Grosse-Brauckmann, 1964, Tafel I, II). Dass jedoch in den meisten Fällen auch an den relativ gut erhaltenen Resten schon mehr oder weniger tiefgreifende Veränderungen stattgefunden haben, lässt sich an mikromorphologischen Bildern zeigen. Dabei handelt es sich in manchen Fällen deutlich um Ergebnisse eines Humifizierungsvorganges, in anderen Fällen dagegen macht der anatomische Zustand der Pflanzenreste den Eindruck, dass ausschliesslich Mineralisierungsvorgänge - ohne gleichzeitige Humifizierung - stattgefunden haben.

Die besten Erhaltungszustände ganzer Pflanzenteile, bei denen noch das mehr oder weniger unveränderte histologische Bild festzustellen ist, findet man in günstigen Fällen bei Laubmoosen (Bryales) (Abb.1, 4) und Sphagnen sowie bei den Wurzeln und Achsen der Holzpflanzen (Abb.7, 9). Auch gewisse Früchte und Samen zeichnen sich durch sehr gut erhaltungsfähige Teile aus. Von den krautigen Pflanzen findet man im Torf dagegen nur selten mehr als einzelne, offenbar besonders gut erhaltungsfähige Teile bzw. Gewebe. Dabei handelt es sich oft nur um Epidermen (Abb.2) - gelegentlich unter Einschluss subepidermaler Schichten - mitunter sogar nur um die Epidermisaussenwände (Abb.2, 20), und dazu treten gelegentlich verholzte Teile der Achsen (z.B. bei *Eriophorum vaginatum*). Eine gute Erhaltung des Zellverbandes von Parenchymen - sei es von Blättern oder Blattscheiden, sei es von Erdsprossen oder von Wurzeln - kommt zwar hin und wieder vor (Abb.1, 5, 6, 1 gehört jedoch zu den Ausnahmen. Der meist nur mässig gute Erhaltungszustand der Mehrzahl der Pflanzenreste im Torf steht im Einklang mit den relativ hohen Stoffverlusten, die offenbar stets mit der Torfbildung verbunden sind.

Parenchyme und andere Gewebe fallen nicht nur bei den oberirdischen und oberflächennahen Organen der torfbildenden Pflanzen - sofern diese überhaupt figurierte Reste hinterlassen - der Zersetzung zum Opfer, vielmehr fehlt auch bei den Resten tief reichender Organe (Wurzeln oder Erdsprosse) im allgemeinen ein Teil der ursprünglichen Gewebe (Abb.4, 13, 16). Derartige Bilder weisen darauf hin, dass auch in etwas tieferen Bereichen eines wachsenden Moores noch Zersetzungsprozesse ablaufen können. Eine Humifizierung findet dort aber wahrscheinlich nicht statt; dafür spricht, dass am Ort der verschwundenen Gewebe in der Regel keine Humifizierungsprodukte festzustellen sind.

Wie weit einzelne Gewebe vollständig verschwunden sein können, ohne dass Humifizierungsprodukte zurückgeblieben sind, lässt sich durch mikroskopische Untersuchung allerdings nicht mit Sicherheit beurteilen; denn im lebenden Moor muss stets mit Vorgängen einer wenigstens geringfügigen Stoffverlagerung gerechnet werden, sei es im gelösten Zustand oder durch die Tätigkeit von Tieren. In der Tiefe der Moore dürften solche Umlagerungen jedoch eine um vieles geringere Rolle spielen als in den oberflächennahen Schichten.

Humifizierungsprodukte in Pflanzenresten

Die Produkte von Humifizierungsvorgängen, die besonders im oberflächennahen Bereich der wachsenden Moore ablaufen dürften, fallen bei der Untersuchung der Torf-Dünnschnitte schon durch ihre dunkelbraune Farbe meist besonders ins Auge. Ihre Erscheinungsbilder sind dabei ausserordentlich verschiedenartig; sie reichen von blossen Bräunungen der Zellwände noch erhaltener Gewebeteile und scharf umschriebenen braunen Ablagerungen im Inneren von Zellen oder Interzellularräumen bis hin zu schlierenartigen oder krümelig erscheinenden Huminstoffbereichen, deren pflanzliche Ausgangsstrukturen nicht mehr vorhanden oder nur noch zu vermuten sind.

Starke Bräunungen der Zellwände findet man besonders auffällig beim Mesophyll von Laubblättern (Abb.5, 6, 17), an dem übrigens immer auch Bereiche mit mehr oder weniger zerstörten Zellen zu erkennen sind. Auch erhalten gebliebene Sklerenchyme krautiger Pflanzen sind meist stark gebräunt. Desgleichen zeigen Laubmoose vielfach eine starke Bräunung; *Sphagnum*-reste sind allerdings meist hell. Dunkelbraun erscheinen in den Schnitten weiterhin meistens die Rinden und Periderme (Abb.7, 9, 10, 11); hier sind jedoch nicht nur die Zellwände gefärbt, sondern die Zellen sind oft auch völlig mit braunen Substanzen angefüllt. Ganz besonders gilt das von den Peridermen (Abb.9, 11). Auch diese Stoffe, bei denen es sich wahrscheinlich um Oxydationsprodukte gerbstoffartiger Primärsubstanzen handelt, sind vermutlich wenigstens teilweise von huminstoffartigem Charakter. Auch die Zellen der Markstrahlen von Holzpflanzen enthalten vielfach braune Ausfüllungen oder, noch häufiger, einzelne oder mehrere braune Tröpfchen (Abb.8) in ihrem Inneren; diese verdanken ihren Ursprung offenbar den primären Zellinhaltsstoffen, denn die den Markstrahlen benachbarten

Leitungselemente und Fasern besitzen im allgemeinen keine derartigen braunen Inhaltsstoffe. In ihrer Farbe und Verteilung äusserlich ähnlich sind bräunliche Bildungen in der interzellularenreichen "Innenrinde" mancher Cyperaceen-wurzeln (Abb.12, 14); auffälligerweise kommen sie jedoch nicht in den Lumina der Zellen sondern gerade in den Interzellularräumen vor.

Einen charakteristischen Fall, in dem die Herkunft humifizierten Materials gerade noch festzustellen ist, obwohl Gewebestrukturen nicht mehr erhalten sind, liefern die Reste der Wollgras-Blattscheiden (Abb.6): sie laufe in typischen, eng schichtigen Bündeln mehr oder weniger wellig zwischen den übrigen Komponenten der Torfe hindurch. Dabei erscheinen die Epidermisreste in der Regel als schmale braune Bänder, an denen kaum noch Strukturen der ursprünglichen Epidermiszellen und nur gelegentlich Reste der subepidermalen Baststränge zu erkennen sind.

Von derartigen Erscheinungsformen gibt es zahlreiche Übergänge zu völlig diffus verteilten Vorkommen von Huminstoffen. Im frischen, wassergesättigten Torf bilden diese dabei vielfach ein die figurierten Pflanzenreste und die wassererfüllten Hohlräume dicht umgebendes, zusammenhängendes Gefüge; im Dünnschnitt erscheinen sie dagegen, bedingt durch den mit der Anwendung des Einbettungsmittels (Cremolan) verbundenen und bezweckten Wasserentzug, meistens in kleinere Teilstücke zergliedert, und nur selten hängen noch grössere Bereiche zusammen (Abb.19).

Die Art und Weise der Schrumpfung, besonders die Grösse, Form und Richtung der Schrumpfungsrisse, stellen dabei zweifellos kennzeichnende Merkmale der Torfe dar, in denen sich spezifische Eigenschaften der (ungeschrumpften) Ausgangssubstanz widerspiegeln. Da im übrigen auch in den obersten Lagen wachsender Moore hin und wieder Zeiten mit trockeneren Bedingungen auftreten, dürfte vielfach schon im fertig abgelagerten Torf eines noch nicht entwässerten Moores eine kleinere oder grössere Anzahl von Schrumpfungsrissen vorgebildet sein, so dass die Cremolananwendung diese Strukturen dann also lediglich sichtbar macht und sie keineswegs neu entstehen lässt.

Verschiedentlich scheint eine feine Verteilung von dunklen, nicht mehr zellulär strukturierten Komponenten jedoch auch von Natur aus vorhanden zu sein; manchmal lässt sich dabei zeigen, dass die dunkelgefärbte Komponente der Torfe nicht völlig zersetzt ist, sondern zahlreiche fein verteilte Fragmente von erst wenig veränderten Pflanzenresten enthält. Hierauf wird noch zurückzukommen sein.

Zur Erhaltungsfähigkeit ursprünglicher Pflanzenstoffe

In morphologisch gut erhaltenen Pflanzenresten lässt sich vielfach noch Zellulose nachweisen, sei es durch Mikroreaktionen, sei es - besonders einfach - auf polarisationsoptischem Wege. Das gilt vornehmlich für die Blätter und Stengel von Moosen, das gilt aber auch für manche Reste von Kräutern (Abb.3, 21) und Holzpflanzen. Jedoch trifft man nicht selten auch auf Strukturen, die - bei guter histologischer Erhaltung - keinerlei Zellulo reaktion mehr zeigen (Abb.2, 3). Auch dabei kann es sich um Reste von krautigen wie von Holzpflanzen handeln. Das ist ein Hinweis darauf, dass der gute Erhaltungszustand von Gewebestrukturen keineswegs als Beweis für noch erhaltene primäre Zellwandbausteine aufgefasst werden darf. (Für das Lignin gilt Entsprechendes wie für die Zellulose). Vielmehr können die ursprünglichen Gewebestrukturen auch nach stärkerer stofflicher Umwandlung des primären Materials unverändert erhalten bleiben. Diese stofflichen Veränderungen, die zweifellos auch mit nennenswerten Stoffverlusten verbunden sind, werden im Dünnschnitt an einer mitunter beträchtlichen Schrumpfung von histologisch noch unversehrt erscheinenden Pflanzenteilen deutlich (so besonders an Holzresten: Abb.10).

Stark humifizierte Bereiche der Schnitte erweisen sich oft, wie zu erwarten, als frei von Zellulose. Das ist aber keineswegs immer so: In dem erwähnten Fall, wo die völlig zersetzt erscheinende Komponente aus fein zerteilten Pflanzenfragmenten besteht, zeichnen sich diese nämlich meist durch deutliche Zellulosereaktion aus; besonders häufig findet man das bei stark zersetzten Hochmoortorfen (Abb.21).

Zum Zersetzungs- und Torfbildungsvorgang

Ebenso wie den Torfschnitten manches über die Zersetzungsprodukte zu entnehmen ist, ergeben sich aus ihnen auch Hinweise auf den Zersetzungsvorgang selbst: Dabei ist die Rolle, die die Pilze hier spielen, besonders auffällig; denn immer wieder sind in den Schnitten Pilzhyphen zu sehen, teilweise im mehr oder weniger zersetzten Material, teilweise in noch figuriert erhaltenen Pflanzenresten (Abb.8, 18). In Hölzern und Moosstengeln, aber auch in anderen Organen verlaufen die Pilzhyphen bevorzugt in der Längsrichtung; sie sind demzufolge besonders auffällig in Längsschnitten (Abb.8, 18), während sie in Querschnitten weniger in Erscheinung treten. Auch Pilzsporen sind häufig in Torfen anzutreffen; sie scheinen dabei in Hochmoortorfen häufiger zu sein als in Niedermoortorfen. Dasselbe gilt auch für die sklerotienartigen Gebilde, die in sehr verschiedenen Grössen vorkommen und vornehmlich in Hochmoor- oder hochmoor-artigen Torfen anzutreffen sind, wenn sie auch in Niedermoortorfen nicht fehlen.

Dass Bakterien bei der Zersetzung neben den Pilzen immer eine Rolle spielen, gilt auch für die Torfbildung und auch im Hochmoor. Ihr Nachweis in Torfschnitten, der vielleicht auf fluoreszenzmikroskopischem Wege denkbar ist, ist jedoch bislang noch nicht versucht worden.

Die Tätigkeit von Boden-Kleintieren lässt sich in Schnittpräparaten durch die vorhandene Losung nachweisen. Dass Tiere in den obersten Schichten entwässerter, nicht zu kalkarmer Moore bei der Torfzersetzung wesentlich mitwirken, ist bereits gezeigt worden (Puffe und Grosse-Brauckmann, 1963). Gelegentlich findet man ihre Spuren jedoch auch in grösseren Tiefen und in unentwässerten Mooren; die Exkremente, die hier offenbar schon aus der Zeit der Torfbildung stammen, sind dabei allerdings im ganzen nur im Inneren von holzigen Teilen oder von Blättern (Abb.17) anzutreffen, also an Stellen, wo die Gefahr einer allmählichen Verschlämmung der Exkremente besonders gering ist. Da es sich offenbar um terrestrische Organismen gehandelt hat, darf man ihre Losung als einen Hinweis dafür ansehen, dass während der Torfbildung zeitweilig verhältnismässig trockene Bedingungen geherrscht haben. Zumindest muss dass für die im Inneren zerfressenen Pflanzenteile selbst gelten, die vielleicht eine Zeitlang an der Oberfläche eines (sonst sehr nassen) Moorstandortes gelegen oder über sie hinausgeragt haben. Exkremente von Wassertieren, die in bestimmten Torfen eigentlich zu erwarten wären, haben sich bislang in Schnitten noch nicht nachweisen lassen; sie sind offenbar besonders schlecht erhaltungsfähig.

Durch diese Befunde bestätigt sich die Vorstellung vom Torfbildungsvorgang als einem zunächst und in der Hauptsache aeroben Zersetzungsprozess: Während der Dauer der aeroben Phase laufen biologisch gesteuerte und mit starken Stoffverlusten verbundene Humifizierungsvorgänge ab. Dafür spricht das völlige Zurücktreten der oberirdischen Organe höherer Pflanzen und, sofern sie überhaupt figuriert erhalten bleiben, ihre deutliche Humifizierung. Der aeroben Phase folgt dann noch eine Phase mit anaeroben Zersetzungsvorgängen. Hierbei scheint es jedoch zu wesentlich kleineren Stoffumsetzungen zu kommen als im aeroben Milieu, und die Vorgänge müssen als reine Mineralisierungsvorgänge aufgefasst werden. Dafür spricht das meist völlige Fehlen brauner Umsetzungsprodukte an der Stelle verschwundener Gewebe in tiefer gehenden Organen wie Wurzeln (Abb.4, 15) oder Rhizomen. Je nach der Tiefe, die derartige Organe erreichen, und je nach den sonstigen Voraussetzungen kann ihr Erhaltungszustand übrigens verschieden gut sein: vom ziemlich unversehrten Querschnitt (Abb.1, 12) bis zu völlig zerdrückten, mehr oder weniger zersetzten Resten einerseits (Abb.14) und bis zu völlig inhaltsleer gewordenen, senkrechten, rundlichen Gängen in der stärker oder schwächer zersetzten Torfgrundsubstanz andererseits (Abb. 16).

SCHLUSSBEMERKUNGEN

In der hiermit vorgelegten Abhandlung haben wir die grosse Vielgestalt der Torf-Pflanzenreste und ihrer Erhaltungszustände eingehend erörtert; denn gerade diese Merkmale liefern ein besonders reichhaltiges Material zum Studium der ersten Zersetzungsphasen, so dass dadurch einiges Licht auf die bisher noch wenig untersuchte Frage der Bildung von Feuchthumusformen fällt. Dabei mögen sich auch manche Ansatzpunkte zum Vergleich mit terrestrischen Humusformen ergeben. Deshalb dürften mikromorphologische Untersuchungen an Torfen auch über den engen Bereich rein moorkundlicher Fragen hinaus von allgemeinerem bodenkundlichem Interesse sein.

Die Untersuchungen wurden dankenswerterweise durch eine Sachbeihilfe der Deutschen Forschungsgemeinschaft ermöglicht.

SUMMARY

Microtome-section investigations of a peat profile from Teufelsmoor near Bremen

Thin sections of peat samples, taken from a peat profile of northwestern Germany, gave some ideas of the decomposition processes involved in peat formation. These processes are concluded to be prevailingly aerobic and biochemical ones: Most of the plant material is wholly destroyed by them and no remains of tissue structure are left. The small number of plant remains, being preserved in peat, as a rule are also more or less affected by destruction, the extent of it being dependent on their histology and histochemistry. In cases of good preservation of tissue the primary cell wall material is not always identifiable. On the other hand, areas of high humification, sometimes contain cellulose within small cell wall fragments.

LITERATUR

Grosse-Brauckmann, G., 1964. Zur Artenzusammensetzung von Torfen (Einige Befunde und Überlegungen zur Frage der Zersetzlichkeit und Erhaltungsfähigkeit von Pflanzenresten). Ber. Deut. Botan. Ges., 76 : 22–31.

Puffe, D. und Grosse-Brauckmann, G., 1963. Mikromorphologische Untersuchungen an Torfen. Z. Kulturtech., 4 : 159–188.

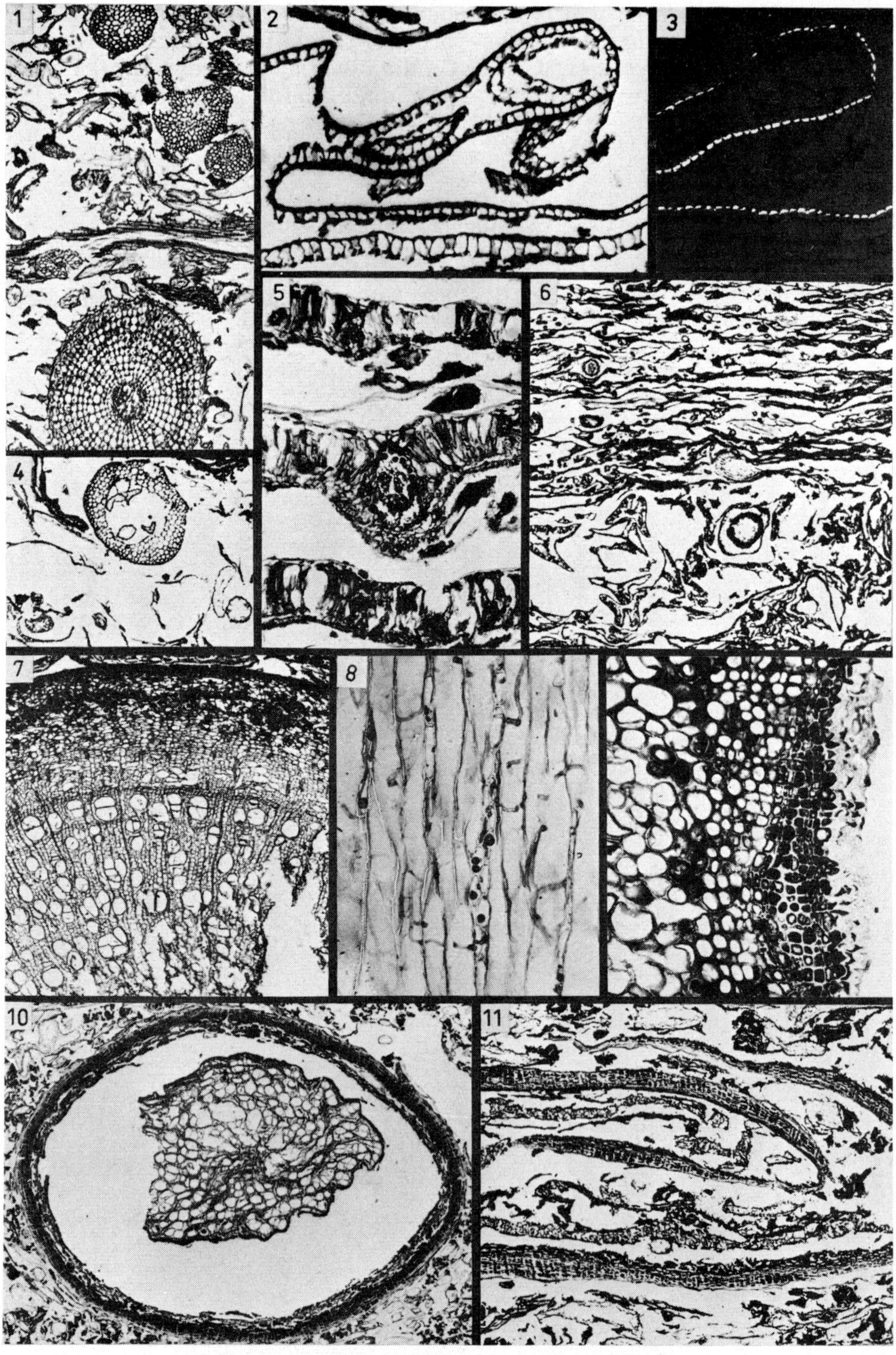
1
2
3
4
5
6
7
8
10
11

ABB.1–11.

Abb.1 (x 32). Oben drei Querschnitte von Laubmoos-Stengeln (cf. *Drepanocladus* spec.), zwei davon sehr gut erhalten. Unten ein Querschnitt durch eine recht gut erhaltene Seggenwurzel (*Carex limosa*): Der Zentralzylinder ist weitgehend zersetzt, auch die Epidermis ist nur in Resten erhalten (einige Wurzelhaare erkennbar); die "Innenrinde" ist histologisch fast vollständig, enthält jedoch in den meisten Interzellularräumen sowie in einem Teil der Zellumina dunkle Ablagerungen. Der Schnitt enthält ausserdem, teils quer, teils längs getroffen, zahlreiche feine Würzelchen und weitere Moosreste.
Abb.2 (x 64). Zwei Epidermen von *Menyanthes* -Blattscheiden, quer getroffen ganz oben Teile einer dritten, mehr längs. Ausserdem fünf Querschnitte von Resten feiner Wurzeln. Innenwände der Epidermis teilweise zerstört, verdickte Aussenwände gut erhalten (wechselnde Wanddicke durch z.T. schiefe Orientierung vorgetäuscht).
Abb.3. Wie Abb.2 (rechter Teil), aber im Polarisationsmikroskop bei gekreuzten Polarisationsebenen: bei der unteren Epidermis keine Zellulosereaktion, bei der oberen nur die Aussenwand (zum grossen Teil) aufleuchtend.
Abb.4 (x 32). Moosstengel (wohl *Drepanocladus* spec.), zentrale (dünnwandige) Teile verschwunden, hier mehrere feine Kräuterwurzeln eingewachsen. Weitere Wurzelreste unten rechts im Bild, weitere Moosreste unten links.
Abb.5 (x 64). Verhältnismässig gut erhaltene Weidenblätter, zu mehreren übereinanderliegend. Hier und da sind Reste der Epidermis und der Kutikula zu erkennen. Das kleinzellige Schwammparenchym ist undeutlicher als das (zweischichtige) Palisadenparenchym, dessen Zellwände sämtlich stark gebräunt und z.T. auch ganz verschwunden sind. Im mittleren Blatt ist ein Seitennerv getroffen, dessen Leitbündel weitgehend zerstört oder humifiziert ist, während die Zellen der Leitbündelscheide sich durch braune, etwas geschrumpfte Füllungen ihrer Lumina gut voneinander abheben.
Abb.6 (x 16). Obere Hälfte: stark humifizierte Wollgras-Blattscheidenreste im Querschnitt; zwischen ihnen - in der Wiedergabe nur undeutlich zu erkennen - einzelne *Sphagnum*-Stengel sowie mehrere quer getroffene Reste von Wollgras-Sklerenchymfaser-Strängen; ausserdem wohl auch feine Wurzeln von Ericaceen. Untere Hälfte: neben Wollgrasresten zwei beblätterte *Calluna*-Zweige im Querschnitt; der eine (links) mit sehr gut erhaltenen Blättern (ihr Mesophyll ist, soweit noch erhalten, stark gebräunt).
Abb.7 (x 16). Birkenwurzel, etwa 5-jährig, zentraler Teil des Holzes stark angegriffen und zerfallend; periphere (jüngste) Zonen aber noch sehr gut erhalten, obwohl leicht geschrumpft (rechts eine Schrumpfungsspalte). Rinde ebenfalls histologisch gut erhalten (mit dunkler Sklereidenzone im äusseren Teil), Periderm schmal und dunkelgefärbt.
Abb.8 (x 128). Birken-Wurzelholz im Längs- (Tangential-)Schnitt. In den Zellen des (zweireihigen) Markstrahls auffällige braune Tröpfchen. In Fasern und Gefässen (diese rechts und links nur noch teilweise im Bild) zahlreiche Pilzhyphen, z.T. unscharf abgebildet.
Abb.9 (x 128). Periderm und anschliessende Rindenzone eines 7-jährigen Weidenstämmchens oder -astes im Längsschnitt; die Lumina zahlreicher Zellen mit auffälligen (teilweise allerdings wohl auch schon im lebenden Zustand vorhanden gewesenen) dunkelbraunen Ausfüllungen.
Abb.10 (x 16). Quer getroffene dünne Birkenwurzel; Periderm gut, Bast mässig gut erhalten; das Holz macht histologisch durch die zahlreichen unversehrten Gefässe einen gut erhaltenen Eindruck (links ist bei der Präparation ein kleines Stück abgebrochen), es ist aber stark geschrumpft und hat sich im Bereich des ursprünglichen Kambiums, wo eine schmale, stark humifizierte Zone zur Ausbildung gekommen ist, von der Rinde gelöst. In der Umgebung sind zahlreiche feine Würzelchen, meist quer geschnitten, und humifiziertes Material zu erkennen.
Abb.11 (x 32). Quer getroffene Birkenreste, nach völligem Verschwinden des Holzes nur noch aus stark veränderten Resten des Bastes sowie dem ziemlich gut erhaltenen Periderm (mit dunklen Zellinhalten) bestehend, zerrissen und stark zusammengefaltet, mit einigen eingewachsenen kleinen Kräuterwurzeln (solche auch ausserhalb: oben und unten) neben humifiziertem Material.

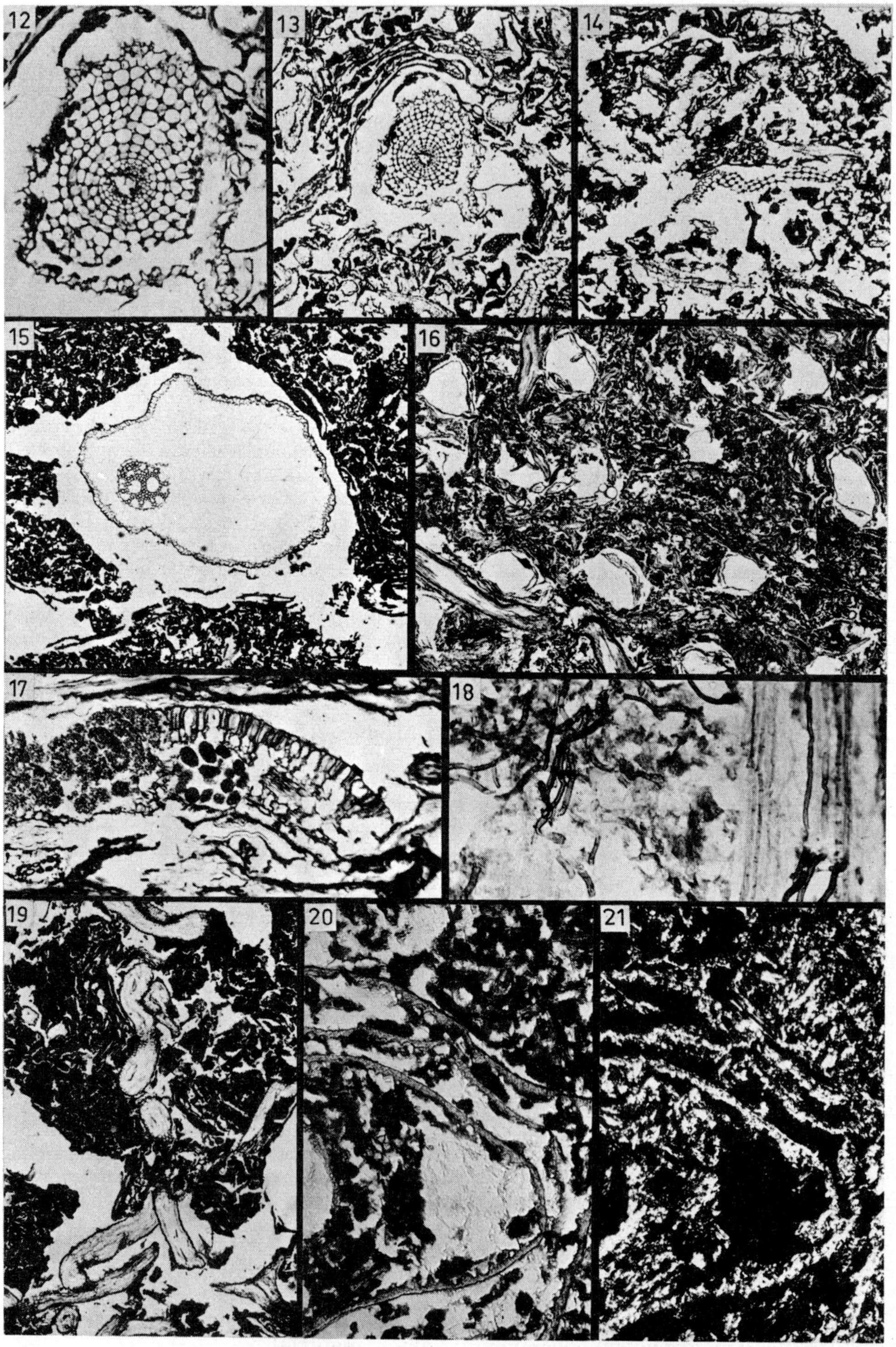
12
13
14
15
16
17
18
19
20
21

Abb.12 (x 64). Rest einer Cyperaceenwurzel, quer getroffen. Zentralzylinder und äussere Rindenzonen weitgehend zerstört, innere Rinde gut erhalten. Hier in ("karoförmigen") Interzellularräumen zum Teil dunkle Ablagerungen.
Abb.13 (x 32). Übersicht zu Abb.12. Der von der Wurzel bei ihrem Eindringen in das sie umgebende Material geschaffene Hohlraum tritt deutlich hervor.
Abb.14 (x 32). Sehr kleiner, gerade eben noch identifizierbarer Rest einer Cyperaceenwurzel in einem weitgehend zersetzten Material.
Abb.15 (x 32). Grosse, in stark humifiziertes Material eingewachsene Monocotylenwurzel mit völlig zersetzter Rinde: Exodermis und Zentralzylinder dagegen gut erhalten.
Abb.16 (x 8). Durch Einwachsen von Wurzeln in stark humifiziertes Material entstandene, sehr dicht angeordnete Löcher; teilweise mit weitgehend zersetzten Resten der ursprünglichen Wurzeln. Oben links und unten links: längs getroffene *Sphagnum*-Stengel.
Abb.17 (x 64). Quer getroffenes Birkenblatt, teilweise im Inneren zerfressen. Epidermis fast überall verschwunden, Kutikula (hell) aber noch teilweise erhalten, desgleichen Teile des Palisadenparenchyms (stark gebräunt). Losung teilweise noch deutlich geformt, teilweise mehr oder weniger zerfallen. In der Umgebung des Blattes stark humifizierte, nicht näher bestimmbare Reste.
Abb.18 (x 128). Pilzhyphen, teilweise (links) in völlig humifiziertem Material, teilweise im Gewebe eines Seggen-Rhizoms (*Carex limosa*, Hypoderm, von der Fläche getroffen; Hyphen dabei in der Längsrichtung verlaufend).
Abb.19 (x 32). Feine Kräuterwurzeln als einzige figurierte Reste in völlig humifizierter Grundsubstanz.
Abb.20 (x 64). Niederblattreste von *Trichophorum caespitosum* (im wesentlichen nur die stark verdickten Epidermis-Aussenwände der Blattaussenseiten) und dunkles, weitgehend humifiziert erscheinendes Material (wohl zum grossen Teil aus den übrigen Geweben von *Trichophorum* hervorgegangen).
Abb.21 (x 64). Wie Abb.20, aber im Polarisationsmikroskop bei gekreuzten Polarisationsebenen. Ausser der Epidermis leuchten auch ausgedehnte Bereiche in dem weitgehend humifiziert erscheinenden Material auf.

ALLGEMEINE CHARAKTERISIERUNG DER FÜR DIE SCHNITTE VERWENDETEN TORFPROBEN

Abb.5, 9: Seggentorf[1] mit Weidenholz, ziemlich stark zersetzt, 242 cm unter der Oberfläche (u.O.).
Abb.2, 3: Seggentorf[1] mit *Menyanthes* , mässig stark zersetzt, 222 cm u.O.
Abb.12, 13, 14: wie vor, 212 cm u.O.
Abb.11: Seggentorf[1] mit einigen Birkenresten, ziemlich stark zersetzt, 192 cm u.O.
Abb.19: Seggentorf[1] stark zersetzt, 187 cm u.O.
Abb.10: Seggentorf[1], birkenholzreich, stark zersetzt, 180 cm u.O.
Abb.1, 4, 18: Moos-Seggentorf, ziemlich schwach zersetzt, 126 cm u.O.
Abb.15: Birkenbruchtorf, stellenweise mit *Sphagnum*, stark zersetzt, 126 cm u.O.
Abb.7, 8: wie vor 124 cm u.O.
Abb.16, 17: wie vor 120 cm u.O.
Abb.6: Hochmoortorf, wollgrasreich, stark zersetzt, 97 cm u.O.
Abb.20, 21: Hochmoortorf mit *Trichoforum caespitosum*-Resten, stark zersetzt, aus einer Tiefe wenig unterhalb des Schwarztorf-Weisstorf-Kontaktes, 180 cm u.O. (Diese Probe stammt nicht von dem beschriebenen Profil des Kollbecksmoores sondern aus einem nicht weit davon entfernten Profil südlich von Bornreihe).

Bei den Schnitten zu den Abb.9, 12, 13, 15, 16, 18, und 19 lag die Schnittebene horizontal; die übrigen Schnitte waren Vertikalschnitte. Den Abb.5, 8, 17, 18, 20 und 21 liegen ungefärbte Schnitte zugrunde, den übrigen mit Gentianaviolett gefärbte Schnitte.

[1] Seggentorf = "Riedtorf".

CLASSIFICATION OF PLASMIC FABRICS OF SOIL MATERIALS

R. BREWER

Commonwealth Scientific and Industrial Research Organization, Division of Soils, Canberra (Australia)

INTRODUCTION

In an earlier paper (Brewer, 1960), the term *matrix* was proposed for the material in which pedological features are usually embedded within peds or in unaggregated soil materials. In sedimentary petrology, however, "matrix" has a more specific meaning. Sedimentary rocks are considered to consist of a framework of mineral grains with finer grained matrix packed between the members of the framework. Framework and matrix are defined by the size range of their members, the grains in the former being greater than 30 μ in diameter and those of the latter being less than 30 μ in diameter (Spencer, 1963). To avoid confusion, therefore, *s-matrix* (that is, matrix plus skeleton grains) of a soil material is defined as the material within the simplest (primary) peds, or comprising apedal soil materials, in which the pedological features occur; it consists of the plasma, skeleton grains and voids that do not occur in pedological features other than plasma separations (Brewer and Sleeman, 1960). Plasma separations are treated as part of the s-matrix because they are not recognizable three-dimensional entities. Pedological features, too, have s-matrices.

S-matrices are not entities with definite boundaries; they merely form the background material within which pedological features occur. Therefore, they have no attributes of size, shape and arrangement as units, so they must be classified solely on their internal characteristics, that is, on the size, shape and arrangement of the plasma (plasmic structure - Brewer, 1960), skeleton grains and voids, and the relationships of the plasma and skeleton grains. These characteristics are the elements of *basic structure* (Brewer and Sleeman, 1960; Brewer, 1960) which is, in effect, the structure of the s-matrix. The *plasmic structure* of soil materials (and pedological features), therefore, is the size, shape and arrangement of the plasma grains, and the voids between them, in the s-matrix without reference to the skeleton grains of larger intrapedal voids.

CLASSIFICATION

The plasmic fabric is the organization of the constituents of the plasma and the associated very small voids, which result from the packing of plasma grains, within the s-matrix of the soil material or of individual pedological features. Individual grains and voids are largely beyond the resolving power of normal optical microscopes and special techniques, such as electron microscopy, are necessary for direct observation of them. Failing this, the voids of the plasmic fabric are assumed to be simply packing voids and receive no further attention. A good deal of morphological description of plasmic fabric, however, is possible by optical methods in thin section on the petrological microscope by interpretation of the extinction patterns of the various kinds and arrangements (distribution and orientation patterns) of plasma grains, provided the optical and physical characteristics of the plasma grains are known. Although the majority of the constituents of the plasma (clay minerals, oxides and hydroxides, soluble salts and rock-forming minerals) are crystalline and anisotropic, they appear isotropic in thin section if the individual grains are unorientated with regard to their neighbours; this is because the grains are very small (less than 2 μ in thickness) compared with the thickness of a thin section (25 μ). Anisotropy of the plasma in thin section, therefore, is due either to translucent, anisotropic plasma grains that have some degree of preferred orientation with regard to their neighbours or to larger crystals of anisotropic plasma; the former may form recognizable units which are apparently identical with the so-called *domains* (Aylmore and Quirk, 1959; Emerson, 1959).

Description and classification of plasmic fabrics, therefore, is based on interpretation of optical properties between crossed nicols, especially extinction phenomena due to:

(1) visible crystals of plasma, (2) the kind and degree of orientation of the plasma grains, including distinctions between opaque, isotropic and anisotropic plasma, (3) the kind and degree of preferred orientation of domains, if they occur, which causes flecked and striated orientation patterns (Brewer and Sleeman, 1960), and particularly:(4) the kind and degree of development of plasma separations, which can be described in terms of size, shape, and basic and referred distribution and orientation patterns (Brewer and Sleeman, 1960).

An important difficulty in classification of plasmic fabrics is the fact that their characteristics form a complete continuity. At any magnification there is a continuous gradation from plasma that appears opaque, or isotropic, to plasma that appears entirely anisotropic, and from fabrics that appear to have an entirely flecked extinction pattern to those which appear to consist entirely of a striated orientation pattern; in the latter, the intermediate stages form a continuity both in degree of development of plasma separations with striated orientation and in the distribution patterns of the plasma separations. Similarly there is a gradual transition from plasmic fabrics that consist entirely of anisotropic domains (whether with striated or flecked extinction patterns), through plasmic fabrics in which the boundaries of the domains are not apparent but the plasma has an indistinct, undulose extinction pattern, to plasmic fabrics that are either opaque or isotropic even at the highest magnifications and with the highest possible light intensity. The best that can be done in the classification of plasmic fabrics, therefore, is to select typical modal fabrics that can be given appropriate names.

The following classification is proposed for the typical plasmic fabrics observed to date. They are classified first into broad groups on their general anisotropy and extinction patterns in thin section between crossed nicols on the petrological microscope; each of these groups can be subdivided further according to the details of their extinction patterns. The classification is summarized in Table I.

Asepic[1] plasmic fabrics

Asepic plasmic fabrics have dominantly anisotropic plasma with anisotropic domains which are unoriented with regard to each other, that is, they have a flecked extinction pattern; there are virtually no plasma separations. They are subdivided into two subgroups:

Argillasepic[2] fabric

The plasma of the argillasepic fabric (Fig.1) consists dominantly of anisotropic clay minerals and exhibits a flecked orientation pattern with recognizable domains.

Silasepic[3] fabric

The soil material of the silasepic fabric has a wide range of particle sizes with relatively high proportions of silt-size grains so that domains are difficult to recognize; the matrix has a flecked extinction pattern.

Sepic plasmic fabrics

Sepic plasmic fabrics have recognizable anisotropic domains with various patterns of preferred orientation, that is, plasma separations with a striated extinction pattern are present. They are subdivided on the characteristics of the plasma separations into:

Insepic[4] fabric

The plasma of the insepic fabric separations occur as isolated patches or islands, within the dominantly flecked plasma which may have a very weak striated orientation at high magnifications.

Mosepic[5] fabric

Mosepic fabric (Fig.2) is an extreme development of insepic fabric; the patches with striated orientation may adjoin each other, or be separated by small areas of plasma with flecked orientation, but have no preferred orientation with regard to each other.

[1] The stem of the adjective "sepic" is derived from "separation" so that "asepic" means "without plasma separations", and "sepic" means "with plasma separations".
[2] "Argill-" from "argillaceous".
[3] "Sil-" from "silt".
[4] "In-" from "insular".
[5] "Mo-" from "mosaic".

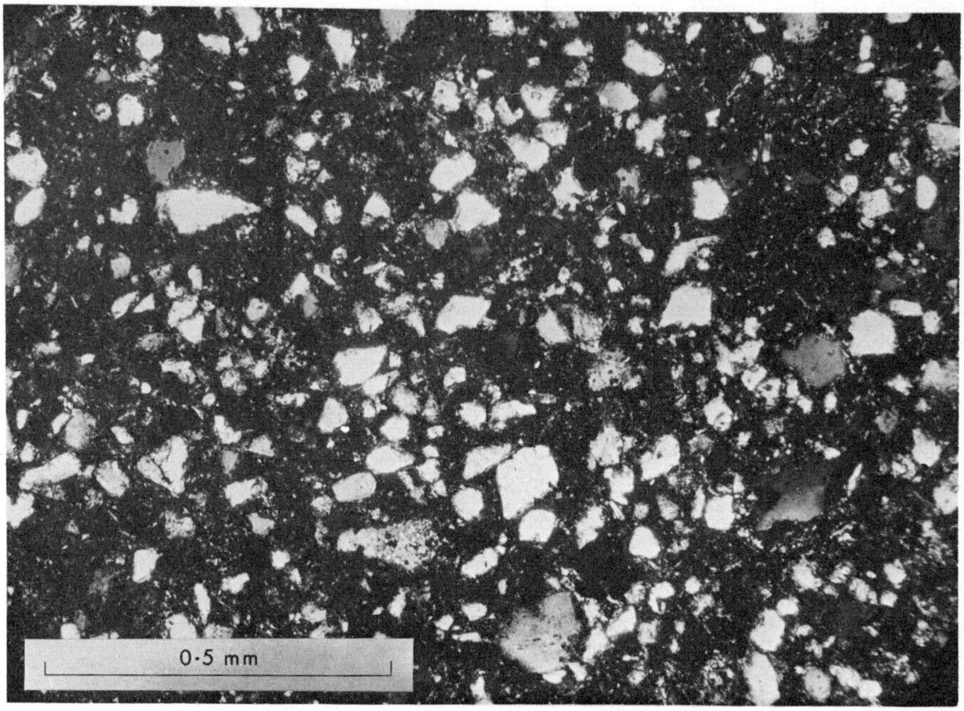

Fig.1. Argillasepic plasmic fabric. Crossed nicols. There are no zones or patches of plasma with striated extinction patterns, but small anisotropic flecks (domains) are observable.

Vosepic[1] *fabric*

The plasma of the vosepic fabric (Fig.3) has a flecked orientation pattern, but plasma separations with striated orientation occur associated with and adjoining the walls of voids; the striated orientation of the plasma separations is dominantly parallel to the walls of the void.

Skelsepic[2] *fabric*

The plasma of the skelsepic fabric (Fig.4) has a flecked orientation pattern, but plasma separations with striated orientation occur adjoining the surfaces of skeleton grains; the striated orientation of the plasma separations is dominantly parallel to the surfaces of framework units.

[1] "Vo-" from "void".
[2] "Skel-" from "skeleton grains".

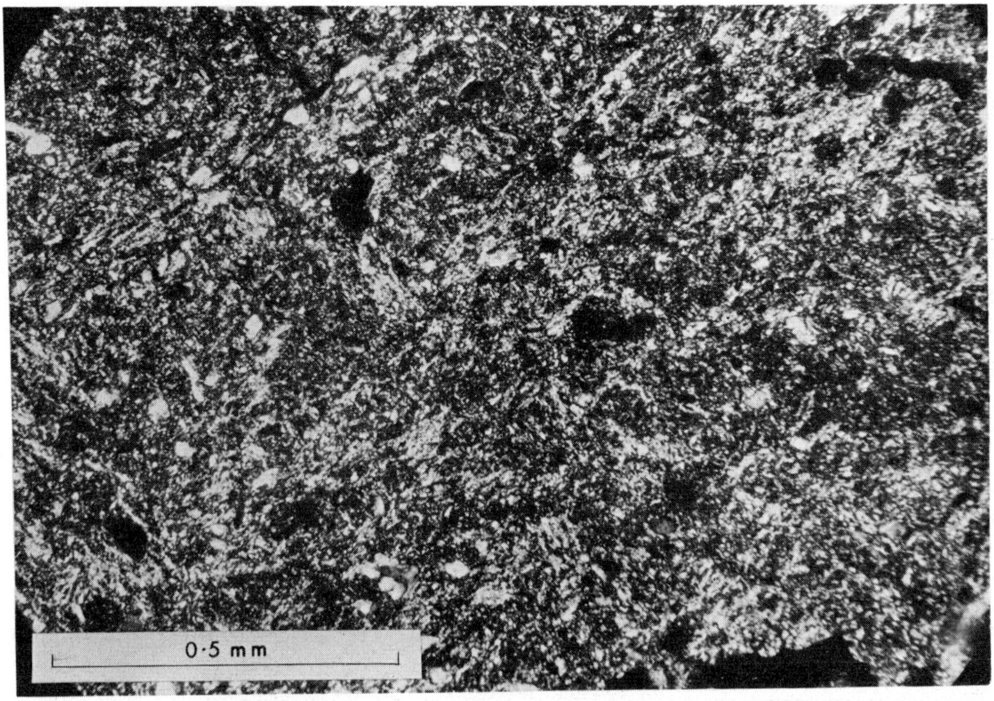

Fig.2. Mosepic plasmic fabric. Crossed nicols. The whole of the plasma consists of relatively large patches with a striated extinction pattern; the patches virtually adjoin each other.

Masepic[1] *fabric*

The plasma of the masepic fabric (Fig.3) has a flecked orientation pattern, but plasma separations occur as zones within the s-matrix apparently not associated with the walls of voids or the surface of skeleton grains; the striated orientation is elongated parallel to the length of the zones. All the zones may be subparallel or there may be two or three sets of subparallel zones at definite angles to each other; the latter may be termed *bimasepic* and *trimasepic* fabric respectively.

Lattisepic[2] *fabric*

The lattisepic fabric is essentially a variety of bimasepic fabric; the plasma generally has a flecked orientation pattern, but acicular and prolate domains occur in a lattice-like pattern, that is, there are two sets of very short, discontinuous plasma separations usually oriented approximately at right angles to each other.

[1] "Ma-" from "matrix".
[2] "Latti-" from "lattice".

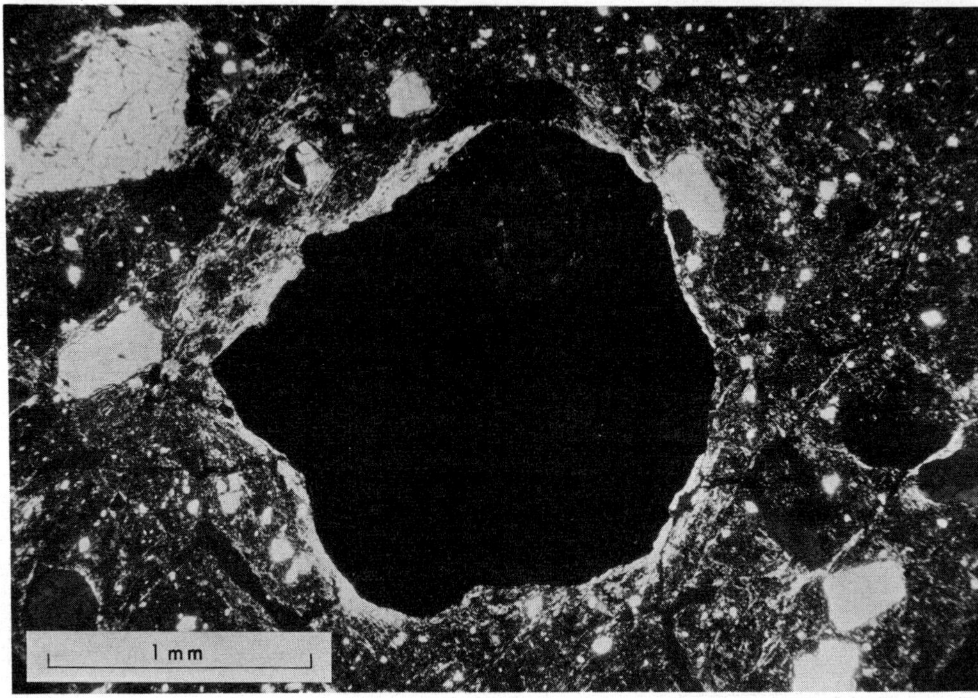

Fig.3. Skelsepic plasmic fabric. Crossed nicols. The large central dark area is a skeleton grain at extinction. The bright zone around it is a plasma separation with a striated extinction pattern whence is derived the plasmic fabric name skelsepic. The other larger grains, which are not at extinction, also have plasma separations associated with their surfaces.

Omnisepic[1] *fabric*

All the plasma of the omnisepic fabric exhibits a complex striated orientation pattern; in thin section the striae have an intricate pattern as regards direction of preferred orientation.

Each of the named sepic–plasmic fabrics is characterized by a single kind of plasma separation; this is an idealised concept. Soil materials usually exhibit combinations of different kinds of plasma separations. These more complex fabrics can be named rather specifically by using combinations of the prefixes used for the simple types, with the convention that the best developed fabric is written last, and qualification by addition of other appropriate prefixes. A commonly occurring plasmic fabric with well developed plasma separations adjoining the walls of planar voids and less well developed plasma separations associated with the surfaces of skeleton grains is adequately named as *skel-vosepic* fabric.

The named fabrics can be subdivided further according to the details of the characteristics of the plasma separations, particularly the relative

[1] "Omni-" meaning "all".

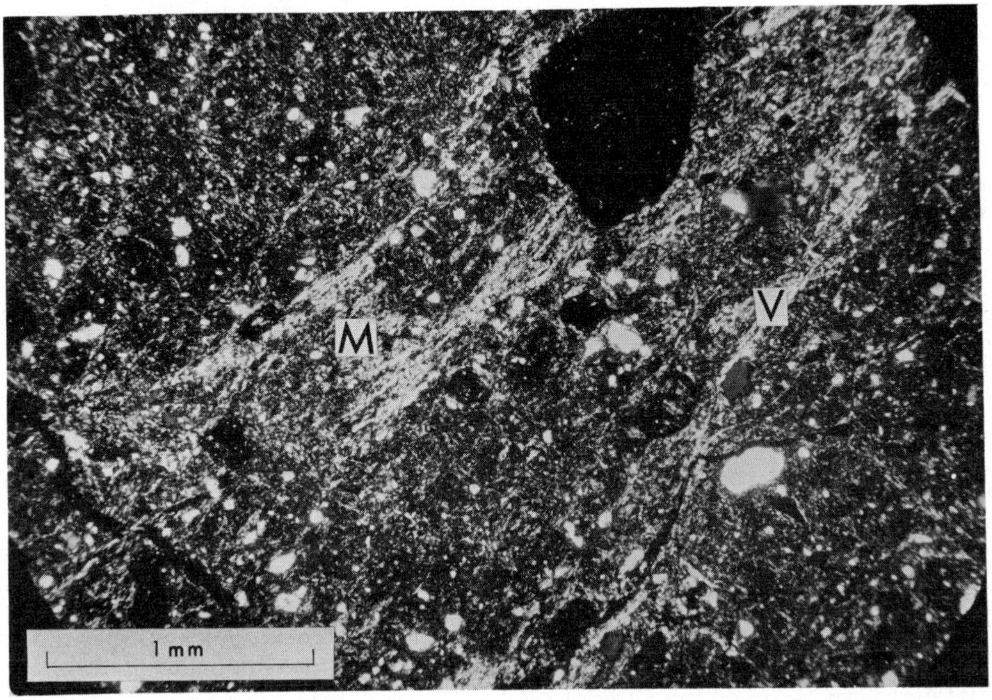

Fig.4. Vo-masepic plasmic fabric. Crossed nicols. M is the centre of a zone of plasma with a striated extinction pattern (plasma separation) not associated with a void (masepic), and V is the centre of a thin zone of plasma with a striated orientation pattern associated with a planar-shaped void (vosepic); this combination of plasma separations defines vo-masepic fabric.

width of the zones of plasma separations in thin section, and the degree of development of the striated orientation pattern within them, that is, the degree of preferred orientation. Much more experience is necessary before subgroups can be named according to these criteria. For the present the named fabrics are qualified for the width of plasma separations (*broad:* wider than 200 μ; *normal:* between 200–25 μ; *narrow:* 25 μ) and degree of preferred orientation within plasma separations (*strong:* >60% of the material exhibits the linear arrangement; *moderate:* 20–60% exhibits the linear arrangement; *weak:* 20% exhibits the linear arrangement); for example, broad, weak vosepic fabric and narrow, strong vosepic fabric.

Undulic[1] *plasmic fabrics*

The plasma of undulic plasmic fabrics is indeterminate, that is, it is practically isotropic at low magnifications and weakly anisotropic with faint

[1] The stem of "undulic" is derived from "undulose" which describes the extinction pattern.

undulose extinction at high magnifications and high light intensities; domains are not recognizable, or are indistinct, even at high magnifications. Observations on undulic fabrics are so sparse that little is known of the range of variation that may occur. One variation has been observed and named *inundulic* fabric in which the normal undulic fabric is modified by the presence of small, rather indistinct domains.

Sepic fabrics have been defined as being dominantly anisotropic with recognizable domains and with various patterns of plasma separations, that is, they are basically asepic fabrics with superimposed plasma separations. Fabrics that are basically undulic or inundulic usually do not exhibit plasma separations, but a number of such fabrics have been observed with plasma separations associated with planar voids. Such a fabric can be classified quite adequately as *vosepic undulic* fabric, and any other such variants can be classified similarly by using the sepic plasmic names as adjectives qualifying undulic and inundulic.

Isotic[1] *plasmic fabrics*

Isotic plasmic fabrics have apparently isotropic plasma, that is, the plasma is indeterminate (Brewer and Sleeman, 1960) at even the highest possible magnifications with high light intensity. It may be possible to subdivide isotic fabrics on the basis of the cause of the isotropic characteristics, that is, true isotropy of the minerals and opacity due to iron oxides, organic matter, and so on.

It is also possible for plasmic fabrics that are basically isotic to exhibit certain kinds of plasma separations that are typical of the sepic plasmic fabrics. These can be classified by the same means used for their counterparts which are basically undulic, for example, *vosepic isotic* fabric is basically isotic with plasma separations associated with voids.

Crystic[2] *plasmic fabrics*

In crystic plasmic fabrics the plasma is usually anisotropic and consists of recognizable crystals, usually of the more soluble fractions of the plasma. Such fabrics can be subdivided on the details of arrangement of the crystals, for example, *random crystic* fabric in which the crystals are arranged randomly, *spherulitic crystic* fabric in which the crystals form spherulites, *concentric crystic* fabric in which the crystals are arranged concentrically, and so on.

---

[1] The stem of "isotic" is derived from "isotropic".
[2] The stem of "crystic" is derived from "crystallized".

TABLE I

Morphological classification of typical modal plasmic fabrics

Fabric class	Extinction pattern	Plasma aggregates	Plasma separations
Asepic	flecked		
argillasepic		recognizable	absent
silasepic		present but difficult to distinguish from silt grains	absent
Sepic	flecked and striated		
insepic			as occasional isolated patches
mosepic			as numerous patches
vosepic		recognizable	associated with voids
skelsepic			associated with skeleton grains
masepic			as elongated zones
lattisepic			as two sets of very short discontinuous plasma separations
omnisepic			strongly developed, complex striated pattern
Undulic	dominantly undulose		
Inundulic		recognizable but rare	rare isolated patches
Isotic	isotropic (or opaque)	absent	absent
Crystic	recognizable birefringent (or isotropic) crystals	absent	absent

[1] Plasma aggregates are recognized as small optically anisotropic units composed of clay minerals; they are apparently identical with domains (Aylmore and Quirk, 1959).

INTERPRETATION

Although any manifestations of optical anisotropy of the plasma of the s-matrix can be confidently attributed to preferred orientation of the plasma grains, the processes involved in producing such preferred orientation are obscure, and interpretation must await accumulation of basic data from experimentation and fundamental studies of the nature of the constituents of the plasma and their behaviour under different conditions, and an accumulation of a bulk of supporting observational data. A number of hypotheses, however, can be proposed for the formation of the broad named groups of plasmic fabrics on the basis of the little that is known.

Meade (1960) pointed out that preferred orientation of clay-mineral grains in sediments can occur by three mechanisms:

(1) as the result of deposition from suspension, (2) by mechanical reorientation due to pressure from an overburden, and (3) by recrystallization of unstable minerals to form more stable clay minerals and micas in the later stages of compaction. He also found that there was a correlation between the size of the clay-mineral grains and the ease of mechanical reorientation by pressure; larger particles, such as illite, kaolinite and chlorite, were reoriented quite readily while smaller particles, such as montmorillonite, were not reoriented by pressure because of the relatively large effects of surface forces. Similar processes are known to operate in soil materials, but the preferred orientation observed within the plasma of the s-matrix of soil materials is not due to deposition from suspension, which would produce pedological features (Brewer and Sleeman, 1960), nor is it likely to be due to recrystallization except under special circumstances. It is probably due to the effect of pressure, not pressure due to a large overburden, but rather pressures and tensions produced by wetting and drying. These are the bases on which the following hypotheses of the formation of the various plasmic fabrics are put forward.

Asepic plasmic fabrics

Flocculation during transport and sedimentation of alluvial materials is the most probable cause of asepic plasmic fabrics. Clay-size minerals may be carried in a completely dispersed state or as floccules. Under either condition some degree of flocculation will occur immediately after dropping of the load of sediment, and some degree of preferred orientation of the individual grains will take place within the floccules as they dry from a saturated state (due to drying tensions), whereas there is no reason to expect the floccules to become oriented with regard to each other. The greater the degree of flocculation and the wider the particle size range the weaker is the degree of preferred orientation as judged by extinction phenomena in thin section (Brewer and Haldane, 1957). Also the exchangeable cations and contaminants affect the final degree of preferred orientation and the size of the floccules (Meade, 1960). It is relevant to note that many of the grey and brown soils of heavy texture (Stephens, 1956) formed on alluvial sediments in the Riverine Plain of western New South Wales have dominantly asepic fabrics.

Domains without preferred orientation may possibly be formed by uneven wetting and drying on a micro-scale within the soil mass, or they may be inherited in colluvial or mass movement material derived from an older soil material, or they may be formed by normal weathering of mineral grains which often produces equant or prolate domains that are unoriented with regard to each other. In all these cases, disturbance of the soil material by soil fauna, heaving, soil creep or mass movement may prevent preferred orientation between domains.

Sepic plasmic fabrics

Sepic plasmic fabrics differ from asepic fabrics in that some proportion of the plasma occurs as acicular to prolate domains that have a preferred orientation with regard to their neighbours to form plasma separations.

There is strong evidence for attributing the striated orientation patterns of some of the plasma separations in sepic fabrics to the effects of pressures and tensions produced by wetting and drying (e.g., Rode et al., 1960). In particular, the voids in vosepic fabrics are the sites of maximum wetting and drying where pressure due to expansion and tension due to shrinkage are at a maximum; this is exemplified in slickensided planar-shaped voids which are indisputably due to pressures of swelling and are always accompanied by vosepic fabric. Masepic fabric probably has a similar origin by uneven wetting and drying causing plastic flow without actual cracking to form recognizable planar-shaped voids; one or more directions of preferred orientation may be developed according to the conditions of wetting and drying. Skelsepic fabric is also probably the result of swelling pressures since the skeleton grains present a solid surface against which pressures can be exerted. Lafeber (1962) suggested that the patterns of plasma separations associated with the skeleton grains in a number of skelsepic fabrics can be explained by rotation and/or translocation of physically anisotropic skeleton grains under pressure. In the soil materials he studied, the plasma separations had "tails" that extended from the surface of the skeleton grains into the plasma of the s-matrix. He attributed parallel tails on opposite sides of the one skeleton grain to translocation of the grain along a line, and somewhat curved tails pointing in opposite directions on opposite sides of the one grain to rotation, the direction of rotation being indicated by the aspect of the tails. Such a mechanism acting on a rather restricted scale could cause lattisepic fabric. It should be noted, however, that plasma separations associated with the surfaces of skeleton grains in other soil materials do not have such tails.

These hypotheses are not so satisfying for the formation of insepic fabric and mosepic fabric because the distribution of the patches of plasma separations has no apparent consistent relationships with locations that are propitious for larger pressures or tensions than the rest of the plasma. The extreme development of omnisepic fabric, however, may be possible under extreme conditions. A more readily acceptable origin of mosepic and insepic fabrics is by inheritance. Many sedimentary rocks have sepic fabrics and physical weathering of such rocks to form soil materials may produce variants of mosepic fabric depending on the original pattern of orientation in the rock, the degree of disturbance, and the degree of alteration of the pattern of orientation due to wetting and drying within the soil material. If the parent material of the soil is an unconsolidated sediment that contains fragments of such fine-grained sediments, weathering, physical softening and disintegration of the fragments may form insepic fabric; this process was observed in some terrace soils from Kempsey in New South Wales. Insepic fabrics, however, probably have other unexplained origins since not all of them can be explained definitely by inheritance.

Undulic plasmic fabrics

Drying of highly flocculated colloids composed of anisotropic clay-mineral grains produces a fabric that is essentially undulic (Brewer and Haldane, 1957) provided the proportion of skeleton grains to clay-size material is relatively low. Undulic plasmic fabrics, therefore, could result from deposition of alluvial materials under appropriate conditions. Such materials, however, usually have an asepic fabric, while undulic fabric is common in such soils as residual krasnozems in which there is no suggestion of the parent material being alluvial in origin. In general there is a high correlation between undulic fabric and relatively high proportions of iron oxides, so the extinction phenomena observed may be due to the effects of high amounts of iron oxides or hydroxides. The mechanism is not understood but iron oxides are known to flocculate and stabilize clay minerals, which would tend to reduce the susceptibility of the individual grains to assume preferred orientation with regard to their neighbours. Also iron oxide and hydroxide grains are generally not platy or acicular so they have no great tendency to assume a preferred orientation themselves, and they tend to prevent preferred orientation of the clay-mineral grains with which they are intimately mixed. Certainly the low degree of anisotropy of many undulic fabrics at low magnifications can be attributed to the presence of high proportions of amorphous or very finely crystalline, unoriented iron oxides or hydroxides. Inundulic fabric is essentially an intermediate between undulic and insepic fabrics.

Isotic plasmic fabrics

Plasmic fabrics that are apparently isotropic at all magnifications and levels of light intensity may be due either to the clay minerals themselves being isotropic or, if anisotropic, very fine grained and unoriented with regard to their neighbours, or to the presence of opaque materials such as organic matter. No information is available as to the proportions, size or degrees of orientation of fine-grained crystalline materials that will cause apparent isotropy in thin section.

Crystic plasmic fabrics

Crystic plasmic fabrics are simply due to crystallization of relatively soluble fractions of the plasma and can be explained by the general laws of nucleation and crystal growth.

SUMMARY

An attempt has been made to classify the plasmic fabrics of soil materials on morphological characteristics, especially the extinction patterns observed in thin section between crossed nicols on a petrological microscope. Although plasma grains cannot be resolved by optical methods when they are of colloidal size, their arrangement can often be interpreted from the extinction patterns provided the broad classes of minerals present are known. In particular, clay-mineral grains commonly assume preferred orientations to form small, recognizable anisotropic domains, and these may have varying degrees of orientation with regard to each other to form plasma separations with striated extinction patterns. The classification is based largely on the degree of anisotropy of the plasma, the presence or absence of anisotropic domains, and the pattern of distribution of plasma separations where they are present. Hypotheses concerning the genesis of the various classes of plasmic fabrics are proposed on the basis of the known properties of the common constituents of the plasma and on processes known to operate in the formation of soil parent materials and profiles.

REFERENCES

Aylmore, L.A.G. and Quirk, J.P., 1959. Swelling of clay-water systems. Nature, 183 : 1752–1753.

Brewer, R., 1960. The petrographic approach to the study of soils. Trans. Intern. Congr. Soil Sci., 7th, Madison, 1 : 1–13.

Brewer, R. and Haldane, A.D., 1957. Preliminary experiments in the development of clay orientation in soils. Soil Sci., 84 : 301–309.

Brewer, R. and Sleeman, J.R., 1960. Soil structure and fabric: their definition and description. J. Soil Sci., 11 : 172–185.

Emerson, W.W., 1959. The structure of soil crumbs. J. Soil Sci., 10 : 235–244.

Lafeber, D., 1962. Aspects of stress-induced differential movements of fabric elements in mineral soils. Proc. Australian Road Res. Board, 1st, Canberra, pp.1059–1067.

Meade Jr., R.H., 1960. Compaction and development of preferred orientation in clayey sediments. Ph. D. Thesis, Stanford Univ., Stanford, Calif., 73 pp.

Rode, A.A., Yarilova, Ye.A. and Rashevskaya, I.M., 1960. Certain genetic characteristics of dark soils. Soviet Soil Sci., 8 : 799–809.

Spencer, D.W., 1963. The interpretation of grain size distribution curves of clastic sediments. J. Sediment. Petrol., 33 : 180–190.

Stephens, C.G., 1956. A Manual of Australian Soils. C.S.I.R.O., Melbourne, 54 pp.

VERWITTERUNGS-MIKROMORPHOLOGIE DER MINERAL-SPEZIES IN MITTELEUROPÄISCHEN HOLOZÄN-BÖDEN AUS PLEISTOZÄNEN UND HOLOZÄNEN LOCKERSEDIMENTEN

B. MEYER und E. KALK

Agrikulturchemisches und Bodenkundliches Institut der Universität Göttingen[1], Göttingen (Deutschland)

ÜBERBLICK

Wenn bisher mineraloptische Untersuchungen an den Sand- und Schluff-Korngrössen-Fraktionen von Böden und Boden-Horizonten, die aus Lockersedimenten hervorgegangen sind, durchgeführt wurden, stand meist eines der folgenden Probleme im Vordergrund:

(1) Nachweis der vertikalen sedimentären Homogenität innerhalb des Solums durch Ermittlung des Verhältnisses der schwer verwitterbaren Mineralarten (z.B., Quarz, resistente Schwerminerale) untereinander - eine der Voraussetzungen für die Bilanzierung bodenspezifischer Verwitterungs-, Stoffwandlungs- und -wanderungs-Prozesse im Profil.

(2) Herkunfts-Bestimmung und Unterscheidung der bodenbildenden Lockersediment-Schichten im Bereich und unterhalb des Solums, z.B., anhand von Leitmineralen (beispielsweise Tuffbestandteilen der Kärlicher oder Laacher Eruptionen in Lössdecken des Jungwürms).

(3) Charakterisierung des pedogenetischen Verwitterungs-Zustandes des Solums durch Ermittlung der Mengen-Quotienten von leichter verwitternden Mineralarten, z.B., Glimmern, Feldspäten, Augiten, Epidoten, Hornblenden, u.a., zu resistenten Bezugsmineralen (s.o.!) sowie der Gradienten dieser Verhältnisse innerhalb des Bodenprofils. Dies ist ein gängiger Weg für überregionale Vergleiche der Böden, Stoffmengen-Bilanzierungen, relative Altersbestimmungen, Ermittlungen der potentiellen Nährstoff-Leistung und Klärung anderer Fragen mehr.

(4) Charakterisierung der pedogenetischen Verwitterungs-Tendenz des Solums durch Ermittlung der bodenspezifischen *Stabilitäts-Reihen* der Mineral-Arten. Dies sind Reihen zunehmender Verwitterungs-Resistenz, die sich aufgrund des unterschiedlich raschen Verschwindens einzelner Mineral-Arten durch Auflösung oder deren verschieden stark ausgeprägten Auflösungs-Formen aufstellen lassen.

Für die Stabilität in Parabraunerden aus mitteleuropäischen Jungwürm-Löss gilt zum Beispiel (Meyer et al., 1962): Calcit < Apatit < Olivin < Biotit < Augit < sonstige Pyroxene < Chlorit < Muskovit < Epidot < Hornblende < Plagioklase < K-Feldspäte < Quarz.

In Bänder-Fahlerden und *Calluna*-Podsolen aus mittelsandigen Flugsanden des nordwestdeutschen Tieflandes (Bosse, 1964): Olivin < Biotit = Augit < Hornblende < Epidot = sonstige Pyroxene < Plagioklase = K-Feldspäte

[1] Direktor: Prof. Dr. Dr. h.c. F. Scheffer.

< Chlorit < Muskovit < Quarz.

Neben den Stabilitätsreihen soll zur Charakterisierung bodenspezifischer Verwitterungs-Tendenzen auch die Ermittlung von *Mineral-Umwandlungs-Folgen* dienen, wie, z.B.: Feldspat ⟶ Opal, Feldspat ⟶ Hydrargillit, Feldspat ⟶ Kaolinit oder Feldspat ⟶ ausschliesslich ionare Abbau-Produkte; in feinkörnigen Fraktionen: Biotit ⟶ Hemiexpandit (Scheffer et al., 1964) + FeOOH ⟶ Expandite oder Al-fixierte aufgeweitete Illite (Scheffer et al., 1961).

PROBLEMSTELLUNG

Die Ermittlung des Verwitterungszustandes der Mineral-Komponente in den einzelnen Bodenhorizonten und die Charakterisierung der herrschenden Verwitterungs- und Umwandlungs-Tendenzen allein aufgrund der üblichen quantitativen Auszählungen von Korngrössen-Fraktionen bleibt häufig ein recht unbefriedigender methodischer Weg. Dies gilt bevorzugt für die Böden unseres Klimaraumes, die erst innerhalb des kurzen holozänen Zeitabschnittes aus den Lockersedimenten des würm-(weichsel-)zeitlichen Glazial- und Periglazial-Raumes hervorgegangen sind, besonders aber für solche, die als noch jüngere Verwitterungs-Bildungen erst aus holozänen Ablagerungen entstanden sind.

Diese Böden, besonders letztere, weisen häufig ein quantitativ noch recht schwach entwickeltes "Verwitterungs-Profil" auf, obwohl qualitativ, d.h., verwitterungs-morphologisch, durchaus schon eine beachtliche Horizont-Differenzierung vorliegen kann.

Vom frischen, unangewitterten Mineralkorn bis zu seinem völligen lösungs-chemischen Verschwinden, seiner restlosen Umwandlung oder seiner mechanisch-lösungschemischen Verkleinerung und damit Anreicherung in feineren Korngrössenklassen alles Grössen, auf die sich bisher meist die quantitative Untersuchung beschränkte, kann sich ein recht langer Weg einer schrittweisen verwitterungs-mikromorphologischen Umwandlung erstrecken. Bei den von uns durchgeführten mineraloptischen Auszählungen interessierte uns daher immer wieder das Problem, wie diese Teilschritte und Stadien der morphologischen Verwitterungs-Umbildung einzelner Mineral-Spezies *nomenklatorisch* und *klassifikatorisch* ordnend erfasst und damit einer quantifizierenden Auswertung zugänglich gemacht werden könnten.

Dieses Vorgehen war auch insofern wichtig, als die Wegrichtung des Verwitterungs-Formenwandels je nach herrschendem Bodenmilieu unterschiedlich ist. Die Verfolgung der Umwandlungs-Sequenz einzelner Mineral-Arten liefert daher weitere Boden-Spezifika, die bereits dann erfassbar werden, wenn noch keine wesentlichen quantitativen Veränderungen des Mineralarten-Bestandes durch Verwitterung eingetreten sind.

Im Folgenden soll dies an den nachstehenden zwei Mineralarten-Gruppen geringer bis mässiger Verwitterungs-Stabilität demonstriert werden:

(1) An solchen Mineral-Arten, die in unseren Locker-Sedimenten mengenmässig bedeutend sind und daher ausschlaggebenden Anteil an der Stoffwandlungs- und -wanderungs-Bilanz der aus diesen Sedimenten hervorgegangenen Böden (Schwarzerden, Braunerden, Parabraunerden, Podsolen,

Pseudogleyen) haben: Biotiten, Chloriten, Muskoviten und Feldspäten.
(2) An einigen mengenmässig weniger, dafür aber als Zeiger-Mineralen bedeutenden Mineral-Arten der Schwermineral-Fraktion: Hornblenden, Epidoten, Augiten.

Dabei soll zunächst auf jene sehr instabilen, abgesehen vom Calcit aber im allgemeinen auch mengenmässig für den Verwitterungs-Prozess weniger bedeutenden Mineral-Arten Erdalkali-Carbonate, Olivin, Apatit, vulkanische Gläser, Gips, die wir z.T. bereits früher bei Untersuchungen an Parabraunerden aus Löss quantitativ mit behandelt haben (Meyer et al., 1962; Fölster et al., 1963), verzichtet werden. Infolge ihres raschen Abbaues bieten sie nur geringe Möglichkeiten der sicheren Erfassung spezifischer Umwandlungs-Folgen.

Die nachstehenden Ausführungen beziehen sich ausdrücklich auf mitteleuropäische Böden aus Locker-Sedimenten und zwar in erster Linie sand- und schluffreichen Ablagerungen des jungpleistozänen Glazial- und Periglazial-Raumes (Löss, glazifluviale und äolische Sande, Grundmoräne, feinkörnigere Frost-Lager- und Wanderschutt-Decken) und der holozänen Akkumulations-Räume (Auenlehm, Binnendünen usw.). Wenn man in diesem Zusammenhang von den grobkörnigen Schuttkörpern absieht, lassen sich nämlich die Prozesse des verwitterungs-bedingten Formenwandels der Mineral-Arten in den genannten Böden als mehr oder weniger gleichgerichtet zusammenfassen. Sie scheinen sich grundsätzlich von den Verwitterungs-Vorgängen des isomorphen Gesteinszersatzes, die bei der Boden-Entwicklung aus anstehenden kompakten Massengesteinen ablaufen, zu unterscheiden (Fölster, mündliche Mitteilung).

BIOTIT

Die holozänen Böden Mitteleuropas aus Lockersedimenten zeigen im Anschluss an den Prozess der Entkalkung die Prozesse der Verbraunung (Eisenoxid-Freisetzung und -Neuverteilung) und Verlehmung, d.h., Erhöhung des Mengenanteiles der Tonfraktion, vorwiegend durch chemisch-physikalischen Zerfall gröberer Mineral-Individuen, speziell der Glimmer, und weniger durch Ton-Aufbau aus ionaren Verwitterungs-Lösungen.

An den letztgenannten beiden Prozessen hat der Biotit mengenmässig einen entscheidenden Anteil: In einer bis 180 cm Tiefe entkalkten Parabraunerde aus Jungwürm-Löss können, z.B., während des Holozäns unter 1 m^2 Oberfläche bis zu 150 kg (6% der festen Bodensubstanz) Dreischichtsilikat-Ton neu gebildet worden sein. Diese Menge stammt zum weitaus überwiegenden Anteil aus dem Biotit-Zerfall (Abschuppung). Im A_3-Horizont können dabei 70–100%, im B-Horizont ca. 30–60% aller Biotite zerfallen sein. Etwa 1/2 bis 2/3 des im Biotit primär gebundenen Eisens geht hierbei in freies Eisenoxid über, das die von den Biotiten "abschelfernden" Ton-Schuppen ummantelt, den Gehalt der Tonfraktion an freiem Fe im B-Horizont auf 2,5% (C-Horizont ca. 1%) ansteigen lässt und dadurch zur Verbraunung beiträgt.

Die morphologisch erfassbare Verwitterungs-Resistenz ist generell sehr gering: > Olivin, ≈ Augit, < Chlorit, Hornblende, Epidot.

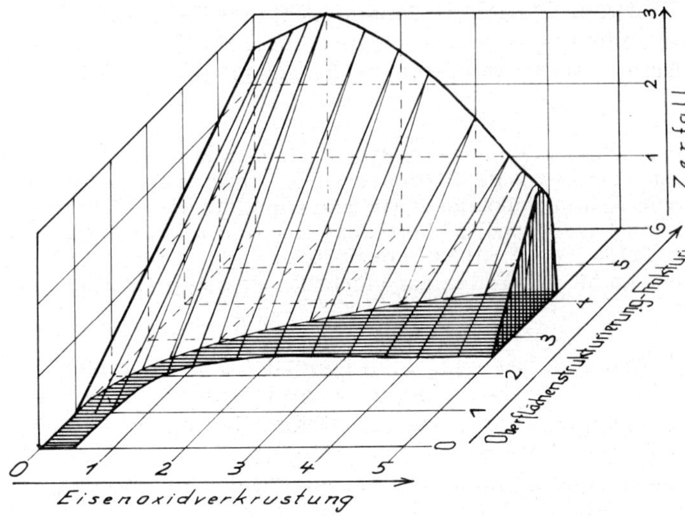
Zerfall
Oberflächenstrukturierung-Frakturierung
Eisenoxidverkrustung
0 1 2 3 4 5
0 1 2 3 4 5 6
1 2 3

1 A

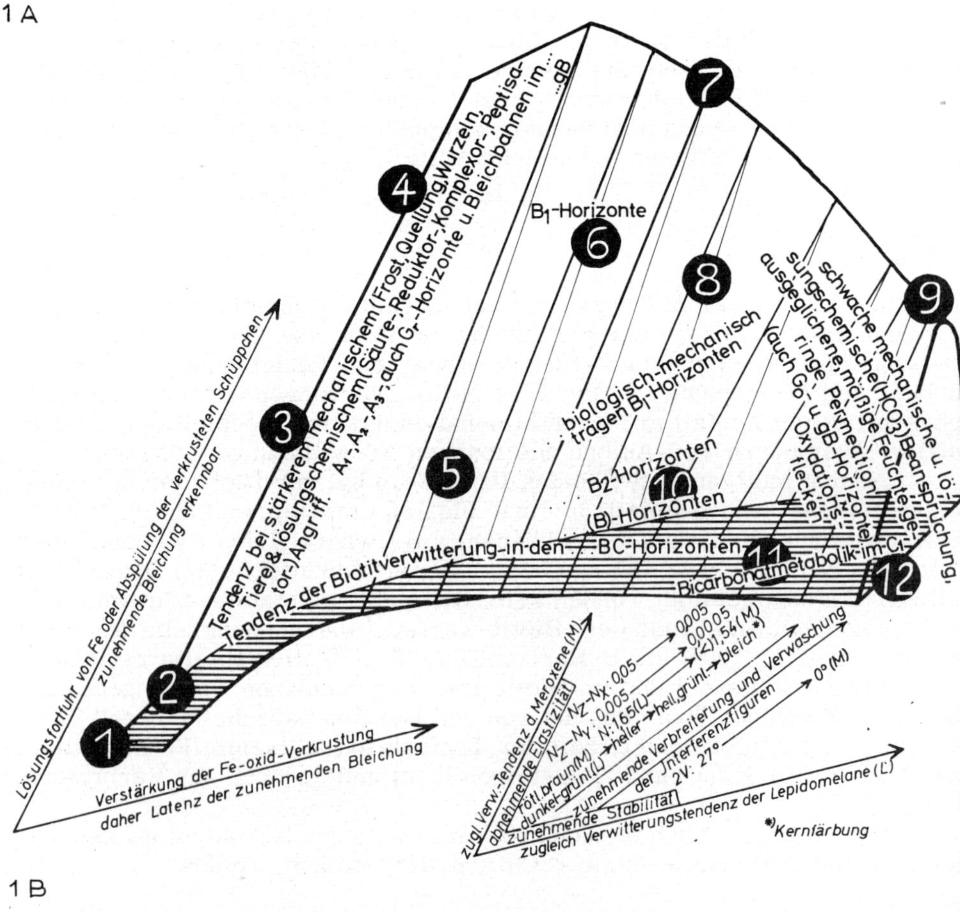
1 2 3 4 5 6 7 8 9 10 11 12
B1-Horizonte
Tendenz bei stärkerem mechanischem (Frost, Quellung, Wurzeln, Tiere) & lösungschemischem (Säure-, Reduktor-, Komplexor-, Peptisator-) Angriff.
A1-, A2-, A3-, auch G1-Horizonte u. Bleichbahnen im ...gB
biologisch-mechanisch trägen B1-Horizonten
B2-Horizonten
(B)-Horizonten
Tendenz der Biotitverwitterung in den ... BC-Horizonten
Bicarbonatmetabolik im C1
schwache mechanische u. lösungschemische (HCO3) Beanspruchung, ausgeglichene, mäßige Feuchte, gering Permeation (auch Go- u. gB-Horizonte) Oxydationsflecken
Lösungsfortfuhr von Fe oder Abspülung der verkrusteten Schüppchen
zunehmende Bleichung erkennbar
Verstärkung der Fe-oxid-Verkrustung
daher Latenz der zunehmenden Bleichung
zugl. Verw.-Tendenz d. Meroxene (M)
abnehmende Elastizität
zunehmende Stabilität
zugleich Verwitterungstendenz der Lepidomelane (L)
zunehmende Verbreiterung und Verwaschung der Interferenzfiguren
2V: 27° → 0° (M)
*) Kernfärbung

1 B

Alle in Böden aus Locker-Sedimenten angetroffenen morphologischen Verwitterungs-Stadien des Biotits lassen sich in die in Abb.1 dargestellten Raumdiagramme einordnen, deren Ordinaten sind die Vektoren *Zerfall*, d.h. Abschuppung, Abschelferung, Abblattung, *Eisenoxid-Umkrustung* und *-Inkrustation* und - als auslösende Ursache beider Vorgänge - die *Frakturierung* und lösungschemische *Strukturierung* (als Sammel-Begriff für alle Veränderungen der Oberfläche).

Beim *Zerfall* lassen sich folgenden Stadien abgrenzen:

(0) Intaktes Mineralkorn mit präsedimentären, geraden oder geschwungenen Abbruch-Kanten.

(1) Geringe Fein-Abschuppung, zu den Randpartien hin etwas verstärkt; beginnendes randliches Einreissen.

(2) Feinschuppiges Abschelfern auf der gesamten Oberfläche, besonders stark in den Randzonen; fransig, dann schartig werdende Ränder und lappiger Korn-Umriss.

(3) Starkes Abschelfern und Abschuppen der gesamten Oberfläche, und Abblattung der dadurch ausdünnenden Randzonen; zackige und dann zerfetzte Ränder, stark gebuchtete Umgrenzung der Biotitblättchen.

Die *Eisenoxid-Abscheidung* lässt sich in fünf Teilschritten verfolgen.

Endstufe ist hier das völlig verkrustete, erste Ansätze zu einer konkretionären Hüllen-Verdickung zeigende Biotit-Teilchen Nr.12 (Abb.2). Setzt die Eisenoxid-Abscheidung an bereits stärker strukturierten Körnern ein oder nimmt das relative Ausmass von fortschreitender Strukturierung und Zerfall im Vergleich zu der Eisenoxid-Ablagerung zu, so tritt die Oxid-Abscheidung weniger in Form einer Umkrustung als vielmehr einer Inkrustation zwischen den oberflächlich aufgelockerten und abgehobenen Biotit-Plättchen und -Schüppchen auf (vgl. Abb.2, Nr.5).

Die Abfolge der *Frakturierung* und *Oberflächenstrukturierung* in Stufen, von denen hier sechs ausgeschieden wurden, lässt sich am besten anhand der Verwitterungs-Sequenz A (Abb.2), Nr.1, Nr.2, Nr.3, Nr.4 und Nr.7 verfolgen, da hier - im Gegensatz zu den anderen Folgen (B, C, D) - rasche Strukturierung und Zerfall über die Eisenoxid-Freisetzung und -Abscheidung dominieren. Dadurch bleiben die wesentlichen Struktur-Phänomene gut erkennbar.

Die glatte, gleichmässig transparente, klare Oberfläche (Nr.1) zeigt als erste Folge-Erscheinung eines lösungschemischen Beanspruchung eine Veränderung in Form eines farblichen Sprenkelung (Maserierung) (Nr.2). Es entstehen trüb-bräunlich-, ring- bis wabenförmig umrandete Flecke vom Durchmesser 1–2 μ und kleiner, die in der Mitte noch die ursprüngliche klare Eigenfarbe des Biotits aufweisen. Die Erklärung hierfür ist vermutlich in der Entstehung eines feinen submikroskopischen, unregelmässig bis wabenartigen Sprungnetzes (Frakturierung) senkrecht zu 001 in

---

Abb.1. Generelles Raumdiagramm der möglichen morphologisch erfassbaren Verwitterungs-Zustands-Stufen von Biotiten in mitteleuropäischen Böden aus Locker-Sedimenten. A. Raumdiagramm mit Ordinaten. B. Trends der Biotit-Verwitterung bei unterschiedlichem Boden-Milieu; Lage-Angabe der in Abb.2 abgebildeten Biotite; Trends der Abwandlung mineraloptischer Merkmale.

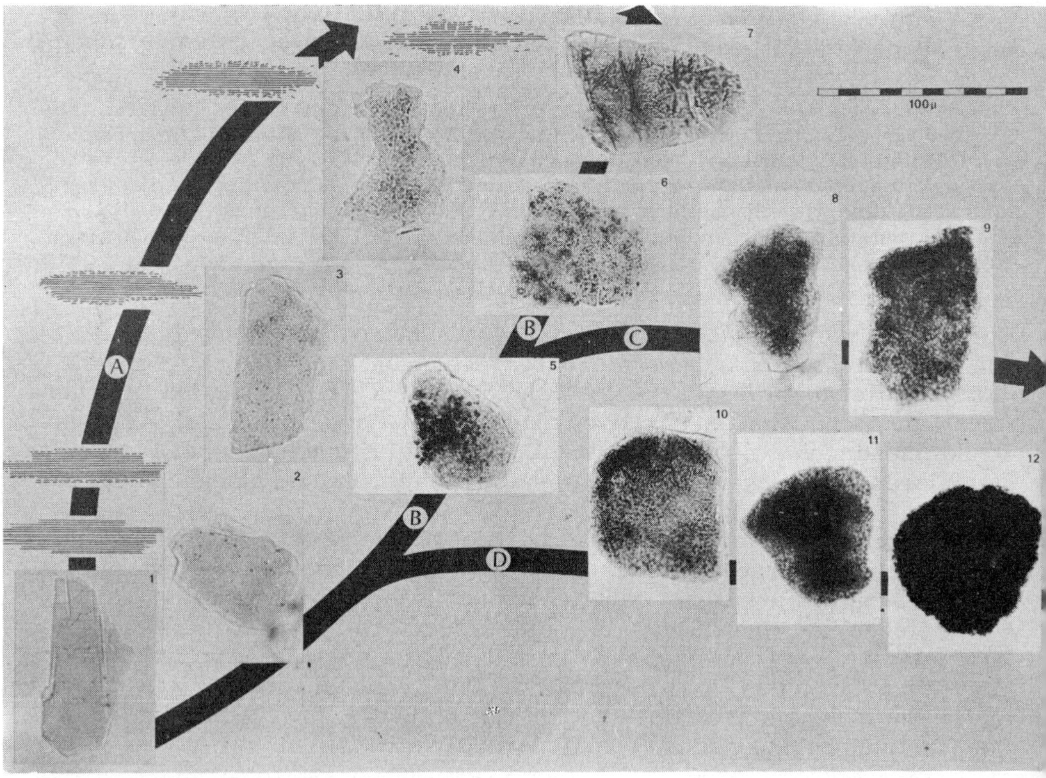

Abb.2. Morphologische Verwitterungs-Stadien der Biotite. Anordnung und Ziffern entsprechen der Anordnung in den Raumdiagrammen in Abb.1. Die Querschnitts-Schemata sollen die Entwicklung der Verwitterungs-Strukturen verdeutlichen. Die Hohlraum-Systeme sind mit Eisenoxyd gefüllt zu denken. Die Pfeile mit Buchstaben kennzeichnen die hauptsächlichen bodenspezifischen Verwitterungs-Morphosequenzen.

der obersten Glimmer-Blättchenlage zu suchen. Von den umgrenzenden Sprüngen aus scheint die Eisenoxidation horizontal zentripetal in die von den Sprungmaschen umgrenzten Tesserae hinein fortzuschreiten.

Die - vermutlich mit der initialen Fe-oxidation im Zusammenhang stehende - Zunahme der Sprödigkeit führt dazu, dass die oberste Glimmer-Lage entlang den sie durchsetzenden Haarsprüngen, die ausserdem einer ständigen lösungschemischen Erweiterung und Vertiefung unterliegen, aufplatzt. Dadurch, dass die Oberfläche in Schüppchen, die sich z.T. randlich aufbiegen und aufstellen, aufspringt, wird sie deutlich rauh (Nr.3). Die isolierten, nunmehr ganz trübbraunen Schüppchen sind von einem Netz heller Lichtsäume umrandet. Dies scheint nicht dadurch bedingt zu sein, dass in den angekippten Randpartien der Schüppchen die höheren Lichtbrechungs-Werte der Y- und Z-Achse in Erscheinung treten. Es zeigt sich nämlich in der Phasenkontrast-Aufnahme der Mineralkörner Nr.3 und Nr.4 in Abb.3, dass die helle Waben-Struktur der Oberfläche bereits durch hochlichtbrechende Eisenoxid-Abscheidungen hervorgerufen ist, die das durch fortgeschrittene Lösungs-Verwitterung erweiterte Rissnetz ausfüllen, wulstig überwuchern und somit eine genarbte Oberfläche erzeugen. Die Abscheidungen müssten, da sie höhere Lichtbrechung besitzen, im Phasenkontrast dunkel erscheinen. Ihre Helligkeit gegenüber dem Biotit-Untergrund wird durch Rand-Beugung und Interferenz und dadurch bedingte Phasenumkehr hervorgerufen.

In den schematischen Querschnitts-Skizzen der Abb.2 ist der Übersichtlichkeit halber nur die fortschreitende Entwicklung, d.h., Vertiefung und Verbreiterung des Sprungnetzes zu Rissen, Klüften und Furchen dargestellt. Alle diese Struktur-Systeme sind mit Eisenoxid ausgefüllt zu denken, das sich oberflächlich oder horizontal-interlamellär zwischen Schüppchen und Spalt-Plättchen ausbreiten kann und schliesslich als Belag und Umkrustung mit den abschuppenden Biotit-Teilchen in die Tonfraktion eingeht.

Bei Nr.4 (Abb.2, 3) entwickelt sich aus der Feinnetz-Struktur ein secundäres tiefer gehendes und durch Anlösung breiter ausgefurchtes Rissnetz, das zunächst noch offen ist, dann aber über die ganze Oberfläche zusammenwächst. Durch Herausfallen einzelner Schuppen wird die Oberfläche rissig-borkig oder - bei relativ stärkerer Oxid-Abscheidung, z.B., bei Nr.6, - schorfartig-warzig. Schliesslich entstehen noch grobmaschigere tiefergehende tertiäre und quartäre Riss- und Spaltensysteme, die sehr breit ausgefurcht und von Eisenoxid wulstig-schildförmig umrandet sind. In diesem zerfurchten und zerklüfteten Zustand ist jedoch der Zerfall, d.h., neben der Abschuppung auch die Abblätterung, bereits so stark, dass sich keine weiteren Strukturierungs-Vorgänge mehr abzeichnen.

Die geschilderten Vorgänge können sich nun - je nach Bodenmilieu - unter einer schwachen oder auch stärkeren Eisenoxid-Verkrustung des Biotit-Teilchens vollziehen. Bei stärkerer Umhüllung äussert sich ein fortgeschrittener Strukturierungs- und Zerfalls-Zustand darin, dass die Kruste nur noch auf den zentralen Partien der Glimmerplatten-Oberfläche intakt ist, während in den Randzonen die Verkrustung meist infolge der hier herrschenden stärker flächigen Abschelferung und -blätterung beseitigt oder von vorn herein hintangehalten ist (vgl. Nr.8). Auf den hier durch Abschuppung exponierten frischen Blättchen-Lagen finden sich - im Gegensatz zu den

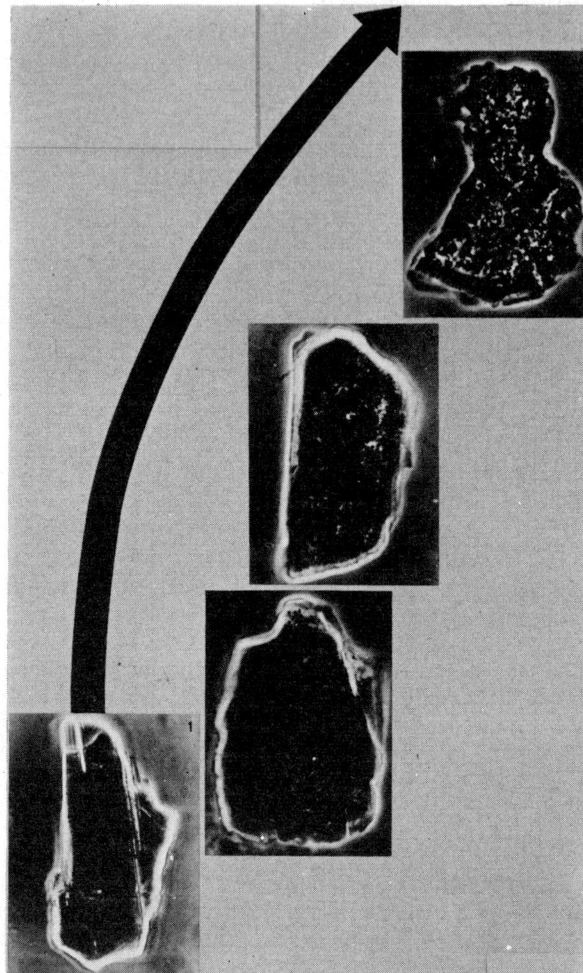

Abb.3. Phasenkontrast-Aufnahme der Biotite Nr.1, 2, 3 und 4 aus der Morphosequenz A in Abb.2 zur Verdeutlichung der wabennetz- und punktförmigen Eisenoxid-Abscheidungen auf der Oberfläche, die sich mit zunehmendem Strukturierungs-Grad entwickeln. Die Oxid-Abscheidungen sind infolge Rand-Beugung und Interferenz hell gegen den dunklen Biotit-Untergrund abgesetzt.

zentralen Teilen - wieder die jungen Anwitterungs-Strukturen.

Die Form des triangulär-dachförmigen Verbreitungs-Gebirges der Biotit-Verwitterungs-Stadien (Abb.1) zeigt, dass die Kombinationsmöglichkeiten der genannten drei morphologischen Verwitterungs-Phänomene in Böden aus Locker-Sedimenten recht begrenzt sind. Im wesentlichen zeigen sich zwei Grund-Tendenzen der verwitterungs-morphologischen Biotit-Umwandlung, die in Abb.2 mit A, B und C, D angedeutet sind.

Im ersten Fall dominieren Strukturierung und Zerfall, die eine visuelle Oxid-Verkrustung hintan halten. Die lösungschemische Eisen-Freisetzung ist dabei zwar nicht schwächer, die Lösungs-Fortfuhr infolge stärkerer Perkolation, sowie Säure-, Reduktor- und Komplexor-Aktivität und die Beseitigung von Krusten-Ansätzen infolge des progressiven Zerfalls unter Beteiligung von Peptisatoren jedoch relativ stärker.

Die Sequenz A läuft hauptsächlich in A_1-, A_2- und A_3-Horizonten von eu- bis oligotrophen Parabraunerde, Fahlerden und Podsolen, Bleichbahnen in B-Horizonten, z.T. auch G_{red}-Horizonten und intrastadialen Nassböden im Löss ab.

In der zweiten Richtung überwiegt die zerfalls-behindernde Eisenoxid-Umhüllung der in Abbau begriffenen Biotite.

Die Sequenz D findet sich besonders in Horizonten mit ausgeglichener Durchfeuchtung, aber oxydativem Milieu und geringerer Lösungs-Beanspruchung bei schwacher mechanischer Einwirkung, wie, z.B., (B)-Horizonten von Parabraunerden aus Löss, gB- und G_{ox}- Horizonten aus schluffigem Material, braunen Vegen, ferner auch würm-interstadialen Böden im Löss.

In B und C halten sich Strukturierung, Zerfall und Oxid-Umhüllungen annährend die Waage. Inkrustationen, die sich, von dem vertieften Risssystem ausgehend, flächenhaft zwischen den aufgelockerten Schüppchen und Blättchen ausbreiten können, sind häufig. Die auffallenden Schüppchen sind stärker als in A mit Eisenoxid besetzt. Die Sequenz ist, z.B., typisch für Parabraunerde-B-Horizonte aus Lössen.

Beide Hauptzüge der Biotit-Umwandlung mit ihren Verwitterungs-Stadien sind charakteristisch für bestimmte Verwitterungs-Milieus von Bodenhorizonten und können durchaus Indikator-Wert im Falle von quartären Paläoböden haben, z.B., bei der frühkaltzeitlichen Pseudovergleyung von interglazialen Parabraunerden und ähnlichen Fällen. Die Korrelation zwischen bestimmten Boden-Horizonten und dem Trend der Biotit-Umwandlung ist in Abb.1B dargestellt.

Verbunden mit dem von Nr.1 ausgehenden Auseinanderfächern der verschiedenen morphologischen Verwitterungs-Abläufe ist eine - in ihrer generellen Tendenz gleichgerichtete, im Falle der verkrusteten Teilchen nur latente - Veränderung der mineraloptischen Eigenschaften des Biotits, die in Abb.1B ebenfalls zusammenfassend dargestellt ist.

Die Beträge der maximalen Doppelbrechung, die wir - da im Hinblick auf die morphologische Untersuchung mit Streupräparaten gearbeitet werden musste - nicht bestimmen konnten, wurden von Rimsaite (1957) übernommen und sind in diesem Zusammenhang vorsichtig zu beurteilen.

Bei den in Abb.2 wiedergegebenen Biotiten handelt es sich vorwiegend um Teilchen von Schluff-Grösse, wie sie besonders in Lössen und den daraus entstandenen Böden auftreten. Eine mengenmässig ausreichende präparative Gewinnung von Mineral-Individuen gleichen Verwitterungs-Grades für die Röntgen-Analyse war uns nicht möglich.

Bei Phasenkontrast-Beobachtung lässt sich unter Einbettung in Kanada-Balsam feststellen, dass in den Verwitterungs-Stufen Nr.4 und Nr.6–9 breite Randzonen-Bereiche der Biotit-Teilchen im Zuge der Bleichung in ihrer Lichtbrechung bis herunter auf die Lichtbrechungs-Werte (1,54) des Einbettungsmittels abgesunken sein können (Helligkeits-Gleichheit). Auf grund von Vergleichen mit ähnlich veränderten gröberen

Biotiten aus isomorphem Gesteins-Zersatz bedeutet dies jedoch noch nicht, dass in diesem Zustand bereits in jedem Falle im Bereich der Kristall-Struktur eine Zwischen schicht-Aufweitung des Biotit-Kristalles in seinen Randzonen stattgefunden haben muss (Fölster, mündliche Mitteilung). Allerdings haben die abgelösten Schüppchen, die in der Ton-Fraktion angereichert werden, bereits überwiegend Hemiexpandit-Charakter (Fölster et al., 1963).

Zusammenfassend zeigt der dargestellte Ablauf der Biotit-Verwitterung, dass der Abbau einerseits direkt zur Anreicherung eisenoxid-umhüllter Teilchen von Tongrösse infolge Abschuppung ohne intermediäre Anreicherung von Fein-Schluff führt. Andererseits ist hervorzuheben, dass auch bei der verwitterungs-bedingten mikromorphologischen Formwandlung der Biotite durch Oxid-Umhüllung Möglichkeiten zu einer langfristigen Stabilisierung gegeben sind, die den auf diese Weise konservierten Teilchen Indikatorwert für verdeckte ehemalige Verwitterungs-Bedingungen (Staunässe, Nassböden) verleihen können.

MUSKOVIT

Während die Biotite heteroklastisch zerfallen, d.h., sich überwiegend auf direktem Wege in Ton-Schüppchen gleicher Maximal-Grösse zerlegen, erfolgt die Muskovit-Verwitterung isoklastisch. Nach Beginn der Verwitterung neigen die frischen Teilchen (Abb.4, Nr.1) dazu, stark aufzublättern und sich dabei in dünne Blättchen von 2–10 μ Dicke zu zerlegen (Nr.4). Gleichzeitig mit der Aufblätterung bilden sich Sprünge (Nr.2), entlang denen die Muskovitblätter in annähernd gleichgrosse Teilstückchen zerfallen, wobei längere geschwungene Bruchkanten und spitze Ecken entstehen.

Diese Prozesse setzen jedoch erst dann ein, wenn die Biotite bereits einen fortgeschrittenen Abbau erkennen lassen.

Das stark gezackte Teilchen Nr.3 zeigt, dass der geschilderte Zerteilungs-Prozess in gleicher Weise weiter fortschreitet. Schliesslich fallen, ähnlich wie beim Biotit, Plättchen von Tongrösse an. Bei diesem Zerfall der Muskovite werden jedoch, abweichend vom Biotit, Muskovit-Teilchen intermediär in allen Korngrössen-Fraktionen bis herunter zum Feinschluff angereichert, die jedoch alle einer, wenn auch sehr langsamen, so doch ständig weiter fortschreitenden Zerkleinerung unterliegen. Die Aufnahme Nr.3 könnte genau so gut aus der Feinschluff- oder Grobton-Fraktion stammen.

Die Aufnahmen Nr.2–4 würden im Phasenkontrast gemacht, um bei der geringen Blättchen-Dicke die Strukturen sichtbar zu machen.

CHLORIT

In ihrem morphologischen Verwitterungs-Verhalten nehmen die Chlorite eine Mittelstellung zwischen den Biotiten und den Muskoviten ein. Dabei zeigen die *eisenärmeren* Chlorite Anklänge an das Verwitterungs-Bild der Muskovite. Nr.13 in Abb.5B ist solch ein dickeres abgeblattetes Chlorit-Teilchen, Nr.14 ein abgelöstes Blättchen mit gezacktem Umriss und kurzen geschweiften (flachgebuchteten) Bruch-Kanten. Demgegenüber zeigen die *eisenreicheren* Chlorite in bestimmten Phasen ihres Verwitterungs-Ablaufes

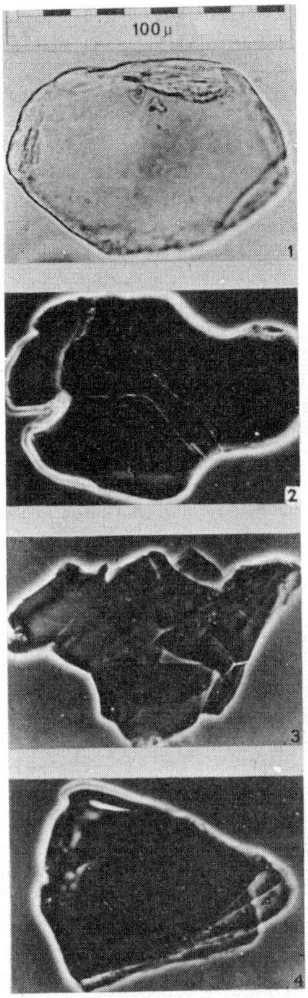

Abb.4. Verwitterungs-Formen am Muskovit.

eine gewisse Ähnlichkeit mit Biotit-Verwitterungsformen.

Der Formenwandel der eisenreicheren Chlorite lässt sich anhand der in Abb.5A dargestellten Sequenzen verfolgen.

Sequenz A : Verwitterung der primären kompakten Chlorit-Teilchen von Schluff bis Feinsand-Grösse bis zum Beginn der Abblätterung.

Nr.1 zeigt ein frisches, präsedimentär mechanisch gerundetes, auf der linken Seite abgebrochenes Korn, das, wie frischer Chlorit meist, dicker ist als frischer Biotit der gleichen Korngrössen-Klasse. Von der Bruchkante gehen Schlagstrahlen aus (einzelne Sprunglinien einer Schlagfigur), Eisenoxid-Abscheidungen fehlen noch. Nr.2 zeigt an den Schnitt-Linien von randlichen Bruchflächen (senkrecht zu 001) und Schichtebenen eine beginnende linienhafte Eisenoxid-Abscheidung. Ähnlich dem Biotit, jedoch weniger intensiv, beginnt sich eine Aufrauhung und Narbung der Oberfläche abzuzeichnen, die bei Phasenkontrast-Beobachtung (Ph) deutlicher wird. Ferner erfolgt eine rasche Ausbreitung eines - im Vergleich zum Biotit grobmaschigeren - Netzes sehr feiner Sprünge, die die oberflächennahen Glimmer-Lagen durchsetzen.

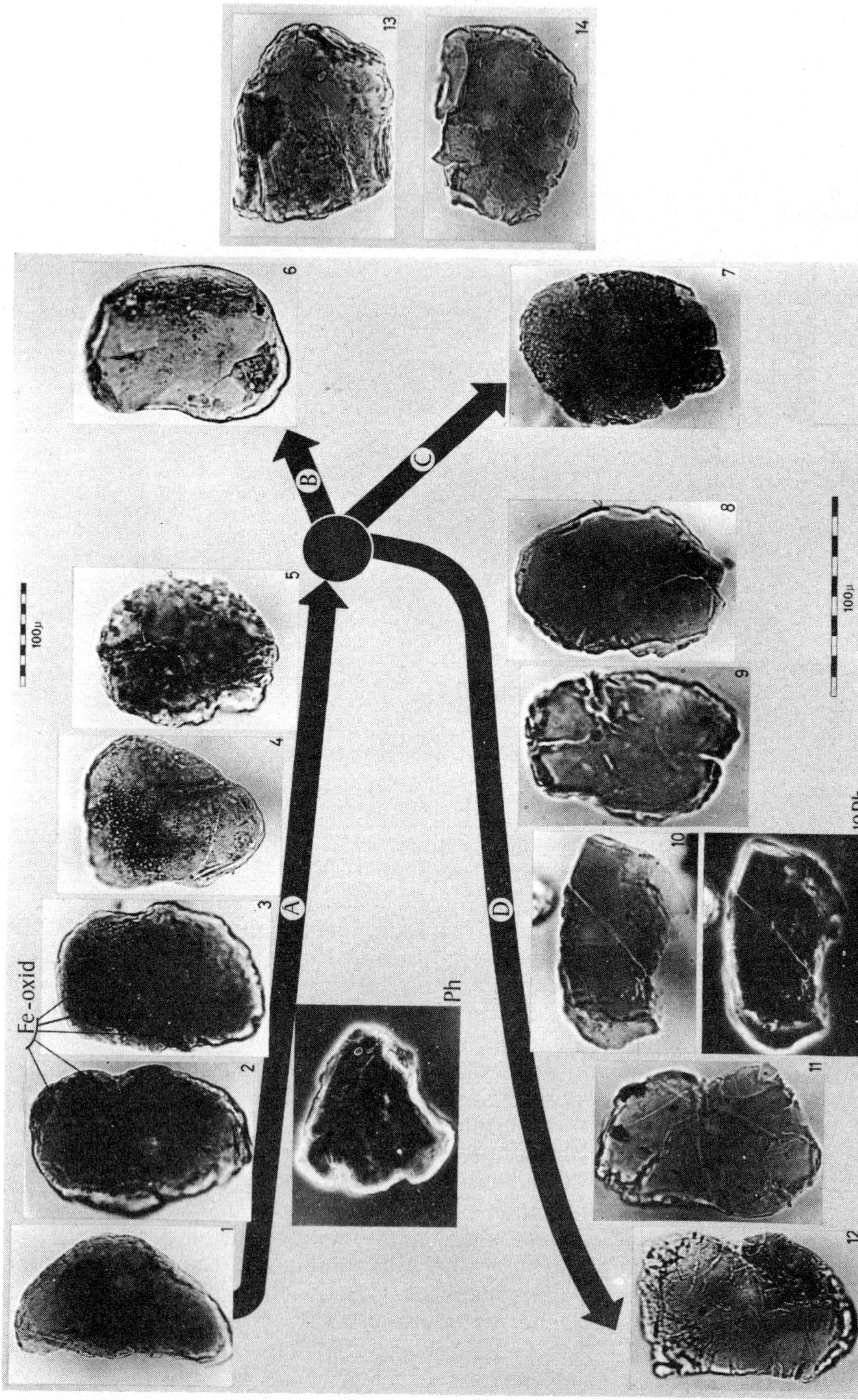

Abb.5. Verwitterungs-Folgen beim Chlorit. A. Eisenreich. B. Eisenarm.

In Nr.3 verstärken sich die genannten Phänomene. Die Eisenoxid-Verkrustung greift von den Rändern her auch auf die Sprungrisse über und vernarbt die Oberfläche ähnlich wie bei den Biotiten.

Nr.4 stellt das potentielle Maximum der Eisenoxid-Verkrustung bei Chloriten dar, denn spätestens im Verwitterungs-Stadium.

Nr.5 setzt - ähnlich wie beim Muskovit - ein intensives Ablösen dünner, grosser Chlorit-Blättchen ein, die entlang den vorgegebenen, nur wenig in die Oberfläche eingetieften Sprüngen abreissen können. Die hierdurch ausgelöste "Teilchen-Vervielfachung" führt oft, z.B., in den B-Horizonten von Parabraunerden aus Löss, dazu, dass dort, wo sich dieser Zerfalls-Prozess vollzieht, der mikroskopisch quantitativ auszählbare Mengen-Anteil der Chlorite in den Schluff-Fraktionen sprunghaft in die Höhe schnellt. Das Zerblättern scheint in diesem Stadium rascher voranzuschreiten als die Oberflächen-Strukturierung und Eisenoxid-Abscheidung.

Aus diesem Zerfalls-Stadium resultieren nämlich *Sequenz B*: (Nr.6) Abgeblattete grobe Chlorit-Teilchen, die nur darum frisch erscheinen, weil die oberflächlichen aufgerauhten, verkrusteten und von Fraktur-Spalten durchsetzten Glimmer-Lagen abgeblättert sind. In den Randpartien (rechts) sind jedoch die Relikte der vorausgegangenen Strukturierung und Oxid-Abscheidung noch deutlich zu erkennen.

Sequenz C : Nr.7 ist ein unmittelbar von der Oberfläche abgelöstes Chlorit-Blättchen, das eine sehr starke biotitähnliche Feinstrukturierung zeigt.

Sequenz D : umfasst die weitere Abbaufolge von Blättchen, die beim Aufblättern des Chlorits aus tiefer liegenden (inneren) Partien frei gelegt worden sind.

Nr.8 stellt solch ein strukturell intaktes Chloritblättchen dar, das ca. 3 μ dick ist. Die Dicken dieser Blättchen schwanken sonst zwischen $<$ 2–10 μ. Wie beim Muskovit entwickeln sich allerdings sehr schnell wieder gestreckte feine Sprünge. Die Abbildungsfolge Nr.9, Nr.10, Nr.11, Nr.12 zeigt die fortschreitende Verdichtung dieses Frakturspalten-Netzes, das die Blättchen in Schüppchen bis herunter zu Tongrösse zerlegt.

Bei dem stark eingerissenen Teilchen Nr.12 lösen sich in den Randzonen bereits solche Schüppchen ab, die dann in die Tonfraktion eingehen.

Im Zuge dieses Prozesses nimmt die Bleichung zu, während gleichzeitig in den Fraktur-Spalten Eisenoxid zur Abscheidung gelangt, was wiederum im Phasenkontrast (10 Ph) deutlich sichtbar wird.

In den Parabraunerden aus schluffigem Material, wie z.B., Löss, selbst sauren Pseudogleyen, findet man das ganze Spektrum der Chlorit-Verwitterungs-Stadien von Nr.1–10 nebeneinander vorliegen, wobei in den unteren Horizonten der Schwerpunkt bei den Stadien Nr.1–4, in den B- und A_3-Horizonten bei Nr.6–10 liegt. Die Endstadien Nr.10–12 wurden aus Bänder-Fahlerden und *Calluna*-Podsolen aus Flugsand-Decken isoliert. Die grosse Verteilungsbreite gegenüber dem Biotit erklärt sich aus der wesentlich geringeren Verwitterungs-Geschwindigkeit des Chlorits.

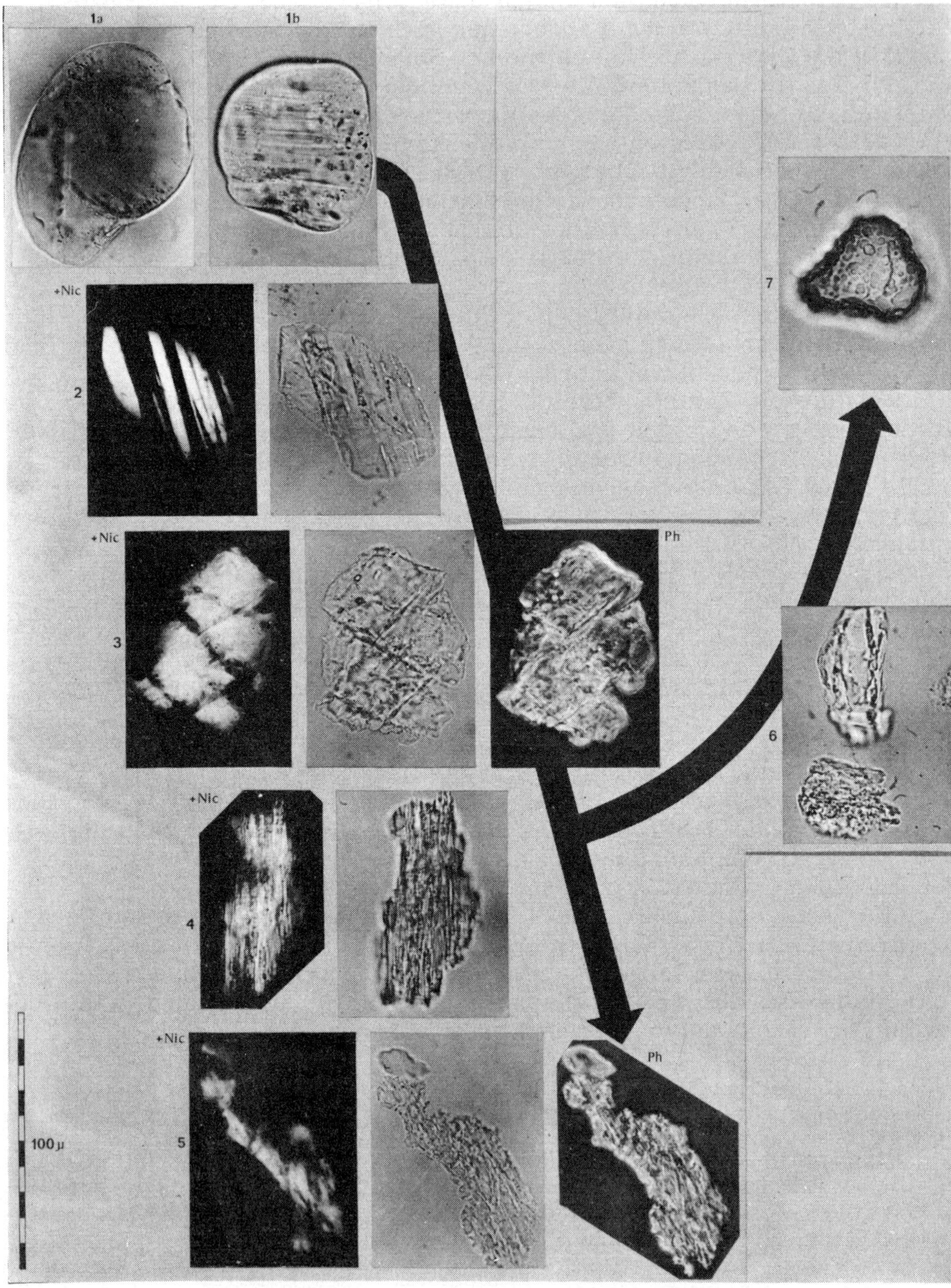
1a
1b
+Nic
2
+Nic
3
Ph
7
6
+Nic
4
+Nic
5
Ph
100 µ

FELDSPAT

Über die standörtliche Abhängigkeit der Feldspat-Verwitterungs-Intensität ist noch relativ wenig bekannt. Im Zuge der holozänen Boden-Entwicklung braucht selbst bei starker Versauerung in feinkörnigen Lockersedimenten keine oder nur eine schwache visuelle Feldspat-Verwitterung stattgefunden zu haben, die dann meist auf die A-Horizonte, d.h., die Zonen mit starker Humus-Metabolik beschränkt ist.

Andererseits stellt man in den Böden, die aus grobkörnigeren Sedimenten (Flugsanden, Sandstein-Fliesserden usw.) hervorgegangen sind (Braunerden, Bänder-Fahlerden, Podsolen) meist einen trotz - oder vielleicht infolge - der gröberen Korngrösse recht intensiven Feldspat-Abbau fest, der in den A_2 - und A_3 -Horizonten bis zu 80% der gesamten primär vorhandenen Feldspat-Substanz ausmachen kann (Von Engelhardt, 1940; Nordmeyer, 1959; Bosse, 1964). Daher sind am ehesten in solchen Böden Verwitterungs-Studien an Feldspäten durchzuführen. Abb.6 stellt, z.B., eine Morpho-Sequenz der Feldspat-Verwitterung dar, wie sie in Podsolen und Bänder-Fahlerden, die im Holozän aus Sanden der nordwestdeutschen Geest entstanden sind, anzutreffen ist. Diese Sequenz ist in ihrer Verbreitung nach dem bisherigen Kenntnisstand wohl zunächst noch auf Böden dieser Art zu beschränken.

Die dargestellte Verwitterungs-Abfolge der Feldspäte, bei denen es sich in den genannten Böden ausschliesslich um die Vertreter der Kalifeldspat-Reihe und von der Plagioklasreihe Albit, Albiklas und nur noch etwas Oligoklas handelt, erfolgt unter Auflösung und Abfuhr der Feldspat-Substanz. Dabei wird intermediär Opal und Allophan-Opal gebildet, wobei wir hier unter Opal optisch isotrope SiO_2 verstehen wollen ohne dass wir dabei bisher nach der kristallographischen Feinstruktur unterscheiden können.

Die Ursache dafür, dass eine intensive Feldspat-Zersetzung in Lockersediment-Böden, sowie der aufgezeigte spezifische Verwitterungs-Ablauf bisher nur in sandigem Material zu finden ist, muss wohl darin gesucht werden, dass nur gröbere Feldspat-Körner zu einer progressiven Intergranular-Verwitterung befähigt sind. Diese schreitet auf den diversen Unstetigkeits-Flächen (Spalt-Ebenen) der Kristalle rasch nach innen fort und breitet sich im Inneren, vermutlich unter Ausbildung eines spezifischen Lösungs-Milieus, aus. Dabei bleibt der Lösungs-Kontakt zwischen Innen- und Aussenlösung bestehen, der die diffusive Ionen-Bewegung und damit die Lösungs-Abfuhr ermöglicht. Ein lösungschemisch selektiver Angriff der Kalifeldspäte und Plagioklase konnte nicht beobachtet werden, d.h. beide verwittern annähernd gleich schnell.

Nr.1 stellt als Ausgangsform "frische" transport-mechanisch gut gerundete Feldspat-Körner des C-Horizontes dar, wobei 1a ein Orthoklas, 1b ein Albit ist.

Nr.2 zeigt ein in seiner Kontur durch Anlösung bereits unregelmässig gewordenes Albit-Teilchen, in dem sich - im Schnittebenen-Bild gestreckt

---

Abb.6. Verwitterungs-Formenfolge beim Feldspat im Zuge der partiellen Umwandlung zu Opal in Podsolen, Braunerden und Bänder-Fahlerden aus mittelsandigen Flugsand-Decken des norddeutschen Tieflandes.

schlauchförmige - heller erscheinende Umwandlungs-Zonen abzuzeichnen beginnen. Zwischen gekreuzten Nicols sieht man diese Zonen als ganz schwach graue (isotrope) Einlagerungen winklig zu den durch Auslöschung dunklen Zwillings-Lamellen verlaufen.

An Nr.3, einem Albiklas-Blättchen, ist der Prozess der intragranulären Feldspat-Verwitterung bereits weiter fortgeschritten. Die parallel zu den Haupt-Spaltrichtungen angeordneten, sich nahezu im rechten Winkel schneidenden Lamellen geben sich zwischen gekreuzten Nicols als isotrop zu erkennen und sind im Phasenkontrast hell. Wenn man die Spaltenfüllungen durch Salz-Sprengung präparativ isoliert, erweisen sich diese als durchschnittlich 1–2 μ starke Opal-Lamellen. Ihre Lichtbrechung hat jedoch eine Streubreite - auch innerhalb ein und desselben Teilchens - die von 1,48 bis 1,56 schwankt. Das bedeutet, dass man mit höheren Al_2O_3-Gehalten in der Opal-Substanz rechnen muss, die den nach Glühen erhaltenen Lichtbrechungs-Werten zufolge etwa bis zu einem Al_2O_3/SiO_2-Verhältnis von 1/2 ansteigen können. Es dürfte daher zweckmässiger sein, von Allophan-Opal zu sprechen.

Nr.3–7 stammen bereits aus den A_2 bzw. A_3-Horizonten der genannten Böden. Auffällig ist, dass hier der Feldspat in Blättchen zerfällt.

Nr.4, ein durch rhythmisches Aufdrücken auf das Deckgläschen hochkant gestelltes Feldspatteilchen, zeigt, wie es zu diesem Zerfall kommt. Es ist von einem System parallel 001 angeordneter Allophan-Opal-Lamellen durchsetzt, entlang denen das Teilchen leicht aufspalten und zu dünnen Plättchen zerfallen kann.

Nr.5, ein ca. 5 μ starkes Feldspat-Plättchen, zeigt das kavernösbrückige Endstadium dieses Zerfalls. Zwischen gekreuzten Nicols erscheinen grosse Flächenteile, die in Allophan-Opal umgewandelt sind, dunkel, während sie im Phasenkontrast als helle Zonen hervortreten.

Bei diesem Feldspat-Zerfall und besonders dann, wenn eine zusätzliche Salzsprengung angewendet wird, werden die einzelnen gröberen Allophan-Opal-Lamellen und -plättchen freigelegt.

Nr.6 stellt stark mazerierte und in Zerfall befindliche Feldspat-Teilchen dar; aus dem oberen ragen bereits die hell erscheinenden Lamellen schräg heraus.

Die freigelegten Allophan-Opal-Täfelchen zeigen im Primär-Zustand häufig einen irregulär zonalen Aufbau, meist in dem Sinne, dass die plättchen inneren Lamellen-Partien niedrigere Lichtbrechungs-Werte aufweisen, also mehr dem Opal angenähert sind. Die äusseren Lagen sind dagegen höher lichtbrechend, und müssen demnach einen höheren Al-Gehalt besitzen. Beide Lagen können mehr oder weniger scharf von einander zu trennen sein, häufig auch mechanisch, sind aber sonst meist irregulär, wulstig-gewunden miteinander verzahnt, gelegentlich auch so, dass Protuberanzen der äusseren Lagen nach innen gehen.

Diese Allophan-Opal-Plättchen unterliegen wie der Feldspat ebenfalls einer - wenn auch anscheinend etwas verzögerten - Auflösung. Es scheint als ob dabei die Al-reicheren Zonen rascher aufgelöst werden als die mehr opalartigen Teile. Zurück bleiben muschelig-kavernöse oder löcherige Opalteilchen niedriger Lichtbrechung.

Nr.7 stellt ein solches residuales Opalplättchen dar. Die Anordnung Al-reicherer und ärmerer Lagen innerhalb der Allophan-Opal-Lamellen in den sich zersetzenden Feldspäten lässt darauf schliessen, dass es sich bei diesen

Bildung weniger um eine Ablagerung als vielmehr um eine von den Spalten ausgehende Feldspat-Umwandlung handelt. Diese erfolgt unter Substanz-Abfuhr (Alkali-, Erdalkali-Ionen, Aluminium, auch SiO_2) und wird durch eine unterschiedliche intragranuläre Diffusion der Lösungs-Partner und zonare Substanz-Differenzierung (z.B., relative oder auch absolute Al-Anreicherung in den spaltenferneren, feldspat-zugewandten Zonen der Opal-Füllungen) während der Auflösung beeinflusst und lässt sich etwa unter dem Gesichtspunkt der Restschicht-Theorie von Correns verstehen. Andere Umbildungs-Produkte wie Kaolinit oder Hydrargillit lassen sich bei dieser Form des Feldspat-Abbaues nicht nachweisen.

AUGITE

Unter den Schwermineralen sind die Augite vor den Hornblenden und Epidoten durch eine sehr geringe Verwitterungs-Stabilität ausgezeichnet, die allerdings grösser ist als die der Olivine.

Die Aufnahme in Abb.7 wurden an Böden (*Calluna*-Podsolen, Bänder-Fahlerden) aus Flugsand-Decken der nordwestdeutschen Geest gemacht. Gleiches gilt für die Hornblenden und Epidote. Es besteht nämlich bei all diesen leicht abbaubaren Schwermineralen nur dann die Möglichkeit für Detail-Studien der Verwitterungs-Morphologie, wenn man von Körnern grösseren Durchmessers ausgeht. In den genannten Böden liegt in den A-Horizonten bereits eine starke prozentuale Lösungs-Verminderung der Gesamtmenge dieser 3 Mineralarten vor; bei den Augiten sind, z.B., 50–70% verschwunden, der Rest ist stark verwittert. Es lassen sich also vom C-Horizont nach oben fortschreitend alle Zonen zunehmenden Verwitterungs-Grades verfolgen. In allen drei Abbildungen (7, 8A und B) beginnen die Folgen mit Nr.1, den frischen Mineralkörnern des C-Horizontes, und enden bei den verwitterungs-morphologischen Endstadien in den A_2- und A_3-Horizonten.

Abb.7, Nr.1 und Nr.2 sind frische Augite, von denen Nr.1 durch prä-sedimentären Transport mechanisch gut gerundet, Nr.2 mechanisch zugebrochen ist, allerdings oben rechts schon Auflösungs-Erscheinungen zeigt. In den Stadien Nr.3 und Nr.4 äussert sich die einsetzende Lösungs-Verwitterung darin, dass sich die glatte, mechanisch zugeschliffene Oberfläche auflöst, indem die Gitterflächen 110 als Flächen höherer Verschiebungs-Geschwindigkeit terrassenartig herauspräpariert werden. Dies wird besonders bei Nr.4 deutlich, wo die ursprünglich glatte Oberfläche über und über wie von Fingernagel-Eindrücken eingekerbt erscheint. In Nr.5 beginnt dann die zinnenförmig-schartige Auflösung der Ränder, die über Nr.7 nach Nr.8 fortschreitet. Bei Nr.6 setzt Eisenoxid-Verkrustung ein, die sich ebenfalls bis Nr.8 verstärkt. Dadurch scheint dem in Zersetzung befindlichen Korn eine gewisse Stabilisierung geboten zu werden. In den A-Horizonten, in denen eine verstärkte Eisen-Mobilisierung herrscht, ist die starke Verminderung der Augit-Körner vom Verwitterungsgrad Nr.8 vielleicht dadurch zu erklären, dass die verwitterten Augite rasch zu kleinen Kristalliten zerfallen, die dann einer schnellen Auflösung unterliegen.

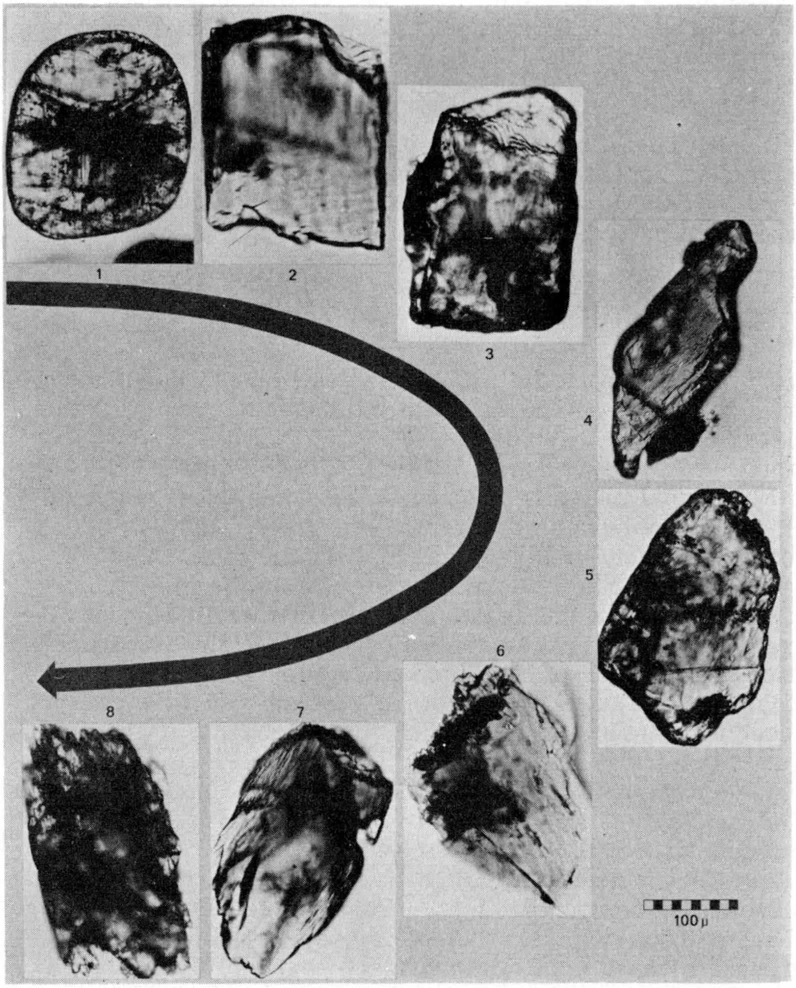

Abb.7. Morpho-Sequenz der Augit-Verwitterung (norddeutsche Sandböden).

EPIDOT

Noch stärker als der Augit zeigt der Epidot (Abb.8A) die lösungschemische Herauspräparierung von Kristall-Strukturen im Zuge der Auflösung, die auch hier wieder von der glatten, beim Transport mechanisch geschliffenen Oberfläche ausgeht und dann auf den kristallographischen Unstetigkeits-Flächen (Spalt-Ebenen) nach innen vordringt. Vom "frischen" Korn (Nr.1) ausgehend, setzt bei Nr.2 randlich die lösungschemische Exponierung der einzelnen Flächen-Elemente ein. Bei Nr.3 (Pistazit) ist die Oberfläche bereits völlig in eine feinkörnige rauhe periphere Zone von lösungschemisch heraus-modellierten Rest-Kriställchen umgewandelt worden. Bei der Epidot-Auflösung ist zwischen dem unterschiedlichen Verwitterungs-Verhalten der

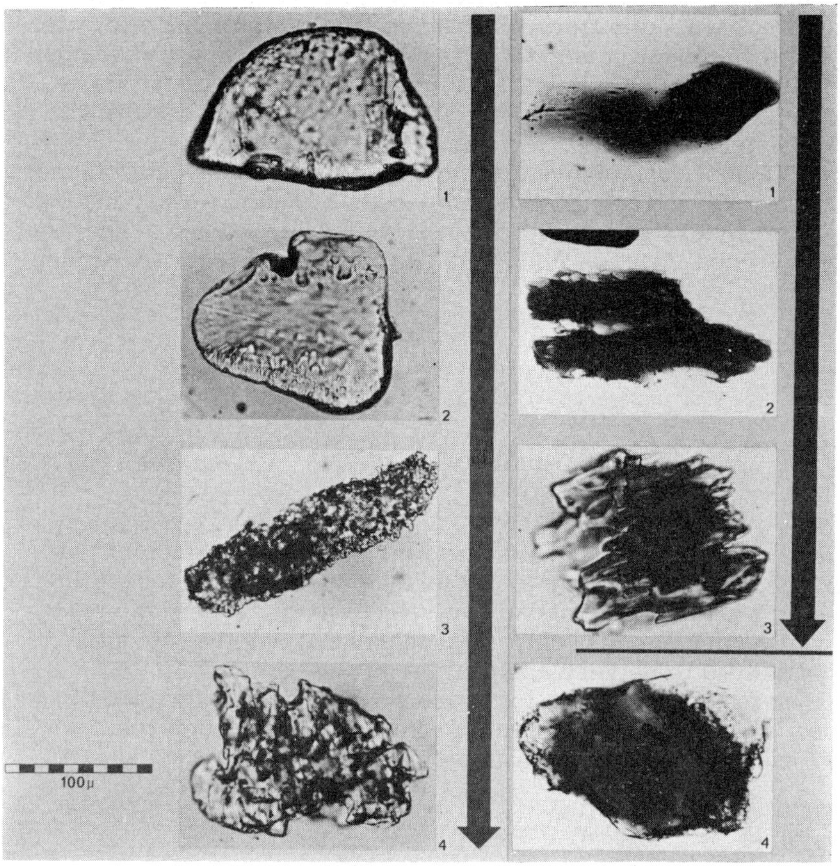

Abb.8. Verwitterungs-Morpho-Sequenzen von Epidot (A) und Hornblende (B) in den Sandfraktionen von Böden aus Flugsand-Decken in Norddeutschland.

eisenreicheren Pistazite und der eisenärmeren Klinozoisite zu unterscheiden. Erstere unterliegen einer rascheren Auflösung, die zu einer relativen Anreicherung der Klinozoisite in den A-Horizonten führen kann. Die Teilchen Nr.3 und Nr.4 gehören nicht einer Sequenz an. Das parallel 001 und 100 abgeblätterte Teilchen Nr.4 stellt ein stark verwittertes Klinozoisit-Teilchen dar, wie es in den A-Horizonten der genannten Böden vorherrscht.

HORNBLENDE

In den Podsolen und Fahlerden aus Flugsand-Decken erweist sich die Hornblende als weniger stabil als der Epidot. In Parabraunerden aus Löss kann aber die Abfolge auch umgekehrt sein.

Wie beim Augit und Epidot erfolgt auch bei der Verwitterung der Hornblenden (Abb.8B) eine lösungschemische Auflösung der Oberfläche unter Herauspräparierung von Kristall-Flächen - allerdings in sehr viel geringerem Masse, denn die schwächere Differenzierung der flächen-spezifischen Anlösungs-Geschwindigkeiten sorgt für rundliche Anwitterungs-formen. In den initialen Verwitterungs-Stadien (Nr.2) sind mandelförmig-rhombische Lösungs-Augen in der Oberfläche, (wie sie schon von Jasmund, 1939–1940, skizziert wurden), sowie entsprechende sichelförmige Rand-Einbuchtungen eine häufige Erscheinung. Diese gibt Anlass zu der später, in Nr.3, hervortretenden steil zinnenförmigen Auflösungs-Struktur. Nr.4 soll nur verdeutlichen, dass besonders bei den eisenreicheren Hornblenden ebenfalls in frühen Verwitterungs-Stadien wieder Eisenoxid-Verkrustungen abgeschieden werden können, die die Verwitterungs-Strukturen verdecken, zumal sie Ablagerungs-Zentren für die in den Podsolen durchwandernden Huminstoffe darstellen.

SUMMARY

The micromorphology of the weathering of mineral species in central-European Holocene soils, developed in loose Pleistocene and Holocene sediments

To get a clear insight into the intensity and direction of development of weathering processes occurring in soils formed during the Holocene in loose sediments, micromorphological investigations of several mineral species - biotites, muscovites, chlorites, feldspars, augites, epidotes, hornblendes - proved to be very adequate.

It appeared to be possible to form one or more characteristic weathering sequences for the different species. Their micromorphological features and formation conditions are dealt with in detail.

LITERATUR

Bosse, I., 1964. Verwitterungs-Bilanzen von charakteristischen Bodentypen aus Flug-sanden der nordwestdeutschen Geest (Mittelweser-Gebiet). Thesis, Georg-Augus Universität, Göttingen, 75 pp.

Fölster, H., Meyer, B. und Kalk, E., 1963. Parabraunerden aus primär carbonathaltige Würm-Löss in Niedersachsen. II. Profilbilanz der zweiten Folge bodengenetische Teilprozesse: Tonbildung, Tonverlagerung, Gefügeverdichtung, Tonumwandlung. Z. Pflanzenernähr., Düng., Bodenk., 100 : 1–12.

Jasmund, K., 1939–1940. Über den Mineralbestand einiger Kaoline unter besonderer Berücksichtigung der kolloiden Grössenordnungen. Chem. Erde, 12 : 508–535.

Meyer, B., Kalk, E. und Fölster, H., 1962. Parabraunerden aus primär carbonat-haltigem Würm-Löss in Niedersachsen. I. Profilbilanz der ersten Folge boden-genetischer Teilprozesse: Entkalkung, Verbraunung, Mineralverwitterung. Z. Pflanzenernähr., Düng., Bodenk., 99 : 37–54.

Nordmeyer, H., 1959. Mineralogische Untersuchungen an drei Bodenprofilen auf Zech-steinletten, unterem Buntsandstein und Zechsteindolomit. Beitr. Mineral. Petrog 7 : 1–4.

Rimsaite, J., 1957. Über die Eigenschaften der Glimmer in den Sanden und Sandsteine Beitr. Mineral. Petrog., 6 : 1–51.

Scheffer, F., Meyer, B. und Fölster, H., 1961. Dreischicht-Tonminerale mit Aluminiu Zwischenschichtbelegung in mitteldeutschen sauren braunen Waldböden. Z. Pflanzenernähr., Düng., Bodenk., 92 : 201–207.

Scheffer, F., Fölster, H. und Meyer, B., 1964. Zur Diagnostik und Systematik von Dreischicht-Tonmineralen in Böden und pedogenen Sedimenten. Chem. Erde, 21 : 210–238.

Von Engelhardt, W., 1940. Zerfall und Aufbau von Mineralen in norddeutschen Bleicherdewaldböden. Chem. Erde, 13 : 1–43.

Weyl, R., 1952. Schwermineraluntersuchungen an Bodenprofilen. Z. Pflanzenernähr., Düng., Bodenk., 57 : 135–141.

MICROMORPHOLOGICAL EFFECTS OF METASOMATIC AND COLLOIDAL PHENOMENA DURING HYPERGENESIS

V.V. DOBROVOLSKY

The Moscow Lenin Pedagogical Institute, Moscow (U.S.S.R.)

INTRODUCTION

Until recently metasomatic phenomena have generally been related to magmatic and metamorphic processes. Lately incidents of metasomatism have been recorded during the formation of sedimentary rocks at the stage of diagenesis. The author has found that metasomatic phenomena are quite common in soil hypergenesis (= alteration of rocks at the surface of continents due to geographic factors; Fersman, 1922).

The author has conducted a systematic investigation of hypergenesis of Quaternary sediments in all the representative natural zones of the U.S.S.R.: tundra, taiga-forest, forest-steppe, steppe and desert. It has been found that the hypergenetic alteration of Quaternary deposits results in the partial decomposition of their components and the development of new formations. The mineralogical composition of new formations is multiform, and comprises nearly fifty names for the area in question. Every zone is represented by certain new formations which remain wellpreserved when buried. Studies of the paragenesis of new formations within Quaternary sediments permits the acquisition of knowledge as to how geographical conditions had been changing in the Quaternary, which is very important for paleogeographical reconstructions.

MORPHOLOGY OF NEW FORMATIONS

Structurally new formations result from the concentration of certain chemical elements in the process of their hypergenetic redistribution.

Three categories of epigenetic concentrations of substances can be distinguished:

(1) Dispersed segregations (crystalline and amorphous).

(2) Isolated forms of concentration (aggregates, fine accumulations, druses, concretions, loess dolls and cements).

(3) Solid formations (crusts, layers, lenticular bodies and bodies of irregular shape).

The process of concentration varies from place to place due to different topographic and geochemical conditions which affect the morphology of new formations.

Characteristic of hypergenesis in *eluvial* environments are: (1) dispersed segregations; (2) aggregates and fine accumulations; (3) concretions;

(4) loess dolls; (5) inwashes.

Hypergenesis in a *superaqueous* environment gives rise to the following forms: (1) pockets and druses; (2) concretions; (3) fumarole incrustation and crusts.

Subaqueous environments produce: (1) concretions; (2) lenticular bodies and layers of precipitation.

The morphology of concretions is fairly multiform. They may be grouped as follows:

A. Isometric (and slightly elongated).

(1) Having an internal cavity (concretions): (a) crystalline (gypsum, calcite); (b) pseudo-amorphous (hydrogoethite, psilomelane).

(2) Having no internal cavity (oolites): (a) crystalline (calcite); (b) pseudo-amorphous (hydrogoethite, pyrolusite, psilomelane, marcasite).

B. Cylindrical.

(1) Having an internal cavity: (a) crystalline (calcite incrustations in pores, gypsum and calcite "akkirshs"); (b) pseudo-amorphous (hydrogoethite rohrensteins, vivianite incrustations in root remains).

(2) Having no cavities - not found.

C. Flattened.

(1) Having a cavity: (a) crystalline (calcite); (b) pseudo-amorphous - not found.

(2) Having no cavities: (a) crystalline - not found; (b) pseudo-amorphous (hydrogoethite, pyrolusite).

Different topogeochemical conditions are characterized by typical forms of concretions. Thus tubular forms, for example, are characteristic of superaqueous conditions, among which humid zones form rohrensteins (ferruginous formations) and arid ones "akkirshs" (gypsum and calcite formations). Subaqueous conditions are generally characterized by oolite-like concretions which in humid zones are mainly concentric ferruginous-manganese concretions with detrital minerals in the centre, and in arid zones calcite concretions of similar structure.

MICROMORPHOLOGY OF NEW FORMATIONS

Micromorphologically new formations can be divided into two groups: (1) cryptocrystalline metacolloidal and amorphous colloidal formations, and (2) crystalline new formations.

Cryptocrystalline metacolloidal and amorphous colloidal formations

Ferruginous and manganese formations have the finest structure. The size of crystalline forms does not exceed 0.02 mm, and is generally measured in μ. In some cases the crystallization of the formations is beyond the power of a microscope, and must be identified either by X-ray or thermal analysis.

The structure of carbonate formations varies from cryptocrystalline to fine-crystalline. The size of the finest forms measures a few μ, and that of the largest 30–50 μ. Formations originating in sub- and super aqueous environments are larger than analogous ones formed in eluvial geochemical environments.

Formations having cryptocrystalline structures are of various metacolloidal microtextures (Fig.1). Cryptocrystalline structure and metacolloidal microtexture indicate that the substance was in a gel state during the development of new formations. In the process of the crystallization of gels the original volume was reduced, resulting in the formation of syneresis fractures. Photographs of syneresis fractures have been published by Dobrovolsky (1955). Here we include a photograph of syneresis fractures in a ferri-oxide microconcentration (Fig.2).

Some new formations are actual gels. Examples are opal and allophanes.

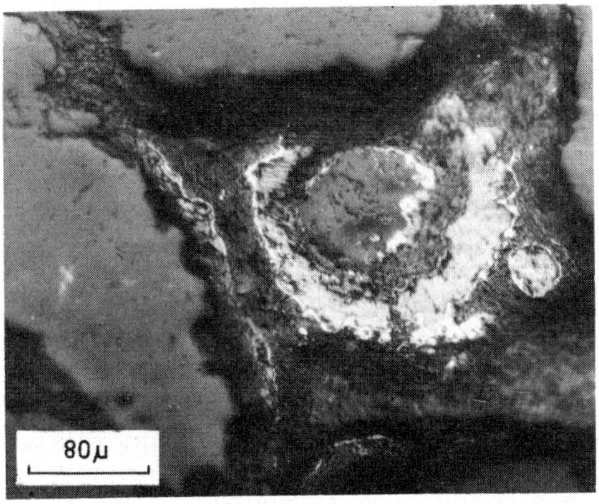

Fig.1. Metacolloidal microtexture of psilomelane (white). Quartz grains (grey) are highly corroded. Manganese ortsand. Polished section.

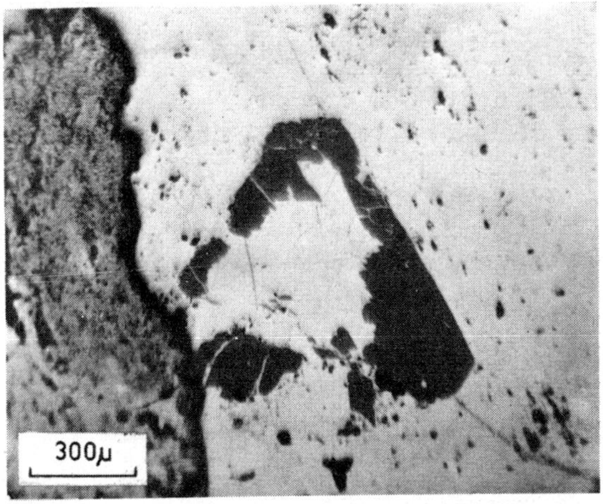

Fig.2. Hydrogoethite microconcretion from bog ore. Radial cracks of syneresis are seen. Polished section.

Crystalline new formations

The formations of this group have relatively large crystalline structures. The size of the individual crystalline forms generally exceeds 100 μ, and frequently amounts to 10–20 cm. Crystals of such a new formation vary in size from place to place, resulting in an irregular granular structure. Most peculiar is a needle-columnar structure of gypsum from old hydromorphous formations of the arid zone. The crystals of this group are translucent and are nearly free of inclusions. The exception is the socalled "Repeteck" gypsum, which has a giant-granular structure whose crystals incorporate many sand grains. No signs of colloidal phenomena have been observed in the formations of this group.

These formations seem to have been calmly crystallized from water-ionic solutions.

The formations of both groups may be found in both arid and humid zones, although arid topogeochemical conditions are more favourable for the latter group.

MICROMORPHOLOGICAL SIGNS OF METASOMATISM

Some of the newly formed minerals do not react with minerals composing Quaternary deposits. Crystals of such minerals are either formed due to high crystallization energy (gypsum) or in cavities (epsomite halite and other water-soluble formations).

Other newly formed minerals, including such well-known minerals as hydrogoethite, psilomelane and calcite, interact with Quaternary minerals in the process of their formation. As a result of this, mineral components of Quaternary deposits are corroded and metasomatically replaced by hypergenetic formations.

The effect of hypergenetic metasomatism depends on the resistance of the rock-forming minerals, also on the replacement capacity of the newly formed minerals. The most resistant rock-forming mineral is quartz, and the least resistant is feldspar. A finely dispersed fraction of Quaternary deposits is metasomatically replaced most easily.

Minerals of ferric and manganese hydroxide groups are particularly capable of metasomatic replacement. These minerals not only powerfully replace a finely dispersed mass, but also highly corrode quartz, the most resistant mineral (Fig.3). Metasomatic action of ferromanganese formations is especially strong under the conditions of hydrogenous hypergenesis of the humid zone. The corrosive capacity of iron and manganese hydroxides is reduced in the arid zone.

Calcite is also capable of metasomatic replacement.

EFFECT OF METASOMATISM ON THE CHEMICAL COMPOSITION OF NEW FORMATIONS

The process of hypergenetic metasomatism involves considerable amounts of chemical elements in migration (including elements from hard, decom-

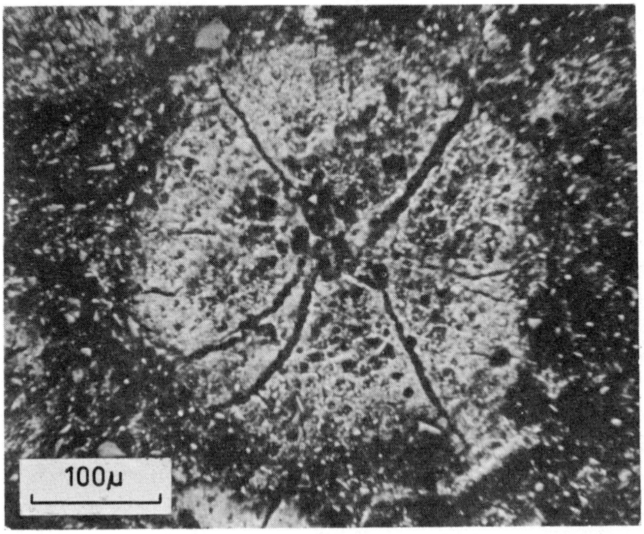

Fig.3. Metasomatic replacement of a quartz crystal (grey) by hydrogoethite (white). Ferric concretion, polished section.

posable compounds). The importance of this phenomenon for geochemistry is only now becoming clear. Corrosion of quartz grains by iron hydroxides was first mentioned by Krotov (1950) for Karelian lakes. In 1955 the author of this paper stated the metasomatic role of calcite new formations, and later (Dobrovolsky, 1960, 1961) discussed the metasomatic phenomena widely present in hypergenesis of Quaternary deposits.

Great amounts of admixtures in new formations are related to colloidal and metasomatic processes. By using the term "admixtures", we mean accessory constituents of a new formation composition, rather than mechanical inclusions of mineral components of Quaternary deposits. These admixtures are in an isomorphous and sorbed state as well as in a specific state of dispersion.

Due to the presence of variable amounts of mechanical inclusions in new formations, it is impossible to use bulk chemical analysis in studying them. A special technique of ratio analysis (Dobrovolsky, 1960, 1961) is recommended when the substance of a new formation is wholly extracted, while mechanical inclusions are removed as unsoluble residue.

Calculation of the results of ratio analyses, per 100% of the newly formed substance, gives an idea of the actual chemical composition of a formation.

New formations resulting from metasomatic and colloidal processes (ferruginous, manganese and carbonate) have a particularly high content of admixtures. The content of admixtures in ferric formations is generally 10–30% of the newly formed substance.

The amount of admixtures is maxima in superaqueous, medium in eluvial, and minima in subaqueous formations.

The results of mineralogical analyses of ferric formations show that when there is a low content of silica and aluminium they are mainly present in a sorbed state. In case of high content these elements are present as amorphous allophanes (complex gels of variable composition). "Pure" silica gel, opal, is found in less amounts, and minerals of free alumina (hydragillite) are quite rare.

Manganese formations also have large amounts of admixtures (25–60% of a newly formed substance). Carbonate formations contain less admixtures, generally not more than 10%.

New formations contain much less admixtures if they crystallized from water-ion solutions and formed without the participation of colloidal and metasomatic processes. Thus, for example, minerals of the vivianite group, widely known in new formations of the humid zone, contain only 1–2% of admixtures. The same amounts of admixtures are found in gypsum formations, which occur widely in Quaternary deposits of the arid zone. It is interesting to note that silica plays the greatest role amongst the admixtures in gypsum.

In addition to principal chemical elements, epigenetic new formations also contain various dispersed chemical elements. The maximum content and variety of dispersed chemical elements have been found in formations greatly affected by metasomatic and colloidal processes.

CONCLUSION

Hypergenesis of Quaternary deposits is accompanied by colloidal and metasomatic processes. Many hypergenic formations (ferric, manganese and carbonate) have cryptocrystalline structures, metacolloidal microtextures, and metasomatically replace some components of Quaternary deposits. Influenced by colloidal and metasomatic phenomena, the chemical composition of these formations acquires considerable amounts of admixtures and a high content of some rare elements.

SUMMARY

Studies of the micromorphology of soil and hypergenic new formations made it possible to establish the wide participation of metasomatism processes in hypergenesis during the Quaternary period. Hypergenetic metasomatism results from interreactions of new formations with mineral components of Quaternary deposits.

Manganese, ferric and carbonate formations corrode and metasomatically replace mineral components of Quaternary deposits.

The effect of hypergenetic metasomatism is dependent upon the resistance of the rock-forming minerals and the replacement capacity of newly formed minerals. The most resistant rock-forming mineral is quartz and the least resistant is feldspar. A finely dispersed fraction of Quaternary deposits is metasomatically replaced easily.

New formations developing from the participation of metasomatic processes possess micromorphological signs of a gel state. These formations are characterized by a cryptocrystalline and metacolloidal microtexture. In the process of gel crystallization the original volume reduces, resulting in the formation of syneresis fractures.

Some new formations are devoid of the micromorphological effects of metasomatism (minerals of the vivianite group, gypsum, epsomite, halite, etc.). These formations have a good crystalline structure; the size of the crystalline forms ranges from 0.1 mm to a few decimetres. No signs of colloidal phenomena have been observed in the new formations of this group.

Micromorphological signs of metasomatic and colloidal processes are confirmed by alterations in the chemical composition and the content of rare elements in new formations.

REFERENCES

Dobrovolsky, V.V., 1955. Mineralogy of carbonate concretions from Quaternary loam soils. Zap. Vses. Mineralog. Obshchestva, 2 : 198–208.

Dobrovolsky, V.V., 1960. On hypergenesis mineralogy in Quaternary deposits of Central Russian forest steppe area. Byul. Mosk. Obshchestva Ispitatelely Prirody, Otd. Geol., 4 : 83–99.

Dobrovolsky, V.V., 1961. Typomorphic new formations in Quaternary deposits of the desert zone of the U.S.S.R. Pochvovedenie, 10 : 44–58.

Fersman, A.E., 1922. Geochemistry of Russia. Nauchnoye Chimico-Technicheskoe Isdatel'stvo, Petrograd, 214 pp.

Krotov, B.P., 1950. Metasomatic replacement in lacustrine metalliferous sediments of different types. Dokl. Akad. Nauk S.S.S.R., 73 (6) : 1263–1266.

GEFÜGE- UND STRUKTURUNTERSUCHUNGEN AN VULKANOGENEN EDAPHOIDEN

MARINA KRESS VOLTZ

Instituto de Edafologia y Agrobiológia, Madrid (Spanien)

Unter *Gefüge* soll die räumliche Anordnung der Bodengemengteile zu einander verstanden werden. Unter *Struktur* soll hier, der ursprünglichen Bedeutung der Bezeichnung folgend, die Eigenschaft des Bodens verstanden werden, bestimmte Formen und Grössen von Aggregaten und Hohlräumen zu bilden. Die Strukturuntersuchung ermittelt Gesamtgehalte und Verhältniswerte zwischen den einzelnen Grössenklassen und Formtypen. Sie dient in erster Linie der Landwirtschaft, dem Waldbau, der Bodenbiologie und der Standortkunde. Unter *Edaphoiden* werden Naturbildungen verstanden, die zwar bodenähnlich sind, deren Entstehung aber nicht auf bodenbildende Prozesse zurückgeht. Die nähere Bezeichnung *vulkanogen* besagt, dass es sich hier um solche Edaphoide handelt, deren Genese ausschliesslich vulkanothermischer Art ist.

Eines der wertvollsten Produkte dieser Art ist der Puzzolano in seinen verschiedenen Formen; Schlackenpartikelchen, die zu grössten Teilen aus vulkanischem Glas bestehen und - durch grossen Gehalt an Gasen zur Zeit ihrer Bildung - von zahlreichen Blasenräumen durchsetzt sind. Sie finden sich verbreitet auf den Kanarischen Inseln. Ähnliche Bildungen sind auch von anderen Vulkangebieten bekannt. Einige werden innerhalb dieser Mitteilung zum Vergleich auch untersucht. Dieser Schlackengrus ist eine Gesteinsart und kann nicht einmal als ein Rohboden bezeichnet werden. Aber er hat ein vollkommenes, hochstabiles Hohlraumsystem wie eine hochentwickelte Bodenbildung und hohe Wasser- und Luftkapazität, die kaum von der einer echten Bodenbildung übertroffen sind; ferner besitzt er, dank seines speziellen Kapilarsystems eine einmalige Schutzvorrichtung gegen Wasserverlust durch Austrocknung, so dass er in Trockengebieten, in denen Böden schwierig zu bewirtschaften sind, besondere Bedeutung erlangt. Sie hängt aber im wesentlichen von drei typischen Strukturmerkmalen ab: (1) von der Korngrösse, (2) von der Hohlraumgrösse, und (3) von der äusseren und inneren Oberfläche der Aggregate.

GEFÜGETYPEN VULKANOTHERMISCHER EDAPHOIDEN

Braunlehmähnliche Edaphoiden (Tafel B2)

Der von uns untersuchte Bimsteingrus konnte sehr verschiedene Farben aufweisen. Aus reinem hellen Glas bestehende Schlacken waren völlig farblos und zeigten im Dünnschliff auch keinerlei Doppelbrechung zwischen gekreuzten Nicols. Solche, die einen reichen Gehalt an Magnetit und anderen dunklen Mineralien enthielten, waren - wie der Puzzolano von Lanzarote - völlig schwarz gefärbt. Unter den farbigen Varianten stehen braune und rote Formen im Vordergrund. Jedoch geben diese rein makroskopischen Merkmale noch keine sicheren Möglichkeiten zur Differenzierung dieser Edaphoide Erst die mikromorphologische Dünnschliffuntersuchung lässt eine wirkliche Typisierung zu. Da es uns zudem darum geht hier diejenigen Gefügetypen aufzuzeigen, die solchen von echten Bodenbildungen analog sind, wird versucht, die charakteristischen Eigenschaften einiger solcher Edaphoiden aus verschiedensten Gebieten zu untersuchen. Zunächst wenden wir uns Gefügebildungen mit braunlehmähnlichen Charakter zu.

Braunlehmähnliches Puzzolano, Pico de Sta. Isabel, Fernando Póo (Probe 54)

Die Probe, die, wie alle übrigen, der Sammlung des Instituto de Edafologia in Madrid entnommen wurde, stammt aus einer Seehöhe von etwa +2.500 m N.N. Die Probestelle gehört bereits der Zwergstrauchstufe des 3.007 m hohen Vulkans an und liegt demnach in einer Zone, in der Gefügebildungen eines unvererdeten Braunlehms nicht mehr entstehen können. Die Probe entstammt dem Untergrund eines braunen Rankers auf Basaltschlacke. Der Schlackengrus ist gelbbraun bis rötlich-braun gefärbt. Im Mikroskop zeigt sich ein Blasenraumgefüge in einer sattgelben bis leicht orangefarbigen ausserordentlich homogenen Grundmasse, in der z.T. Phänokristalle, in erster Linie von Titan-augiten, eingebettet sind. Die Grundmasse zeigt eine Farbtiefe, wie sie bei Bodenbildungen sonst nicht vorkommt. Unter dem Lichtmikroskop erweist sich die Grundmasse selbst als völlig amorph. In der vorliegenden Probe scheinen zwischen gekreuzten Nicols kleine Stengelchen von Plagioklas, daneben kleine Augitkriställchen auf. Die Gegenwart dieser völlig unverwitterten Phänokriställchen unterscheidet dies Schlacken von vornherein von (B)/C-Horizonten von Braunlehmen, da sie in solchen nur in verwittertem Zustand vorkommen.

Braunlehmähnliche Bolus in Blasenräumen, Pico de Sta. Isabel (Probe 40)

Die Probe stammt aus dem C-Horizont eines erdigen Braunlehms aus einer Seehöhe von ca. 1.500 m. Hier kann ein bolusartiger unvererdeter Braunlehm durch blosse Verwitterung nicht gebildet werden. Die rötliche Schlackenmasse zeigt ein dichtes Blasenraumgefüge mit ausserordentlich verschieden Hohlraumgrössen. Die verfestigte glasige Grundmasse enthält neben kleinen Plagioklasleisten und Augitkriställchen dichte Ausscheidungen von grellroter bis orangegelben Eisenoxydhydratkriställchen die im Durchlicht opak sind. Die Blasenräume enthalten zum grössten Teile eine weiche grellgelbe, amorphe Bolusmasse, in der das Eisenoxydhydrat vollständig peptisiert ist. Zwischen gekreuzten Nicols zeigt sie keine Doppelbrechung und enthält auch keine Mineralkörnchen. Sie ist fliessbar und kann sich in vorhandene Spalten hineinbewegen. Sie füllt die Blasenräume mehr oder weniger aus. Ihr Braunlehmgefüge unterscheidet sich von dem in Braunlehmen die durch Bodenbild entstanden sind nur durch den um vieles grelleren Farbton.

Braunlehmähnliches Edaphoid aus dem Trumao-Gebiet von Mittel-Chile

Dieses Material wurde von Herrn Arnaldo Mella, Santiago de Chile, gesammelt und untersucht. Wir verdanken die Möglichkeit, seine Angaben hier zu zitieren, seinem freundlichen Einverständnis, wofür ihm bestens gedankt sei. Der Trumao ist ein Boden, der sich nach den Untersuchungen Mellas aus einem bolusartigen Material gebildet hat, das in Form von Detritus als vulkanische Asche abgelagert wurde und durch sekundäre Ver-erdung das Gefüge eines erdigen Braunlehms erlangt hat. Die Böden zeigen eine Ansammlung von verschlämmtem, unvererdetem Bolus, der etwa den Füllmassen in den Blasenräumen der Probe 40 entspricht. Dieser Bolus unterscheidet sich von einem durch Bodenbildung entstandenen Braunlehm dadurch, dass er keinerlei Mineralkörner als Einsprenglinge enthält. Dies kommt bei Bodenbildungen fast nur bei bestimmten Terra-Fusca-Formen vor, bei denen die Mineralkörner, soweit sie nur aus Kalzit bestanden haben, sich infolge der starken Verwitterung in Feuchtklimaten völlig gelöst haben.

Die vulkanogenen Bolusvorkommen in Chile und vielen anderen Gebieten unterscheiden sich von echten Bodenbildungen in der Regel dadurch, dass nicht nur die Kieselsäure, die Eisenoxydhydrate und die Aluminiumoxydhydrate, sondern auch die Tonsubstanz in amorpher Form vorliegen. Die Allophantone gehen erst durch die Bodenbildung in kristalline Tone wie Kaolinite über (Besoain und Garcia–Vicente, 1962). Sehr häufig unterscheiden sich vulkanogene Bolusarten von Bodenbildungen schon durch weitaus grellere Farben im Dünnschliffbild.

Pseudogleyähnliche Edaphoiden (Tafel B3)

Braunlehmähnliche Gefüge mit peptisiertem und mit durch Diffusion hochbeweglichem Eisenoxydhydrat können durch Eisenabwanderung gebleicht werden. Die auffallend gelbe Farbe der sonst unveränderten Matrix verwandelt sich in graue Farbtöne oder wird völlig weiss; im Dünnschliff ist sie bei Durchlicht farblos. Dies kann bei Bodenbildungen überall dort eintreten, wo Stoffwanderungen durch Diffusion möglich sind, vor allem bei der Pseudovergleyung und bei der Laterisierung. Bei der Pseudovergleyung wird die Enteisung der Grundmasse dadurch erzeugt, dass das amorphe hochbewegliche Eisenoxydhydrat durch Wanderung fleckenförmige Ausscheidungsstellen entstehen lässt. Dort fällt es durch Kristallisation (in Form von Goethit oder Lepidokrokit) aus. Diese Enteisung in einer für Diffusion gut leitbaren dichten Matrix ist auch bei vulkanothermischen Bildungen häufig zu beobachten.

Gefleckter Puzzolano, Pico de Sta. Isabel (Probe 57)

Die Probestelle zeigt erdigen Braunlehm über Puzzolano in einer Seehöhe von ca. 2.700 m. Aus dem Untergrund des Profils stammt die Probe, die Schlackengrus von rund 3–6 cm Grösse enthält. Sie zeigt Fragmente von hellgrauer, gelblicher bis braunroter Farbe. Die Zersatzmasse ist zwischen den Fingern leicht zerreiblich. Im Dünnschliff sieht man unter dem Polarisationsmikroskop Blasenraumgefüge mit auffallend grellgelber bis orangefarbiger optisch isotroper Grundmasse, in der sich Phänokriställchen von Augiten, kleine Plagioklas-Stengelchen und Magnetitkörner zeigen. In der Grundmasse treten ferner grellrote, durchscheinende Aggregatkomplexe von auskristallisiertem Eisenoxydhydrat auf, die als scharf abgegrenzte Flecken die Matrix durchsetzen. Diese Tendenz zur Auskristallisation hat eine Enteisung der Grundmasse zur Folge. Die orangerote bis sattgelbe Farbe hellt sich auf, wird leicht gelb, dann grau und verschwindet bei starker Enteisung völlig.

Schluffiges pyroklastisches Sediment aus pseudogley-artigen Gefügefragmenten, Pico de Sta. Isabel (Probe 52)

Die Probe stammt aus dem Untergrund eines braunen Mullrankers aus einer Seehöhe von 2.625 m. Das Material stellt makroskopisch ein braungelbes, im feuchten Zustande völlig unplastisches dichtes Aschensediment mit rötlichen Flecken dar. Mikroskopisch zeigt sich, dass es gänzlich aus einem losen Gemenge von Gefügefragmenten von Schluffgrösse besteht. Unter ihnen lassen sich zunächst farblose kantig-splitterige Bruchstücke von vulkanischem Glas feststellen, ferner rundliche Aggregate von farbloser, grauer, hellgelber sattgelber, orangeroter bis grellroter Farbe. Sie stellen Teile eines ehemals zusammenhängenden pseudogley-ähnlichen Gefüges dar, das in seiner Gesamtheit immer noch Pseudogleyschichten vortäuscht, was durch Vorhandensein zusammenhängender grellroter Kristallaggregate von Grobsandgrösse und die stellenweise Häufung von rötlichem eisenreichen Detritus noch verstärkt wird. Der Standort zeigt keinerlei Neigung zu Staunässe und hat ein ungewöhnliches Aufsaugevermögen für Wasser. Das Material lässt einen ausserordentlich leichten Gefügezerfall in unregelmässig geformten rauhwandigen Bröckeln erkennen.

Lateritähnliche Edaphoiden (Tafel B4)

Rote Basaltschlacke bei Chillon, Ciudad Real (Ci 3)

Die schwarze, stark abfärbende Basaltschlacke zeigt eine grellrote Oberflächenschicht, die ins Innere des Gesteins hineinreicht und grössere Erstarrungsräume ausfüllt. Im Mikroskop sieht man ein Gefüge, das durch opake, im Auflicht grellrote, Eisenleisten zerteilt ist, die Querschnitte von Spaltausfüllungen darstellen. Die Räume zwischen ihnen waren früher von einer grellgelben, dichten, braunlehmähnlichen Masse ausgefüllt, in der sich schwebend kleine, im Lichtmikroskop noch gut erkennbare, durchscheinend grellrote, rundliche bis fast deutlich sechseckige Kristalliten von Eisenoxydhydrat, wahrscheinlich von Hämatit, gebildet haben. Die Kriställchen vereinigten sich bei weiterem Nachwachsen und bilden nun eine Art Netz, wie es bei Lateritstrukturen häufig zu finden ist. Die grellgelbe Matrix wurde bis auf wenige Reste ausgewaschen. Weitere Lateritähnliche Merkmale, sowie Andeutungen von Lateritprofilen fehlen. Diese Gefügebildungen sind bei vulkanogenen Umwandlungen häufig zu finden und geben leicht zu Verwechslungen mit Lateriten Anlass.

Vulkanothermischer Fleckenton, Pico del Teide, Gipfel Fumarole, Teneriffa (Te 2)

Die Probe, die vom Gipfel des Teides (3.780 m) stammt, lässt noch die frühere Schlackenstruktur der Lava, aus der das Material entstanden ist, erkennen. Letztes ist aber so stark zersetzt dass es bereits bei leichtem Fingerdruck zerbricht. Es zeigt eine weissliche enteisente Grundmasse, in der dunkelrote Flecken von Kristallisationsaggregaten von Eisenoxydhydrat aufscheinen. Der Zerfall des Materiales ist staubartig; dieser Feindetritus-staub erlangt durch Vermischung der Teilelemente einen rosafarbigen Ton. Schlackenbruchstücke, die noch Reste einer wenig enteisenten auffallend grellgelben Grundmasse aufweisen, zeigen einen geringeren Zerfall. Sie lassen ferner erkennen, dass die frühere Grundmasse, ähnlich wie bei der Laterisierung bei Bodenbildungen, von einer Art Braunlehmgefüge mit peptisiertem amorphen Eisenoxydhydrat ihren Ausgang genommen hat. Die staubartig zerfallene, rosafarbige Masse nimmt Wasser gierig auf, bildet aber einen unplastischen Teig, der nach dem Austrocknen wieder zerfällt.

Das Dünnschliffbild zeigt eine farblose bis graue Matrix mit im Auflicht grellroten Kristallisationsaggregaten von Eisenoxydhydrat. Zum Teil zeigen sich Felder mit kleinen rundlichen Schwebekriställchen (Hämatit) wie in Probe Ci 3.

Das Gefüge ist den lateritischen Fleckentönen ("mottled clays") ausserordentlich ähnlich, auch bezüglich der Tatsache, dass sich die Kristallisationsaggregate leicht zu opaken Ausscheidungen verdichten, im Gegensatz zu den fast immer durchscheinend bleibenden Ausscheidungsaggregaten in Pseudogleyen.

MIKROMORPHOMETRISCHE STRUKTURUNTERSUCHUNG AN EINEM PUZZOLANO VON LANZAROTE (OSTKANAREN)

Als Beispiel für die quantitativen Strukturverhältnisse werden in folgenden einige mikromorphometrische Untersuchungsergebnisse über den Puzzolano von Lanzarote angeführt. Bestimmt wurden der Gesamthohlraumgehalt, der mittlere Durchmesser der Intra-Aggregat-Hohlräume, sowie die innere Oberfläche dieser Hohlräume und darüber hinaus auch die äusseren Oberflächen der Aggregate.

Dazu wurden sowohl fluoreszierende Anschliffe nach Werner (1962) hergestellt, als auch einige Dünnschliffe. Die Messungen erfolgten an Strukturphotogrammen von Anschliffen und Dünnschliffen. Der Gesamthohlraumgehalt wurde photometrisch an Strukturphotogrammen der Anschliffe bestimmt. Die Hohlraumdurchmesser sind am Teilchengrössenanalysator (Zeiss) gemessen worden; die innere und äussere Oberfläche wurde an Strukturphotogrammen verschiedener Vergrösserung mit Hilfe eines Kurvimeters ermittelt (Geyger 1962; Kubiena et al., 1963).

Abb.1 zeigt, dass sich zwischen diesen Puzzolano-Aggregaten in jener Lagerung welche man in Tabakpflanzungen von Lanzarote antrifft, schon ein sehr stark wechselndes Hohlraumsystem bildet. Durch die komplizierte Umrissform der Aggregate finden sich nur selten Hohlraumformen, die über längere Strecken einen gleichmässigen Durchmesser aufweisen. Vielmehr

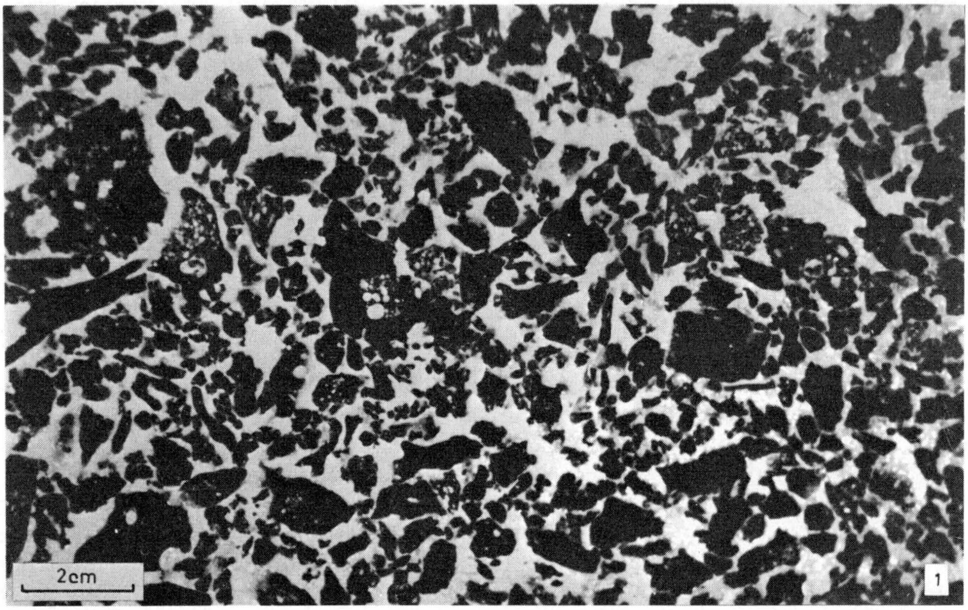
2cm
1

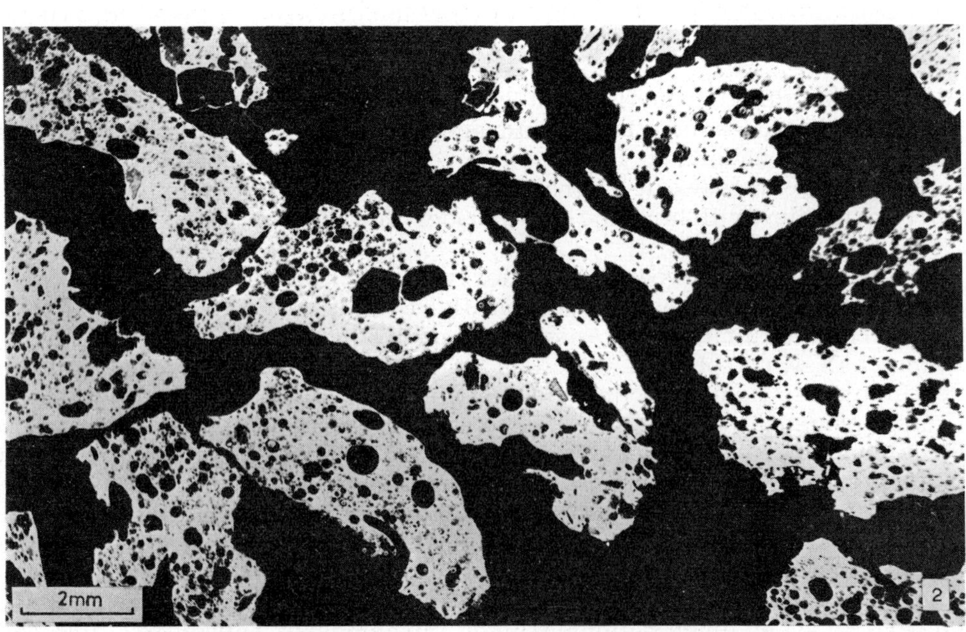
2mm
2

ist zu erkennen, dass der Hohlraumdurchmesser sehr stark schwankt. Es wechseln also häufig - schon in den Zwischen-Aggregat-Hohlräumen - über und unter kapillare Hohlräume oft ab (vgl. Abb.1, 2). Von grosser Bedeutung für den Wasserhaushalt dieser Puzzolano-Auflage ist darüberhinaus zunächst die äussere Oberfläche, die die Aggregate bieten. Durch die kompliziert gelappten Umrissformen dieser Bröckel ergibt sich der relativ hohe Wert von 10,7 cm^2/cm^3. Abb.3 zeigt, dass die Hohlräume in sehr mannigfalten Formen tief in die Aggregate hineingreifen. Dieses Bild, sowie das in Abb.4 und dazu die direkte Untersuchung unter dem Auflichtmikroskop, lassen erkennen, dass die Hohlräume im Inneren der Puzzolano-Aggregate nur in seltenen Fällen abgeschlossene, sog. "Blasenporen" sind, sondern dass vielmehr auch im Inneren weitverzweigte Hohlraumsysteme vorhanden sind. Auch hier wechseln die mittleren Hohlraumdurchmesser auf geringstem Raum sehr stark. Auf grund dieser Verhältnisse ergibt sich der sehr hohe Wert von 25,0 cm^2/cm^3 innerer Oberfläche der Aggregate. Von besonderer Bedeutung für den Wasserhaushalt in dieser Puzzolano-Auflageschicht ist darüberhinaus natürlich das Grössenverhältnis dieser Intra-Aggregat-Hohlräume. Die Messungen ergaben, dass Hohlraumdurchmesser von 50–100 μ am stärksten vertreten sind, jedoch überwiegen die Mittelporen (Odén, 1957) ganz allgemein. Grobporen von 200–500 μ sind nur noch mit 10% des Gesamtanteiles vertreten. Zu beachten ist jedoch, dass die Hohlraumdurchmesser schon auf kleinstem Raum sehr schnell wechseln, so dass auch im Inneren dieser Aggregate nur selten ein durchgehendes Kapillarsystem vorhanden ist; vielmehr wird die kapillare Wasserbewegung stets durch Strecken mit überkapillaren Hohlraumdurchmessern unterbrochen (vgl. dazu Abb.3, 4). Die Messergebnisse dieser Hohlraumdurchmesser-Bestimmungen gibt die Abb.5. Alle mikromorphometrischen Auswertungen sind in der Tabelle I zusammengestellt.

Abb.1. Strukturphotogramm eines fluoreszierenden Anschliffes. Puzzolano-Aggregate (Lanzarote, Kanarische Inseln). Die Abbildung zeigt die Aggregate in natürlicher Lagerung. Man erkennt, dass ihre Formen ausserordentlich vielgestaltig sind: nur wenige der Bröckel sind rundlich, die meisten stark gelappt. Auf diese Weise ergeben sich zwischen ihnen Hohlräume von sehr verschiedener und stark wechselnder Breite. Das Innere der Aggregate ist sehr hohlraumreich. Man kann schon bei direkter Betrachtung erkennen, dass auch die Hohlräume im Innern sehr vielgestaltig sind.

Abb.2. Strukturphotogramm eines Dünnschliffes. Puzzolano-Aggregate, Lanzarote (Kanarische Inseln). Weiss: feste Substanz; schwarz: Hohlräume. Bei stärkerer Vergrösserung ist jetzt zu erkennen, dass die inneren Hohlräume der Bröckel sehr verzweigt mit einander verbunden sind und - räumlich betrachtet - Netzräume bilden. Der Durchmesser der Hohlräume ist ausserordentlich verschieden, nur selten finden sich solche, die über längere Strecken gleiche Breite haben; ihr Querschnitt ist meist rundlich.

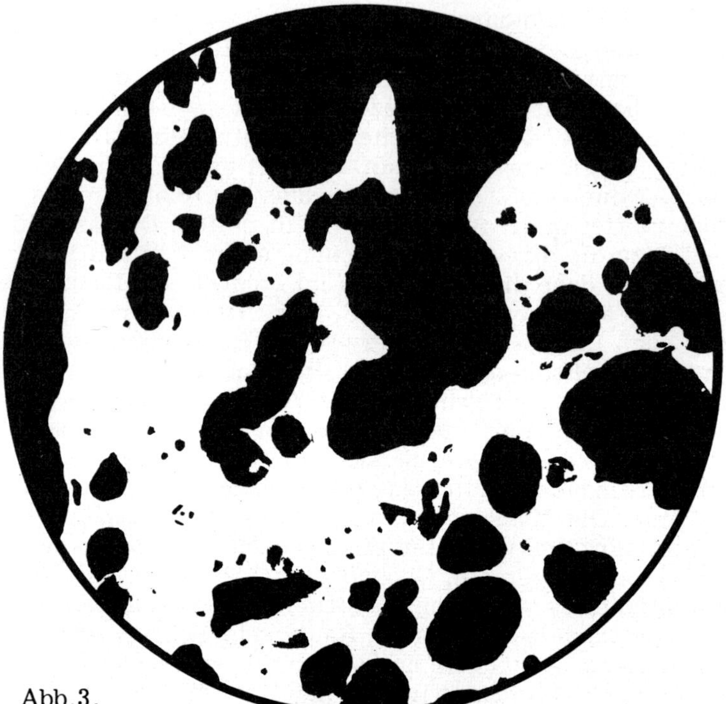

Abb.3.

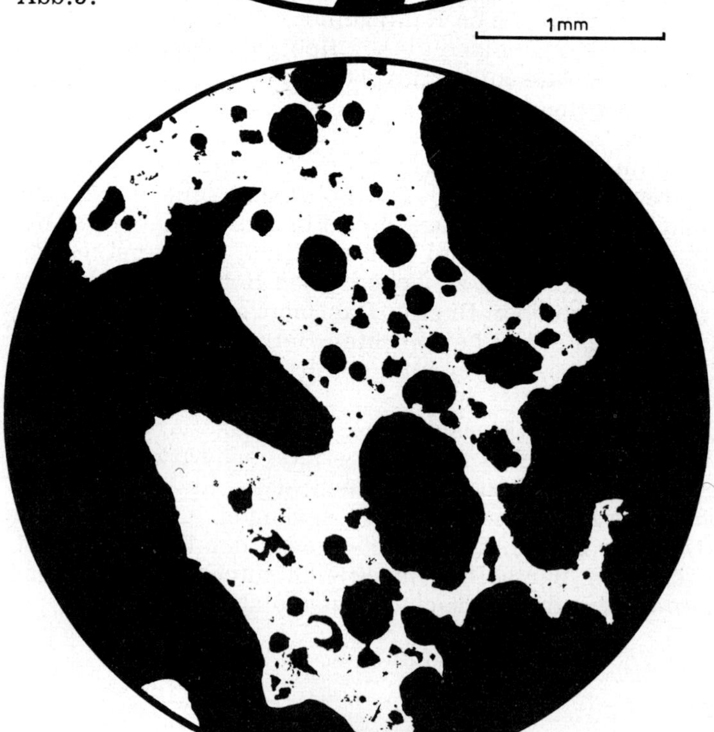

Abb.4.

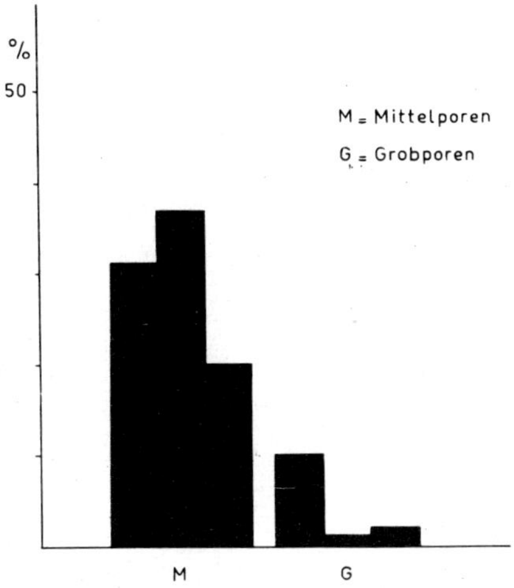

Abb.5. Mittlere Hohlraumdurchmesser im Inneren der Aggregate (Einteilung nach Odén, 1957).

Abb.3. Strukturphotogramm eines Dünnschliffes. Puzzolano-Aggregat, Lanzarote (Kanarische Inseln). Weiss: feste Substanz; schwarz: Hohlräume. Das Bild zeigt einen Ausschnitt aus einem grösseren Bröckel mit einem Teil seines Randes. Man erkennt, dass von aussen die Hohlräume in lappigen Buchten tief in das Innere der Aggregate eindringen, sich oft verbreitern und verengen. In der rechten Hälfte des Bildes sind einige Querschnitte durch Hohlräume zu erkennen: sie sind rundlich, manchmal kreisförmig. Oberhalb und unterhalb der Schnittebene des Dünnschliffes hängen alle Hohlräume wie in einem Schwamm miteinander zusammen, sodass sich ein weitverzweigtes Hohlraumsystem ergibt.

Abb.4. Strukturphotogramm eines Dünnschliffes. Puzzolano-Aggregat, Lanzarote (Kanarische Inseln). Weiss: feste Bodensubstanz; schwarz: Hohlräume. Wie in Abb.1 wird hier ein Ausschnitt aus einem grösseren Aggregat gezeigt. Es ist eine besonders hohlraumreiche Partie: die feste Bodensubstanz bildet nur noch dünne Wände zwischen den grösseren und kleineren Hohlräumen. Man erkennt auch hier, dass die Querschnitte der Hohlräume meist rundlich sind. Wie die direkte stereoskopische Betrachtung ergibt, handelt es sich aber nur in seltenen Fällen um abgeschlossene sog. Blasenporen.

TABELLE I

Mikromorphometrische Untersuchungen von Puzzolano-Aggregaten Lanzarote (Ostkanaren)

Gesamthohlraumgehalt, äussere Hohlräume	40%
Äussere Oberfläche der Puzzolano-Aggregate	10,7 cm^2/cm^3
Innere Oberfläche der Puzzolano-Aggregate	25,0 cm^2/cm^3
Mittlere Durchmesser der Intra-Aggregat-Hohlräume	24– 50 μ: 31,1%
	50–100 μ: 36,9%
	100–200 μ: 19,8%
	200–500 μ: 10,3%
	>500 μ: 1,8%

DISKUSSION DER ERGEBNISSE

Die mikromorphologischen Untersuchungen haben gezeigt, dass bodenähnlich Gefüge auch in vulkanogenen Produkten vorkommen können; besonders solch mikromorphologische Bildungen, die von feuchten oder wechselfeuchten Tropenböden bekannt geworden sind, wie von Braunlehmen tropischen Pseud gleyen und Lateriten. Das ist leicht verständlich, denn die Bildungsbedingun der vulkanogenen Zersatzprodukte ähneln gewissermassen jene der Tropenböden nach bezw. sie steigern in einer Weise, wie sie durch blosse Verwitterung auch an den heissesten und zugleich feuchtesten Standorten der Tropen nicht gefunden werden können. Dieser Umstand hilft uns, vulkanogen Edaphoide von echten Bodenbildungen durch bestimmte Gefügemerkmale, durch ihre Farbe in einigen Fällen auch durch ihre besondere Mineralführu unterscheiden zu können.

Es ist sicherlich häufig möglich, vulkanogene Produkte durch die Feld untersuchung festzustellen und von echten Böden, die deutlich edaphische Profile bilden, unterscheiden zu können. Die Erkennung durch die Standortaufnahme und die blosse chemische Untersuchung ist jedoch nur dann schwierig, wenn solche Edaphoide durch Erosion verlagert wurden und als tiefe Sedimente vorliegen. Solche Sedimente sind in Vulkangebieten häufig und wurden wiederholt als Paleoböden (z.B., Tertiär-Rotlehme) angesproch Aus diesem Grunde ist es wichtig, die mikromorphologische Untersuchung, der bei der Diagnose der vulkanogenen Edaphoide eine besondere Rolle zufällt, zu einem sicheren, auch in komplizierteren Fällen untrüglichen Bestimmungsmittel auszubauen.

Durch die mikromorphometrische Untersuchung von Puzzolano der Insel Lanzarote (Ostkanaren), der zum Anbau von Tabakpflanzungen verwendet wird, ist zum ersten Male deren Hohlraumsystem auf Korngrössen, Hohlraumgrössen sowie innere und äussere Oberfläche untersucht worden. Es zeigt sich ein zu dem Verwendungszweck sehr vollkommenes Kapillarsystem, das bei sonst keiner Bodenbildung gegeben ist. Es kann einerseits grosse Wassermengen aufnehmen, es gibt aber diese infolge des Wechsels von Feinkapillaren und Grosskapillaren und die dadurch bedingte Verhinderung eines an die Bodenoberfläche führenden Kapillarstromes nicht oder nur wenig wieder ab.

ANERKENNUNG

Herrn Prof. Dr. Kubiena möchte ich für seine freundliche Anleitung, Herrn Dr. Beckmann für seine Hilfe bei den mikromorphometrischen Messungen und Herrn Dr. Alonso für die elektronenmikroskopischen Untersuchungen, verbindlich danken.

SUMMARY

Fabric and structure investigations on volcanogenetic edaphoids

Micromorphological investigations are of special significance for the recognition of volcanogenetic edaphoids. As a result of their formation conditions, they resemble tropical soils to a certain degree, viz. braunlehms, tropical pseudogleys and laterites.

Volcanogenetic edaphoids differ from soils chiefly by distinct fabric features, as well as by a particular mineralogical composition.

Micromorphometrical investigations have proved that a Puzzolano used for tobacco culture in Lanzarote (East Canary Islands) has, for this purpose, an ideal capillary system which has not yet been found in soil.

LITERATUR

Besoain, M. and Garcia–Vicente, J., 1962. Clay mineralogy of some volcanic ash soils of Chile. Neues Jahrb. Mineral, Abhandl., 98 : 349–366.

Beckmann, W., im Druck. Möglichkeiten der stereoskopischen Untersuchung im Innern der natürlichen Bodenstruktur. Zeiss-Mitt.

Geyger, E., 1962. Zur Methodik der mikromorphometrischen Bodenuntersuchung. Z. Pflanzenernähr., Düng., Bodenk., 99 : 118–129.

Kubiena, W., Beckmann, W. und Geyger, E., 1963. Mikromorphometrische Untersuchungen an Hohlräumen im Boden. Anales Edafol. Agrobiol. (Madrid), 22 : 551–568.

Odén, S., 1957. Förslag till klassifikation av markporer. Kgl. Skogs- Lantbruksakad. Tidsskr., 96 : 297–313.

Werner, J., 1962. Über die Herstellung fluoreszierender Anschliffe. Z. Pflanzenernähr., Düng., Bodenk., 99 : 144–150.

FARBENBILDER

Tafel B2. Gefüge eines Puzzolanos des Pico de Sta. Isabel Fernando Póo (Probe Fe 57) mit braunlehmähnlichem Gefüge. Die amorphe Grundmasse von verwitterndem vulkanischem Glas ist durch peptisiertes hochbewegliche amorphes Eisenoxydhydrat gefärbt. Rechts oben Kristallisationsaggregate.

Tafel B3. Puzzolano vom Pico de Sta. Isabel (Probe Fe 54) mit zum Teil enteisenter graufleckiger Grundmasse. An der Oberfläche des Aggregates durch Kristallisation von Eisenoxydhydrat gebildeter, im Durchlicht opaker, bei Auflicht grellroter Saum.

Tafel B4. Lateritähnliche vulkanogene Ausscheidungen von kristallisiertem Eisenoxydhydrat in der Grundmasse und an den Wänden der Blasenräume. Probe Fe 53. Pico de Sta. Isabel.

DIE KLIMAXBILDENDEN RANKER SPANIENS, IHRE MIKROMORPHOLOGIE UND GENESE[1]

JOSE M. ALBAREDA HERRERA

Institut für Bodenkunde und Agrobiologie, Consejo Superior de Investigaciones Cientificas, Madrid (Spanien)

Ranker sind nach Kubiena (1948) kalkarme terrestrische Bodenbildungen auf Silikat-oder Kieselgesteinen, die nur aus dem Humushorizont bestehen, der unmittelbar auf dem Muttergestein aufliegt. Bei Übergangsbildungen (braunen Rankern) schiebt sich ein mehr oder minder bandförmig ausgebildeter, geringmächtiger, mässig verwitterter brauner B-Horizont zwischen Humushorizont und Muttergestein ein. Die Verwandtschaft mit den Rendsinen (terrestrische AC-Böden auf Kalkgesteinen) ist nicht nur durch den ähnlichen Profilaufbau und durch eine gewisse mikromorphologische Ähnlichkeit gegeben, sondern auch dadurch, dass sich Rendsinen, besonders solche auf Kalksilikatgesteinen oder Kalksandsteinen, durch starke Entkalkung und zunehmende Azidität in Ranker oder rankerähnliche Bildungen umwandeln können. Diese Verwandtschaft lässt einige Probleme der Genese besser verstehen, worauf später zurückgekommen werden soll.

Im Unterschied zu den Rendsinen sind die Ranker auf Gebiete mit speziellen Umweltverhältnissen beschränkt. Solche Umweltverhältnisse sind in Spanien in besonderem Masse gegeben, so dass in Europa bis heute kein anderes Land bekannt geworden ist, in dem ein gleicher Formenreichtum oder in dem so hochentwickelte Ranker, wie jene in Nordwestspanien, mit ihren metertiefen schwarzen Humushorizonten, festgestellt werden konnten. Die Beantwortung der Frage nach der Genese dieser Ranker und deren Abhängigkeit von den Umweltverhältnissen ist wiederholt in Angriff genommen worden (vgl. Huguet del Villar, 1937; Kubiena, 1938, 1953; Franz, 1956). Immer blieben aber noch Einzelheiten zu klären. Eine Anzahl von Exkursionen des Verfassers, besonders in die Rankergebiete Galiziens und in der Sierra de Gata im Jahre 1963, haben einige neue Gesichtspunkte zur Genese dieser Rankerformen erkennen lassen. Diese in das Gefüge des bisher Bekannten einzubauen, soll mit dieser Studie versucht werden.

Zunächst ist zu sagen, dass man unter Ranker streng terrestrische Bildungen versteht. Anmoorige oder torfige Böden scheiden darum, wie bereits Franz (1955) betonte, von vornherein aus, auch als Vorstufen der Entwicklung. Die Mikromorphologie aller Ranker, die sich aus wenig zersetzten Pflanzenresten und mehr oder minder stark humifizierter Lösung einer terrestrischen Kleinfauna zusammensetzt, deutet auf eine durch die Tätigkeit von Landtieren geprägte Humusform hin. Die Ableitung der nordwestspanischen Ranker von semiterrestrischen Vorstufen, wie sie von Huguet del Villar (1937) angenommen wurde, lässt sich mikromorphologisch nicht bestätigen.

[1] Die hier behandelten alpinen und subalpinen Böden sind laut der U.S.D.A. Soil Classification, 7th approximation (1960) wahrscheinlich: Psammonte, die atlantischen Ranker: Umbrepte (Red.).

Die tiefgründigen nordwestspanischen Ranker wurden zuerst von Huguet del Villar (1937) beschrieben. Eine Arbeit über die subalpinen bis alpinen Ranker der Sierra de Guadarrama wurde erstmalig von Claver Aliod (1947) veröffentlicht. Er verwendete für sie die Bezeichnung "Suelos silico-húmicos" in Anlehnung an den von Pallmann (1933) für alpine AC-Böden auf festen Silikatgesteinen, die er als erster beschrieben hat, gebrauchten Namen "Humussilikatböden". Claver hat eine grosse Anzahl von Proben analysiert und auch die ersten Photographien von dem mikromorphologischen Aufbau dieser Böden zu geben versucht. Seine mikroskopischen Untersuchungen wurden bereits an Dünnschliffen durchgeführt und ergaben die Feststellung eines Modergefüges mit zahlreichen losen, unverwitterten Mineralkörnern, in erster Linie Quarzen, Feldspäten und Glimmern mittlerer Korngrösse. Die Bedeutung des Begriffes Ranker und seiner Spielarten für Spanien wurde auch von De Cañedo–Argüelles Quintana (1949) betont.

Tiefgründige Ranker in subalpinen Bereichen wurden wiederholt auch von russischer Seite, besonders im Kaukasus, beschrieben, aber nicht als eigene, von den bisher bekannten Bodentypen verschiedene Bildung bezeichnet. In diesem Zusammenhang scheint die Bemerkung von Glinka (1914) zu stehen, dass eine Bildung von Tschernosemen auch auf Granit als Ausgangsmaterial festgestellt werden kann. Die nordspanischen Ranker auf Granit (das hier fast allein für sie in betracht kommende Muttergestein) haben aber ausser ihrer Tiefgründigkeit und ihrer relativ geringen chemischen Verwitterung nichts mit den Tschernosemen gemeinsam; sie sind Produkte einer völlig anderen Umwelt, einer anderen Vegetation und einer völlig anderen Standortbiologie. Die von Tschernosemen grundverschiedene Mikromorphologie lässt sich durch eine einfache Dünnschliffuntersuchung sofort überzeugend erkennen. Dies dürfte auch für die dem Steppenklima entrückten südrussischen, insbesondere kaukasischen Formen zutreffen.

Eine umfassende Einteilung aller Rankerformen zu geben, wurde von Kubiena (1948, 1953) versucht. Sie trägt dem Umstand Rechnung, dass sich die Ranker, ähnlich wie die Rendsinen, in erster Linie durch ihre Humusform unterscheiden. So scheinen in seiner Systematik Protoranker (Humusform Silikatmoder), mullartige Ranker (Humusform mullartiger Silikatmoder), Mullranker, und dystrophe Ranker (Humusform dystropher Grobmoder bis Rohhumus) auf. Nach dem Entwicklungsgrad werden zusätzlich unterschieden: graue Ranker (mit grauen C-Horizonten, charakterisiert durch weitgehenden physikalischen Zerfall), Podsolranker (mit beginnender Durchschlämmung), C_m-Ranker oder Ganter (mit chemisch stark verwittertem, aber nicht durch Oxydation des freien Eisenoxydhydrats gebräuntem C-Horizont) und braune Ranker (mit einem den A-Horizont bandartig säumenden, beginnenden (B)-Horizont). Starken Einflüssen durch besondere Standortverhältnisse wurden durch die Begriffe Xeroranker oder Trockenranker (Kubiena, 1953, S.210), Tundra-Ranker, Eilag-Ranker, Tangel-Ranker (durch den Einfluss besonderer Tangelpflanzen hervorgerufen, Kubiena 1948, S.126) Rechnung getragen. Da auch die Humussilikatböden Pallmanns (1933) mikromorphologisch als zu dieser Bodengruppe gehörig erkannt werden konnten, wurden sie der höheren Einteilungskategorie Ranker zugeordnet und als alpine Ranker bezeichnet.

Dem Problem der Klimabedingtheit der klimaxbildenden Ranker hat Franz eine eigene Veröffentlichung gewidmet, in der er neben ostösterreichischen Xerorankern (vgl. Franz, 1955) vor allem spanische Rankerformen berücksichtigt hat. Es ist sein Verdienst, auf die Einmaligkeit der Ranker im Nordwesten Spaniens hingewiesen und sie mit dem feuchtwarmen Klima ihrer Verbreitungsgebiete in Zusammenhang gebracht zu haben. Er hatte Gelegenheit, diese Rankervarietät, die er als atlantischen Ranker bezeichnet hat, auf zahlreichen Exkursionen in Galizien und in der Sierra de Estrella (Portugal) kennen zu lernen und festzustellen, dass sie auf ihren Standorten als Endbildungen der gegenwärtigen Bodenentwicklung auftreten.

Neben den von Franz behandelten atlantischen Rankern und den alpinen Rankern der spanischen Hochgebirge, die den Humussilikatböden Pallmanns entsprechen, weist Spanien noch eine Fülle von anderen Rankervarietäten auf.

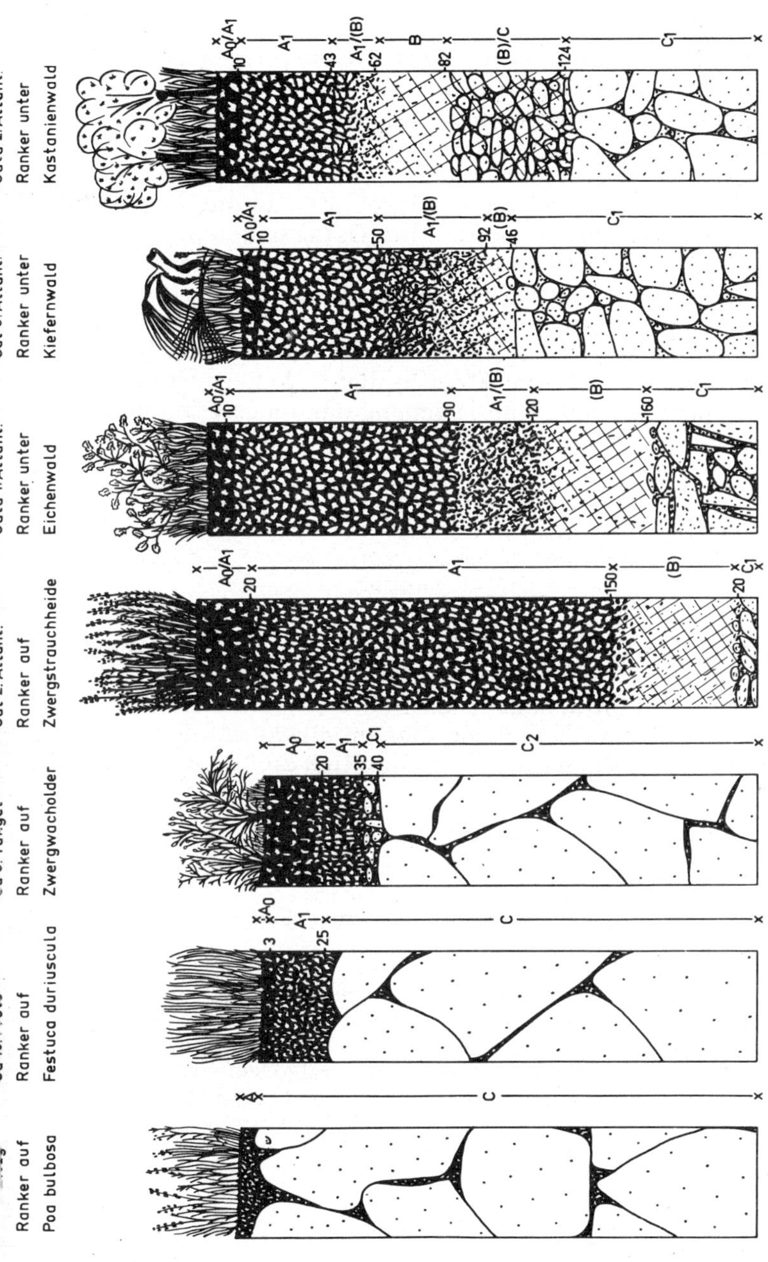

Abb.1. Profilübersicht spanischer Ranker.

Als die wichtigsten Ranker Spaniens sind ausser den obigen zu nennen: die subalpinen Tangelranker, die trockenen Protoranker, die mullartigen und Mullranker der Páramos, die Protoranker und mullartigen Ranker der Stranddünen der andalusischen Küstengebiete (die in Nordeuropa fehlen und in Spanien als Endstadien der gegenwärtigen Bodenentwicklung auftreten können), die Podsolranker Asturiens, die z.T. grössere Flächen einnehmen die Waldranker der semihumiden Braunerdegebiete und die besondere Varietät der Launa-Ranker (Kubiena, 1953) in Ostandalusien. Der dieser Mitteilung gesteckte Rahmen würde in keiner Weise ausreichen, alle diese Formen zu behandeln. Sie wird sich auf solche beschränken, die klimaxbildend sind und deren mikromorphologische Untersuchung neue Schlüsse bezüglich ihrer Genese zulässt. Zu diesem Zwecke soll zunächst die Beschreibung einiger hierfür wichtiger Profile, nebst Anführung der Ergebnisse der mikromorphologischen Untersuchung der ihnen entnommenen Rahmenproben, vorausgenommen werden.

ALPINE UND SUBALPINE RANKER

Probe Gd 5

Alpiner Eilag-Ranker auf Granit unter Rasen von *Poa bulbosa;* Guarramas 2.240 m, Sierra de Guadarrama. Ebener, stark windgefegter Gipfelstandort. Der Rispengrasrasen ist ausserordentlich dicht, entwickelt sich auf den Felsenflächen des Gipfels oder nahe diesem direkt auf der Oberfläche des Gesteins, ohne dessen Verwitterung wesentlich zu beeinflussen (Abb.1).

Profilaufbau[1]

0–4 cm: A_0, dichter, torfähnlich zusammenhängender Horizont von toten, wenig zersetzten und frischen Wurzelresten, die ein dichtes Geflecht bilden, das auch in die Gesteinsspalten hineinreicht. pH = 5,9.

>4 cm: C, unverwitterter Zweiglimmergneis mit glatten Oberflächen.

Mikroskopische Untersuchung

Probe aus 0–4 cm: A_0, lockerer Grobmoder, hauptsächlich aus braun gefärbten Wurzelresten mit fast völlig unzerstörten Zellstrukturen bestehend. Rote Zersetzungsfarben. Neben den Pflanzenresten finden sich nur kleine schwärzlichbraune Losungsstücke in Grösse und Form von Oribatidenlosung. Mineralkörner (Quarze, Muskovite, Biotite, Orthoklase) treten mengenmässig stark zurück und beschränken sich auf solche von Feinsand- und Schluffgrösse. Alle sind völlig unverwittert, haben keine Kornbeläge und tragen nur dünnen Organismenaufwuchs oder feinsten Losungsdetritus an der Oberfläche.

Anmerkung

Dieser Boden wurde auch von Claver (1947) untersucht. Es wurde hier mit erweiterter mikroskopischer Analyse zu den vergleichenden Untersuchungen der spanischen Ranker herangezogen, weil er zum Verständnis der Genese der Rankerformen an feuchtwarmen Standorten wesentlich beiträgt. Die geringe Zersetzung der Organismenreste und das rohhumusähnliche Aussehen ist nicht durch seine Dystrophie, sondern vor allem durch den kalten Trockenstandort (Windeinfluss) bedingt.

[1] Horizontbenennung nach Kubiena (1953).

Probe Gd 16

Protoranker auf Granit; unter Rasendecke von *Festuca duriuscula*; El Santo, 2.200 m, bei Piedrahita, Puero de Villatoro, Sierra de Guadarrama. Die Umweltverhältnisse bewirken fast keine chemische Verwitterung des Gesteins, obwohl ein starker mechanischer Zerfall im Bereiche der Bodenbildung festzustellen ist (Abb.1).

Profilaufbau

0– 3 cm: A_0, dieser Horizont besteht fast ganz aus wenig zersetzten und wenig zerkleinerten Stengel-und Wurzelresten von Gramineen, die aber keinerlei Merkmale von Rohhumus erkennen lassen. Eine Blattbräunung ist nicht sichtbar.

3–25 cm: A_1, lockerer, im trockenen Zustande staubender brauner, feucht schwärzlich brauner, leicht zerfallender Boden mit grossem Sand- bis Grusgehalt (Gesteinsbruchstücke). Verhältnismässig stark durchwurzelt, doch nicht in dem Grade wie bei Gd 5.

>25 cm: C, unverwitterter, in der Übergangszone lediglich mechanisch zerfallener Granit. Mineralführung: Quarz, Orthoklas, Albit, Biotit und Muskovit.

Mikroskopische Untersuchung

0–3 cm: A_0, die Pflanzenreste, z.T. unzerkleinert, z.T. in kleine Splitter aufgelöst, zeigen zwar mehr oder minder starke Bräunung, aber keinen Zerfall des Zellgefüges. Ein grosser Teil der Rückstände erscheint völlig frisch. Zwischen ihnen zeigt sich Losung von Kleinarthropoden, die aber gegenüber den unzersetzten Pflanzenresten mengenmässig stark zurücktreten. Die Mineralkörner erreichen Grobsandgrösse und zeigen fast nur mechanischen Zerfall. Die zahlreichen Biotite weisen zu grossem Teile noch olivgrüne Farbe auf, die Orthoklase starken Zerfall in Spaltstücke; nur die Albite lassen teilweise beginnende Verwitterung in Serizit innerhalb der Mineralkörner erkennen (Abb.2).

Probe aus 3–4 cm und aus 15–19 cm: A_1, der Losungsanteil von Kleinarthropoden (Hornmilbengrösse) nimmt zu, doch ist der Anteil an gebräunten, doch strukturell gut erhaltenen Pflanzenresten in der oberen Probe immer noch gross. Sie erfahren lediglich eine stärkere Zerkleinerung und liegen hauptsächlich in Form von kleinen Splittern vor. In der tieferen Probe nimmt der Losungsanteil wesentlich zu, die Farbe der Losungsstücke ist jedoch mehr braun als schwarz, was mit der relativ geringen Humifizierung zusammenhängt. Der Gehalt an Grob- und Feinsand nimmt gegen den C-Horizont zu, die geringe chemische Verwitterung bleibt bestehen.

Probe Gd 3

Tangelranker auf Granit unter Buschwerk von Zwergwacholder (*Juniperus communis nana alpina*); Guarramas, ca. 2.000 m. Dieser Boden, der sich durch einen mächtigen F-Horizont mit wenig zersetzten und wenig zerkleinerten Pflanzenresten charakterisiert (Tangelschicht), ist in seinem Charakter von den Umweltverhältnissen stark von der Vegetation abhängig. Das Profil gehört zu einem (subalpinen) Tangelranker, dessen Standort sich an der Grenze zu den alpinen Rasenböden befindet (Abb.1).

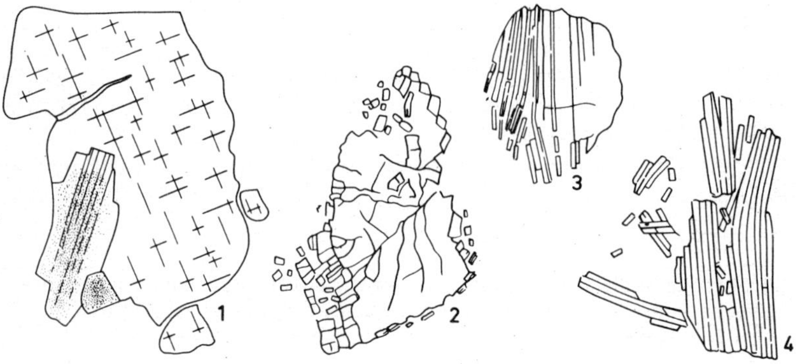

Abb.2. 1 = Mikroklin und Plagioklas mit Serizit; 2 = Orthose; 3 = Albit; 4 = Biotit.

Profilaufbau

0–20 cm: A_0, Pflanzenreste fast ausschliesslich gebräunte Rückstände von Nadeln und Zweigstückchen von Zwergwacholder, die torfartig miteinander verflochten sind.

20–35 cm: A_1, losungsreicher mullartiger Moder von dunkelgrauer, in feuchtem Zustande schwärzlich braungrauer Farbe und lockerer Beschaffenheit.

>35 cm: C_1, vornehmlich physikalisch verwittertes grobsandig bis grusiges Material mit eingelagertem Mehlsand.

Mikroskopische Untersuchung

Probe aus 0–4 cm: Pflanzenreste mit Zellstrukturen von hellgelber, zum Teil braunroter bis grellroter Farbe (zumeist Nadel- und Zweigreste von Wacholder). In den roten Rückständen häufig Frasskavernen mit Milbenlosung von gleicher Farbe. Ausserhalb der Pflanzenreste schwärzlichbraune Losung von Arthropoden. Von Mineralkörnern sind fast nur solche in Schluffgrösse festzustellen.

Probe aus 25–29 cm: A_1; Pflanzenreste treten fast völlig zurück; im Vordergrunde stehen dunkelbraune bis schwärzliche Losungsstücke verschiedener Grösse und Form. Der Mineralgehalt hat stark zugenommen, die Mineralkörner zeigen geringe Verwitterung und sind in erster Linie durch mechanischen Zerfall freigelegt worden.

ATLANTISCHE RANKER[1]

Probe Gal 1

Brauner mullartiger Ranker auf Granit unter Zwergstrauchheide; Alto de Martelo, Seehöhe 425 m, Hangneigung 5–10°N. Die Pflanzendecke ist im wesentlichen ein *Erica-Calluna*-Heide, die sich wahrscheinlich aus einem Eichenwald entwickelt hat. In der gegenwärtigen Pflanzendecke findet sich: *Erica ciliaris, Erica cinerea, Erica umbellata, Calluna vulgaris, Ulex europeus, Agrostis setacea, Arrhenaterum Thoree*. Die Reaktion im A_u-Hor. ist pH = 5,4.

[1] Einige chemische Analysendaten für die folgenden Profile sind in Tabelle I zusammengestellt.

TABELLE I

Chemische Analysenwerte

Horizont	pH in H_2O	H^+	Ca^{2+}	Mg^{2+}	K^+	Na^+	S	T	S/T x100	C	N	C/N	Orgische Substanz
Gal. 1 A_0	4,5	57,25	8,82	0,50	1,46	0,36	11,14	68,39	16,28	27,72	1,21	22,99	47,67
A_1	4,79	61,29	1,28	-	0,63	0,05	1,96	63,25	3,09	7,50	0,92	8,1	12,90
A/(B)	4,9	50,80	1,12	-	0,67	0,09	1,88	52,68	3,56	6,67	0,68	9,81	11,74
(B)	4,95	34,68	2,57	-	0,65	0,06	3,28	37,96	8,64	3,72	0,34	10,94	6,39
Gal. 2 A_1	4,8	22,58	2,57	-	0,64	0,05	3,26	25,84	12,62	8,90	0,56	15,88	15,20
(B)	5,28	49,50	2,38	-	0,73	0,09	3,20	52,70	6,07	2,21	0,16	13,83	3,80
Gal. 3 A_0/A_1	4,71	41,53	0,92	-	0,65	0,12	1,69	43,22	3,91	6,14	0,46	13,35	10,56
A	4,95	53,63	1,28	-	0,62	0,08	1,98	55,61	3,56	6,06	0,75	6,96	10,42
(B)	5	28,22	0,73	-	0,65	0,09	1,47	29,69	4,98	1,22	0,46	4,69	2,09
C	5,2	17,34	1,84	0,27	0,68	0,08	2,87	20,21	14,20	1,25	0,27	4,63	2,15
Gata 1 A_0/A_1	4,9	62,4	2,39	0,07	2,05	3,05	7,56	69,96	1,08	10,39	0,81	12,82	17,87
A_1	5,05	59,20	3,67	-	2,00	3,25	8,92	68,12	1,30	4,33	0,63	6,87	7,45
(B)	5,40	12,4	1,83	-						3,94	0,39	10,10	6,77
Gata 2 A_1	4,80	15,60	1,10	-	1,90	3,25	6,25	21,35	2,86	6,92	0,65	10,64	11,90

Profilaufbau

0– 12 cm: A_0, vorwiegend braun gefärbtes Geflecht von wenig zersetzten Pflanzenresten, durchsetzt von schwärzlichen koprogenen Aggregaten. pH = 4,5.

12– 70 cm: A_1, im feuchten Zustande tiefschwarzer, trocken schwärzlich grauer, gut durchwurzelter mullartiger Moder, der gute Zersetzung und Humifizierung zeigt und beim Graben in 1–5 mm grosse, erdige Krümelaggregate zerfällt. Die trockene Bodenmasse benetzt sich anfangs schwer, nimmt aber dann grosse Mengen Wasser auf, das sich gleichmässig im Gefüge verteilt. pH = 4,8.

70– 74 cm: A/(B), geringmächtige humose, dem A-Hor. ähnliche, doch etwas bräunlicher gefärbte Übergangsschicht. pH = 4,9.

74–102 cm: (B), in feuchtem Zustande dunkelgraubraune, trocken grauockerfarbige, erdige, leicht zerfallende, mineralreiche, aber bindesubstanzarme Bodenmasse. pH = 4,95.

102–150 cm: C_1, in feuchtem Zustande hellolivgrüne, schneidbare, im trockenen Zustande bröckelig zerfallende, weissliche Zersatzschicht.

>150 cm: C_2, weisslichgrauer Zweiglimmergranit.

Mikroskopische Untersuchung

Probe aus 0–4 cm: A_0, überwiegend braunrot bis grellrot gefärbte, wenig zersetzte, doch zerbissene Pflanzenreste mit gut erhaltenen Zellstrukturen. Daneben kleine, dunkelbraune zylindrische bis unregelmässig geformte Losung von Kleintieren. Als weiterer Gemengteil zeigen sich lose Mineralkörner von vorwiegend Feinsandgrösse, in erster Linie Quarze, Muskovite und braune, z.T. olivgrüne Biotite mit geringer Verwitterung. Feldspäte treten stark zurück.

Probe aus 20–24 cm: A_1, der Boden besteht hier fast ganz aus kleinen, schwärzlichen, in dünner Schicht dunkelbraunen, unregelmässig rundlich geformten Losungsstücken. Unzersetzte Pflanzenreste mit konservierten Zellstrukturen treten stark zurück. Die Losung ist fast mineralfrei. Mineralkörner von Feinsandgrösse (zum Teil Grobsandgrösse) ausserhalb der Kleintierlosung sind zahlreicher. Auch hier finden sich fast unverwitterte Biotite; Feldspäte sind relativ selten.

Probe aus 60–64 cm: A_1, die Losung ist dichter aneinander gerückt, teils zu komplexen Losungsstücken grösserer Arthropoden, teils zu leicht kohärenten Humusmassen vereinigt, die bei Austrocknung Sprungrisse zeigen und in unregelmässig-rauhwandige Absonderungsstücke zerfallen. Mineralkörner sind zahlreich, solche von Grobsandgrösse nehmen zu. Die Bodenmasse ist schwer benetzbar.

Probe aus 80–84 cm: (B), die Aggregate von gleicher Grösse und Form wie obige Kleinlosungsstücke, nur durch freies Eisenoxydhydrat braun gefärbt. Dieselbe Masse findet sich in Form von mehr oder minder dicken Belägen an der Oberfläche von Mineralkörnern angelagert (besonders an solchen der Grobsandfraktion).

Probe Gal 2

Brauner mullartiger Ranker auf Granit unter Zwergstrauchheide; oberhalb der Talsperre von Barrié de la Maza, La Coruña, Galizien. Der Standort ist auf einem Nordhang von 20–25° Neigung, seine Seehöhe ist 320 m. Die Vegetation unterscheidet sich von der vorigen dadurch, dass *Ulex europeus* (Stickstoffsammler) weit mehr in den Vordergrund tritt und auch Gräser mengenmässig stärker vertreten sind. Dadurch ist die Zersetzung und Humifizierung besser als in Gal 1. Dies drückt sich auch in der Horizontierung aus, indem der A_0-Hor. in Gal 1 hier durch einen A_0/A_1 ersetzt ist und der A_1 gegenüber von Gal 1 eine ausserordentliche Tiefe von 150 cm erreicht. In der Nähe des Profils findet sich ein flaches wasserreiches Gerinne, das den Boden feucht hält. Dennoch zeigt sich hier keine Anmoorbildung (Abb.1).

Profilaufbau

0– 20 cm: A_0/A_1, die wenig zersetzten Pflanzenreste treten weit stärker zurück, der Losungsanteil stärker hervor als in dem gleichen Horizont des Profils Gal 1.

20–150 cm: A_1, gut humifizierter, in feuchtem Zustande kohlig schwarzer, in trockenem Zustande dunkelgrauer, feinaggregatreicher, lockerer Moder, der beim Graben bei genügendem Wassergehalt erdig zerfällt. Der Humus ist stark färbend, staubt in trockenem Zustande ungewöhnlich stark und hat an Händen, Papier und Wäschestücken, die sich leicht schwärzen, eine auffallend hohe Färbekraft. Bei Wiederbefeuchtung zeigt sich nach anfänglich schwerer Wiederbenetzbarkeit ein ungewöhnlich hohes Wasseraufsaugevermögen und eine hohe Wasserkapazität (78,9 Vol.%). Mit der Tiefe nimmt der Mineralgehalt zu. Während die Biotite fast unverändert sind, zeigen die Feldspäte, die gegenüber Quarz und Glimmer zurücktreten, fast immer Spuren chemischer Zersetzung. Bei den Plagioklasen ist ein serizitischer Zersatz fast die Regel. pH = 4,8.

150–200 cm: (B), dieser Horizont ist mächtiger als in anderen Profilen, macht aber immer noch nur ein Drittel des A-Horizontes aus, weshalb der Boden (der ausserdem zu geringe chem. Verwitterung zeigt) noch als brauner Ranker und nicht als Braunerde bezeichnet werden muss. Die Feuchtfarbe ist braun, die Trockenfarbe bräunlichocker; die Bodenmasse zeigt ebenso leichten erdigen Zerfall wie der A_1-Horizont. pH = 5,3.

200–230 cm: C_1, schneidbare, im trockenen Zustande weissliche, bröckelig zerfallende Masse.

und darunter: C_2, dieser wurde im Aufschluss nicht erreicht, das frische Gestein (Zweiglimmergranit) war aber in Form zahlreicher Lesesteine in Profilnähe leicht zugänglich.

Mikroskopische Untersuchung

Probe aus 8–12 cm: A_0/A_1, stark durchwurzelte, Wurzel- und Pflanzenreste enthaltende, doch zugleich sehr losungsreiche Schicht. Die Losung zeigt sehr verschiedene Formen und Grössenklassen; auch zusammengesetzte Losung, aus Kleinarthropoden-Exkrementen, zerbissenen Pflanzenresten und Mineralkörnern bestehend, ist häufig. Die Pflanzenreste zeigen zu grossem Teile auffallend braunrote bis grellrote Farben. In ihnen sind auch Frasskavernen mit Losung von minierenden Milben zu finden. Die Mineralkörner sind blank, wenig verwittert, und treten mengenmässig zurück; die Kleinarthropodenlosung ist mineralkornarm bis mineralkornfrei (Tafel B5).

Proben aus 44–48 cm und 65–69 cm: A_1, Gefügebild ähnlich wie beim A_1-Hor. der Probe Gal 1. In der unteren Probe nehmen Gesteinsfragmente von Grusgrösse zu. Die Mineralkörner sind fast unverwittert. Auffallend ist, dass unverwitterte Kalifeldspäte relativ sind.

Probe aus 160–164 cm: (B), Bodenmasse hauptsächlich aus ockerbraunen Kleinaggregaten bestehend, die sich zu grösseren Aggregatkomplexen vereinigen können. Gesteinsfragmente von Grusgrösse neben Einzelkörnern sind zahlreich. Auffallend ist auch hier die geringe Verwitterung der Mineralien: Biotitplättchen können noch in olivgrüner Färbung neben schwach gebräunten Exemplaren vorkommen, gebleichte Formen fehlen fast völlig; fast unverwitterte Orthoklase sind häufig.

Probe aus ca. 210 cm: C_1, diese Zersatzschicht des Muttergesteins ist nicht homogen, sondern zeigt Partien mit verschiedenem Verwitterungsgrad. Die Verwitterung zeigt sich vor allem an den Feldspäten, die, unter Auflicht in der unpräparierten Probe gesehen, Trübung bis erdigen Zerfall zeigen. In unmittelbarer Nähe von Biotiten, die durchwegs ihre schwärzliche Farbe behalten haben, bemerkt man Bräunung durch freies Eisenoxydhydrat. Neben den verwitterten Partien zeigen sich andere, in denen alle Gemengteile einschliesslich der Feldspäte völlig unverwittert geblieben sind, da sie offenbar von den Sickerwässern nicht erreicht worden sind.

C_2, als Lesestein vorgefundenes Handstück. Mineralführung: Quarz, Orthoklas, Mikroklin, Albit, Biotit und Muskovit; mit relativ hohem Biotitgehalt.

Probe Gal 3

Brauner mullartiger Ranker auf Granit unter Kiefernwald; Alto de Catoira, Provinz Pontevedra, Galizien. Der Standort liegt auf der fast ebenen Gipfelfläche eines Hügels in einer Seehöhe von 145 m. Der Wald ist eine Neuaufforstung von *Pinus pinaster* (Alter 10–12 Jahre), die Kraut- und Strauchschicht setzt sich in erster Linie zusammen aus *Erica cinerea*, *Erica umbellata*, *Calluna vulgaris*, *Alimium occidentalis*, *Daboisia polyfolia*, *Agrostis setacia* und *Thymus caespiticius* (Abb.1).

Profilaufbau

0–10 cm: A_0/A_1, ähnlich wie im vorigen Profil treten unzersetzte Pflanzenreste zurück; gut vererdeter Humus reicht bis fast an die Oberfläche des Bodens. pH = 4,7.

10–50 cm: A_1, in feuchtem Zustande braunschwarzer, trocken dunkel sepiabrauner, lockerer, leicht in Bröckel bis Kleinkrümel zerfallender Boden, der im trockenen Zustande eine auffallend schwere Benetzbarkeit hat. An die trockene Oberfläche aufgebrachtes Wasser sammelt sich, ohne einzudringen, in Vertiefungen an. Einmal benetzt, zeigt er ein gutes Kapillarsystem, das grosse Wassermengen festzuhalten vermag. Dieses merkwürdige Verhalten der Schwerbenetzbarkeit wurde weder im A_0/A_1-Horizont noch in den folgenden Horizonten, dem A(B) und (B) festgestellt. pH = 4,95.

50–92 cm: A(B), wie der vorige nur durch geringeren Humusgehalt und höheren Gehalt an freiem Eisenoxydhydrat brauner gefärbt. Die Aggregatbildung ist mehr rundlich knollig.

92–96 cm: (B), hellockerfarbige, in feuchtem Zustande sattbraune, lockere Masse mit deutlichem Tongeruch und bröckelförmigem Zerfall in Kleinaggregate. pH = 5.

C_1, wurde nur auf etwa 10 cm aufgeschlossen. Weiche, schneidbare bis bröckelig zerfallende Zersatzschicht von weisslicher Farbe. pH = 5,2.

Mikroskopische Untersuchung

Probe aus 0–4 cm: A_0/A_1, in den Pflanzenresten sind braunrote bis grellrote Farben häufig zu sehen. Wesentlich ist, dass solche nur in diesem Horizont vorkommen, während sie im A_1-Hor. und im A/(B)-Hor. völlig aufgearbeitet und humifiziert werden. Zu den Pflanzenresten treten aber reichlich Losungsstücke von Kleinarthropoden von sehr verschiedener Grösse und unregelmässig rundlicher Form, so dass sie stellenweise überwiegen. Die Mineralkörner treten mengenmässig zurück. Biotite zeigen z.T. noch olivgrüne Farbtöne, Feldspäte lassen stärkere mechanische, z.T. chemische Verwitterung erkennen.

Probe aus 15–19 cm: A_1 (Tafel B6), dieser Horizont, der sich durch eine besonders langsame Benetzbarkeit charakterisiert, zeigt eine dichte Aneinanderlagerung von humusreicher, schwärzlicher, in dünner Schicht dunkelsepiabrauner, Kleintierlosun[illegible] deren Einzelstücke sich innerhalb weiter zusammenhängender Partien berühren. M[illegible] dieser dichteren Aneinanderlagerung scheint auch die Schwerbenetzbarkeit bei star[illegible] Austrocknung zusammenzuhängen.

Probe aus 50–54 cm: A(B) (Tafel B7), auch hier ist die Lagerung der Losungsaggre[illegible] ziemlich dicht; die Mineralkörner zeigen zudem starke Oberflächenbeläge, zum Te[illegible] sogar Kornbrücken zerfallener, dichtgeschlämmter Aggregatmasse. Trotz dieser Häufung hat die Schwerbenetzbarkeit des Bodens abgenommen, was ohne Zweifel mi[illegible] dem grösseren Mineralgehalt der Schlämmassen zusammenhängt, die sich auch du[illegible] stärkere Eisenbräunung grösseren Feinglimmergehalt anzeigt.

Probe aus 94–98 cm: (B), ähnliches Material wie im vorigen Horizont, doch humusa[illegible]

und mit stärkerem Aggregatzerfall. Verdichtete Schlämmassen zeigen sich nicht nur an den Kornoberflächen, deren Beläge ausserordentlich verstärkt sind, sondern auch innerhalb der Intergranular-Räume. Trotz diesem sichtlich stärkeren Mineralzerfall zeigen sich noch zahlreiche, wenig verwitterte Biotite und Feldspäte.
Probe aus etwa 100 cm Tiefe: C_1, ähnliches Bild wie im gleichen Horizont des vorangehenden Profils.

Probe Gata 1

Brauner mullartiger Ranker auf Granit unter Eichenwald; oberhalb San Martin de Trevejo, Puerto de Sta. Clara, Sierra de Gata, Extremadura. Die Profilstelle liegt auf einem leichtgeneigten Westhang in einer Seehöhe von 1.092 m. Die Vegetation besteht aus: *Quercus pyrenaica*, *Genista cinerea*, *Sarothamnus scoparius*, *Erica lusitanica*, *Lavandula pedunculata*, *Pteridium aquilinum*, *Asphodelus albus* (Abb.1).

Profilaufbau

0– 10 cm: A_0/A_1, in feuchtem Zustande pechschwarzer, trocken bräunlich schwarzer, wenig aufhellender Humus, der erdigen Zerfall in Kleinaggregate zeigt. Gute Benetzbarkeit. pH = 4,9.

10– 90 cm: A_1, im feuchten Zustande schwarzer, trocken nach unten zu heller werdender, bis bräunlich dunkelgrauer Horizont mit starkem erdigen Zerfall, guter und tiefreichender Durchwurzelung, zunehmendem Mineralgehalt und Tongeruch. Gute Benetzbarkeit. pH = 5,05.

90–120 cm: $A_1/(B)$, Strukturbildung wie voriger, nachlassender Humusgehalt, zunehmende Eisenoxydhydratfärbung. Gute Benetzbarkeit.

120–160 cm: (B), guter krümeliger Zerfall, Material stark grusig bis steinhaltig, die Bodenmasse ist dunkelgraubraun im feuchten, hellockerfarbig im trockenen Zustande. pH = 5,4.

160–200 cm: C_1, heller, feucht schneidbarer, trocken erdig zerfallender Granitzersatz. C_2, heller Granit, der in einzelnen Blöcken in dem weichen Zersatz ansteht.

Mikromorphologie
Probe aus 3–7 cm: A_0/A_1, der Boden zeigt mit Auflicht und ohne Präparierung mikroskopiert ein lockeres, ausserordentlich hohlraumreiches Gefüge, das nicht nur durch ein dichtes Netz von Wurzelhaaren, sondern auch von Pilzhyphen in seiner Struktur befestigt ist, durch eine vollkommene Lebendverbauung stabilisiert erscheint. Die erweiterten Hohlräume sichern eine erleichterte Benetzbarkeit. Der Dünnschliff zeigt dunkel gefärbte Losungsstücke verschiedener Grösse und Form mit guter Humifizierung. Das Innere der Losung ist mineralarm. Unter den Pflanzenresten sind grellrote Farben häufig.
Proben aus ca. 30–34 und 60–64 cm: A_1, mit Auflicht wird ein ähnlicher hohlraumreicher Gefügeaufbau und eine ähnliche Stabilisierung der Struktur zufolge einer vollkommenen Lebendverbauung durch ein Netz von Wurzelfasern und Pilzhyphen sichtbar. Der Dünnschliff zeigt gute Aggregatbildung, die roten Pflanzenumwandlungsstoffe sind durch gute Humifizierung völlig aufgearbeitet.
Probe aus ca. 96–100 cm: $A_1/(B)$, gute Aggregatbildung, stärkeres Hervortreten der Bräunung durch freies Eisenoxydhydrat.
Probe aus 125–129 cm: (B), hier zeigt sich bereits eine Tendenz zu Aggregatzerfall, die sich durch eine Zunahme von breiten Anlagerungen von Zerfallmasse an der Oberfläche der Mineralkörner und an verschwemmter Bodenmasse in den Intergranularräumen anzeigt. Der Horizont ist humusfrei und zeigt Eisenfärbung. Doch sieht man immer noch unzersetzte olivgrüne Biotite und unverwitterte Kalifeldspäte.

Probe aus ca. 180 cm: C_1, der Granit besteht im wesentlichen aus Quarz, Orthoklas, Mikroklin, Albit, Biotit und Muskovit. In diesem Horizont lässt sich ein Zerfall in Bruchstücke von 2–3 cm Durchmesser erkennen. Dazwischen zeigt sich Gefügezerfall, verursacht durch die chemische Verwitterung der Albite, während die Orthoklase, Mikrokline und Biotite fast unverwittert geblieben sind.

Probe Gata 2

Brauner mullartiger Ranker auf Granit unter Kastanienwald; San Martin de Trevejo, Strasse zum Puerto de Sta. Clara, etwa 1.000 m, unterhalb Gata 1. Das Profil liegt auf einem Westhang mit 28% Neigung. Die Vegetation besteht aus: *Castanea sativa*, etwas *Quercus pyrenaica*, *Sarothamnus purgans*, *Hedera helix*, *Genista cinerea*, *Rubus* sp. (Abb.1).

Profilaufbau

0– 5 cm: A_0/A_1, losungsreiche gut zusammenhängende, z.T. krümelig, z.T. in 3–5 cm grosse Bruchstücke zerfallende, im feuchten Zustande schwärzlichbraun, in trockenem dunkel sepiabraun gefärbte Schicht. Nach Austrocknung tritt schwere Benetzbarkeit ein. Nach Befeuchtung gut leitbares Kapillarsystem.

5– 40 cm: A_1, dunkelbraune, im trockenen Zustande sepiabraune Masse, die im feuchten Zustande erdig, im trockenen fast staubartig, doch immer noch in Kleinaggregate zerfällt. Der Horizont ist wesentlich mineralreicher als der vorangehende und ist auch nach starker Austrocknung leicht benetzbar. pH = 4,8.

40– 62 cm: A_1/(B), ähnlich wie A_1, nur humusärmer.

62– 82 cm: (B), lockerer, erdig zerfallender Boden mit nur geringer Bräunung. Farbe trocken hellocker, feucht braunocker. Die Masse ist reich an Gesteinsgrus.

82–124 cm: (B)/C, Farbe weisslich hellocker, leicht in Bruchstücke und Feindetritus zerfallend.

124–200 cm: C_1, fast weiss gefärbt, leicht in Bruchstücke von 3 - 50 cm zerfallend. C_2, lag in Blöcken an der Bodenoberfläche vor.

Mikroskopische Untersuchung

Probe aus 1–5 cm: A_0/A_1, das Auflichtbild des unpräparierten Bodens zeigt wie bei Gata 1 eine Festigung des Feinaggregatbaues zufolge einer vollkommenen Verflechtung durch ein Netz von Wurzelfasern und Pilzhyphen. Im Dünnschliff dominieren unzusammenhängende Losungsstücke sehr verschiedener Grösse und Form, in denen die braune Farbe der Pflanzenreste dominiert und die zum Teil einen Aufbau aus zerkleinerten, aber sonst wenig zersetzten Pflanzenresten erkennen lassen. Die Verspinnung durch Pilzhyphen ist selbst im Dünnschliff noch zu erkennen; die Farbe der Hyphen ist fast durchwegs dunkelbraun. Unter den Losungsarten sind auch solche von grösseren Arthropoden vertreten. Die Schwerbenetzbarkeit scheint mit der Art und Zusammensetzung der Losung in engem Zusammenhang zu stehen. Daneben finden sich nackte, unverwitterte Mineralkörner bis zu Grobsandgrösse. Auch grüne, bzw. nur leicht gebräunte Biotite sind häufig.

Probe aus 44–48 cm: A/(B), hier beginnt sich bereits der Aggregatzerfall deutlich auszuwirken. Er zeigt sich in dicken Belägen aus verschlämmter Aggregatmasse an der Oberfläche der Mineralkörner sowie an Einlagerungen des gleichen Materials in den Intergranularräumen. Letztes ist humusärmer, zeigt mehr Eisenfärbung und lässt zwischen gekreuzten Nicols zahlreiche feine Glimmersplitter im Innern erkennen.

Probe aus 70–74 cm: (B), ähnliches Bild wie bei der vorangehenden Probe; Aggregate und Schlämmassen nur frei von Humusbeimengung.

Probe aus ca. 140 cm: C_1, diese Zersatzschicht lässt eine etwas stärkere chemische Verwitterung erkennen als bei den vorigen Profilen. Sie betrifft ausser den Plagioklasen in Form von mehr oder minder starken Anwitterungsspuren auch die Orthoklase, Mikrokline und z.T. auch die Biotite. Doch lassen sich daneben immer noch zahlreiche unverwitterte Mineralkörner feststellen.

AUSWERTUNG ZUR ERKLARUNG DER GENESE DER KLIMAXBILDENDEN RANKER SPANIENS

Ranker sind Anfangsphasen der Bodenbildung. Sie kommen darum, wie die Rohböden, in allen Teilen der Welt vor, auch dort, wo sie nur ein kurzes Zwischenglied in einer Entwicklung sind, die zu Böden mit intensiver Verwitterung und kompliziertem Profilaufbau führt. Sie treten aber, da sie dort nur Übergangsbildungen sind, mengenmässig stark zurück und zeigen sich nur dem, der zu Studienzwecken nach ihnen sucht. Hingegen treten Ranker überall dort in den Vordergrund, wo sie als Endglieder einer Entwicklung aufscheinen, wo die Bildungsbedingungen für Bildungen mit stärkerer Verwitterung, Tonbildung, Tonverlagerung u.a. nicht ausreichen. Wir sind gewohnt, Klimaxbildungen von Böden geringer Entwicklungshöhe dort zu suchen, wo eine weitere Entwicklung durch zu geringe Temperaturen (Kältesteppe) oder durch zu geringe Durchfeuchtung (Trockensteppe) gehemmt ist. Auf Grund dieser Erfahrungen ist jedoch schwer zu verstehen, wieso AC-Böden, wie die atlantischen Ranker Galiziens und der Sierra de Gata Extremaduras in einem ausgesprochenen feuchtwarmen Klimagebiet Endstadien bleiben und sich nicht in kurzer Zeit zu Braunerden oder Semipodsolen mit starker Verwitterung und Durchschlämmung umwandeln.

Umwelteinfluss bei alpinen und subalpinen Rankern

Wenn wir uns zur Erklärung dieser Zusammenhänge zuerst den alpinen und subalpinen Rankern der spanischen Hochgebirge zuwenden, so wollen wir uns vor allem den Umweltverhältnissen zuwenden, die zur Ausbildung der in vorangehendem angeführtem Beispielprofile geführt haben. Leider sind klimatische Daten der Standorte selbst unbekannt. Die nächste meteorologische Beobachtungsstation liegt am Navacerrada-Pass in einer Seehöhe von 1.440 m. Sie liegt also 800 m tiefer als das Profil Gd 5 des Eilag-Ranker auf dem Gipfel des Guarramas, 760 m tiefer als das Protorankerprofil Gd 16 El Santo und 560 m tiefer als das Profil Gd 3 des Tangel-Rankers auf dem Guarramas. Die Messungen der Beobachtungsstation betragen für die mittlere Jahrestemperatur 5,9° C, für die mittleren Maxima 9,7° C und die mittleren Minima 2,2° C. Die mittlere Monatstemperatur für Juli ist +15,7° C, die mittlere Monatstemperatur für Januar -0,8° C, das absolute Minimum in der Beobachtungszeit (1951–1960) betrug -15° C. Der mittlere Jahresniederschlag ist 1.268,3 mm; ein grosser Teil fällt jedoch als Schnee im Winter, der Sommer ist niederschlagsarm. Die Höhenunterschiede besagen, dass die Temperaturwerte im Januar an den Profilstandorten wesentlich niedriger liegen werden. Verwendet man den Höhengradient von Lautensach (1962) von 0,60° je 100 m Seehöhenzunahme für den Winter und von 0,45° je 100 m für den Sommer so kommt man zu den in Tabelle II wiedergegebenen Werten:

TABELLE II

Mittlere Januar-, Juli- und Jahrestemperaturen

Profile	Seehöhen	Bodentypus	Mittlere Temperaturen (°C)		
			Januar	Juli	pro Jahr
Gd 5	2.240 m	Eilagranker	-5,6°	12,1°	1,5°
Gd 16	2.200 m	Protoranker	-4,8°	12,3°	1,7°
Gd 3	2.000 m	Tangelranker	-4,0°	13,2°	1,8°
Gata 1	1.092 m	Brauner mull-artiger Ranker	2,7°	19,1°	11,1°
Gata 2	1.004 m	Brauner mull-artiger Ranker	3,0°	19,3°	11,4°
Gal 1	425 m	Brauner mull-artiger Ranker	7,3°	18,0°	12,4°
Gal 2	325 m	Brauner mull-artiger Ranker	7,7°	18,5°	12,9°
Gal 3	145 m	Brauner mull-artiger Ranker	7,6°	18,5°	12,9°

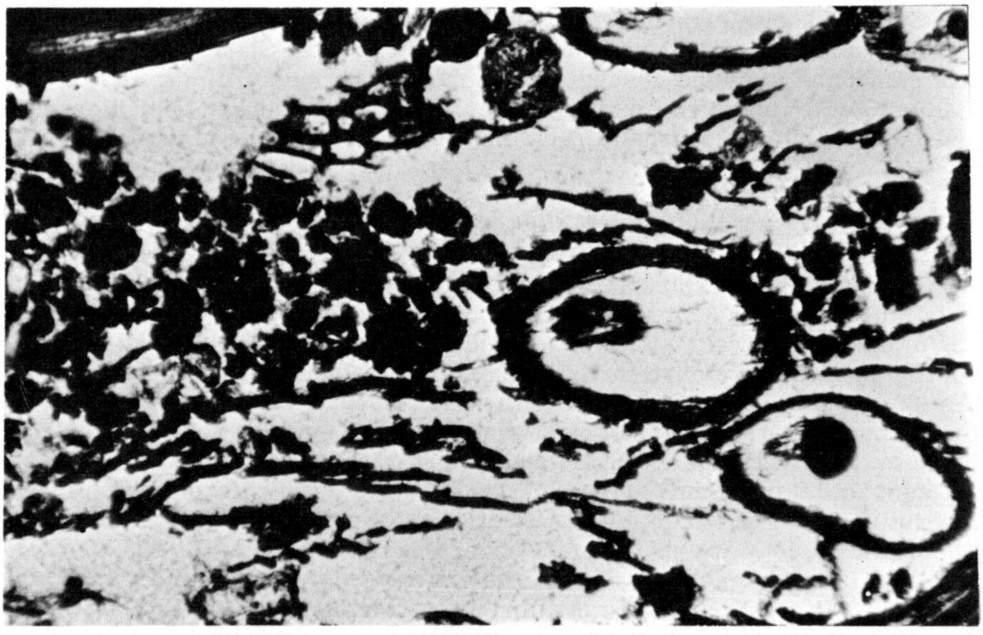

Abb.3. Dünnschliff eines alpinen Protorankers, Guarramas, Sierra de Guadarrama. Sehr lockeres, primitives Gefüge aus splitterigen Pflanzenresten, Kleintierlosung und lose Mineralkörner.

Die Beobachtungsstation für die ersten drei Profile ist Puerto de Navacerrada (1.444 m), für Gata 1 und 2, Navasfrias (975 m) in der Provinz Salamanca, für die Profile 1 und 2 von Galizien, Santiago de Compostela

(250 m) und für den Profil 3 von Galizien, Pontevedra (4 m).

Die langsame Bodenentwicklung, die geringe chemische und die starke mechanische Verwitterung, das Fehlen jeder Tonbildung und jeder Abscheidung von freiem Eisenoxydhydrat im Profil, der Vorrang der Tätigkeit der Kleinarthropoden, (die auch bei starkem Winterfrost noch tätig sein können) gegenüber jener der pflanzlichen Mikroorganismen, die Umsetzung in schwärzliche Losung, das lockere, leicht verwehbare und leicht verschwemmbare Bodengefüge, die geringmächtigen Humushorizonte, das alles sind Folgen der niedrigen Temperaturen, des Winterfrostes, der langen Schneedecke, die über gefrorenem Boden auftaut und der raschen Austrocknung des Bodens (vgl.Abb.3). Im Falle des Eilagrankers Gd 5 kommt die starke Windwirkung hinzu, im Falle des Tangelrankers Gd 3 der Einfluss der Pflanzendecke (Tangelpflanzen), die indes zwar eine Vertiefung der Humushorizonte und eine Verbesserung der biologischen Verhältnisse mit sich bringt, aber die Intensität des chemischen Mineralzerfalles nicht wesentlich beeinflusst.

Umwelteinfluss bei den atlantischen Rankern

Wenn wir zur Charakterisierung der Umweltverhältnisse der atlantischen Ranker das Klima von Santiago de Compostela (Seehöhe 269 m) heranziehen, so finden wir sehr im Gegensatz zu den Bedingungen in der Sierra de Guadarrama eine mittlere Wintertemperatur von 10,0° C und eine mittlere Sommertemperatur von 17,7° C bei einem mittleren Jahresniederschlag von 1.480 mm. Die Flächen ausserhalb der Rankerstandorte zeigen als Endbodenbildung humide Braunerden bis Semipodsole (podsolige Braunerden) . Es scheint demnach nur das Vorhandensein der tiefen terrestrischen Humushorizonte der atlantischen Ranker (vgl. Abb.4) der Grund zu sein, dass die Entwicklung des Bodens über dieses Stadium nicht hinausgeht. Wir haben in der Bodenkunde Paralellen. Auch die Tangelrendsina, die einen mächtigen Humushorizont ausbildet, übt mit ihm eine hemmende Wirkung auf die weitere Bodenentwicklung aus. Eine Ähnlichkeit besteht auch bei der alpinen Pechrendsina, die in einer besonderen Varietät auch in Spanien zu finden ist (Sierra de Grazalema). Wenn diese durch Auswaschung stark entkalkt wird, nähert sie sich in ihrem Charakter und ihrer Mikromorphologie wesentlich den atlantischen Rankern. Aber auch die braunen Tangelranker der tieferen Lagen der Sierra de Guadarrama (in eta 1.800 m, besonders unter *Sarothamnus purgans*) können mächtige A_0/A_1 -Horizonte ausbilden und die Weiterentwicklung zur Braunerde wesentlich verzögern. Was für eine Erklärung gibt es hierfür?

Eine sehr bemerkenswerte Eigenschaft der atlantischen Ranker ist ihr hohes Wasserfesthaltvermögen. Für den A_1-Horizont des Profils Gal 2 wurde eine Wasserkapazität von rund 80 Vol.% gemessen. Dies bedeutet, dass bei der Humustiefe von 150 cm des Profils Gal 2 ohne Unterbrechung 1.100 L auf eine Fläche von 1 m^2 fallen müssten, um den Boden so zu sättigen, dass er Wasser an den Untergrund abzugeben beginnt. Das sind 1.100 mm Niederschlag ohne Unterbrechung gegenüber der mittleren Niederschlagsmenge von 1.480 mm Santiagos für das ganze Jahr. Aber selbst für eine Humustiefe von 1 m (wie bei Gata 1) könnte der Boden ohne

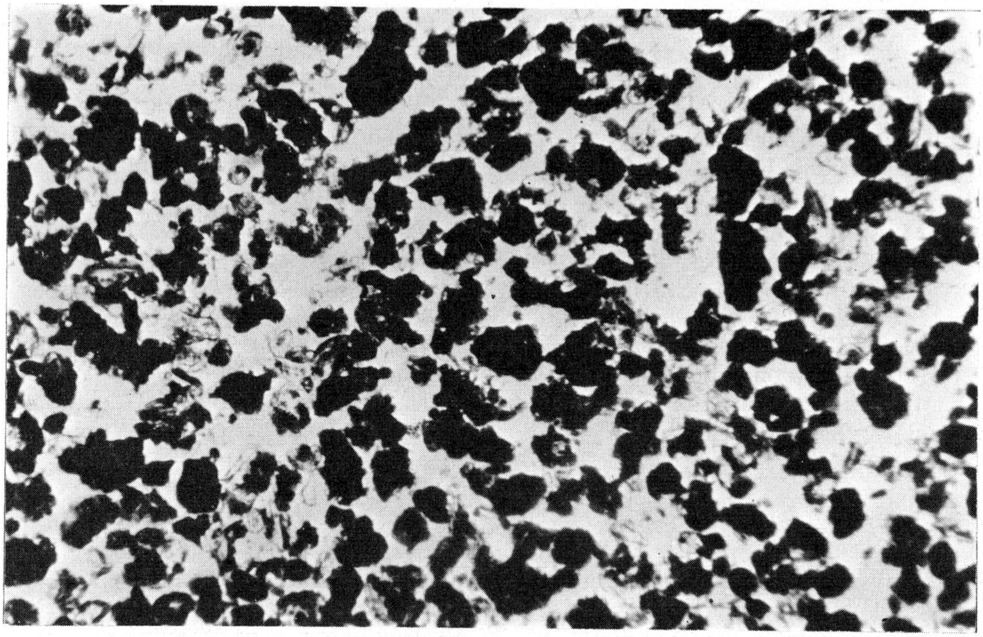

Abb.4. A_1-Horizont eines atlantischen Rankers (Profil Gal 5). Während bei dem Protoranker in Fig.3 eine Scheidung in A_0 und A_1-Horizonte nicht auftritt, zeigt sich hier in der eigentlichen Humusstoffschicht ein Gefüge von ausschliesslich schwärzlichen Losungsstücken (neben lockeren Mineralkörnern), das in dem vorliegenden Beispiel besonders gleichmässig und weitmaschig entwickelt ist, eine gute Benetzbarkeit und eine hohe Kapillarit Wasserleitfähigkeit und Wasserkapazität aufweist.

Unterbrechung mit 780 L beregnet werden, ohne dass der Untergrund auch nur einen Tropfen Wasser erhält, und bei einer Humustiefe von 60 cm wären es immer noch 450 L. Bei dem grossen Aufsaugevermögen des einmal angefeuchteten Bodens sind auch Spalten, Sprünge und Wurzelröhren wenig leitfähig, ebensowenig als ein Spalt in einer Lage von Filterpapieren leitfähig für Wasser sein könnte, das man in sie einzugiessen versucht, bevor die saugende Papiermasse noch nicht gesättigt ist. Bei einem so ausser ordentlichen Speicherungsvermögen ist es begreiflich, dass trotz bedeutende Niederschlagmengen, der Boden nicht mehr als einen braunen Saum unterhalb des A-Horizontes entwickeln konnte und dass auch dieser nur eine sehr mangelhafte Verwitterung zeigt, bei der nicht nur die Kalifeldspäte sondern auch ein grosser Teil der Biotite fast unverwittert blieben.

Wesentlich erschwert werden diese Gegebenheiten dadurch, dass die Hauptmenge der Niederschläge in der kühlen Jahreszeit fällt, während die Sommer niederschlagsarm sind. Zu dieser Zeit zeigen die Humusdecken Austrocknung und staubigen Zerfall (siehe auch Franz, 1956). Das Gestein ist an dieser Austrocknung beteiligt und bekommt so gut wie keine Feuchtigkeit nachgeliefert, die seine Verwitterung intensivieren und die Entwicklung nach einem Braunerdeprofil befördern könnte. An der schnellen Austrocknur ist das stark leitfähige Kapillarsystem des befeuchteten Bodens wesentlich beteiligt. Jede an der Oberfläche verdunstende Wassermenge wird sofort aus dem Innern des Bodens bis zu seiner völligen Erschöpfung nachgeliefert.

Von wesentlicher Bedeutung ist, dass der Boden mit zunehmender Reife, da mit dieser auch die Mächtigkeit des Humushorizontes zunimmt, die entwicklungshemmende Wirkung verstärkt. Während geringmächtige Humushorizonte noch einige Wassernachlieferung für die (B)/C-Horizonte gewähren, halten tiefe A-Horizonte so gut wie alle Feuchtigkeit für den eigenen Ausbau zurück.

Es bleibt noch zu klären, wie es zu der Bildung der tiefen Humushorizonte kommt. Wir sehen, dass für sie bestimmte Pflanzendecken in betracht kommen, in ähnlicher Weise wie für die Tangelrendsinen, Tangelranker und Rohhumusbildungen. Immer sind es Pflanzendecken, die grosse Mengen von Bestandesabfall liefern. Bei ungünstigen dystrophen Verhältnissen für die Zersetzung und Humifizierung, entstehen aus diesen grossen Rückstandsmengen Rohhumusformen. Bei den Tangelrendsinen und Tangelrankern sind die klimatischen Verhältnisse noch so, dass mitunter mächtige A_0-Horizonte (Tangelschichten) entstehen. Bei den atlantischen Rankern werden zufolge der optimalen Verhältnisse für die Zersetzung und Humifizierung die A_0-Horizonte weitgehend abgebaut und rascher als bei verwandten Bodentypen in A_1-Horizonte übergeführt.

Die atlantischen Ranker sind, so gut erhalten sie sich heute auch unserer Untersuchung darbieten, keine progressiven Bildungen, ihre Areale werden deutlich kleiner. Die zunehmende Trockenheit in unserer Zeit, der gegenüber sie besonders empfindlich sind, wird wahrscheinlich von ihren Standorten am ersten Besitz ergreifen. Sie sind in erster Linie Waldböden und ihre ursprüngliche Vegetation scheinen vor allem zwergstrauchreiche (besonders ginsterreiche) Eichenwälder gewesen zu sein. Die Zwergstrauchheide und die Neuaufforstungen von Kiefern und Kastanien sind als sekundär entstandene, auf menschliche Einwirkung zurückgehende Pflanzendecken anzusehen. Dennoch muss diesen eine erhaltende, aufbauende Wirkung zugerechnet werden. Bei Kastanienwäldern ist dies weit mehr der Fall, als es durch das hier angeführte Beispiel gezeigt werden konnte.

SUMMARY

The climax forming Spanish rankers; their micromorphology and genese

The rankers in Galicia and Sierra de Gata (Extremadura) prove to be not initial stages of soil formation which turn soon into braunerde or semipodsol, but climax formations of a pronounced humid and warm climate.

In the alpine and subalpine rankers, low temperatures, winter frost and prolonged covering with snow blankets cause a slowly proceeding soil development with a weak chemical and a relatively strong mechanical weathering.

One of the most outstanding properties of the Atlantic rankers is their high water capacity. These rankers are not progressive formations; their areas are clearly getting smaller, mainly because of their susceptibility to the increasing dryness of the present time. Their original (oak forests) and secundary vegetation (afforeststations of pine and chestnut trees) have a protecting and improving effect.

LITERATUR

Albareda Herrera, J.M. y Hoyos de Castro, A., 1955. Edafologia, 2 ed., S.A.E.T.A., Madrid, 368 pp.
Claver Aliod, M., 1947. Contribución al estudio de los suelos silicico-humicos de la Sierra de Guadarrama. Anales Edafol. Agrobiol. (Madrid), 6 : 5–52.
De Cañedo–Argüelles Quintana, E., 1949. El "ranker" de Kubiena, en nuestra Sierra de Guadarrama y en las Montañas Rocosas. Montes, 5 : 189–192.
Franz, H., 1955. Zur Kenntnis der "Steppenböden" im pannonischen Klimagebiet Österreichs. Bodenkultur, 8 (1955) : 125–132.
Franz, H., 1956. Drei Klimabedingte Ranker-Subtypen Europas. Trans. Intern. Congr. Soil Sci., 6th, Paris, 1956, 5 : 135–141.
Glinka, K., 1914. Die Typen der Bodenbildung, ihre Klassifikation und geographische Verbreitung. Bornträger, Berlin, 365 pp.
Huguet del Villar, E., 1937. Los suelos de la Peninsula Luso-Ibérica. Madrid, 416 pp.
Kubiena, W., 1938. Micropedology. Collegiate Press, Ames, Iowa, 243 pp.
Kubiena, W., 1948. Entwicklungslehre des Bodens. Springer, Wien, 215 pp.
Kubiena, W., 1953. Claves sistemáticas de suelos. Consejo Super. Invest. Cient., Madrid, 388 pp.
Lautensach, H., 1962. Caracteristicas y ritmo anual de la temperatura en la Peninsula Ibérica. Estud. Geograf., 23 : 259–292.
Pallmann, H., 1933. Die Bodentypen der Schweiz. Mitt. Gebiete Lebensm. Hyg., 24 : 8–20.
Pallmann, H. und Haffter, P., 1933. Pflanzenphysiologische und bodenkundliche Untersuchungen im Oberengadin. Ber. Schweiz. Botan. Ges., 42 : 357–466.

FARBENBILDER

Tafel B5. Dünnschliff des A_0/A_1-Horizontes eines atlantischen Rankers (Profil Gal 2, aus 8–12 cm). Braunrote Pflanzenreste mit unzerstörten Zellstrukturen, schwärzliche Kleintierlosung und einzelne Mineralkörner (losungsarme Partie; x 75).

Tafel B6. Gefügebild des A_1-Horizontes eines atlantischen Rankers (Profil Gal 3, aus 15–19 cm). Die roten Pflanzenreste sind verschwunden, die schwärzliche Kleintierlosung dominiert. Daneben unverwitterte Mineralkörner (x 75). Die Losung ist hier dicht aneinandergedrängt (schwere Benetzbarkeit), die losungsfreien Lücken sind mit Mineralkörnern ausgefüllt.

Tafel B7. A/(B)-Horizont eines atlantischen Rankers (Profil Gal 3, aus 50–54 cm). Erdarmes Gefüge mit überwiegend unverwitterten Mineralkörnern. Zunehmender Aggregatzerfall. Die verschlämmte humusarme Aggregatmasse findet sich in erster Linie in Form von Anlagerungen an den Kornoberflächen; x 75.

SOIL DEVELOPMENT IN SANDY MATERIALS OF THE BELGIAN CAMPINE

F. DE CONINCK and J. LARUELLE

Soil Survey Centre, Ghent and Geological Institute, State University, Ghent (Belgium)

INTRODUCTION

During the cartography of the sandy soils of the Belgian Campine - generally characterized by a pronounced podzolisation - a great complexity was found in their morphology.

This paper is an attempt towards genetic classification of the sandy podzols. In an earlier paper (De Coninck and Laruelle, 1960), some preliminary results of this study have already been published. The present work is based mainly on detailed field observations and on micromorphological studies of thin sections.

PRELIMINARY REMARKS

In order to make the text easier to understand, we give here the definitions of some of the terms used:
The terms *skeleton* and *plasma* of the soil are used in the sense of Kubiena (1938): *Skeleton* includes principally the slightly or unaltered mineral grains and the undigested organic remnants; *plasma*, which forms the essential element of soil dynamics, includes humic matter, clay minerals, and free oxides or hydroxides, especially of iron.

Regarding terminology for the organic matter, we refer to Jongerius (1957): *Moder* is composed of excretions of small soil fauna, with the exception of the earthworms; *mull-like moder* is composed of moder and fine mineral particles; *dispersed humus* is colloidal humic matter liberated at the top of the profile and deposited at some depth in the form of humus skins or "cutans" on the skeleton grains (the term *cutan* is used in the sense of Brewer, 1960).
For the denomination of the genetic horizons, we follow mainly the Soil Survey Staff, U.S.D.A. (1960):

A2 : a mineral horizon in which the feature emphasized is loss of clay, iron or aluminium, with resultant concentrations of quartz or other resistant minerals in sand or silt sizes.

B2 : that part of the horizon where the properties, on which the "B"-horizon is based, are without clearly expressed subordinate characteristics, indicating that the horizon is transitional to an adjacent overlying A- or an adjacent C-horizon.

B2t : B-horizon of (illuvial)[1] clay.

---

[1] We use "illuvial" between brackets to indicate that in this case B-horizons do not necessarily contain illuviated matter.

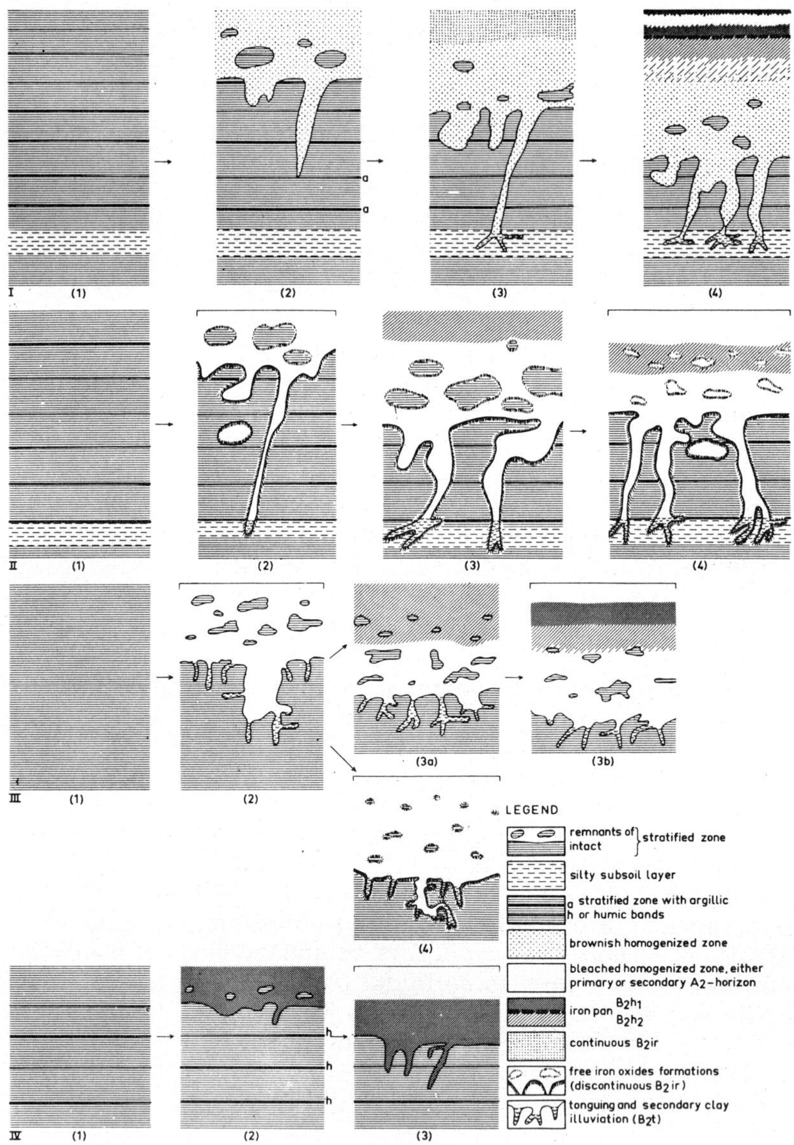
I (1)
(2)
(3)
(4)
a
a
II (1)
(2)
(3)
(4)
III (1)
(2)
(3a)
(3b)
(4)
IV (1)
(2)
(3)
h
h
h
LEGEND
remnants of intact } stratified zone
silty subsoil layer
a stratified zone with argillic h or humic bands
brownish homogenized zone
bleached homogenized zone, either primary or secondary A2-horizon
iron pan B2h1 B2h2
continuous B2ir
free iron oxides formations (discontinuous B2 ir)
tonguing and secondary clay illuviation (B2t)

Fig.1.

I. Sequence of dry soils with brownish plasma

(1) Primary clay enrichment.

(2) Homogenization: brownish homogenized zone.

(3) Formation of a diffuse B2ir-hor. in the brownish homogenized zone; secondary clay illuviation.

(4) Formation of a podzol with B2h1- and B2h2-hor.

II. Sequence of wet soils with brownish plasma

(1) Primary clay enrichment.

(2) Homogenization : formation of a bleached A2- and discontinuous B2ir-hor.; secondary clay illuviation.

(3) Beginning illuviation of organic matter.

(4) Formation of a B2h-hor. around mottles and concretions.

III. Sequence of soils with greenish plasma

(1) Primary clay enrichment.

(2) Homogenization : formation of a bleached A2; secondary clay illuviation.

(3a) Formation of B2h2-hor. in bleached A2-hor.

(3b) Evolution to a profile with an A2-, a bleached B2h1- and a brownish B2h2-hor.

(4) Formation of mottles and concretions.

IV. Sequence in soils with a primary podzol

(1) Primary organic matter enrichment (B2h2-hor.).

(2) Homogenization : formation of a B2h1-hor.

(3) Evolution to a profile with a bleached A2-, a blackish B2h1- and a brownish B2h2-hor.

B2ir : B-horizon of (illuvial) iron.
B2h : B-horizon of (illuvial) humus.
B2h1 : upper B-horizon of humus with blackish or grayish color.
B2h2 : lower B-horizon of humus with brownish or reddish-brown color.

The term *homogenization* is used to indicate the process involving disturbance of the original stratification. It is mainly due to the activity of burrowing fauna and of roots (Hoeksema, 1953; Edelman, 1960). The phenomenon is predominantly located in the upper part of the soil profile.

By the term *clay* is meant the fraction less than 2 μ, as determined by particle size distribution analysis. It does not necessarily mean that this fraction is composed of clay minerals.

Soil consistency is named in accordance with the terminology of the Soil Survey Manual (Soil Survey Staff, U.S.D.A., 1951).

The term *podzolization* is used in a rather limited sense: it indicates, the illuviation of organic matter and the formation of B2h-horizons.

The phenomena described in this paper include all soils of the region except the alluvial soils with a permanently high water table.

The distinction between dry and wet soils is based on the absence or presence of a periodic or permanent water table in the profile. In dry soils, the profile never undergoes the influence of a water table. In wet soils, influence of a water table may be permanent or periodical.

ORIGIN AND COMPOSITION OF THE PARENT MATERIAL

The sandy deposits dating from the last glaciation and the postglacial period show a more or less horizontal stratification. The texture varies from coarse sand with a mediane around 200 to very sandy loam with 35–50% silt.

The mineralogical composition exhibits a striking difference between south and north. In the southern Campine, where the substratum is composed of glauconitic sands of the Upper Tertiary, the Quaternary deposits always contain a variable percentage of glauconite; in the northern Campine, where the substratum consists of non-glauconitic Lower Pleistocene deposits, the parent material is almost free of glauconite.

STAGES OF DEVELOPMENT

The following stages of development have been distinguished:
Stage A : formation of a primary colloidal enrichment;
Stage B : homogenization of the profile and formation of secondary clay;
Stage C : formation of an iron or humus B-horizon;
Stage D : formation of mottles and concretions in soils with greenish plasma.

An overall picture of the soil sequences which can originate in this way, is given in Fig.1.

Stage A : Primary colloidal enrichment

The first stage of development seems to consist of a formation of a cutan of plasma around every skeleton particle. This plasma may be organic or mineral.

The organic cutan occurs in specific wet soils, namely those located in small isolated depressions. These soils thoroughly show cutans of dispersed humus, especially expressed as subhorizontal darker bands in case of strong accumulation. In this case, the original stratification of the sediment becomes very obvious. The depth of humus illuviation may be considerable (several meters). The consistency is normally firm, even very firm in some cases. In thin sections, a continuous isotropic brownish cutan of dispersed humus is striking. It may sometimes overlie a very thin birefringent cutan. This humus illuviation forms a B2h2-hor.

In all other soils, an oriented birefringent clay cutan is found around the skeleton particles. There may be successive translocation of clay, giving rise to the well-known horizontal bands or lamellae of clay accumulation.

The color of this clay cutan differs strikingly. In dry non-glauconitic soils, the plasma is brownish; under the influence of a water table, the color turns rusty, although the difference with that of the dry soils may be small.

In dry glauconitic soils, the color is generally comparable to that of dry non-glauconitic soils. However, it may be distinctly greenish, especially when the glauconite content is high. In wet conditions, there is a general tendency towards a greenish color, although many soils keep a brownish color. The tendency towards a rusty color, as in non-glauconitic soils, is much less pronounced, at least in this first stage.

The most important feature of this first stage is the presence in all considered soils of a cutan around every skeleton particle, even between the bands of clay accumulation, which is normally considered as the C-hor. of the profile. However, this phenomenon is highly divergent. In coarse non-glauconitic sands, the clay cutans are hardly visible in thin sections. In very glauconitic materials, the plasma may be so abundant that it fills an important part of the intergranular space. In the lamellae, one distinctly recognizes concentric clay accumulations, clearly indicating the phenomenon of successive migration.

It is remarkable that in the stratified part of the profile roots are almost entirely lacking. This is in accordance with the idea of Edelman (1960) that the pores in sandy soils with original stratification are too small for roots without the preliminary burrowing of soil fauna.

Stage B : Homogenization of the profile and formation of secondary clay

These two phenomena belong to the same process. They will, however, be treated separately for clarity.

Homogenization

This phenomenon presents itself in a very heterogeneous way. The lower limit of the homogenized zone is very irregular and discontinuous : non-homogenized patches are always left in the homogenized zone. It is characteristic that the roots adapt themselves completely to the homogenized zone.

The process of homogenization can take on different aspects according to the nature of the enveloping cutan.

(1) In soils with a cutan of dispersed humus, the homogenization gives rise to a zone of grayish color, often with a pinkish tinge. Consistency is considerably diminished. During the process, dispersed humus is replaced for the most part by moder humus scattered among bleached sand and silt grains. We call these profiles primary podzols.

Up to now, we have had no clear evidence that at the moment of formation of the B2h2-hor. there was already a bleached A2-hor., but we could establish without doubt that the homogenized zone was changing into a bleached horizon.

(2) In soils with a mineral cutan, the homogenization may produce the formation of a brownish or a bleached homogenized zone.

The formation of a *brownish homogenized zone* occurs in dry soils with brownish plasma. The homogenized zone gets a slightly lighter color, the consistency appreciably diminishes and, if the plasma is abundant, the clay content decreases considerably.

In thin sections, a partial disappearance of the continuous oriented cutan from the first stage is observed; parts of the clay cutans remain on the sand grains, so that these grains are partially bleached.

The diminishing consistency seems to be due to the disappearance of the continuous cutan which acts as glue or cement.

The formation of a *bleached homogenized zone* takes place in wet soils with brownish plasma and in all soils with greenish plasma. The homogenization gives rise to a strongly bleached zone, with an ever-decreasing consistency.

In wet soils with brownish plasma, another process takes place at the same time : rusty strips are formed at the margin between the bleached and the stratified zone. In thin sections, it appears to be due to a partial destruction of the clay minerals in the stratified zone and to a translocation of free oxides, especially of iron. This is analogous to the formation of a profile with destroyed textural B-hor., described by Ameryckx (1960) for the soils of the Flemish Sandy Region.

In the upper part of the profile, a few non-homogenized patches are left. These patches are relatively small. Surrounded by a rusty, iron-enriched band, they form indurated mottles and, in extreme cases, iron concretions. Thus, these mottles and iron concretions are remnants of the stratified zone, and are enriched with iron, especially at their peripheries. These iron accumulations consist of intergranular non-birefringent plasma, which owes its micromorphological characteristics to the fact that seemingly unoriented clayish material, mixed with humic and ferric substances, has replaced intergranular filling of a former birefringent kind. The same unoriented clayish material may also fill other intergranular spaces. The problem of either replacement or simple filling still remains open for discussion. Birefringent, well-oriented colloidal material (obviously clay-ferric complexes) is still present in the form of thin cutans around all of the skeleton particles. In the extreme case of concretions, the intergranular space is almost entirely filled. No plasma is left in the bleached part.

In the soils with greenish plasma it seems that free oxides are not produced at this stage, so that the bleached zone passes directly into the greenish stratified zone.

As a conclusion, it may be said that the homogenization brings about differentiation in the profile. The homogenized zone forms an impoverished horizon compared to the stratified zone; this impoverishment presents important differences according to the cases. In the coarse non-glauconitic sands with little plasma in the stratified zone, the impoverishment is small. In the glauconitic materials, the difference in clay content may reach several percents and in this case, the profile can be considered as having an A2-hor. overlying a B2t-hor.

In the case of the wet soils, mottling often forms a zone between the bleached A- and B2t-hor. which can be assimilated to a discontinuous B2ir-hor. This B2ir-hor. does not form in the A2-hor., but in the upper part of the B2t-hor. by accumulation of free iron oxides from the A2-hor.

Formation of secondary clay

We have established that intense translocation of clay from the homogenized zone often produces an accumulation of clay in the root holes at the bottom of the homogenized zone. We have called this "secondary clay", since this form of translocation and accumulation is completely different from the clay enrichment during the first stage. The secondary clay is either grayish in non-glauconitic materials or greenish in glauconitic materials, even when the primary clay is brownish, probably indicating a reduction process.

This secondary accumulation is especially pronounced in soils with greenish plasma and silty parent material (30–50% of silt). In this case, the accumulation takes place just below the bleached A2-hor. and accentuates the pre-existing difference in clay content between the A2-hor. and the stratified B2t-hor. In such profiles, differences in clay content reaching 20% can be found. (Plate C1.)

In the soils with brownish plasma, the secondary clay accumulation is mainly pronounced when silty horizons are present in the subsoil. These layers are thoroughly interwoven with very small channels, formed by a multitude of rootlets and covered or filled up with clay cutans (Plate C8). Such silty horizons always have a mottled aspect and the hard or firm consistency of a fragipan horizon (Deckers and Baeyens, 1963).

In thin sections, the secondary clay is present in the shape of bulky and localized accumulations, often with concentric forms of superposed layers, and with strong birefringence and distinct extinction bands. All these criteria clearly indicate clay migration (Brewer and Haldane, 1957).

The present state of the profile determines the further development.

Stage C : Formation of an iron or humus B-horizon

Soils with unbleached homogenized horizon

In this horizon, a brown diffuse B2ir-hor. is formed. This is probably due to oxidation or alteration in situ of remnants from the primary clay and of alterable minerals which are still present in the horizon at this stage of development. It is the well-known B2ir-hor. of the brown podzolic soils of the American authors (Frei and Cline, 1949) or of the "Podsol-Braunerde" of German authors (Altemüller, 1962). We normally find this profile in the glauconitic dry soils with brownish plasma where the oxydation of the grains of glauconite, not yet altered in the first stage, is greatly responsible for the brownish color and the relatively high free iron content.

The non-glauconitic dry soils probably had the same stage of development. Yet, presently, they always show a distinct podzol, characterized by a differentiation of the humus accumulation in two distinct horizons, an upper blackish one (B2h1-hor.) and a lower brownish one (B2h2-hor.); there is an abrupt transition between the two horizons, often accentuated by a very thin, very hard iron pan. The B2h1-hor. is strongly impoverished in iron and calcinates white; the B2h2-hor. is not impoverished in iron, but on the contrary often presents a slight accumulation of free iron (De Coninck 1954). This is probably the remnant of a diffuse B2ir-hor.

In thin sections, the B2ir-hor. of the brown podzolic soil indicates the presence of small floculations of a brown non-birefringent plasma on and between the skeleton grains. It seems to be composed of ferro-organic compounds. We also find much mull-like moder formed by root decomposition.

In the B2h1-hor. of the humus podzol, the organic matter is composed of moder and dispersed humus (Edelman, 1960). The moder is formed by decomposition of the roots, the development of which is very intense in this horizon. These are the pellets of the American classification (1960).

The thin iron pan presents a very abundant, dark brown, non-birefringent plasma (Plates C6 and C7), which fills most of the pore space. The great content of organic matter in this horizon points out that it probably consists of ferro-organic compounds.

The organic matter of the B2h2-hor. is chiefly composed of dispersed humus; however, there is always an appreciable amount of moder humus.

Soils with bleached homogenized horizon

In these soils, B2ir-hor. formation is impossible in the bleached A2-hor. due to a lack of alterable substances. A brown podzolic soil with B2ir-hor. is never found. There can only be translocation of organic matter, which always forms a very diffuse illuviation. First, a brown or reddish-brown horizon of dispersed humus originates; in a second stage, a grayish-brown horizon appears in the upper part of the B2h-hor. by replacement of the dispersed humus by root-moder. The color of this horizon evolves into pinkish gray. The transition between the two horizons is very gradual. In some profiles, we could even find the whole B2h-hor. chiefly composed of root moder.

In the soils with rusty mottles and concretions (discontinuous B2ir-hor the B2h-hor. can be formed just over and between these iron accumulations. Such profiles can be considered as having a B2h- and a B2ir-hor., the last one being discontinuous. But strictly speaking, the B2ir-hor. formation is not involved in the process of podzolisation.

Chemical analyses of the iron concretions give a relatively high organi matter content (2–3%). We suppose that this organic matter originated from root decomposition and formed compounds with the iron oxides during the homogenization.

A similar process may occur in soils with greenish plasma; the B2h-hor. then develops between and around the remnants of the greenish B2t-hor.

Stage D : Formation of mottles and concretions

In many soils with greenish plasma - and only in these soils - we find numerous mottles and concretions in the A2-hor. in the zone of contact between the A2- and the B2t-hor.

The study of thin sections shows that this is not a question of iron translocation and accumulation, but alteration in situ, probably an oxydation, of the greenish plasma. During the transition, the color gradually passes from greenish to brownish or to reddish-brown, while the birefringence decreases in the same way and disappears almost completely in the mottles and concretions (Plates C2–5). This phenomenon, however, is not going on in all soils with greenish plasma; up to now, we have not found an explanation for this difference.

DESCRIPTION OF SOIL PROFILES[1]

The descriptions are to be an illustration of the preceding chapter. These profiles represent the most important in the described sequences[2]:

(1) A dry profile on non-glauconitic sand with a podzol having a B2h1-hor. and a B2h2-hor. separated by a thin iron pan, and a C-hor., differentiated into a homogenized C1-hor. and a stratified C2-hor.

(2) A dry profile on non-glauconitic sand with slight evidences of free oxide formation and strong secondary clay illuviation in a silty subsoil layer.

(3) A wet profile on non-glauconitic loamy sand with a discontinuous B2ir-hor., composed of indurated rusty mottles and concretions, and with an incipient formation of a B2h-hor.

(4) A toposequence on glauconitic loamy sand going from a dry to a wet profile. The dry profile has a brown podzolic B2ir-hor., which is formed in the homogenized zone and shows typical primary clay illuviation bands in the stratified zone. The wet profile has a B2h-hor. in the bleached homogenized A2-hor. and shows a discontinuous B2ir-hor. overlying the stratified zone, and secondary clay illuviation in the more silty subsoil.

---

[1] According to the Soil Survey Staff, U.S.D.A. (1960), profiles 1 and 2 are orthumods, 3 and 4A ferraquods, and 4B and 5 humaquods (Ed.).

[2] Another paper dealing with "Pedogenèse dans des sols sableux glauconifères" was presented at the VIIIth International Congress of Soil Science, held in Bucarest (1964). This deals mostly with similar profile descriptions, emphasizing those of strongly glauconitic materials.

Profile 1: Dry profile on non-glauconiferous coarse sand

Locality: Sheet of Iepenrooy, municipality of Meer (northern Campine). Vegetation: Natural vegetation of *Calluna vulgaris*, *Pinus silvestris*, *Betula*, *Quercus*, mosses. (Fig.2).

Macromorphology

<2–0 cm:	01, little altered plant remains of *Calluna* and many bleached sand grains;
0–3/4 cm:	A1, very dark gray (10 YR 2/1[1]) bleached sand[2]; humic; structureless, massive; very friable;
3/4–15/20 cm:	A2, grayish-brown to dark grayish-brown (10 YR 3/2–4/2) bleached sand; little humic; structureless, single grain: loose to very friable; with few, faint to distinct, irregular, very dark gray (10 YR 3/1) humic patches;
15/20–23/30 cm:	B2h1, black (5 YR 2/1) sand; very humic; structureless, massive; firm; with many thin rootmats and bleached sand grains in the mats; thick, dark reddish-brown (5 YR 3/2) rootmat and iron pan at the lower boundary;
23/30–30/35 cm:	B2h2, dark reddish-brown to black (5 YR 2/2–2/1) sand; humic; without bleached sand grains; structureless, massive; friable to firm; very few roots;
30/35–50/55 cm:	B3h, dark reddish-brown (5 YR 3/5–3/2) sand; humic; structureless, massive; friable; with many, very irregular, darker humic lamellae and spots;
50/55–65/75 cm:	C1, variegated sand, dark brown (7.5 YR 3/2 and 10 YR 4/4) and yellowish-brown (10 YR 5/4) irregular spots; structureless, massive; loose to very friable; with few, thin, humic, slightly firmer lamellae and some round, dark reddish-brown (5 YR 3/3) spots, filled up with dead roots and organic matter;

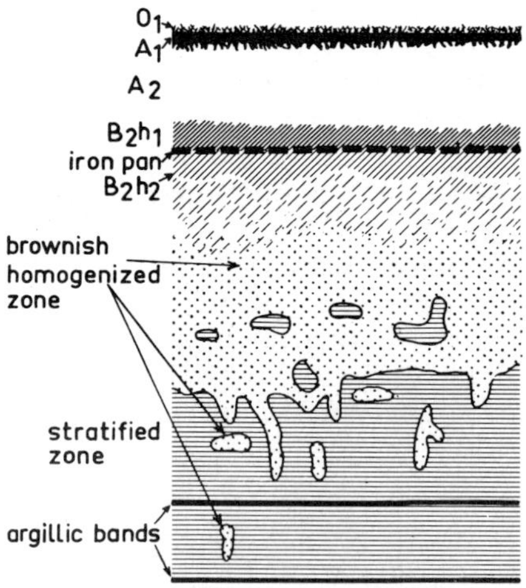

Fig.2. Profile 1.

---

[1] Color notations according to Munsell Soil Color Charts. All colors are given for moist soil materials.

[2] Textural class is given according to the Belgian texture classification.

>65/75 cm: C2, light yellowish-brown to brownish-yellow (10 YR 6/4–6/6) sand; structureless, massive; very friable, firmer than the C1-hor.; stratified with (1) medium and large, distinct(especially when dry), yellowish brown (10 YR 5/4) (when dry, color shows lighter than in the matrix) spots, which are distinctly looser; the extent of these spots gradually diminishes to the bottom; all the roots are present in these loose spots; (2) very irregular humic lamellae and humic spots; (3) brown to dark brown (7.5 YR 4/4) B2t-hor. lamellae, firm when wet, very hard when dry, discontinuous in the upper part (breakdown of the B2t-hor.) and forming "double bands" with the humic lamellae.

Micromorphology

In the horizons A1, A2 and B2h1, the organic matter is chiefly composed of moder humus between bleached sand grains. Especially in the B2h1-hor., the accumulation of rootmoder humus is very important. Just above the boundary with the B2h2-hor., a mineral skeleton is lacking almost entirely; we have only unaltered root remains and root-moder. Often the excrements are present around and in the root remains, indicating undoubtedly that the moder is formed by small soil fauna from the root remains. The excrements show a strong tendency to form clusters and to pass into dispersed humus.

In the iron pan, the intergranular space is entirely filled up with a non-birefringent plasma, consisting primarily of free oxides and of organic matter.

In the B2h2-hor., mineral skeleton prevails, while organic plasma fills most of the pores, especially in the upper part. Around the sand grains we find thick continuous cutans of dispersed humus, forming bridges at contact points of the grains. In the intergranular space, great quantities of moder, illuviated or formed in place, are present.

The argillic lamellae show two forms of plasma: there is a thin, strong birefringent cutan on the skeleton, while an important part of the intergranular space is filled with birefringent plasma, often with the typical indication of successive layering.

Profile 2: Dry profile on non-glauconitic sand with silty layer in the subsoil

Locality: Sheet of Iepenrooy, municipality of Meer (northern Campine). Vegetation: Natural vegetation of *Calluna vulgaris, Molinia coerulaea, Quercus, Betula, Pinus silvestris,* little *Erica* and *Genista,* mosses. (Fig.3.)

Macromorphology

<1–0 cm: 01, little altered plant remains and many bleached sand grains;

0–25 cm: A2, dark gray to very dark gray (10 YR 4/1–3/1) sand, humic; structureless, massive; loose; with very dark gray (10 YR 3/1), common, distinct, diffuse, medium and coarse patches and strips;

25/30–30/40 cm: B2h1, very dark gray (5 YR 3/1) sand, very humic; many bleached sand grains; structureless, massive; firm; with few bleached A2-hor. patches; abrupt, irregular boundary, coinciding with a very hard, reddish-brown (5 YR 2/2), thin iron pan, underlying a distinct rootmat;

30/40–50/60 cm: B2h2, variegated sand: matrix very dark brown (7.5 YR 2/2); yellowish-red (5 YR 4/6) little spots and horizontal strips just underneath the iron path (remains of an earlier B2ir-hor.); dark yellowish-brown (10 YR 3/4); brown (10 YR 5/3); dark brown to very dark gray (7.5 YR 3/3); structureless, massive; very friable (dark parts), loose (light parts); light patches form irregular cavities and holes in the dark matrix;

50/60–75 cm: B2ir, mottled sand : matrix yellowish-brown (10 YR 2.5 Y 5/4);

structureless, massive; friable; non-stratified; with many, distinct, medium and coarse, diffuse to clear, very hard mottles, reddish-brown to dark reddish-brown (7.5–5 YR 4/4), with many black spots and faint stratification; remark : mottles form a discontinuous B2ir hor.;

75–80 cm: C, yellowish-brown (10 YR–2.5 Y 5/4) sand; structureless, massive friable; non-stratified; with (1) common, faint, medium and coarse, diffuse, yellowish-brown (10 YR 5/4) patches, which are the continuation of the pale patches of the B2h2-hor., with still more friable consistency than matrix; locally brown to dark brown (10 YR 3/3–3/4), slightly humic material around these patches; (2) brown to dark brown (10 YR 3/3–3/4) humic patches; (3) common, faint, little to coarse, diffuse, firm, dark yellowish-brown (10 YR 4/4) stratified patches;

80–90/95 cm: B2t, pale brown (10 YR 6/3) sand; structureless, massive; loose; with many, prominent, medium and coarse, sharp, firm, strong brown (7.5 YR 5/6), stratified patches, locally occupying the largest part of the horizon; the strong brown patches form the remnants of a former continuous B2t-hor.;

90/95–100/105 cm: II B2t, variegated light sandy loam : light gray (2.5 Y 7/2; type 1); light brownish-gray (2.5 Y 6/2; type 2); strong brown (7.5 YR 5/8; type 3); these three components show a regular arrangement : the upper part consists of 1 and 2, with 1 above and 2 beneath; next a discontinuous 3 band, again a 1 band, a thick 2 band, an irregular 3 band, again a (1+2) band with 2 prevailing in the central and bottom part; the whole horizon is structureless and massive; the three components exhibit a difference in composition and consistency : 1

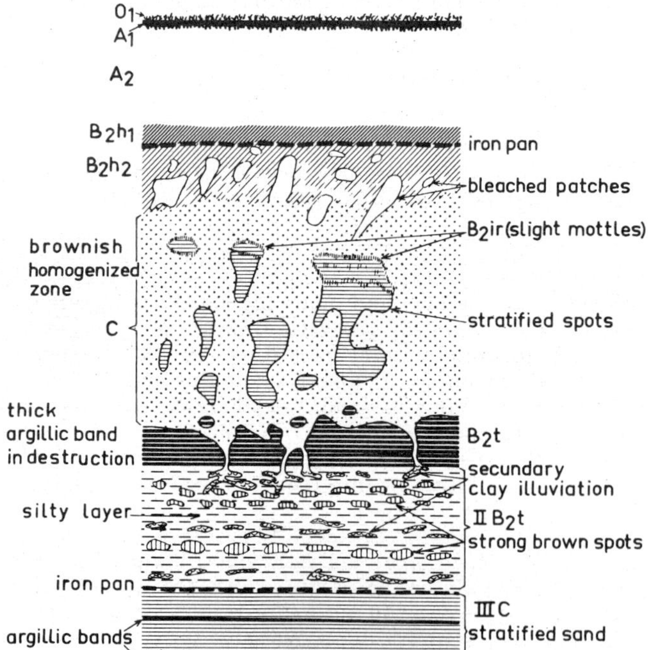

Fig.3. Profile 2.

silty, friable, non-sticky and non-plastic; 2 is more clayey because of thin irregular clayey channels, friable to firm, slightly sticky and slightly plastic, with many dead roots; 3 is silty, stratified, friable to firm, non-sticky, non-plastic and more brittle; the lower boundary is abrupt, accentuated by a very hard iron pan formed by successive horizontal thin yellowish-red (5 YR 4/8) layers of free oxides;

>100/105 cm: III C , yellowish-brown (10 YR 5/4) sand, stratified, structureless, massive; very friable; with irregular strong brown (7.5 YR 5/8) argillic bands.

This profile is in the first place an illustration of the primary B2t-hor. destruction and of the illuviation of secondary clay in a silty layer. This silty layer causes a slight water retention, probably causing the formation of little mottling in the profile (discontinuous B2ir-hor.)

Micromorphology
The A1-hor. has moder humus with a tendency to formation of clusters (first stage of dispersed humus). All the skeleton grains are bleached. In the A2-hor., we find the same skeleton with scattered moder humus between and on the skeleton grains, with a tendency to grow together to compact clusters (disappearance of the original excrements structure). In a few places, fresh excretions are present in cavities of not-yet-altered plant remains.

In the B2h1-hor., the intergranular spaces are for the most part filled with rootmoder and illuviated moder. A large part of this moder has lost its original shape to form compact black clusters and shapeless aggregates enclosing the most part of the skeleton.

In the B2h2-hor., the skeleton is fully enveloped with a brown continuous cutan of dispersed humus and with root moder (few little iron concretions were also found). The most striking differences between the B2h1-hor. and B2h2-hor. are the color (black in the B2h1-hor. and brown in the B2h2-hor.) and the way in which the skeleton is coated (more complete in the B2h2-hor.)

In the B2ir-hor., the skeleton is fully enveloped with thin cutans, with more or less distinct birefringence. There are also a number of brown birefringent plasma concentrations but of amygdalus type, which are an indication of pseudogley conditions.

The micromorphology of the C-hor. indicates a loose structure, composed of grains more or less enveloped with birefringent plasma, but practically without intergranular braces. A few little-iron concentrations are present.

The destroyed B2t-hor. shows localised strong birefringent pore fillings often with layered structure. Little slightly birefringent or opaque iron concentrations show pseudogley conditions.

In the B2t-hor., we find bulky, localised clay illuviation. The color is grayish, with high birefringence, indicating reducing circumstances. Their localisation follows large conducting channels, but penetrates also deeply between the skeleton grains. Around the skeleton, a thin light-colored birefringent plasma is present.

The iron pan at the lower boundary has abundant non-birefringent plasma, filling up a large part of the porespace. It seems to be built up of goethite and hematite.

Profile 3: Wet profile on non-glauconitic loamy sand

Locality: Sheet of Brecht, municipality of Brecht/Overbroek (northern Campine). Vegetation: Plantation of *Pinus silvestris*. (Fig.4.)

Macromorphology

<2–0 cm: 01, accumulation of needles of *Pinus silvestris;*

0–25/35	cm:	Ap, variegated plow layer;
25/35–35/60	cm:	A2+B2h, variegated loamy sand, pale brown (10 YR 6/3), brown to dark brown (7.5 YR 4/2) and dark grayish-brown (10 YR 4/2); structureless, massive; friable, with few distinct, common, friable to firm, strong brown (7.5 YR 5/6) mottles and concretions;
35/60–50/70	cm:	B2ir + B2h, composed of three different materials : (1 strong brown (7.5 Y 4/6) and dark reddish-brown (5 YR 3/4 and 5 YR 3/2), structureless, massive, extremely firm, large concretions, the number of which increases from the top to the bottom; (2) dark reddish-brown (5 YR 3/2), structureless, massive, friable to firm, irregular, humic patches and horizontal, thin, humic lamellae; (3) brown (10 YR 5/3), structureless, massive, friable patches;
>50/70	cm:	Cg, stratified horizontal layers, consecutively strong brown (7.5 YR 5/6) and pale gray to pale yellow (2.5 Y 6/2–6/4),structureless, massive, friable; locally pale brown (10 YR 6) patches and tongues.

In this profile the stratified zone has a rusty color; at the top of this zone, a discontinuous B2ir-hor. composed of extremely firm concretions is present. Between these concretions, a faint B2h-hor. is formed.

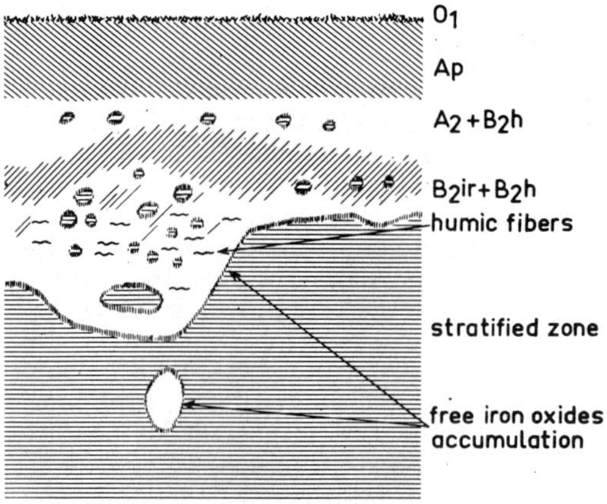

Fig.4. Profile 3.

Micromorphology

The humic part of the A2+B2h-horizons is composed of rather indistinct, scattered agglomerations of mullic moder. In the A2 part, localized cutans of more or less birefringent plasma are present. This is probably a remnant of a primary clay cutan.

The B2ir-hor. concretions show two distinctly different forms of plasma : (1) most of the skeleton grains are coated with birefringent, yellowish-brown cutans; (2) many intergranular spaces are filled with non-birefringent, yellowish-brown to brownish-red plasma, composed of ferro-organic compounds. This difference can only be explained by two subsequent stages of formation.

In the Cg-hor. the birefringent cutan around the skeleton grains is present everywhere, while non-birefringent intergranular plasma is lacking almost entirely.

Profile 4: Toposequence on glauconitic loamy sand

Locality: Sheet of Geel, municipality of Olen (southern Campine).

Dry profile

Vegetation: Plantation of *Pinus silvestris;* undergrowth of *Festuca rubra, Descampsia flexuosa, Agrostis stolonifera, Luzula, Genista, Calluna vulgaris, Betula, Quercus.* (Fig.5.)

Macromorphology

<4–0 cm: 01, unaltered and slightly altered plant remains;

0–25/27 cm: Ap, brown to dark brown (10 YR 5/3–4/3) loamy sand, humic; structureless, massive; very friable; small pieces of charcoal and bricks;

25/27–45 cm: B2ir, dark yellowish-brown (10 YR 4/4) loamy sand; structureless, massive; friable;

45/75 cm: C1, light yellowish-brown (2.5 Y 6/4) sand; structureless, massive; loose to very friable; non-stratified; with common, faint, clear, light olive brown (2.5 Y 5/4–5/6) stratified patches with slightly firmer consistency; on breaking, these patches form angular fragments, while the matrix forms rounded ones; roots are present only in matrix;

>75 cm: C2, light olive brown (2.5 Y 6/4–5/6) sand; structureless, massive; very friable; stratified; with (1) horizontal, prominent, stratified, sharp, yellowish-brown to dark yellowish-brown (10 YR 5/6–6/4), structureless, massive, friable to firm, (very hard when dry), 0.5–2 cm thick argillic bands, partially destroyed in the upper part; (2) faint, many, fine to coarse, clear, light yellowish-brown (2.5 Y 6/4), loose to very friable, irregular pale olive (5 Y 6/3) spots that are bordered by strong brown (7.5 YR 5/6–5/8) strips and patches with firmer consistence; where these spots cut the argillic bands, the latter get a strong brown color and firm to very firm consistency; roots are only present in these looser spots.

Wet profile

Vegetation: Plantation of *Pinus silvestris;* undergrowth of *Festuca rubra, Descampsia flexuosa, Agrostis stolonifera, Luzula, Genista, Calluna vulgaris, Betula, Quercus.*

Macromorphology

0–45/55 cm: Ap, dark brown (10 YR 3/3) loamy sand, humic; structureless, massive;

45/55–70/75 cm: B2h, variegated loamy sand; structureless, massive; (1) matrix dark reddish-brown (5 YR 3/2), very friable; (2) dark brown (10 YR 3/3), loose to very friable, forming distinct, sharp, very irregular patches; (3) little, faint, sharp, rusty spots and a few dark reddish-brown (5 YR 3/3) very firm concretions, forming discontinuous horizontal bands (remnants of argillic lamellae); the lighter patches pass into the A2-hor. material;

70/75–95/100 cm: A2, variegated loamy sand; structureless, massive : (1) yellowish-brown to light olive brown (10 YR 2.5 Y 5/6), friable; (2) light olive brown (2.5 Y 5/4), more friable; the two materials form faint, diffuse patches among each other; with many rusty spots and dots and a few firm concretions;

95/100–110/115 cm: B2ir, composed of two materials : (a) variegated loamy sand; structureless, massive; friable to firm according to the darkness of the color; stratified; (1) dark yellowish-brown (10 YR 4/4); (2) yellowish-red (5 YR 4/6); (3) locally reddish-brown to dark reddish-brown (5 YR 4/4–3/4), forming thin horizontal layers; (b) idem as the A2-hor. with more rusty spots : at the boundary of (a) and (b) thin reddish-brown to yellowish-red (10 YR 4/4–4/8) layers or strong brown (7.5 YR 5/6) non-stratified, more friable material;

110/115–150/160 cm: Cg, light olive brown (2.5 Y 5/6–5/4) loamy sand; structureless, massive; friable; stratified; with (1) many, prominent, coarse, sharp to diffuse, irregular, loose, pale olive to olive (5 Y 6/3–5/3) non stratified patches; locally these spots feel a little more clayey and are more olive in color (5 Y 6/4–5/4); (2) dark yellowish-brown (10 YR 4/4), yellowish-red (5 YR 4/6), reddish-brown to dark reddish-brown (5 YR 4/4–3/4) mottles, strips and circlets; (3) in the whole horizon, but especially in the rusty part many black and dark brown dots; the thin silty layers grow thicker and more numerous downwards;

>150/160 cm: IIB2t, yellowish-brown (10 YR 5/6) loamy sand to light sandy loam; structureless, massive; friable; stratified; with distinct, many, fine to coarse, diffuse, green, clayey, slightly plastic, non-stratified patches; in the core of these patches, olive (5 Y 7/4), loose, non-stratified material may be present.

This toposequence illlustrates very well on the one side the formation of a brownish homogenized zone with a diffuse, continuous B2ir-hor. in the dry soil. On the other side, the presence of a bleached A2-hor. with a B2h-hor. and a discontinuous B2ir-hor. at the top of the stratified zone are found in the wet soil; a silty layer in the subsoil with green secondary clay has to be pointed out. In the zone of transition between the two described profiles, the homogenized zone grows gradually more bleached from dry to wet with a disappearance of the diffuse B2ir-hor. and the appearance of a discontinuous B2ir-hor.

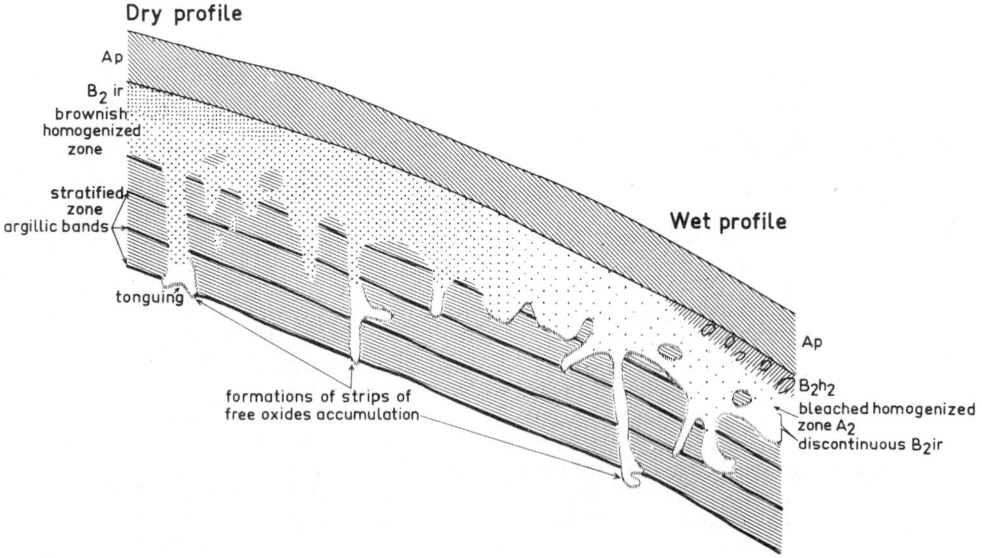

Fig.5. Toposequence (4).

Micromorphology

Dry profile: In the B2ir-hor. the skeleton is partially enveloped by a dark brown plasma cutan in which the birefringence, present in underlying horizons, has mostly disappeared. Some non-birefringent intergranular fillings are found. This opaque plasma is formed by alterations of the birefringent plasma. Locally it has a loose structure, probably indicating organo-ferric compounds.

In the homogenized and stratified part of the C-hor., little difference is visible in the nature and volume of the birefringent plasma.

In the B2t bands, an abundant birefringent plasma envelops the whole skeleton, often forming successive layering.

Wet profile: In the B2h-hor. the skeleton is partially enveloped with dispersed humus and loose rootmoder accumulations.

In the A2-hor., many grains are still surrounded by a thin, slightly birefringent cutan. The glauconite grains show different colors, according to the stage of alteration.

In the B2ir-hor. matter, a partially birefringent cutan is present around a part of the skeleton grains. Locally, non-birefringent plasma fills the intergranular space. The rusty dots are formed by brown or dark brown altered opaque glauconite grains.

In the stratified Cg-hor., the cutan is more distinct and more birefringent than in the B2ir-hor., indicating that in the last horizon an alteration of the primary cutan is going on.

In the silty II Cg-hor. the plasma fills most of the intergranular space. It is highly birefringent with clear extinction bands. Locally, the birefringence diminishes, while the color turns a darker brown, indicating the beginning of an alteration.

Profile 5: Wet profile on glauconitic sand with primary podzol

Locality: Sheet of Grobbendonk, municipality of Grobbendonk (southern Campine). Vegetation: Plantation of *Pinus silvestris*, with *Molinia coerulaea*, *Erica tetralix*, *Descampsia flexuosa* and a few mosses. (Fig.6.)

Macromorphology

<2/3–0 cm:	01, little altered needles;
0–20/25 cm:	A2, very dark gray (10 YR 3/1) sand, humic; structureless, massive; very friable to loose; with (1) many, very dark gray to black (5 YR 3/1–2/1) spots and vertical strips (rootholes); (2) many, faint, diffuse, looser, dark grayish-brown to dark gray (10 YR 4/2–4/1) patches, especially in the bottom part;
20/25–35/40 cm:	B2h1, very dark to black (5 YR 3/1–2/1), rubbed dark reddish-brown (5 YR 3/2), passing downwards to dark brown (5 YR 3/2), sand, very humic; structureless, massive; friable; with (1) many, faint, diffuse, looser and less humic, dark brown (7.5 YR 3/2) patches; (2) many dark spots and strips like those in A2;
35/40–45/65 cm:	B2h2, dark reddish brown (5 YR 3/3–3/2) passing downwards to dark reddish-brown (5 YR 3/3) sand, humic, without bleached sand grains; structureless, massive; locally friable, locally very firm stratified lumps, separated by planes with plenty of roots; with irregular, darker humus fibers and rootmats; many roots in the loose part, no roots in the firm lumps;
45/65–75/80 cm:	B3h1, dark yellowish-brown (10 YR 4/4) sand; structureless, massive; friable; stratified; with (1) little and coarse, faint, diffuse, looser, yellowish-brown to dark yellowish-brown (10 YR 5/6–4/4) patches, especially in the upper part in which only roots are present; (2) humic fibers closely following the stratification; (3) firmer, dark brown (7.5 YR 3/1) humic band at the lower boundary;

75/80–100 cm: II B3h2, olive brown to dark yellowish-brown (2.5 Y–10 YR 3/4) sand, structureless, massive; firm; stratified; with a few, faint, humic fibers, especially in the upper part; roots are lacking entirely; many little cobbles at the lower boundary;

>110 cm: III Cg, green stratified very glauconitic sand.

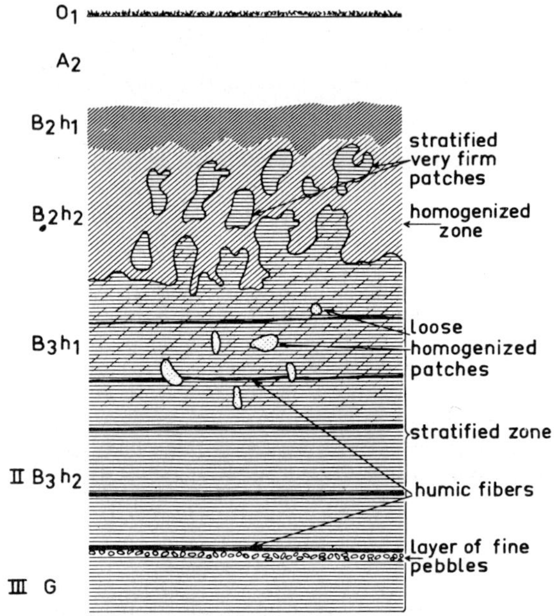

Fig.6. Profile 5.

Micromorphology

The A2-hor. shows residual discontinuous, brown, non-birefringent, humic plasma.

In the B2h2-hor., important cutans and intergranular fillings of dispersed humus are present. Locally, rootmoder is found. The numerous glauconite grains are freshly green and show no alteration at their periphery. In the humic bands, the cutans of dispersed humus are still more continuous.

The micromorphology of the homogenized and stratified parts of the B3h-hor. shows little or no differences. The glauconite grains can show a transition to a brown color, especially at their periphery. Nowhere do they pass into a colloidal form.

ACKNOWLEDGEMENT

We would like to acknowledge our thanks to Prof. Dr. R. Tavernier, Director of the Soil Survey Center at Ghent, for the very real help he has provided us in these studies. Without his active support and restless interest this research would never have been possible. Our thanks are also addressed to several members of the staffs of the Soil Survey Center and of the Geological Institute, State University of Ghent.

SUMMARY

A genetic classification of sandy soils of the Belgian Campine based on field observation and micromorphology is attempted. The southern part of this region has a glauconitic parent material which has a great influence on the soil development.

Analogous stages of development can be distinguished. The intensity of these stages determines the final composition of the profile.

In all considered soils, a colloidal cutan is formed around the skeleton particles. This cutan can be organic or mineral, it is a "primary podzol" in the case of an organic cutan, and a "primary clay" in the case of a mineral cutan (stage A).

Stage B coincides with the homogenization of the profile (disappearance of the original stratification) and a translocation of clay (secondary clay) which brings about a textural differentiation in the profile.

Stage C shows the development going on in the homogenized zone. When it has a brownish color, the sequence of development is first a diffuse B2ir-hor., then a humic podzol with iron pan and iron accumulation in the bottom part of the B2h-hor. When the homogenized zone is completely bleached, only organic accumulation is present.

Stage D is restricted to some profiles on glauconitic material, where the alteration of the birefringent cutan gives rise to mottles and iron concretions.

Four profiles and a toposequence are described. The most important micromorphological features are discussed.

REFERENCES

Altemüller, H.–J., 1962. Beitrag zur mikromorphologischen Differenzierung von durchschlämmter Parabraunerde, Podsol-Braunerde und Humus-Podsol. In: Arbeiten aus dem Gebiet der Mikromorphologie des Bodems. Chemie Verlag, Weinheim/Bergstrasse.

Ameryckx, J., 1960. La Pédogénèse en Flandre sablonneuse. Pédologie, 10 (1) : 124–190.

Brewer, R. and Haldane, A.D., 1957. Preliminary experiments in the development of clay orientation in soils. Soil Sci., 84 : 301–309.

Brewer, R., 1960. Cutans: Their definition, recognition, and interpretation. J. Soil Sci., 11, 280–292

Deckers, J. en Baeyens, L., 1963. Polysequumprofielen van de Hoge Kempen. Pedologie, 13 (1) : 120–154.

De Coninck, F., 1954. Différences dans la morphologie des podzols suivant l'humidité (Campine Anversoise). Trans. Intern. Congr. Soil Sci., 5th, Leopoldville, 4 : 412–417.

De Coninck, F. et Laruelle, J., 1960. L'influence de la composition minéralogique sur la formation des podzols. Trans. Intern. Congr. Soil Sci., 7th, Madison, 1960, 5 (22) : 157–164.

Edelman, C.H., 1960. Podzols forestiers et podzols de bruyère. Pédologie, 10 (2) : 229–249.

Frei, E. and Cline, M.G., 1949. Profile studies of normal soils of New York. Micromorphological studies of the gray-brown podzolic-brown podzolic soil sequence. Soil Sci., 76 : 333–344.

Hoeksema, K.J., 1953. De natuurlijke homogenisatie van het bodemprofiel. Boor Spade, 6 : 24–30.

Jongerius, A., 1957. Morfologische onderzoekingen over de bodemstructuur. Mededel. Sticht. Bodemkartering, Bodemkundige Studies, 2 : 156 pp.
Kubiena, W.L., 1938. Micropedology. Collegiate Press, Ames, Iowa, 243 pp.
Soil Survey Staff U.S.D.A., 1960. Soil Classification, a Comprehensive System. 7th Approximation. Soil Conservation Service, U.S. Dept. Agr., Washington, D.C., 265 pp.
Soil Survey Staff U.S.D.A., 1951. Soil Survey Manual. U.S. Dept. Agr., Washington, D.C., 503 pp.

COLOURED ILLUSTRATIONS

Plate C1. Greenish B2t-hor. in very glauconitic material: abundant strongly birefringent plasma, filling up all the pore space. At the lower boundary concentric secondary clay accumulations.
Thin section 223, crossed nicols; x 45.
Plate C2. Formation of an iron concretion in non-glauconitic sand. All the pore space has been filled with non-birefringent dark brownish plasma (left part). Between the non-birefringent plasma and the skeleton grains, and in the right part, the primary birefringent cutan is visible.
Thin section 206, plain light; x 27.
Plate C3. Same view as Plate C2. Crossed nicols; x 27.
Plate C4. Formation of a concretion in soils with greenish plasma by alteration of the greenish birefringent plasma (right) to brownish non-birefringent plasma (left).
Thin section 238, plain light; x 45.
Plate C5. Same view as Plate C4. Crossed nicols; x 45.
Plate C6. B2ir of brown podzolic soil on glauconitic sand with thin non-birefringent plasma of organo-ferric compounds. Glauconitic grains in different stages of alteration: from greenish to brownish, showing a gradual lowering of the birefringence.
Thin section 232, plain light; x 45.
Plate C7. Same view as Plate C6. Crossed nicols; x 45.
Plate C8. Strongly birefringent secondary clay illuviation along a root hole.
Thin section 239, crossed nicols; x 45.

MEDITERRANEAN BROWN FOREST SOILS OF THE SIERRA MORENA (SPAIN), THEIR MICROMORPHOLOGY AND PETROGRAPHY[1]

G. PANEQUE and N. BELLINFANTE

Centro de Edafologia y Biologia Aplicada del Cuarto, Sevilla (Spain)[2]

INTRODUCTION

The Sierra Morena, situated in the southwest of Spain north of the Quadalquivir river, is a peneplain renewed by fluvial erosion. It shows a labyrinthic landscape, with a residual relief of lithological and structural origin. In the western region, where this study has been carried out, the highest altitudes do not reach 900 m.

The lithology corresponds to archaic and paleozoic formations. The former is represented by gneiss, mica-schist, crystalline limestones, and chloritic and micaceous slates, and the latter by clayey or micaceous shales and crystalline limestones. There are some plutonic intrusions, mainly formed by granites and syenite in this region, as well as some volcanic ones with diabase, diorite, and serpentine (Macpherson, 1879; Barras de Aragón, 1899; Mingarro Martin, 1961).

The climate of the western Sierra Morena is temperate, with moderate rainfall (750 mm) and generally cloudless skies. The summer is long, dry and very hot, while the other seasons are mild and somewhat humid. The rainfall is highly influenced by the Atlantic Ocean. Winds, generally proceeding from the southwest, move a large amount of clouds, which are held by the mountains of the Sierra Morena after crossing the Guadalquivir Valley. Thus the mountains receive the precipitation. Maximum temperatures rise from about 12°C in January to nearly 35°C at the beginning of August. Minimum temperatures rarely fall below 0°C.

The phytosociology varies with altitude, from the typical species of a Mediterranean sclerophylous forest to those of an Atlantic çentro-European caducifolius cold forest. Below 400 m altitude is a climax vegetation of *Oleo-Ceratonion*, defined by the sub-association *Asparageto-Rhamnetum cistetosum*. Between 400–700 m the climax vegetation is *Quercion rotundifoliae*, divided into three sub-associations: *Pireto-Quercetum suberetosum*, *Pireto-Quercetum ilicetosum* and *Pireto-Quercetum faginetosum* (Estudio Agrobiológico de la Provincia de Sevilla, 1962).

The most frequent soil in the Sierra Morena is the Mediterranean brown forest soil, soil climax of this region. This soil is developed mainly

---

[1] The here treated soils are - according to the U.S.D.A. Soil Classification, 7th approximation (1960) - ochrepts, probably in particular ustochrepts (Ed.).

[2] Director: Prof. Dr. F. Gonzâlez.

on granite and shale, as well as on red soil sediment ("rotlehm"), which represent the three most outstanding varieties of Sierra Morena Mediterranean brown forest soil. The similarity of their humus horizons allows one to classify them as analogous soils according to Pallmann. In particular cases due to topography, lithology, microclimate, vegetation, etc., there are also other soils more or less related to the above: xeroranker, ferritic "Braunerde", relict red soils, etc.

The present work[1] aims to complete and relate the micromorphological study to the main chemical, mechanical and mineralogical determinations of the Mediterranean brown forest soil (Meridional "Braunerde") of the Sierra Morena.

SOIL DESCRIPTION

Profile I

Situation: El Ronquillo, altitude 250 m, mountainous relief, medium slope. Parent material: sienitic granite. Vegetation: climax vegetation *Oleo-Ceratonion*; sub-association *Asparageto-Rhamnetum cistetosum*.

Soil Horizons

0–10 cm: A, very dark grey brown (10YR 3/2), sandy, crumbly to granular, medium organic matter content; good permeability and root penetration;
10–30 cm: (B), dark brown (10YR 4/3), clayey sand, crumbly; medium aeration, somewhat plastic;
30–50 cm: (B)/C, yellowish brown (10YR 5/6), sandy, compact; poor aeration, moist;
50 cm: C_1, decomposed granite.

Micromorphology

A-hor., skeleton of medium sized, somewhat altered, sand grains; the matrix contains brown flocculated iron hydroxides, brownish grey clay-humus complexes, and a little yellow dense plasma. Soil structure shows micro-aggregates and compact fabric, and not much evidence of good biological activity.

(B)-hor., skeleton of small and medium sized, altered grains, which are immersed in the yellow and dense clayey matrix. The latter contains finely flocculated iron hydroxides and possibly brown clay–humus complexes which determine the formation of some aggregates.

(B)/C-hor., skeleton of large grains which have the structure of altered granite; yellowish-brown clayey plasma interposed in the skeleton.

Profile II

Situation: Cala, altitude 530 m, relief flat. Parent material: sienitic granite. Vegetation: climax vegetation *Quercion-Rotundifoliae*; sub-association *Pireto-Quercetum suberetosum*.

[1] This work forms a part of the Doctoral thesis of Mr. Bellinfante, which will be submitted to the Faculty of Science of the University of Seville (Spain).

Soil Horizons
0–40 cm: A, brown (10YR 5/3), loamy sand, crumbly, low organic matter content; good permeability and root penetration;
40–80 cm: (B), yellowish brown (10YR 5/4), sandy, crumbly rather compact; good permeability and penetration;
80 cm: decomposed granite.

Micromorphology
A-hor., skeleton mainly formed by large sand grains, and a few practically unaltered plant remains. The matrix contains flocculated iron hydroxides, scarce non-uniform reddish-yellow and disorientated plasma, and clay–humus complexes. In thin sections soil shows reddish-brown colour and spongy microstructure.

(B)-hor., soil in thin section is similar to the A-hor. in colour, mineral skeleton and matrix. Some irregular microaggregates are also present amongst the sand grains.

C_1-hor., mainly mineral skeleton of medium sized grains; there is a reddish-brown plasma of fluid contexture, somewhat interspersed in the altered granitic.

Profile III

Situation: Cañaveral del León, altitude 580 m, mountainous relief. Parent material: violaceous shales. Vegetation: association climax *Mirteto-Quercetum rotundifolii.*

Soil Horizons
0–10 cm: A/(B), grey-brown (10YR 5/2), sandy-loam, crumbly; some shale fragments;
>10 cm: C_1, violaceous grey (5YR 5/1) altered shales.

Micromorphology
A/(B)-hor., skeleton of small shale fragments, very fine grains and scarce plant remains in good decomposition. The matrix contains clay minerals, various iron oxides and clay-humus complexes. Well-developed microspongy structure with pores and microaggregates of very different sizes; the microaggregates are linked to the firm mass by fine bridges.

C_1-hor., clay-shales rich in very fine opaque grains of iron oxides.

Profile IV

Situation: Cortelazor, altitude 700 m, mountainous relief. Parent material: mica-schists. Vegetation: association climax *Mirteto-Quercetum rotundifolii*.

Soil Horizons
0–10 cm: A, reddish grey (10YR 5/2), loamy sand, crumbly; medium content of organic matter in different degrees of decomposition;
10–20 cm: A/C_1, very pale brown (10YR 7/3), sandy loam, crumbly; rock fragments frequently present;
>20 cm: C_1, very altered mica-schists.

Micromorphology
A-hor., skeleton formed by some altered fine plants remains; few rock fragments and some sand grains. Matrix is formed of clay–humus complexes, flocculated iron hydroxides and dense plasma. The latter encloses the decomposed rocks with red iron oxides which are turning brown. There are also excrements of worms and other little animals which promote the formation of aggregates, cavities and pores.

Profile V

Situation: Cortelazor, altitude 490 m, mountainous relief. Parent material: slates. Vegetation: association climax *Mirteto-Quercetum rotundifolii.*

Soil Horizons
0–15 cm: A/(B), light yellowish-brown (10YR 6/4), sandy loam, crumbly, medium organic matter content; good permeability and root penetration;
15–45 cm: (B_1), brownish-yellow (10YR 6/6), loamy, polyedric crumbs; low permeabilit
45–70 cm: (B_1)/C, yellow (10YR 7/6), clayey loam, polyedric to laminar; slates at different stages of degradation;
>70 cm: C, chloritic slates.

Micromorphology
A/(B)-hor., skeleton formed of small fragments of very altered chloritic rock with red iron oxides and few plant remains. The clayey or silty matrix contains very homogenously distributed, flocculated, reddish-brown hydroxides, opaque crystallites of iron oxides, and clay–humus complexes, coloured by iron oxides. Irregular micro-structure of biogenic aggregates, "worked" through the slate fragments and linked with them in a spongy structure.

(B_1)/C-hor., very altered slates with a dense, slightly birefringent, red plasma, infiltrating and surrounding the rock mass. The plasma contains very fine crystallites of iron (and possibly manganese) oxides.

Profile VI

Situation: Cortelazor, altitude 650 m, mountainous relief. Parent material: chloritic slates. Vegetation: association climax *Mirteto-Quercetum rotundifolii.*

Soil Horizons
0–15 cm: A/(B), dark reddish-brown (5YR 3/3), loamy, crumbly; high organic matter content;
15–50 cm: (B_1), red (2,5YR 5/6), clayey loam, polyedric crumbs; medium permeability and root penetration;
>50 cm: (B_1)/C, very altered slates.

Micromorphology
A/(B)-hor., skeleton formed of small slate fragments, various mineral grains and decomposed plant remains. The silty clay matrix contains brown iron hydroxides and brown clay–humus complexes; when observed in reflected light it shows a red colour due to various iron oxides. Well-developed micro-aggregate structure.

Profile VII

Situation: El Pedroso, altitude 500 m, mountainous relief. Parent material: mica-schists. Vegetation: olive tree plantation (climax vegetation *Quercion-rotundifoliae*).

Soil Horizons

0–20 cm: A_p, reddish-yellow (7,5YR 7/6), sandy, crumbly; low organic matter content;

20–40 cm: (B_1), reddish-yellow (5YR 7/8), sandy loam, compact granular; low permeability and root penetration;

40–65 cm: $(B_1)/C$, reddish-yellow (5YR 6/6), loamy sand, granular; medium permeability.

Micromorphology

A_p-hor., skeleton formed of coarse sand grains and very few plant remains. Soil in thin sections is reddish-brown. It contains flocculated iron hydroxides, very scarce red dense plasma, and probably organo-mineral complexes. There are some small microaggregates between large grains of the skeleton.

(B_1)-hor., skeleton of altered small and large grains, immersed in a red matrix or enclosed by cutans. This is due to the retraction of the fundamental soil mass, which has a more fluid and birefringent contexture than the remaining soil. Dense microstructure with few aggregates, which are principally formed by retraction.

$(B_1)/C$-hor., skeleton of large grains, immersed in a matrix which mainly consists of a dense mass of yellowish-red plasma of fluid contexture; this is caused by infiltration into the altered rock (mica-schist). The matrix also contains minerals from the mica-schists in various stages of alteration. The microstructure is dense with few retraction and conduction voids.

ANALYTICAL DATA

The micromorphological study of the Sierra Morena brown forest soils can be completed and related to the chemical, mechanical and mineralogical analyses of these soils (Kundler, 1959; Brewer, 1960; Paneque, 1961; Lotti; 1962). Tables I, II and III contain the analytical results.

DISCUSSION

Profiles I and II of Mediterranean brown forest soil ("Braunerde") on granite are best known, and their analytical results more easily understood, by thin section study. Microscopic examination of (B)/C-horizons reveals the existence of a dense (probably braunlehm) plasma (Kubiena, 1953; Paneque, 1961), infiltrated in the altered granitic rock. Consequently the structure is compact, the aeration poor, and the texture more sandy in deeper layers. Horizon-A does not show large micromorphological differences. However it seems clear that horizon-(B) of profile I differs from that of profile II because of its higher clay content, compaction, and other facts confirming the genesis and development of profile I under shrubs on ondulate land, and the aluvial origin of profile II.

Profile I shows an almost neutral pH, a medium organic matter content, and a C/N quotient higher than 15. The exchange capacity and the base content increase with depth, reaching a maximum in horizon-(B) which is more humid, less sandy and more clayey (Tables I and II). The sand fraction contains a large proportion of heavy minerals, which consist mainly of hornblende, which increases towards the rock, a sienitic granite. Light minerals (quartz and alkaline feldspars) are also abundant, while medium minerals (plagioclases) are scarce (Table III). Kaolinite is common throughout the

TABLE I

Chemical analysis of Mediterranean brown forest soils of Sierra Morena

Profile	Horizon	Depth (cm)	pH	O.M. (%)	C (%)	N (%)	$\frac{C}{N}$	B.E.C. m.e.q./100	Ca^{2+} Exchangeable	Mg^{2+} Exchangeable
I	A	0–10	6.15	2.33	1.35	0.08	16.8	6.94	163.5	49.6
	(B)	10–30	6.05	0.83	0.48	0.04	12.0	15.90	247.7	108.1
	(B)/C	30–50	6.95	0.29	0.17	0.02	8.5	12.66	204.4	73.0
II	A	0–40	5.30	1.23	0.72	0.05	14.4	4.37	92.2	43.8
	(B)	40–80	5.85	0.55	0.32	0.03	10.6	9.87	120.2	2.1
	C_1	>80	6.07	0.22	0.13	0.02	6.5	-	370.3	140.2
III	A/(B)	0–10	5.35	2.47	1.43	0.10	14.3	6.58	91.4	71.5
IV	A	0–10	5.10	4.03	2.34	0.14	16.7	6.58	103.4	83.2
	A/C_1	>10	5.00	1.83	1.06	0.09	11.7	-	62.5	49.6
V	A/(B)	0–15	5.75	1.81	1.63	0.12	13.6	6.29	76.9	44.8
	(B_1)	15–45	5.60	0.84	0.49	0.07	7.0	13.19	79.4	64.2
	(B_1)/C	45–70	6.05	0.19	0.11	0.01	11.0	22.00	101.0	202.9
VI	A/(B)	0–15	6.00	11.72	6.80	0.52	13.0	8.77	132.2	61.3
	(B_1)	15–50	5.25	0.50	0.29	0.02	14.5	-	29.0	44.0
	(B_1)/C									
VII	A_p	0–20	5.95	0.43	0.25	0.03	8.3	6.65	120.2	48.2
	(B)	20–40	4.95	0.69	0.40	0.04	10.0	14.30	120.2	92.0
	(B)/C	40–65	4.85	0.40	0.23	0.02	11.5	14.15	96.2	173.7

TABLE II

Mechanical analysis of Mediterranean brown forest soils of Sierra Morena

Profile	Horizon	Depth (cm)	Moisture (%)	Coarse sand (%)	Fine sand (%)	Silt (%)	Clay (%)
I	A	0–10	2.32	37.18	35.52	4.90	22.50
	(B)	10–30	5.74	19.36	30.65	5.20	44.80
	(B)/C	30–50	3.11	47.22	28.16	8.10	16.40
II	A	0–40	1.58	34.96	33.40	16.80	13.60
	(B)	40–80	1.45	44.67	29.82	7.20	19.30
	C_1	>80	4.88	54.90	16.70	10.70	16.80
III	A/(B)	0–10	1.51	32.03	10.33	19.20	28.80
IV	A	0–10	1.95	37.85	23.89	8.60	29.90
	A/C_1	>10	2.20	31.55	13.28	12.80	42.10
V	A/(B)	0–15	1.60	37.54	12.30	22.60	27.10
	(B_1)	15–45	4.39	9.39	3.93	76.30	12.20
	(B_1)/C	45–70	4.68	16.51	15.87	38.30	29.20
VI	A/(B)	0–15	5.54	16.60	22.00	47.40	13.80
	(B_1)	15–50	1.70	9.20	29.30	32.60	28.40
	(B_1)/C						
VII	A_p	0–20	1.29	44.85	31.18	8.70	17.60
	(B)	20–40	2.75	26.63	11.05	48.04	13.20
	(B)/C	40–65	2.79	33.19	16.06	18.10	33.20

whole profile, illite is highest in horizon-A, and montmorillonite is included in (B) and (B)/C (Table III).

In profile II the pH and organic matter content are lower. The C/N quotient and base exchange capacity are smaller, and exchangeable Ca and Mg are lower in horizon-A and (B) (Table I). Coarse sand increases towards the rock, while the percentage of fine sand and silt is highest in the upper horizons (Table II), possibly because of fluvial deposition. In the sand fraction heavy minerals increase towards the rock, a sienitic granite with much hornblende. Horizons-A and (B) contain accessory minerals (staurolite and titanite) foreign to the granitic rock; light minerals are abundant and increase in the upper horizons. Illite and kaolinite are the most abundant clay minerals; in horizon-A there is some montmorillonite and in hor.-(B) chlorite (Table III).

The Sierra Morena Mediterranean brown forest soil, on shale and slate, is generally a shallow soil, permanently young because of erosion except in privileged sites. Profiles III and IV belong to this type of soil. Micromorphological study of them adds much information to the field description, and also explains the results of chemical and mineralogical analyses satisfactorily.

Microscopic observation of the rock from profile III in thin section shows that the slate is rich in fine grains of iron oxides (hematite and magnetite) which, together with clay minerals, determine the brown reddish grey colour of the original material. The well humified organic matter in horizon A forms organo-mineral complexes and leads to the formation of

TABLE III

Mineralogical analysis of Mediterranean brown forest soils of Sierra Morena

			Sand minerals (0.4–0.04 mm)				
Profile	Horizon	Depth (cm)	Light minerals D = 2.68	Medium minerals D = 2.68–2.9	Heavy minerals D = 2.9	Relative frequency of main heavy minerals[1]	
I	A	0–10	48.25	2.75	49.00	+++ Hrn	++ He
	(B)	10–30	49.34	6.10	44.56	++++ Hrn	+ He
	(B)/C	30–50	42.40	12.70	44.90	+++++ Hrn	
II	A	0–40	97.37	-	2.63	++ Hrn	+ St
	(B)	40–80	86.90	4.30	8.80	+++ Hrn	+ St
	C_1	>80	78.25	7.05	14.70	+++++ Hrn	
III	A/(B)	0–10	30.39	69.50	0.11	++ Hrn	+++ He
IV	A	0–10	29.35	28.05	42.40	+++++ Cl	
V	A/(B)	0–15	43.10	48.60	8.30	++++ Cl	+ Ep
	(B_1)	15–45	78.49	19.40	2.11	+ Cl	+++ Ep
	$(B_1)/C$	45–70	94.15	4.30	1.55		++++ Ep
VI	A/(B)	0–15	42.30	41.00	16.70	++++ Cl	
	(B_1)	15–50	72.60	24.30	3.10	+++ Cl	+ Ep
	$(B_1)/C$	>50	85.06	14.40	0.54	+ Cl	++++ Ep
VII	A	0–20	75.36	11.45	2.92	++++ Cl	+ Ep
	(B)	20–40	90.99	8.78	0.23	+++ Cl	+ Ep
	(B)/C	40–65	90.91	8.90	0.19	+ Cl	++ Ep

[1] +++++ Predominant = 75–100%
++++ Dominant = 50– 75%
+++ Abundant = 25– 50%
++ Common = 10– 25%
\+ Present = 0– 10%

[2] For the identification the following methods have been used: DTA, X ray, (%)K_2O E.G. retention and B.E.C.

a well-developed spongy micro-structure. However a reddish colour, from the rock iron oxides, can be observed in thin section.

The soil has a low pH, medium organic matter content, a C/N quotient near 15, a low base exchange capacity and base content (Table I), and a loamy texture (Table II). The sand fraction contains few light and heavy minerals, and a large amount of medium ones. Heavy minerals consist mainly of hematite, hornblende, chlorite, zircon, and magnetite; clay minerals are fundamentally illite and chlorite (Table III).

Profile IV is a ranker or young Mediterranean brown forest soil on mica-schist. It is from a higher (700 m) and more hilly land than that of profile III. Microscopic observation of the soil in thin section shows a rathe complex organization, as described above. As in profile III the pH is low, the organic matter content high, the C/N quotient near 17, the base exchange

Relative frequency of main heavy minerals[1]			Clay minerals[2] Relative frequency[1]				Profile
			+++ Ill	+++ K			I
			+ Ill	+++ K		++ Mont	
				+++ K		+++ Mont	
+++ Ti			+++ Ill	+++ K		+ Mont	II
++ Ti			+++ Ill	+++ K	+ Cl		
++ Cl	+ Zr	+ Mg	+++ Ill		+++ Cl		III
			+++ Ill	++ K	+ Cl		IV
+++ He	++ Li		+++ Ill	+ K	+++ Cl		V
+++ He	++ Li		+ Ill	+++ K	+ Cl		
+++ He	++ Li		+ Ill	++++ K			
+++ He	++ Li		+++ Ill	+ K	+++ Cl		VI
+++ He	++ Li		+ Ill	++++ K			
+++ He	++ Li		+ Ill	++++ K			
+ He	+ Zr	+++ Mg	+++ Ill	+++ K			VII
+ He		++ Mg	++ Ill	++++ K			
	+ Zr		++ Ill	++++ K			

Hrn = Hornblende
He = Hematite
St = Staurolite
Ti = Titanite
Cl = Chlorite

Zr = Zircon
Mg = Magnetite
Li = Limonite
Ep = Epidote

Ill = Illite
K = Kaolinite
Mont = Montmorillonite
Cl = Chlorite

capacity low, the base content low in horizon-A (Table I), and the texture is silty-sandy (Table II).

The more permanent sand minerals are heavy ones, almost all being chlorites. Light and medium minerals occur in equal quantities (Table III). Clay minerals are illite, kaolinite and chlorite (Table III).

Another variety of Mediterranean brown forest soil in the Sierra Morena shows the characteristics of being developed on "red loam (Rotlehm) sediment", principally shale. Profiles V, VI and VII represent those soils which normally are located in areas higher than 500 m, with a humid microclimate. Under these conditions the "browning" of the red soil can be important. The intensity of this process allows the establishment of the following genetical series; red loam sediment (paleosoil)→"Ranker" or browned red loam→ Mediterranean brown forest soil on shale.

In these profiles there is a contrast between the upper reddish-brown horizon-A, and the underlying eroded red or reddish yellow paleosoil. In thin section the distinction is very clear; especially obvious are the differences between the skeletons, their iron oxide and hydroxide contents (Kubiena, 1956, 1962), and their contents of brown clay–humus complexes. The microstructure of brown and red horizons also differs greatly; micro-aggregate formation is well-developed in the A-horizon, while in the red paleosoil birefringent plasma and compact structure occur.

The analytical data corresponding to these soils are very interesting. Red horizons are the most acid and have the lowest organic matter content, and the C/N quotient in general is near to ten. The base exchange capacity increases with depth (Table I), together with the clay content of these generally loamy soils, and coarse sand increases towards the soil surface (Table II).

The mineralogical composition of the sand and clay is very similar in the three profiles (Table III). An increase in the proportion of heavy minerals can be observed in the A-horizon due to the abundance of chlorite, which shows a certain "antipathy", in the profile, for epidote. Iron minerals (hematite, magnetite and limonite) are relatively abundant. There is a decrease in the content of medium minerals (plagioclases) in the red horizons, and an increase in light minerals (quartz and alkaline feldspars).

Clay minerals are fundamentally illite, kaolinite and chlorite; illite + chlorite increase in brown horizons, while kaolinite increases in red ones (Table III).

SUMMARY AND CONCLUSIONS

The micromorphological and petrographical study carried out on the three main varieties of Mediterranean brown forest soil of the Sierra Morena (Spain) can be summarized in the following conclusions:

(1) The Mediterranean brown forest soil (meridional "Braunerde") on granite (sienitic) shows an abundant skeleton, mainly formed by alkaline feldspars, quartz and hornblende. They exist in various proportions, according to the degree of substitution of micas by amphibol in the rock.

The matrix contains flocculated iron hydroxides in horizons-A and (B), organo-mineral complexes and clay. The clay minerals are illite (more abundant in horizon-A), kaolinite (uniformly distributed through the profile), montmorillonite (associated with a compact structure and almost neutral reaction of the medium), and a little chlorite, possibly contained in the rock or proceeding from the transformation of micas.

The micro-structure of horizon-A shows micro-aggregates, and is very different from that of horizons-(B)/C and C. The latter shows birefringent dense plasma, infiltrated through the altered rock or from the structure of horizon-(B), which is more or less porphiritic.

(2) The Mediterranean brown forest soil on shale and slate has a mineral skeleton largely formed by very small fragments of altered rock. These microfragments play an important role in the soil structure, together with the organo-mineral complexes of the matrix. The matrix also contains finely flocculated iron hydroxides of actual formation, and fine-grained

crystallized iron oxides either contained in the sedimentary rock (shale) or proceeding from a red paleosoil ("Rotlehm"), which can be traced in thin section. Clay minerals, illite and chlorite, are mainly a heritage from the rock; the kaolinite proceeds from the red soil. The microstructure is formed by irregular, well-developed aggregates linked to the skeleton.

(3) Finally, the Mediterranean brown forest soil on red soil on shales shows a complex profile. The skeleton of A-horizon is similar to that of actual Mediterranean brown forest soil on shale, but has a larger content of rock altered by an old pedological process. The matrix contains flocculated iron hydroxides, reddish-brown organo-mineral complexes, and crystallized iron oxides proceeding from the old red soil and the rock. Clay minerals, illite and chlorite, are of secondary formation, as well as inherited from the original material.

The micromorphology of the red sediment shows characteristics typical of a relict red loam ("Rotlehm"), more or less red in colour. The mineral skeleton is formed by fine grains and shale fragments; red plasma of fluid structure with rather fine birefringent crystallites; peptized iron hydroxides and iron oxides (goethite and hematite).

The sand fraction is mainly formed by light minerals (alkaline feldspars, quartz, etc.), a few medium minerals (plagioclases, etc.), and scarce heavy minerals (epidote, hematite, limonite, etc.). The clay fraction contains mainly kaolinite and a little illite.

REFERENCES

Barras de Aragón, F., 1899. Apuntes para una descripción geológica-mineralógica de la Provincia de Sevilla. Alonso e Hijos, Palencia, 303 pp.

Brewer, R., 1960. The petrographic approach to the study of soils. Trans. Intern. Congr. Soil Sci., 7th, Madison, 1 : 1–13.

Estudio Agrobiológico de la Provincia de Sevilla, 1962. Consejo Super. Invest. Cient., Centro de Edafol. Biol., 137pp.

Kubiena, W., 1953. Bestimmungsbuch und Systematik der Böden Europas. Enke, Stuttgart, 392pp.

Kubiena, W., 1956. Rubefication and laterisation (their morphological distinction). Trans. Intern. Congr. Soil Sci., 6th, Paris, 1956. Comm. 5 : 247–250.

Kubiena, W., 1962. Taxonomic importance of the type and formation in tropic soils. Z. Pflanzenernähr., Düng., Bodenk., 98 : 205–213.

Kundler, P., 1959. The sod podsols and grey forest soils of Central Russia as compared with the leached soils of Western Europe. Z. Pflanzenernähr., Düng., Bodenk., 86 : 16–36.

Lotti, G., 1962. L'impiego delle sezioni sottili nello studio della microestructura del terreno. Agrochimica, 6 (3) : 265–274.

Macpherson, J., 1879. Estudio geológico–petrográfico del N. de la Provincia de Sevilla. Comm. Mapa Geol. España, Madrid, 6 : 97–268.

Mingarro Martin, F., 1961. Estudio del Carbonífero del Norte de la Provincia de Sevilla. Tesis Doctoral, Universidad de Madrid, 125pp.

Paneque, G., 1961. Estudio micromorfológico de algunos sedimentos béticos. En: Reunión de Sedimentologia, 2. Consejo Super. Invest. Cient., 9–21.

MICROMORPHOLOGY AND CHEMISM OF HUMUS-CLAY NEW-FORMATIONS IN GREY FOREST SOILS

E.I. PARFENOVA, E.F. MOCHALOVA and N.A. TITOVA

Dokuchaev Soil Institute, Moscow (U.S.S.R.)

The grey forest soils[1] are developing under broad-leaved forests of the forest-steppe zone of the U.S.S.R. They form a transitional type between chernozems and podzolic soils. These grey forest soils are characterized by various features similar to the soils mentioned above as well as by definitely different ones. In addition they have peculiar morphological and chemical characteristics (Rubilin et al., 1964). The mobile humus-clay new formations, which are the object of our studies, reflect their specificity most clearly.

In the profiles of the loamy grey forest soils we find the following clearly outlined horizons (horizon designation after Nocin and Petrov, 1959):

0–45 cm: A, humus horizon, usually showing three sub-horizons:
0–20 cm: A_1, grey or dark-grey, loose, crumb structure, containing large amount of thin roots.
20–35 cm: A_1 A_2, darker brownish grey, fine granular structure; usually defined as second humus horizon.
34–45 cm: A_2 B, heterogeneously coloured greyish brown with yellow patches and grey tongues, bleached by podzolization; the nutty structural units are covered with a whitish powder.
45–150 cm: B, illuvial horizon; in its upper part (45–70 cm) it is dark brown, prismatic to nutty, with dark cutans and whitish bleached powdering on the surface of the structural units; below 70 cm the colour fades out and changes to brownish-grey and brownish-yellow; prismatic units become larger but are found to be less stable.
>150 cm: C, the horizon consists of yellow structureless loam with or without carbonates.

From the zone of transition, between humus horizon and illuvial horizon, we find clearly outlined longitudinal nearly parallel cracks penetrating to 2–2.5 m below surface. The walls of the cracks are covered with black brown cutans (in thickness varying from 0.5–2 mm). In hollows, when the cracks cross each other, the cutan becomes thicker (up to 5 mm). The surface of the cutan shows the imprints of a branching network of hair-roots 1 mm in diameter. These roots have penetrated into the cracks where they have been pressed and decomposed in situ; sometimes we find slightly decomposed or still living hair-roots here. The fine roots do not penetrate

[1] In all probability the here treated soils belong to the udalfs, according to the 7th approximation of the U.S.D.A. Soil Classification (1960). (Ed.)

the dense mass of the soil. They favour the cracks in order to avoid obstacles and to find nutritive matter (in the previously accumulated humus) and probably better humidity conditions (during humid periods the water flows along the walls of the cracks).

In the dense part of the deposit we find new-formations of another kind, namely "clayified roots" - the clay-humus pseudomorphs for plant roots (Fig.1). Many profiles show layers of light coloured loam at a depth of 60–70 cm. They have a distinct root system of black-brown colour with a well-preserved main root (diameter 1–2 cm) and branching thin roots which are completely transformed into a clay substance.

Fig.1. Humus–clay pseudomorph for root. A little less than natural size.

The root system seldom preserves its natural appearance and usually only shows its fragments. At points of root penetration the surface of structural units is covered with a thin humic cutan. These pseudomorphs are outlined particularly clearly in soils having a high level (3–5 m below surface) of ground water rich in calcium bicarbonate.

In a profile out on a slightly sloping plain (ground water level below 5 m), under a mixed forest (oak, birch, maple, elm) with platyphylous grass the following samples were taken: (1) Bulk soil samples of horizons, (2) dark humus-clay cutans, separated from the wall of a deep vertical crack and (3) clayifield root at a depth of 110 cm, stretching horizontally. The samples were studied in thin sections. Their bulk chemical composition and fractional humus content were determined.

MICROMORPHOLOGY

Horizon-A_1 consists of micro-aggregates (1–2mm). They are composed of angular particles of the primary minerals quartz, feldspars, hornblende, mica, epidote and glauconite and rarely zircon, garnet, tourmaline and titanite (predominantly 5–10 μ). Among them we see loosely distributed plant residues in various stages of decomposition, phytolithes[1] and micro-ortsteins. By comparing the fabric of various states of organic residue one may obtain successive data on their decomposition; first one observes still fresh residues, then transitional forms with a partly decomposed cell structure, and then a yellow, brown and black-brown colloformic[2] substance.

In fresh tissues the cells are filled with a light yellow, almost colourless amorphous substance. Often the walls are strongly anisotropic, showing light-grey or golden colours between crossed nicols.

With progressing decomposition the walls of the cells lose birefringence and the cell content changes its colour to brownish. In many cases this transparant brown substance is weakly or strongly birefringent, showing similarities with the "Braunlehmteilplasma" after Kubiena (1956). At the next stage the walls of the cells decompose, but the brown colloidal substance retains the square shape of the cells. Further observations show that the outlines of the cells become diffused and brown, shapeless colloidal concentrations are formed; they contain tiny fragments of still intact tissues. Often one can see how this mass contracts, forming coagels which show birefringence because of the orientation of the particles.

Besides this brownish-yellow substance, black and brown condensates of humus are seen (diameter 10 μ) forming micro-aggregates of the second and third order.

In the following horizon, A_1 A_2, the smallest amount of organic remains can be found. Black-brown humified micro-aggregates are also fewer. Here the humus is peptized and becomes mobile. We observe independent concentrations of dispersed ("punctual") black particles in streaks which are accumulated inside the pores, and are more or less noticable in the yellow strong birefringent clay material. This is a characteristic micromorphological of illuviation. This substance forms cutans on the walls of micropores and microcracks, or even fills them with conchoidal layers. This fact determines the dark subhorizon A_1 A_2 as humus-clay illuvial. In podzolized microspots of this horizon fine clay substance is almost absent; primary minerals without cutans can be seen with particles of peptized humus amidst them.

In the fabric of transitional horizon A_2 B the following micro-spots heterogeneous coincide: (a) Dark brown, enriched with mostly peptized humus, much less in aggregates; (b) brownish-yellow, with deposits of colloidal anisotropic clay in microcavities, with or without dark humus; (c) light greay micro-spots, bleached by podzolization.

In horizon-B the deposits of colloformic anisotropic clay intermixed with dispersed particles of humus are especially abundant. The amount of the humus decreases with depth and in the lower part of the horizon it is almost completely absent. The accumulation of dark humus and clay takes place in this horizon, along the cracks of the walls and the surfaces of pris-

[1] Phytolithes: siliceous particles of vegetable origin. Cf. for instance: F. Smithson, 1956. Silica particles in some British soils. J. Soil Sci., pp. 122–129; F. Smithson, 1958. Grass opal in British soils. J. Soil Sci., pp. 148–155. (Ed.)

[2] The author uses the word "colloformic" instead of "colloidal", because the first means only that the substance has a colloidal appearance. (Ed.)

matic structural units.

In thin sections, prepared of cutans taken from the walls of the cracks, we see that the cutan has a flow fabric (Fig.2) and consists of birefringent homogeneous colloformic material; sometimes it has abundant though uneven concentrations of black particles of peptized humus. This fact permits us to state that a mechanical mixture of humus and clay has taken place here without any evidence of chemical combination. In these cutans the remains of strongly decomposed tissues are noticed. In some micro-spots a yellow amorphous substance is present. Macromorphology and micromorphology of the cutans show a mobile state of clay and dark humus, both migrating from the top to the bottom of the profile. Moreover we cannot exclude the possibility of the formation of humus (and possibly clay minerals) in situ from decomposing roots.

The fabric of clayifield roots is very peculiar and repeats anatomical particularities of the tissue of living root. In a living root we can see parallel arrangements of woody fibres with vascular bundles separating them; in a cross-cut the latter show circular openings. In clayifield roots we see in longitudinal section parallel layers. The layers consist of two kinds of material; some layers are filled with a yellow anisotropic colloformic substance, while in others we notice an admixture of angular particles (~ 5 μ) of primary minerals (Fig.3). In cross-cuts we cannot see the distinct texture of the material (possibly this is due to the fact that these roots were located horizontally and were deformed by the pressure of the soil above), but we see a round hole (diameter 200 μ) similar to the cross-cuts of vessels in living root in some places (Fig.4).

THE ORIGIN OF HUMUS-CLAY NEW-FORMATIONS

The observations show that we deal with a more complicated process than a simple infillment of the root channel by a mixture of material mechanically transported from above. Micromorphological studies of these new-formations permit us to state that here the impregnation of decayed tissues of the root by a fine colloidal solution, still retaining its anatomical structure, took place. Particles of primary minerals found their home in cavities of vascular bundles, moved by the flow of the solution during this process. It is most probable that the products of decomposition in situ of the root tissues participate in the formation of this colloformic substance. This is proved by the marked enrichment of layers where primary minerals are absent, by dark fine humus and by the composition of the humus itself. We shall discuss this later on. Such formations, apparently, were described by Brewer and Sleeman (1963) as striotubules with lamellary texture.

The fact that intact clayifield roots are met comparatively rarely (usually we only find fragments of them) leads to the conclusion that they cannot survive in these soils for a long time. As they are rich in organic substances they are destroyed by micro-organisms. We assume that destruction takes place swiftly, for otherwise we would be able to find an enormous amount of these pseudomorphs as a result of the long period of forest growth. Particularities of the fabric of pseudomorphs and the views stated above, point to the fact that these new-formations are recent formations.

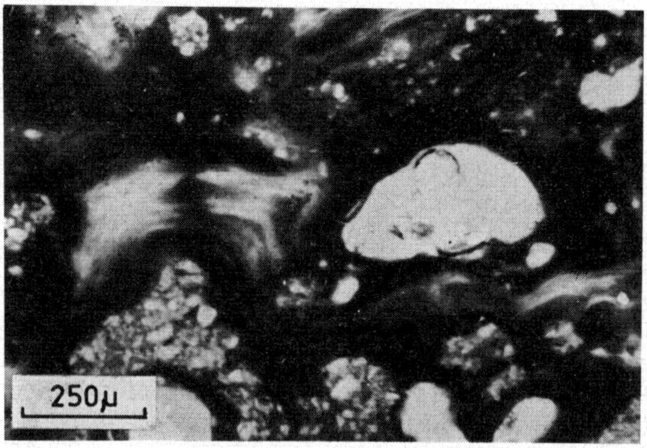

Fig.2. Flow fabric of a cutan on the wall of a crack.

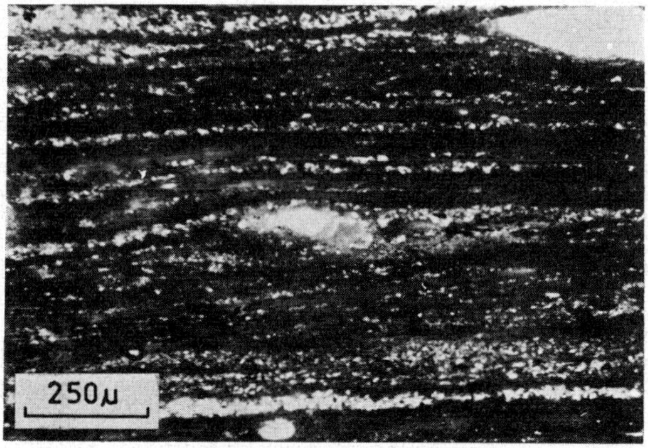

Fig.3. Clayifield root; longitudal cut.

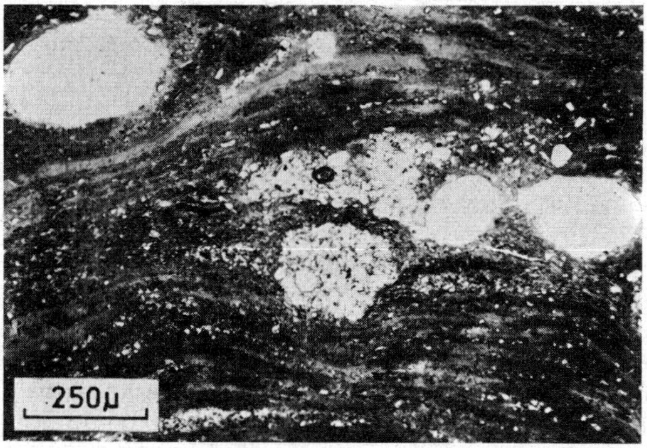

Fig.4. Clayifield root; cross cut.

The humus–clay cutans on the walls of the cracks are also recent new-formations, developed under forest. This is proved by the gradual disappearance of the cutans in cultivated soils. In but recently cultivated soils the cutans are still found; however they become lighter in colour (light grey) and much thinner. In soils which have been cultivated for longer time they are almost absent. By then micromorphological features of A_1-horizon resemble those of chernozem soils. Only the specific fabric of the illuvial horizon proves that the forest has grown on these soils. In the opposite case, when forests are being grown on chernozem soils which have no illuvial features, these soils shown humus–clay streaks under a microscope after a time.

We consider these conclusions to be rather important for they demonstrate that the process of migration of humus–clay substance from top horizon to lower parts, a process of lessivage, is a recent one. The question posed is what kind of substance is the mobile substance? Does it appear as a result of natural removal of fine colloidal parts of the soil-forming deposit itself? Or is it synthesized in the top horizon and then moved downward? What is the mechanism or the phenomena in this case?

The solution of this problem first involves overcoming great difficulties. In the case of the development of soils on continental deposits we have no direct data for deciding if the present clay substance is a new-formation or may be inherited from the deposit. Clay minerals distributed on the surface of the earth are not very different. We have thousands of names for primary minerals, while for clay minerals we have less than several dozens names all together, and even not more than three main groups: hydromicas, montmorillonites, kaolinites. The predominance of a group can only be seen in broad latitudinal climate zones. In the moderate zone, to which the soils taken for our studies belong, we found that minerals of hydromica and montmorillonite types predominate (mixed-layered di–trioctahedral minerals with non-expanding and expanding layers in the lattice).

The bulk chemical composition of fine fractions is very close to small variations in zonal soil types in all horizons. Thermogramms and X-ray data also show a rather close resemblence (Dobrovolsky, 1962; Kundler, 1959; Parfenova and Yarilova, 1960; Yarilova and Parfenova, 1960). We found several suggestions in other publications pointing out admixtures of other minerals - vermiculite, chlorite, kaolinite, etc. But their amount is small and we have not found any clear differentiations of them along the profiles.

In dealing with the quantitative distribution of clay substance along the profile, we find that in grey soils developed under forests data of mechanical analyses constantly show higher presence of fraction $<1\ \mu$ in the top part of the A_1, as compared with the underlying layer; a further accumulation of this fraction is observed in the illuvial horizon. Especially high accumulations of small particles take place in cutans and pseudomorphs of roots. Supercentrifugal fractioning of cutans and of the soil they are covering, was done by R.H. Aidinyan (personal communication). His experiments revealed that cutans consists of fractions $<0.05\ \mu$ to 70-80%. In soil it amounts to 25%.

Table I shows the bulk composition of the upper part of the horizon A_1 (0–5 cm), the parent material (110–115 cm), the cutans and the pseudomorphs as well as the composition of their colloidal fractions ($<0.2\ \mu$).

TABLE I

Composition of A_1 and B-horizon, cutan and clayified root of forest soil and of their colloidal fraction <0.2 μ (% of the ignited sample)

Sample	Samples in total								
	loss on ignition	SiO_2	Al_2O_3	Fe_2O_3	P_2O_5	CaO	MgO	K_2O	Na_2O
A_1,0–5 cm	16.95	76.69	12.26	3.88	-	1.85	1.38	2.38	0.83
B,110–115 cm	5.71	74.00	13.03	5.25	-	1.63	1.92	2.62	0.99
Cutan	14.24	63.41	20.41	7.49	-	2.23	2.53	2.50	0.47
Clayified root	13.92	61.52	20.77	9.63	-	2.08	2.53	2.49	0.46

Sample	Colloidal fraction <0.2 μ								
	loss on ignition	SiO_2	Al_2O_3	Fe_2O_3	P_2O_5	CaO	MgO	K_2O	Na_2O
A_1,0–5 cm	23.61	53.24	28.70	9.85	0.04	2.77	2.82	2.78	0.19
B,110–115 cm	9.42	54.49	25.48	12.35	0.01	2.51	3.26	2.64	0.13
Cutan	19.42	54.16	29.92	10.59	0.15	2.57	3.32	2.64	0.19
Clayified root	18.10	54.73	26.96	10.37	0.15	2.09	3.02	2.90	0.13

In the samples analysed as a whole, the difference in the composition of the soil itself and the separated new-formations can be clearly seen. There are high amounts of SiO_2 and Na_2O in the soil, mainly on account of stable quartz and albite, and high quantities of R_2O_3, CaO and MgO in new-formations, which are connected with the accumulation of fine fractions and their exchangeable capacity.

In colloidal fractions these differences are wiped out and we see that the compositions of all the samples are almost indentical, excluding obvious accumulations of phosphorus in the films and the clayified roots and certain accumulation of Fe_2O_3 in deeper horizons. We can mention the tendency for the Al_2O_3 content to increase in the upper layer of horizon-A and in the new-formations, as compared with the deposit. Comparing the data of losses on ignition at the top of Table I and the lower ones, we can clearly see the connection between the organic substance and the colloidal fractions.

The composition of the organic substance in the studied samples is shown in Table II.

TABLE II

Composition of humus of A_1 and B- horizon, cutan and clayified root

Sample	Initial sample (%C)	Organic matter, extracted with 0.1 N H_2SO_4[1] (%C)	Humic acids[1] (%C)	Fulvic acids[1] (%C)	$\frac{C_{h.ac}}{C_{f.ac}}$	Humic acids extracted with 0.1 N NaOH without decalcification[1]	Residue[1] (%C)	$\frac{E_4}{E_6}$
Soil 0–5 cm	6.01	0.21 3.5	1.29 21.5	1.20 20.0	1.08	1.29 100	3.52 58.6	4.72
Soil 110–115 cm	0.22	0.06 27.3	0.02 9.1	0.07 31.8	0.29	-	0.13 59.1	-
Cutan	1.40	0.12 8.6	0.25 17.9	0.29 20.7	0.86	traces	0.86 61.4	4.05
Clayified root	1.37	0.10 7.3	0.34 24.8	0.25 18.2	1.36	traces	0.78 56.9	4.05

[1] Top row of figures indicates the carbon content of the different substances as percentage of the soil weight; the lower line gives the figures as percentage of organic matter.

As the micromorphological data show, humus of the top part of horizon A_1 consists of plant residue in different degrees of decomposition, up to shapeless yellow, brownish-grey or black-brown formations of a colloidal nature (amorphous and weakly anisotropic) accumulating in loose micro-aggregates. Chemical analysis shows rich amounts of organic matter in this layer. Humus contains almost equal amounts of humic acids and fulvic acids. The humic acids are rather unstable and pass completely, without preliminary decalcification, into 0.1 *N* NaOH extract. This composition of the humus explains its capacity for migration; it flows along the cracks, together with the clay and is deposited on the crack walls as cutans. The cutans contain more fulvic acids than the upper soil layer, but we cannot detect fulvic acids under the microscope, since they are camouflaged by the brownish yellow colouring of the clay substance.

Dark dispersed and unevenly distributed humus in cutans consist of humic acids of more complicated composition. They have a more clearly developed aromatic structure, as may be judged by the fact that only slight traces of these acids pass without decalcification into 0.1 *N* NaOH extract and also by the degree of optical density. The different composition of the humic acids in this case can probably be explained by the fact that hair-roots decaying in situ take part in the formation of humus. This idea is proved by the composition of humus of root pseudomorphs, where a high $C_{h.ac}/C_{f.ac}$ ratio is developed on account of the rising amount of humic acids completely bound with calcium.

Humus does not penetrate into soil-forming rock, except for cracks and clayified roots. The sample of the lower part has been taken at the same depth where the cutans and clayified root were detected, it contains a very negligible amount of humus in the form of extremely mobile fulvic acid, extracted by 0.1 *N* H_2SO_4. The research carried out, and the data found in publications, allow us to state now certain views on clay formation.

As far back as 1945 Polynov (1956) based his hypothesis of the biogenic origin of synthetic clays (of montmorillonite and illite types) on his studies of primary stages of soil formations on volcanic rocks. He pointed out that the most probable original products are organo-mineral compounds of aluminium and wadozic silica, liberated by the process of humification and mineralization of organic residues. Since that time science has acquired various experimental data which support Polynov's idea.

Kononova's monograph (1961) reflects the contemporary situation in studies of soil organic matter. It shows that the main forms of humus compositions in soil are the metal-organic compounds-complex, chelates and adsorption forms, bound with clay ("argyllites"). The latter may form more than 50% of the general content of humus in heavy soils. Kononova points out that complex compositions of the chelatic type (ferro-humus, complexed compositions of silica and aluminium) were obtained in the laboratory experiments. We have seen dry remnants of Fe-organic complex, studied by Titova (1962), under the microscope. Amidst amorphous particles of weakly anisotropic substance, they reveal the material similar to clay substances which is seen among decaying organic residues in thin sections.

Other publications describe numerous experiments showing the synthesis of clay minerals from ionic solutions at normal temperature and pressure; some point out the fact that the presence of oxalic acid provides beneficial conditions for artificial synthesis (Siffert and Wey, 1961). The sam conditions are created in the top part of the soil horizon because of the soluti of crystals of calcium oxalates, which, as is generally known, is present in abundance in plant residues. Considering the fact that organo-mineral (Al,Fe, Si) complexes are present in the top soil horizon, and that possibilities for the synthesis of clay minerals out of elements exist in natural solutions, we may take for granted that interaction can easily take place between them with the formation of absorptive compounds of biogenic character ("argyllite" type). It is probable that the incredibly homogeneous transparent anisotropic clay substance revealed under the microscope in recent new-formations (cutans, clayified roots, streaks in microcavities) is of a similar type of formation.

In various bio-climatic zones the conditions for synthesis may become difficult or more or less favourable, depending on relations between the content in organic residues of carbon in one case and ash bases and nitrogen in the other (Ponomareva, 1956; Duchafour, 1957). It is difficult in sand podzolic soils of high latitudes because of the prevalence of fulvic acids in humus composition. In this case they determine the high migrating capacity of almost all the elements, except Al and Fe, which are part of the content of Al and Fe-fulvates. That is the reason why microscopic observation of these soils either do not show birefringent clays or only show very slight traces.

The described form of synthesis of organic clays does not exclude the process of gradual transformation of primary minerals into clay minerals after the schemes, for example, of Jackson and Sherman (1953) and De Vore (1960). De Vore also took synthesis of clay minerals into consideration. In his interpretation this mechanism is described as the extrication of fragment of colloidal size (tetrahedrons, their chains and hexagonal rings) from crystalline lattice frame of primary minerals, followed by their resynthesis to a structure of layer type.

However, this process takes place mainly in subsoil horizons, where synthesis occurs without direct interference from a vegetative cover (but with the participation of micro-organisms and percolating organic solutions). Micromorphological data confirm this peculiarity: clay minerals are found in the form of tiny scales of intangled-fibrous fabric, uniformly distributed inside the soil material. Clay minerals are also found in the form of clay pseudomorphs for primary minerals which keep the original crystallized shape.

The overall distribution of vegetation, the sorting out of chemical elements of primary minerals in the process of mineral nutrition, and the synthesized way of clay formation, accepted by majority of scientists as a basic one (Antipov-Karataev and Kader, 1956; Keller, 1957; Lerz and Borchert, 1962,and many others) probably determine the existence of so few clay mineral species on the surface of the earth. Predominance of this or that group of clay minerals is connected with the various migrational capacity of chemical elements, the participants in the synthesis, and the stability of their compounds under given bio-climatic conditions.

SUMMARY

The article deals with the study of grey forest soils of the European part of the U.S.S.R. Their mobile humus–clay new-formations, effecting the specificity of the development of the soils of this type, are the main subject of the studies described.

New-formations (the cutans on the walls of the cracks, streaks in the illuvial horizon, humic–clay pseudomorphs for the plant roots) are studied by micromorphological method in comparison with the chemical bulk and humus composition of the soils. The authors came to the conclusion that new-formations are a product of recent processes in soil formation. They assume that it is an adsorption compound of organo-mineral complexes (chelates) and clay substance synthesized from ionic solutions. Original material for this compound is found in decomposing organic residue and ions of soil solution. The data give some additional experimental explanation to the problem of the biogenic character of synthesized soil clays - the hypothesis which was originally stated by the Soviet scientist Polynov.

REFERENCES

Antipov-Karataev, I.N. and Kader, G.M., 1956. The question of the genesis of clay minerals during weathering of primary minerals. Tr. Pochv. Inst. Akad. Nauk. S.S.S.R. 51 : 98–157. (Russ.)

Brewer, R. and Sleeman, J.R., 1963. Pedotubules: their definition, classification, and interpretation. J. Soil Sci., 14 : 156–166.

De Vore, G.W., 1960. The surface chemistry of feldspars as an influence on their decomposition products. Clays Clay Minerals, Proc. Natl. Conf. Clays Clay Minerals, 6 (1959) : 26–41.

Dobrovolsky, V.V., 1962. Geochemical peculiarities and genesis of fine dispersed masses of Quaternary formations. Dokl. Akad. Nauk. S.S.S.R., 147 (2) : 466–468. (Russ.)

Duchaufour, P., 1957. Sur le rôle de la matière organique dans les phénomènes de lessivages et de podsolisation. Compt. Rend., 245 : 1154–1157.

Jackson, M.L. and Sherman, G.D., 1953. Chemical weathering of minerals in soils. Advan. Agron., 5 : 219–318.

Keller, W.D., 1957. The principles of chemical weathering. Columbia, Mo., 111 pp.

Kononova, M.M., 1961. Soil organic matter - its nature, its role in soil formation and in soil fertility. Pergamon, Oxford, 450 pp.

Kubiena, W.L., 1956. Zur Mikromorphologie, Systematik und Entwicklung der rezenten und fossilen Lössböden. Eiszeitalter Gegenwart, 7 : 102–112.

Kundler, P., 1959. Zur Kenntnis der Rasenpodsole und Grauen Waldböden Mittelrusslands im Vergleich mit den Sols lessivés des westlichen Europas. Z.Pflanzenernähr., Düng., Bodenk., 86 : 16–36.

Lerz, H. und Borchert, W., 1962. Verwitterung von Microcline unter atmosphärischen Temperatur-Druckbedingungen. Chem. Erde, 22 : 386–429.

Nocin, V.A. and Petrov, B.F., 1959. Methodics and technics of pedological field survey. In: Pedological Survey. Akad. Nauk. S.S.S.R., pp. 45–105. (Russ.)

Parfenova, E.I. and Yarilova, E.A., 1960. The problem of lessivage and podzolization Pochvovedenie, 9 : 1–15. (Russ.)

Polinov, B.B., 1956. First formation stages on massive crystalline rocks. Selected papers, Akad. Nauk. S.S.S.R., pp. 395–407. (Russ.)

Ponomareva, V.V., 1956. Humus-formation, its nature and geographical distribution. Congr. Intern. Sci. Sol, 6e, Rappt. pp. 221–240.

Rubilin, E.V., Vernander, N.B., Parfenova, E.I., Suslova, E.V. and Troitzki, A.I., 196- Grey forest soils of the European part of U.S.S.R. Intern.Congr. Soil Sci., 8th, Rept. Soviet Soil Scientists, in press. (Russ.)

Siffert, B. et Wey, R., 1961. Sur la synthèse de la kaolinite à température ordinaire. Compt. Rend., 253 : 142–144.

Titova, N.A., 1962. Iron-humus complexes of some soils. Pochvovedenie, 12 : 38–43. (Russ.)

Yarilova, E.A. and Parfenova, E.I., 1960. Investigation of clay minerals of soil colloi- Kolloidn. Zh., 22 : 237–242. (Russ.)

ZUR MIKROMORPHOLOGIE LESSIVIERTER BÖDEN IN VERSCHIEDENEN KLIMAGEBIETEN

G. REUTER

Universität Rostock, Rostock (Deutschland)

EINFÜHRUNG

In früheren Untersuchungen an lessivierten Böden auf jungpleistozänem Geschiebemergel Norddeutschlands konnte festgestellt werden, dass hier bei zahlreichen Standorten die profil- und mikromorphologisch deutlich vorhandenen Lessivé-Merkmale (Horizonte-El und It) mit mehr oder weniger stark ausgeprägten Braunerde-Merkmalen (Horizont-Be) interferieren bzw. sogar von diesen übertönt werden (Reuter, 1958, 1962a). Es wurden in den Böden dieses maritim getönten gemässigten humiden Klimagebietes ausserdem in starkem Masse Interferenzen des Lessivé mit Merkmalen der Pseudovergleiung (marmorierter Horizont-P) und Podsolierung (Horizonte-Ep und Ih) festgestellt:

231 Podsoliger pseudovergleiter Braunerde-Lessivé (Doberan, Norddeutschland) (siehe Reuter, 1962a, Abb.2);
Om–As(Ep)–AsBe–Be–El–ItP–M (ab 210 cm). (Für die gebrauchte Horizontbenennung siehe Reuter, 1957).
426 Podsol-Pseudoglei-Lessivé (Sassnitz, Norddeutschland) (siehe Reuter, 1962b, Dia 29);
Or–Orm–As–Ep–(Ih)–El–ItP–Mv–M (ab 170 cm).
503 Lessivé-Braunerde (Rostock, Norddeutschland) (siehe Reuter, 1962a, Abb.1);
pAn–Be–It–Mv–M (ab 110 cm).

Von zwei möglichen Hypothesen zur genetischen Deutung dieser Befunde wurde der einer durch Klima-änderung bedingten disharmonischen Entwicklung der Vorzug gegeben und das Optimum der Lessivierung für das norddeutsche Gebiet in den gegenüber den rezenten Verhältnissen deutlich wärmeren Abschnitt des Atlantikum gestellt, während die Vorgänge der erdigen Verbraunung, Podsolierung und Pseudovergleiung als rezente Bodenbildungsprozesse angesehen wurden (Reuter, 1958, 1962a). Ähnliche Vorstellungen wurden teilweise von Manil (1962) und Altemüller (1962) entwickelt. Auch von Ehwald (1961) und Lieberoth (1962) wird auf die Möglichkeit einer rezenten Verbraunung von Lessivés hingewiesen, allerdings nur unter Ackerkultur. Zur Überprüfung unserer Hypothese von einer optimalen Lessivierung mitteleuropäischer Böden im Atlantikum erschien es zweckmässig Standorte zu untersuchen, die sich heute unter einigermassen vergleichbaren Klimabedingungen, d.h. vor allem bei etwas höheren Mitteltemperaturen, entwickeln. Mit freundlicher Unterstützung der Kollegen des Puschkarov-Instituts für Bodenkunde in Sofia und des Instituts für Bodenkunde

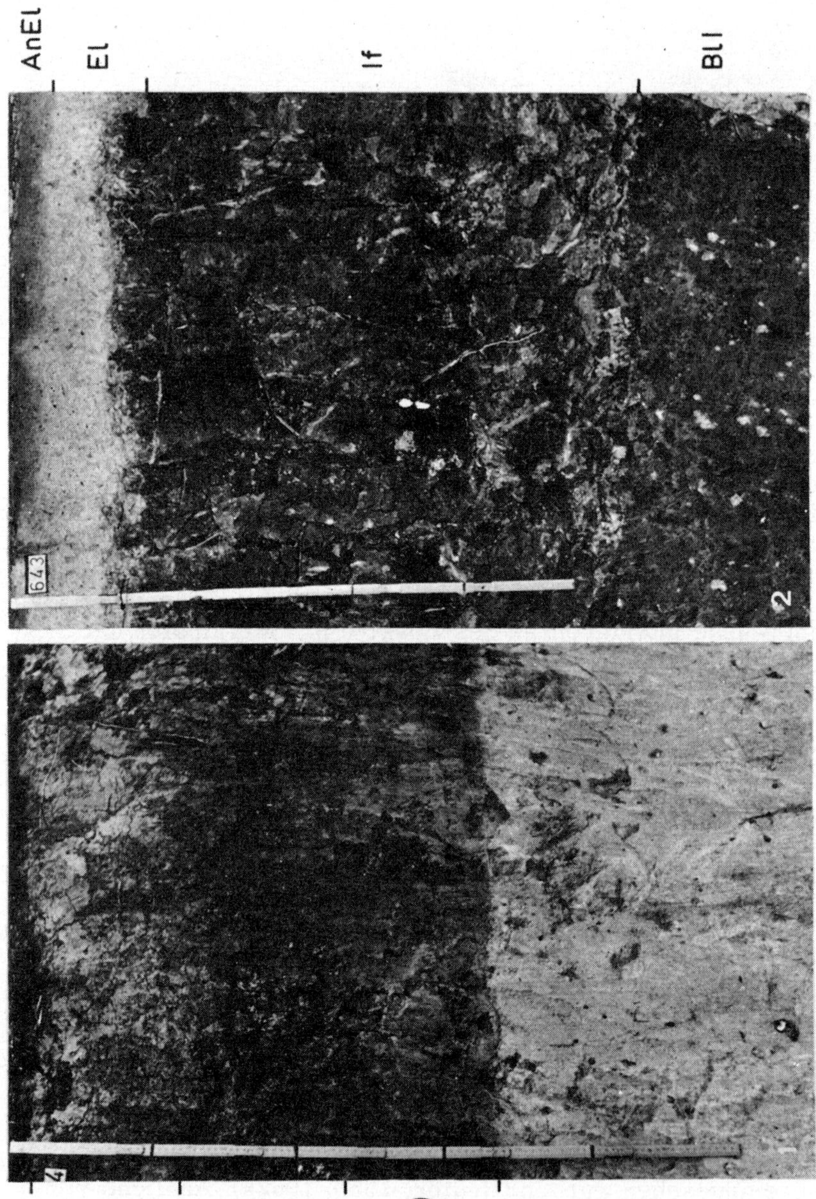

Abb.1. Lessivé in Ungarn (Profil 674, Vergleichsmass 1 m).

Abb.2. Lessivé in Südbulgarien (Profil 643, Vergleichsmass 1 m).

und Agrochemie in Budapest konnte in Bulgarien und Ungarn eine grössere Zahl von lessivierten Böden aufgenommen werden, deren analytische Daten und tonmineralogischen Befunde an anderer Stelle veröffentlicht werden (Reuter, 1964; Reuter und Menning, 1964). Weiterhin wurden Beobachtungen und analytische Ergebnisse von Böden aus Polen und Nordamerika in den Vergleich mit einbezogen.

PROFILMORPHOLOGISCHE UNTERSUCHUNGEN

Die profilmorphologische Auswertung dieses Materials lässt bereits deutliche Unterschiede zwischen den verschiedenen Klimagebieten erkennen. Während in Norddeutschland (Jahresmittel 550–650 mm bzw. 7–8° C, Juli 16–17° C)[1] die Profile vorherrschend starke Interferenzen zwischen Lessivierung, Verbraunung, Pseudovergleiung und Podsolierung erkennen lassen (s.o.), fanden wir in Ungarn und Südbulgarien unter wärmeren, mehr sommertrockenen und stärker kontinental geprägten Bedingungen diese Interferenzen kaum vor, dafür in viel stärkerer Verbreitung typische Lessivés. Beispiele aus Ungarn (Jahresmittel 570–730 mm bzw. 10,7–10,8° C, Juli 21,5–21,7° C)[1].

674 Lessivé (Kincses) auf Lössmergel (siehe Abb.1);
As–El–It–Mv(It)–M (ab 70 cm).
680 Lessivé (Hollad) auf Lössmergel;
(Om)–As–El–It–Mv–M (ab 90 cm).
678 Lessivé (Savoly) auf Sandmergel mit periglazialem Eiskeil (siehe Stefanovits, 1963, Abb.1, Tafel III);
As–El(It)–It–M (ab 335 cm).

Die A-Horizonte sind schwach bis mässig sauer. Braunerdemerkmale oberhalb des Ton-Illuvialhorizontes fehlen hier völlig, bemerkenswerterweise auch bei dem nur relativ schwach entwickelten Lessivé 674 von Kincses (siehe Abb.1). Es konnte hier auch keine lokale Vergesellschaftung mit Braunerde festgestellt werden, obwohl in diesem Gebiet (Umgebung des Balaton-Sees) nach Stefanovits (1963) echte Braunerden verbreitet sind. Dies war indessen der Fall bei Standorten in Gebirgslage (Jahresmittel 700 mm bzw. 7° C, Juli 17° C):

687 Lessivé (Matragebirge, Nordungarn) auf Andesittuff;
(Om)–An–El–It–MvIt (ab 60 cm).
686 Saure Braunerde (Matragebirge, Nordungarn) auf Hydroandesit;
(Om)–As–B–Mv (ab 70 cm).

Auch in Südbulgarien fanden wir typische Braunerden nur in Gebirgslagen (z.B. oberhalb Batak, Rhodopengebirge, auf Silikatgestein), dagegen im Bereich der thrazischen Ebene (Jahresmittel 500–560 mm bzw. 11,8–12,6° C, Juli 22,6–24,4° C) typische Lessivés, vergesellschaftet mit Braunlehmen, die häufig auch Lessivierungserscheinungen erkennen lassen:

---

[1] Sämtliche Klima-angaben beschränken sich auf die untersuchten Standorte.

643 Lessivé (Madjereto) auf kalkhaltigem Lehm (siehe Abb.2);
AnEl–El–It–BlIt–Mv (ab 160 cm).
Pseudovergleiung und Mn-Konkretionen vorhanden.
648 Lessivé, rötlich-braun (Selo Sbor) auf kalkhaltigem Lehm;
pAsEl–It–Bl–Mv (ab 125 cm).
Der Mv enthält reichlich Karbonatkonkretionen.
638 Lessivé (Swesditza) auf Sandmergel;
pAn–El–It/El–It–M (ab 145 cm).
647 Lessivierter rötlicher Braunlehm (Chaskovo) auf Kalkphyllit;
pAn–BlIt–Bl–Mv (ab 60 cm).

Bei der mikromorphologischen Auswertung wurden die aus Rahmenproben gewonnenen Dünnschliffe der einzelnen Bodenhorizonte bzw. Teilhorizonte auf ihren prozentualen Anteil an Orientierungsdoppelbrechung (%D) innerhalb der karbonatfreien Tonsubstanz durchgemustert, um auf diese Weise zu einem Mass für den anteiligen Gehalt an Braunlehmteilplasma (nach Kubiena, 1956) zu kommen. Die Ergebnisse %D wurden zusammen mit anderen Analysendaten bereits tabellarisch mitgeteilt (Reuter, 1964). Obwohl diese %D-Zahlen mit einem Fehler belastet sind, der zwischen 2–20% liegen kann, lassen sie doch angesichts der teilweise recht grossen Unterschiede deutliche Tendenzen erkennen: In sämtlichen Lessivé-Böden aller Untersuchungsgebiete liegt das Maximum der Orietierungsdoppelbrechung im Ton-Illuvialhorizont-It. Dieses gemeinsame Merkmal ist von grosser diagnostischer Bedeutung, vor allem auch bei erodierten und bei petrographisch uneinheitlichen Profilen, die keine Bilanzierung der Tonverlagerung gestatten. Die höchsten Werte wurden mit %D = 90 in den stark pseudovergleiten Illuvialhorizonten-ItP der Profile in Norddeutschland festgestellt, jedoch stehen die südosteuropäischen Standorte kaum nach; und selbst der relativ schwach entwickelte Lessivé 674 von Kincses, Ungarn (siehe Abb.1), weist im It-Horizont 50% Orientierungsdoppelbrechung auf. Wurde der Ton-Illuvialhorizont aufgegliedert, so nimmt im allgemeinen der Anteil %D mit zunehmender Tiefe ab.

Auffallend niedrig sind die %D-Zahlen im Eluvialbereich der typischen Lessivés (meist < 10%) und auch in den podsolierten Horizonten (< 5%). Durchweg hohe Werte (70–80%) zeigt der lessivierte Braunlehm in Südbulgarien. Auf die sehr geringe bzw. fehlende Orientierungsdoppelbrechung in den Braunerde-Horizonten bzw. die Übergangserscheinungen bei Lessivé-Braunerde-Interferenzen Norddeutschlands war bereits früher hingewiesen worden (Reuter, 1962a).

Eine spezielle mikromorphologische und mineralogisch-chemische Bearbeitung erfordern noch die rötlich-braunen Lessivés im Zusammenhang mit den vergesellschafteten rötlichen Braunlehmen in Bulgarien, die von Gerassimow (in Gürov und Ninov, 1963) als "zimtfarben" bezeichnet wurden. Da es sich in den meisten Fällen mit hoher Sicherheit um Standorte der rezenten Lessivierung handelt, interessiert in besonderem Masse die Frage, ob die hier auftretenden rötlichen Eisenoxyde dabei rezent gebildet werden, oder nach den Vorstellungen von Koinov (1963), Gürov und Kinov (1963) vorwiegend als Relikte paläopedologischer Prozesse aufzufassen sind. Für die erstere Version könnte das Profil 648 sprechen, da hier die rötlichen Substanzen ausser in Konkretionen auch ziemlich diffus verteilt auftreten, was nach Kubiena (1962) als Merkmal der Rubefizierung einer Braunlehmgrundmasse anzusehen ist. Dagegen beschränken sich die rötlichen Komponenten bei Profil 643 (siehe auch Abb.2) ausschliesslich auf isoliert erscheinende Konkretionen. Dies spricht mehr für einen Reliktcharakter der Rubefizierungserscheinung.

Eine gewisse Zwischenstellung hinsichtlich der Profilausbildung und Mikromorphologie nehmen mehrere von uns in Polen aufgenommene Böden ein. Es wurde hier im gemässigten humiden jedoch etwas kontinental getönten Klima in den Lessivés verschiedentlich ein Verbraunungshorizont gefunden, der profilmorphologisch den Interferenzen Norddeutschlands ähnelt (A–B–E1–It), dessen mikromorphologische Eigenschaften (%D > 30) und niedriges Eisenverhältnis (Reuter, 1964) aber nicht der typischen Braunerde entsprechen.

Soweit die anlässlich der Exkursionen in Nordamerika (I.S.S.S., 1960) vorgeführten lessivierten Böden von uns mikromorphologisch untersucht werden konnten, ergab sich ein relativ häufiges Auftreten von Lessivé-Braunerde-Interferenzen in den nordöstlichen Gebieten (z.B. sog. *'Typudalf'* von Erie, Pa. und sog. *'Fragudalf'* von Wooster, Ohio) gegenüber den in dieser Beziehung mehr typischen Lessivés in den etwas wärmeren Gebieten (z.B. sog. *'Albaqualf'* bei Vandalia im südlichen Illinois). Mikromorphologisch deutliche Merkmale der Tonverlagerung fanden wir in verschiedenen Brunizems, deren Bezeichnung als *'Argudoll'* diese Tendenz auch zum Ausdruck bringt. Gewöhnlich ist bei diesen Böden jedoch die Lessivierung profilmorphologisch nicht offensichtlich, da kein El-Horizont auftritt und vor allem die oberen Horizonte mehr den Gefügeeigenschaften einer Braunerde entsprechen. Wir möchten diese Verhältnisse deshalb als starke Lessivé-Braunerde-Interferenzen bezeichnen, ohne dass die betreffenden Böden mit den untersuchten europäischen Böden direkt identifiziert werden können. Demonstration von Beispielen aus Illinois, Wisconsin und Iowa.

SUMMARY

Micromorphology of lessivé soils in different climate regions

There are fundamental differences between the lessivé soils of northern Germany and the comparatively investigated lessivé soils of Hungaria and southern Bulgaria; the former have a frequent occurrence of interference marks of browning (Verbraunung), pseudogleization and podsolization, while the latter generally do not possess such marks. The sites of southeastern Europe with 3–5° C higher average temperature are obviously better qualified for recent formation of simple lessivé profiles. "Braunlehm", partly in a reddish variation, appears instead of "Braunerde" as a preceding stage of development of these soils. These statements support our hypothesis of optimum lessivation in central Europe during the Atlantikum period.

The so-called %D-number has been found suitable as a half-quantitative measure for the degree of orientation and double refraction of carbonate-free clay material in thin-sections. All lessivé soils show a maximum in the clay-illuvial It-horizon.

LITERATUR

Altemüller, H.J., 1962. Beitrag zur mikromorphologischen Differenzierung von durchschlämmter Parabraunerde, Podsol-Braunerde und Humus-Podsol. Z. Pflanzenernähr., Düng., Bodenk., 98 (143) : 247–258.

Ehwald, E., 1961. Aufgaben des Instituts für Bodenkunde der Deutschen Akademie der Landwirtschaftswissenschaften zu Berlin. Sitz. ber., Deut. Akad. Landwirtschaftswiss. Berlin, 10 : 10.

Gürov, G. und Ninov, N., 1963. Die zimtfarbenen Böden in Bulgarien. Vortragsmanuskript, Sofia.

Intern. Soc. Soil Sci., 1960. Excursion guide Congr. Intern. Soc. Soil Sci., 7th, Tour 1., 104 pp.

Koinov, W., 1963. Die Besonderheiten der wichtigsten Böden Bulgariens und die bei ihnen verlaufenden genetischen Prozesse. Vortragsmanuskript, Sofia.

Kubiena, W.L., 1956. Zur Mikromorphologie, Systematik und Entwicklung der rezenter und fossilen Lössböden. Eiszeitalter Gegenwart, 7 : 101–112.

Kubiena, W.L., 1962. Die taxonomische Bedeutung der Art und Ausbildung von Eisenoxydhydratmineralien in Tropenböden. Z. Pflanzenernähr., Düng., Bodenk., 98 (143) : 205–213.

Lieberoth, I., 1962. Über den Einfluss der Ackerkultur auf die Bodenentwicklung im sächsischen Lössgebiet. Albrecht-Thaer-Archiv, 6 : 3–30.

Manil, G., 1962. Diskussion über den Ausdruck "Lessivé" auf Grund mikromorphologischer Beobachtungen. Z. Pflanzenernähr., Düng., Bodenk., 98 (143) : 214–21

Reuter, G., 1957. Beitrag zur Nomenklatur der Bodenhorizonte. Wiss. Z. Univ. Rostoc 2 : 207–211.

Reuter, G., 1958. Tendenzen der Bodenentwicklung im Küstenbezirk Mecklenburgs. Habilitationsschrift Univ. Rostock. Auch: Wiss. Abhandl., Deut. Akad. Landwirtschaftswiss. Berlin, 1962, 49 : 28 S.

Reuter, G., 1962a. Lessivé-Braunerde-Interferenzen auf Geschiebemergel. Z. Pflanze ernähr., Düng., Bodenk., 98 (143) : 240–246.

Reuter, G., 1962b. Bodenkundliche Farbdia-Serie A1–50, 2. Aufl. Institut für Agrikult chemie und Bodenkunde der Universität Rostock.

Reuter, G., 1964. Vergleichende Untersuchungen an lessivierten Böden in verschieder Klimagebieten. Proc. Intern. Congr. Soil Sci., 8th, Bucharest, in Vorbereitung.

Reuter, G. und Menning, P., 1964. Tonmineralogische Untersuchungen an lessivierten Böden verschiedener Klimagebiete, in Vorbereitung.

Stefanovits, P., 1963. Magyarország talajai. Akadémiai Kiadó, Budapest, 442 S.

ZUR MIKROMORPHOLOGIE UND MIKROMORPHOGENESE DER LÖSSBÖDEN NEUSEELANDS

W.L. KUBIENA

Bundesforschungsanstalt für Forst- und Holzwirtschaft, Abteilung Bodenkunde, Reinbek bei Hamburg (Deutschland)

GLOBALE EINORDNUNG DER NEUSEELÄNDISCHEN LÖSSBÖDEN

Weltsystem der Lössböden

Die Kenntnis der Lössböden Neuseelands vermittelt uns durch ihre besondere Eigenart neue wertvolle Erkenntnisse über die komplexe Mikromorphogenese der Lössböden überhaupt. In typischer Weise tritt auch hier wieder ihre grosse Variabilität der Formbildung unter dem Diktat der heutigen sowie auch aller vorzeitlichen Umweltverhältnisse in Erscheinung. Wie kaum in anderen Bodenbildungen erscheinen fast alle Einflüsse, frühere und rezente, in dem Gefügebild der einzelnen Horizonte mit solcher Klarheit verzeichnet. Um die Lössböden Neuseelands in ihrer Eigenart besonders herausstellen zu können, soll zunächst versucht werden eine vereinfachte Weltübersicht der rezenten und vorzeitlichen Lössböden und ihrer Entwicklungstendenzen zu geben (Tabelle I). Da sich in den verschiedenen Zeitperioden die Grenzen der einzelnen Lössbodentypen, wie jene der Bodentypen überhaupt, erheblich verschoben haben, vor allem ihrer planetarischen, hypsometrischen und kontinental–maritimen Lage nach, handelt es sich mit der beigegebenen Tabelle nicht um eine Übersicht in Hinsicht auf ihre gegenwärtige Verbreitung, wenn auch die Abhängigkeit ihres Vorkommens von der geographischen Lage darin zum Ausdruck gebracht werden soll, sondern, um eine Übersicht über die genetischen Beziehungen zueinander und damit über die auf diese Weise entstehenden Formenkreise.

Allgemeine Charakteristik

Die neuseeländischen Lössböden, deren Hauptverbreitungsgebiet im Osten der Südinsel liegt, die jedoch auch im Südteil der Nordinsel zu finden sind, fallen mikromorphologisch hauptsächlich unter die Gruppe D der Weltübersichtstabelle. Sie sind in Europa gefügekundlich den in Lössprofilen des östlichen Niederösterreichs, des Burgenlandes, Ungarns, Jugoslawiens u.a. als mehr oder minder rötlich gefärbten Varianten vorkommenden Böden am ähnlichsten, die dort mehr oder minder nur als fossile Bildungen vorkommen. Die Tatsache, dass sie die bereits wohlbekannten Lössbodenmerkmale erkennen lassen und dass ihnen die gleiche Variabilität, die gleiche Tendenz zur Ausbildung bestimmter Varianten, vor allem Pseudogleys, Lessivés, erdige Lessivés und Löss–Braunerden eigentümlich ist, zeigt, unter anderen

TABELLE I

Vereinfachte Weltübersicht der rezenten und vorzeitlichen Lössböden und ihrer Entwicklungstendenzen

A. Ehemalige arktische Bodenprovinz (nur fossile Böden)

Arktischer Rohboden (Rämark)

Arktischer Gley

Arktischer Pseudogley

Tundra–Pararendsina→Arktische Lockerbraunerde (durch Vererdung von Lessivé- und Pseudogleyrelikten entstanden)

B. Feucht-temperierte Bodenprovinz

Lössrohboden

Pararendsina

Stagnogley←Pseudogley→Feuchttemperierter Lessivé (gebleichte Parabraunerde)

Anmoor

Pseudo-Tschernosem

erdiger Lessivé

Lössbraunerde

podsolige Braunerde (Semipodsol)

C. Steppenbodenprovinz

Lössrohboden

Anmoor

Smonitsa

Tschernosem

degradierter Tschernosem

Pararendsina

Braunerde

D. Submediterrane bis subtropische Bodenprovinz (in Europa nur fossile Böden)

Lössrohboden

Pararendsina

braunlehmreicher Lessivé

Tirsifizierter Pseudogley ←südlicher (rötlicher) Pseudogley

leicht rubefizierter braunlehmreicher Lessivé

xeromorph verhärteter Pseudogley (Hardpanbildung)

xeromorph verhärteter braunlehmreicher Lessivé (Hardpanbildung)

Merkmalen, dass es sich auch bei den neuseeländischen Ausgangsmaterialien um wirkliche, den Vorkommen auf der Nordhalbkugel vollkommen entsprechende Lösse handelt. Als Bodenvarianten stehen in Neuseeland braunlehmreiche Lessivés und südliche Pseudogleys im Vordergrunde. Gegen den wärmeren Norden nimmt der Braunlehmgehalt zu, nach dem Süden und in höheren kühleren Lagen die sekundäre braune Vererdung, so dass einige dieser südlichen Formen bereits den deutschen, belgischen und nordfranzösischen Lössböden nahkommen. Die deutliche Zunahme der Xeromorphie, die besonders in den sommertrockenen Gebieten des Otago- und Canterbury-Distrikts zum Ausdruck kommt, entspricht der gleichen Zunahme der Xeromorphie im südlichen Europa. Alexandra, der Hauptort in central Otago, hat mit 330 mm die niedrigste jährliche Niederschlagsmenge in Neuseeland. Bildungen eines ursprünglich feucht-warmen Klimas, wie die braunlehmreichen Lessivés, unterliegen unter ihrem Einfluss nicht der braunen Vererdung, sondern einer fast irreversiblen Verhärtung ohne wesentliche Gefügeveränderungen, wie dies die tertiären Braunlehmsedimente im Stadtgebiet von Madrid zeigen. Der stete Wassermangel unter Erwärmung führt zu den als xeromorph verhärteten braunlehmreichen Lessivés und Pseudogleys bezeichneten Varianten der Südinsel, die vor allem der Gruppe der "yellow-grey earths" nach neuseeländischer Nomenklatur angehören (s. Raeside, 1956). Standorte rezenter Vernässung zeigen im Gegensatz hierzu Anmoorbildung über Gley oder Pseudogley (z.B. der "Temuka-silt loam"), extrem wechselfeuchte Standorte im sommertrockenen Bereich tirsoide Pseudotschernoseme, wie die Profile an den Steiluferabbrüchen im Löss am Rakaia River bei Barrhill Village.

MIKROMORPHOLOGISCHE UNTERSUCHUNG EINIGER PROFILTYPEN

Felduntersuchung

Zu den Feldstudien und zur Aufsammlung von Rahmenproben für die Laboratoriumsuntersuchung hatte der Verfasser durch seine Teilnahme an einer Bereisung der Südinsel anlässlich der Internationalen Konferenz der IV. und V. Kommission der Internationalen Bodenkundlichen Gesellschaft in Palmerston-North in Neuseeland Gelegenheit. Von allen Rahmenproben wurden Dünnschliffe hergestellt. Von den untersuchten Bodenprofilen sind hier nur die für diese Abhandlung wichtigsten angeführt.

Profil NS14: Braunlehmreicher Lessivé-Pseudogley, Gleniti, Timaru, Canterbury. Neuseeländische Bezeichnung: "Moderately leached yellow-grey earth" bzw. "moderately enleached gammate pallic soil from moderately argillised loess" (S. Taylor und Pohlen, 1962; Pohlen, 1962).

Das Profil liegt auf einem konvexen 5°S geneigten Hang der Canterbury Downlands, die gegenwärtige Vegetation besteht aus Grasland (vorwiegend *Agropyron* spp.), die natürliche Vegetation war "silver tussock" (*Poa caespitosa*). Das Ausgangsmaterial ist mässig verwitterter Löss von Grauwacke, die Seehöhe ist 60 m; das Klima charakterisiert sich durch 600 mm Jahresniederschlag, eine mittlere Januartemperatur von 16° C und eine mittlere Julitemperatur von 5° C

Profilaufbau[1].

0– 30 cm	: A-Hor., dunkelgraubrauner Schlufflehm, gut durchwurzelt, locker, krümelig. pH = 4,1–5,4.
30– 60 cm	: A/g-Hor., hellockergrauer, sehr dichter Schlufflehm mit dunklen Humusröhren. pH = 5,8.
60–100 cm	: B_1/g-Hor., sehr dichte, überwiegend hell gefärbte Pseudogleyschicht mit eisenreichen roten Tonfragmenten, deren Zwischenräume eine weissliche Tonfüllung zeigen (Teba-Netz). pH = 5,3.
100–150 cm u.m.	: B_2/g-Hor., dichter, verhärteter, ockerfarbiger Schlufflehm, in diesem Eiskeile mit weisslicher Tonfüllung.

Mikromorphologie.
A-Horizont, Probe aus 2–6 cm: Der Dünnschliff zeigt eine dichte hellgraue Grundmasse, die in ineinander gepressten knollig-nierigen Aggregatformen vorliegt. Letzte deuten auf intensive Tätigkeit von Regenwürmern im Humushorizont. Daneben zeigen sich aber auch freiliegende, locker gefügte Aggregatformen (Arthropodenlosung). In der hellen Grundmasse treten mehr oder minder rundliche Kristallisationsfelder von Eisenoxydhydrat oder Bruchstücke von solchen auf, die neben der eisenarmen Matrix auf eine ehemalige Pseudogley-dynamik hindeuten. Es zeigt sich so gut wie kein ockergelbes Braunlehm-Teilplasma bzw. nur sehr spärliche Rückstände von solchem. Ebenso fehlt Braunerde-Teilplasma vollständig. Zersetzung und Humifizierung der Pflanzenreste sind gut, als Humusform liegt Grasland-Mull vor.

A/g_1-Horizont, Probe aus 20–24 cm: Zusammenhängende (weniger aus aneinandergepressten Aggregaten bestehende) dichte, weisslichgraue Grundmasse mit braunroten, im Auflicht grellroten Bruchstücken von Pseudogley-Kristallisationsaggregaten, daneben einzelne tiefbraune knollige Braunlehmkonkretionen von Eisenoxydhydrat. Die Grundmasse ist sehr reich an Feinsand- und Schluffkörnern, deren Dichte und Zusammensetzung in gleicher Weise auch in den roten Kristallisationsfeldern von Eisenoxydhydrat zu finden ist.

A/g_1-Horizont, Probe aus 32–36 cm: Ähnliches Gefügebild wie in voriger. Hier sind kleine Braunlehmflitter, auf die später noch zurückgekommen werden soll, gut erhalten. Daneben zeigen sich Glaukonitkörner (vgl. S.231).

A/g-Horizont, Probe 40–44 cm: Auch hier flitterförmiger Braunlehmdetritus. In der Bodenmasse treten gelbe Flecken auf, die Stellen andeuten, die reich an stark peptisiertem Braunlehm-Teilplasma sind. Diese Anreicherungen zeigen sich in Form von Leitbahnfüllungen, aber auch in der Weise, dass sich die Intergranularräume ganzer Gefügepartien völlig mit Braunlehm-Teilplasma ausfüllen. Dieses ist in dünner Schicht eigelb, in dicker Schicht fast orangefarbig.

B_1/g-Horizont, 95–100 cm: Die Anreicherungen von gelbem bis orangefarbigem Braunlehm-Teilplasma, besonders in Form von, z.T. stark verbreiterten Leitbahnfüllungen, haben sich verstärkt. Die Pseudogleymerkmale treten zurück.

B_2/g-Horizont, 145–150 cm: weiterhin starke Anreicherung von Braunlehm-Teilplasma, Zurücktreten von Pseudogleymerkmalen.

Diskussion der Ergebnisse

Der Pseudovergleyung ging eine Lessivierung voraus, die, da sich in rezenten Sprungrissen kein Braunlehm-Teilplasma mehr ansammelt, nicht mehr aktiv zu sein scheint. Die nachfolgende Pseudovergleyung ist durch Tagwasseranstau entstanden und reichte im wesentlichen bis zu 1 m Tiefe. Im Hinblick auf die Verhärtung des Unterbodens durch anhaltende

[1] Abweichende Horizonttiefen erklären sich daraus, dass Proben häufig etwas abseits der vom Soil Survey von Neuseeland vorbereiteten Profilstellen genommen wurden. Eine Reihe von Profilen wurden vom Verfasser allein aufgesucht und in die vorliegende Studie einbezogen.

Austrocknung und die terrestrische Humusbildung scheint sie heute nicht mehr wirksam zu sein. Die starke Neigung zu Staunässe in einem früheren feuchteren Klima ist für die braunlehmreichen neuseeländischen Lessivéprofile typisch. Es gibt kaum eines unter ihnen, das nicht zugleich Pseudogleymerkmale zeigt. Der Braunlehmreichtum und Pseudogleycharakter hat, wie aus dem Profilaufbau hervorgeht, bereits vor der periglazialen Überformung und der Entstehung der Eiskeile bestanden.

Profil NS22: Braunlehmreicher Lessivé-Pseudogley, Greytown-Martinborough road, southern Wairarapa District, Nordinsel.
Neuseeländische Bezeichnung: "Wharekaka silt loam". "Yellow-grey earth" bzw. "moderately enleached, gleyed, net gammate pallic soil".

Das Profil liegt auf einem 2° W geneigten Hang in flachhügeliger Landschaft, die gegenwärtige Vegetation besteht aus dem Grasstreifen des Strassenrains, die natürliche Vegetation war broadleaf forest. Das Ausgangsmaterial ist ein lössartiger Schluff über Schotter. Die Seehöhe ist 43 m, das Klima charakterisiert sich durch 900–1.000 mm Jahresniederschlag und eine mittlere Jahrestemperatur von 14° C.

Profilaufbau.

0– 9 cm	: A-Hor., dunkelgraubrauner dichter Schlufflehm mit bröckeliger Absonderung. pH = 5,9.
9– 15 cm	: A/g-Hor., dichte, tonige,graue Grundmasse mit kantiger Absonderung. In der Grundmasse braune eisenreiche Flecken, im oberen Teile dunkelbraune, gut gerundete Braunlehmkonkretionen. pH = 5,9.
15– 26 cm	: g-Hor., dichte, tonige, stark verhärtete, hellgraue Grundmasse mit braunen Eisenflecken und kantiger Absonderung. pH = 6,3.
26–100 cm u.m.:	B/g-Hor., kantige bis leicht gerundete, braune eisenreiche Gefügefragmente mit Eisensäumen in hellgrauer Grundmasse. Kantige bis prismatische Absonderung, starke Verhärtung. pH = 6,5 im oberen Horizontteil.

Mikromorphologie.
A-Horizont. Die Humusform ist ein Mull, der sich aus einem Anmoor entwickelt. Die enteisente, dichte Grundmasse und die Ausbildung von im Auflicht roten Kristallisationsfeldern von Eisenoxydhydrat deutet auf Pseudovergleyung. Reste von Braunlehm- und Braunerde-Teilplasma fehlen. Das Gleiche gilt für den g-Horizont.

B/g-Horizont. In der weisslichen, dichten Grundmasse eisenreiche Gefügepartien, in denen die Intergranularräume völlig von starkpeptisierten Braunlehmplasma ausgefüllt sind. Dieses ist in dünner Schicht eigelb und wird bei stärkerer Konzentrierung intensiv orangefarbig. In dieser Braunlehm-Matrix entstehen dichte, durchscheinende Kristallisationsaggregate von Eisenoxydhydrat (Goethit) in tief orangeroter bis blutroter Farbe (Abb.1). Das Braunlehmplasma tritt auch in Form von Wandanlagerungen und Füllmassen in Spalten auf. Wandanlagerungen von konkretionärem Eisenoxydhydrat sind opak, im Auflicht grellrot.

Diskussion der Ergebnisse

Das Profil gibt ein der vorigen Bodenbildung entsprechendes, obwohl auch gegenwärtig noch unter dem Einfluss von Staunässe stehendes Beispiel auf der Nordinsel wieder, das in den gegenwärtigen Umweltverhältnissen den ursprünglichen näherkommt. In Mitteleuropa zeigt es sich mikromorphologisch einer fossilen Bodenbildung in einem Bodenprofil bei Ebersbrunn, Niederösterreich, am ähnlichsten. Die orangefarbige Tönung des Braunlehmplasmas nimmt in der Nähe der Kristallisationsfelder zu, steht demnach mit dem Pseudogleycharakter in Zusammenhang.

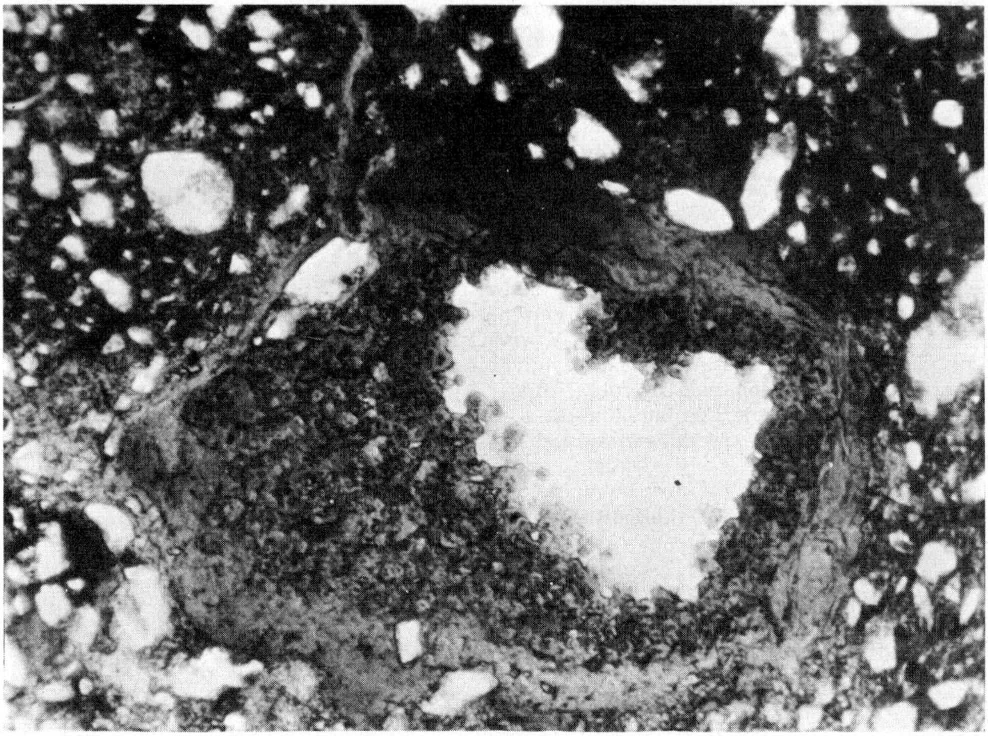

Abb.1. Braunlehmreicher Lessivé-Pseudogley NS22, Greytown–Martinboroug road, Nordinsel; B/g-Hor. Die Intergranularräume sind von einer stark peptisierten, eigelben bis orangefarbigen Braunlehm-Matrix erfüllt in der sich dichte durchscheinende Kristallisationsaggregate von Goethit von grellroter Farbe bilden.

Profil NS3: Vererdeter, pseudogleyiger Braunlehm-Lessivé, Otautau, Südende der Südinsel.
Neuseeländische Bezeichnung: "Southern yellow-grey earth to southern yellow-brown earth (luvi-fulvic soil) intergrade".

Das Profil, das bei der Grundaushebung zu einem Hausbau freigelegt worden war, liegt auf einer ebenen Schafweidenfläche, die zu der ebenen bis leicht welligen, aus pleistozänen, z.T. holozänen alluvialen Terrassen, entstandenen, zu grossem Teile von einer 1–2 m mächtigen Lössschicht bedeckten Southland Plain gehört. Letzte gilt als eines der fruchtbarsten Gebiete Neuseelands. Die Seehöhe ist ca. 20 m, der Jahresniederschlag um 1.100 mm, die mittlere Januartemperatur um 14°C, die mittlere Julitemperatur um 4°C, das Jahresmittel um 9,9°C, die Anzahl der Frosttage bewegt sich um 250.

Profilaufbau.

0– 30 cm	: A-Hor., dunkelgraubrauner Schlufflehm mit lockerem erdigem Gefüge.
30– 70 cm	: A/g-Hor., weisslich ockergrauer dichter Schlufflehm.
70–100 cm	: g/B_1-Hor., weisslich ockergelber sehr dichter, doch feuchter und grabbarer Schlufflehm.
100–120 cm	: B_2/g-Hor., dichter ockerfarbiger, eisenfleckiger Schlufflehm.
120–180 cm	: B_3/g-Hor., sehr schwerer tief braunocker gefärbter toniger Lehm.
180–184 cm	: E_1/B-Hor., sehr dichtes toniges tiefbraunes Fliesslehmband.
184–200 cm u.m.	: E_2/B-Hor., sehr dichter sattockerfarbiger toniger Fliesslehm.

Mikromorphologie.
A-Horizont, Probe aus 2–6 cm (Tafel B8). Starkvererdetes, hohlraumreiches, flockiges Braunerdegefüge, das sich aus grossen, z.T. knolligen Bodenaggregaten zusammensetzt, die auf intensive Regenwurmtätigkeit hindeuten. Neben letzten finden sich auch längliche Formen mit kreisrunden Querschnitten (Losung grösserer Arthropoden). Unter den Feinsand- und Schluffkörnern, die das Mikroskelett zusammensetzen finden sich auch wenig verwitterte Biotite, Kalifeldspate, Chlorite, Augite und Hornblenden.

A/g-Horizont, Probe aus 40–44 cm. Die weissgraue Grundmasse zeigt Spuren starker Bleichung, doch zugleich auch deutlicher Durchschlämmung, die Reste von braunem flockigem Braunerde-Teilplasma zurückliess. Daneben sind aber auch scharfkonturierte Kristallisationsreste von grellroter Farbe vorhanden. Diese zeigen im Feinskelett die gleiche Mineralzusammensetzung wie die Grundmasse, haben sich darum aus dieser entwickelt.

g/B_1-Horizont, Probe aus 76–80 cm. Die Grundmasse, die zum Teil noch weisslich, zum Teil hell ockergelb gefärbt ist, wird dichter und braunlehmreicher. Die Mikromorphologie dieses Horizontes charakterisiert sich durch ein ausgesprochenes Leitbahngefüge mit sehr homogenen dichten Braunlehm-Füllmassen, die zum Teil geschichtete Anlagerungsgefüge zeigen. Die Eisenoxydhydrat-Abscheidungen kommen nicht nur in Form von Flecken und Kristallisationsnestern, sondern auch von roten, durchscheinenden Wandanlagerungen von auffallender Doppelbrechung vor.

B_2/g-Horizont, Probe aus 100–104 cm. Sehr dichter Horizont, der fast völlig aus eigelber Braunlehmgrundmasse mit Leitbahngefüge besteht. In dieser zeigen sich zugleich die oben genannten Pseudogleymerkmale (Abb.2). Dasselbe gilt auch für den folgenden B_2/g-Horizont. Als hauptsächliches Tonmineral wurde ebenso wie für die darunterliegende Fliesserde Illit gefunden.

Die E/B-Horizonte bestehen aus der dichten, stark peptisierten Braunlehm-Grundmasse mit starker, Teilchendoppelbrechung zeigender Schlierenbildung (Tafel D1).

Diskussion der Ergebnisse

Die stark südliche Lage, das rauhere Klima mit seinen Winterfrösten haben in dem ursprünglich braunlehmreichen und pseudovergleyten Oberboden nach dem Nachlassen der Bodenvernässung eine weitgehende Vererdung hervorgerufen, die sich durch die starke Durchschlämmung (unter Zurücklassung des flockigen Braunerde-Teilplasmas) noch verstärkt hat. Diese Vererdung hat aber nur bis zu einer Tiefe von etwa 50 cm gereicht. Die unteren braunlehmreichen Profilteile zeigen Spuren starker Schlämmstoffeinlagerung. Da die Austrocknung nicht so wirksam war wie in den Profilen NS14 und NS22, ist die extreme und irreversible Verhärtung nicht in dem Masse eingetreten wie dort.

Profil NS12: Lessivé-Braunerde, Pigroot, Central Otago.
Neuseeländische Bezeichnung: wahrscheinlich "Upland yellow brown earth".

Das Profil, das nicht näher untersucht werden konnte, stammt aus einer Seehöhe von ca. 200 m. Das Ausgangsmaterial ist Löss und Verwitterungsmaterial aus kristallinem Schiefer. Der Jahresniederschlag ist 650–750 mm, die Winter sind hart und weisen mehr als 200 Frosttage auf.

Profilaufbau.
0–20 cm: A-Hor., dunkelbraungrauer Schlufflehm mit lockerem erdigem Gefüge.
20–53 cm: B-Hor., braunockerfarbiger Schlufflehm, dichter als A-Hor. Die tieferen Profilteile wurden nicht erfasst.

Mikromorphologie.
A-Horizont, Probe aus 0–4 cm. Die Bodenmasse zeigt gute braune Vererdung, gute Aggregatbildung durch die ein lockeres hohlraumreiches Schwammgefüge entstanden ist. Peptisiertes Braunlehm-Teilplasma fehlt völlig, nur in bezug auf die Farbgebung zeigen sich in der braunen Lockermasse runde gelblich gefärbte Aggregate, wahrscheinlich Regenwurmlosung von Unterboden-Material. Das Feinskelett zeigt in auffallender Weise Biotite, Plagioklase, Augite mit geringer Verwitterung und kleine Bruchstücke von Gneis. Daneben zeigen sich wiederholt grüne und gelbliche Glaukonitkörner, auf deren Bedeutung für die Genese der Lösse später zurückgekommen werden soll. Auffallend sind auch die kleinen Braunlehmflitter, auf die bereits in Profil NS14 Bezug genommen wurde.

B-Horizont, Probe aus 30–34 cm. Dieser Horizont zeigt noch leichte Vererdung ist jedoch in der Farbe gelblicher, in den Aggregatformen eckiger. Doch ist die Verschlämmbarkeit auch hier nicht gross, peptisiertes Braunlehm-Teilplasma in Hohlräumen und Spalten fehlt so gut wie völlig.

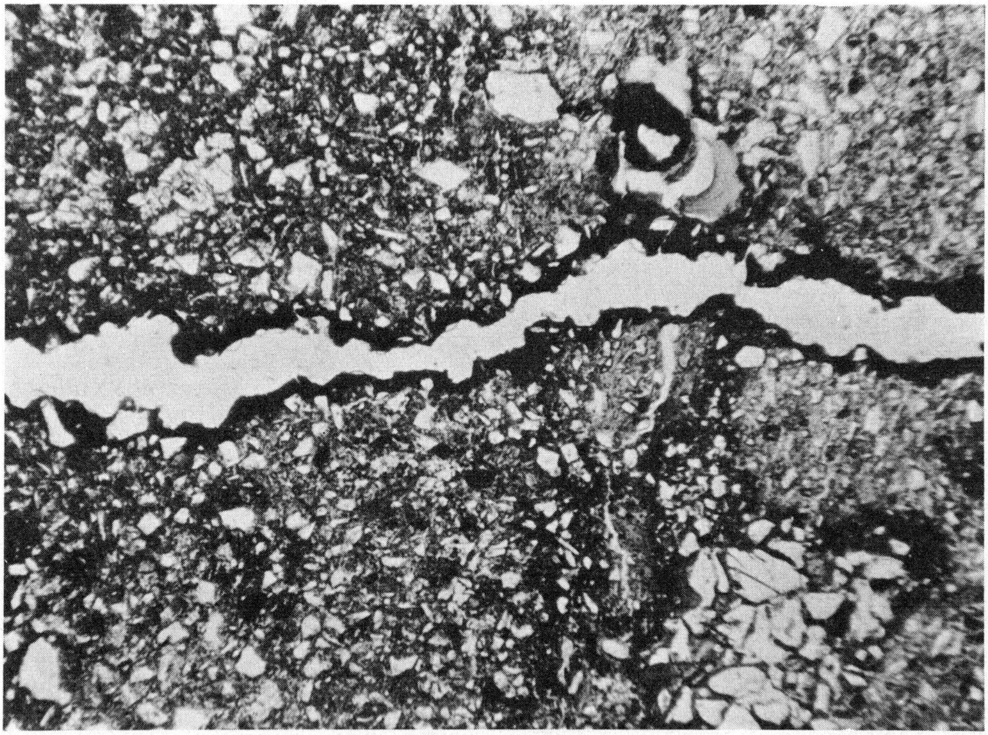

Abb.2. Untergrund von vererdetem Braunlehm-Lessivé, Otautau, E_2/B-Hor. Sehr dichtes, feinsandig-schluffiges Braunlehmgefüge mit Pseudogley-Merkmalen. Im Bilde randständige, der Bodenspalte folgende Ausscheidung von grellrotem Eisenoxydhydrat durch Auskristallisation in Form von Goethi

Diskussion der Ergebnisse

Auch hier wird der Einfluss des feucht-wechselwarmen Klimas mit häufigen Winterfrosten sichtbar, das in unvernässten Böden eine starke Vererdung oder wenigstens eine wesentliche Abstumpfung der Peptisierbarkeit des

Braunlehmplasmas mit sich bringt. Neben einem stark verwitterten Bodenanteil zeigt sich jedoch ein anderer, der, sichtbar in dem Zersetzungsgrad des Feinskeletts, eine verhältnismässig geringe Verwitterung aufweist. Er geht im wesentlichen auf detritischen Material von den kristallinen Schiefern der Otago Mountains zurück.

BODENENTWICKLUNG UND KLIMAWANDEL

Polygenese

Dass die Eigenart der hochentwickelten Lössböden zufolge des starken Klimawandels in den Zeiträumen, in denen sich ihre Entstehung vollzog, nur durch eine sehr komplexe Genese erklärt werden kann, hat der Verfasser bereits in einem Beitrag auf der Jahrestagung der Deutschen Quartärvereinigung im Jahre 1955 ausgeführt und zur Diskussion gestellt. Es wurde versucht, vor allem der Herkunft des Braunlehm-Teilplasmas in den Lessivéformen und Pseudogleyen nachzugehen. Bezugnehmend auf die damaligen Deutungsversuche (Kubiena, 1956) soll an der Hand des neuseeländischen Materials die Frage nach der Morphogenese wieder aufgenommen werden. Hierfür sollen auch neue Ergebnisse, erzielt an Untersuchungen von Lössböden Chinas und Argentiniens, verwendet werden. Über die Mikromorphologie und Entwicklung der "yellow-grey earths" Neuseelands, einer Bodengruppe, der ein grosser Teil der neuseeländischen Lessivés und Pseudogleys angehört, wurde am Internationalen Bodenkundigen Kongress in Paris von Raeside (1956) ein sehr beachtenswerter Beitrag vorgelegt, in dem dieser Autor, unabhängig vom Verfasser, auf Grund von Profilstudien, mineralogischen und mikromorphologischen Untersuchungen zu der gleichen Schlussfolgerung gelangt ist, dass die Entstehung dieser Böden nicht allein dem gegenwärtigen Klima zugeschrieben werden kann, sondern dass sie das Resultat von abwechselnd feuchtwarmen und trockenkühlen Klimaten gewesen sein musste. Besonders setzt die Bildung der Hartschichten im Unterboden der "yellow-grey earths" (vgl. Profil NS14 und Profil NS22) eine viel grössere jahreszeitliche Feuchtigkeit voraus, als sie dem gegenwärtigen trockenen und kühlen Klima ihrer Verbreitungsgebiete entspricht. Die gegenwärtigen Umweltverhältnisse könnten ferner niemals die Tonverlagerung hervorrufen, die sich im Profil in der Richtung der Hartschichten vollzogen hat (Leitbahnfüllungen und Leitbahn-anlagerungen). In vier von fünf Jahren zeigt gegenwärtig der Boden während der Sommer- und Herbstmonate nur eine Durchfeuchtung nahe dem Welkepunkt und während des übrigen Jahres wird der Untergrund nur kurze Zeit feucht (Raeside, 1956).

Braunerde- und Braunlehm-Komponente

Wenn wir die neuseeländischen Lössböden mit jenen anderen Gebieten vergleichen, ist wohl der ausserordentliche Reichtum an Braunlehmteilplasma bei den meisten Formen der Unterschied, der mikromorphologisch am stärksten ins Auge fällt. Mit ihm hängt auch die Neigung zu Pseudovergleyung und Hardpan-Bildung und wohl auch die starke Erodierbarkeit der Lössgebiete zusammen. Auf letzte weist auch Raeside hin und bringt mit ihr die durch Reduzierung und neuerliche Auflagerung entstandenen, ausserordentlich verschiedenen Profilmächtigkeiten in Zusammenhang.

Die Lössböden Neuseelands bestehen im wesentlichen aus zwei verschiedenen Komponenten, einer, die durch Braunlehmverwitterung, und einer, die durch Braunerdeverwitterung entstanden ist. Gegen den kälteren Süden zu und mit steigender Seehöhe nimmt die Braunerdekomponente zu, gegen den Norden und mit fallender Seehöhe die Braunlehmkomponente. In der erdigen Braunerdekomponente zeigt sich eine relativ geringe Verwitteru mit wenig angegriffenen Biotiten, Plagioklasen, Kalifeldspaten und Augiten, in der Braunlehmkomponente ist die Verwitterung hoch, alle Silikate sind in stark peptisierte, ockergelbe, durch amorphes Eisenoxyd gefärbte Tonmassen umgewandelt. In den nördlichen und küstennahen Lessivés kann die Braunlehmkomponente so stark zunehmen, dass der Boden fast ganz aus ihr besteht, in den südlichen und hochgelegenen Varianten kann sie auf mengenmässig stark zurücktretende und wenig inkorporierte (durch Alterung verhärtete) Braunlehmflitter beschränkt bleiben.

Vergleich mit monogenetischen Böden in situ

Dieses wechselnde Verhältnis zwischen den beiden Aufbaukomponenten mach die Untersuchung des Gefügeaufbaus in streng autochthonen Bodenbildungen interessant, die in situ auf bestimmten Muttergesteinen entstanden sind. Der Verfasser hatte Gelegenheit, Proben von Böden auf Grauwacke und Glimmer schiefer auf dem Old Man Range und dessen Umgebung zu nehmen und die Bodenentwicklung bis auf dessen Gipfel zu verfolgen. Ein Profil einer subalpinen Braunerde in einer Höhe von ca. 1.800 m entwickelte unter *Dracophyllum prostratum* und anderen Polsterpflanzen einen A-Horizont vor 15 cm und einen B-Horizont mit geringer chemischer Verwitterung von 30 cm Mächtigkeit unmittelbar über dem Gestein. An anderer Profilstelle zeigt sich im 50 cm Tiefe Permafrost. Alle untersuchten Böden zeigten ein ausgesprochenes Braunerdegefüge. Aus den Bereichen mit grösserer Seehöhe sind keine Braunlehmrelikte bekannt. Auch die wenig verwitterten Ando-Böden bei Waioulu auf ca. 700 Jahre alter rhyolitischer Asche und ca. 2.700 Jahre alter andesitischer Asche bei Waioulu und bei Ohakuni im Bereiche des Ruapehu-Vulkans, Nordinsel, zeigen mikromorphologisch Braunerde-Elemente. Hingegen sind in tieferer Lage auf Grauwacke auf der Nordinsel ausserordentlich intensiv verwitterte, schwere, völlig bolusartige Braunlehme zu finden, wie der Taita clay loam auf den Hügeln der Ostseite des Hutt Valley bei Wellington (Profil NS21). Das vom Soil Survey vorgeführte Profil oberhalb seiner Forschungsstation ist ein vorzügliches Beispiel dieser Entwicklung und kann von uns zum Ausgangspunkt unserer vergleichenden Untersuchungen gemacht werden. Sein ausserordentlich dichter, ockergelber B-Horizont liegt mit einer Mächtigkeit von 110 cm auf einem braunroten, pseudogleyigen B–C-Horizont auf. Mikromorphologisch zeigt er ein stark peptisiertes, hohlraumfreies Braunlehmgefüge, wie es extrem humiden tropischen bis subtropischen Umweltverhältnissen entspricht (Tafel D2). Starke Schlierenbildung, die zwischen gekreuzten Nicols in Erscheinung tritt, deutet auf hohe Beweglichkeit der gesamten Bodenmasse und auf das Vorhandensein illitischer bzw. vermikulitischer Tone, die durch Analysen des Soil Survey nebst hohem Feinglimmer und Kaolinitgehalt bestätigt wird. Die in den Befunden aufscheinende Zunahme von Eisenoxydhydrat in den unteren Profilteilen und die Feststellung von Hämatit deutet auf die mikromorphologisch deutlich in Erscheinung tretende Neigung zu Pseudovergleyung subtropischer bis tropischer Prägung hin. Klimatisch zeigt der Standort eine Seehöhe von 115 m, eine Neigung von 16°N, eine jährliche Regenmenge von 1.300 mm, eine mittlere Januartemperatur von 16°C und eine mittlere Julitemperatur (Winter) von 4° C. Die gegenwärtige Vegetation ist Wald von *Pinus radiata* mit *Leptospermum manuka*, *Erica lusitanica* und *Ulex europaeus*. Die natürliche Vegetation war Wald von *Nothophagus truncata*, die geographische Lage ist 41°S 174° E.

Planetarische Lagevergleichung

Vergleicht man Umweltverhältnisse und Bodenbildungen obiger geographischer Lage in Neuseeland mit denen jener Punkte, die ihr planetarisch in anderen Teilen des Erdglobus entsprechen, so schneidet der entsprechende Meridian der antipodischen Lage (6° E 41° N) eine Linie im Mittelländischen Meer, die Barcelona mit der Stadt Bonifacio im südlichen Korsika verbindet. In Spanien zeigen sich nächst Barcelona auf Silikatgesteinen gut erhaltene (wenig verėrdete) Relikte von Braunlehm und Rotlehm, daneben grosse Flächen wenig veränderter Braunlehm- und Rotlehmsedimente. Die rezente Bodenbildung hat den Charakter einer Trockenbraunerde mit geringer chemischer Verwitterung und geht erst in den Pyrenäen mit steigender Seehöhe in humide Braunerden über. Lösse wurden kürzlich im Bereich von Barcelona festgestellt (Solé Sabaris u.a., 1957). Es scheint, dass auch sie z.T. braunlehmreiche und rote pseudogleyige Bodenbildungen enthalten, doch liegen noch keine mikromorphologischen Untersuchungen in dieser Richtung vor. In der antözischen Lage (41° N 174° E) schneidet der neuseeländische Meridian von Taita im pazifischen Ozean eine Linie, welche die Stadt Hachinohe auf der japanischen Insel Hondo mit der Stadt Bureka in Kalifornien verbindet. Im kalifornischen Bereich kommen gut konservierte Braunlehmrelikte vor; die rezente Bodenbildung begünstigt jedoch in dieser nördlichen Lage Bildungen, in denen braunerdeähnliche Gefüge-Elemente überwiegen. Die Lage in Japan ist rauher und charakterisiert sich durch harte Winter. Da hier die Schneegrenze besonders tief ist, werden die auch in Neuseeland vorkommenden Andoböden mit ausgesprochener Braunerde-Mikromorphologie in Nord Hondo weit tiefer in die küstennahen Regionen reichen und Braunerden bis Semipodsole die nicht alpinen und subalpinen Lagen und die nicht vernässten Küstengebiete der Insel vielleicht völlig beherrschen. Die der antözischen Lage am nächsten liegenden bekannten Lössböden beginnen westlich von Peking und erstrecken sich auf einer breiten Fläche mit extrem strengen kontinentalen Wintern gegen Südwesten. Der Verfasser hatte Gelegenheit, in das von Herrn Prof. Dr. Kowalinski bei seiner Forschungsreise im chinesischen Lössgebiet aufgesammelte Probematerial und in die davon hergestellten Dünnschliffe Einblick zu nehmen. Hier sind ausgesprochene Braunerden mit völlig fehlendem Braunlehm-Teilplasma häufig (Abb.3), doch auch die Lessivéformen zeigen ein gut entwickeltes Braunerde-

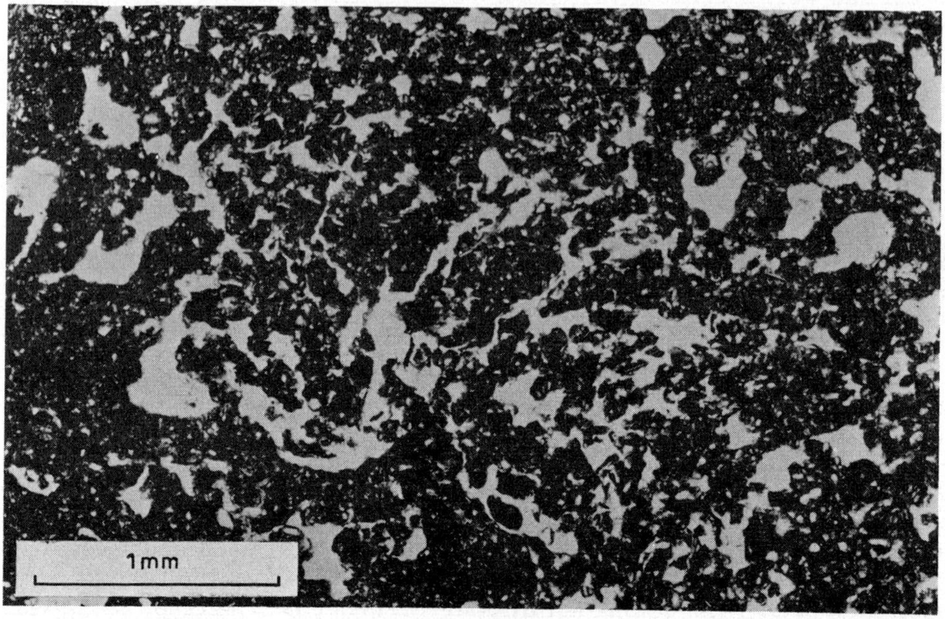

Abb.3. Durch Klima-einfluss stark vererdeter, hohlraumreicher, Lössboden aus China (Probenahme Prof. Dr. Kowalinski).

Grundplasma und neigen im Braunlehm-Teilplasma häufig zu brauner Granulierung und sekundärer brauner Vererdung. In der periözischen Lage schneidet der ihr entsprechende Meridian (6° E) den Parallelkreis von Taita (41° S) in einem Punkt des südlichen Atlantischen Meeres $7\frac{1}{2}°$ östlicher und $\frac{1}{2}°$ südlicher als die Insel Gough. Über deren Bodenbildungen ist auch im Schrifttum nichts bekannt. Die nächstliegenden Böden finden sich in der Kap-Provinz Südafrikas fünf Breitengrade weiter nördlich. Hier fand der Verfasser vorzüglich erhaltene Relikte von Braunlehmen und Rotlehmen und grosse Flächen von Bodensedimenten gleichen Charakters vor. In 500 m Seehöhe können sich humide Braunerden bilden. Nur auf den extrem nährstoffarmen Tafelsandsteinen entwickeln sich auf der Gipfelfläche des Tafelberges (1.082 m) Ranker und auf dem Kap der Guten Hoffnung unter üppigen Erikabüschen sogar Humuspodsole. Die etwa auf gleicher Breite, aber wesentlich nördlicher reichenden Lössgebiete Argentiniens weisen Lössböden auf, die der Gruppe D der Weltübersicht (Tabelle I) angehören und dem streng kontinentalen Steppenklima entsprechend, schwarzerdeähnliche Varianten mit einem gewissen Gehalt an nicht aktivierten Braunlehmflokken, aber ohne wesentliche Bildung von Lessivé-Gefügen und Leitbahn-Einlagerungen zeigen.

Diskussion der Lagevergleichung

Obige Gegenüberstellungen ergeben, dass die neuseeländischen Braunlehme im Bereiche von Taita nicht nur die beste Entwicklung, sondern auch unter allen planetarisch entsprechenden Lagen, die für eine Braunlehmbildung günstigsten Umweltbedingungen zeigen. Auch dann, wenn die heutigen Verhältnisse eine weitere Braunlehmbildung in dieser Form nicht mehr hervorbringen könnten, so wird sie doch so, dass die Böden in ihrem Braunlehmcharakter erhalten bleiben und keinen allzugrossen Veränderungen (ausser jenen durch Staunässe) unterworfen sind. Dies wird auch dadurch verstärkt, dass das Niederschlagsmaximum hier auf den Januar d.i. auf den wärmsten Sommermonat fällt, aber reiche Niederschläge im Laufe des Winters zu verzeichnen sind. Die bisherigen Erfahrungen zeigen, dass Gebiete, die reich an Braunlehmen und Braunlehmsedimenten sind, auch in den Lössböden einen grösseren Gehalt an Braunlehm-Teilplasma zeigen. In gleicher Weise kann auch das Umgekehrte für braunlehmarme Bereiche behauptet werden. Neuseeland ist besonders reich an braunlehmreichen Lessivés und Pseudogleyen und steht in dieser Beziehung unter allen bekannten Lössgebieten an der Spitze.

UNTERSUCHUNGEN ZUR MIKROGENESE

Herkunft des Braunerde-Teilplasmas

Darüber, dass ein Teil des in den Lössböden enthaltenen Ausgangsmateriales aus den verwitternden Gesteinen stammt, welche die Inseln aufbauen, besteht kein Zweifel. Die Herkunft ist nicht nur aus den überaus resistenten Silikaten wie Epidot, Klinozoisit, Zirkon und Turmalin nachweisbar, sondern kann auch an sonst leicht verwitterbaren Mineralien wie Biotit, Augit, Chlorit, Plagioklasen und Orthoklasen erkannt werden. Ihr verhältnismässig geringerer Zersetzungsgrad lässt vermuten, dass es sich in erster Linie um Material von Gebirgsböden und deren Sedimenten handelt, in denen das in Betracht kommende Ausgangsmaterial zu suchen ist. Diese werden auch in Bezug auf ihr lockeres Gefüge am leichtesten durch Winderosion zerstört und aus ihnen die feinsten Kornarten zum Transport aussortiert. Ihre weitere Verwitterung führt heute, besonders in den rauheren Lagen, zur Vermehrung der Braunerdekomponente.

Herkunft des Braunlehm-Teilplasmas

Mit Sicherheit kann angenommen werden, dass das Vorkommen von Braunlehmen in situ auf Silikatgesteinen in der Vorzeit in Neuseeland z.T. wesentlich grösser war als in der Gegenwart, dass sie in unvererdeten Formen auch höher an den Berghängen hinaufreichten. Sie und ihre Erosionssedimente sind wahrscheinlich die Hauptquelle des Braunlehm-Teilplasmas gewesen. Doch kommen zusätzlich auch andere Quellen in Betracht. An den Küsten von Braunlehmgebieten sind die Schelfablagerungen in der Regel braunlehmreich. Die Braunlehm-Sinkstoffe können völlig frei (in kleinen Aggregaten) abgelagert oder in organogenen Kalkabscheidungen eingeschlossen vorliegen (Tafel D3). In beiden Fällen werden sie aus den in den Kaltzeiten wasserfrei gewordenen Schelfgebieten durch den Wind landeinwärts geweht. Ein wichtiges Leitmineral für diesen Einfluss sind die häufigen Glaukonitkörner, die sich nicht nur im Rohlöss, sondern auch in den meisten Lessivés und sogar in höher gelegenen Lössbraunerden wiederfinden. Sie sind neben Braunlehmdetritus ein wesentlicher Gemengteil in fast allen tertiären und vortertiären Schelfsedimenten Neuseelands. Ein anderer Gemengteil von besonderer Tragkraft sind die von Raeside erstmalig festgestellten Seeigel-Stacheln, die mit Küstennähe zunehmen und gleichfalls die Herkunft eines Teiles des Ausgangsmateriales aus marinen Sedimenten bestätigen.

Eigenschaften des Rohlöss

Dieser unterscheidet sich mikromorphologisch in zwei Merkmalen von den europäischen Lössen: (1) Dass sie nur geringe Kalkgehalte zeigen und (2) dass ihr Anteil an Braunlehmelementen stellenweise ziemlich hoch sein kann. Dieser Braunlehm-Anteil ist auch schon im Rohlöss u. zw. noch im Zustande des lockeren hohlraumreichen Gefüges (das im ursprünglichen Aufbau auch erweiterte Intergranularräume zeigen kann) hoch beweglich und kann sich in wandständigen Anlagerungen in den Leitbahnen und grösseren Kornzwischenräumen ablagern. Dennoch stehen auch solche Lösse in steilen Wänden an. An den Steilabfällen zum Rakaia-River (Südinsel) findet sich ein Rohlöss mit Wandanlagerungen von auffallend orangeroter Farbe. Plasma-Ansammlungen dieser Art werden auch makroskopisch in Form von horizontalen rötlichen Bänderungen sichtbar. Die Pseudotschernosemprofile, die sich auf ihm entwickelt haben (von einem rezenten Rohlöss überlagert), deuten auf ehemalige Vernässung, das rote Plasma auf pseudogley-ähnliche Vorgänge hin.

Rasche Verlehmung im Pleistozän

Auffallend ist, dass hellgefärbte Ablagerungen von Rohlöss gegenüber verlehmten bzw. vererdeten Lössen stark zurücktreten. In dem Verhältnis zwischen Rohlöss und Verlehmungszonen zeigt sich gegenüber Europa und den mächtigen Rohlösswänden in China eine völlige Umkehrung. Dies lässt vermuten, dass die Lösse in den Interglazialzeiten verhältnismässig rasch zu tiefgründigen Lessivés und Pseudogleys sowie Fliesserden und Erosionssedimenten von solchen umgewandelt worden sind. Die vielen Pseudogleyspuren, Anmoorformen und Gleye deuten darauf hin, dass sie um vieles feuchter waren als die Gegenwart. Unter dauernder Feuchthaltung und Staunässe wird nicht nur eine Vermehrung des Braunlehm-Teilplasmas und ein völliger Abbau vorhandenen Braunerde-Teilplasmas bewirkt, die Braunlehmkomponente wird auch weitgehend von sekundärer Vererdung in den Kaltzeiten geschützt, die sich nur unter ausgesprochen terrestrischen Umweltverhältnissen vollzieht. Die starke Entkalkung scheint gleichfalls eine Folge des feuchtwarmen Klimas der Interglazialzeiten Neuseelands gewesen zu sein. In immerfeuchten Braunlehmgebieten über Kalkstein erscheinen nicht nur die Verwitterungsdecken völlig entkalkt, die Entkalkung greift nach erfolgter Verkieselung der oberen Gesteinschichten auf diese selbst über, so dass sie, wie im Gebiet der tropischen Terra fusca von Mvuasi Kokosi im unteren Kongo, mit Salzsäure kaum noch aufbrausen.

Rezente Zunahme der Xeromorphie des Bodens

In fast allen Teilen der heutigen Subtropen (einschliesslich der wechselfeuchten Tropen) zeigt sich in den Bodenbildungen eine starke Zunahme der Xeromorphiemerkmale, die weitgehend auf die heutigen Mediterrangebiete übergreift (vgl. Kubiena, 1962a, 1962b, 1964). Auch in all den Gebieten, die in S.229–230 zur planetarischen Lagevergleichung zitiert wurden, tritt diese Xeromorphiezunahme deutlich in Erscheinung. Wie überall so wirkt sich auch in Neuseeland der Einfluss des Menschen durch seine Wirtschaftsweise sowie durch seine Eingriffe in den natürlichen Landschaftshaushalt den Einfluss des natürlichen Klimawandels verstärkend und beschleunigend auf die Entwicklung der Xeromorphiemerkmale des Bodens aus. Sie hat in Neuseeland jedoch nur eine erhöhte Austrocknung und einen verstärkten Gefügezerfall mit einer erhöhten Erosierbarkeit durch den Wind bei den vererdeten Formen und eine zunehmende Verhärtung der Unterböden der braunlehmreichen Varianten zur Folge gehabt.

ANERKENNUNG

Der Verfasser dankt Herrn Dr. McCraw für die vorzügliche Führung und die zahlreichen Auskünfte während der Durchführung der Reise, Herrn Dr. Norman H. Taylor und dem Soil Survey Neuseelands für zahlreiche Hilfen im Lande selbst und die freundliche Übernahme des Transportes der Bodenproben nach Europa.

SUMMARY

Micromorphology and micromorphogenesis of the loess soils of New Zealand

According to our nomenclature, the loess soils of New Zealand can be designated as lessivés and pseudogleys with high braunlehm content. In the southern locations and those of higher elevation characterized by greater frost frequency and alternate humidity, a more or less strongly developed tendency to the formation of earthy varieties ("braune Vererdung") can be observed. This leads partly to the genesis of typical braunerde. The loess varieties rich in braunlehm tend to water stagnation, pseudogleyization and gleyization. The respective humus forms vary from dystrophic forest moder to grassland-mull, from anmoor varieties to pseudochernozem humus.

In a world system of the loess soils, the varieties rich in braunlehm of New Zealand and their development tendencies range in a particular group of formations which are associated with submediterranean to subtropical environment conditions. They are characterized by an extreme density, by a tendency to hardpan formation and by a particular reddish pseudogleyization very frequently in warm climates. They are, however, only partly products of the recent climatic conditions.

The monogenetic soil formations in situ on silicate rocks in the region of the loess soils and their erosion sediments are either braunlehm or braunerde varieties. The different loess soils vary most typically in their ratio of braunerde and braunlehm plasma content. The lessivé soils rich in braunlehm plasma are primarily to be found in braunlehm areas. In typically earthy loess soils, the braunlehm plasma is in most cases only present in the form of inactive and irreversibly hardened fabric fragments.

A comparative investigation of the planetary position shows that the loess soils of New Zealand, as varieties rich in braunlehm, possess not only their utmost development, but that also their locations present the best conditions for their conservation among all corresponding latitudes.

The loess varieties of New Zealand are mostly low in calcium carbonate content, but in spite of this, they have a porous fabric rich in cavities and partly considerably widened intergranular spaces. In humid environments, the admixture of braunlehm plasma becomes activated already in the raw loess, in which it may be peptized in its percolation water and precipitated in the form of depositions on the space walls.

In opposition to European loess occurrences, raw loess deposits are less frequent in New Zealand than highly decomposed lessivé and pseudogley formations or their respective erosion sediments. Under the present environment conditions, a marked increase of xeromorphology characters can be observed in wide sections of the loess soil area.

LITERATUR

Excursion Guide, Southern Tour. International Soil Conference, New Zealand, 1962, 115 pp.

Kubiena, W., 1956. Zur Mikromorphologie, Systematik und Entwicklung der rezenten und fossilen Lössböden. Eiszeitalter Gegenwart, 7 : 102–112.

Kubiena, W., 1962a. Bodenkunde. In: H. Abel (Redakteur), Deutsche Afrika-Wissenschaft – Stand und Aufgaben. Wissenschaftlicher Ausschuss Deut. Afrika Ges., Köln, pp.7–17.

Kubiena, W., 1962b. Die Böden des Mediterranen Raumes. In: Kalium Symposium, Bern, pp.167–190.

Kubiena, W., 1964. Welche Aufgaben hat die Planung der Bodennutzung in den Tropen. Ber. Tagung Projektleiter, Feldafing, Deut. Stift. Entwicklungsländer, Bad Godesberg, im Druck.

Pohlen, I.J., 1962. Soil classification in New Zealand. Trans. Joint Meeting Comm. IV and V Int. Soc. Soil Sci., pp.440–452.

Raeside, J.D., 1949. Ecological investigations in the Nothofagus-forests in New Zealand. New Zealand J. Forestry, 5 : 401–410.

Raeside, J.D., 1956. Yellow-grey earths of South Island, New Zealand, an example of polygenesis in soil development. Congr. Sci. Sol, 6e, Paris, V 107.

Solé Sabaris, L., Virgili, C. et Ripoll Perelló, E., 1957. Environs de Barcelona et Montserrat. Livret guide de l'excursion B, Intern. Assoc. Quaternary Res., Congrès Intern., 5e.

Taylor, N.H. and Pohlen, I.J., 1962. Soil survey method. New Zealand, Dept. Sci. Ind. Res., Soil Bur. Bull., 25 : 242 pp.

Woldstedt, P., 1962. Die Vergletscherung Neuseelands und die Frage ihrer Gleichzeitigkeit mit den europäischen Vereisungen. Eiszeitalter Gegenwart, 12 : 18–24.

FARBENBILDER

Tafel B8. Dünnschliff vom A-Hor. des vererdeten Braunlehm-Lessivés, Otautau. Probe aus 2–6 cm. Durch Klima-einwirkung entstandenes stark vererdetes, flockiges Braunerdegefüge; x 75.

Tafel D1. Gleicher Boden, Probe aus dem Untergrund, 180–184 cm. Dichtes, unvererdetes Braunlehmgefüge. Auf der linken Bildseite oben und schräg unten gefügefremde gealterte Braunlehmfragmente; x 75.

Tafel D2. Braunlehm in situ auf Grauwacke, NS21, Taita, Probe aus dem B-Hor. Sehr dichtes Braunlehmgefüge mit beginnender Ausscheidung von roten Kristallisationskomplexen (Neigung zu Pseudovergleyung); x 75.

Tafel D3. Im Schelfgebiet abgelagerter tertiärer Mergel mit Füllmassen von Braunlehm in Hohlräumen. Invercargill, Südinsel. Fragmente dieser Füllmassen finden sich im gealterten Zustande in den Lössen und Lössböden (vgl. Tafel D1); x 75.

OBSERVATIONS ON THE ORIGIN OF A CUTAN IN THE YELLOW-BROWN SOILS OF THE HIGHLANDS OF NEW GUINEA

G.K. RUTHERFORD

Department of Geography, Queen's University, Kingston, Ont. (Canada)

INTRODUCTION

Plasma concentrations and plasma separations associated with natural surfaces in soil materials are common pedological features. Some of these phenomena have been recognised as coatings of clay minerals and they have been referred to as "clay skins", "Tonhäutchen" etc., terms which imply that surfaces of peds or voids have been coated with clay by deposition after illuviation in suspension. As many phenomena have been observed which are neither "clay" nor depositional in origin, Brewer (1960) coined the term cutan and defined it as:

> "a modification of the texture, structure or fabric at natural surfaces in soil material due to concentration of particular components or in situ modification of the plasma; cutans can be composed of any of the component substances of plasma."

During the course of soil survey in the Highlands of New Guinea, the author noticed what appeared to be very strongly developed cutans in the yellow-brown soils which are the zonal soil of the area. The object of this paper is to indicate the nature of the cutans compared with the surrounding soil matrix and the origin of them.

PHYSICAL SETTING

The soil profile investigated occurs near Nipa patrol post in the Southern Highlands District in the Territory of Papua and New Guinea. The present rugged and extremely complex landscape is largely the product of Plio-Pleistocene orogeny and dissection and Pleistocene vulcanism which has blanketed large areas of the landscape with ash showers of andesitic composition. The annual precipitation of 2,250–3,400 mm is well distributed throughout the year. Annual variations in rainfall are small, and there are no prolonged dry periods. The mean percentage cloud cover is 80–95% and the relative humidity approximately 70% at 3 p.m. The potential evapotranspiration is estimated to be 750–1,000 mm. Thus the climate can only be described as perhumid.

The mean annual temperature is approximately 11–20° C depending on the altitude and monthly fluctuations are small. The mean annual temperature at Nipa is estimated to be approximately 18° C. Thus the effect of the climate on soil formation is continuous without any periods of dormancy due to drought or low temperatures.

THE SOIL

Rutherford (1962) first drew attention to the widespread occurances of morphologically similar humic yellow-brown to strong brown clay soils on a variety of parent rocks in the highland districts of Papua and New Guinea. He briefly discussed their affinities to the yellow-brown earths and yellow-brown loams in New Zealand and the inceptisols of the United States Department of Agriculture seventh approximation (1960) and tentatively suggested they should be known as yellow-brown soils. Haantjens and Rutherford (in preparation) have assessed in more detail the zonal aspects of these soils.

Soil profile description

On gently sloping ash covered ridge crest with 3° slope; *Themeda imperata* garden regrowth succession.

0–3 inch. [1] 5YR3/2 (7.5YR4/2) Dark reddish brown silty clay loam; weak medium angular blocky structure; very friable when moist, moderately hard when dry; abrupt, smooth boundary; grass roots form tangled mat.

3–9 inch. 7.5YR5/6 (7.5YR4/6) Strong brown clay; medium angular blocky structure; friable when moist, slightly hard when dry; gradual, regular boundary; moderate amounts of dark minerals; common, fine roots.

9–>50 inch. 7.5YR4/6 (5YR3/4) Strong brown clay; "structureless"; well developed 7.5YR5/6 (10YR7/6) cutans up to 1/8 inch thick; some 2.5YR3/6 dark red cutans; friable when moist, slightly hard when dry; soil matrix high in dark minerals.

EXPERIMENTAL

Soil samples with cutans at 16 and 44 inches from the surface were selected, impregnated with "Plastrene 47" and slides for inspection with petrographical microscope were prepared. The soil samples were selected as far as possible to include the boundary between the soil matrix and a cutan. The cutans were carefully scraped from the dry soil matrix. The cutan and the soil matrix samples from both depths were sent for X-ray analysis to Mr. S.A. Forman, Head Mineralogy Section, Soil Research Institute, Canada Department of Agriculture, Ottawa.

RESULTS OF X-RAY ANALYSES

Initial X-ray patterns showed no significant differences between the matrix of the 16 inch and the 44 inch sample. Neither could any significant difference be detected between the cutans in the 16 and 44 inch samples. The soil

[1] Dry Munsel colours in brackets.

matrix and cutan in the 44 inch sample were then subjected to closer analysis.

X-ray analysis of the 0.002 mm fraction gave the following results:

(1) Cutan; consists mainly of gibbsite, moderate amounts of amorphous material assumed to be allophane and minor magnetite.

(2) Matrix; predominantly magnetite with minor gibbsite and allophane with a trace of quartz and actinolite.

(3) One X-ray line appeared in both the cutan and the matrix at 4.05Å. This could not be related to any specific compound but it could be the strongest line for ramsdellite (MnO_2). Chemical tests for Mn in the matrix showed 395 p.p.m.

(4) The X-ray patterns indicated that all fragments of the soil material appeared to be composed of a series of cores surrounded by clay size material. A likely explanation is that the intense leaching in such a perhumid climate has caused clay mineral formation to be centered around a resistant non-reactive core such as magnetite. There is no evidence from the slides to indicate any such concentric arrangement in the soil fabric.

(5) Although hornblende is very common in the fine sand fraction of the matrix it could not be identified in the clay size fraction of the matrix or cutan.

MICROMORPHOLOGY OF THE CUTAN

It is not the scope of this paper to describe the plasmic fabrics of the soil matrix but relevant information on the nature of the cutan is presented. Brewer (1960) suggested that the most useful sequence for considering the differentiating characteristics of cutans is:(1) characteristics of surfaces affected; (2) mineralogical nature of the cutanic material; (3) fabric of the cutans; (4) additional features.

With respect to the soil materials under discussion the following observations were noted:

(1) The cutans occur on both channel and joint and skew planes.

(2) The cutanic material in reflected light has an uneven ropy appearance with waxy lustre. In the field they have a distinct "fatty" feel. X-ray analysis shows that the cutanic material is predominantly gibbsite with some amorphous material (allophane?) and magnetite.

(3) It is suggested that the fabric of the cutan should be described by (a) the sharpness of the boundary and (b) the orientation pattern (Brewer and Sleeman, 1960). The cutanic material was found to be (a) sharply divided from the matrix and (b) at 200 magnifications, appears moderately oriented and striated. The orientation is parallel to the cutanic surface.

(4) Additional characteristics: (a) Separation; the cutanic material is strongly separated from the matrix so that its characteristics are entirely due to the characteristics of the concentrated component of the plasma. (b) Adhesion; the cutanic material is weakly adhesive to discrete when gently crushed in the dry state.

CLASSIFICATION OF THE CUTANIC MATERIAL

The material is a simple cutan composed of a uniformly intimate mixture of magnetite, gibbsite and allophane. The discrete nature of the cutans, their lack of adhesion, their strong orientation and their position on channels and joint and skew planes indicate a probable illuvial origin. In addition they are composed mainly of materials which readily form suspensions and undergo reorientation. However, whilst no one characteristic is completely diagnostic for the process of formation, the perhumid environment and the conclusions reached by Haantjens and Rutherford (in press) that the yellow-brown soils probably never have been dry is further evidence on the illuvial nature of these cutans.

ACKNOWLEDGEMENTS

The author wishes to express his deepest gratitude to Messrs. S.A. Forman and N.M. Miles of the Canada Department of Agriculture for carrying out X-ray analysis.

SUMMARY

A cutan occurring in the deep B-horizon of a yellow-brown soil formed from andesitic ash in the Highlands of New Guinea is described and compared with the matrix material. X-ray analyses of the clay-sized fractions of the cutan and matrix show that the matrix is composed mainly of magnetite with some gibbsite and amorphous material (supposed to be similar to allophane). The cutan is composed mainly of gibbsite with moderate amounts of amorphous material and minor magnetite. In the matrix the intense leaching appears to have caused clay formation to be centered around non-reactive cores such as magnetite. The position of the cutans on channels, and joint and skew planes together with their discrete nature and lack of adhesion indicates a probable illuvial origin.

REFERENCES

Brewer, R., 1960. Cutans: their definition, recognition, and interpretation. J. Soil Sci 11 (2) : 280–292.

Brewer, R. and Sleeman, J.R., 1960. Soil structure and fabric: their definition and description. J. Soil Sci.,11 (1) : 172–185.

Haantjens, H.A. and Rutherford, G.K., (in preparation). Soil zonality and parent rock in a perhumid tropical mountain region. Trans. Intern. Congr. Soil Sci., 8th, Bucharest, 1964.

Rutherford, G.K., 1962. The yellow-brown soils of the highlands of New Guinea. Trans. Joint Meeting Comm. 4 and 5, Intern. Soc. Soil Sci., Palmerston North, pp. 434–439.

A CONTRIBUTION TO THE MICROMORPHOLOGICAL INVESTIGATIONS OF PSEUDOGLEYS IN THE NORTH-WESTERN PART OF YUGOSLAVIA

Z. RACZ

Department of Soil Science, Faculty of Agriculture, Zagreb (Yugoslavia)[1]

In the opinion of Škorić (In: Škorić et al., 1962), Cirić (In: Filipovski and Cirić, 1963), and other Yugoslav pedologists who are concerned with the study of pseudogleys, the morphology of these soils is characteristic. On its basis they can easily be distinguished from other soil types. However, when one wishes to estimate the dynamics of the development of pseudogley one encounters difficulties. It is frequently difficult to bring the existing morphological properties into accord with recent pedogenetical processes. Therefore the authors suggest that in working with these soils the criterion of relictness, as well as other criteria, should be introduced for easier subdivision into lower systematic units.

The aim of this work is to study the same problems on the basis of the micromorphological properties of pseudogley. By the application of microscopic technique we can expect detailed knowledge of the morphology of these soils as well as a better possibility for the mutual delimitation of specific pedodynamic formations which came into being under the influence of recent or relict processes.

The applied procedure for the preparation of thin sections is basically similar to the one recommended by Altemüller (1962b). The only difference is that an unsaturated polyester resin of Yugoslav manufacture, "Chromoplast M 9851" (Racz, 1963), was used.

BASIC DATA ON INVESTIGATED SOILS

In the northwestern part of Yugoslavia pseudogleys are found mainly on deep strata of non-calcareous Pleistocene loams. This area has a moderately continental climate with a mean annual temperature of about 11° C, and an average annual rainfall of about 900 mm. The topography is from undulating to hilly, and the typical pseudogley occurs on terraced sites. The duration of the stagnation of atmospheric water is characterized by the middle period (medium time interval) of excess moistening. In the past a forest of sessile oak and hornbeam (*Querceto-Carpinetum croaticum*) was located here. The major part of the forest was cleared and today the area is used for the cultivation of field crops.

The first investigations on pseudogleys in Yugoslavia were also related to these soils developed on Pleistocene loams (Neugebauer, 1958). One of

[1] Head of the Institute: Prof. Dr. A. Škorić.

Fig.1. Vertical section of the pseudogley profile on Pleistocene loams. Transition part of g-horizon and marmorized B1g-subhorizon.

Fig.2. Vertical section of the pseudogley profile on Pleistocene loams. B2g-subhorizon; first and commencement of second horizons of fossil soil.

these profiles (in the Maksimir Forest, near Zagreb) was presented and explained by Neugebauer and his collaborators to the participants at the International Conference for the Genesis, Classification and Cartography of Soils of South-European Countries (Zagreb–Beograd, 1958), and was the subject of lively discussion as to the delimitation of concepts of lessivé soils, pseudogley and podzols. On the basis of available data on the same profile, which were published later (Neugebauer et al., 1961), as well as on the basis of the works of Škorić et al. (1962) and Filipovski and Cirić (1963) mentioned in the introduction, a layered structure of the profile occurs frequently in these soils. Horizons of fossil soil follow the upper layer of profiles with an Agh-g-Bg [1] sequence of horizons (Fig.1 and 2). According to Janeković (1960

[1] Horizon designation after Neugebauer et al. (1961) (Ed.).

the non-calcareous Pleistocene loams are of aeolian origin, and a gradual deposition of these materials during the Würm stage conditioned their layered structure. He believes that only the surface horizons correspond to the Holocene, while the marmorized Bg-horizon represents a fossil "tundra soil" belonging to the last great Würm stadial (Würm III). One or several preserved layers of fossil pseudogleys and fossil tundra soil follow up to the Pliocene substratum.

Wishing to study the question of recent and relict micromorphological properties as completely as possible, comparative investigations of the described soils with pseudogleys on other substrata of neighbouring areas were also carried out. One of them was a transitional subtype lessivé-pseudogley. It developed on carbonate loess, and possesses a plain formed structure of the profile and the horizons Apg–A2g–gB–Cca. However, a layered structure occurs again in the second comparative profile, which developed on deeper deposits of terra rossa. Signs of gleyification can be seen here in the upper, namely the former A3 and Bt-horizons of lessivé terra rossa, and in the lower parts of the profile, actually in the fossil D-horizons.

RESULTS OF MICROMORPHOLOGICAL INVESTIGATIONS AND DISCUSSION

Characteristics of superficial horizons

In regards to the more detailed study of micromorphological characteristics of individual horizons, it should be noted that in the investigated soils the Ao-subhorizon rarely occurs. In so far as it is actually found in forest soil varieties, it is fresh litter which will decompose in the course of the year.

A1-horizon is well-developed, but its thickness and colour vary depending on the vegetation and the amount of organic matter. The organic matter is well dispersed and exhibits features of mature humus. Sometimes small concretions already appear in it, so that it may also be designated as Agh-horizon. Here, as well as in the following g-horizon skeleton elements prevail within the soil fabric. For the most part they are quartz, micaschists and chlorite, whose grain size corresponds to the dimensions of fine sand and coarse silt. The skeleton particles are densely distributed; their interstices are small and filled with colloidal substance.

In comparison with the foregoing horizon, g-horizon possesses a lighter yellowish-grey colour because of the smaller amount of organic matter. This light colour of uniform elementary fabric on the one hand, and the presence of dark brown concretions on the other, condition a special appearance of thin sections of this horizon (Fig.3). The investigations of the typical pseudogley g-horizon, of A2g-horizon of transition subtype lessivé-pseudogley, and of A3-horizon of the lessivé terra rossa with the initial signs of pseudogleyification, are indicative of an increased amount of iron-hydroxide concretions which is an important characteristic of recent pseudogleyification. Because the segregation of concretions is the result of definite reduction and oxydation processes, their form, size and colour can serve as a measure of intensity of pseudogleyification (Blümel, 1962). Progressing from the lessivé soils towards pseudogley the concretions become more numerous and larger. Their distribution within the soil mass is significant, with an obvious

FIG.3–10

Fig.3. g-horizon of pseudogley, on substratum of Pleistocene loam (x 25). Within the homogeneous fabric of this bleached-out horizon the concretions of iron hydroxide are particularly conspicuous. Their distribution within the basic soil mass is significant, while the walls of pores and small canals are completely free.

Fig.4. Coloured parts of marmorized B1g-horizon of the same profile (between crossed nicols, x 80). In the middle part is a pore filled with colloidal substance to a greater extent. In contrast to similar accumulations and over lays in the illuvial horizons of lessivé soil, here the colloidal substance has partly lost its fluid character and birefringence.

Fig.5. The first horizon of the fossil soil, the lower part of profile on Pleistocene loams (x 25). The formations similar to concretions which appears here have an entirely different shape and distribution than the concretions in the g horizon of the upper part of the profile. Strongly compacted elementary fabric of fossil horizons, as well as the concentrations of the mentioned formations round the pores and small canals, are probably the main reason for the poor permeability of the entire profile.

Fig.6. The second horizon of the fossil soil of the same profile (between crossed nicols, x 250). Between the coloured and colourless parts of this marmorized horizon there are no larger differences as to their fabric. Within the light grey patched (5 Y 7/1) accumulations of a peptized colloidal substance also appear.

Fig.7. A2g-horizon of transition subtype lessivé-pseudogley, parent material carbonate loess (x 80). Occurrences of iron hydroxide concretions in former eluvial horizons of lessivé soils are closely bound to the process of pseudogleyification. Completely formed concretions frequently have a regular shap They are from brown to black in colour, and are sharply delineated.

Fig.8. gB-horizon of the same profile (between crossed nicols, x 80). Under the influence of pseudogleyification the changes of elementary fabric of illuvial horizons occur gradually. In the individual parts of the horizon the fabric of lessivé soil, with characteristic descending flows of fluid plasma on the walls of small canals is still well preserved.

Fig.9. gB-horizon of transition type, on substratum of deep deposit of terra rossa (x 17). In bleached-out parts of marmorized horizons the loss of colo is conditioned by displacement of the whole plasma or only of iron.

This is an example of loss of colour due to the increased displacemen of the plasma. Coloured groups of aggregates exhibit a yellowish-red colou (5 YR 5/8), while the bleached-out parts are from reddish yellow to brownis yellow in colour (7.5 YR–10 YR 6/6).

Fig.10. Marmorized fossil, within the deeper deposits of terra rossa (x 17). Here such differences in fabric do not occur, although the contrasts of colours are much greater. The coloured parts are red (2.5 YR 4/6), while the patches and irregular streaks are from light grey to pale yellow in colour (10 YR 7/2–2.5 Y 7/4). In this case the changes are effected by the loss of iron, probably under entirely other conditions of pseudogley formati

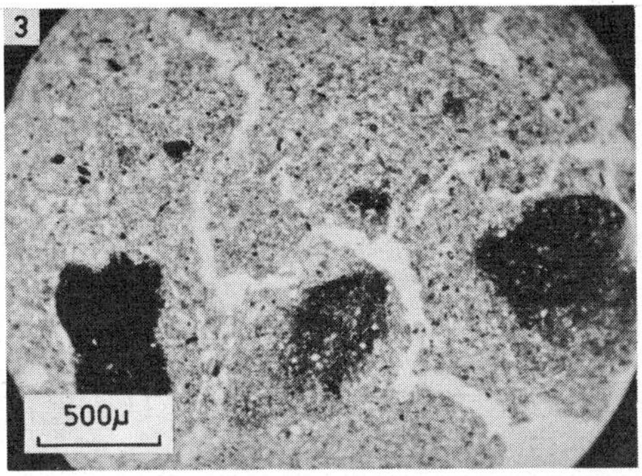
3
500μ

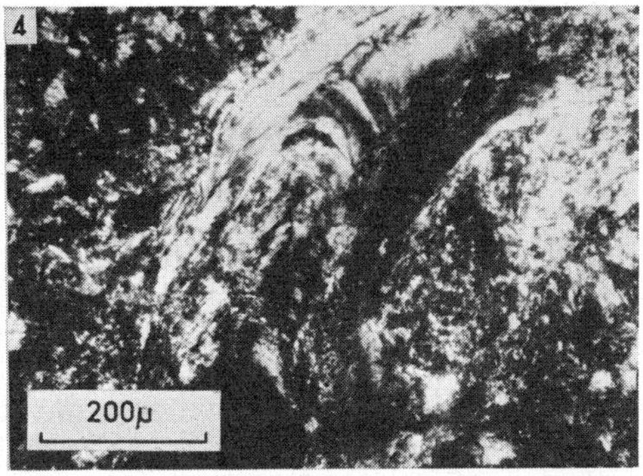
4
200μ

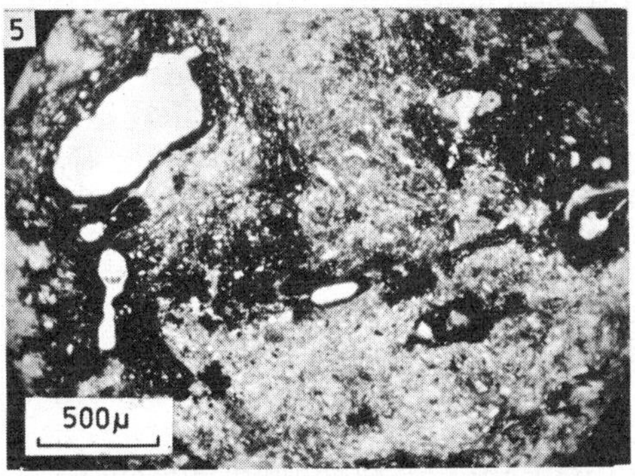
5
500μ

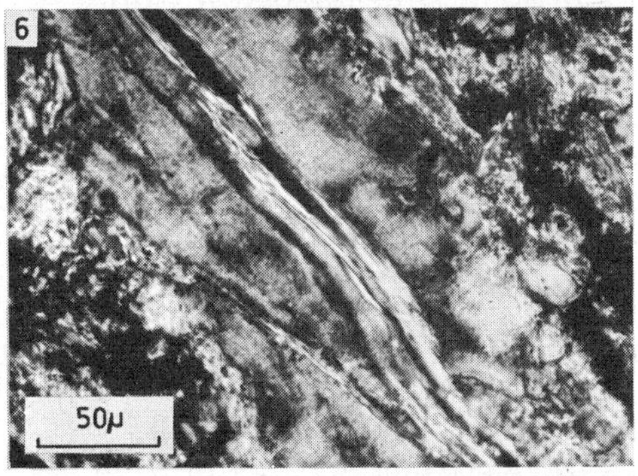
6
50μ

7
200μ

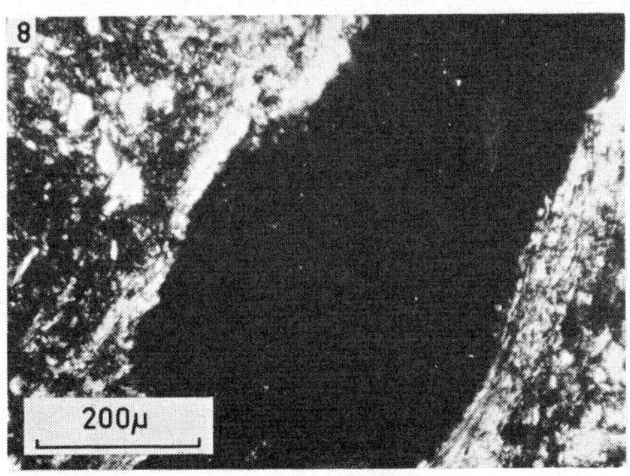
8
200μ

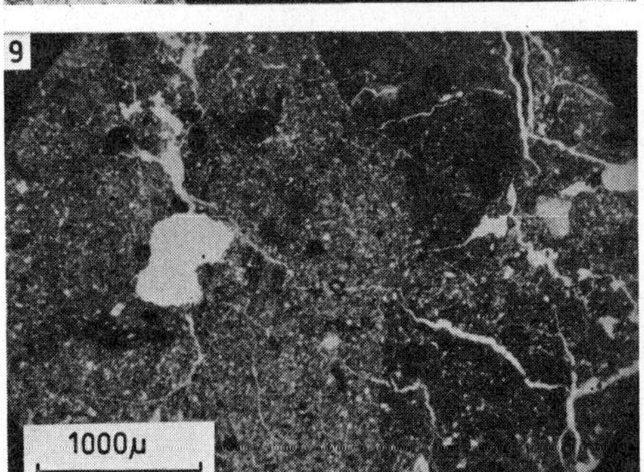
9
1000μ

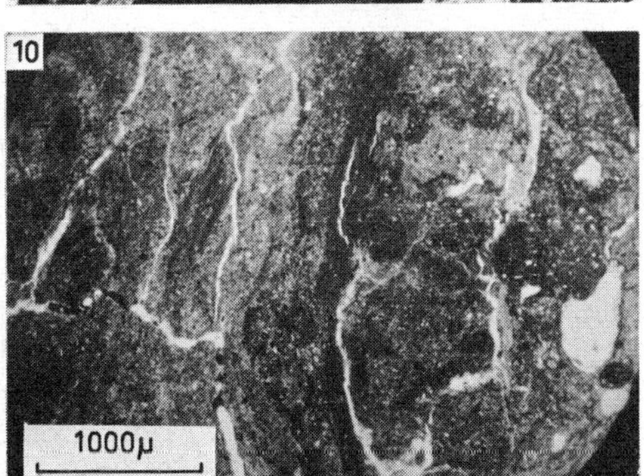
10
1000μ

tendency towards the formation of regular, more or less spherical, forms (Fig.7). With a further increase of humidity towards stagnogley, however, the concretions either change their shape or completely fail to appear.

Characteristics of marmorized Bg-horizon

As distinguished from the upper horizons, which essentially possess a uniform colour and fabric throughout, Bg-horizon is non-uniform in fabric and colour. This accounts for the specific appearance of marmorized horizons by which the pseudogley morphologically distinguishes itself from other soil types.

In samples which were separated from the brownish-yellow (10 YR 6/6 parts of horizons the existing colloidal substance is mainly coagulated, and only in some places does it exhibit a fluid character (Fig.4). Comparative investigations show (Fig.8) that the higher the intensity of surface-water excess, the fewer the occurrences of this peptized substance (similar to "Braunlehm"). However the other components of the fabric are also involved in the changes, especially in the upper part of Bg-horizon. Phenomena of various colours occur within the accumulations of colloidal substance because of the variable concentration of iron compounds. Within the transmitted light of the microscope one discerns a whole series of transitions from bright yellow to brown-black. Examining this heterogeneity of colour and fabric, Zimmermann (In: Mückenhausen, 1962) concluded that it is conditioned by the differences within the component parts of the clay substance and its various gel structures.

Streaks and patches of light grey colour (5 Y 7/2) appear among the coloured groups of aggregates. These colourless sections are made up of different fabrics depending on the depth of the profile in which they appear. In the upper part of B1g-subhorizon, within these light places, grains of skeleton prevail so that the loss of colour is actually conditioned by the loss of colloidal substance. In B2g-subhorizon, however, the colloidal substance is better preserved and in places it accumulates because of the descending flows as vertical streaks. Therefore the loss of colour is conditioned only by the loss of iron here. Both examples, i.e., the loss of colour along with the displacement of the whole plasma and the shifting of the iron, are also stated by Kubiena (1956). But as the former case occurs more frequently in marmorized horizons of the upper parts of the profile (Fig.9), and the latter in the lower parts and in the marmorized horizons of the fossil soil (Fig.10) the question arises as to whether the loss of iron occurs before the general loss of plasma, or whether they are two different processes occurring in various conditions of pseudogley formation.

It is difficult to state exactly what conditions favour the formation of marmorized pseudogley horizons. In the preceding pages the opinion of Janeković (1960) is mentioned, according to which the Bg-horizon of the upper part of profile is also already the result of relict pedogenetic processes Mückenhausen (1962) mentions that a similar standpoint is supported by other European pedologists. On the other hand, the American researchers (Grossman et al., 1959) connect the occurrence of the layered structure of profile and the marmorized horizons ("fragizones") only with the existing conditions of the natural drainage of the soil. On the basis of these investigations it is only possible to conclude that the marmorized horizons are neither exclusively relict, nor exclusively recent formations.

Characteristics of fossil horizons

Further horizons, designated as component parts of the fossil soil, are distinguished from the horizons of the upper part of the profile by their fabric. In general, the share of colloidal substance is a larger one here and it increases more and more with depth. The differentiation of the horizons is sharper, and the contrasts of colours are greater. Besides, in the marmorized horizons which also appear here the mottlings do not display a vertical direction of stretching, as they do in the upper horizons. According to Janeković (1960) these irregularities are conditioned by solifluction displacements in the past.

Within the first horizon one finds almost black formations (resembling concretions), which actually permeate the whole soil mass (Fig.5) more frequently. Under the existing conditions of the environment, the colloidal substance which builds them is coagulated. They have an indefinite shape, and their prevalent disposition round the fissures and the previous small canals indicates that they could have been produced by a fluid movement. Similar formations resembling concretions are described by Mückenhausen et al. (1959) as frequent forms occurring in the lehm fabric.

A marmorized horizon of strong brown colour (7.5 YR 5/8) follows next. The mentioned formations occur in a smaller number, but there is more plasma in a peptized condition. In transmitted light the colour of this colloidal substance is yellow or light grey, while between crossed nicols it distinguishes itself by a high birefringence (Fig.6). In regard to the increased presence of the well-preserved substance in a mobile condition, these horizons are easily distinguished from the Bg-horizons of the upper part of the profile. The compactness of the fabric is also much greater here. There are very few voids, and they too are completely or largely filled with a colloidal substance.

On the basis of the foregoing discussion we can conclude that the micromorphological properties of these horizons also point to their fossil character. The increased presence of elements of lehm fabric are not in harmony with today's conditions of pseudogley formation. According to various authors (Kubiena, 1956; Altemüller, 1962a, 1962b; Dalrymple, 1962), this type of fabric could have been produced only under conditions of intensive weathering. Whether such conditions existed in the course of Würm, which would assume that these materials had been deposited then, or whether these were older materials or admixtures of other materials included in the course of displacement, remains for the moment an open question.

CONCLUSIONS

These investigations, as well as the previous experiences of other authors (Kubiena, 1956; Blümel, 1962; Mückenhausen, 1962), lead to the conclusion that not only the appearance of the profile, but also the elementary fabric of pseudogley, is specific. In this soil type the microscopic pictures of g-horizon and marmorized Bg-horizon (or g1 and g2-horizons according to the nomenclature used by Mückenhausen) are particularly conspicuous. In our case, pseudogley on Pleistocene, the differentiation of the profile is still more evident because of the appearance of fossil horizons.

The presence of fossil horizons in these soils is not only of theoretical but also of practical importance. The poor permeability of these horizons, due to strongly compacted elementary fabric, aggravates the unfavourable water-air conditions of the pseudogley still more. As the horizons of fossil soil occur at different depths within the profile because of a varying intensity of erosion in the past, they also exert different influences on the recent dynamics of the development of these soils. Therefore the proposal of Škorić (1962), that the depth at which these horizons commence should also be a criterion for the sub-division of pseudogley into the lower systematic units may be a worthwhile one.

ACKNOWLEDGEMENTS

The author wishes to thank all those who helped him in the course of his work on this occasion. He is especially indebted to Prof. Dr. A. Škorić for the useful advice and the numerous discussions in connection with the material which is presented in this work, as well as to Dr. M. Malez, with whose assistance the taking of microscopic photographs was carried out.

SUMMARY

The results of micromorphological investigations on pseudogley occurring on Pleistocene loams are discussed. This soil type is widely distributed in northwestern Yugoslavia, and it is characterized by the layered structure of the profile. Because of the inbalance of the morphological properties of the individual horizons and the recent conditions of formation, some Yugoslav pedologists propose that one should also introduce the criterion of relictness in the subdivision of lower systematic units. The observed differences in the elementary fabric between the horizons of the upper part of the profile, and the fossil soil of the lower part, are in agreement with these proposals and confirm their correctness.

REFERENCES

Altemüller, H.J., 1962a. Beitrag zur mikromorphologischen Differenzierung von durchschlämmter Parabraunerde, Podsol–Braunerde und Humus–Podsol. Z. Pflanzenernähr., Düng., Bodenk., 98 : 247–258.

Altemüller, H.J., 1962b. Verbesserung der Einbettungs und Schleiftechnik bei der Herstellung von Bodendünnschliffen mit "Vestopal". Z. Pflanzenernähr., Düng., Bodenk., 99 : 164–177.

Blümel, F., 1962. Formen der Eisenoxydhydrat-Ausscheidungen in Gleyen und Pseudogleyen. Z. Pflanzenernähr., Düng., Bodenk., 98 : 258–264.

Dalrymple, J.B., 1962. Some micromorphological implications of time as a soil forming factor, illustrated from sites in southeastern England. Z. Pflanzenernähr., Düng., Bodenk., 98 : 232–239.

Filipovski, G. and Cirić, M., 1963. Zemljista Jugoslavije (Soils of Yugoslavia). Jugoslovenskog Drustva za Proucavanje Zemljista, Beograd, 498 pp.
Grossman, R.B., Stephen, I., Fehrenbacher, J.B., Beavers, A.H. and Parker, J.M., 1959. Fragipan Soils of Illinois. Soil Sci. Soc. Am. Proc., 23 : 65–75.
Janeković, D., 1960. Starost i geneza pseudogleja na beskarbonatnom praporu jugozapadnog oboda Panonskog bazena. (Age and genesis of pseudogley on a non-carbonate loess of the southwestern border of Panonnian Basin). Thesis for the habilitation, Agric. Fac., Sarajevo, 92 pp.
Kubiena, W.L., 1956. Zur Mikromorphologie, Systematik und Entwicklung der rezenten und fossilen Lössboden. Eiszeitalter Gegenwart, Ohringen, 7 : 102–112.
Mückenhausen, E., Gerkhausen, W. und Kerpen, W., 1959. Entstehung und Eigenschaften der Böden auf den fossilen Verwitterungsdecken der Eifel. Z. Acker Pflanzenbau, 108 : 201–222.
Mückenhausen, E., 1962. Entstehung, Eigenschaften und Systematik der Böden der Bundesrepublik Deutschland. Deut. Landwirtsch. Ges., Frankfurt am Main, 148 pp.
Neugebauer, V., 1958. Pitanje pseudogleja u Hrvatskoj. (Über die Frage der Pseudogleys in Kroatien). Zemljiste i Biljka, 1–3 : 209–211.
Neugebauer, V., Cirić, M. i Živković, M., 1961. Komentar pedološke karte Jugoslavije. (Soil map of Yugoslavia). Jugoslavenskog Drustva za Proucavanje Zemljista, Beograd, 107 pp.
Racz, Z., 1963. Prilog mikromorfološkom istraživanju naših tala. (A contribution to the micromorphological investigations of our Soils). Zemljiste i Biljka, 1–3 .
Skorić, A., Kovačević, P., Racz, Z. i Modrić, A., 1962. Detaljno proucavanje glavnih tipova tala Jugoslavije. (A detailed study of the main Yugoslav soil types). Rept. Federal Res. Develop. Board, Beograd, 56 pp.

ZUR KENNTNIS DER ROTLEHME VON DER CHALKIDIKI (GRIECHENLAND) UND IHRER VERÄNDERUNG DURCH BODENEROSION

ZLATKO GRAČANIN

Institut für Bodenkunde und Bodenerhaltung der Justus Liebig-Universität, Giessen (Deutschland)

Über die Anwesenheit der Rotlehme auf der Chalkidiki wurde bereits von geologischer Seite berichtet (Maravelakis 1933, S.11). In neuerer Zeit hat Alexiadis (1960) ihre tonmineralogische Zusammensetzung untersucht. Doch über ihre Morphologie und ihr Verhältnis zu der Umwelt ist bis jetzt wenig bekannt. Mikromorphologisch wurden sie meines Wissens bis jetzt noch nicht untersucht. Bei meinen Bodenuntersuchungen in Nordgriechenland hatte ich Gelegenheit im östlichen Teil der Halbinsel Chalkidiki verschiedene Rotlehm-Profile zu untersuchen; über einen Teil dieser Beobachtungen sei mir gestattet hier einen ersten vorläufigen Bericht zu geben.

Die von mir untersuchten Rotlehme befinden sich im östlichen Teil der Chalkidiki, im Bereich der kristallinen Schiefer, auf denen sie aber keineswegs der einzige Bodentyp sind. Vielmehr kann man entlang der frischen Strasseneinschnitte oft die rasche Abwechslung der Rotlehme mit anderen Bodentypen verfolgen. Dies ist aber kaum durch die rezenten bodenbildenden Prozesse, sondern vielmehr durch die geologische und tektonische Geschichte dieses Gebietes zu erklären. Solches Vorkommen der Rotlehme auf der östlichen Chalkidiki spricht für die These von Kubiena (1962, S.173), nach der die Rotlehme des Mediterrans Reste alter Bodendecken darstellen, die unter dem Einfluss vorzeitlicher Klimate (vorwiegend im Tertiär, zum Teil aber auch im Pleistozän) entstanden sind. Die kristallinen Schiefer von Chalkidiki gehören der alten Rodopenmasse an, so dass bei ihnen die alten Bodenbildungen durchaus zu erwarten sind. Dies um so mehr, da nach den bisherigen Untersuchungen die Chalkidiki von der Vereisung nicht betroffen wurde, so dass eine mehr oder weniger kontinuierliche Bodenbildung mindestens seit Tertiär anzunehmen ist.

Andererseits ist aber hervorzuheben, dass sich manche Rotlehme wegen des bergigen Reliefs nicht mehr in der ursprünglichen Lagerung befinden und dass sie vielfach durch Zerstörung ihrer Vegetationsdecke der Bodenerosion ausgesetzt wurden.

Es bestehen leider keine meteorologischen Messungen, die das Klima, unter dem sich die Rotlehme auf Chalkidiki gegenwärtig befinden, näher charakterisieren könnten. Die nächste meteorologische Station, über die ich die Angaben habe, ist Thessaloniki. Die dort ermittelten mittleren jährlichen Niederschläge und Temperaturen (Mariolopoulos, 1937) sowie die aus ihnen berechneten Regenfaktoren sind aus der Tabelle I ersichtlich. Dazu sind noch Bezeichnungen für die Humidität und den Wärmecharakter des Klimas im Sinne der Klassifikation von M. Gračanin (1950) beigegeben worden.

TABELLE I

Die mittleren Niederschläge und Temperaturen[1] sowie Regenfaktoren in Thessaloniki

Monat	Niederschläge (mm)	Temp. (0° C)	Regenfaktor	Humidität des Klimas	Wärme-charakter
Jan.	35,4	5,6	6,32	semihumid	mässig kalt
Febr.	31,6	6,7	4,72	semiarid	mässig kalt
März	38,1	10,4	3,66	semiarid	mässig warm
Apr.	42,1	14,6	2,88	arid	warm
Mai	52,9	19,3	2,74	arid	warm
Juni	34,6	23,6	1,47	arid	heiss
Juli	25,2	26,3	0,96	arid	heiss
Aug.	23,5	25,9	0,91	arid	heiss
Sept.	32,6	22,3	1,46	arid	heiss
Okt.	58,5	17,2	3,40	semiarid	warm
Nov.	64,4	11,4	5,65	semihumid	mässig warm
Dez.	47,1	7,6	6,20	semihumid	mässig kalt
Jahr	486,1	15,9	31,83	arid	warm

Wegen des bergigen Reliefs muss man annehmen, dass die in Thessaloniki gemessenen Werte auf Chalkidiki in Abhängigkeit von der Höhenlage, Exposition und Geländeform weitgehenden Modifikationen unterworfen sind. Wesentlich feuchtere Mikro- und Lokalklimate müssen die Buchenwaldstandorte haben. Noch am geringsten werden diese Abweichungen im Bereich der immergrünen mediterranen Vegetation sein, wo die hier darzustellenden Bodenprofile 15 und 16 geöffnet wurden.

Profil 15 (20. Oktober 1961)
Strivoti, oberhalb der Strasse, etwa 350 m über N.N. Südexponierter Hang oberhalb des breiten Tales Megali Panaghia–Gomati. Neigung etwa 30%. Muttergestein: grüner Schiefer.

Vegetationsdecke: offenes Buschwerk, Degradationsstadium von *Andrachno-Quercetum ilicis*, dass hier an der kontinentalen Grenze seiner Verbreitung ist (vgl. auch Horvat, 1962a, 1962b). Die Zerstörung des Bestandes durch Abholzung, Beweidung und Brand ist so weit fortgeschritten, dass der Kronenschluss der Sträucher nur 0,5–0,6, die Höhe vorwiegend 0,2–2,10 m beträgt; nur einzelne Sträucher von *Arbutus andrachne* und *Quercus coccifera* zeichnen sich durch grössere Höhen aus. In der Strauchschicht befinden sich: *Arbutus andrachne* L. 2.3, *A. unedo* L. 2.3, *Phillyrea media* L. 2.3, *Erica verticillata* Forsk. 2.3, *Erica arborea* L. 2.3, *Calycotome villosa* (Poir.) Lk. 2.3, *Cistus villosus ssp. eu-villosus v. incanu.* (Rchb.) Freyn. 1.3, *Cistus salviaefolius* L. 1.3, *Anthyllis Hermanniae* L. 1.3, *Quercus ilex* L. 1.3 und *Pistacia terebinthus* L. + Krautschicht, nur etwa 10% der Fläche deckend: *Chrysopogon gryllus* Trin. 2.2, *Stipa bromoides* (L.) Brand. 1.2, *Thymus striatus* Vahl. + 2, *Teucrium polium* L. +, *Brachypodium pinnatum* (L.) Beauv. +, *Briza maxima* L. +.

Skelett bedeckt etwa 30–40% der Bodenoberfläche, und zwar in Form von Steinen verschiedener Grösse, von Kies und von einzelnen kleineren

[1] Nach Mariolopoulos, 1937.

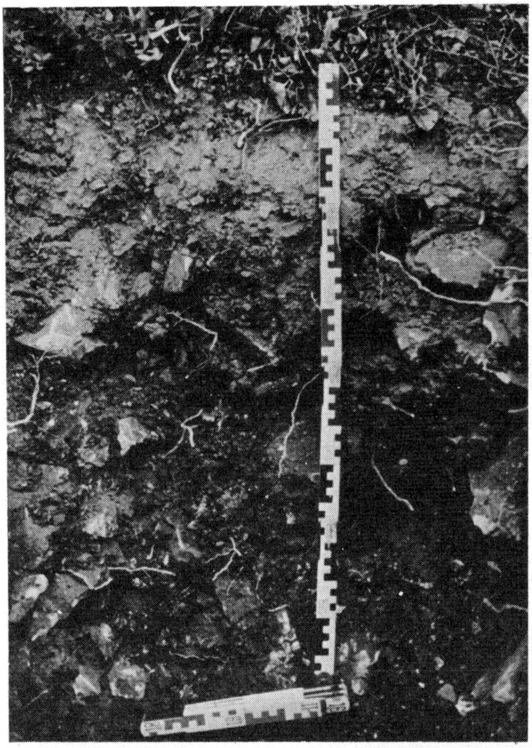

Abb.1. Verbraunter relikter Rotlehm. Strivoti (Chalkidiki).

Blöcken des anstehenden Gesteins. Ungeschützte Feinerde bedeckt etwa 10–15%.

Profil (Abb.1), geöffnet unter einem Strauch von *Arbutus andrachne*:

0– 3 cm: A_0', sehr lose gelagerte Laubstreu;

3–3,5 cm: A_0'', nur etwa 0,5 cm mächtiger Horizont der halbzersetzten organischen Substanz;

3,5– 16 cm: A_1, skelettreicher, kiesiger und steiniger, staubschluffiger Lehm, staubiger bis bröckeliger Struktur, ziemlich reichlich durchwurzelt. Dünnschliff (Abb.2) weist braune Vererdung, ein auch innerhalb der Aggregate hohlraumreiches Gefüge, viele Gesteinsbruchstücke, aber im Vergleich zum untersten, typischen Rotlehmhorizont auch Fremdminerale auf. Die Hohlraumformen sind sehr mannigfaltig, meist abgerundet, keine Sprungrisse. Allmählicher Übergang in:

23– 35 cm: A_3B_1, skelettreichen, steinigen Horizont mit dunkelbraunen staubschluffigem Ton, schwach rötlicher Nuance. Die Struktur ist ausgesprochen bröckelig, es herrschen Aggregate mit 0,5–1,5 cm Durchmesser vor. Im Dünnschliff (Abb.3) ist eine sehr deutliche Rubefizierung und ein im Vergleich zum untersten typischen Rotlehmhorizont aufgelockertes Gefüge zu erkennen. Es kommen nur einzelne Sprungrisse vor, die meisten Hohlräume sind vielgestaltig. Allmählicher Übergang in:

40– 80 cm: B_2C, skelettreichen, steinigen und kiesigen Horizont mit dunkelrotbraunen staubschluffigem Ton, bei der Probeentnahme trocken und in diesem Zustand gut ausgeprägter bröckeliger Struktur. Es herrschen

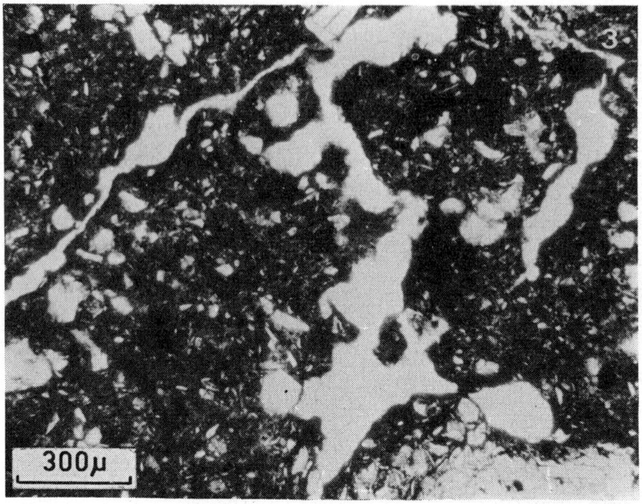

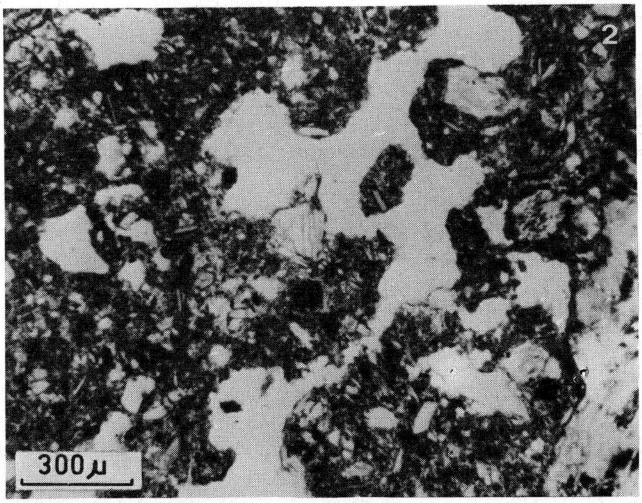

Abb.2. Dünnschliff aus dem A_1-Horizont des verbraunten relikten Rotlehms von Strivoti. Hohlraumreiches erdiges Gefüge. Ein Gesteinsbruchstück am rechten Bildrand.

Abb.3. Dünnschliff aus dem A_3B_1-Horizont des verbraunten relikten Rotlehms von Strivoti (Chalkidiki). Gelockertes Rotlehmgefüge: Sprungrisse nur vereinzelt, andere Hohlraumformen wiegen vor.

nussförmige Aggregate von 2–3 cm Durchmesser vor. Auch dieser Horizont enthält noch manche Wurzeln. Dünnschliff (Abb.4) zeigt ein dichtes Rotlehmgefüge, von Sprungrissen durchsetzt, in welchem Eisenkonkretionen bzw. -ausscheidungen sichtbar sind. Das Rotlehmgefüge ist noch sehr typisch, weist nur wenig Vererdung auf.
Reaktion auf $CaCO_3$ ist im ganzen Profil negativ.

TABELLE II

Analysenergebnisse Profil 15

Horizont	A_1	A_3B_1	B_2C
Skelett in der Bodenprobe (%)	54,1	56,9	48,7
Feinerde enthält (%)			
<2 μ	20,7	46,6	54,4
2– 20 μ	40,9	28,6	24,6
20– 60 μ	13,8	7,6	5,6
60– 200 μ	14,5	10,0	7,8
200–2000 μ	10,1	7,2	7,6
Strukturaggregate (%) - Trockensiebung			
<2 mm	34,2	10,5	6,1
2–3 mm	7,5	7,5	6,5
3–5 mm	14,5	16,2	14,5
>5 mm	43,8	65,8	72,9
Einige chemische Eigenschaften			
pH (H_2O)	6,35	6,36	6,38
pH (N KCl)	5,58	5,23	5,07
Organische Substanz (%)	3,94	2,96	1,41
C (%)	2,29	1,72	0,82
N (%)	0,21	0,17	-
C:N	10,90	10,12	-
Minerale in der Tonfraktion (2 μ) - durch Röntgenanalyse identifiziert			
Vermikulit	+	+	+
Chlorit (in B_2C Bodenchlorit)	+	+	+
Kaolinit	+	+	+
Illit	+	-	+
Quarz	+	+	+

Die Eigenschaften dieses Profils seien durch Analysenergebnisse näher charakterisiert (Tabelle II). Dieselben Tonminerale wurden auch von Alexiadis (1960) in Rotlehmen aus "greenstones" und Phylliten der Chalkidiki verzeichnet, doch in den letzteren ist der Illit vorherrschendes Mineral, was in unserem Profil sicherlich nicht der Fall sein wird, weil hier die Illit-Linien sehr schwach sind und im A_3B_1-Horizont ganz fehlen.

Auffällig ist die grosse Zunahme des Tongehaltes mit der Tiefe. Der mikromorphologische Befund spricht jedoch dafür, dass sie nicht - oder im wesentlichen nicht - auf eine Tonverlagerung im Profil, sondern auf die Erhaltung des alten tonreichen Rotlehms im B_2C-Horizont zurückzuführen ist.

Auf Grund der Eigenschaften dieses Profils muss man ihn im Sinne von Kubiena (1953, 1962) als verbraunten relikten Rotlehm bezeichnen.

Unter gegenwärtiger Nutzung als Weideland, ist der untersuchte Rotlehm der Bodenerosion ausgesetzt. Im Zusammenhang mit der

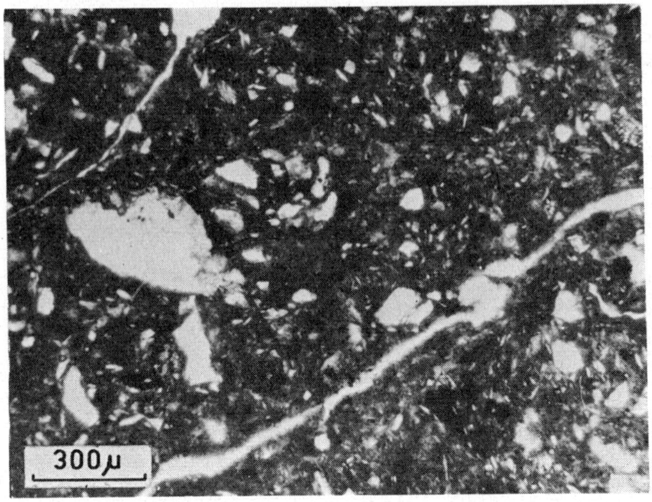

Abb.4. Dünnschliff aus dem B_2C-Horizont des verbraunten Rotlehms von Strivoti. Dichtes Rotlehmgefüge mit zwei Sprungrissen. Dunkel: Eisenanhäufungen.

fortschreitenden Zerstörung der Macchia, erfasst der Abtrag immer grösser Teile der Fläche. Um festzustellen, welche Bodenveränderungen dabei stattfinden, wurde das Profil 16 untersucht.

Profil 16 (20. Oktober 1961)
Strivoti, unbewachsener Teil der Fläche, auf der Profil 15 geöffnet wurde. Die Wirkung der Bodenerosion durch Wasser ist deutlich sichtbar. Wenn auch der oberflächliche Abtrag vorherrscht und keine tiefen Gräben entstehen, so kommt hier - ebenso wie auf den Karsthutweiden in Kroatien (vgl. Z. Gračanin 1962, S.192, 284) - keine gleichmässige Schichterosion zustande, sondern es entsteht ein Mosaik von erodierten und wenig oder nicht erodierten Stellen nebeneinander am Hang. Ein grosser Teil der Bodenoberfläche ist von Kies und Steinen bedeckt (Abb.5), die den Überrest des abgeschwemmten Teiles der Profils darstellen. Solche Skeletthorizonte an der Bodenoberfläche schützen den Rest der Feinerde und die tieferen Horizonte von weiterem Abtrag. Für sie habe ich die Bezeichnung A_{sk}-Horizont vorgeschlagen (Z. Gračanin, 1963). Zum Teil kommt aber auch ungeschützte Feinerde zum Vorschein:

Abb.5. Erodierter verbraunter Rotlehm vom Strivoti (Chalkidiki). Bildung des A_{sk}-Horizontes an der Bodenoberfläche durch Ansammlung des Skelettes aus abgeschwemmten Horizonten.

Abb.6. Strukturphotogramm der dicht zusammengeschwämmten Oberfläche des erodierten verbraunten Rotlehms von Strivoti. Plattige Struktur parallel zur Oberfläche. Hohlräume und Mineralkörner schwarz.

5

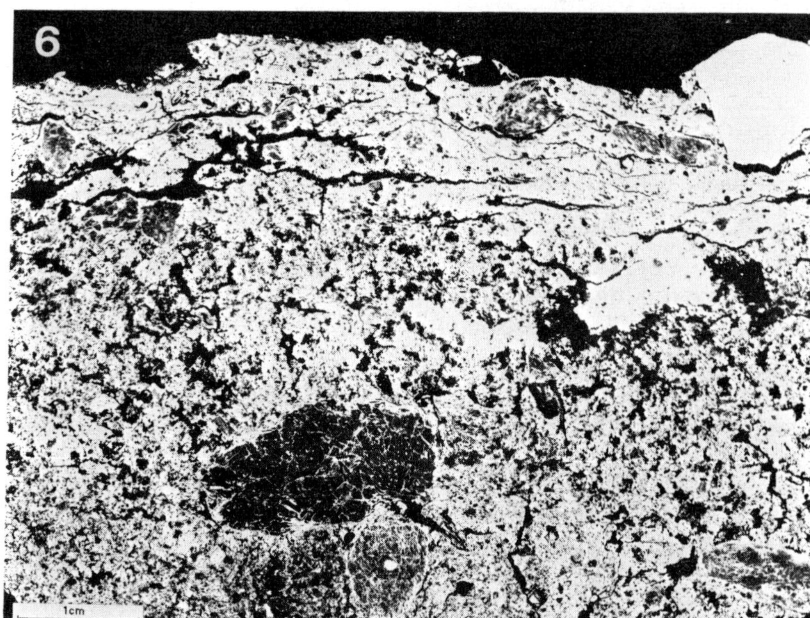
6
1cm

0– 5 cm: A_1, brauner, stark skeletthaltiger, staubschluffiger Lehm, ausgesprochen plattiger Struktur (Abb.6). Das Material dieses Horizontes stammt wohl grösstenteils aus dem A_1, ist aber umgelagert und dicht zusammengeschwemmt worden, worauf die Bildung der plattigen Struktur zurückzuführen ist. Folgt:

5–16 cm: A_3B_1, skelettreicher, steiniger Horizont mit braunem staubschluffigem Lehm schwach rötlicher Nuance. Reaktion auf $CaCO_3$ ist negativ.

TABELLE III

Analysenergebnisse Profil 16

Horizont	A_1	A_3B_1
Skelett in der Bodenprobe	45,4	56,9
Feinerde enthält (%)		
<2 μ	21,9	30,4
2– 20 μ	34,9	33,7
20– 60 μ	12,9	14,7
60– 200 μ	16,7	12,6
200–2000 μ	13,6	8,6
Strukturaggregate (%) - Trockensiebung		
<2 mm	38,0	29,4
2–3 mm	8,8	11,0
3–5 mm	13,6	19,2
>5 mm	39,6	40,4
Einige chemische Eigenschaften		
pH (H_2O)	6,60	6,60
pH (N KCl)	5,84	5,62
Organische Substanz (%)	3,84	3,16
C (%)	2,23	1,84

Ein Vergleich der beiden Profile lässt erkennen, dass infolge der Bodenerosion durch Wasser: (1) A_0', A_0'' und ein grosser Teil des A_1-Horizontes verlorengegangen sind; (2) der A_{sk}, ein Skelett-Horizont als Überrest der abgeschwemmten Horizonte auf dem grössten Teil der Bodenoberfläche gebildet wurde und (3) zur Verkrustung der Bodenoberfläche gekommen ist, wobei durch die Prall- und Plantschwirkung der Regentropfen, vermutlich aber auch durch die Trittwirkung des weidenden Viehs, das ursprüngliche Gefüge der Reste des A_1-Horizontes so weit verschlechtert wurde, das sie zusammengeschlämmt werden konnten und so zur Ausbildung der plattigen, für die Infiltration äusserst ungünstigen Struktur kam.

ANERKENNUNG

Die Untersuchungen auf Chalkidiki wurden im Rahmen einer längeren Forschungsreise in Nordgriechenland durchgeführt, die auf Anregung von Herrn Prof. Dr. Ing. H. Kuron mit der Unterstützung der Deutschen Forschungsgemeinschaft zustande kam, wofür ich ihnen sehr zu Dank verbunden bin. Herrn Forstmeister N. Harissiadis (Arnaia) möchte ich für die Führung im Gebiet seines Forstamtes und Herrn Prof. Dr. W. Kubiena für die Ermöglichung der mikromorphologischen Untersuchungen meinen verbindlichsten Dank sagen.

SUMMARY

A contribution to the knowledge of the red loams of the Chalkidiki (Greece) and their alteration under influence of soil erosion

Under present environmental conditions, the relictic red loams in the area of the Mediterranean vegetation (degradation stages of the *Andrachno-Quercetum ilicis*) of the Chalkidiki are exposed to changes which lead to the formation of a brunified red loam ("verbraunter Rotlehm"). The microscopic picture shows "braune Vererdung" and loosening of the structure in the upper horizons.

By excessive pasturing, cutting and burning, the vegetation cover of the red loams in gradually being destroyed, starting soil erosion by water. As a result, the following changes of their profiles have been observed: (1) a loss of the A_0', A_0'' and great parts of the A_1-horizons; (2) the formation of the skeletal A_{sk}-horizon on a large part of the soil surface; and (3) the formation of the compact platy structure on the unprotected fine soil.

LITERATUR

Alexiadis, K.A., 1960. I oryktologiki systasis tis argillou ton erythropilon tis Chalkidikis (Mineral composition of the clay of red-loam soils of Chalkidiki). Sonderdruck aus: Parartimatos tis Epistimonikis Epetiridos tis Geoponikis kai Dasologokis Scholis, Thessaloniki, 45 S.

Gračanin, M., 1950. Mjesečni kišni faktori i njihovo značenje u pedološkim istraživanjima (Monthly rain-factors and their significance for pedological investigations). Poljoprivredna Znanstvena Smotra, 12 : 51–67.

Gračanin, Z., 1962. Verbreitung und Wirkung der Bodenerosion in Kroatien. Osteuropastudien der Hochschulen des Landes Hessen. Schmitz, Giessen, 1 (21) : 335 S.

Gračanin, Z., 1963. Zur Nomenklatur der Bodenhorizonte erodierter skeletthaltiger Böden. Z. Pflanzenernähr., Düng., Bodenk., 101 : 42–48.

Horvat, I., 1962a. Die Grenze der mediterranen und mitteleuropäischen Vegetation. Botan. Ges., 75 : 91–104.

Horvat, I., 1962b. Die Vegetation Südosteuropas in klimatischem und bodenkundlichem Zusammenhang. Mitt. Österr. Geograph. Ges., 104 : 136–160.

Kubiena, W., 1953. Bestimmungsbuch und Systematik der Böden Europas. Enke, Stuttgart, 392 S.

Kubiena, W., 1962. Die Böden des mediterranen Raumes. Kalium Symposium 1962 : 167–190.

Maravelakis, M.I., 1933. Oi geologikoi kai makroseismikoi charaktires ton seismon tis Chalkidikis Septembrios 1932 (Les caractères géologiques et macrosismique des séismes de Chalcidique, Septembre 1932). Panepistimion Thessalonikis, Ergastirion Oryktologias–Geologias–Petrologias, Thessaloniki, 33 pp.
Mariolopoulos, E.G., 1937. La distribution des éléments méteorologiques en Grèce. Athènes.

MICROMORPHOLOGICAL INDICATIONS OF THE PROCESSES ASSOCIATED WITH THE FORMATION OF THE KRASNOZEMS (RED EARTHS) AND THE RED-COLOURED CRUST OF WEATHERING IN THE TRANSCAUCASUS

A.I. ROMASHKEVICH

Geographical Institute of the U.S.S.R. Academy of Sciences, Moscow (U.S.S.R.)

INTRODUCTION

The thick crusts of weathering of the Transcaucasus, and the soils which developed on them, have invariably attracted the attention of scientists over a period of many years. They are the most ferrallitised formations in the U.S.S.R.

Dokuchayev placed them among laterites. Later on, these formations were given the name of krasnozems and red-coloured crusts of weathering. These names owe their origin to the fact that free aluminium hydroxide, a criterion of laterite weathering, was found in small amounts in the weathering crusts of the Transcaucasus against a background of predominantly kaolinite-halloysite composition, in the presence of large amounts of iron oxides and hydroxides.

Recently, however, geologists (Yakovleva, 1958; Lisitsina, 1962) have essentially amended the characteristics of the red-coloured formations of the Transcaucasus. This was due to the discovery of considerable amounts of free aluminium hydroxide, in the form of gibbsite which forms during both the first and the last stages of profile development, in a number of districts. In this connection, the predominantly clayey (kaolinite–halloysite) composition of the weathering crust can be regarded as stage in the process of laterisation.

CONTEMPORANEOUS WEATHERING

To a certain degree this problem can be elucidated by a study of the course taken by the modern weathering in the territory. With this aim in view, we have investigated those fragments of fresh basalts and andesite-basalt porphyrites which have a weathering incrustation on the surface.

Morphologically, the incrustations differ from thick red-coloured crusts of weathering by having lighter, brownish hues, better homogeneity and a coarser mechanical composition. In thickness they range from several cm to several mm. The incrustation does not always retain the structure of fresh rock. Most of the feldspars are leached and voids have developed in their place. Some of the decomposed feldspars have yielded products which are unidentifiable under the microscope. Iron hydroxides, in the form of cords and granular agglomerations, are segregated around the crystals of pyroxenes which are sometimes also leached. The chloritised fine-grained mass, and the chlorite matter in the form of aggregates, are ferruginated. No clay minerals have been detected under the microscope. The magnetite in the incrustation has a somewhat obscure form, though no distinct formation of iron hydroxide due to it has been observed.

Thermal analysis (Fig.1D, E) of the incrustations has shown that the

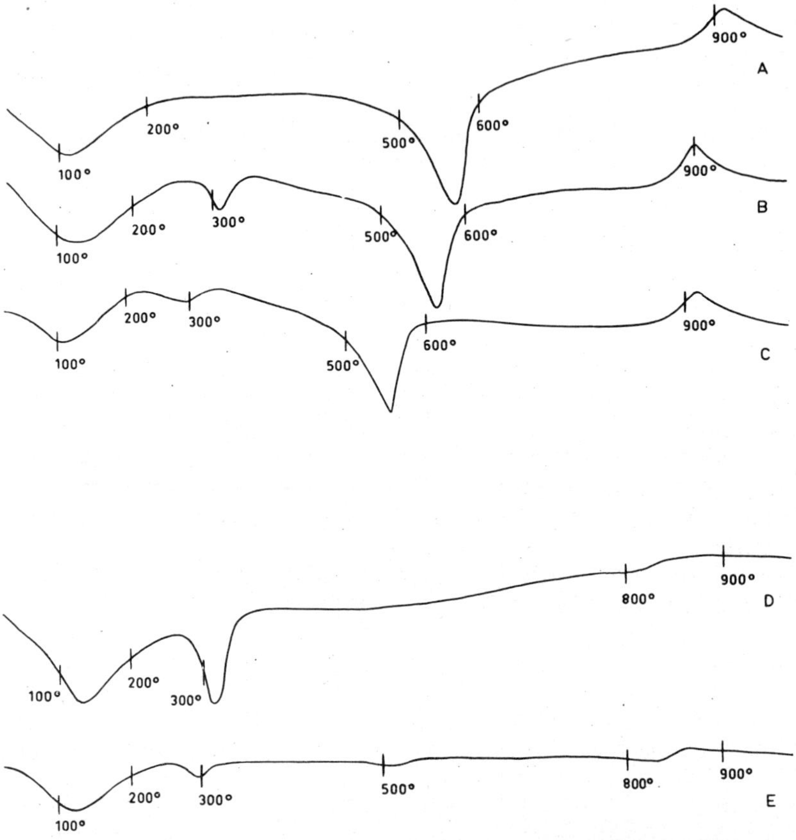

Fig.1. Thermograms of krasnozem (B, C), crust of weathering (A) and incrustations on the fragments of porphyrite (D) and basalt (E).

material of which they are made possesses high hydrophility, and the samples contain varying but noticeable amounts of crystallized sesquioxides. Apparently hydrogoethite–goethite predominates among the crystallized sesquioxides.

The thermal curves show no effects which might indicate the presence of clay minerals. Most probably silicic acid, aluminium hydroxide and, to a certain extent, iron hydroxides are amorphous in the incrustation. It is possible that they form compounds of allophane.

These data suggest that the modern weathering processes of the primary minerals reach the state of amorphous material, attended by a considerable hydration of this material and a vigorous leaching of the silicic acid and part of the aluminium. The external effect of these processes take the form of clearly expressed ferrugination. These data are corroborated by chemical analyses testifying to the outwash of much silicic acid, and of less aluminium with the accumulation of iron (Table I).

TABLE I

The total of SiO_2, Al_2O_3 and Fe_2O_3 in basalt, andesite-basalt porphyrite and the weathering incrustations (% ignited material)

Object	Hygroscopic moisture (%)	Ignition loss (%)	SiO_2	Fe_2O_3	Al_2O_3
Fresh basalt	11.67	5.08	44.70	15.94	23.27
Weathered incrustation	19.68	7.13	30.85	36.24	27.62
Fresh andesite-basalt porphyrite	5.00	2.12	52.28	9.83	18.20
Weathered incrustation	11.35	4.64	44.54	25.32	18.13

In this way the contemporaneous weathering of the Transcaucasus possesses all the features of laterisation. Somewhat out of place here is the accumulation of the weathering products in an amorphous state, rather than in the form of crystallized oxides, as in the tropics, and an outwash of some aluminium.

We do not consider it possible to draw a complete analogy between the incrustations of the weathering and the thick crusts. Moreover, if we compare the material of thick crusts of weathering with the modern incrustations we might conclude that they have very few features in common. Kaolinisation is typical in one case, and the accumulation of amorphous material in the other.

At the same time, the processes which formed the thick red-coloured crusts, and which are now forming the incrustations, have a common tendency to wash out the silicic acid and accumulate sesquioxides. Apparently it can be assumed that the accumulation of amorphous material is the first stage of the weathering process, followed by kaolinite formation. As far as the second stage is concerned, the formation of kaolinite is a process developing in time during the course of which the environment should obviously remain the same. This course could be expected if we recall the available data concerning the weathering processes and the artificial synthesis of clay minerals (Noll, 1936; Kanno, 1956; Green, 1959).

SOIL FORMATION AND WEATHERING

The upper part of the crusts of weathering terminates in a zone where the processes of soil formation carry on most actively. On the basis of his studies of the laterite formations of Brazil, China and Guinea Gerasimov (1957, 1961; Gerasimov and Man–Yun–Ji, 1958; Gerasimov and Romashkevich, in press) suggested the name "active soil-forming layer" for this upper zone with intensive manifestations of biological activity, good aeration, and maximum wetting with atmospheric precipitation.

Morphologically, this zone usually differs from the crust of weathering by having lighter (yellowish) hues and a loss of the texture of the weathering crust. Sometimes the boundary between them is very difficult to establish morphologically because of the bright hues in the colouring brought about by abundant mottles of iron oxides and hydroxides.

The concept of an "active soil-forming layer" is extremely important for the further study of laterite formations. As a matter of fact, we have too little knowledge so far of the weathering and soil-forming processes in the tropical and subtropical regions to be able to differentiate and identify these processes. Therefore, it is extremely important, both from the genetic and practical points of view, to establish characteristic features of the laterite crust of weathering and the active soil-forming layer which develops on such crust.

The application of micromorphological and mineralogical methods to the study of the laterite profiles of Guinea (Gerasimov and Romashkevich, in press) has been found to produce interesting results. In particular, first evidence was obtained as regards the microstructure and mineralogy of both the "active layer" and the zone of lithomarge (crust). We continued these investigations on the red-coloured formations of the Transcaucasus.

The active layer of soil formation, or the krasnozem profile proper, is about one metre thick here, and turns into the red-coloured crust of weathering of the main effusive rocks of Middle Eocene or the shingles of Pliocene in its bottom part. We investigated the profiles of krasnozem and crusts of weathering on the basalts and andesite-basalt porphyrites in the most elevated hilly part of Ajaria, where the denudation processes repeatedly eroded the soils and the upper parts of the weathering crust. In this case one can clearly observe the superimposition of the modern soilforming process on the already formed crust of weathering, as well as the indications which separate the "active layer" from the crust of weathering.

The crust of weathering

Plate D4 shows that the structure of the initial hard rocks is well-preserved in the crusts of weathering. The phenocrysts of feldspars are frequently leached or replaced by kaolinite and halloysite. Upon weathering, the phenocrysts of pyroxenes produce segregations of iron hydroxides. The ground mass of the rocks, composed of microcrystals of feldspars, pyroxenes, volcanic glass and chlorite, is completely kaolinised (Fig.1A). In this way the crust of weathering here, as in other cases, retains the structure of compact rocks and spotted segregations of iron hydroxides, i.e., the crust of weathering under study does not differ in this respect from the lithomarge zone in the laterites (Gerasimov and Romashkevich, in press). It is very important that during the process of weathering of compact rocks, due to the leaching out of a number of substances, the forming crust of weathering acquires a considerable porosity and, hence, the ability to rapidly let the abundant precipitations through.

In this crust of weathering gibbsite occupies a subordinate position, and manifests itself only locally on feldspars and pyroxenes, in the form of streaks in the rock fissures and as sinters in the voids. However, there is apparently more gibbsite than can be observed with the microscope, as we have certain indications to the effect that a considerable part of it may have evolved into an amorphous or metacolloidal form (Ostryakov, 1915, 1917).

The upper parts of the crusts of weathering undergo essential transformations in places due to the substances percolating from above. The voids are filled with sinter of oriented clay, and banded deposits of amorphous iron hydroxides can be observed; the structure of the weathering crust is noticeably disturbed here. The clay mass can change the orientation of its particles and assume a new orientation under the effect of the percolating solutions.

Active (soil) layer

The crust of weathering passes gradually to the bottom of the krasnozem active layer. Plate D5 shows the boundaries of the areas of the weathering crust, and the areas transformed by the soil-forming processes. It is interesting to note that the clay mass of the upper parts of the weathering crust becomes more dispersed, disaggregated as it were, and loses the structure of the rock. Humus, which uniformly impregnates the clay mass of the weathering crust, causes the spotted segregations of iron hydroxides to disperse and distribute uniformly in the clay mass. Impregnation of the clay mass by humus and iron hydroxides decreases the birefringence of the clay mass and, apparently, disorients it. A considerable amount of sinter of oriented clay can be observed in the lower portion of the active layer.

The humus horizons of the active layer contain a clearly defined microstructure in the form of fragments and spherical microaggregates. The latter own their origin to the soil fauna whose role in the formation of the structure of the active layer can hardly be overestimated. Investigations show that there are no ferruginous films on the aggregates. The structural jointings owe their strength to the close links inside the microjointings of elementary clay particles due to the thin ferruginous and humus-ferruginous films which cover them.

The humus horizons of the active layer are characterized by a further disorientation of clay particles and by reduced birefringence of the soil mass. The humus horizons contain a considerable amount of partially decomposed vegetative remains, especially when the active layer is below a fern cover. Phytolites[1] are very few in number. Apparently, we observe a process of their rapid dissolution and washout by soil waters here.

It follows from the above that krasnozem (active layer) differs noticeably in microstructure from the crust of weathering on which it has developed, but it has an identical mineralogical composition. It should be noted that the thermal curves yield the same picture both for the soil and the crust of weathering, taken without separation into fractions, and for the clay fraction. The thermograms (Fig.1A, B, C) reveal the halloysite-kaolinite composition of the soil and the crust of weathering and the presence of free sesquioxides, with iron hydroxide predominating in the latter.

---

[1] See foot-note on page 201 (Ed.).

The principal features of the soil-forming process

The profile under investigation clearly illustrates the superimposition of the soil-forming processes on the already formed crust of weathering. As we saw, the clay material and spotted segregations of iron hydroxides were formed during the weathering process at the expense of the primary minerals while retaining the structure of the initial hard rock. The weathering and washout of all substances of separate minerals produced porous rocks. The soil-forming processes superimposed on the crust disturb its structure, and disperse and disorient the clay mass. The organic matter plays the main role in these processes. It causes the spotted segregations of iron hydroxides to dissolve and redistributes iron in the clay mass. As it interacts with the iron hydroxides, the organic matter forms humus-ferruginous films around clay minerals. The low absorption capacity of the active layer can be attributed to the presence of ferruginous films on elementary clay particles, which act as isolators. At the same time, this layer is characterized by the washout of clay followed by the formation of sinters of oriented clay in the lower part of the active layer and at the top of the crust of weathering. Thus, we observe the process of lessivage, whose degree of intensity depends on the age of the active layer.

The manifestations of the soil-forming processes also affect the upper parts of the weathering crust. Clay particles percolate along pores and cracks into the crust together with soil water, settle on its walls, and form sinters of oriented clays as they dry. The soil-forming processes also result in banded deposits of iron hydroxides in the upper part of the weathering crust.

In this way, the process of soil formation is here distinguished by the dispersion of all matter inherited from the crust of weathering, its redistribution along the profile, and a high mobility of clay and iron hydroxides whose leaching is limited mostly by the top of the weathering crust.

SUMMARY

The study of the crusts of weathering and the overlying soils in the Transcaucasus, with the aid of micromorphological and mineralogical methods, makes it possible to reveal a number of characteristic features acquired by krasnozem profiles during the processes of weathering and soil formation. This study is based on the concept of an "active soil-forming layer", introduced by Gerasimov. Analysis of the microstructure and mineralogical composition of the profiles has shown that the crusts of weathering and the active soil-forming layer clearly differ in microstructure, while possessing the same mineralogical composition. The crusts of weathering and the contemporaneous weathering have all the features of laterisation. The processes of soil formation are distinguished by dispersion and redistribution of the components inherited from the crusts of weathering, by the disorientation of the clay mass, and by lessivage phenomena.

REFERENCES

Gerasimov, I.P., 1957. Geographical Observations in Brazil. Izv. Akad. Nauk S.S.S.R., Ser. Geograf., 2 : 52–66.

Gerasimov, I.P. and Man–Yun–Ji, 1958. Genetic Types of Soils in the Territory of the People's Republic of China and Their Geographical Distribution. Izd. Akad. Nauk S.S.S.R., 86 pp.

Gerasimov, I.P., 1961. Modern Laterites and Lateritic Soils. Izv. Akad. Nauk S.S.S.R., Ser. Geograf., 2 : 3–12.

Gerasimov, I.P. and Romashkevich, A.I., in press. Genetic Profile of Modern Laterite. Trans. Intern. Congr. Soil Sci., 8th, Bucharest, 1964.

Green, R.E., 1959. Mineralogy of Clays. Publ. Office Foreign Lit. U.S.S.R., Moscow, 452 pp.

Kanno, Y.A., 1956. A Pedological Investigation of Japanese Volcanic-ash Soils. Trans. Intern. Congr. Soil Sci., 6th, Paris, 1956, Comm. 5, E : 105–109.

Lisitsina, N.A., 1962. On the gibbsite-bearing crust of weathering on the Batumi coast of the Caucasus. In: Crust of Weathering. Izdatel'stvo Akad. Nauk S.S.S.R., 4 : 96–120.

Noll, W., 1936. Über die Bildungsbedingungen von Kaolin, Montmorillonit, Serecit, Pyrophyllit und Analcim. Mineral. Petrog. Mitt., 48 : 210–246.

Ostryakov, A., 1915. Introduction to Laterite Soils, 1. Kazan, 199 pp.

Ostryakov, A., 1917. Introduction to Laterite Soils, 2. Kazan, 127 pp.

Yakovleva, M.N., 1958. On the allite crust of weathering in humid subtropics of the Transcaucasus. In: Investigation and Use of Clays. Izd. Lvov Univ., Lvov, pp. 209–229.

COLOURED ILLUSTRATIONS

Plate D4. Microstructure of the weathering crust of andesite-basalt porphyrite; x 45, nicols II. 1 = voids at the sites of leached feldspars; 2 = sinters of oriented clay in voids at the site of leached feldspars; 3 = aggregates of chlorite matter; 4 = finegrained basal mass of rock.

Plate D5. Microstructure of krasnozem, 110–120 cm. The photo shows a clearly delineated boundary of the soil sections (1) and the sections of the crust of weathering unaffected by soil formation (2); x 45, nicols II.

MICROSCOPIC STUDIES OF LATERITE FORMATIONS

R. HAMILTON

Huizen, N.H. (The Netherlands)

INTRODUCTION

The word laterite covers a rather large succession of tropical weathering products. Yet all of these products - varying from earthy to rocky materials and coloured red or brown - do form one single order. In the U.S.A. these soils are called oxisols (U.S.D.A., 1960). This large group includes all the members of one genetical chain, starting with fresh rock and ending with the iron crust, i.e., the "indurated dead" soil. Between these extremes are the intermediary latosols, i.e., tropical "red" earths. The nomenclature of these member "laterites" will be discussed in detail elsewhere, in this article "laterite" means iron crust only.

In central west Africa, viz. the Volta River basin and some adjacent areas, the two most important, striking large iron crusts occupy distinct topographical sites. These sites are the uplands or high grounds, and the lower slopes or low grounds. Accordingly, such laterites are called high and low level laterites.

THE GENESIS OF HIGH AND LOW LEVEL LATERITES

The genesis of both these laterites is quite different. High level laterites are formed by normal residual weathering in tropical regions. They are therefore residual or eluvial laterites. They are formed in the uplands because weathering and drying conditions are optimal here: high and exposed positions favour oxidation, dehydration, evaporation and lateral drainage.

Low level laterites are formed by the ferruginization of colluvial deposits. These ferruginizating iron colloids come from the uplands, and flow downward into these deposits. Accordingly, these laterites are also called colluvial or seepage laterites. They are formed in the lower slopes, or sometimes in the bottoms of narrow valleys, because here the change in gradient and soil texture retards the viscous iron colloids sufficiently to fix them on the spot.

Originally the uplands must have been rich in iron, for otherwise low level laterites could not been formed. Very often these uplands are iron-crusts, iron gravels or very ferruginous latosols. In a few other cases, however, metalliferous veins, ore bodies (gossans) or iron rich rocks appear to be the sources of the iron.

The whole theory can be represented beforehand in summary as visualized in Fig.1.

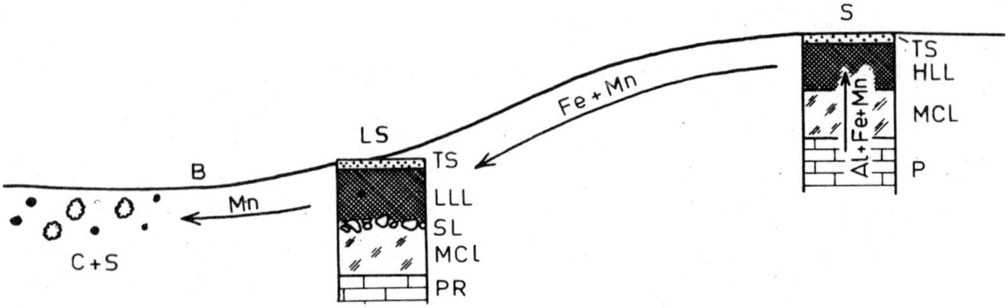

Fig.1. Site and schematic representation of high and low level laterites. Arrows indicate the origin of materials. The ultimate distribution of these materials is: Al in **HLL**, Fe in **LLL** and Mn in B. B = valley bottom; C + S = concretions and stains of Mn (+ Ca); **HLL** = high level laterite (bauxitic); **LLL** = low level laterite (ferruginous); **LS** = lower slope = foot slope; **MCl** = mottled clay; **PR** = parent rock; **S** = upland = summit; **SL** = stone line; **TS** = top soil.

Both of the above genetical processes have been studied in hundreds of thin sections made from a great many profiles in West Africa. In this way the course of both processes has been gradually disclosed.

Although a few details, e.g., concerning the status of manganese and titanium, are still being investigated, the main points are now clear. To explain the theory some of the pictures made during these studies are represented here (see also the coloured pictures of Kubiena, 1943, 1962).

The formation of high level laterites

Fig.2–8 show the formation of a high level laterite. It is a series of pictures starting with the fresh underlying rock (Fig.2) and ending with the overlying end product, or iron crust (Fig.8).

The rock is cracked by physical weathering, i.e., a long succession of heatings by the sun and coolings by rains or nights. Along these cracks water and air, above all CO_2 and O_2, enter the rock and chemical weathering is started (Fig.3). Accordingly, the very soluble bases are leached away. Only calcium is sometimes partially recovered as stains and concretions in the valley bottoms. The less soluble oxides like Al_2O_3, Fe_2O_3, and SiO_2 are left behind as hydrous colloids. What happens to Mn_2O_3 and TiO_2 has not yet been established. In the investigated cases all the parent rocks were poor in these oxides. Consequently their concentration was never high enough to produce observable results. Some resistant minerals like quartz also remain. Gradually all these remains concentrate, and finally the colloids of alumina and silica interact to form kaolinite. Colloidal iron oxide, however, remains free and mobile. The red-brown mass filling the cracks at this stage is therefore nothing but iron-stained kaolinite with a few quartz grains. It represents the very first beginnings of a latosol. In other words, these red-brown cracks are very young but genuine latosols on the smallest possible scale (Fig.4). Gradually the whole parent rock becomes pierced with an increasing number of widening and reddening cracks. At long last the whole rock is changed into a latosol (Fig.5). In these latosols the still mobile iron colloids, protected by silica, are generally easy to observe. They form clear birefringent streams and fields of a pale yellow-brown colour. In the course of time, with increasing age and the leaching of the last bases, these latosols dry out (Sherman et al., 1953) and acidify (down to pH 4). With increasing acidity there is a slight formation of iron and aluminium solutions which are leached. The over-concentrated iron colloids flocculate. Afterwards they continue to dehydrate and then start to crystallize. The role of daily and seasonal changes in temperature, and of earth or thermo currents, in these "ageing processes" is not yet solved. In any case electric currents do exist in iron crusts and bauxites. In some bauxites potential differences up to 400 mV have been measured.

This entire process can be inferred from X-ray and thin section studies. Diffractograms of the crack materials, i.e., of unfinished young latosol, indicate only some kaolinite and quartz. Those of the mature latosols, however, show several crystalline iron oxides in addition to these minerals. Yet kaolinite remains dominant (up to 80%). With increasing age these soils show (successively): first some diffuse limonite, i.e., cryptocrystalline goethite, then distinct goethite, and finally goethite with hematite.

This "ageing" process (i.e., dehydration + concentration + crystallization), which gradually changes the dilute amorphous iron colloids into dense crystalline iron minerals (MacKenzie, 1959), is called "siderization" (sideros = iron) or ferruginization. (In some cases, e.g., in the low level laterites, it includes the original influx of iron colloids too). It can be seen in all older latosols as distinct dark patches with irregular or round outlines (Fig.6). Such patches form the well-known concretions or iron gravel when removed from a latosol. The patches grow because the "ageing process" gradually involves the whole latosol. In addition, there is also some supply of new iron colloids from the surroundings, that is, upward from the rotting rock below and laterally from the adjacent soils. Finally these patches meet and coalesce. The intergrowth of irregular patches is very gradual and commonly imperceptible. They form dark, rather homogeneous laterites. On the other hand, the intergrowth of round patches (concretions) is often quite evident. They can be seen meeting and then being cemented together by the iron colloids. These colloids fill the angular cavities between the meeting concretions, and afterwards wholly overgrow them (Fig.7). In this way concretionary laterites are formed with clearly visible concretions (Fig.8).

The mineralogy of various iron crusts, as revealed by X-rays, is rather simple, and only few differences are observed. Their chief minerals are kaolinite, goethite and hematite. Sometimes a little magnetite and gibbsite can be noticed too, but neither manganese nor titanium minerals can be determined. Dr. R. Schmidt–Lorenz (personal communication, 1961) found traces of lepidocrocite and maghemite in some laterites. Alexander (Alexander and Cady, 1962) found anatase!

Hematite and magnetite are the normal and predictable end members of a dehydration and ageing series. The presence of gibbsite, however, is something new. It points to the breakdown of kaolinite and the possible

FIG.2–12

Fig.2. The fresh rock, a granite. It is chiefly composed of biotite, orthoclase, quartz and some magnetite; x 20.

Fig.3. The cracked rock. Along these cracks weathering enters into the rock and starts to change it wholly; x 20.

Fig.4. An enlarged picture of a crack. This crack is filled with iron-stained kaolinite, i.e., with a very young latosol; x 20.

Fig.5. A young but fully developed latosol; x 20.

Fig.6. An old latosol with several dark patches. These ferruginization patches gradually increase in size and number as the latosol grows older; x 20.

Fig.7. Two round concretions have met. They are now cemented together by iron colloids, which fill an angular cavity. When the cavity is wholly filled, the colloids start to overgrow the concretions; x 40.

Fig.8. A concretionary laterite, i.e., the final product of coalescing round concretions; x 20.

Fig.9. Colloidal silica. It is quite evident in the colourless borders; x 20.

Fig.10. A latero-bauxite. This is a transitional stage between laterites and bauxites. The original laterite pattern is still obvious. Pale parts are almost pure gibbsite with some diaspore; however, the dark parts are ferruginous and lateritic; x 20.

Fig.11. A low level laterite (young stage). Some vigorous flow structures of iron colloids can be noticed like ramifying streams; x 20.

Fig.12. A low level laterite (old stage). (Sub)-angular quartz from colluvial deposits is cemented together by iron colloids. Flow structures are almost absent and indistinct. The dark parts are saturated by iron colloids. In the paler parts these colloids are still infiltrating and of lower concentration; x 40.

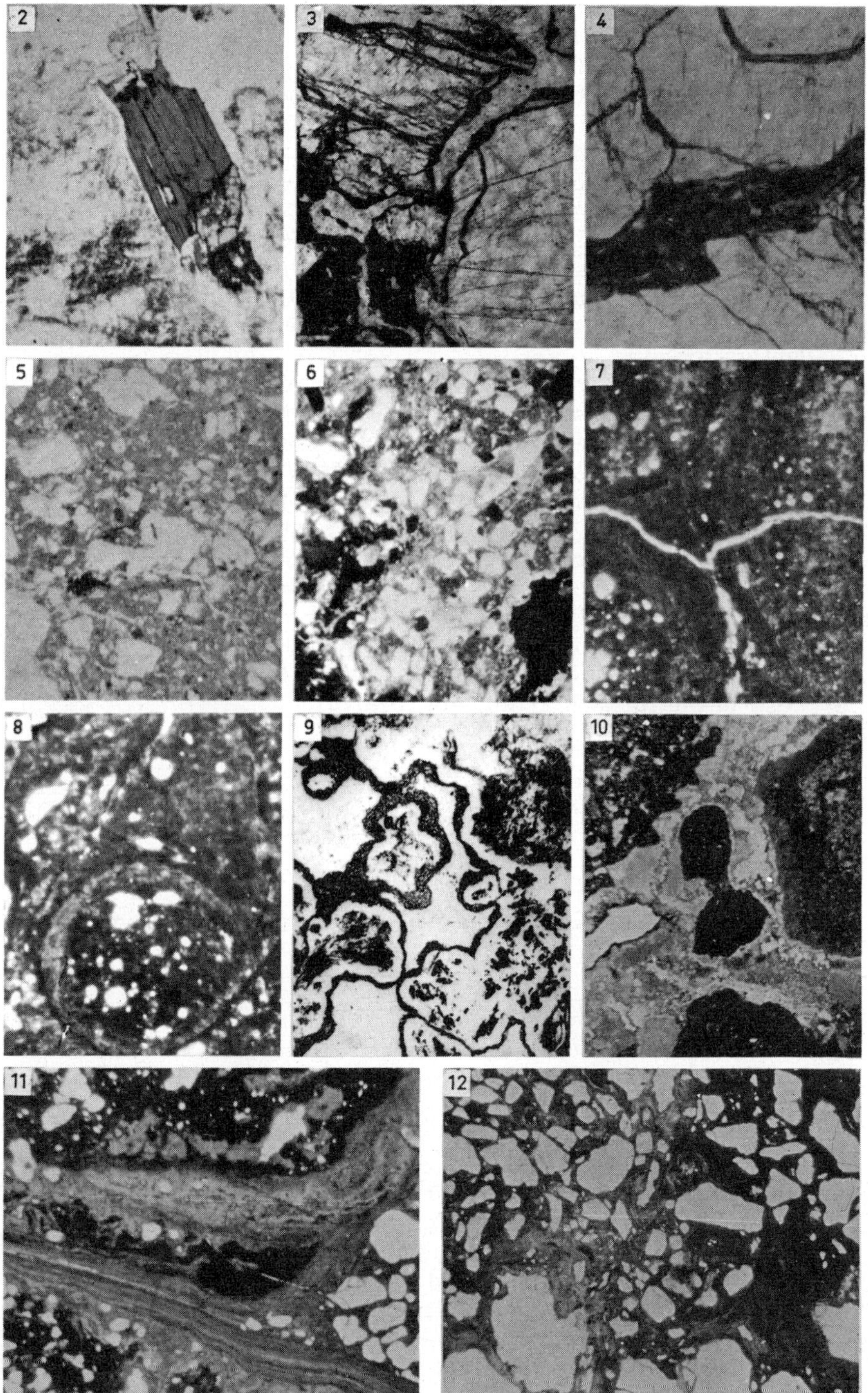
2
3
4
5
6
7
8
9
10
11
12

(differential) leaching of some iron. This announces the final stage of laterite formation in West Africa, i.e., its transformation into bauxites. The gibbsite is probably formed from alumina derived from the breakdown of kaolinite. This is quite likely, for both the defective kaolinite and the remaining silica have been found in some very old laterites. Defective kaolinite was inferred from its somewhat different diffractogram, and colloidal silica was seen in some thin sections (Fig.9). Therefore it appears quite possible that a laterite with sufficient aluminium can be transformed into an aluminous bauxite in the course of time. Many, if not all, the bauxites in West Africa are therefore refined high level laterites (Cooper, 1936). Whether these bauxites are also improved afterwards by the removal (differential leaching) of some iron and silica is not yet definitely known. Along with some other details, this is still under investigation.

In any case, a complete transitional series from ferruginous laterites to aluminous bauxites does exist. It can be observed both in the field and under the microscope. A good example of such a transitional stage between laterites and bauxites is represented in Fig.10. The white dots in this picture are small rosettes of gibbsite in a lateritic milieu (Lacroix, 1913). However, the pale areas, i.e., high grade bauxite, are chiefly gibbsite with some diaspore. Both these minerals were determined by X-rays. Traces of boehmite, but no bayerite, were sometimes observed by Dr. R. Schmidt–Lorenz (personal communication, 1961) in a few bauxites.

No pictures of pure bauxites are included, for they are almost colourless in thin sections and thus do not show off well. In summary, one can show the whole process in the following diagram (Fig.13).

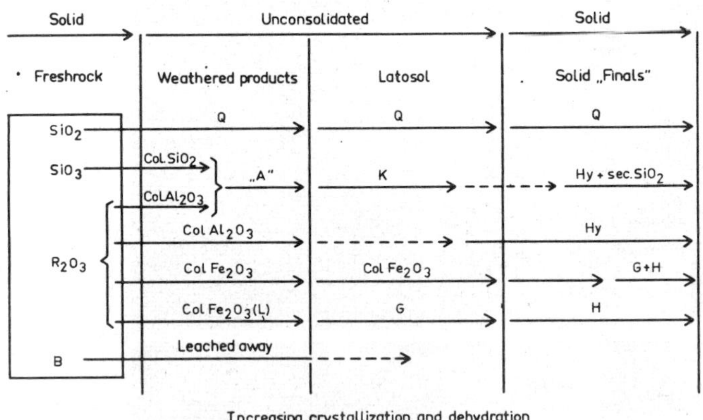

Fig.13. Diagram of formation of high level laterite. A = allophane = colloidal "kaoline"; B = bases; Col = colloidal; Finals = new-formations = ferruginous laterites (gravel, pans, crusts) and aluminous laterites (bauxites); G = crystalline goethite; H = crystalline hematite; Hy = crystalline hydrargillite = gibbsite; K = crystalline kaolinite; L = limonite = mixture of colloidal + crystalline Fe_2O_3; Q = free quartz (free silica); R_2O_3 = sesquioxides = Fe_2O_3 + Al_2O_3; Sec = secondary colloidal silica; SiO_2 = free silica (quartz); SiO_3 = combined silica = silicates.

The formation of low level laterites

So far we have only discussed the upland laterites. As to the low level laterites, their formation is much simpler and easier to investigate. They are formed simply by the filling and saturating of colluvial deposits with mobile, silica protected iron colloids (a little manganese oxide is present too!). These iron colloids are supplied laterally from the surrounding ferruginous uplands, from whence they flow down the slopes (Lamotte et al., 1962). This supply goes on until the whole colluvial mass is soaked with it. Afterwards these colloids start to flocculate, age and crystallize in the same way as the high level laterites. When the uplands are poor in iron no low level laterites are found. On the other hand, uplands very rich in iron form large slope laterites, sometimes even to the upper slopes. Such large slope laterites are the consequence of a continuous supply, which does not end until the source of iron is nearly exhausted.

The mineralogy and micrography of these laterites are therefore not different from that of the upland laterites. Thin sections of low level laterites (Fig.11, 12) are sometimes difficult to distinguish from those of certain upland laterites. Genetically the colluvial laterites are somewhat younger. Accordingly, some of them still show vigorous colloidal streams (Fig.11) when the upland laterites are already consolidated. Chemically they are mostly richer in manganese, for manganese is very mobile and easily removed from the uplands.

How these manganese oxides move has not yet been undisputably settled. This problem is still being examined. It seems, however, that at least one part moves as true solutions, and the remainder as colloids. Probably the colloids remain wholly or partially in the low level laterites, whilst the solutions move on to the valley bottoms. Here they form manganese stains and concretions. Many valley soils contain free manganese, sometimes more than 200 p.p.m. Even valley streams may have free manganese (up to 0.2 p.p.m.) in their water. Some valley concretions, effervescent with H_2O_2, probably enclose the mineral pyrolusit. Total MnO can be high as 3.1% in valley soils, whereas rocks and laterites rarely contain more than 1.0%.

The presence of manganese often gives the low level laterites a peculiar dull brown colour, strongly contrasting to the bright red brown colour of the upland laterites. In the field, however, the sites are characteristic and the differences clear (cf. Fig.1).

ACKNOWLEDGEMENTS

Prof. Dr. W. Kubiena and his staff are very gratefully noted for all their aid, and for the fine thin sections they made. These thin sections facilitated the investigations enormously.

Dr. R. Schmidt–Lorenz has helped me very much with his enviable X-ray equipment. For the valuable diffractograms made I am much obliged to him.

Finally, I am greatly indebted to Dr. B.J. Harris. The time and devotion he spent has made the micrographs a real success.

SUMMARY

The weathering and disintegration of a rock in the tropics is described. In the formed decomposition products a soil gradually develops. Still later, solid iron stone concretions and pans are formed in this soil.

By the leaching of these residual upland laterites a secondary laterite is built up in the colluvial mass of the foot slopes.

REFERENCES

Alexander, L.T. and Cady, J.G., 1962. Genesis and hardening of laterite in soils. U.S. Dept. Agr., Tech. Bull., 1282 : 90 pp.

Cooper, W.G.G., 1936. The bauxite deposits of the Gold Coast. Gold Coast, Geol. Surv Bull., 7 : 35 pp.

Kubiena, W.L., 1943. Gefüge Untersuchungen an tropischen und subtropischen Rotlehmen. Beitr. Kolonialforsch., 3 : 48–58.

Kubiena, W.L., 1962. Structure and Genesis of Tropical Soils. Bundesanstalt Forst-Holzwirtsch., Reinbek bei Hamburg, 3 pp.

Lacroix, A., 1913. Les latérites de la Guinée et les produits d'altération qui leur sont associés. Nouvelles Arch. Musée Hist., T5 : 255–356.

Lamotte, M. et Rougerie, E.G., 1962. Les apports allochtones dans la génèse des cuirasses ferrugineuses. Rev. Géomorphol. Dyn., 13 : 145–160.

MacKenzie, R.C., 1959. The ageing of sesquioxide gels. I. Iron oxide gels. Mining Mag. (London), 32 : 153–165.

Sherman, G.D., Kanchiro, Y. and Matsusaka, Y., 1953. The role of dehydration in the development of laterite. Pacific Sci., 7 : 438–446.

Soil Survey Staff U.S.D.A., 1960. Soil Classification, a Comprehensive System, 7th Approximation. Soil Conservation Service, U.S. Dept. Agr., Washington, D.C., 265 pp.

A SHORT NOTE ON DROPLET-FORMATION IN IRONCRUSTS

R. HAMILTON

Huizen, N.-H. (The Netherlands)

In the course of ironcrust investigations in West-Africa, a few times some curious phenomena were observed, i.e., droplet-formations. They are by no means of frequent occurrence: during the investigation of several hundreds of thin sections the phenomena were seen less than half a dozen times. Yet they are of some universality. They were noted e.g., both near Monrovia and near Accra i.e., more than 760 miles apart. Further they occurred both on mountains up to 2,500 ft. high and in lowlands.

The only thing the phenomena have in common is that they seem to occur merely in old ironcrusts (laterites). It is therefore that they are considered as ageing or as ex-solution phenomena. It is emphasized however that this is only a working hypothesis and not at all a settled fact. Positively these "droplets like dots" are not colonies of iron bacteria. Neither they are faecal pellets, because these iron crusts are hard and not nourishing.

Owing to their rare occurrence it is very difficult to get sufficient material for a conclusive investigation. So it was decided to publish this note. It is hoped that it will draw the attention of other investigators to these phenomena and also to learn their comments. May be with some collaboration it will be possible to solve the problem.

These "droplets like dots" are both dark or light coloured. Further the dark droplets are more uniform in size (Fig.1), whereas the pale drops are clearly of varying size (Fig.2). The pale drops do suggest something of a (normal) growing process i.e., increasing size. The dark droplets increase enormously in number only and seem to coalesce at last.

From comparative studies with other ironcrusts (laterites) and bauxites it was tried to learn something about their composition. The dark droplets are probably iron oxides i.e., concentrated haematite and magnetite. The pale drops however may consist of alumina i.e., of gibbsite and diaspore. This is quite in accordance with the ultimate transformation of "ferruginous" laterites in "aluminous" bauxites (Hamilton, 1964).

For the present it is impossible to say more about these interesting and curious things.

SUMMARY

In some old laterites small striking dots-like droplets were noticed. Their possible importance and composition are discussed.

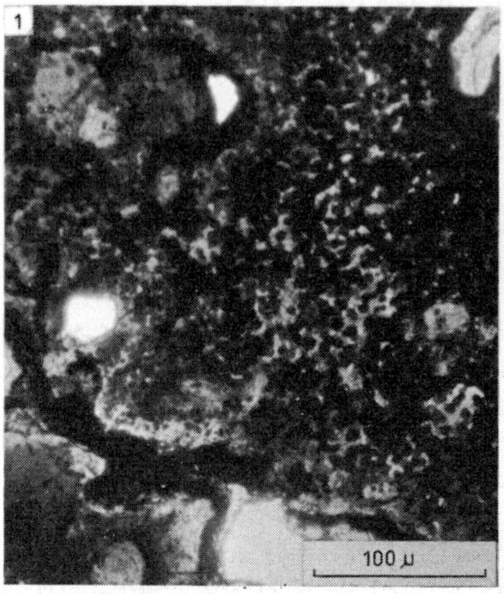

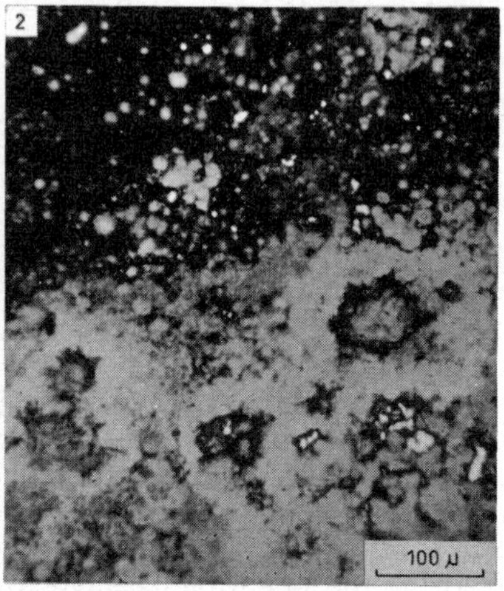

Fig.1. Dark droplets or dots of uniform size in a bouldery ironcrust along the coast (45 miles west of Accra). Ordinary light.

Fig.2. Pale drops or dots of varying size in a massive ironcrust-block (5 miles north of Accra). Ordinary light.

REFERENCE

Hamilton, R., 1964. Microscopic studies on laterite formation. In: A. Jongerius, (Editor), Soil Micromorphology. Elsevier, Amsterdam, pp.269–276.

ZUR MIKROMORPHOLOGIE DER EISEN- UND ALUMINIUMOXYDANREICHERUNG BEIM TONMINERALABBAU IN LATERITEN KERALAS UND CEYLONS

R. SCHMIDT–LORENZ

Bundesforschungsanstalt für Forst- und Holzwirtschaft, Reinbek bei Hamburg (Deutschland)

EINLEITUNG

Bei der Diskussion über die Entstehungsweise der Laterite ist die Frage, in welchem Masse Eisen- und Aluminiumoxydanreicherungen[1] das Produkt absoluter oder relativer Akkumulation (D'Hoore, 1954) sind, von grundsätzlicher Bedeutung. In neueren Arbeiten (Prescott und Pendleton, 1952; Alexander et al., 1956; Alexander und Cady, 1962; Sivarajasingham et al., 1962) wird vornehmlich die Anreicherung durch Stoffzufuhr von aussen behandelt (absolute Akkumulation). Dagegen wird die Lateritentstehung durch relative Anreicherung, d.h., durch Abfuhr anderer Substanzen (neben Basen bes. Kieselsäure, ggf. auch Aluminiumoxyd) ohne Stoffzufuhr meist nur verhältnismässig kurz abgehandelt.

Nach Kubiena (1954) geht die Lateritbildung in mehreren Phasen vor sich, wie an Vorkommen Spanisch Guineas demonstriert wurde. Dabei erfolgt eine Eisenoxydanreicherung durch Ausfällung des peptisierten, amorphen Eisenoxydhydrats innerhalb einer dichten Braunlehm-Grundmasse in kleinen schwebenden Hämatitkriställchen und- kristallaggregaten und durch deren Zusammenschluss im Verlaufe der Auswaschung der restlichen eisenfreien Grundmasse (siehe auch Kubiena, 1956, 1957, 1962). Auch Alexander und Cady (1962) fanden in westafrikanischen Lateriten Eisen-Mikroaggregate ("tiny spherical aggregates"), ohne auf deren Entstehungsweise näher einzugehen. R. Hamilton (briefliche Mitteilung, 1964) beobachtete in Dünnschliffen aus westafrikanischen Eisenkrusten "pale and dark droplets like dots of uniform und varying size". Bereits 1943 erwähnt Kubiena Eisen-Mikroaggregate in einer in Hydrargillit (syn. Gibbsit) umgewandelten Grundmasse, also die nebeneinander vorkommende, voneinander unabhängige Ausscheidung von Eisen- und Aluminiumoxyd (Demonstration anhand eines Dünnschliffs aus Lateritpanzer von St. Thomas, Indien). Diese für die Lateritforschung bedeutsamen Erkenntnisse schliessen an die älteren, vornehmlich aus dem Vergleich chemischer Analysen hergeleiteten Ansichten über die Entstehungsweise der Laterite an (Newbold, 1844; Holland, 1903; Glinka, 1914; etc.).

Neuere, bodenkundlich ausgerichtete Arbeiten über Lateritvorkommen im Bereich der Malabarküste, insbesondere aus dem Gebiet der ersten Fundstellen des Materials "Laterit" (Buchanan, 1807) sind spärlich (Satyanarayana und Thomas, 1961, 1962); detaillierte mikromorphologische Untersuchungen fehlen bisher.

[1] Eisenoxyd steht für die im Boden vorkommenden Verbindungen des Systems Fe_2O_3/FeO/H_2O; entsprechend Aluminiumoxyd.

UNTERSUCHUNGSMATERIAL

Das Probenmaterial entstammt einer grösseren Anzahl von Lateritvorkommen, die in Kerala, in den Distrikten Palghat, Alleppey, Quilon und Trivandrum, ferner im Südwesten Ceylons im Herbst 1962 aufgesucht wurden. Besonders zu erwähnen ist der Einbezug von Material aus Lateriten der "type locality" Buchanans im Raum von Angadipuram (Fox, 1936; Stephens, 1961).

Das aus einer grösseren Anzahl von Laterithorizonten entnommene Material besitzt unterschiedlichen Aufbau, vorherrschend jedoch die für den Buchanan-Laterittyp charakteristische Vermikularstruktur, d.h.,schwammartige Anlage des sesquioxydreichen Gerüstes. Dessen gewundene röhrchenförmige Kavernen sind mit einem eisenärmeren Ton von Braunlehmcharakter (i.f. "Füllton") ausgefüllt.

Folgende Probenahmeorte mit Lateriten ausgeprägter Vermikularstruktur sind besonders zu nennen:

(a) Angadipuram, Distrikt Plaghat; Aufnahme eines "low level laterite" unweit Bahnstation und mehrerer Einzelproben aus "high level laterite" einige km west des Ortes; Gestein: Hornblendegranulit.

(b) Pattambi, nahe Angadipuram; Profil eines "low level laterite"; Einzelproben aus pisolith-haltigen Böden; Probe eines verfestigten Pisolithlaterits aus Bachbett.

(c) Varkala, Trivandrumdistrikt; stark aluminiumoxydhaltiger Laterit an der Steilküste; aus jüngeren Sedimenten (Braunkohlenschicht in ca. 30 m Tiefe).

(d) Lunawa (einige km ost des Ortes), ca. 15 km süd-südost von Colombo über Biotitgneis.

FRAGESTELLUNG

Im Dünnschliff führen alle aufgesammelten Vermikularlaterite grosse Gefügebereiche mit den von Kubiena (1954) aufgezeigten Eisen-Mikroaggregaten und zwar umso ausgeprägter, je weniger Fremdbeimengungen und Gesteinsstrukturen am Aufbau der Gerüstsubstanz beteiligt sind. Pisolithe erweisen sich häufig als gerundete Bruchstücke aus solchen Vermikularlateriten.

Die starke Verbreitung der Eisen-Mikroaggregate spricht dafür, dass die zu ihrer Bildung führenden und damit verbundenen Prozesse ein wesentliches Merkmal des Buchanan-Laterittyps sind.

Auf der Grundlage der vergleichenden mikromorphologischen Analyse wurde angestrebt, Näheres über den Aufbau und die Bildungsbedingungen der Eisen-Mikroaggregate und über die mit ihrem Auftreten verbundenen Veränderungen in der sie umgebenden Grundmasse zu erfahren. Resultate der ergänzend durchgeführten chemischen und röntgenographischen Analyse, u.a. von mikrotechnisch isolierten Gefügepartien, werden hier nur teilweise genannt.

UNTERSUCHUNGSERGEBNISSE

Zur Entstehung des sesquioxydreichen Gerüstes typischer Vermikularlaterite aus Braunlehmmaterial

Die Art des inneren Aufbaus des Gerüstes der untersuchten Vermikularlaterite hängt weitgehend davon ab, in welchem Bereich die Laterisierung eingesetzt hat, d.h., im wesentlichen, ob in braunlehmartig vorverwitterten Gesteinszersatzzonen oder in Braunlehmdecken. Letztere führen je nach Entstehungsweise wechselnde Anteile an Fremdbeimengungen, nämlich sesquioxyd-inkrustierte Gesteinsbruchstücke und Pisolithe unterschiedlicher Herkunft und Bildungsweise.

Dass wesentliche Anteile des Gerüstes der aufgenommenen Vermikularlaterite aus Braunlehmmaterial hervorgegangen sind, liess sich teilweise bereits makroskopisch feststellen und zwar am ehesten bei Material nicht zu hohen Alters, Ausbildungs- und Verhärtungsgrades.

Charakteristische Merkmale des Gerüstes eines fast ausschliesslich aus Braunlehmmaterial hervorgegangenen Vermikularlaterits (z.B., Laterit der Lokalität Lunawa, Ceylon) bei Lupenvergrösserung sind (vergl. dazu Tafel E2).

(a) Ausgeprägte Differenzierung in Aufbau und Färbung.

(b) Aus einer Vielzahl in sich dichter, überwiegend glattflächig begrenzter scharfkantiger Gefügekörper von Sand- bis Erbsengrösse zusammengesetzt; diese sind meist dunkel gesäumt und brekzienartig ineinander gefügt; die feine Felderung wird erst im Bruch sichtbar.

(c) Neben verschiedenen gelblichen, braunen und roten Farbabstufungen sind Gefügekomplexe mit violettem Farbstich besonders auffallend und verbreitet; daneben kommen fast farblose vor; teils diffuse Farbübergänge, jedoch nicht so häufig wie im angrenzenden, nicht laterisierten Füllton.

(d) Die Härte der Strukturkörper läuft i.a. mit der Farbtiefe parallel; nicht verhärtete Anteile des Gerüstes besitzen tonige, teilweise gelartige Konsistenz.

Die sesquioxydreichsten Bereiche des Gerüstes einerseits und der unveränderte Füllton in den Kavernen des Vermikularlaterits andererseits sind grundverschieden aufgebaut. Berücksichtigt man jedoch die Gefügezonen zwischen diesen beiden Extremen, ferner den mikromorphologischen Aufbau verschiedener Laterithorizonte unterschiedlicher Organisationshöhe, so ergibt sich aus dem Vergleich eine lückenlose Folge mikroskopischer Gefügeeinheiten, an deren Anfang ein feinkornreiches Braunlehmmaterial ohne Laterisierungsmerkmale, an deren Ende eine dichte, eisenoxydreiche Masse steht, deren Aufbau aus einzelnen Eisen-Mikroaggregaten nur aus dem Zusammenhang hervorgeht.

Im Dünnschliff lässt sich unschwer erkennen, dass die Anlage der brekzienartigen Felderung des Gerüstes auf Quellungs- und Schrumpfungsvorgänge unter Bildung von Feinpolyedern im Braunlehmstadium zurückzuführen ist. Entsprechend dem halloysitisch-kaolinitischen Tonmineralbestand ist die Schrumpfung wenig intensiv, so dass die Polyeder zumeist nur schmale Risse zwischen sich frei lassen. Durch Einfliessen von Feinsubstanz erhalten die Polyeder auf ihren Begrenzungsflächen filmartige, vielfach rhythmisch angelegte Beläge aus orientiertem Feinton mit diffus

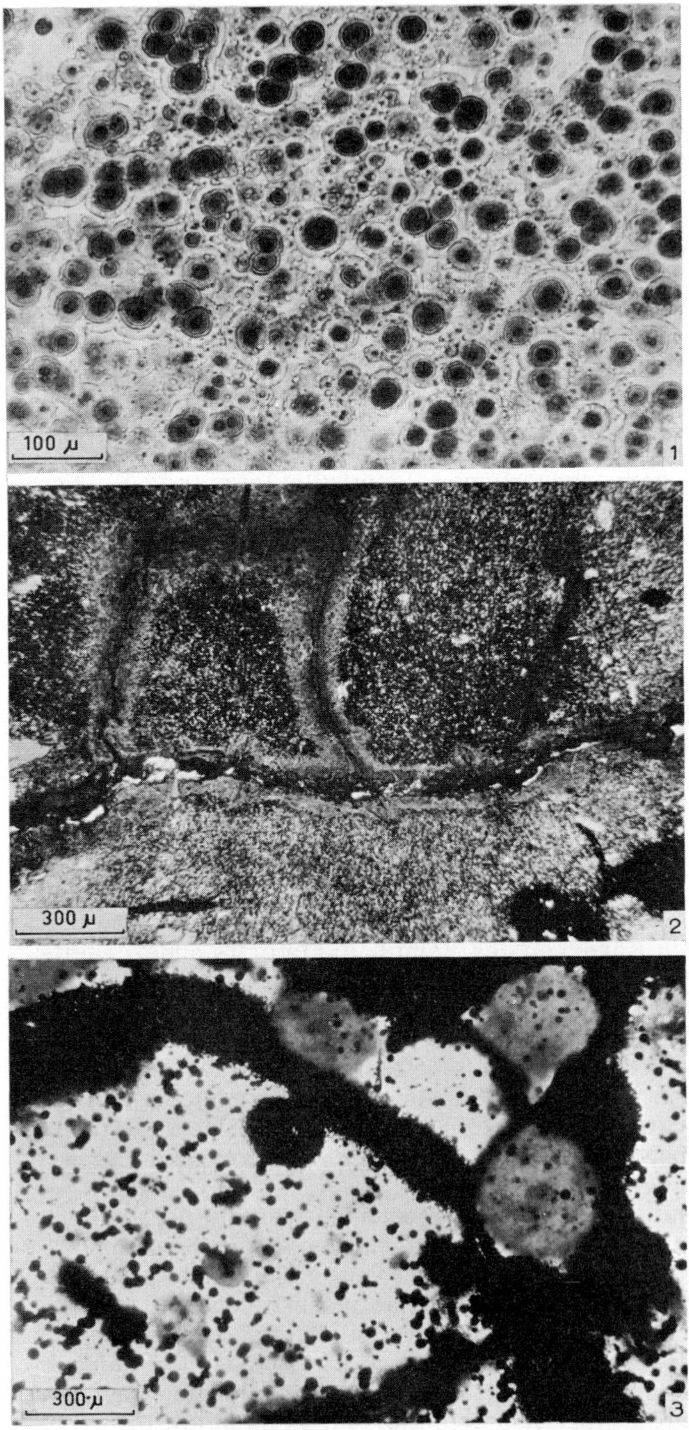
100 μ
1
300 μ
2
300 μ
3

feinverteiltem Eisenoxyd. Diese Fliessplasmafilme besitzen offenbar grundsätzliche Bedeutung bei den nachfolgenden Abbauprozessen und Stofftrennungen innerhalb der Polyeder im Zuge der Laterisierung. Kubiena spricht von "abgekapselten" Gefügeelementen.

Umwandlungen durch Laterisierung

Aufbau und Ausscheidungsorte der Eisen-Mikroaggregate

Eisen-Mikroaggregate der nachfolgend beschrieben Art sind innerhalb dichter, homogener, feinstkornreicher Grundmasse und zwar vorherrschend in abgekapselten Gefügekomplexen ausgeschieden. Zwei verschiedene Formen werden beobachtet, die wahrscheinlich prinzipiell gleichen Bildungsbedingungen unterliegen.

(1) Grössere Sphärokörper mit konzentrischem Aufbau (im Auflicht erkennbar) und glatter Begrenzung; Grösse zwischen ca. 1 μ und 10 μ $\emptyset$; Ausscheidung dieser Form vornehmlich in den Innenbezirken der Polyeder (Abb.3).

(2) Kleinere, teils rundliche, meist jedoch andeutungsweise hexagonale Mikrokörper ohne erkennbare Innenstruktur; Grösse um 1 μ und kleiner; Ausscheidung teilweise im Inneren dichter Gefügekomplexe, vorherrschend jedoch in deren Randbezirken (Abb.2; auch Abb.3).

Diese beiden Ausscheidungsformen sind nicht immer eindeutig voneinander zu unterscheiden. Sie zeigen weiterhin folgende Merkmale:

(a) Sehr wechselnder gegenseitiger Abstand; im Innengefüge von Polyedern entspricht der Abstand i.a. dem 1–5 fachen Mass ihres Durchmessers; in Polyederrandbezirken oft recht dichte Packung.

(b) Die Ausscheidung des diffus feinverteilten Eisenoxyds erfolgt offenbar grundsätzlich durch engräumige zentripetale Strömung und nachfolgende Ausfällung, nicht durch Verdrängung der Grundmasse, sondern durch deren Einschluss.

(c) Ausscheidung auch innerhalb orientierter Tonmineralfliessfilme; dabei regelmässige Anordnung der Eisenmikrokörper parallel zur Tonmineralbasisfläche, aber auch in grösseren, fleckenförmigen Konzentrationen; ferner sehr feine Ausscheidungen in Randzonen solcher Tonmineralanhäufungen (Abb.6).

Abb.1. Durchlichtaufnahme (x 120); Eisenoxyd-Sphärolithe mit konzentrischen Fällungszonen in verwitterndem Hypersthen in Charnockit (Lok.: Elpitiya, südwest Ceylon).

Abb.2. Durchlichtaufnahme (x 48) aus jüngerem Laterit (Lok.: Pattambi, Palghat); gleichmässig verteilte Hämatit-Mikroaggregate in isotroper Grundmasse; Schwundrisse mit regelungs-doppelbrechendem Fliessplasma.

Abb.3. Durchlichtaufnahme (x 48); aus Lateritbruchstück (Lok.: Nähe Kayangulam, Alleppey) mit besonders grossen Eisen-Mikroaggregaten (Sphärokörper) in isotroper Grundmasse; rundliche Gefügekomplexe in der Grundmasse noch doppelbrechend.

(d) Bei sehr wechselnder Doppelbrechung erscheinen die Eisen-Mikroaggregate meist opak; bei Auflicht sind sie rot gefärbt; jüngere Ausscheidungen sind seltener und gelblich-braun.

(e) Gefügebereiche mit nicht zu hoher Konzentration an Eisen-Mikroaggregaten sind am Handstück recht häufig violett gefärbt.

(f) Nach der röntgenographischen Analyse handelt es sich bei der im Auflicht rot gefärbten Form wahrscheinlich ausschliesslich um Hämatit mittleren bis guten Kristallisationsgrades.

Umwandlungen in der Grundmasse, in der die Eisen-Mikroaggregate ausgeschieden sind

Der kleinsträumigen Konzentrierung des diffus feinverteilten Eisenoxyds entspricht eine nahezu vollkommene Bleichung der restlichen Grundmasse. Deren Doppelbrechung nimmt über den Betrag hinaus ab, der aus dem Verlust an feinverteiltem Eisenoxyd resultiert. Häufig erscheint die verbleibende Grundmasse isotrop. Auch orientierte Tonmineralfilme werden von diesen Veränderungen betroffen. Nach Verlust ihrer Regelungsdoppelbrechung sind sie schliesslich nur noch auf Grund ihrer Morphologie identifizierbar.

In älteren, besonders *eisenoxydreichen Lateriten* werden darüber hinaus folgende Gefügeveränderungen und neuen Eigenschaften festgestellt:

(a) Grössere Gefügezonen mit erhöhter Konzentrierung von Eisen-Mikroaggregaten, die ineinander verwachsen erscheinen.

(b) Hierdurch Zunahme des Eisengehaltes pro Volumeneinheit.

(c) Im Zuge der Konzentrierung Zunahme des Hämatit-Kristallisationsgrades und der Härte.

(d) Intensiver, unregelmässig rauher Feinbruch des dichten Gefüges, oft quer durch Zonen mit hoher Eisenoxyd-Konzentration und durch eiseninkrustierte Polyeder-Abkapselungsschichten.

In den *aluminiumoxydreichen Lateriten* treten zusätzlich folgende Merkmale auf:

(a) Die restliche Grundmasse ist lokal oder über grössere Gefügebereiche hinweg in Gibbsit umgewandelt.

(b) Überwiegend feinkristalline Ausscheidung des Gibbsits; gröbere Auskristallisierung um in der Grundmasse eingebettete grössere Quarze und andere verwitterungsresistente Minerale.

(c) Idiomorphe Gibbsitausbildung innerhalb kleiner Hohlräume; lokal grössere Gibbsit-Sphärolithe in ehemaligen Hohlräumen.

(d) Die Gibbsitisierung lehnt sich u.a. an Fluidalstrukturen des ehemaligen Braunlehms an; sie erfasst auch Tonmineralfilme.

(e) Ausheilung von Schwundrissen durch reinen Gibbsit.

(f) Soweit die gibbsitisierten Gefügezonen eisenhaltig sind, liegt Eisenoxyd gewöhnlich in Form von in der Gibbsitgrundmasse eingeschlossenen Mikroaggregaten vor (vergl. Kubiena, 1943; Abb.17, 18).

(g) Mikroskopisch weitgehend rein erscheinende, nahezu eisenfreie, harte Gibbsitknollen führen nach röntgenographischem Befund geringe, aber doch deutliche Anteile an Kaolinit.

SCHLUSSFOLGERUNGEN UND DISKUSSION

Unter "real laterization" versteht D'Hoore (1954) einen extremen Fall der relativen Anreicherung. Im Gegensatz zur absoluten Anreicherung genügt zur Betrachtung des Vorgangs der relativen Akkumulation die Berücksichtigung des Umwandlungsraumes (D'Hoore, 1954). Gesicherte Aussagen über die abgeführten Stoffe setzen die Kenntnis des ursprünglichen Zustandes dieses Raumes voraus. Diese ergibt sich bei dem hier untersuchten Material aus indischen und ceylonesischen Lateriten aus dem nebeneinander von Produkten unterschiedlichen Umwandlungsgrades. Bei der Umwandlung werden die Tonminerale des Braunlehms in ihre Hauptbestandteile Kieselsäure und Aluminium zerlegt. Die freigesetzte Kieselsäure wird nahezu vollkommen ausgewaschen. Sekundäre Verkieselungen wurden nicht beobachtet. Häufig wird die Entfernung der Kieselsäure jedoch offenbar stark verzögert (Entstehung von Kieselsäure-Gel), was u.a. durch die Abkapselung der Gefügekomplexe bedingt sein kann. Das freigesetzte Aluminiumoxyd verhält sich demgegenüber anders: entweder kristallisiert Aluminiumoxyd in situ aus oder es wird aus dem Gefügekomplex ausgewaschen. Gleichzeitig mit der Destruktion der Tonminerale wird das ursprünglich diffus feinverteilte Eisenoxyd des Braunlehms im wesentlichen in Eisen-Mikroaggregaten innerhalb der dichten Grundmass ausgeschieden.

Entsprechend dem unterschiedlichen Verhalten von Kieselsäure und Aluminiumoxyd lassen sich prinzipiell zwei verschiedene, ineinander übergreifende Prozesse unterscheiden, die man als Intensitätsstufen der Laterisierung durch relative Anreicherung auffassen kann.

In der ersten Intensitätsstufe verbleibt das freigesetzte Aluminiumoxyd im wesentlichen in situ und kristallisiert zu Gibbsit (Abb.4, 5, Tafel E1), vermutlich zumindest teilweise als Pseudomorphose nach dem ursprünglich vorliegenden Tonmineral (Erhaltung von Fliessstrukturen aus dem Braunlehmstadium). Infolge Fixierung der eingeschlossenen Eisen-Mikroaggregate unterbleibt deren weiterer Zusammenschluss. Die Mikroaggregate sind hier häufig auffallend klein. Oft sind die gibbsitisierten Gefügekomplexe recht eisenoxydarm. Diese Eisenarmut kann in einem primär eisenärmeren Ausgangsmaterial begründet sein, möglicherweise steht sie aber auch mit geringem Abkapselungsgrad der Polyeder in Verbindung (bei geringer Einschlämmung von unveränderter Feinsubstanz). In diesem Fall ist eine weitreichende Entfernung des durch Kieselsäure geschützten Eisenoxyds aus dem Gefügekomplex in Betracht zu ziehen. In ähnlicher Weise könnte man die oft beobachteten feinen Eisenoxyd-Ausscheidungen in den Randzonen von abgekapselten Gefügekomplexen mit Konzentrationsabnahme in Richtung Innengefüge erklären. Zarte, gelblichbraune isotrope Beläge auf den Oberflächen von Polyedern kann man als eisenoxydreiche Gele deuten, deren Substanz dem Gefügeinneren entstammt (Tafel E3). Möglicherweise gehören einige der von Harrassowitz (1926) und Fox (1936) analysierten, stärker aluminiumhaltigen Lateritvorkommen aus Indien dieser Gruppe an.

In der zweiten Intensitätsstufe der Laterisierung durch relative Anreicherung wird ausser Kieselsäure auch das freigesetzte Aluminiumoxyd aus dem Gefügekomplex ausgewaschen. Hieraus folgt eine gegenüber dem ersten Fall intensivere Eisenkonzentrierung. U.a. aus dem häufigen Vorkommen von gut gerundeten, relativ grossen Eisen-Mikroaggregaten in

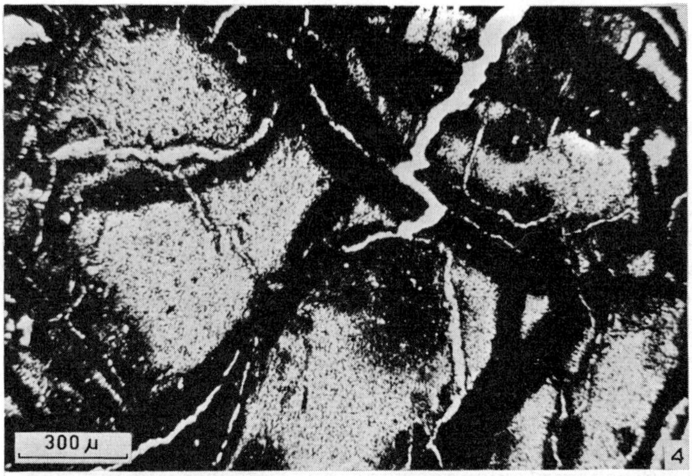
300 μ
4

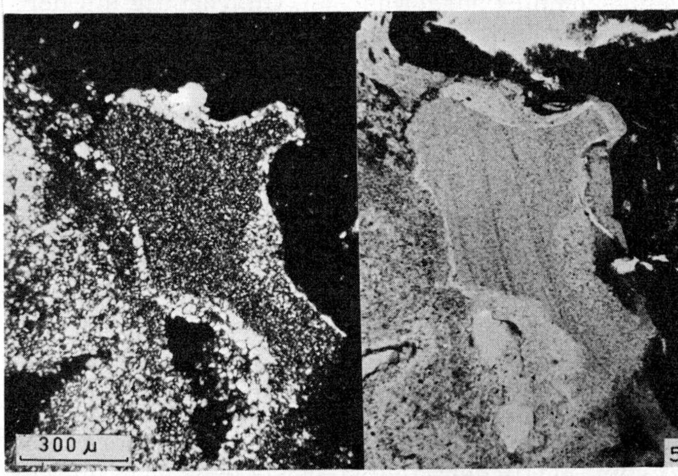
300 μ
5

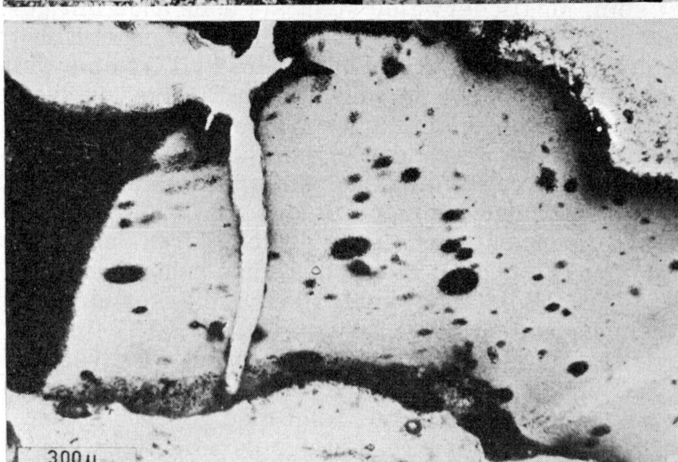
300μ
6

isotroper Grundmasse (Abb.3) wird geschlossen, dass bei frühzeitiger Aluminiumoxyd-Entfernung und gleichzeitig verzögerter Kieselsäureabfuhr besonders günstige Bedingungen zur Ausfällung des feinverteilten Eisenoxyds in mehr oder minder grossen Sphärokörpern gegeben sind. Deren Ausscheidung erfolgt offenbar durch Strömung des feinverteilten Eisenoxyds und dessen rhythmische Ausfällung nach Art Liesegangscher Ringe im Kieselsäure-Gel.

In diese Richtung weist auch folgendes Beispiel der Verwitterung eines besonders Aluminiumarmen primären Minerals, nämlich von Hypersthen in einem Laterit aus Charnockit (Lok.: Elpitiya, südwest Ceylon). Dicht neben noch unverwitterten Mineralen werden stark veränderte Hypersthene beobachtet, in deren nahezu isotroper Grundmasse gelblichbraune Eisen-Sphärolithe (im Sphäriten Kreut) polarisierten Licht mit konzentrischen rhythmischen Fällungszonen ausgeschieden sind (Abb.1). Daneben finden sich alle Übergangsstadien bis zu Opaken, im Auflicht rot gefärbten Sphäroaggregaten innerhalb des selben Mineralindividuums. Die morphologische Ähnlichkeit der Endprodukte lässt den Schluss zu, dass die Bildung von kugelförmigen Eisen-Mikroaggregaten bei der Laterisierung von Braunlehmmaterial in prinzipiell gleicher Weise vor sich geht. (Die aufgezeigte Art der Hypersthenverwitterung demonstriert ausserdem, dass die Laterisierung durch relative Anreicherung auch an noch nicht verwitterten primären Mineralen direkt einsetzen kann.) Wahrscheinlich bestehen Beziehungen zwischen der Kieselsäurefreisetzung aus den Tonmineralen und der von verschiedenen Faktoren abhängigen Kieselsäureabfuhr aus dem Gefügekomplex, insbesondere deren Geschwindigkeit, einerseits und den beobachteten Verschiedenheiten von Grösse, Form und Verteilung der Eisen-Mikroaggregate andererseits.

In der Endphase der Laterisierung erscheinen die einzelnen Eisen-Mikroaggregate über grössere Gefügebereiche hinweg ineinander verwachsen. Ihren Zusammenschluss setzt die Auswaschung der Kieselsäure voraus, der die Hydrolyse des Kieselsäuregels vorausgeht.

Mit der Konzentrierung der verbleibenden Stoffe durch Substanzverlust sind ausgeprägte Schrumpfungsvorgänge durch Verdichtung infolge Dehydration und Durchkristallisation der Residualprodukte verbunden. Sie haben eine teilweise Regenerierung des Leitbahnsystems zur Folge als Voraussetzung für neue Stoffzufuhr in Form von Schlämmstoffen. Auf diese Weise, zusätzlich durch Mitwirkung von Bodenkleintieren wird eine Wiederholung (Kubiena, 1954)

Abb.4. Durchlichtaufnahme (x 48); weitgehend eisenoxydfreie, gibbsitisierte Grundmasse mit zahlreichen, zu verschiedenen Zeiten entstandenen Schwundrissen, teils mit Gibbsit verheilt (Lok.: Varkala).

Abb.5. Aufnahme im Durchlicht und polarisiertem Licht (x 48); gibbsitisiertes Tonmineralschichtpaket mit perlschnurartig angeordneten Eisenoxyd-Ausscheidungen auf den Schichtflächen (aus Laterit der Lok.: Nuvahad, Alleppey).

Abb.6. Durchlichtaufnahme (x 48) aus dem gleichen Dünnschliff wie Abb.5; noch nicht so weit umgewandelte Tonsammlung in Hohlraumnische; das ehemals diffus feinverteilte Eisenoxyd liegt am Rande des Komplexes und in rundlich begrenzten Ansammlungen in der isotropen Grundmasse vor.

der demonstrierten Vorgänge eingeleitet. Diese währen so lange, bis infolge Alterung und Härtung ein Endzustand erreicht ist und/oder eine weitere Stoffzufuhr oder -abfuhr unterbunden wird.

Sichere Anzeichen der lateralen Stoffzufuhr über die Lösungsphase in den Gefügebereichen der relativen Anreicherung fehlen. Über den Verbleib der Abbauprodukte (bzgl. teilweise Eisenoxydabfuhr, siehe oben) ergeben sich aus der Mikromorphologie nur mittelbare Schlussfolgerungen. Migration von Fliessplasma aus den Akkumulationsbereichen heraus liess sich nicht nachweisen. Die Fortfuhr der Abbauprodukte erfolgt vermutlich allein in echter und kolloidaler Lösung.

Die erwähnte, bereits im Gelände auffallende Violettfärbung ("lilac" nach Newbold, 1844) ist an die Gefügezonen mit noch nicht zusammengeschlossenen Eisen-Mikroaggregaten gebunden. Nach Mahlen violett gefärbten Materials, auch nach dessen Kochen in Lauge tritt bisweilen ein Verlust der Violettfarbkomponente ein. Sie resultiert offenbar aus dem Nebeneinander von Hämatit-Mikroaggregaten und noch nicht abgeführten Tonmineralabbauprodukten. Deren Alterung ist vermutlich wesentlich beteiligt am Laterithärtungsprozess, der bei der Lufttrocknung frisch entnommenen, zur Ziegelherstellung geeigneten Materials vor sich geht.

Bezüglich der in Lateriten recht verbreitet vorkommenden Vermikularstruktur ist zu betonen, dass makroskopisch ähnliche Formen grundsätzlich verschiedenen Innenaufbau haben können, wie sich aus dem Vergleich mit westafrikanischem Material ergibt. Die eindeutige gegenseitige Abgrenzung dieser Konvergenzerscheinungen ist am ehesten mit Hilfe des Dünnschliffs möglich.

ANERKENNUNG

Folgenden Herren dankt der Verfasser für Organisation der durchgeführten Exkursionen bzw. für Führung im Gelände: Dr. K.V.P. Menon, Direktor der Central Coconut Research Station, Kayangulam, Kerala, und dessen Mitarbeiter Prof. E.J. Verghese, soil chemist, ferner Dr. M.L.M. Salgado, Direktor der Coconut Research Institute, Lunuwila, Ceylon, und dessen Mitarbeiter Dr. D.A. Nethsinghe, soil chemist.

Die Arbeit wurde mit Mitteln der Deutschen Forschungsgemeinschaft, Bad Godesberg, und der Hamburgischen Forschungsgemeinschaft durchgeführt.

SUMMARY

On the micromorphology of FeO and AlO accumulation by destruction of clay minerals in laterites in Kerala state and in Ceylon

A relative accumulation of FeO and AlO by a process involving tiny FeO-sphero-aggregates with haematite as a final product occurs in laterites in Kerala, including Buchanan's type locality, and also in southwest Ceylon. This process takes place within a matrix which is rich in colloids. The exportation of silica as a product of destruction of clay minerals appears to be combined with this process of segregation of the FeO, but it is probably independent of the fate of the AlO which may crystallize in situ to gibbsite.

LITERATUR

(Bezüglich der älteren Literatur siehe unter J.A. Prescott and R.L. Pendleton, 1952.)

Alexander, L.T., Cady, J.G., Whittig, L.D. and Dever, R.F., 1956. Mineralogical and chemical changes in the hardening of laterite. Trans. Intern. Congr. Soil Sci., 6th, Paris, 1956, E : 67–72.

Alexander, L.T. and Cady, J.G., 1962. Genesis and hardening of laterite in soils. U.S. Dept. Agr., Tech. Bull.,1282 : 90 pp.

D'Hoore, J.L., 1954. The description and classification of free sesquioxide accumulation zones. Trans. Intern. Congr. Soil Sci., 5th, Leopoldville, 1954, 4 : 39–44.

Kubiena, W.L., 1943. Gefügeuntersuchungen an tropischen und subtropischen Rotlehmen. Beitr. Kolonialforsch., 3 : 48–58.

Kubiena, W.L., 1954. Micromorphology of laterite formation in Rio Muni (Spanish Guinea). Trans. Intern. Congr. Soil Sci., 5th, Leopoldville, 1954, 4 : 77–84.

Kubiena, W.L., 1956. Rubefizierung und Laterisierung (zu ihrer Unterscheidung durch mikromorphologische Merkmale). Trans. Intern. Congr. Soil Sci., 6th, Paris, 1956, E : 247–249.

Kubiena, W.L., 1957. Neue Beiträge zur Kenntnis des planetarischen und hypsometrischen Formenwandels der Böden Afrikas. Stuttgarter Geograph. Studien, 69 : 50–64.

Kubiena, W.L., 1962. Die taxonomische Bedeutung der Art und Ausbildung von Eisenoxydhydratmineralien in Tropenböden. Z. Pflanzenernähr., Düng., Bodenk., 98 : 205–213.

Prescott, J.A. and Pendleton, R.L., 1952. Laterite and lateritic soils. Bur. Soil Sci. (Gt. Brit.), Tech. Commun., 47 (1952) : 51 pp.

Satyanarayana, K.V.S. and Thomas, P.K., 1961. Studies on laterites and associated soils, I. Field characteristics of laterites of Malabar and South Kanara. J. Indian Soc. Soil Sci., 9 : 107–118.

Satyanarayana, K.V.S. and Thomas, P.K., 1962. Studies on laterites and associated soils. II. Chemical composition of laterite profiles. J. Indian Soc. Soil Sci., 10 : 211–222.

Sivarajasingham, S., Alexander, L.T., Cady, J.G. and Cline, M.G., 1962. Laterite. Advan. Agron., 14 : 1–60.

Stephens, C.G., 1961. Laterite and the type locality, Angadipuram, Kerala, India. J. Soil Sci., 12 : 214–217.

FARBENBILDER

Tafel E2. Auflichtaufnahme (x 28) aus Vermikularlaterit (Lok.: Lunawa, Ceylon); mit allen Übergangsstadien vom unveränderten Braunlehm bis zu Gefügezonen mit hoher Eisenoxyd-Konzentration; ausgeprägtes Polyedergefüge, intensive Abkapselung der Gefügeelemente; nur wenige Bruchstücke aus Gestein und primären Mineralen; teils Spuren von Kleintiertätigkeit.

Tafel E3. Durchlichtaufnahme (x 150) aus "high level laterite" (Lok.: Angadipuram, Palghat); umgewandelte Braunlehmkomplexe mit dicht gelagerten Hämatit-Mikroaggregaten in isotroper Grundmasse; daneben weitgehend enteisente Partie; Schwundrisse mit orientiertem Fliessplasma gefüllt, teils mit beginnender Eisenoxydausscheidung; teils gelartige Eisenoxyd-Beläge.

Tafel E1. Aufnahme bei polarisiertem Licht (x 150) aus Laterit mit lokal ausgeprägter relativer Aluminiumoxyd-Anreicherung (Lok.: Varkala, Trivandrum); gibbsitisierte Grundmasse; zu verschiedenen Zeiten gebildete Schwundrisse; die älteren, bereits früh mit Fliessplasma verheilten Risse sind dicht von Hämatit-Mikroaggregaten umgeben; jüngere Risse quer zu früheren angelegt mit unverwittertem Fliessplasma mit Regelungsdoppelbrechung.

EISENOXYDKONKRETIONEN UND SCHLIERENBILDUNG IN EINIGEN TROPENBÖDEN ECUADORS

ERWIN FREI

Eidgenössische Landwirtschaftliche Versuchsanstalt, Zürich–Oerlikon (Schweiz)

Die chemische Verwitterung karbonatfreier Muttergesteine ist in den feuchten, warmen Tropen sehr intensiv. Die Bildung der verschiedenen Sekundärmineralien wird von der Zusammensetzung des Muttermaterials und durch die klimatischen Verhältnisse gesteuert. Das Feuchtigkeitsregime im Boden ist ausschlaggebend für die Wanderung und die Umkristallisierung der amorphen Eisen- und Aluminiumhydroxyde, sowie der Alumosilikate und der freien Kieselsäure. Aber die völlige Klärung der Genetik der Latosole und ihrer verwandten Bodenbildungen steht noch aus, die vorliegende Arbeit behandelt anhand einiger Dünnschliffuntersuchungen ecuadorianischer Tropenböden nur ein Fragment dieses Problems.

Für die meisten Latosole Ecuadors ist der feinkörnige, rot–gelb–weiss maserierte Unterboden (oxic subsurface horizon) typisch. Der pastellrote Ton ergibt sich wahrscheinlich durch die Umhüllung der roten Eisenoxyde mit weisslichen Mineralien oder Gelen. Je nach Zusammensetzung des Muttermaterials wiegen die roten oder die weisslichen Farbkomponenten vor. Mikromorphologisch (Abb.1; Tafel D6) besteht der Boden aus dunkelroten,

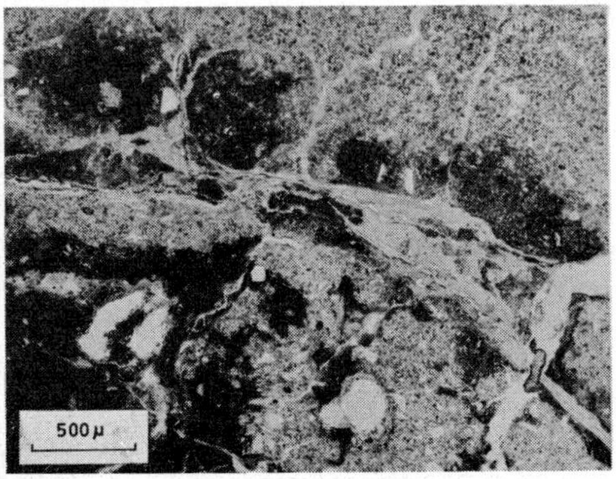

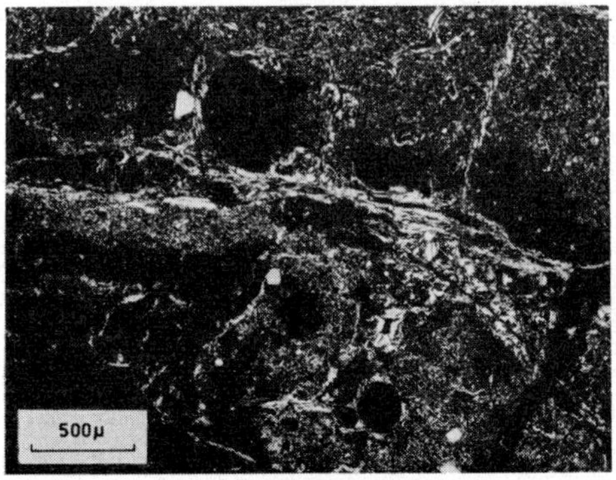

Abb.1. Gelbroter Latosol Ecuador; Dünnschliff. "Oxid horizon", flockige, opake Eisenoxydkonkretionen treten vorwiegend in Nähe der Hohlräume auf.

Abb.2. Wie Abb.1, jedoch zwischen gekreuzten Nicols. Die weissliche Grundmasse weist stark doppelbrechende Schlieren auf, die vorwiegend dem Hohlraumsystem parallel laufen.

flockigen Eisenoxydkonkretionen in einer feinkörnigen, weisslichen Grundmasse, wahrscheinlich Aluminiumoxyd und Kaolinit. Besonders auffallend für dieses Stadium ist die scharfe Separierung der dunklen und der hellen Komponente, ausgenommen die häufige Einwachsung von Primärmineralien in die Konkretionen. Die flokkigen Eisenoxydkonkretionen sind deutlich umgrenzt, sehr unregelmässig geformt, einigermassen dem Hohlraumsystem des Bodens angeglichen und meist optisch opak. Ihre Entstehung dürfte mit dem Feuchtigkeits- und Durchlüftungsregime des Bodens in Zusammenhang sein. Da die Böden feinporig und reich an stark hydratisierten Hydroxyden sind, kann bei starken Regenfällen auch bei normaler Durchlässigkeit eine vorübergehende völlige Porensättigung vorkommen. Dies führt jedoch zu periodischem Sauerstoffmangel, der erst nach Abzug des Sickerwassers wieder behoben ist. Bei der Bildung von Konkretionen könnte somit die Wanderung von reduziertem Eisen eine Rolle spielen. In Latosolen, die einer kurzen Trockenzeit ausgesetzt sind, findet eine allmähliche Entwässerung und Umkristallisierung der gewanderten Gele statt. Das optisch opak Verhalten deutet auf kristalline Eisenoxyde die jedoch noch weich und schneidbar sind wie Käse. Viele Latosole Ecuadors befinden sich erst in einem initialen Entwicklungsstadium, da sie nur wenig Konkretionen enthalten, sodass die Hauptmasse des Bodens eine homogen gelborange Färbung aufweist (Tafel D7).

In normalem Licht erscheint der weissliche Anteil des "oxic horizon" im Dünnschliff feinkörnig porös, aber im übrigen strukturlos. Zwischen gekreuzten Nicols (Abb.2), treten jedoch Schlieren sehr hoher Doppelbrechung hervor. Die hellsten Streifen sind deutlich mit dem sichtbaren Porensystem korreliert. Die Ursache dieser schon früher von Kubiena (1962) beschriebenen Erscheinung, könnte thixotropischem Verhalten der feinen Kaolinit-, Boehmit- oder Gibbsitkristalle zugeschrieben werden. Da die optischen Achsen mit den Kristallachsen mehr oder weniger zusammenfallen, ist bei physikalischer Ausrichtung der Kristallite eine Formdoppelbrechung zu erwarten. Wie entsteht diese gleichmässige Ausrichtung? Für ein allgemeines Fliessen fehlen alle erforderlichen Indizien. Es ist unmöglich, dass in einem stabilen Bodengefüge ein grosser Teil der Bodenmasse in Bewegung sein könnte. Die sehr feinen, tafeligen Kriställchen sind stark hydratisiert und möglicherweise durch anhaftende Kieselsäure entladen. Dem Spannungsfeld des Absorptionswassers ausgesetzt, vermögen sie sich thixotropisch nach ihren Kristallachsen auszurichten. Die Formdoppelbrechung ist deshalb am grössten in Nähe der Hohlräume und rings um stabile Materialoberflächen, weil dort die Wassertensionen am kräftigsten wirksam sind. Die allgemein bekannte schwierige Dispergierbarkeit der Kolloidfraktion der Latosole steht nicht im Widerspruch zu dieser Thixotropietheorie, aber sie erklärt vielleicht warum die doppelbrechenden Schlieren im Bodengefüge erhalten bleiben.

Latosole die sich in einem Klima mit längerer Trockenzeit entwickeln zeichnen sich meist durch einen braunen Oberhorizont aus, der dispergierte Tone ins Perkolationswasser abgibt. Die Art der Wiederanlagerung solcher Tone zeigt Abb.3; es entstehen geschichtete Tonhäute von starker Formdoppelbrechung und mit deutlichen Interferenzlinien.

Auf Kalkgestein werden im warmen Tropenklima Ecuadors mit langer Trockenzeit roterdeähnliche Böden gebildet. Die rundlichen, körnigen Eisen

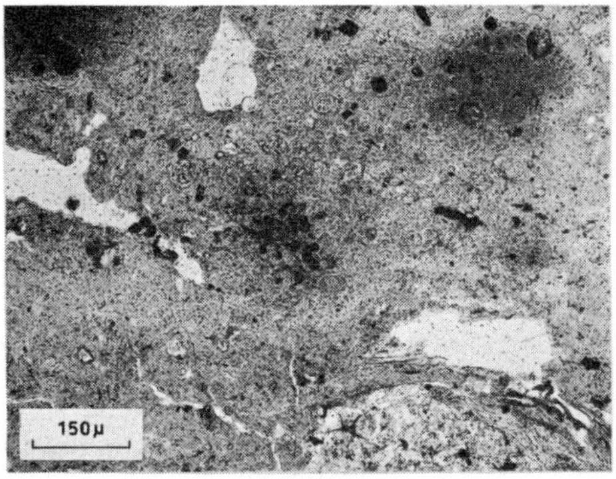

Abb.3. Brauner Latosol, Ecuador; Dünnschliff, zwischen gekreuzten Nicols. Geschichtete, stark doppelbrechende Tonhäute in Hohlraum.

Abb.4. Roterde, Ecuador; Dünnschliff. B-Horizont mit zahlreichen, körnigen Eisenoxydkonkretionen von 10 µ Durchmesser, sowie beginnende Verkittung und Wachstumsstadien der Konkretionen.

oxydkonkretionen, das krümelige Aggregatgefüge und der harte rotbraune Farbton des B-Horizontes, sind für die Roterden allgemein typisch. Unter 10 µ kleine Eisenoxydkonkretionen, überschwemmen die ganze Bodensubstanz des meist ziemlich scharf begrenzten B-Horizontes. Abb.4 erlaubt einen Einblick in die Entstehungsart grösserer Konkretionen um einen oder mehrere kleine Kristallisationskerne. Es bilden sich vorerst diffuse Flecken, die allmählich zu konzentrisch aufgebauten, makroskopischen Konkretionen auswachsen. Einzig der Kristallisationskern kann opak sein, im übrigen bleiben die Eisenoxyde in der Roterde jedoch transparent (Tafel D8). Die Grundmasse des Bodens ist sehr tonreich. Was im Vergleich zu den "oxic horizons" der Latosole besonders auffällt, ist die allgemein braunrote Farbe der Grundmasse. Die Aluminiumsilikate bleiben in der Roterde mit Eisenhydroxyden imprägniert. Man muss unter diesen Umständen annehmen, dass die mässig alkalische Reaktion und die Kalziumsättigung dafür verantwortlich sind.

SUMMARY

Iron oxide concretions and formation of cutans in some tropical soils of Ecuador

Flake-like, sharply bordered, opaque concretions of iron oxide resting in a whitish, very fine textured matrix are characteristic of the oxic horizons of some investigated latosols. The formation of these ferric concretions depends on the chemistry of the parent material, but also very much on the hydrology and the aeration of the soil. The pale matrix shows very strong birefringence, particularly along pronounced streaks supposed to be of thixotropic origin. The minute aluminium oxide and aluminium silicate crystals undergo an orientation according to their crystal axes. Real clay cutans are rather scarce in brown-latosols of Ecuador. Granular ferric concretions of concentric structure are characteristic for terra rossa-like soils on calcareous parent material in a warm tropical climate with strong alternation of dry and humid periods. In these soils, only a part of the iron is concentrated in the concretions. The rest impregnates the abundantly present clay; therefore, the matrix is of reddish-brown color.

LITERATUR

Guennelon, R., 1951. Étude préliminaire à la caractérisation des oxydes de fer des sols rouges méditerranéens. Ann. Agron., Sér. A, 3 (4) : 506–509.

Kubiena, W., 1962. Die taxonomische Bedeutung der Art und Ausbildung von Eisenoxydhydratmineralien in Tropenböden. Z. Pflanzenernähr. Düng. Bodenk., 98 (3) : 203–213.

Shermann, G.D., Matsuraka, Y., Jkawa, H. und Uehara, G., 1964. The role of the amorphous fraction in the properties of tropical soils. Agrochimica, 8 (2) : 146–163.

Soil Survey Staff, U.S.D.A., 1960. Soil Classification, a Comprehensive System, 7th Approximation. Soil Conservation Service. U.S. Dept. Agr., Wahington, D.C., 265 pp.

FARBENBILDER

Tafel D6. Gelbroter Latosol Ecuador; Dünnschliff; x 72, Durchlicht mit Auflicht. Weitgehende Separierung des Eisens in Eisenoxydkonkretionen, die in eine gelbweissliche Grundmasse von Aluminiumoxydkristallen und allenfalls Kaolinit eingebettet sind. Primärmineralien sind noch vorhanden.

Tafel D7. Initialer Latosol, Ecuador; Dünnschliff; x 72. Das Eisen ist noch weitgehend diffus in der ganzen Bodenmasse verteilt. Die Konkretionsbildung ist erst im Anfang.

Tafel D8. Roterde, Ecuador; Dünnschliff; x 72. Durchlicht mit Auflicht; B-Horizont mit rotbrauner, toniger Grundmasse und zahlreichen körnigen, Eisenoxydkonkretionen mit Kristallisationskernen.

MIKROMORPHOLOGISCHE UNTERSUCHUNGEN ÜBER DIE ENTWICKLUNG VON PODSOLEN IM AMAZONASBECKEN

H.-J. ALTEMÜLLER und H. KLINGE

Institut für Bodenbearbeitung der F.A.L., Braunschweig–Völkenrode; Hydriobiologische Anstalt der Max-Planck-Gesellschaft, Plön (Deutschland)

EINLEITUNG

Bei pedologischen Studien im nicht überschwemmten Areal am unteren Rio Negro (brasilianisches Amazonasgebiet) wurden von Klinge an Talhängen Catenen aufgenommen, die Braunlehme und Podsole sowie Übergangsformen zwischen beiden Typen umfassen.

Nach einer im Druck befindlichen Arbeit (Klinge, 1964), in der die Zusammenhänge zwischen mineralischem Ausgangsmaterial, Topographie und Vegetation sowie Böden diskutiert werden, sind in praktisch noch ebener Lage an den Talrändern auf stärker sandigen, fluviatilen Sedimenten Braunlehme ausgebildet; in den meisten Fällen zeigen diese Böden Andeutungen von Bleichhorizonten und werden dann als "schwach gebleichte Braunlehme" (Kubiena, 1953) bezeichnet. Ist bei stärkerer Geländeneigung ein makroskopisch erkennbarer, durchlaufender Bleichhorizont ausgebildet, liegt nach Kubiena (1953) ein "sandiger Braunlehm" vor. Hangabwärts stellt sich unterhalb des immer mächtiger werdenden Bleichhorizontes ein Anreicherungshorizont für humose Schlämmstoffe (B_{h_2}) ein; solche, von Kubiena nicht beschriebenen stark gebleichten Braunlehme werden "eluvierte Braunlehme" genannt. Im weiteren Verlauf der Bodenentwicklung greift der Bleichhorizont taschenförmig in den Unterboden aus und zerstört den früher gebildeten B_{h_2} Horizont teilweise; dabei entsteht ein zweiter, schmaler Humusanreicherungshorizont (B_{h_1}), der der Untergrenze des A_e-Horizontes folgt und dem B_{h_2} Horizont aufliegt. Böden mit zwei humosen Anreicherungshorizonten werden als Braunlehm-Podsole bezeichnet. Ist im Zuge der weiteren Bodenentwicklung am Unterhang der B_{h_2}-Horizont restlos aufgelöst und nur noch der B_{h_1}-Horizont vorhanden, liegen Podsole vor, deren A_e-Horizonte einige Meter betragen können, während der B_{h_1}-Horizont nur wenige Zentimeter misst. Die Bleichhorizonte der Podsole und Braunlehm-Podsole sind stets von zahlreichen, parallel verlaufenden dünnen Humusbändern durchsetzt.

Die Vegetation der Podsole ist eine rachitische Waldform (Campina), die der Braunlehme der typische Regenhochwald (Mata da terra firme).

Die mikromorphologische Bearbeitung der Böden aus Catenen mit den entwicklungskundlich interessanten Übergangsformen von Braunlehmen zu Podsolen hat Altemüller übernommen. Den Ergebnissen dieser Untersuchungen werden einige wichtige Gefügemerkmale der Entwicklungsreihe Parabraunerde - Podsol aus dem gemässigten Klima vorausgeschickt.

GEFÜGEKUNDLICHE MERKMALE DER ENTWICKLUNGSREIHE PARABRAUNERDE - PODSOL IM GEMÄSSIGTEN KLIMA

Die Begriffe Podsol und Podsolierung wurden in den vergangenen Jahren vor allem mit Unterstützung durch mikromorphologische Untersuchungen enger gefasst und schärfer abgegrenzt, indem die Prozesse der Tonverlagerung (Lessivierung), die ohne wesentliche chemische Veränderung der Schlämmstoffe ablaufen, von der Podsolierung getrennt und besonderen Bodentypen zugeordnet wurden. Podsole selbst wurden allerdings noch relativ wenig mikroskopisch untersucht.

Die Trennung wurde auch dadurch vereinfacht, dass Podsole, z.B., in der norddeutschen Glaziallandschaft ziemlich substratgebunden auf verarmten Sanden vorkommen, während Parabraunerden für die Lössgebiete typisch sind. Die vor der eigentlichen Podsolierung häufig anzutreffende Ausbildung von braunerdeartigen Gefügen unterstützt die Trennung von Tonverlagerungsprozessen.

In manchen Fällen gehen aber auch auf sandigen Substraten der Entwicklung von Humus- oder Eisenhumuspodsolen Tonverlagerungen voraus, die den Verlagerungsprozessen in den Parabraunerden vergleichbar sind (Altemüller, 1962). Hierbei kommt es zur Ausbildung von Hüllengefügen aus Tonteilchen, die den Skelettkörnern als geschichtete und vorzüglich geregelte stark doppelbrechende Filme anhaften. Die Ortsteine der Podsole sind zwar in guter Ausbildung ebenfalls durch schichtig gebaute Hüllengefüge charakterisiert, doch bedingt ihre stoffliche Zusammensetzung aus kolloidalen Humusstoffen und organo-mineralischen Verbindungen andere Färbung, typische netzartige Schrumpfung und optische Isotropie. Auf Grund dieser Gegebenheiten scheint die gefügekundliche Trennung von Lessivierung und Podsolierung einfach. Allerdings ist - wie schon erwähnt - über die Variationsbreite der Podsole noch wenig bekannt.

MIKROMORPHOLOGISCHE KENNZEICHNUNG DER ENTWICKLUNGSREIHE BRAUNLEHM - PODSOL IM FEUCHT TROPISCHEN KLIMA

Die Abgrenzung Ton- (Humus-) Verlagerung–Podsolierung verdient bei den tropischen Böden des unteren Rio Negro besondere Aufmerksamkeit, weil die Podsole sich auf sandigem Substrat unmittelbar aus Braunlehmen verschieden starken Bleichungsgrades entwickeln. Hohe Verwitterungsintensität, starke Stoffbewegungen und tiefgründige Bleichung spiegeln die Bildungsbedingungen des feucht tropischen Klimas wider. Anreicherungshorizonte sind nur wenig mächtig entwickelt und vor allem durch die dunkle Färbung gekennzeichnet; die ermittelten Kohlenstoffgehalte dieser Horizonte entsprechen allerdings nicht ihrer intensiven Färbung.

Die auffallend geringe Mächtigkeit der Podsol-B-Horizonte kann auf seitliche Entwässerung der Böden zurückgeführt werden, für die die Voraussetzungen (dichter Untergrund, Lage der Böden an Talhängen) gegeben sind. Auch die durch humose Stoffe verursachte Färbung der die Podsolgebiete entwässernden Bäche spricht für diese Annahme.

Nachstehend werden die gefügekundlichen Verhältnisse der untersuchten Böden beschrieben.

Braunlehm

Ein Braunlehmprofil, das makroskopisch keine Anzeichen für Bleichung oder Stoffverlagerung aufweist, entspricht der von Kubiena gegebenen allgemeinen Kennzeichnung des typischen Braunlehms (Kubiena, 1953) in wesentlichen Merkmalen.

Unter einer dünnen Streudecke und schwach entwickelten Humuszone von etwa 6 cm Mächtigkeit folgt eine einheitlich gelbbraune dichte Bodenmasse. Im Dünnschliff sind vorwiegend zwei Komponenten zu bemerken (Abb.1): (a) gelber, ziemlich gleichmässig feiner Ton als dichte, teilweise etwas aggregierte Grundmasse und (b) eingelagerte Sandkörner, die ein teils offenes, teils auch enger gefügtes Skelett bilden.

Dieses Sandskelett besteht fast ausnahmslos aus ungleich geformten, gerundeten, auch kantigen Quarzkörnern, die starke Lösungserscheinungen in Gestalt von Aushöhlungen, und tief eingefressenen Kanälen aufweisen, in denen Tone und teilweise auch dunkelbraune Huminstoffe eingelagert sind. Manche kleinen Quarzsplitter sind Reste grösserer Körner, die nahezu vollständig der Auflösung anheimgefallen sind. Diese intensiven Verwitterungserscheinungen sind in allen untersuchten Profilen gleichermassen zu finden. Zwischen gekreuzten Polarisatoren löschen die Quarze in ihrer Mehrzahl uneinheitlich undulös aus. Sie enthalten ausserdem zahlreiche Einschlüsse, wie Flüssigkeitsbläschen und Minerale, häufig Nadeln von Rutil. Unter den grösseren Körnungen sind ausserdem nur vereinzelt vorkommende hochlichtbrechende Minerale, vor allem Zirkon, und wenige opake Erzbröckchen zu nennen. Die dichte Grundmasse besteht vorwiegend aus sehr feinen Körnungen unter 1 μ Grösse (Abb.3), von denen man bei mittlerer bis starker Vergrösserung vor allem massenhaft vorhandene Körnchen hoher Lichtbrechung von runder, teilweise auch stäbchenförmiger Gestalt erkennen kann. Teilweise sind sie von gelblicher Farbe und stellen möglicherweise Goethitkriställchen dar. Eisenhydroxyde sind darüberhinaus auch in feinkolloidaler Verteilung vorhanden, doch fehlen Konkretionen, die in den gleichmässig leitfähigen Braunlehmgefügen sonst meist vorhanden sind (Kubiena, 1953), hier vollkommen.

Bei gekreuzten Polarisatoren bleibt die Tonmasse fast vollkommen dunkel (Abb.2). Die einzelnen Gemengteile weisen selbst nur geringe Doppelbrechung auf, die wegen der Kleinheit der Kriställchen nur bei höchster Vergrösserung sichtbar wird. Anzeichen für eine Teilchenregelung, wie sie durch Fliessbewegungen oder andere Verlagerungsvorgänge entstehen, sind ebenfalls nicht sichtbar.

In dem stark verwitterten Material sind nach mikroskopischer Feststellung keine glimmerartigen Tonminerale vorhanden, die durch ihre hohe Doppelbrechung und leichte Orientierungsfähigkeit sonst immer auffallen. Allenfalls sind Kaolinit und möglicherweise auch amorphe Tone vorhanden.

Eine Neigung des Braunlehms zum Zerfliessen ist trotzdem an mancherlei Erscheinungen festzustellen. So bleiben, z.B., Aggregate aus Kleintierkot offensichtlich nicht lange erhalten, sondern schliessen sich zu einer einheitlichen Masse zusammen. Feine rundliche Poren, die von eingeschlossener Luft (Luftbläschen) herrühren, bleiben aber lange erhalten und durchsetzen auch andere Teile des Braunlehms, denen man die ehemals vorhandenen Kleintieraggregate nicht mehr ansieht. Schrumpfungsrisse sind

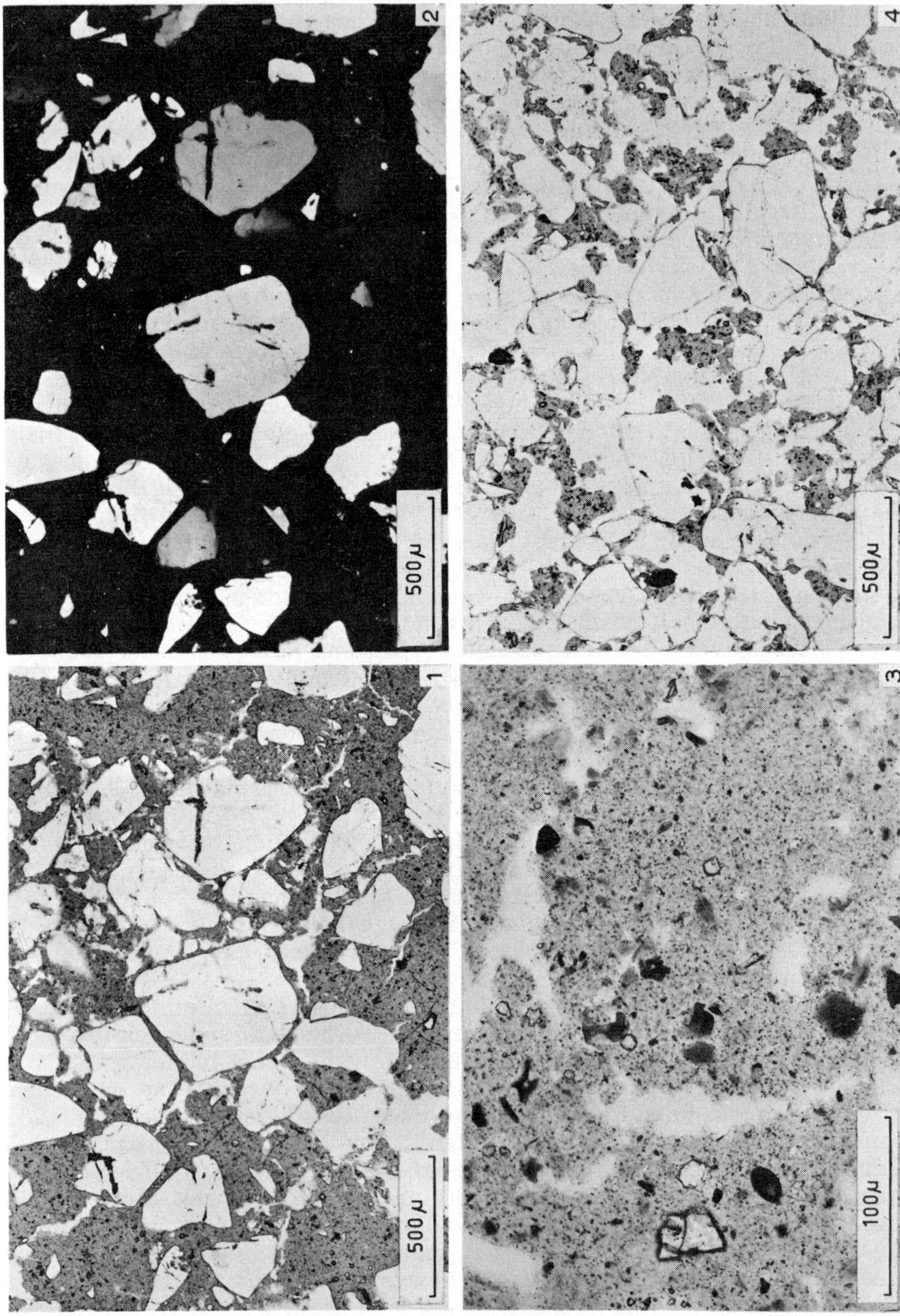
1
500 μ
2
500 μ
3
100 μ
4
500 μ

in den dichteren Partien stets ausgebildet.

Andererseits ist die Ausrundung der Poren wenig vollkommen, und die den Skelettkörnern anhaftenden Beläge sind häufig mehr aggregiert als glatt angelagert; meniskusartige Verbindungen zwischen den Körnern sind selten. So sind also auch Merkmale vorhanden, die eine gewisse Stabilität des Braunlehms anzeigen.

Zwei Beobachtungen sind noch anzuschliessen. Organische Stoffe sind als dunkelbraune stark zerteilte Pflanzenreste stellenweise und nach der Tiefe abnehmend in die Tonmasse eingelagert. Ferner sind hier und in sämtlichen folgenden Profilen in allen Tiefen kleine kristallartige, vieleckige Körnchen von 5–15 μ Grösse eingelagert. Es handelt sich wahrscheinlich um Pflanzenopal, der noch näher untersucht werden muss (vgl. Abb.3).

Gebleichter Braunlehm

Gebleichte Braunlehme kommen in mehreren Beispielen vor. Sie wurden je nach Ausbildung am Profil als schwach gebleichter Braunlehm, sandiger Braunlehm und eluvierter Braunlehm bezeichnet.

Der Übergang vom typischen Braunlehm (siehe p.295) ist in einer Catena an einem kleinen Talhang zu studieren, zu der die beiden folgenden Profile gehören, welche am Oberhang (nächstes Profil) und am Unterhang (letztes Profil dieses Abschnittes) ausgebildet sind.

Allgemein gleichen die gebleichten Braunlehme in ihrer Tonmasse dem typischen Braunlehm weitgehend. Die mineralische Zusammensetzung scheint sich nicht zu ändern. Das dichte Innengefüge der Tonmasse bleibt erhalten.

Am Oberhang ist jedoch der Tongehalt des Oberbodens deutlich geringer. Die tonigen Anteile sind nicht mehr als dichte Grundmasse, sondern als aggregierte Beläge den Skelettkörnern angeheftet. Die Farbe ist etwas mehr bräunlichgelb. Ab etwa 30 cm Tiefe ist wieder hellgelber Braunlehm als dichte Grundmasse vorhanden. Bei gekreuzten Polarisatoren ist die Doppelbrechung ebenfalls sehr gering, so dass bei schwacher Vergrösserung keine Teilchenregelung angelagerter Tone wahrzunehmen ist. Erst bei mittlerer bis starker Vergrösserung wird schwache Regelungsdoppelbrechung deutlich. Die sehr geringe Doppelbrechung der Tonanteile erschwert die Beurteilung der Tonanreicherung im B-Horizont sehr. Es kommt hinzu, dass im Hellfeld auch keinerlei Schichtung feststellbar ist, die sonst ein untrügliches Kennzeichen für nach und nach erfolgte Anlagerung ist.

Abb.1. Braunlehm (B 37 K), (B)-Horizont, 15–19 cm. Objektiv Leitz Pl 4.

Abb.2. Wie Abb.1, jedoch gekreuzte Polarisatoren. Grundmasse ist optisch isotrop.

Abb.3. Braunlehm (B 37 K). Tonmasse mit dunklen punktförmigen Partikeln durchsetzt. Keine Schichtung. Ausserdem Pflanzenteile, Opal und ein Feinsandkorn hoher Lichtbrechung. Objektiv Leitz P 25.

Abb.4. Eluvierter Braunlehm (B 35 K) mit aggregiertem Gefüge im A_e-Horizont.

Trotzdem muss gefolgert werden, dass bei etwa 30 cm Tiefe eine Tonanreicherung vorliegt.

In dem Profil, das hangabwärts folgt und am unteren Hangabschnitt liegt, ist die Verarmung des Oberbodens noch ausgeprägter. Dabei ist der unter einer mehr moderartigen Humusdecke folgende Braunlehm noch stärker aggregiert und im Dünnschliff dunkelgelb bis gelbbraun gefärbt (Abb.4). Man kann hier von einer erdigen Natur des Braunlehms sprechen. Regelungsdoppelbrechung fehlt völlig. Pflanzenteile sind als stark zerteilte Gewebestücke weit verbreitet eingemischt.

Der im Profil erkennbare kaffeebraune humose B-Horizont zeigt im Dünnschliff keine sichtbaren Merkmale für Tonanreicherung. Der Tonanteil ist zwar deutlich grösser, doch bleibt er verhältnismässig gut aggregiert, zeigt keine Schichtung von Anlagerungen und nur äusserst schwache Regelungsdoppelbrechung an wenigen Aggregaträndern.

Eluvierter Braunlehm

Diese Braunlehme, zu denen schon das vorstehend beschriebene Profil gehört, stammen aus einer anderen Catena, die in schwach hängigem Gelände am Rande eines kleinen Tales gefunden wurde. Die Böden sind sämtlich tonärmer und bilden die Übergangsstadien zum Humuspodsol derselben Catena. Skelett und Tonmaterial gleichen in wichtigen Merkmalen den voraus beschriebenen Böden. Folgende Änderungen sind hervorzuheben:

(a) Die Humusformen sind dystropher Moder bis Rohhumus.

(b) Unter der Humusschicht zunehmende Tonverarmung und Bleichung.

(c) Die Tonanteile sind vor allem in den oberen Horizonten, aber auch in tieferen Teilen aggregiert und nicht gerichtet. Regelungsdoppelbrechung fehlt völlig.

(d) Tonanreicherungshorizonte können im Dünnschliff nicht erkannt werden. Möglicherweise erfolgt der Stofftransport noch in tiefere Zonen.

Die Tonverlagerung tritt möglicherweise auch stärker zurück gegenüber der Tonzerstörung, die dann zu einer Verlagerung von gelösten Verwitterungsprodukten führt. Dies muss mit den Ergebnissen der Tonmineraluntersuchungen weiter verfolgt werden. Die hochlichtbrechenden Tonanteile. die man stets und überall im Ton findet und die schwer verwitterbar sind, müssten, da sie in den Bleichhorizonten ebenfalls abnehmen, irgendwo angereichert sein. Dafür sind bis 2 m Tiefe jedoch keine Anzeichen vorhanden. Erst in den folgenden Podsolen ist diese Schwermineralanreicherung der Tonfraktion im Ortstein festzustellen.

(e) Die im Profil mit B_h bezeichneten Horizonte in 40–90 cm Tiefe erweisen sich mit zunehmender Profilentwicklung als Anreicherungshorizonte für Humussole, besonders in dem Profil, das nach den Dünnschliffbefunden schon als Podsol anzusehen ist und als Braunlehm–Podsol bezeichnet wurde.

Die Humussole durchtränken teilweise die vorhandenen Tonmassen und geben ihnen eine schmutzigbraune Färbung. Daneben bilden sich jedoch auch mehr oder weniger deutlich abgesetzte dunkelbraune, optisch isotrope Beläge über den Tonaggregaten und Skelettkörnern oder über beiden. Sie sind den bislang beschriebenen Belägen aus Humusortstein wohl gleichartig und bilden - besonders auf blanken Quarzoberflächen - teilweise die bekannten netzartigen Schrumpfrisse.

Podsole

Unmittelbar an die eluvierten Braunlehme mit dem Profil des Braunlehm-Podsols als Übergang schliesst ein ausgeprägter Humuspodsol mit tiefem Bleichhorizont die Catena zum Talgrund hin ab. Die Merkmale des Humuspodsols sind:

(a) Rohhumus als Humusform.

(b) Stark ausgeprägter, bis 3,5 cm Tiefe reichender Bleichhorizont, der von Humusbändern durchzogen ist. Da die Proben sich nicht ungestört erhalten haben, findet man nur an einzelnen Körnern im Dünnschliff dunkelbraune, etwas schorfige Beläge von angetrockneten Humussolen. Die übrigen Körner sind völlig gebleicht.

(c) Der B_h-Horizont ist deutlich mit schwarzbraunen und rötlichbraunen Fe-haltigen Humussolen angereichert (Abb.5) und enthält ausserdem noch Tonanteile des vorausbeschriebenen Braunlehms, so dass sich beide Komponenten überlagern bzw. vermischen (Abb.6). Durch Glühen der Proben bilden sich rote und blassrosa Farbtöne. In einem Ortsteintopf ist das Gefüge sehr ähnlich.

Nach der Tiefe tritt der Braunlehmcharakter, trotz des relativ geringen Tonanteils, wieder in den Vordergrund. Hier finden sich gut ausgebildete Beläge an den Sandkornoberflächen und meniskusartige Anlagerungen in den Kornwinkeln. Das Tonmaterial ist eisenärmer und im Dünnschliff fahl gelblich. Von oben eingebrachte Anlagerungen sind dunkler und dadurch hier gut erkennbar. Anreicherungen von hochlichtbrechenden Mineralen der Tonfraktion sind hier an mehreren Stellen sichtbar. Neben der eigentlichen Podsolierung findet also noch ein Transport von Tonmaterial statt.

Das hellere Tonmaterial ändert bei näherer Untersuchung mit starken Vergrösserungen im Polarisations- und Phasenkontrastmikroskop seine Eigenschaften gegenüber dem vorausbeschriebenen Braunlehm augenscheinlich auch in der Mineralzusammensetzung. Die hellere Färbung wurde erwähnt. Zudem erkennt man im Phasenkontrast neben den in gleicher Weise vorhandenen Mineralteilchen hoher Lichtbrechung besonders zwei Anteile in Form von Plättchen und Schüppchen bis etwa 2 μ Grösse. Die einen haben etwa gleiche Lichtbrechung wie das Einbettungsmedium, die anderen sind etwas höher lichtbrechend. Im Hellfeld-Durchlicht sind keine Konturen wahrnehmbar. Bei gekreuzten Polarisatoren sehr schwache Doppelbrechung. Nähere Aussagen über die Natur dieser Mineralteilchen können nicht gemacht werden. Für die Deutung sei jedoch auf das nächste Profil verwiesen.

Humuspodsol am Steilhang des Rio Negro

Unter der Rohhumusdecke ist eine moderartige Zone ausgebildet, die sehr zahlreiche kohlige Brandreste enthält. Der folgende Bleichhorizont war am Profil mit Humusbändern durchsetzt, die sich während des Transportes nicht erhalten haben. Bei 23 cm Tiefe beginnt der schwarze B_h-Horizont mit einem taschenartig vertieften Ortsteintopf. Im Dünnschliff findet man sehr gleichmässig feinaggregierte Huminstoffe von kaffeebrauner Farbe um die Sandkörner angelagert (Abb.7), die beim Glühen zu weisser Asche verbrennen.

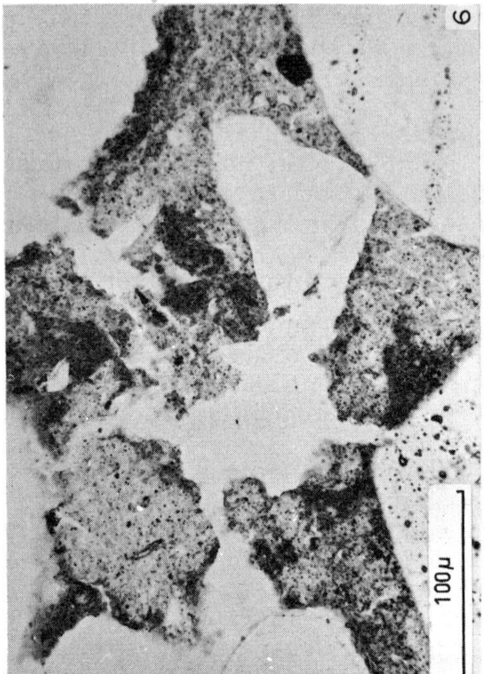
6
100μ

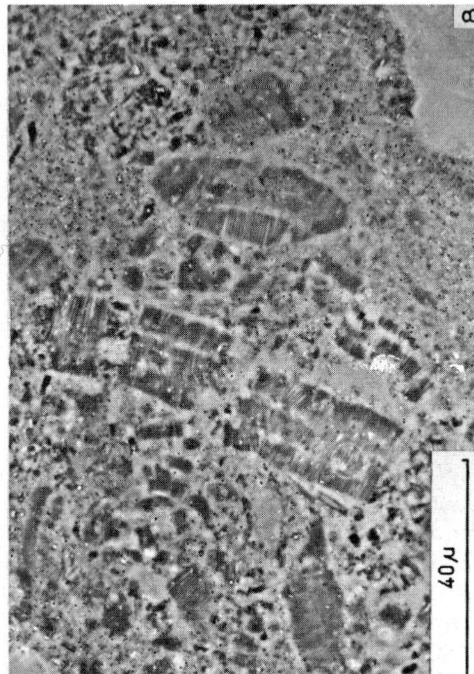
8
40μ

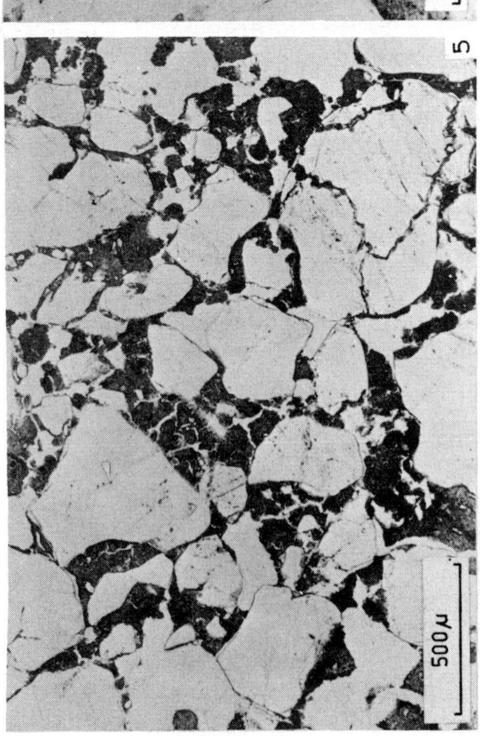
5
500μ

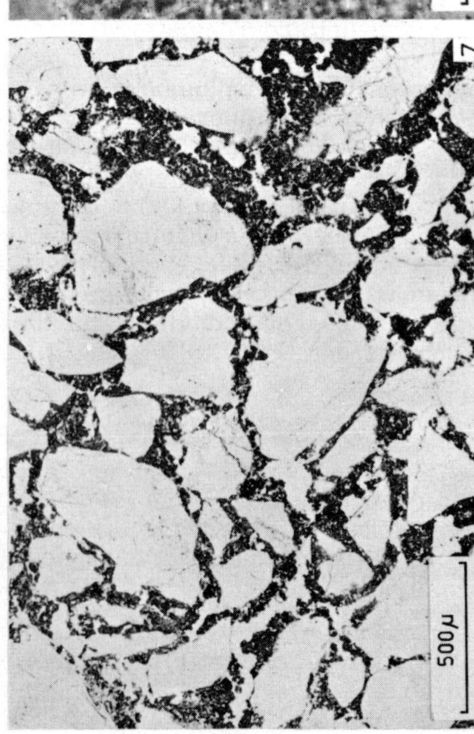
7
500μ

Kolloidale Eisenhydroxyde sind nicht vorhanden. Sie hätten sonst durch Oxydation Rotfärbung verursacht. In der Asche erkennt man jedoch massenhaft hochlichtbrechende Minerale der Tonfraktion, die vordem durch die dunklen Huminstoffe weitgehend verdeckt waren. Über die sonstige Mineralzusammensetzung kann wenig gesagt werden. Irgendwelche Regelungen sind nicht festzustellen. Auch die ungeglühte Probe ist optisch völlig isotrop. Das feinaggregierte Gefüge dieses Ortsteins ist in dieser Form noch nicht bekannt.

Besondere Aufmerksamkeit verdient weiter das Material aus dem B_s-Horizont, das im Dünnschliff ein dem vorausbeschriebenen Braunlehm ähnliches Gefüge erkennen lässt. Die Tonbeläge auf den Sandkörnern sind hier jedoch deutlich geregelt und bilden ein, wenn auch undeutlich geschichtetes Anlagerungsgefüge von blass gelblicher Farbe. Sehr geringe Regelungsdoppelbrechung ist gleichmässig verbreitet festzustellen und deutet auf Anlagerungsvorgänge. Die zwischen Körnern und Tonbelägen verbleibenden Porenräume sind von sehr feinen kristallinen Massen unregelmässig ausgefüllt. Bei Vergrösserungen ab etwa 500 : 1 kann man im Polarisations- und Phasenkontrastmikroskop deutlich zwei Komponenten unterscheiden, die schon im vorigen Profil undeutlich beobachtet worden waren (Abb.8).

Hier sieht man nun in verhältnismässig guter Ausbildung wurmförmige Aggregate, die deutlich aus feinen Plättchen aufgebaut sind. Die Lichtbrechung liegt ganz in der Nähe der Brechzahl des Einbettungsmittels (n_D = 1,558), was bei Beobachtung im Phasenkontrast an der leicht bläulichen Dispersionsfarbe erkennbar ist. Bei gekreuzten Polarisatoren können Gangunterschiede von γ = 25–36 μ festgestellt werden. Die Dicken wurden unter der Annahme, dass die Schnitte nach der Basis annähernd isodiametrisch sind, aus der gemessenen Breite der Aggregate mit 4–6 μ zugrundegelegt. Das ergibt eine Doppelbrechung von Δ = 0,006, dieselbe des Kaolinits.

Die zweite Komponente hat unregelmässig schüppchenförmige oder kornförmige Gestalt und etwas höhere Lichtbrechung und Doppelbrechung als die Kaolinitaggregate. Messungen sind hier schwieriger durchzuführen. Bei ausgewählten Partikeln von etwa 3 μ Grösse ergeben sich Gangunterschiede von rund γ = 70 μ. Das entspricht einer Doppelbrechung von 0,02.

Die Angaben lassen vermuten, dass es sich um kristallisiertes Aluminiumhydroxyd (Hydrargillit) handelt. Beide Minerale sind auf Grund der starken Verwitterung in diesem Bodenmaterial zu erwarten.

Abb.5. Eisenhumus-Ortstein eines Podols (B 64 K). Starke Schwundrisse sind erkennbar.

Abb.6. Gleiches Präparat wie in Abb.5. Ungleichmässige Verteilung von dunklen, humosen und hellen, humusarmen Tonanteilen. Objektiv Leitz P 25.

Abb.7. Humus-Ortstein eines Podsols (B 40 K). Im Gegensatz zu dem Ortstein der Abb.5 feinaggregierte humose Anlagerung.

Abb.8. Neubildung von Kaolinit in wurmförmigen Aggregaten, daneben kleinere dunkle Aggregate vermutlich von Hydrargillit. Dünnschliff, Phasenkontrast. Objektiv Leitz Pr Fluorit Oel 70.

Tonmineralogische Untersuchungen sind zur Kontrolle und Erweiterung dieser Feststellungen notwendig.

Sollten die Beobachtungen Bestätigung erfahren, so handelt es sich um die erste mikroskopische Bestimmung dieser Minerale in Podsolen. Die allgemeinen klimatischen und bodengenetischen Gegebenheiten und die besonderen mikromorphologischen Verhältnisse lassen annehmen, dass Neubildungen vorliegen, die unter dem Einfluss der Podsolierung entstanden sind. Bei der Länge der Aggregate und der Art ihrer Ausbildung ist zumindest vom Kaolinit nicht anzunehmen, dass er im Profil transportiert und wieder angelagert wurde. Beide Minerale, Kaolinit und der vermutete Hydrargillit, sind deutlich in den vordem freien Raum hineingewachsen und einem Material aufgelagert, das schon vorher einen Transport und Anlagerung durchgemacht hat. Es ist allerdings anzunehmen, dass auch das feinere Tonmaterial Kaolinit enthält - wie schon eingangs erwähnt; dieser ist jedoch feindispers verteilt und mikroskopisch nicht sichtbar. Die Schlussfolgerung ergibt sich allein aus der schwachen Regelungsdoppelbrechung.

Erwähnt sei jedoch, dass im Untergrund dieses Profiles weisse Tonbänke auftreten, deren Obergrenze durch eine Zone knolliger, nuss- bis eigrosser Tonaggregate gebildet wird, welche im Profil bei 80 cm angeschnitten ist.

Interessant ist weiterhin die Frage, warum das flachgründige Podsolprofil eine intensivere Kaolinitbildung aufweist als das tiefgründige, vorher beschriebene. Dieser Umstand lässt sich vielleicht aus dem Wasserhaushalt erklären. Klinge nimmt bei den tief ausgebleichten Profilen - wie schon gesagt - einen starken seitlichen Wasserabzug an. Das würde bedeuten, dass von den Verwitterungslösungen nur wenig in den Untergrund eindringt. Der schwach ausgebildete Anreicherungshorizont spricht für diese Annahme, ebenso, dass sich unter dem Ortstein nur wenig Veränderungen des Braunlehms beobachten lassen. Bei dem Humuspodsol am Steilufer des Rio Negro sprechen die mikromorphologischen Befunde für eine geringere Stoffabfuhr; dem entspricht die topographische Situation, die keinerlei Neigung des Geländes erkennen lässt.

SUMMARY

Micromorphological investigations on the development of podsols in the Amazon Basin

On the slopes of the Rio Negro Valley, a braunlehm-tropical topsequence occurs. The authors compared the fabric features of this sequence with those of a "parabraunerde"–podsol genosequence (genetic sequence) of the temperate climate zone.

It appeared that the investigated soils possess some typical features which do not occur in the comparable soils of the moderate climate. However, on the other hand, the sequences show certain morphological parallels.

Bleached "Braunlehm" is the first intermediate member of the tropical toposequence. It is very probable that the mobility of "Braunlehm" material is reduced in these soils, for they show aggregated, more or less earthy fabrics, which are comparable with those in the earthy pre-phases of temperate region podsol. Characteristic clay-accumulations are found only in faintly bleached "Braunlehms". Perhaps they occur also in the other investigated profiles of this type, but at a greater depth, so that they were not found during field studies and therefore were not sampled.

In transitional to podsol soils the accumulation horizons become darker with humus sols, which form plain cutans in extremely developed variants.

The fabrics of podsol ortsteins differ in the investigated soils. Underneath the ortsteins, clay accumulations as well as new formations of minerals are found.

Because of microscopical determinations, the occurrence of kaolinite and hydrargillite is presumed.

LITERATUR

Altemüller, H.J., 1962. Beitrag zur mikromorphologischen Differenzierung von durchschlämmter Parabraunerde, Podsol–Braunerde und Humus–Podsol. Z. Pflanzenernähr., Düng., Bodenk., 98 (143) : 247–258.

Klinge, H., 1964. Podzol Soils in the Amazon Basin. J. Soil Sci., im Druck.

Kubiena, W.L., 1953. Bestimmungsbuch und Systematik der Böden Europas. Enke, Stuttgart, 392 S.

MICROMORPHOLOGY OF SOME TROPICAL MOUNTAIN SOILS

ERWIN FREI

Swiss Experiment Station for Agriculture, Zürich–Oerlikon (Switzerland)

INTRODUCTION

This paper deals only with the soils of the central Andeans (Altiplano) of Ecuador, quite a mountainous country, situated at 2,500–4,000 m above sea level under the equator. In contrast to other tropical mountain regions, the central Andeans receive little precipitation. The rain loaded clouds are losing their water mainly on the outside slopes of the Andean chains. Therefore the local distribution of the rainfall ranges from 150 cm annually to only 20 cm in some low situated basins. Average annual temperature varies from about 5° C on high elevations to over 20° C in the low sites; seasonally there hardly is any variation in temperature. On account of the ecliptic near the equator, the transparent air in the mountains and the scarce cloudiness, a strong global radiation causes a continental character of the climate. As a result there are large temperature gradients between day and night, rapidly drying after the showers and high evapotranspiration rates.

Most of the described area is covered with thick, often alluvial displaced andesitic volcanic ash deposits. The sediments are of Quaternary age, but in historical times only slight ash precipitations have occurred; they did not interrupt the soil formation seriously. The rapid weathering processes in this material guarantee deep and well developed soil profiles in a relatively short time.

The black Andean soils or "Paramo" soils are widespread in the cool, moderately humid high sites, of rather extreme continentality. They can hardly be compared with other soils outside the tropics. In a way some similarity exists to the ando soils of Japan. But for other parts of Asia the ando soils have been described as regosolic and independent of climatic conditions. On the same volcanic ash in the warmer, semihumid or semiarid sites of the central Andeans there are brunizem and chestnut soils. Compared with the black Andean soils they are lighter in colour and have typical steppic character. Few really aridic parts of the central Andreans are covered with a semidesertic vegetation and obviously aridic soils on volcanic ash.

MORPHOLOGICAL AND PEDOGENETICAL FEATURES OF BLACK ANDEAN SOILS

The black, very deep mollic or umbric epipedon of the black Andean soils is very conspicuous. Crumbs of high stability form a spongy and a very porous surface horizon (Fig.1). The lower part of the epipedon is usually more compact, cloddy to coarse prismatic, and even darker than the surface horizon. Destroying the black cutans on the surfaces of the aggregates by rubbing between the fingers makes the striking black appearance fainting. Black cutans and coagulates finer than 200 μ in diameter result probably from the precipitation of dispersed chiefly organic matter migrating downwards by percolation. Also below the epipedon in the B/C-horizon, organic cutans occur on all the primary minerals (Fig.2). It is not known, what

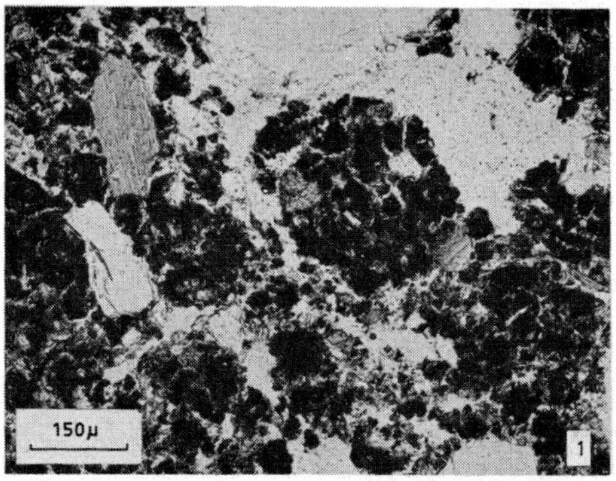

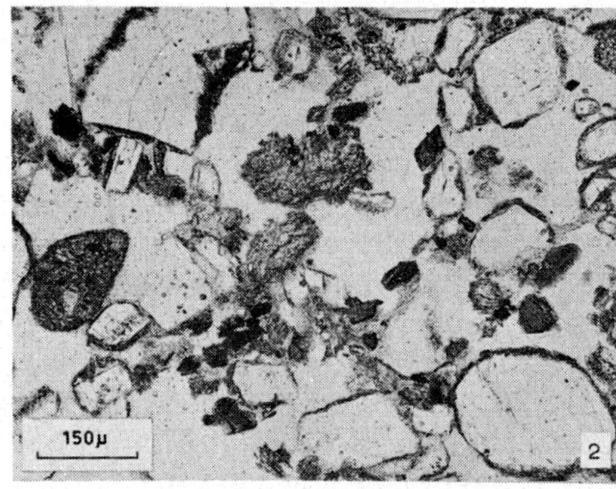

Fig.1. Black Andean soil; thin section. Mollic epipedon with many coagulates of organic matter but without cutans on mineral surfaces.

Fig.2. Black Andean soil; thin section. Transitional subsoil. All mineral surfaces are surrounded by organic cutans; there are still few coagulates.

causes the dispersion of the organic colloids and their precipitation at the mineral surfaces. Perhaps colloidal silica may act as a complexing and a mobilizing agent; in the laboratory very much silica can be dispersed after the destruction of the organic matter. Doubtlessly the steppic grass vegetation with a vigorous matting of root and the high biological activity in the soil are important for the abundant humus production.

These processes take place under slightly acid conditions, a base saturation of the exchange capacity of about 50%, and a sodium saturation of 1–2%. It is supposed that the frequent drying and moistening are also important factors as well as the low soil temperature.

MORPHOLOGICAL AND PEDOGENETICAL FEATURES OF BRUNIZEMS

In contrast to the black Andean soils, the brunizems of the central Andeans have a dark, greyish brown mollic epipedon, with a relatively low content of organic matter. The fabric of the surface horizon seems to be unstable and somewhat powdery, because of impoverishing from colloids. But below this horizon, the epipedon appears darker and more consistent. On cleavage planes and surfaces of the coarse aggregates quite dark cutans occur (Fig.3 Plate F1) while the interior of the aggregates shows a lighter colour. It is interesting to see the organic cutans more or less restricted to the lower part of the epipedon, while the clay moves deeply into the subsoil. Below the mollic epipedon there is an argillic horizon with diffuse boundaries and of a very consistent and primitive fabric (Fig.4; Plate F2). The clay

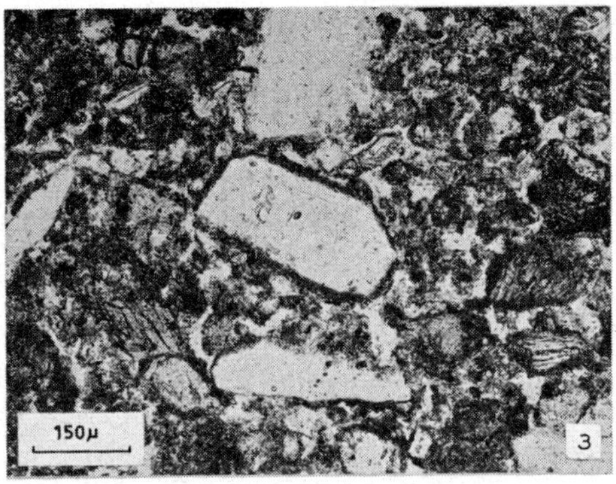

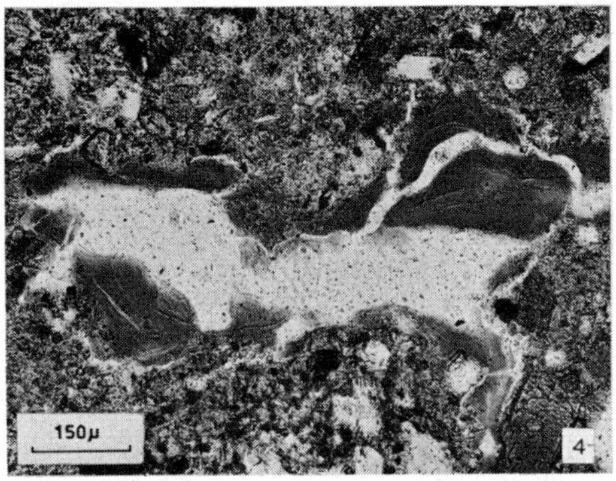

Fig.3. Brunizem of central Andeans; thin section. Mollic epipedon; lower part with mixed organic and clay cutans and a rather compact fabric.

Fig.4. Brunizem of central Andeans; thin section. Argillic subsoil with stratified clay cutans and accumulations in soil pores.

accumulations show the features of migrated cristalline colloids; high birefringence, extinction bands, layering parallel to the ped surface and unmixed with coarse soil material except single grains. Light and dark stratification of some cutans indicate that clay minerals and organic compounds migrate together, and may perhaps be complexed. In very fine textured volcanic ash the clay cutans act as cementing agens and may produce a duripan in the deeper subsoil (Plate F3).

Important factors of the pedogenesis are: the moderate warm soil temperature with frequent moistening and rather severe drying, the absence of calcium carbonate, a base saturation of about 80% of the exchange capacity and a sodium saturation of 2–4%, the increase of the adsorbed sodium with profile depth, the complete absence of soluble salts within the solum and the presence of weatherable aluminium silicates, and perhaps also an excess of silica in duripan soils. The migration of clay and organic matter in these brunizems may be similar to the solodization of some solonetz or solod soils. But such proceedings have also been described from brunizems and chernozems of other countries.

SUMMARY

The black Andean soils of the central Andeans of Ecuador are of high permeability, and characterized by regular percolation. Because of the continental climate, the moderate rainfall and the parent material rich in basic aluminium silicates, the acidification of the solum is very limited. Very much humus and a considerable, but rather modest amount of clay minerals constitute the solum, containing in its upper part still weatherable primary minerals. Free sesquioxydes are quite unimportant, but free silica gels seem to be abundant. Organic colloids (perhaps in combination with silica) are transported downwards and precipitated on the ped surfaces in the lower part of the solum.

The brunizems and chestnut soils of the central Andeans are permeable but by climatic reasons somewhat imperfectly percolated. Clay minerals are very important constituents of the solum. There is not very much humus in these soils, but it is deeply and thoroughly mixed into the epipedon, determining largely its fabric. Weatherable aluminium silicates are less abundant in the upper part of the solum overlying the noncalcic parent material. The percolation stream carries clay minerals and organic colloids from the surface horizons downwards. These colloids are forming skins in the B- and C-horizons of the brunizems.

REFERENCES

Bramao, L.D., 1962. Considerations on the potentialities of the soils of the world. Accad. Ital. Sci. Forest., 11 : 333–349.

Dudal, R. and Soepraptohardjo, M., 1961. Some considerations on the genetic relationship between latosols and andosols in Java (Indonesia). Trans. Intern. Congr. Soil Sci., 7th, Madison, 1961, 4 : 229–237.

Frei, E., 1958. Eine Studie über den Zusammenhang zwischen Bodentyp, Klima und Vegetation in Ecuador. Plant Soil, 9 : 215–236.

Soil Survey Staff U.S.D.A., 1960. Soil Classification, a Comprehensive System, 7th Approximation. U.S. Dept. Agr., Washington, D.C., 265 pp.

White, E.M. and Papendick, R.I., 1961. Lithosolic solodized Solonetz soils in southwestern South Dakota. Soil Sci. Soc. Am. Proc. 25 (6) : 504–506.

COLOURED ILLUSTRATIONS

Plate F1. Brunizem of central Andeans; thin section; x 225. Mixed organic and clay skins. Lower part of the mollic epipedon, transition to the argillic subsoil.

Plate F2. Brunizem of central Andeans; thin section; x 225. Clay skins on ped surfaces, of the upper part of the argillic subsoil.

Plate F3. Brunizem of central Andeans; thin section; x 225. X-nicols + mica plate. "Cangahua" duripan on volcanic ash, cemented by clay skins and probably also by silica.

COMPARATIVE MICROMORPHOLOGICAL CHARACTERISTICS OF SOME SOLONETZ SOILS OF THE STEPPE AND SEMI-DESERT ZONES[1]

E.A. YARILOVA

Dokuchaev Soil Institute, Moscow (U.S.S.R.)

Solonetzs belong to soil types whose micromorphological aspects have not been sufficiently investigated. Such data would provide an opportunity for a more detailed characterisation of these soils in different natural zones (Kubiena, 1962), especially for the purpose of classifying them (Ivanova, 1963a, b). This paper deals with the results of investigations on some solonetz soils of two zones.

SOLONETZ SOILS OF SEMI-DESERT ZONE

Both meadow-steppe saline solonetz of the lowland near the Caspian Sea and a medium-columnar steppe solonetz from Ergeni highland in combination with zonal light chestnut soil, were studied. These soils are formed on a loess-like loam of nearly the same texture (Zaitsev and Zvorykina, 1955; Rode and Polski, 1960) and mineral composition (as shown by our data). The lime content of this loam is about 10%. The distribution of carbonates throughout the profile of both solonetz soils is similar: horizons with a lower lime content and those with carbonate accumulation are well distinguished, the boundary between them lying at a depth of about 50 cm.

Some differences in the natural conditions are reflected in the character of the upper part of the profiles. Light soluble salts and carbonates are more intensively leached in the highland solonetz and the illuviation of different soil components becomes more apparent here.

Meadow-steppe saline solonetz (Dzhanybek, West Kazakh region)

Vegetative cover: *Kochia prostrata*, *Artemisia pauciflora*, less *Poa bulbosa*, *Lepidium perfoliatum*, *Salsola*. The morphology and some chemical and physical properties of this soil have been studied by Rode and Polski (1960). Following horizons have been distinguished:

[1] According to the 7th approximation of the U.S.D.A. Soil Classification (1960), these soils are natrargids (Ed.).

A-horizon (mean thickness 10 cm): Light-grey, fine lamellar, easily crushed into powder; dry.

B-horizon (mean thickness 23 cm): Greyish chocolate brown; crushes into angular fragments 1–2 cm in diameter; salt efflorescences in the lower part of the horizon; dry.

Two saline horizons: The first horizon is pink-brownish (mean depth 50 cm) and of sand-like structure. The second is pale-brownish (mean depth 67 cm) and of nutty-prismatic structure which becomes coarser downwards. Sodium and calcium sulphate accumulations are present.

Gypsum saline horizon (150–300 cm): Pale-brownish of cloddy-prismatic structure with gypsum accumulations.

Saline transitory horizon (300–600 cm): Relatively wet, compact, with gradually diminishing salt content.

Parent rock: Non-saline loess-like loam.

Effervescence with acid from the surface horizon or from the lower part of the B-horizon. During the spring moisture penetrates no lower than the upper part of the first saline horizon. Sodium sulphate sharply predominates among salts. The salt content in the A and B-horizons does not exceed 4–6 m.e./100 g. Exchangeable sodium in the A-horizon makes up 5–12% (0.8–2.1 m.e./100 g), in the B-horizon 30–58% (8.0–13.3 m.e./100 g) of C.E.C. The presence of readily soluble salts suppresses the unfavourable properties of this horizon. In the upper part of the A-horizon the humus content is generally less than 2% and diminishes downwards. The brown colour of the humus brings its relatively simple molecular structure in evidence (Ponomareva, 1956).

Micromorphology

The laminated fabric of the A-horizon represents its macroscopic features in miniature (Fig.1). One millimeter thick laminae are formed by indistinctly shaped micro-aggregates (peds) of the first and second order. Some of them are covered by thin oriented clay cutans. Clay cutans on mineral grains are observed throughout the profile. The greatest part of the clay is flocculated, but there are many very small spots of relatively homogeneous oriented clay which exhibits wavy extinction. Numerous iron hydroxide accumulations are present (50–100 μ), partly with dark-coloured humus. They also form cutans on the surface of part of the mineral grains and plant remains. Round-shaped concretions of iron enriched clay about 200 μ in diameter are sparse.

The following forms of organic matter have been distinguished:

(1) brown-coloured fragments of plant tissue show various degrees of decomposition up to single fibres dissipated in the matrix. Many such fibres show birefringence giving the impression that they are impregnated with oriented clay (a kind of pseudomorph).

(2) Charred fragments of plant tissue disintegrating into black particles with sharp, chiefly rectilinear, contours.

(3) Amorphous humus occurs in the forms of: (a) dark-brown, round-shaped flakes up to 4 μ in diameter which accumulate in places where plant residues decompose; they apparently represent the most condensed calcium-bearing humic acids; (b) brown-coloured dispersed mobile humus intimately bound with the clay. This form predominates.

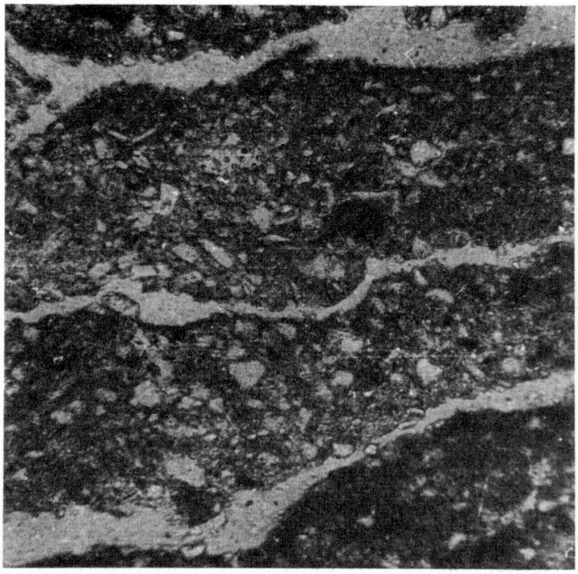

Fig.1. Laminar fabric of A-horizon of meadow-steppe saline solonetz of the semi-desert zone.

Among the plant remains, brown and partly black-coloured droppings of mites 20–50 μ in diameter can be seen as well as fungal hyphae and spores. Also phytolites[1] and single diatom skeletons are present.

This horizon as well as the entire profile is intersperced with microcrystalline calcite efflorescences which are not quite uniformly distributed. Palmate crystals of calcium carbonate occur on some pore walls. Larger but not numerous crystals of calcite are often corroded and partially recrystallized into microcrystals. Single rhomboidal dolomite grains are observed, as well as fragments of mollusc and foraminifer shells (Kaspian fauna).

The B-horizon has a compact fabric with cleavage blocks 1–2 mm in diameter. The clay is generally flocculated, but small spots and flows of oriented brownish-yellow clay can be observed (Fig.2). They include small grains of different minerals, particularly uniformly arranged mica leaflets, dark-brown humus particles and frequently microcrystalline calcite.

Iron oxide accumulations and concretions with concentric oriented clay are not numerous. Some pores in the lower part of the horizon are filled with irregularly formed gypsum crystals up to 40 μ in diameter.

The sandy-like saline horizon is composed of mineral grains up to 100 μ in diameter with cutans of a lime-carrying clay substance, and aggregates, similar in size, of the same substance (Fig.3).

In the second saline horizon this kind of aggregate is much less pronounced and it disappears downwards. Segregations of clay substance are not observed. Chambers tightly filled with gypsum crystals of lens or

[1] See foot-note on page 201 (Ed.).

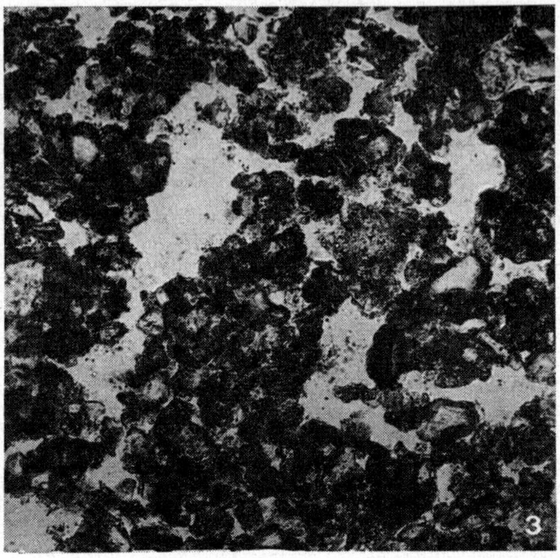

Fig.2. Oriented clay in B-horizon of meadow-steppe saline solonetz of the semi-desert zone.

Fig.3. Sandy-like fabric of the first saline horizon of meadow-steppe saline solonetz of the semi-desert zone.

irregularly shape can be seen. In gypsum-saline and transitional horizons small flows of oriented clay, as well as concretions which include it, again appear.

The C-horizon has a compact fabric heavily interspersed with fine lime. Many carbonate grains of larger sizes are present while gypsum is less abundant. Small patches of oriented gel-like clay can be detected up to the depth of 640 cm (the bottom of the soil profile). They often have lime in them, and may be considered as relicts of a previous meadow stage of the solonetz formation. (Antipov–Karataev, 1953; Rode and Polski, 1960).

Medium columnar solonetz (Ergeni highland)

As the micromorphology of this soil has been reported recently (Yarilova, 1963), only those of its main features which differ from those of saline solonetz will be considered here. *Kochia prostrata, Artemisia incana, Art. pauciflora* and *Camphorosma monospeliatum* predominate in the plant cover. Salt accumulations can be found at the same depth as in the saline solonetz. The upper one, which is poorly expressed at a depth of 50 cm chiefly consists of sodium chloride and sulphate; the lower essentially consists of sodium sulphate. Gypsum crystals have been found here too. In A and B-horizons both carbonates and readily soluble salts are absent.

The content of exchangeable sodium in A-horizon reaches up to 18% (1.8 m.e. /100 g), and to 23% (5.5 m.e.) of C.E.C. in B-horizon. The humus content in A-horizon is 1.9, in B_1-horizon 1.6, and in B_2-horizon 0.9%. The $C_{humic\ acid} : C_{fulvic\ acid}$ ratios are 1.4, 1.3 and 0.4 respectively (Kononova, 1961).

Micromorphology

The lamellar A-horizon is formed by angular to round and elongated peds with rugged surfaces; some of them are considerably humified, apparently representing droppings of soil fauna larger than mites. Humus is presented by similar forms as in the saline solonetz, but it contains relatively more dark-coloured particles. The mite's droppings among plant remains are somewhat larger (up to 80 μ), and are blackish in the upper horizon. The clay is flocculated together with the humus substances. Only few small patches of oriented clay, including dispersed black humus, and cutans around some mineral grains, are distinguished. In the lower part of the horizon small strips of eluviated material can be seen which mainly consists of quartz and feldspars (whitish powder). The concretions are small-sized (100 μ in diameter) and scanty, and are composed of iron oxide and dark coloured humus particles. Phytolites and diatom skeletons are more numerous then in the saline solonetz.

In the B-horizon with a cleavage-bloc fabric the fine flow orientation of the clay suggests its peptized state. Cutans of oriented clay along cracks and pore walls are much more abundant. Mineral grains have thin clay cutans. Some larger cracks are filled with somewhat coarser material.

Carbonates appear in the lower part of B-horizon, some of the larger grains being corroded and recrystallized. This may be attributed to the dissolving action of fulvic acids which penetrate into this horizon in great amounts.

The underlying calcareous horizon shows a slight salt accumulation. The clay is generally flocculated. Some streaks of relatively uniform clay are interspersed with micro-crystalline calcite. Single micropores are filled with gel-like clay without lime apparently of more recent origin. Aggregate formation becomes evident with the increased salinisation in the C-horizon. The second layer of salt accumulation has a sandy-like structure and includes gypsum. In the lower part of the profile chambers with gypsum and calcite crystals appear.

In order to compare the biogenic products in the solonetz soils more exactly we examined the immersion preparations of 10–100 μ and particles 1–10 μ of s.g. < 2.20. (The larger ones are chiefly root fragments). The obtained data (see Table I) show the following differences between the two soils: The steppe solonetz contains much more phytolites, owing to the greater participation of graminae in the vegetative cover, and opaline skeletons of diatoms. In the saline solonetz the diatoms are found only as single exemples; at a certain depth amorphous porous opal particles are encountered. The medium columnar solonetz contains more products of small animal activity (mostly of mites) which concentrate somewhat deeper than the dry surface layer.

Among the components of light fraction, the afore mentioned anisotropic particles having the appearance of fibrous plant root tissue are found.

TABLE I

Composition of the fractions of s.g. < 2.20

Soil	Depth (cm)	Plant residues (%)		Charred particles (%)		Clay aggregates (%)		Phytolites (%)	
		a[1]	b[1]	a	b	a	b	a	b
Meadow-steppe saline solonetz	0–6	15	22	8	13	71	40	5	15
	15–25	33		7	7	45	33	1	
	50–60	19		6	3	37	25		
Medium columnar steppe solonetz	2–10	17		18	20	15	20	34	32
	10–16	9		12		24	100	10	traces

[1]a = fraction 10–100 μ; [1]b = fraction 1– 10 μ.

Examination after moderate ignition confirms the supposition that we deal with clay pseudomorph for plant remains, the clay being oriented more perfectly along the fibres (in the Table they are denoted as clay aggregates). Besides impregnation of decomposing plant tissue by soil solutions containing colloidal clay particles, clay formation from the interaction of molecular soluble components (among them the products of plant residue decomposition is possible. The clay is tightly bound with the brownish humus, which is likely to promote its peptization. This can be seen from the correlation between the amount of clay aggregates and that of brownish humus: in the saline solonetz they are most numerous in the upper horizon, while in the steppe solonetz at the depth of 10–16 cm. In this connection it should be noted that Ponomareva (1956) has found the brown-coloured humus to be chiefly bound with sesquioxides and the clay fraction of desert zone soils.

From thin section examination the similarity of the fabrics of both solonetz soils is obvious. The following are the most characteristic features:

(1) A part of the clay is peptized and mobile. Pockets and cutans of oriented clay on pores walls and cracks are seen throughout the soil profile, with the exception of the sandy-like salt accumulation horizon. This phenomenon is less pronounced in the meadow-steppe saline solonetz and stronger in the steppe solonetz in which eluvial microspots have been formed in the lower part of the A-horizon poor in clay.

(2) There are only small amounts of concretions and of small sizes, indicating low iron mobility.

(3) The surface of the mineral grains is covered with clay cutans partly protecting them from destruction.

(4) Several humus forms can be distinguished; fine brownish humus plays an active role in clay mobilization and its migration down the profile.

Skeletons of diatoms (%)		Sponge spicules (%)		Amorphous opal (%)		Mite droppings (%)		Minerals (%)		Soil
a	b	a	b	a	b	a	b	a	b	
1	3					traces			7	Meadow-steppe
tra-	tra-	1	5	13		traces			55	saline solonetz
ces	ces		5	38					67	
7	8					traces		9	20	Medium
9		traces				27		9		columnar steppe solonetz

SOLONETZ SOILS OF THE STEPPE ZONE

In the Chrenovskaja steppe (Voronesh region) solodized and residual solonetzic solod, formed on calcareous loess-like clay, have been investigated.

Solodized crust-columnar meadow solonetz

This soil found under grass near a sink occupied by aspen groves ("kolok").

0– 1 cm: A_0, moss cover with a small amount of mineral material.
1–3(6) cm: A_2, grey, powdered to dusty, loose.
3(6)– 15 cm: B_1, brownish dark grey, columnar, very compact. Whitish powder on the tops and vertical faces of the columns.
15– 31 cm: B_2, dark grey, with prismatic structure, compact. Effervesces with acid from a depth of 20 cm. Whitish powder on crack walls.
31– 58 cm: B_3, brownish dark grey with cloddy-prismatic structure, compact. Fair effervescence.
58– 78 cm: B_4/C_1, heterogeneously coloured, predominantly brownish, prismatic, compact, with flecks of carbonates. Feeble effervescence.
73– 144 cm: C_1, yellowish-brown with humus accumulations along rootchannels. No carbonates.
144– 200 cm: C_2, yellowish-brown, containing carbonates. Numerous hard iron-manganese concretions, gley spots.

The ground water level was found at the depth of 3 m. The solonetz belongs to the sulphate soda type. The exchangeable sodium in A, B_1 and B_2-horizons makes up 1.5, 33 and 44% C.E.C.

Micromorphology

The fabric of A_2-horizon has a poorly expressed laminarity. Microaggregates are loose, composite, and have ragged surfaces; some of them are strongly humified. Most of the clay is flocculated together with blackish humus particles. There is only a small amount of peptized clay as well as opaque parts rich in humus concretions. Root-channels, with brownish plant tissue residue showing partly birefringent cell walls, are numerous. Among them singular dark-coloured mite droppings 600 μ in length can be observed. Mineral grains are bare and form small accumulations (eluvial microspots from which clay was leached out).

B-horizon has a compact, indistinct cleavage block fabric. The clay shows fine flow structure. Many pores and cracks are filled with oriented clay including dark coloured peptized humus. Oriented clay pseudomorphs for plant fibres like those described above can be observed. A part of the sand grains and pore walls have thin clay cutans. Single wider cracks are filled with clay. Shapeless accumulations of blackish humus particles and iron oxides are scanty, as are concretions also containing brightly birefringent clay in the center.

In the B_2-horizon, as well as in lower horizons, compact calcite precipitations in some pores are covered with thin clay cutans; thus they are of later origin, and were probably called forth by recent soil forming processes. In B_3 and B_4/C-horizons concretions are smaller and less compact.

The C-horizon is strongly interspersed with calcite microcrystals. The soil matrix is incompletely segregated into roundish micro-aggregates about 200 μ in diameter rich in clay. Compact concretions reach several millimeters. Accumulations of iron oxides often take the form of broad patterns.

Residual solonetzic solod

Vegetation: a young aspen grove. The soil contains very small quantities of exchangeable sodium. It shows many similar features, but there are also some essential differences. It contains concretions throughout the profile, and they are particularly numerous in A_2 and B_1-horizon, in which iron and humus have been accumulated as well as a great deal of clay. Their sizes range from several tenths of a millimeter up to one millimeter or more; in B_2-horizon they become less distinct. In the B-horizons the amount of oriented clay cutans in pores and cracks increases as compared with the solonetz. The fabric is rather heterogeneous because of local concentrations of different components.

Meadow-soda sulphate solonetz soils, Kamennaja steppe (Voronezh region)

The micromorphology of this soil has been reported in one of our papers (Yarilova, 1953). Field descriptions are given by Antipov–Karataev and Kader (1963). Vegetation: *Sueda maritima*; ground water level: 1 m. Here we shall stress only some pecularities of the fabric of the calcareous solonetz-solontchak. The exchangeable sodium content makes up 37% of C.E.C. in A-horizon and 68% in B-horizon. The greatest amounts of salt (up to 2.7%) are accumulated in the surface layer 1 cm thick. According to

Kononova and Belchikova (1963), the solonetz has the same humus content as the surrounding chernozems. No differences in the amount of humic and fulvic acids, or in nonhydrolizable residue are observed. C_{humic}/C_{fulvic} ratio is 2.2 up to a depth of 40 cm; at a depth of 40–55 cm it is 1.3. However a greater mobility of humic acid is characteristic of the solonetz. These acids accumulate in B-horizon.

Micromorphology

The fabric of A-horizon is rather heterogeneous (Fig.4). Loose accumulations of blackish humus particles, flocculated with clay, intermix with compact

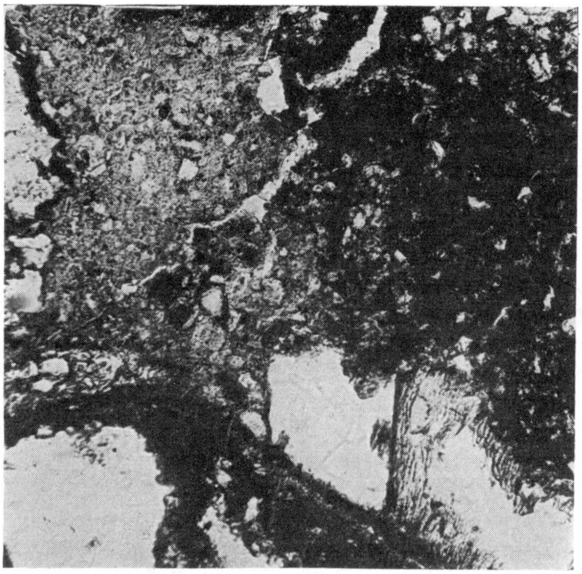

Fig.4. Heterogeneous fabric of B-horizon of the meadow-soda sulphate solonetz. Dark-coloured humus accumulations, grey-coloured clay spots and plant residues in pores can be seen.

brown-coloured spots of clay oriented in the form of fine streams. Single pores have thin clay cutans. Plant remains have the same appearance as in the solonetz of the Chrenovskaja steppe. Strongly birefringent iron oxide accumulations are not numerous. There are many compact roundish aggregates of fine calcite crystals, impregnated with brownish humus. Some of the mineral grains, among them quartz, are cracked into small fragments.

The fabric of B-horizon is still more heterogeneous. Fine lime accumulations are much more abundant. In the lower part of the horizon diffusion rings consisting of the same substances appear besides concretions which grow larger (up to 1 mm).

From the micromorphological descriptions the following characteristic features of the solonetz soils of the chernozem zone may be pointed out:

(1) Heterogeneity of the fabric due to the irregular distribution and isolation of humus, clay and iron especially in the meadow soda-sulphate solonetz soils and solods.

(2) Predominance of partly peptized dark-coloured humus of mull type which is found in concretions, clay pore fillings and crack linings.

(3) Considerably greater mobility of clay and iron than in the solonetz soils of the semi-desert zone. The formation of clay-humus-iron concretions increases with the development of solodization.

(4) Clay cutans around mineral grains appear beginning from the B-horizon; here they include dark humus particles.

(5) In the meadow soda solonetz soils destruction of mineral grains, including quartz, is observed.

SUMMARY

The following micromorphological features of chestnut and chernozem solonetz soils are: (1) Inherent in zonal soils; (2) connected with local conditions of solonetz development (high ground water table, the mode of salinization and others); (3) common to solonetz soils of different zones.

Among the most notable features are: some similarity in fabric of solonetz and light chestnut soils (laminarity of the upper horizon, sandy-like structure of the salt accumulation horizon), a small amount of iron concretions (a low mobility of iron), and a great deal of brown-coloured humus and clay cutans around mineral grains throughout the profile, etc.

Meadow soda solonetz soils located among chernozems have fewer common features with these soils. Most evident is the predominance of dark mull humus, but is in a more disperse and mobile state in the solonetz.

The most characteristic feature of all solonetz soils is the mobility of a part of their clay fraction intimately bound with the fine humus and iron. Oriented clay patches, streaks, cutans along the cracks, on pore walls and around mineral grains and peds, as well as concretions, are seen throughout the profile beginning from the surface. They are most sharply expressed in B-horizon. The clay orientation is controlled by several factors (the amount of exchangeable bases, especially sodium, salt content, percentage of organic matter, its forms, etc.).

REFERENCES

Antipov–Karataev, I.N., 1953. The problems of genesis and geographical distribution of the solonetz soils in U.S.S.R. In: Reclamation of the Solonetz Soils of the U.S.S.R. Akad. Nauk S.S.S.R., Moscow, pp.11–141. (Russ.).

Antipov–Karataev, I.N. and Kader, G.M., 1963. On the development of solonetz processes in the soils of Kamennaja steppe. In: Problems of Grassland Farming. II.Akad. Nauk S.S.S.R., Moscow, pp.514–658. (Russ).

Ivanova, E.N., 1963a. A trial of classification of the steppe solonetz soils. Pochvovedenie, 3 : 14–26. (Russ.).

Ivanova, E.N., 1963b. A trial of the classification of the desert and semi-desert solonetz soils and solods. Pochvovedenie, 4 : 20–28. (Russ.).

Kononova, M.M., 1961. Soil Organic Matter. Pergamon, Oxford, 450 pp.

Kononova, M.M. and Belchikova, N.P., 1963. Organic matter transformation in the ordinary chernozem. In: Problems of Grassland Farming. II. Akad. Nauk S.S.S.R., Moscow, pp.303–360.
Kubiena, W.L., 1962. Wesen, Ziele und Anwendungsgebiete der mikromorphologischen Bodenforschung. Z. Pflanzenernähr., Düng., Bodenk., 97 : 1–13.
Ponomareva, V.V., 1956. Humus of the takyrs. In: The Takyrs of Western Turkmenia. Akad. Nauk S.S.S.R., Moscow, pp.411–438. (Russ.).
Rode, A.A. and Polski, M.N., 1960. Soils of the Dzhanybek station. Tr. Pochv. Inst., Akad. Nauk S.S.S.R., 56 : 3–214. (Russ.).
Yarilova, E.A., 1953. Mineralogy of the solonetz soils of the chernozem zone. Tr. Pochv. Inst., Akad. Nauk S.S.S.R., 56 : 131–142. (Russ.).
Yarilova, E.A., 1963. Micromorphology of the Ergeni Highland Soils. Pochvovedenie, 2 : 33–39. (Russ.).
Zaitsev, N.M. and Zvorykina, K.V., 1955. Soils and vegetation of the first experimental plot of the Arshan-Zelmen experimental station. Tr. Inst. Lesa, Akad. Nauk S.S.S.R., 28 : 108–150. (Russ.).

THE MICROMORPHOLOGICAL METHOD IN PALEOPEDOLOGY AND PALEOGEOGRAPHY[1]

T.D. MOROZOVA

Geographical Institute of the U.S.S.R. Academy of Sciences, Moscow (U.S.S.R.)

INTRODUCTION

Today the micromorphological method is assuming ever greater importance in the profound study of soils and their genetic types. As applied to the study of fossil soils, this method is also very promising. By comparing the microstructure of buried and modern soils, we can make well-grounded conclusions as to the nature of ancient soil-forming processes and the environment under which they have developed.

Two principal aims are pursued by the micromorphological study of old soils in thin sections:

(1) The revelation of survived indications of ancient soilforming processes

(2) The determination of secondary changes in fossil soils which occurred after burial and which markedly conceal their original properties.

In our research we used the micromorphological method to study buried soils in loess deposits of the Russian plain and central Asia. The buried soils and the encompassing loess stratum were investigated concurrently in thin sections, for a number of secondary features of old soils, such as carbonisation, depend on the properties of the loessial roof. The study of the Quaternary stratum, layer by layer, enables us to easily identify the levels of fossil soils which differ from both loess and other loose deposits in thin sections. This is possible because the buried soils effectively preserve a number of characteristics testifying to their original microstructure. Such characteristics are their aggregates and pores, the structure and distribution of organic matter and its relation to the mineral components and the microfabric of clays and ferruginous and salt new formations.

FEATURES OF ANCIENT SOIL-FORMING PROCESSES

Aggregates and pores

The aggregates and pores are comparatively well-preserved in old soils after their burial in loess, with the exception of cases where the soils were subjected to violent secondary mechanical dislocations. They have different structures in conformity to the type and genetic horizon of the old soil.

[1]This paper has been prepared under the guidance of Academician I.P. Gerasimov.

Thus, podsolic horizons of old podsolic soils clearly show a lamellar structure, typical of the A_2-horizons of modern soils of this type. Humus horizons of fossil chernozem expose a distinctive complex aggregation, with the complex aggregates separated by meandering intra-aggregate pores. We have also found numerous spherical biogenic pores here. Humus horizons of fossil grey forest soils are remarkable for their large-sized (1–2 mm) aggregates with rounded edges. Illuvial horizons of old podsolic and grey forest soils, enriched by highly dispersed clay matter, are usually of a compact structure, sometimes composed of large angular aggregates separated by porous cracks. Another aggregate structure has been discovered in the fossil brown and grey-brown soils of central Asia. Large aggregates have been found here in argillised horizons, sometimes with rounded edges, very often angular, and separated by intra-aggregate pores and thin cracks. The grains are tightly packed inside the aggregates.

Organic matter

After burial the organic matter in the soils changes both quantitatively and qualitatively. However it has been shown in the works of Gerasimov (1961) and Glazovskaya (1956) that the relationship between the main humus components nevertheless reflects the leading soil-forming process. In this way we can determine, for example, whether the humus is of forest or herbaceous origin.

The microstructure of humus in buried soils has undergone considerable transformations, however. At the same time, the connection between the organic matter and the mineral mass, and the nature of the humus distribution along the soil profile, have been preserved. Along with other micromorphological features, these data can be utilised to determine the genetic types of fossil soils. Thus in fossil chernozems humus has a finely dispersed structure and dark colour, and uniformly permeates and conceals the clay. The clay humus mass is fixed and shows no indications of shift. Brown and grey-brown soils buried in the central Asian loess contain brownish, highly dispersed humus, evenly distributed and connected with the ferruginated clay mass.

In buried soils, as a rule, we find negligible amounts of vegetative remains. They are usually strongly humified and charred, and seldom preserve their original structure.

Clay

The clay matter of the soil is highly sensitive to soil-forming processes, and is therefore widely utilized today for the study of the genesis and classification of modern soils (Brewer, 1956; Feofarova, 1960; Kubiena, 1938; Parfenova and Yarilova, 1960; Stephen, 1960; Velichko, 1961; Yarilova and Parfenova, 1957). The microstructure of the clay is closely connected with the hydrothermal conditions of pedogenesis, the mechanical composition, and some other factors such as the presence of water-soluble salts.

The microstructure of optically oriented clays is one of the most persistent pedogenetic indications of buried soils. Many old soils in loess deposits of the Russian plain reveal a sinter form of optically oriented clays. The presence of clay matter having such a microstructure, distributed over pores and cracks in the illuvial horizons, is today regarded as an objective sign of processes connected with the shift of the organo-mineral mass along the soil profile. The fluid microstructure of optically oriented clays has been found in many modern soils which are distinguished by illuviation processes to some degree.

It can be assumed that a different sinter micromorphology must correspond to different forms of illuviation (in solonetz, solodised and podsolic soils, etc.). At the present-day level of knowledge these differences are still rather vague. Nonetheless, the first steps have already been taken to clarify this problem.

For example, Stephen (1960) points out that the fluid form of optically oriented clays in solonetz soils has a comparatively weak optical otientation of clay matter (as exemplified by the solodised solonetz soil of southern Australia). Stephen accounts for this fact by the presence of watersoluble salts which promote rapid coagulation and, hence, hamper the optical orientation of clay matter. Parfenova and Yarilova (1960) have noticed certain zonal changes in the morphology of fluids of optically oriented clays. For instance, clay sinters in sod podsolic soils are light and transparent, whereas in grey forest soils they are enriched by humus, which gives them a dark-brown hue and makes them slightly transparent. Some of these peculiarities in sinter morphology (colour, transparency, humus content) were taken into account in the micromorphological study of buried soils, and were utilised for the determination of genetic types.

Sandy and sandy-loam fossil soils of the podsolic type reveal a filmy form of optically oriented clays, in the shape of yellowish-brown transparent coatings around mineral grains and "collars" between the sand grains. This clearly illustrates the dependence of the microstructure of optically oriented clays on mechanical composition. However, in this case too a small amount of clay sinters can be found confined to large pores.

Fossil soils may also contain yet another, rather widespread, form of optically oriented clays: scaly fibrous clays, inherent in soils subjected to processes of intrasoil weathering and argillisation without the displacement of soil formation products (brown, grey-brown and dark brown) (Minashina, 1958, 1960). The scaly fibrous microstructure of clays has been detected in thin sections from fossil soils buried in central Asian loess.

Soil life

Like modern soils, old soils contained abundant organic life. In studying thin sections from fossil soils one can observe numerous traces of soil fauna-dislocations of the soil mass, tracks and fossilised excrements of worms and insects. Mineral inclusions of biogenic origin, such as skeletons of diatoms, spiculae of Porifera and phytolithes, are quite frequent.

Carbonates

The form of carbonates, and their quantitative distribution within the limits of the genetic profile of fossil soils, are extremely important for the solution of problems related to the genesis of soils and loess. Analysis of old soils in thin sections enables more detailed study of the specific features of the composition and the origin of carbonates in soils. We studied the form and distribution of carbonates with a view to determining their soil origin or diagenesis. Inasmuch as the features of old soils acquired afterwards, particularly the distribution and form of carbonates, largely depend on the properties and composition of the loess roof, it is necessary to dwell briefly on the forms of carbonates widespread in loess. As to their origin, these carbonates can be arbitrarily divided into primary and secondary ones. Primary carbonates, brought in with the main mass of sediments, are represented by fragmental calcite and dolomite, frequently diffused on the edges, and uniformly distributed among the mineral grains.

Secondary carbonates exhibit unequivocal indications of uneven distribution and dislocation in the loess stratum. Such carbonates arise during the acquisition of loess appearance by sediments. The mass of secondary carbonates is represented in the loess of the Russian plain by microcrystalline secondary calcite (grain size 1–4 μ), which can be encountered both in a dispersed form and in accumulations confined to the walls of pores, cracks, and holes made by roots.

Like loess, microcrystalline secondary calcite is the most frequent element in thin sections from the buried soils of the Russian plain. It is natural to assume that this form of secondary carbonates in fossil soils is originally of soil and diagenetic origin. For this reason, we determined the genesis of carbonates in our investigations by comparing the forms of carbonates, and their quantitative distribution in buried soils and in the loess roof, in thin section. Due attention was also given to the location of carbonate salts among the other elements of microstructure. A thorough micromorphological study of thin sections from buried soils shows that concentrations of microcrystalline calcite can frequently be detected in the pores above the fluids of optically oriented clays. Since the formation of sinters of optically oriented clays, which prove the illuviation processes in the organo-mineral mass, and the deposition of secondary carbonates cannot be placed simultaneously in time, such a location of microcrystalline calcite can be conceived as corroborating its diagenetic origin.

However, carbonates of ancient soil origin can also be found in thin sections. This, lublinite[1], which is not met in loess, has been discovered in the fossil soils of central Asia. The modern soils display deposits of lublinite in the lower horizons of chernozem, grey-brown and brown soils (Kubiena, 1938; Minashina, 1960). The occurrence of lublinite in fossil soils serves along with other features, therefore, as an indication of soil origin, and can be utilised for the determination of the genetic type of old soil.

---

[1] Lublinite = needle crystals of calcite.

MICROMORPHOLOGICAL INVESTIGATION OF PALEOGEOGRAPHICAL CONDITIONS IN THE MOSCOW–VALDAI (EEM) INTERGLACIAL

The micromorphological method has been applied (together with field morphological and chemical studies) to special research on the stratigraphically persistent horizon of fossil soil in the Upper Pleistocene loess stratum on the Russian plain (Morozova, 1963; Velichko and Morozova, 1963). Exposed buried soils (river banks, ravine sides) were studied at about every 30–50 km along the meridional profile, starting from the regions of the Moscow Glaciation, via Smolensk, Roslavl, Bryansk, Novgorod–Seversky, Putivl, Priluki and Kremenchug. A clearly defined level of supermorainic fossil soil has been traced everywhere within the boundaries of this area. Its stratigraphic location was determined by Velichko (1961; Velichko and Morozova, 1963).

In all the investigated sections, the buried soil occurs at the base of the Valdai (Würm) loessial deposits above the glacial deposits of the Dnieper (Riss) Glaciation in the south, and of the Moscow (apparently, Warthe stage) Glaciation in the north of this territory, and dates from the Mikulino (Eem) Interglacial.

In their structure, the interglacial (Eem) soils resembled present-day soils in the middle part of the Russian plain (sod podsolic, boggy, chernozem and grey forest soils). In their distribution, the interglacial soils were governed by the law of latitudinal zoning. Paleoedaphic investigations, and particularly use of the micromorphological method, brought to light certain regularities in the soil alterations throughout the Interglacial. The data obtained in this manner on the nonuniformity of the climate during the Moscow-Valdai (Eem) Interglacial are in good agreement with paleobotanical data (Grichuk, 1950). The structure of the interglacial soils indicates two stages in their development; beyond any doubt the earlier podsol stage (apparently coinciding with the phase of coniferous and coniferous broad-leaved forests) and the sod stage (the phase of the climatic optimum). This is borne out by the peculiar structure of a number of investigated soils; a thick sod horizon (50–70 mm) is present, together with podsolic and illuvial horizons with well-defined morphological and microstructural features. To all appearances, the podsolic soils were widely developed in the early Interglacial, because the relics of podsolic processes manifest themselves even in the structure of fossil chernozem to the south of this area.

In comparing modern and interglacial soils we cannot fail to notice that the thickness of humus horizons was far greater in old soils, and that they were initially completely leached of all carbonates, whereas modern soils crowning the loessial stratum in the same sections frequently contain a carbonate illuvial horizon. These features of Eemian soils apparently testify to the high humidity of the interglacial climate as compared with today, which on the one hand, gave rise to a luxurious herb vegetation with a highly productive organic mass, and on the other to the leaching of the carbonates beyond the limits of the soil profile.

These features of the interglacial soils apparently owe their origin to the peculiar physical and geographical conditions of the Eemian, and to the history of their formation. These soils represent the final result of continuous soil formation during the interval between the two glaciations, while the modern soils are at but one of the stages of their development.

CONCLUSIONS

(1) The micromorphological study of old soils in loess deposits of the Russian plain shows that this method can be used to great advantage by paleogeography.

(2) After burial, the soils retain many specific features of their microstructure (aggregates, microstructure of clays, microstructure of organic matter and its connection with the mineral part, etc.), which help to restore the original appearance of old soils.

(3) The study of old soils in thin sections makes it possible to determine the secondary features of the soils brought about by diagenesis in a number of cases.

(4) The spatial study of the stratigraphically persistent horizons of fossil soils in a series of consecutive sections enables us to restore the original structure of the soil mantle, and the regularities in distribution of soils during the various stages of the Quaternary.

SUMMARY

This paper describes the results of the micromorphological study of fossil soils buried in loessial deposits of the middle part of the Russian plain. The data obtained testify to the fact that despite prolonged burial the fossil soils retain many specific features of their original microstructure (aggregates and pores, organic matter, microstructure of clays), the analysis of which brings to light the nature of ancient soil-forming processes. The paper describes the paleogeographical conditions in the Moscow–Valdai (Eem) Interglacial on the basis of the study of the horizon of the interglacial fossil soil, in a series of consecutive sections in loessial deposits exposed along the meridian profile.

REFERENCES

Brewer, R., 1956. Optically oriented clays in thin sections of soils. Trans. Intern. Congr. Sci. Sol, 6^{e}, Paris, 1956, Comm., 1–2B : 871–874.

Feofarova, I.I., 1960. Micromorphological characteristics of clay matter in soils. Rept. Soviet Pedologists Intern. Congr. Soil Sci., 7th, Madisōn. U.S.S.R. Acad. Sci., Moscow, pp.461–464.

Gerasimov, I.P., 1961. Buried soils and their paleogeographical significance. Conf. Study Quaternary Period, 1 : 224–232.

Glazovskaya, M.A., 1956. Buried soils, methods for their study and their paleogeographical significance. Collection Articles Intern. Geograph. Congr., 18th, pp. 59–69.

Grichuk, V.P., 1950. Vegetation in the Russian Plain during the Lower and Middle Quaternary. Trans. Geograph. Inst. U.S.S.R. Acad. Sci., 46 : 5–202.

Kubiena, W., 1938. Micropedology. Collegiate Press, Ames, Iowa, 243 pp.

Minashina, N.G., 1958. Optically oriented clays in soils. Pochvovedenie, 4 : 91–96.

Minashina, N.G., 1960. Micromorphological investigation of loess and its changes in soil formation. Rept. Soviet Pedologists Intern. Congr. Soil Sci., 7th, Madison. U.S.S.R. Acad. Sci., Moscow, pp.465–475.

Morozova, T.D., 1963. Structure of old soils and regularities in their geographical distribution during the various stages of soil formation in the Upper Pleistocene (from the materials on the study of buried soils in the loess of the middle part of the Russian plain). Pochvovedenie, 12 : 26–37.

Parfenova, E.J. and Yarilova, E.A., 1960. On the lessivage and podsolidation. Pochvovedenie, 9 : 1–16.

Stephen, I., 1960. Pedology. Clay orientation in soils. Sci. Progr. (London), 48 : 190.

Velichko, A.A., 1961. Stratigraphy of loessial deposits in the periglacial zone of the Valdai and Moscow glacials. In: Paleogeography of the Quaternary Period. State Univ. Publ., Moscow, pp. 93–108.

Velichko, A.A. and Morozova, T.D., 1963. Mukulino fossil soil, its features and stratigraphic significance. In: Anthropogeny of the Russian Plain and its Stratigraphic Components. Akad. Nauk S.S.S.R., Moscow, pp.100–147.

Yarilova, E.A. and Parfenova, E.I., 1957. Newly formed clayminerals in soils. Pochvovedenie, 9 : 37–49.

MICROFABRIC OF MODERN AND OLD MORAINES

N.A. KORINA and M.A. FAUSTOVA

Geographical Institute of the U.S.S.R. Academy of Sciences, Moscow (U.S.S.R.)

The object of our investigations of some of the regions of modern and old glaciation in the U.S.S.R. was the fabric of glacial deposits. We used the micromorphological method, which enables a detailed study of the structural features of these deposits which are otherwise inaccessible to the naked eye.

The samples with an undisturbed structure came from the ground moraines of the Valdai (Würm) Glaciation in the Pskov and Smolensk regions, and the recent moraines of the Tien Shan (central Asia). From the samples thin sections in three mutually perpendicular planes were made and studied with a petrographic microscope.

The microstructure of the moraines has been found to consist of a silty clay mass and inclusions of sand fragments unevenly distributed in the mass. The thin sections obtained from the moraine exhibit the preferred orientation of elongated sand particles, a reticulate and criss-cross fibrous texture of optically oriented clays in clay varieties and scaly in sand varieties. By way of example, we shall describe a thin section of a sandy silt moraine of the Valdai (Würm) Glaciation in the Russian plain near Velikiye Luki. The sample was taken from a depth of 2.5 m.

A sandy silt moraine; particles $> 500\ \mu$ - 1.2%, 200–500 μ - 4.6%, $< 200\ \mu$ - 94.2%, some gravel grains. The thin section clearly reveals the preferred orientation of sand grains. The rock fragments are represented mainly by quartz, feldspars and calcite. Ore minerals, epidote, hornblende, zircon, rutile and mica can also be found. The feldspars, on the whole, are fresh and unaltered, and sometimes weakly sericitised and saussuritised. Individual fragments of limestone are coated by a clay-iron cutan with diffused edges. The filling clay–silt mass consists of fine calcite and hydromica. The texture of the clay is scaly, occasionally reticulate fibrous. Flow structures of optically oriented clay can be encountered in the pores and fissures and around the mineral grains. The pores are closed and spherical. There are spherical ferruginous clay and ferruginous concretions of a loose aggregate structure, 400 μ and smaller in size.

The moraines of various grain sizes have different microfabrics. The orientation of sand particles is the most important structural feature in sandy moraines. The microfabric of clay moraines is determined by the distribution of the aggregates of clay minerals, and their optical orientation. For this reason we shall dwell on the textures of optically oriented clays in the moraines, and on the orientation of the rock fragments, in more detail.

THE TEXTURES OF OPTICALLY ORIENTED CLAYS IN MORAINES

As has been noted above, the microfabric in the clay moraines is determined by the texture of optically oriented clays. Perpendicular fibrous (reticulate) and criss-cross fibrous textures (Fig.1) are most frequent. In the first case, the large fibres are located approximately at right angles to each other, rarely the angle amounts to 50–60°. Inside them the clay particles are oriented similarly and show extinction in one direction. The intersection of the fibres produces a reticulate microtexture with meshes of 100–200 μ. The criss-cross fibrous texture is formed by the fibres of optically oriented clays, haphazardly arranged.

In separate parts of the thin section the fine reticulate texture is sometimes superseded by fields of an oriented clay, 400–800 μ in size. In the moraines containing numerous sand and silt particles, the spaces between the sandy grains are filled with haphazardly arranged elongated scales of optically oriented clays and silt, 50–100 μ, and rarely 200 μ, in size. This texture is called a scaly texture.

A concentric orientation of clay particles, forming a distinctive cutan, can be observed around the sand grains. Such cutans can also be found around the pores. In the pores fibres consisting of optically oriented clay form commissures. A flow structure of optically oriented clay, along the cracks and pores, can be observed in thin sections of moraines taken from a shallow depth (up to 1.5 m). The width of the cutans along small cracks (70–400 μ) ranges from 30–200 μ, and along large cracks (800–1,000 μ) it reaches up to 400 μ. On the edges of the pores the size of the cutans ranges from 30–80 μ. The texture of the plasma concentrations is conchoidal.

It should be noted that the soils and clay rocks of nonglacial origin, which are mostly fine clays, display a perpendicularly fibrous texture of optically oriented clay similar to the one encountered in moraines. It is possible that this texture is connected with recrystallization processes here, and is formed as a result of minute cracks being filled with clay matter (Dmitrieva et al., 1962). Parfenov and Yarilova (1962) relate the formation of such a texture in soils to the processes of intrasoil weathering, and to the shift of clay matter inside individual microsections.

The criss-cross and perpendicularly fibrous textures were observed in all the thin sections of the clay moraines at different depths from the surface (from 0–7 m). No such textures have been found in the varved clays, which have a mechanical composition similar to that of the clay moraines.

To all appearances, the mechanism of the formation of oriented textures of fine particles is rather complex. Some of them may arise during the process of sedimentation, as was shown by Keller (1946), Brewer (1956), Finko (1957) and Lambe (1958). A number of textures are formed by pressure, a fact confirmed by the experiments of Williamson (1947), Popov (1949) and, Seed and Chan (1959). The measure of orientation of clay particles depends greatly on the content of loosely bound water (Denisov, 1951; Seed and Chan, 1959).

Apparently, the various textures of optically oriented clays in moraines are syngenetic. Their origin can be explained as follows. The moraine deposits could have given rise to the formation of textures of optically oriented clays under ice pressure and the presence of water. The reticulate texture

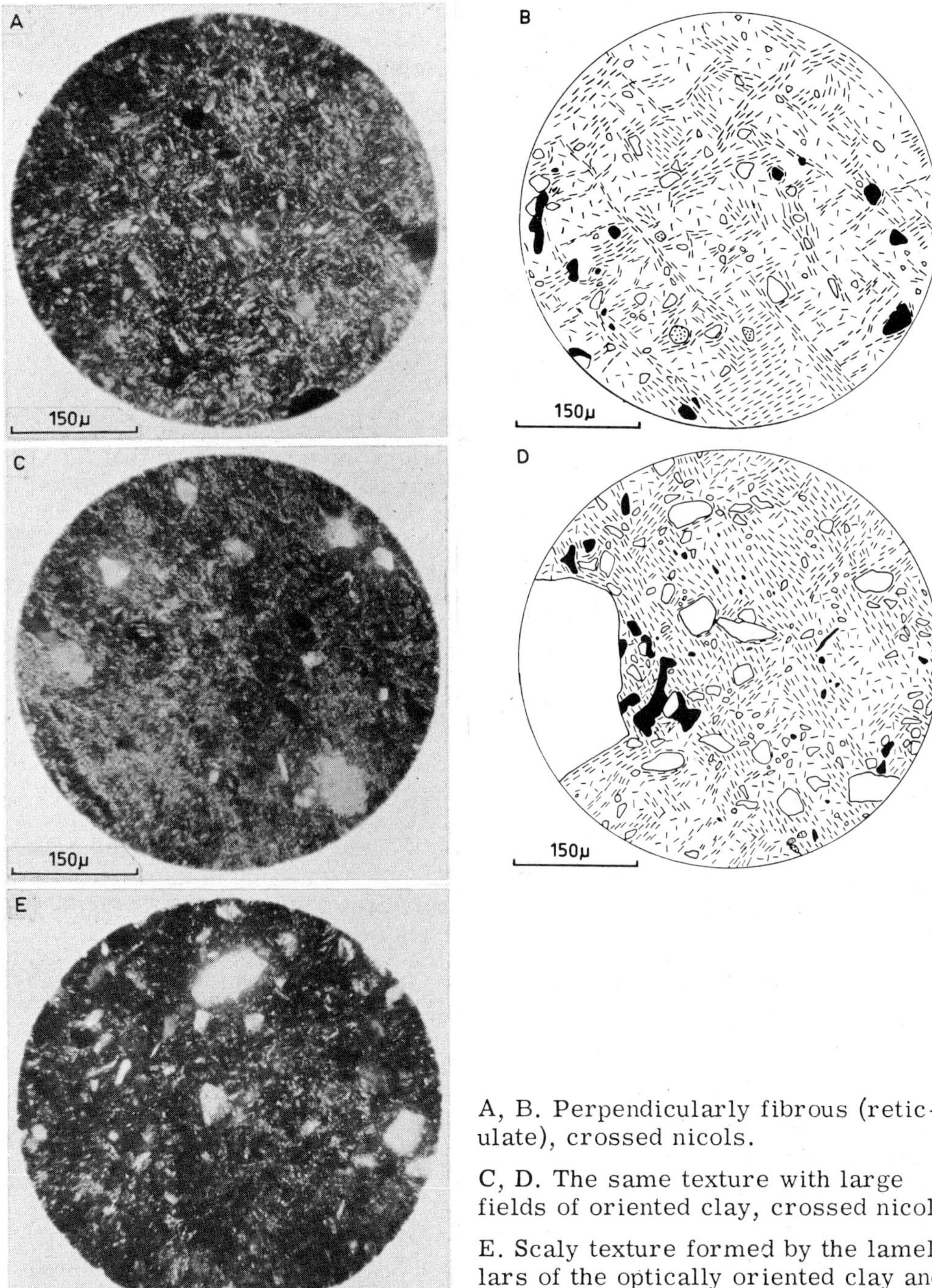

A, B. Perpendicularly fibrous (reticulate), crossed nicols.

C, D. The same texture with large fields of oriented clay, crossed nicols.

E. Scaly texture formed by the lamellars of the optically oriented clay and silt, crossed nicols.

Fig.1. Texture of optically oriented clays in the moraines of the Valdai (Würm) Glaciation of Smolensk region.

in the clay ground moraines, which were formed under tranquil conditions of thawing, might have survived from the structure of the moraine-containing ice, in which the aggregates of clay particles were distributed over the basal planes of ice crystals (Evteyev, 1961). In this case the water from the melting ice was completely or partly removed without causing the material to shift.

We believe, therefore, that the main features of the microfabric of moraines should be regarded as inherited, just as the characteristic features of their macrofabric are assumed to be inherited from the ice fabric (Richter 1932, 1936; Flint, 1957; Yakovleva, 1957). The origin of the textures of optically oriented clay in the moraines should also be considered from this standpoint.

ORIENTATION OF SAND GRAINS IN MORAINES

The question of the orientation of the rock fragments in the moraines has a history of its own. Numerous investigations have established that the direction of the long axes of the boulders usually coincides with the direction of the flow of the glacier (Richter, 1932; Holmes, 1941; Yakovleva, 1955; Harrison, 1957). Simultaneously, data have been obtained to the effect that the disposition of sand grains is governed by the same regularity as that of larger fragments (Harrison, 1957). It has been shown that the shape of the sand grains in no way affects their orientation, but the coefficient of the grain elongation is still very important (Seifert, 1954).

We have studied the orientation of sand grains in thin sections from the ground moraine and the kames of Valdai age (Würm) in the vicinity of Vitebsk and Velikiye Luki, and in the modern ground moraine of the Semyonov glacier (Tien Shan). The orientation of the sand grains was measured on the universal table designed by Fyodorov, in thin sections arranged in three mutually perpendicular planes. The results were depicted on stereograms as a projection of the lower semisphere. All the thin sections of the old moraines of the Russian plain showed a good orientation of sand grains (Fig.2). The horizontal thin sections displayed a rather correct orientation of sand grains from north-northwest to south-southeast.

In the modern ground moraine of the Semyonov glacier the sand grains were oriented from west to east, which coincides with the orientation of pebbles and corresponds with the movement of the glacier. In this case grains which have a long–short axes-ratio of 2 : 1 or more conform most closely to the orientation direction. Grains with a ratio below 2 : 1 are frequently located irregularly.

The clearly expressed orientation of the sand grains in the modern ground moraine of the Semyonov Glacier is explained by a steady thawing of the moraine from under the ice base, and its deposition in place without subsequent displacement of the material. The fact that the orientation direction of the sand material in the moraine coincides with the direction of the glacier movement shows that the structure of the moraine-containing ice has been inherited by the ground moraines. This conclusion can be applied to many old ground moraines which were formed under similar thawing conditions.

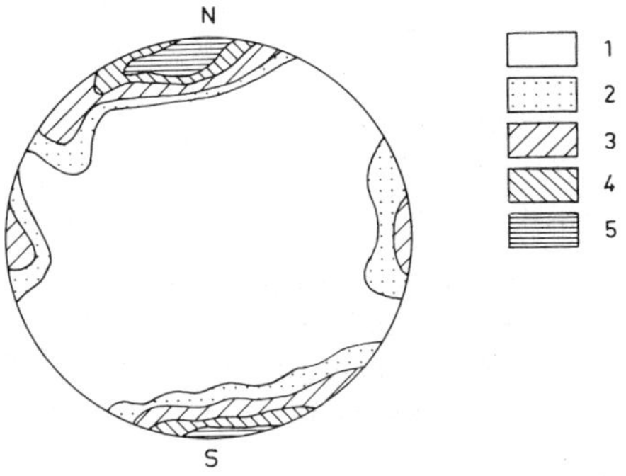

Fig.2. Stereogram of orientation of sand grains in the moraine of the Valdai (Würm) glaciation near Velikiye Luki (depth of 5 m). Symbols: 1 = < 2%; 2 = 2–4%; 3 = 4–6%; 4 = 6–8%; 5 = 8–10%.

THE ROLE OF SECONDARY PROCESSES IN THE ALTERATION OF THE MORAINE FABRIC

The soil-forming and weathering processes, as well as the processes of congelation and slope formation, are vitally important for the development of the microfabric of sedimentary rocks. These processes are apparently capable of changing the original microfabric of the moraine.

The upper part of all old moraines displays formations of pedogenetic origin: clayey flow structures, clay-iron and iron concretions, and segregations of secondary microcrystalline calcite. The traces of clay and iron movement found deep in the moraines can serve as an indication for separating this stratum into horizons of different ages.

Sections of optically oriented clays formed during the weathering of large-size fragments have also been found in old moraines. Various stages of this process have been observed, from bands of optically oriented clays on the edges of the crystal and the cleavage planes, to the reticula formed of optically oriented clays at the site of an almost completely disintegrated fragment. The clay textures of such an origin can usually be seen clearly in thin sections.

The congelation and slope-forming processes create a characteristic macrofabric. These processes cause the fragmental product to shift and reorient itself. It can be assumed, apparently, that they also shape the microfabric of the deposits.

Thus, the microfabric of the moraines possesses primary and secondary features. The former include the location of sand grains and the definite texture of optically oriented clays. They help to restore the conditions under which the moraines were formed. The latter testify to the processes which affected the moraines from the moment they had been deposited.

SUMMARY

(1) Primary and secondary features should be singled out in the microfabric of the moraines.

(2) The primary features include the orientation of the sand grains in the direction of the ice movement, and the definite textures of optically oriented clays. They help to restore the conditions under which the moraines were formed.

(3) The secondary features are acquired by the microfabric of the moraines after their deposition, under the effect of slope-forming, solifluction and congelation processes, as well as pedogenetic processes. They describe the conditions under which the moraines existed from the moment of their deposition.

REFERENCES

Brewer, R., 1956. Optically oriented clays in thin-sections of soils. Congr. Intern. Sci. Sol, 6E, Paris, B : 21–25.

Denisov, N.Y., 1951. On the nature of deformation of clay rocks. River Fleet Ministry Publ., Moscow, 200 pp. (Russ.).

Dmitrieva, E.V., Ershova, G.J. and Oreshnikova, E.J., under the direction of M.F. Vikulova and A.V. Khabakow, 1962. Atlas of textures and structures of sedimentary rocks. I, Gosgeoltekhizdat, Moscow, 1 : 574 pp. (Russ.).

Evteyev, S.A., 1961. Geological activity of the ice sheet in eastern Antarctic. Thesis, Moscow, 399 pp. (Russ.).

Finko, V.I., 1957. Mineralogy and genesis of refractory clays of the Zeisk-Bureinsk Depression. In: Investigation and use of clays. Lvov Univ., Lvov, pp.697–710.

Flint, R.F., 1957. Glacial and Pleistocene geology. Wiley, New York, N.Y., 553 pp.

Harrison, P.W., 1957. A clay-till fabric: its character and origin. J. Geol., 55 : 275–307

Holmes, D.C., 1941. Till fabric. Bull. Geol. Soc. Am., 52 : 1299–1354.

Keller, W.D., 1946. Evidence of texture on the origin of the Cheltenham fire clay of Missouri and associated shales. J. Sediment. Petrol., 16 : 63–71.

Lambe, T.W., 1958. The structure of compacted clay. Proc. Am. Soc. Civil Engrs., J. Soil Mech. Found. Div., 84 : 1–34.

Parfenova, E.I. and Yarilova, E.A., 1962. Mineralogical research in pedology. Akad. Nauk S.S.S.R., Moscow, 205 pp. (Russ.).

Popov, I.V., 1949. Mineralogical investigations of clay rock structures. Problems of Soviet Pedology, 15 : 174–210 (Russ.).

Richter, K., 1932. Die Bewegungsrichtung des Inlandeises, rekonstruiert aus den Kritzen und Langsachsen der Geschiebe. Z. Geschiebeforsch., 8, 1.

Richter, K., 1936. Gefügestudien im Engebrae, Fondalsbrae und ihren Vorlandsedimenten. Z. Gletscherk., 24 : 22–30.

Seed, H.B. and Chan, C.K., 1959. Structure and strength characteristics of compacted clays. Proc. Am. Soc. Civil. Engrs., J. Soil Mech. Found. Div., 85 : 87–128.

Von Seifert, G., 1954. Das mikroskopische Korngefüge des Geschiebemergels als Abbild des Eisbewegung, zugleich Geschichte des Eisabbaues in Fehmarn, Ostwagrien und dem danischen Wohld. Meyniana, 2 : 124–190.

Williamson, W.O., 1947. The fabric, water-distribution, drying-shrinkage and porosity of some shaped discs of clay. Am. J. Sci., 245 : 645–662.

Yakovleva, S.V., 1955. On regulated location of boulders in the strata of basal moraine Materialy Vses. Nauchn. Issled. Geol. Inst., 9 : 40–46.

Yakovleva, S.V., 1957. The orientation of boulders in basal moraines and its importance for determining the direction of the glaciers movement. Tr. komis. po Izuch Chetvertichnodo Perioda Akad. Nauk S.S.S.R., 13 : 300–303.

THE APPLICATION OF SOIL MICROMORPHOLOGY TO THE RECOGNITION AND INTERPRETATION OF FOSSIL SOILS IN VOLCANIC ASH DEPOSITS FROM THE NORTH ISLAND, NEW ZEALAND[1]

J.B. DALRYMPLE

Department of Geography, University of Auckland, Auckland (New Zealand)

INTRODUCTION

In the North Island of New Zealand the classical area for the study of volcanic ash deposits of andesitic composition is the Lower Waikato Valley and the closely adjacent areas. Here, their mode of occurrence on the ground surface varies from thick landscape blankets to additions of a few inches to the top of the soil present at the time of deposition. Grange and Taylor (1931) first distinguished at a reconnaissance level between two major groups of ash deposits; the Mairoa Ash and the Hamilton Ash. This difference was further substantiated when the soil mapping of the area was made by Grange et al. (1939). They found that yellow-brown loams occur on the Mairoa Ash and brown granular loams on the Hamilton Ash. This has recently been questioned by Ward (1961) who considers that both soil groups occur on the topmost member - the Hamilton Shower of Grange and Taylor (1931) - of the Hamilton Ash. Which soil develops on this particular ash member, Ward considers, depends primarily on the position of the site with respect to slope and drainage conditions. Even so, this does not invalidate the general premise that, over a wide area, each one of the two groups of ash deposits is characterized by very different soils.

All workers in this area consider that the Mairoa and the Hamilton Ash are both composed of a number of ash deposits together with a number of phases of weathering. Only Ward (1961), however, has attempted to differentiate in any detail between the ash beds and to trace their distribution. One of the major difficulties here is to distinguish between the individual ash deposits and the subsequent alterations resultant on pedogenesis. This largely results from the fact that few of these ash deposits have a paroxysmal origin. As Taylor (1933) first described, most had an intermittent origin and were deposited over a sufficiently long period of time for weathering and soil formation to have kept pace with deposition. Such fossil soil horizons have not previously been described in detail as, even in Ward (1961), the emphasis is on the ash bed rather than on the zones of weathering.

[1] As the soils treated in this paper all possess a high content of allophanes, it seems that they - according to the 7th approximation of the U.S.D.A. Soil Classification (1960) - are andepts (Ed.).

AREA AND NATURE OF STUDY

The aim of this study is to assess the value of soil-micromorphology as a tool for recognizing and interpreting the fossil soil zones associated with volcanic ash deposits of andesitic composition. This first required the field recognition and distinction between such geological and pedological materials Thus, a reconnaissance survey of the classical area and sites was completed and followed up by more detailed investigations to the north and particularly in the southern and central parts of the North Auckland land district. It is from this area that this study has been made, where in situ deposits of these widely distributed ash deposits have not previously been recorded or studied in detail.

The deposits of ash were recognized in the field on morphological and mineralogical grounds, whilst the criteria used to identify and distinguish between the various phases of soil formation include the depth of oxidative weathering; the depth and character of the B-horizon; the nature, intensity, and phases of clay translocation; the presence of clay formation in situ; and the degree and character of gleying. The micromorphological results were then subsequently compared with the field interpretation of the macro-morphological evidence.

Provided that comparable sites are selected for study, the thickness and nature of these zones of pedological alteration should be significant in interpreting the duration of these various phases of landscape stability and for indicating something of the character of the environment at the time of their formation. Thus, eluvial sites have been chosen where the slope angle is less than 4° and their aspects are similar. Such sites are least likely to undergo secondary changes, as a result of either fluvial or colluvial erosion, subsequent to the deposition of an individual ash member. Thus, both fossil and modern soils from such sites exhibit mature profiles.

PRELIMINARY RESULTS OF THE FIELD STUDY

The preliminary results of the field study in the southern and central parts of the north Auckland province indicate that there are four major groups of ash deposits: groups I, II, III, and IV; each group containing at least two ash facies. Detailed morphological and field descriptions of the four groups of ash deposits and their associated soils will be given elsewhere. Numerals, rather than names have been chosen in order to clearly indicate the local stratigraphical relationships of the ash deposits.

At a site on the main Papakura–Hunua highway (526/334) the red, reddish yellow, and the white and pinkish white ash deposits of group I together with their associated zones of weathering and pedological alteration form the base of the ash deposits. These are overlain by the brownish-yellow ash deposits and their associated soils of group II; which in turn are overlain by the yellow ash deposits with white concretions and their associated soil horizons of group III. At a site on the Bombay–Paparata highway (532/209) similar ash deposits and soil zones of groups I and II occur, but here they are overlain by a series of strong brown ash deposits with yellow patches and tubular concretions of group III. At a site in Gellings Road, near Hunua township (590/291), similar group III ashes and their associated soil zones occur. Here, they are underlain by group I deposits and overlain by yellowish-brown granular ash deposits of group IV.

Although no site has yet been found in this area where both of the two types of group III are present, there is sufficient evidence from these eluvial sites to establish a local stratigraphy for these ash deposits; with group I the oldest and group IV the youngest.(No formal attempt has yet been made to correlate in detail these locally recognized groups of ash deposits with the Mairoa and Hamilton Ash from the lower Waikato Valley). This immediately establishes the relative chronology of the fossil soil horizons at these and other comparable sites.

Most of the ash members within the groups I–IV are associated with zones of pedological alteration. These may be either only weakly developed and represent rather immature soils, as within the ash groups II, III, and IV; or well developed, like those associated with the older ash members of group I. In the field some of the latter resemble planosols; and a few, latosols. However, the major zones of pedological alteration are found in the form of heterogeneous soil mantles between each of the major groups of ash deposits; fossil soil mantle I (formed after the deposition of the last ash member of group I and before the first of group II), fossil soil mantle II (between groups II and III), fossil soil mantle III (between groups III and IV), and fossil soil mantle IV (the modern soil forming on group IV). In comparable eluvial sites, each of these are associated with specific soil types, namely; ando soils with fossil soil mantle IV; minimal grey-brown podzolics with III; modal grey-brown podzolics with II; and red-yellow podzolics with I. Thus, all the fossil soil mantles except IV are characterized by the translocation of clay; and there is evidence for a considerable period of landscape stability between the deposition of each ash group.

THE PLACE OF SOIL MICROMORPHOLOGY IN THIS STUDY

The recognition of ash deposits of andesitic composition

It is possible to recognize these deposits of andesitic ash in thin section because of the character of the clay minerals which were formed concurrently with their slow and intermittent deposition. Thus, little fresh volcanic glass has been observed, most being in the form of either angular shards which are lunar in shape or as cryptocrystalline dust. Usually 40% or more of their ground mass is made up of clay-sized material which in ordinary light is yellow (10YR 7/6, 7/8, 8/6) in colour. Within this area are many discrete, rounded, patches which are very smooth in appearance, are devoid of shrinkage or strain cracks, and are very pale brown (10YR 8/4); (Fig.1). In contrast, smaller areas do show evidence of strain with well developed cracks. These occur less frequently than the very pale brown areas and are brownish-yellow (10YR 6/8). All facies, however, are isotropic between crossed nicols. Both subsiduary areas can occur in all the ash deposits of groups II–IV, but there is a marked tendency for the brownish-yellow areas to occur in the older deposits of group II and for the very pale-brown areas

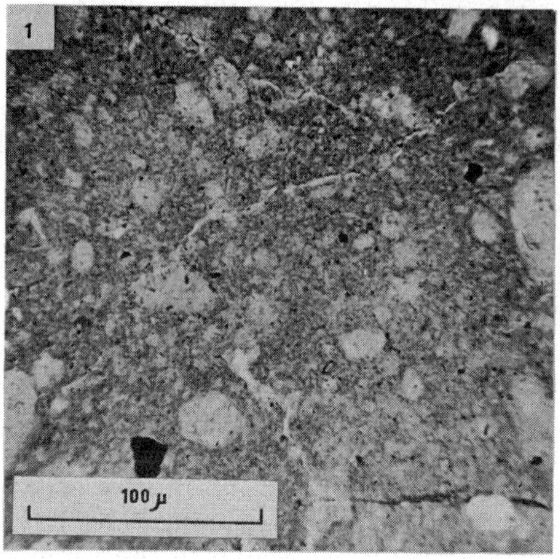

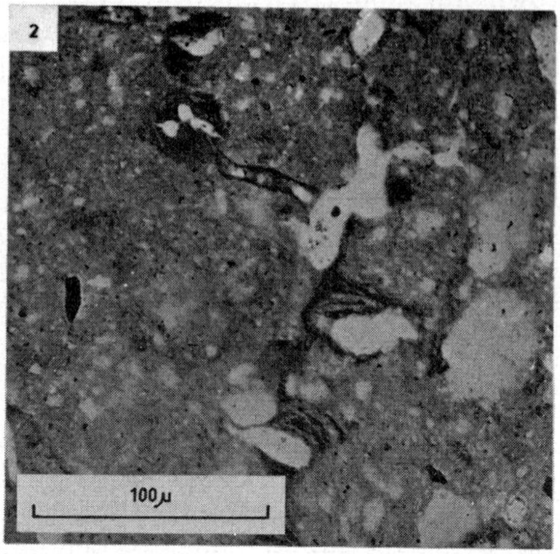

Fig.1. Group III ash deposit of andesitic composition. Its ground mass is composed of yellow (10YR 7/8) clay-sized material with many, discrete, very pale brown (10YR 8/4), and rounded patches which are devoid of shrinkage or strain cracks. Site in the main Papakura–Hunua highway, (526/334).

Fig.2. B_t-horizon of fossil soil mantle II. Additions of strong brown (7.5YR 5/ clay-sized material in the form of thin, optically continuous, lamellae to many of the micro and very fine tubular pores to the yellow (10YR 7/8) groundmass. Site in the main Papakura–Hunua highway (526/334). Under ordinary light.

in the youngest deposits of group IV. Thus, it is not yet possible on micromorphological grounds to definitely distinguish between the ash deposits within groups II–IV. However, it is thought that the number and shape of the voids and cavities; and of the size range, shape, and type of light and heavy mineral grains may well be diagnostic for each of the ash deposits. For example, micas appear to be most frequent in ash deposits of group II, whilst large and fresh mineral grains of andesine are characteristic of group III. So far, however, no detailed quantitative study along these lines has been completed.

As in the field, so in thin section, the ash deposits of group I are very variable and they are easily distinguished from those of groups II–IV. In ordinary light the colour of their ground mass varies from very pale-brown (10YR 8/3, 8/4) to yellowish-red (5YR 5/8) and red (2.5YR 5/8), whilst between crossed nicols this material is mainly isotropic but does contain areas showing slight birefringency and the initial stages of orientation. This is particularly the case in the older members of group I. Even so, the general isotropic character of the clay-sized material making up the matrix of the ash deposits of group I is consistent throughout.

The distinction between ash deposits of andesitic composition and other commonly occurring parent materials

On low-lying sites the processes of colluviation and river action can give rise to very mixed soil parent materials, but in most eluvial sites, such as on the main Papakura–Hunua highway, the deposits of ash directly overlie deeply weathered Tertiary and greywacke beds. In the field, it is not easy to distinguish the base of the red ash deposits of group I from the top of the deeply weathered sedimentary rocks. However, in thin section, the deep yellow and red weathering associated with these rocks all show micromorphological features typical of lehms. Their groundmass, or plasma, is markedly birefringent and shows a strong degree of preferred orientation. Where the ash deposits directly overlie such material there is evidence of a mixed zone with distinct and discrete areas showing either isotropic or lehm characteristics. In addition, fresh angular feldspars and micas occur which are similar to those found in the overlying ash, but which are absent from the weathered Tertiaries and greywackes. This mixed zone, therefore, is considered to represent the results of the first additions at this site of volcanic ash to the already deeply weathered Tertiaries.

Likewise, none of the other commonly occurring parent materials contain clay-sized material which, in thin section, is isotropic. One exception is certain of the basic tephra and volcanic rocks, but these have a distinctive mineral suite. Thus, except in the last case, the ash deposits of andesitic composition can be distinguished on micromorphological grounds from the other commonly occurring parent materials in this area.

The recognition of fossil soil mantles associated with the ash deposits of andesitic composition

Since most of the ash deposits in this area had an intermittent origin, they were all weathered to a greater or lesser degree concomitant with their deposition and before the time unit when the respective soil mantles formed. Thus, it is not surprising that, in thin section, the matrix of the fossil soil mantles associated with the ash deposits of groups II–IV are essentially similar. Between crossed nicols all are isotropic and under ordinary light they have the same colour and form as their respective ash deposits, but they show a more marked development of patches, concentrates, and concretions. Patches are considered as the initial stages in the accumulation of colloidal material, whilst concentrates can be distinguished from concretions by the occurrence of visible mineral grains within the area of accumulation. All these areas are much denser than the remainder of the matrix; and darker in colour, generally brownish-yellow (10YR 6/6, 6/8). As this material becomes more prominent in the ground mass of such fossil soil horizons, the very pale-brown areas of clay-sized material decrease.

However, there is one distinguishing micromorphological feature which occurs both in the fossil soil mantles (except for soil mantle IV) and in many of the less well developed pedological zones within each of the major groups of ash deposits. This is the addition to their groundmass of clay-sized material which has not been formed in situ, but translocated in the soil water from overlying horizons and redeposited as fillings or linings to many of the conducting channels and cavities within the structure units (Fig.2). This material also occurs in the form of either thin and discontinuous, or as thick and continuous, cutans to structure faces. Between crossed nicols, these colloidal additions are very unusual. Unlike any of the otherwise similar material described by many workers from such soil types as grey-brown and red-yellow podzolics, these here are isotropic. Thus, such colloidal additions can only be distinguished from the surrounding matrix under ordinary light.

In fossil soil mantle III colloidal additions take the form of strong brown (7.5YR 5/6, 5/8) linings to a few of the small rounded cavities (root pores) which occur within the blocky structure. This material shows optical continuity and no banding. This is also the case for the thin and discontinuous cutans to some of the structure units.

In fossil soil mantle II, many more cavities and conducting channels are filled with this colloidal material which also tends to permeate into the structure units away from such lines of movement. In the larger conducting channels, banding from strong brown (7.5YR 5/8) to reddish-yellow (7.5YR 7/ is common. These bands are still optically continuous but, between crossed nicols, they show slight birefringency from yellowish-brown (10YR 5/8) to dark yellowish-brown (10YR 4/4). However, the groundmass and the rest of the non-banded colloidal additions remain isotropic.

The micromorphology of fossil soil mantle I differs more from that of its underlying ash than do any of the younger soil mantles from their respective ash deposits. Here, the ground mass of colloidal material (which on the average comprises 68% of the structure units as compared with 20% colloidal additions, 9% mineral grains, and 3% empty cavities) is under ordinary light brownish-yellow (10YR 6/8) and, between crossed nicols, approximately half is isotropic and half shows varying intensities of yellow (2.5Y 7/8, 8/6, 8/8). In the latter areas, the clay-sized materials are markedly orientated and show other features typical of lehms. Most conducting channels and cavities within their ground mass are filled with strong brown (7.5YR 5/8) to reddish-yellow (5YR 5/8) colloidal material which is frequently banded. These bands are not optically continuous; and these layers result from more than one phase of additions of clay-sized material to the groundmass. This is the case both within, and around, structure units. In addition, colloidal material permeates into, and in some cases throughout, the structure units and also coats many of the mineral grains. These additions, in ordinary light, tend towards dark reddish-brown (5YR 3/4) and remain much the same colour between crossed nicols.

Thus, in terms of soil-micromorphology, the four fossil soil mantles (I–IV) can all be recognized and distinguished both from one another and from their respective, underlying, ash deposits. This is possible, largely, as a result of differences in the amounts and character of the additions of clay-sized material to their fabrics.

Within each of the groups II–IV, the less well developed zones of pedological alteration all show micromorphological features resultant on the accumulation of colloidal material within their groundmass, whilst others show features of colloidal additions but which are somewhat less intensely developed than in their respective major soil mantles. In most of the fossil soil horizons occurring beneath soil mantle I, however, the reverse is the case. Here, the secondary additions of colloidal material are so great that the matrix of many of the structure units of these horizons are predominantly composed of translocated clay-sized material. There are areas showing well developed banding and layering; and there is every evidence for successive phases of movement of this clay-sized material. Under ordinary light, it varies in colour from red (10YR 4/6, 4/8; 2.5YR 5/8) to reddish-yellow (5YR 7/6, 7/8) and strong brown (7.5YR 5/6, 5/8). The reddish-yellow and strong brown material is markedly birefringent, whilst the red material is isotropic. The remaining areas of the ground mass are also isotropic, except where such old and deeply buried soil horizons have been strongly gleyed. Then, between crossed nicols, their groundmass shows features of preferred orientation and marked birefringency.

The distinction between the fossil soil mantles associated with the ash deposits of andesitic composition and other commonly occurring soils

The micromorphology of the gleyed fossil soil horizons, which are associated with some of the older ash deposits of group I, can be distinguished from those associated with modern surface water gleys occurring on unweathered Tertiary clays. In the latter case there are no additions of colloidal material to what is otherwise a similar ground mass; and there is an absence of angular mineral grains. The micromorphology of the modern grey-brown and red-yellow podzolic soils, which are formed from the unweathered Tertiary silts and fine sands and from the upper part of the deeply weathered Tertiary and greywacke profiles, is different from those of the similar soils of the fossil soil mantles I, II, and III. In none of the modern soils are their groundmasses isotropic, or are their colloidal additions either isotropic or only slightly birefringent. Red-yellow podzolic soils are also associated with certain of the basalts and basic tephra materials in this area. Again, their colloidal additions are birefringent and show well developed parallel streaks; but some examples have been observed where their groundmass is not birefringent but isotropic. The latter have a rather similar micromorphology to that of fossil soil mantle I, but the characteristic size range, shape, and suite of their mineral grains are diagnostic. This would also be the case for the latosolic soils found on certain other basic rocks. Thus, the fossil soil mantles associated with the ash deposits of andesitic composition contain diagnostic features of soil-micromorphology which enable them to be distinguished from the other commonly occurring modern soils in this area.

Interpretation from the evidence of soil micromorphology of fossil soil mantles associated with the ash deposits of andesitic composition

The recognition of any fossil zone of pedological alteration and the identification of its nature enables the character of the soil forming processes to be identified; it gives an indication of the amount of time available for weathering and soil formation; and it enables something to be inferred of the character of the environment at the time of its formation.

Soil forming processes; and soil types

The overriding and diagnostic micromorphological feature occurring in all the fossil soil mantles is the isotropic character of much of their colloidal material. This reflects the nature and origin of these fossil soils in that allophane is the first clay mineral formed from the weathering of ash deposits of andesitic composition (Birrell and Fields, 1952; Fields and Swindale, 1954). This does not mean that the matrix of these soils is composed solely of such clay minerals, since hydrous feldspars and hydrated ferric oxide may occur in close association with them. Thus, until further information is available from these specific sites, the use of the terms colloidal material or clay-sized material is to be preferred even though allophane is almost certainly the dominant constituent of such material. This micromorphological characteristic is sufficiently diagnostic that, provided the soil is not formed on a basic volcanic parent material, the addition of even small amounts of such ash to the top part of any modern or fossil soil can be recognized from thin sections. This has been shown to hold true even though the soil material containing the ash is subsequently eroded and incorporated into colluvial or fluvial deposits.

It has been shown from thin sections how colloidal material can be concentrated in the groundmass (soil mantle IV); how, after initial mobilization, this material can be translocated in the soil moisture and be redeposited in the conducting channels and cavities within structure units as well as in the form of cutans to structure faces. Such material may be deposited in discrete areas and in small amounts (soil mantle III); it may line many of the conducting channels and cavities, occasionally show banding which remains optically continuous, and it may begin to permeate into the groundmass and form thin and discontinuous cutans to structure faces (soil mantle II); or it may completely fill all the lines of movement within the structure units, many of which show the development of layers which are not optically continuous, and it may permeate throughout the groundmass and form thick and continuous cutans to structure faces (soil mantle I). Thus, apart from soil mantle IV, the major soil process occurring in these fossil soil mantles is the translocation of clay and the formation of B_t-horizons. In each case this pedological process happened subsequent to the complete weathering of the whole ash deposit. Thus, the soil micromorphological evidence supports the contention that each of these soil mantles must represent a period of landscape stability when no further ash was accumulating at these sites and when the processes of soil building were dominant over those of erosion, transportation, and deposition.

Even in the oldest of the fossil soil mantles (I), the colloidal additions are only slightly birefringent and so they are still presumably largely composed of allophane. Thus, the mobilization, translocation, and redeposition of these clay minerals hardly alters their micromorphological characteristic. However, in the groundmass of this fossil soil mantle (I), there are areas which show between crossed nicols preferred orientation and marked birefringence. Thus, the clay mineral fraction of these areas is not allophane; possibly meta-halloysite or montmorillonite (Fields, 1955; Fields and Taylor, 1961). Furthermore, the nature of the micromorphology of this groundmass suggests that clay formation in situ, as opposed to clay translocation, is the process involved here. This is also considered to be the case in many of the fossil soil horizons which occur beneath, and are older than, soil mantle I.

Except for the lack of birefringence, the soil micromorphological characteristics of the colloidal additions to the groundmass of the soil mantles III, II, and I are comparable, and almost identical, to features found in soils formed from loess and other silt-sized deposits in southeastern England (Dalrymple, 1960, 1962a, b, 1964). These micromorphological features have been recognized as characterizing grey-brown podzolic and red-yellow podzolic soils; and interpreted as a time sequence in the development of their B_t-horizons. Thus, the soil micromorphological evidence from eluvial sites in the southern and central part of the North Auckland province confirms the field interpretation of minimal grey-brown podzolics for fossil soil mantle III; of modal grey-brown podzolics for fossil soil mantles II; and of red-yellow podzolics for fossil soil mantle I.

Many of the less well developed fossil soil horizons associated with the individual ash members within group II–IV have micromorphological features which show the development of concentrates comparable to those in soil mantle IV. These fossil soil horizons are interpreted, therefore, as ando soils. Others contain features of colloidal additions comparable to, but less intensely developed than, those in their respective fossil soil mantles. Some of these features result from the superposition of more than one phase of pedogenesis. This happens when the total accumulation of an ash member is small, such that the subsequent phase of soil formation continues into the upper part of, or throughout, the underlying soil. Similarly some, but not all, of the older fossil soil horizons associated with the older ash deposits of group I represent composite profiles. Their complex micromorphological features are very similar to those found in soils formed on the older and higher-lying brickearths in southeastern England and result from the processes of clay translocation; clay formation in situ; and, less frequently, gleying. These soils (Dalrymple, 1960, 1964) are considered to be planosolic, and in a few cases latosolic, soils. Thus, again, the soil micromorphological evidence supports the field interpretation.

Relative lengths of time available for the formation of the fossil soil mantles, and a time, or development, sequence of soils

The micromorphology of the ash deposits of group IV are very similar to those within group III; yet soil mantle IV is an ando soil and fossil soil mantle III a grey-brown podsolic. Thus, it would seem reasonable to infer that insufficient time has been available since the deposition of ash IV for the translocation of clay to develop, particularly since some of the minor fossil soil horizons within groups II–IV have also been recognized as ando soils. Again, the ash deposits within groups III and II have fairly similar micromorphological characteristics, but fossil soil mantle II contains much greater amounts of additions of clay-sized material. Once again, the inference is that a relatively longer period of time is involved during which fossil soil mantle II was formed. The differences between fossil soil mantles II and I are even more marked. In fossil soil mantle I there are not only much greater and more intense additions of clay-sized material, but also there is more than one phase of clay translocation both within and around the structure units. This suggests that a relatively longer period of landscape stability occurred between the deposition of the last ash member of group I and the first of group II than during any of the subsequent phases represented by soil mantles II, III, and IV. This also suggests that fossil soil mantle I represents a major stratigraphical break within the deposition of these ash deposits.

Since the fossil soils have only been studied from directly comparable eluvial sites (see p.341), the only pedogenetic variable other than time is climate. It is, however, very difficult to attribute these micro- and macromorphological differences observed, both amongst the fossil soil mantles and the minor fossil soil horizons, to variations in climate; particularly when the total time span, both for the deposition of all the ash deposits and for the formation of all the soil horizons, is only from the Late Glacial to Recent; and for groups II–IV considerably less than this. Thus, it is considered that, from eluvial sites, variations in the relative lengths of time available for soil formation are primarily responsible for the differences in the nature of those fossil zones of pedological alteration which are formed from ash deposits of andesitic composition in the southern and central parts of the North Auckland province. Furthermore, it is postulated that, under such specific conditions, the sequence of soils, ando, through grey-brown podzolic and red-yellow podzolic, to planosols (and possibly latosols) represents a time, or development, sequence of soils.

ACKNOWLEDGEMENT

The author wishes to acknowledge a research grant from the New Zealand Universities Grants Committee towards the cost of field research.

SUMMARY

From the southern and central parts of the North Auckland province, a local chronology of four groups of ash deposits of andesitic composition, with major zones of pedological alteration between them, has been established in the field from closely comparable eluvial sites. Detailed soil micromorphological studies of these ash deposits and of their associated soils have shown that the nature, and especially the isotropic character, of their clay-sized fraction enables them to be distinguished in thin section from almost all other parent materials and soils in this area. The soil micromorphological features of accumulation of colloidal material in patches, concentrates, and concretions; of the translocation of colloidal material and its redeposition in conducting channels, cavities, and as coatings; of clay formation in situ; and of gleying, enable the many fossil soil horizons to be recognized and distinguished from their respective ash deposits. Furthermore, each of the four major soil horizons can be differentiated one from another.

Many of the minor fossil soil horizons have composite profiles, whilst the four major zones occur as heterogeneous mantles; each associated with a specific soil type. From the soil micromorphological evidence, inferences are made as to the relative lengths of time available for soil formation during the periods of non-deposition of ash. It is considered that time is the major factor for the differences observed in these fossil soils; and a development sequence of soils is postulated from ando, through grey-brown podzolic and red-yellow podzolic, to planosols (and possibly latosols).

REFERENCES

Birrell, K.S. and Fields, M., 1952. Allophane in volcanic ash soils. J. Soil Sci., 3 : 156–166.

Dalrymple, J.B., 1960. The nature, origin, age, and correlation of some of the brickearths and associated soils of southeastern England. Ph.D. Thesis, Univ. London, London, 397 pp.

Dalrymple, J.B., 1962a. Some micromorphological implications of time as a soil forming factor, illustrated from sites in southeastern England. Z. Pflanzenernähr. Düng. Bodenk., 98 : 232–239.

Dalrymple, J.B., 1962b. Some aspects of the alteration of loess by processes of weathering and soil formation and its application to Quaternary chronology. In: Symposium on Geochronology and Land Surfaces in Relation to Soils in Australasia. Div. Soils, C.S.I.R.O., Adelaide, pp.186–188.

Dalrymple, J.B., 1964. The application of pedology to the identification and interpretation of erosion surfaces, illustrated from southeastern England and the North Island, New Zealand. Intern. Geograph. Congr., 20th, London, 1964, Abstr., in press.

Fields, M., 1955. Clay mineralogy of New Zealand soils. 2. Allophane and related mineral colloids. New Zealand J. Sci., B37 : 336–350.

Fields, M and Swindale, L.D., 1954. Chemical weathering of silicates in soil formation. New Zealand J. Sci., B36 : 140–154.

Fields, M. and Taylor, N.H., 1961. Clay mineralogy of New Zealand soils. 5. Mineral colloids and genetic classification. New Zealand J. Sci., 4 : 679–706.

Grange, L.I. and Taylor, N.H., 1931. Reconnaissance soil survey of the central part of the North Island. New Zealand Dept. Sci. Ind. Res., Geol. Surv., Ann. Rept., 25 : 7–8.

Grange, L.I., Taylor, N.H., Sutherland, C.F., Dixon, J.E., Hodgson, L., Seelye, F.T., Kidson, E., Cranwell, L.M. and Smallfield, P.W., 1939. Soils and agriculture of part of Waipa County. New Zealand Dept. Sci. Ind. Res., Bull., 76: 85 pp.

Taylor, N.H., 1933. Soil processes in volcanic ash beds, New Zealand J. Sci., 14 : 193–202.

Ward, W.T., 1961. Geology of Glen Murray and Rotongaro Districts. M. Sc. Thesis, Univ. Auckland, Auckland, 99 pp.

SOIL FABRIC AND SOIL MECHANICS

D. LAFEBER

Soil Mechanics Section, Commonwealth Scientific and Industrial Research Organization, Syndal, Victoria (Australia)

INTRODUCTION

A change in the stress conditions acting on a material system generally results in some form of deformation of the system in order to attain a new state of equilibrium in respect to the changed stress pattern. Sometimes a deformation may become macroscopically observable as a change of shape or volume or of both of the whole system. As far as soils are concerned, the study of such "macroscopical" deformations and their relations to the applied stress configurations comprises an appreciable part of the traditional field of soil mechanics. The principal aim of these studies is to predict the behaviour of soils in response to specified stress conditions. Serious difficulties may arise where the deformations are not simple to observe directly or even indirectly, or where the interest is focused on a problem such as the mode of failure.

In all instances, however, a change in stress conditions resulting in deformation of any compound material system must inevitably involve movements or displacements (strain) of some or all of the individual components of the system.

STRAIN ANALYSIS AND FABRIC ANALYSIS

Strain of kinematical analysis in its widest meaning comprises the recognition and recording of stress-induced movements (or their traces) of the individual elements of a compound material system, and the subsequent construction of the appropriate strain pattern for the whole system. In this sence strain analysis is an indispensable and important tool in the assessment of the stress-strain relations of any material.

In a complex material such as soil, the individual elements (e.g., skeleton grains, clay platelets, pores, etc.) of the system are spatially arranged in a continuous, threedimensional pattern (fabric or fabric pattern). Fabric analysis is concerned with the description, classification, and genetical analysis of the geometrical characteristics of a fabric pattern. This includes the study and analysis of changes in these geometrical data due to the development of a fabric pattern out of an earlier, different one, irrespective of the cause of the development. In other words, strain analysis can then be considered as that branch of fabric analysis which is interested especially in stress-induced fabric changes of macroscopical or microscopical dimensions.

The present position of soil fabric analysis in particular can be described in the following way. Soil fabric studies as envisaged originally by Kubiena (1938) and subsequently developed by other investigators (e.g., Altemüller, 1962), are still considered essentially a branch of pedology. As a result they are very closely associated with, and therefore, to some extent restricted by the taxonomy and genesis of natural soils. For this reason their course and aim run in many respects parallel to those of the fabric analysis of the non-deformed rocks or non-tectonites. Soil fabric analysis as an ally to soil mechanics, on the other hand, although still in its infancy, has to be considered as a development which will have much in common with fabric analyses of deformed rocks or tectonites.

Strain observations as made in traditional soil mechanics testing procedures have the advantage that: (1) they are direct and produce instantaneous data on changes of length or volume ; (2) they are continuous, i.e. the rate of strain can be compared at any moment with the rate of change of the stress conditions.

As soon as more detail - usually of microscopical or submicroscopical dimensions - is required, however, these advantages are immediately lost, and the examinations become generally of a "post mortem" character, largely as a result of preparation procedures. This appears to be particularly serious as strain analysis is supposed to study and analyse movements or displacements as an expression of fabric changes. In other words, the assessment of movements or displacements in a "frozen state" then becomes largely a matter of interpretation.

Fortunately, the same problem occurs in the fabric analysis of deformed rocks, and, in the course of time, has been successfully approached in that field in many instances. It seems logical, therefore, to attempt to develop similar or indentical methods and techniques in the kinematical analysis of deformed soils.

The fundamental principles of the petrological approach are described by Knopf and Ingerson (1938) as follows:

> "Any movement of a natural object is reflected in the position that the object assumes under the influence of that movement"; and: " ..... objects in motion tend to abandon a random scattering of position in favour of a significant arrangement in space that is determined by the movement. Spatial arrangement of this kind is one type of the definitely ordered state of matter that is known as preferred orientation."

These principles are basically correct, but incomplete in so far as the authors consider orientation only. In fact, if a number of objects constituting a compound system have been brought into position as a result of movement, whatever the cause of this movement is, there are two independent geometrical aspects to be taken into account for each and every one of these objects. These are the location within the system or *spatial distribution* and the *orientation*. The importance of both aspects becomes particularly obvious where stress-induced movements affect a system composed of objects whose properties in regard to stress show a wide range of magnitude.

If the movements considered are restricted to those which are due to stress, the previously mentioned statements should be formulated thus: The pattern of stress-induced movements in an assembly of natural objects is reflected in the spatial distribution and orientation of these objects within the particular assembly. Such movements tend to result in either preferred spatial orientation, or in a particular spatial distribution pattern of the objects within the system, or in a combination of both. The relation between spatial distribution and orientation of the objects in the system is completely covered by the term *topotropy* (Sander, 1950).

PREFERRED ORIENTATION AND SPATIAL DISTRIBUTION

I. The traces of both rotational and translational movements have been encountered in deformed rocks. In attempting to find evidence for similar for identical phenomena in soils, it should be realized, however, that there is a very marked difference in mechanical characteristics between rocks and soils. In the majority of rocks the mechanical properties, particularly the strength of the various mineralogical constituents, are of approximately the same order of magnitude. Secondly, the stresses to which rocks are subjected in the process of metamorphism are often of the same order of magnitude as the strength of the constituting minerals. Even if the applied stresses are lower, their effect is still frequently reinforced by the simultaneous action of appreciably higher temperatures. As a result, fracture and/or recrystallization of the constituents are commonly observed metamorphic features. It can very often be inferred, therefore, that the combined and usually simultaneous action of stresses and elevated temperatures affects even the crystal lattices of the minerals of metamorphosed rocks, by either deforming or rebuilding them.

In soils, on the other hand, the mechanical properties, particularly the strength of the various constituents (skeleton grains, clay matrix, and water- or air-filled pores) are orders apart. Secondly, the stresses - either due to natural causes or to engineering measures - acting on soils are extremely small compared with those occurring in rock metamorphism, whereas the influence of high temperatures is negligible. As a result it can be expected that the stresses responsible for the usual type of soil deformation will not directly affect the lattices of the crystalline constituents of the soil.

In other words, stress-induced movements in metamorphic rocks will often result in preferred lattice orientation of one or more minerals, and only occasionally in preferred dimensional orientation (especially when the shape of the mineral grains is strongly controlled by the crystal lattice, as e.g.,with the micas). Stress-induced movements in soils, however, will often reveal themselves in the preferred dimensional orientation of one or more constituents. Complications may arise in the case of the soil clay minerals, which similar to the micas, show a strong control of the external shape by the crystal lattice.

II. *Preferred orientation* of fabric elements is often the most obvious of the two previously mentioned geometrical aspects of many natural and artificial fabric patterns. As a result, it has been studied extensively, particularly in its relations to stress-induced movements in rock and in metal fabrics. These studies have clearly shown that there are, in fact, a number of preferred orientation patterns, closely associated with the different types of movements (e.g., Knopf and Ingerson, 1938). Fundamentally two simple types of preferred orientation can be distinguished.

A. The simplest type is the one in which an external shape vector or a crystallographic vector of a certain type of fabric element - e.g.,the longest axis of a pore, the crystallographic c-axis of a quartz crystal, or the pole to the basal plane of a mica individual - displays the same spatial orientation (statistically) for the individual representatives of that fabric element in the fabric concerned. This basic type is termed the *point maximum orientation pattern*. In a fabric possessing this orientation pattern the mica flakes, for instance, would be (statistically) parallel. Often this type can be recognized readily, in particular where flaky or elongated, rod-shaped fabric elements are involved.

In rock fabric analysis it is customary to demonstrate the orientation of shape or crystallographic vectors of fabric elements (usually mineral grains) in the following way. Each fabric element which possesses the particular vector is thought to be moved *without rotation* to a reference sphere, in such manner that the origin of the particular vector coincides with the centre of the sphere. The intersections of the directions of the vectors with the (usually lower half) of the sphere are then projected on the horizontal equator plane of the sphere, using the equal area spherical projection method (Knopf and Ingerson, 1938). Areas in this plane showing the same density of projected intersections per unit area are then surrounded by contour lines. Using this method, a simple point maximum orientation pattern, e.g., of the optical axes of a number of quartz individuals in a tectonite, would appear as shown in Fig.1. In this diagram, the black area represents the maximum density of projected intersections per unit area, and the surrounding contours show the location of zones of decreasing density.

B. Fabric studies of rocks, however, have revealed that there is another very common orientation pattern which, unfortunately, is not so evident. It can, as a rule, only be recognized by the application of specialized optical and/or X-ray techniques. In this pattern the previously mentioned vectors are (statistically) parallel to a specific plane. The particular orientation type is termed the *girdle orientation pattern*. (Fig.2).

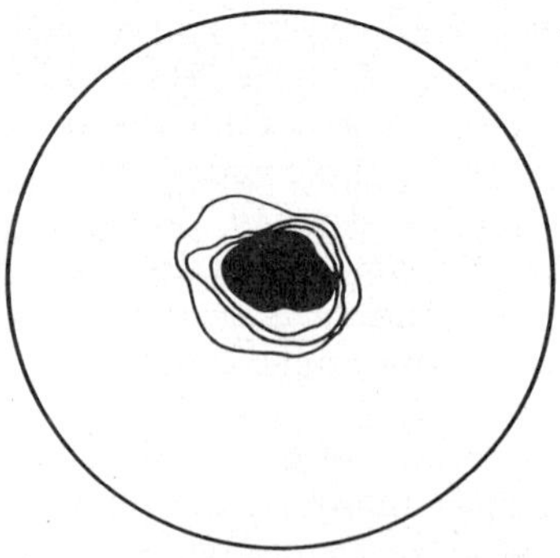

1.

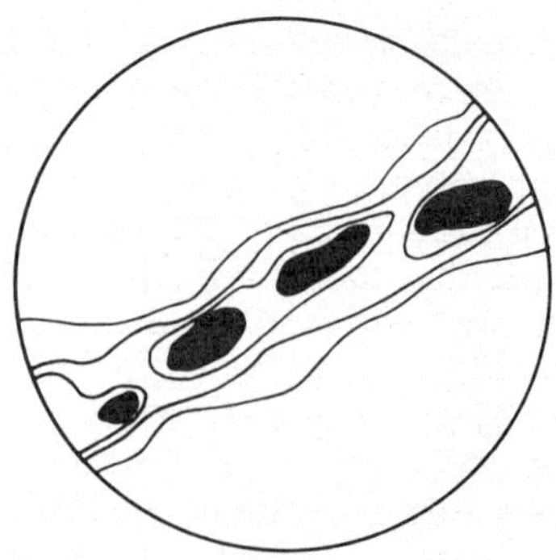

2.

Fig.1. Equal-area spherical projection of a point maximum orientation pattern.

Fig.2. Equal-area spherical projection of a girdle orientation pattern.

In actual materials the conditions are often much more complicated, as more than one point maximum may be present, or one or two point maxima may occur in combination with a girdle (Fig.3), or two differently oriented girdles may be present (Fig.4), or one type of fabric element in a pattern may be characterized by a point maximum, and another type in the same fabric pattern may show a girdle orientation, etc. A number of these possibilities has been encountered in the analysis of rock fabrics. The mutual spatial relations of point maxima and girdles in such complex situations are usually described in rock fabric analysis by means of certain symmetry notations. This method, which has actually a semi-quantitative basis also simplifies in many instances the study of the spatial relations between deformation patterns and the resulting orientation patterns.

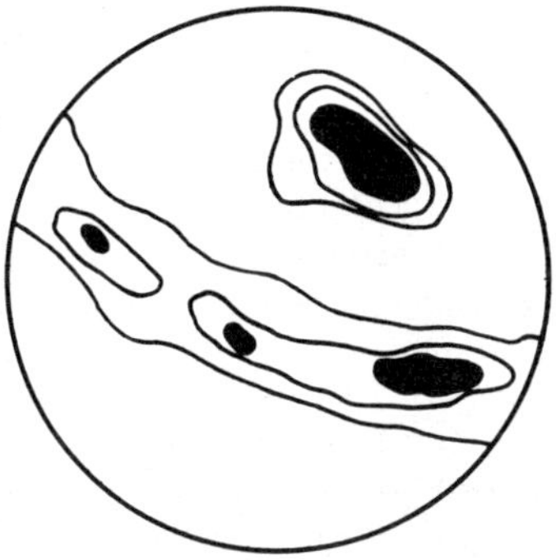

3.

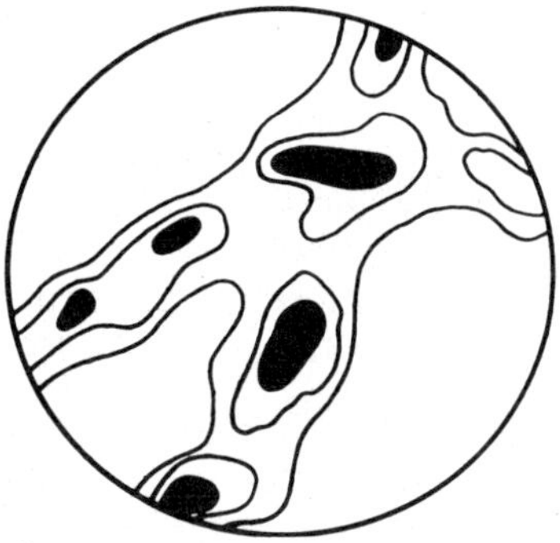

4.

Fig.3. Equal-area spherical projection of the combination of a point maximum and a girdle orientation pattern.

Fig.4. Equal-area spherical projection of two intersecting girdle orientation patterns.

Simple point maximum orientation patterns, e.g.,of elongated quartz (skeleton) grains or flaky clay minerals, have been noticed in soils, although they have not been described as such. They have often been explained as depositional features. In other instances, they have been considered as a result of stress, usually simple compaction (e.g.,Buessem and Nagy, 1954). Simple or complex girdle orientation patterns are presumably the most common type of orientation pattern in deformed rock fabrics. They are often associated with rotational movements. The occurence of girdle orientation patterns in soils has not been assessed definitely at present, although evidence of stress-induced rotational movements (Fig.5) has been found in some natural soils (Lafeber, 1962) and in artificially prepared clay samples (Williamson, 1954).

It should be pointed out, however, that the use of X-ray methods to assess the presence of such orientation patterns, particularly of the clay minerals in soils, may encounter difficulties as a result of the nearly universal occurrence of oriented clay cutans (Brewer, 1960) around skeleton grains. (This type of preferred orientation pattern has no counterpart in the orientation patterns of rock fabrics.)

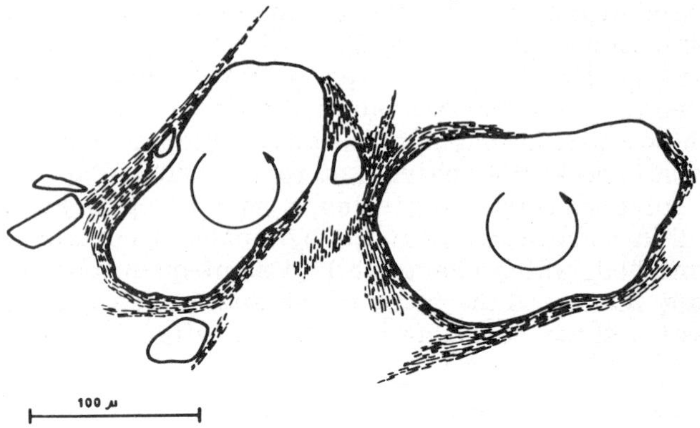

Fig.5. Partial stripping of oriented-clay cutans and formation of cutan tails as a result of stress-induced rotational movements of skeleton grains in a black earth, Adelaide, South Australia. (The thin lines represent the traces of the basal planes of the clay minerals, the curved arrows indicate the direction of rotation of the skeleton grains.)

It is felt, on the other hand, that the discovery of more complex orientation patterns in artificially or naturally deformed soils will be only a matter of time, depending mainly on whether suitable preparation, observation and recording techniques can be developed.

III. The *spatial distribution* of fabric elements in deformed rocks, unfortunately, has not been studied extensively as yet. The interest in the subject originates from Sander and his co-workers and some work in this field has been carried out by them ("Achsenverteilungsanalysen". Sander, 1950). As previously stated, however, spatial distribution analysis can be expected to be of major importance in materials such as soils, in which there is characteristically a very wide range of magnitude in the strength and deformation properties of the various constituents (Lafeber, 1963).

Recent studies by the author on the spatial relations between skeleton grains and pores of a number of Australian soils have shown, for instance, that in the majority of the soils the frequency of contacts between skeleton grains mutually is significantly less than that to be expected if these grains were arranged at random. This appears to be valid in these soils even for skeleton grain contents as high as 50% by volume, and is in agreement with the findings of Mitchell (1956) in certain American soils. It could also support a statement made by Trollope and Chan (1960) that by loading a soil specimen the stress has to be carried, in the first place, by the clay matrix. In other instances, however, there appears to be a faint suggestion that the frequency of the particular contacts may be increased by remoulding the soil.

The frequency of contacts between skeleton grains and pores in these soils has been studied at the same time, and reveals also some intriguing aspects which will be reported elsewhere.

The above studies, however, have also stressed the possible significance of the almost universal occurrence of higher order fabric elements in natural and, possibly even artificially prepared soil specimens. These higher order fabric elements, which are usually of mm or even cm size, are complex units consisting of skeleton grains, as well as clay matrix and pores. Their occurrence seems to be independent of the depth in the soil profile, but their size, shape, and composition apparently vary with depth. Their role in and their reaction on deformation of the soil are not known at present, but there are definite suggestions that in many instances they are hard to destroy (e.g.,Ingles, 1963), unless there is a very appreciable excess of water present (at least the liquid limit water content, possibly much more).

SOIL MECHANICS AND SOIL FABRIC

The interest of soil mechanics in soil fabric is certainly not a new development. Terzaghi (1925) discussed a limited number of largely hypothetical soil fabric types ("Strukturarten") in the introductory chapter of his classical work and, in subsequent chapters, their possible influence on the mechanical behaviour of soils. The subject was taken up again and somewhat enlarged by Casagrande (1932), who attempted to explain the well-known and often striking difference in strength of soils in the undisturbed and in the remoulded state as due to changes in the (hypothetical) fabric pattern. Fig.6 shows the fabric of an undisturbed marine clay, as envisaged by Casagrande, whereas Fig.7, by way of contrast, shows the actual but simplified pattern of a krasnozem from Mt. Tamborine, Queensland. Hvorslev (1937, 1960) noticed and discussed a difference in strength of soils, depending upon whether the load was applied normal to the bedding plane(= plane of preferred orientation of clay mineral platelets) or in another direction.

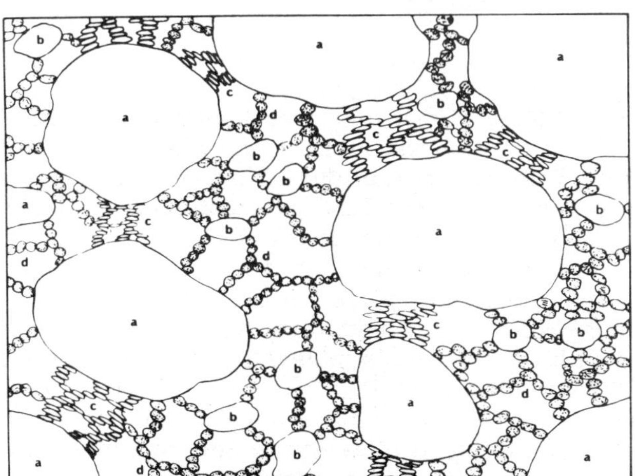

Fig.6. The structure of an undisturbed marine clay (after Casagrande, 1932): (a) silt grain, (b) clay particle, (c) flocculated colloidal particles of high degree of consolidation due to local concentration of pressure, (d) flocculated colloidal particles of low degree of consolidation.

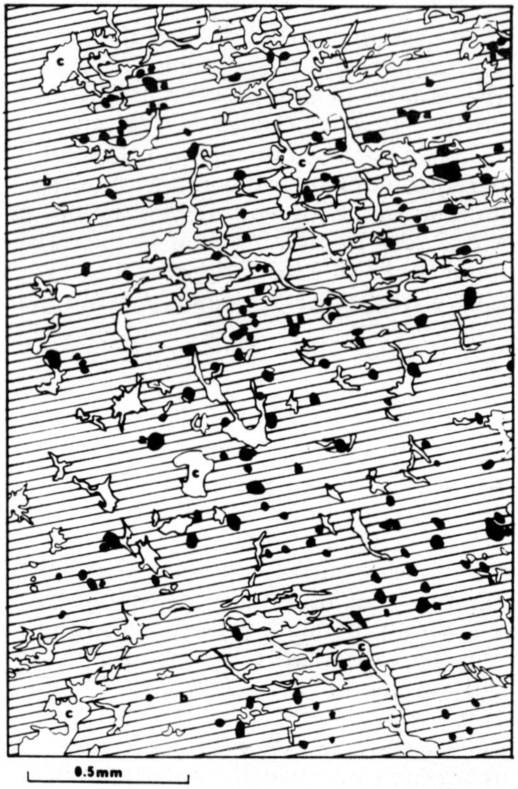

Fig.7. Fabric pattern of krasnozem, Mt. Tamborine, Queensland: (a) skeleton grains (predominantly quartz), (b) Fe-stained clay minerals and submicroscopical pores, (c) micro- and macropores.

Sander (1939) discussed the possible application of methods and techniques developed in rock fabric analysis, on soil fabric patterns.

More recently, Buessem and Nagy (1954) prepared kaolinite samples of various water contents, in which they claim the kaolinite platelets to be largely in a random orientation. They report that during subsequent loading, the kaolinite platelets tended to arrange themselves perpendicular to the direction of the compressive stress. Williamson (1954), on the other hand, prepared kaolinite samples of various water contents, a number of which had a known preferred orientation of the clay platelets prior to testing. Deformation of these samples by rolling produced very distinct fabric changes, which in accordance with the terminology developed in rock fabric analysis were described by Williamson as the development of R-tectonites. Rosenqvist (1955) studied a number of Norwegian clays and demonstrated the preferred orientation of the clay mineral platelets or the lack of such orientation to be related to environmental factors during deposition and, in some instances, also to the particle size. Mitchell (1956) examined a number of North American clays and silty clays, and came to the conclusion that the orientation of the clay particles in relation to the orientation of the applied stress distribution had a marked influence on the strength of the soils.

Several features stand out amongst the results of these studies and a number of other similar ones not mentioned here. Firstly, there appears to be a distinct suggestion of a relation between the strength and certain fabric anisotropies of soils. Secondly, such fabric anisotropies may be depositional features or they may be stress-induced. Thirdly, the observed fabric anisotropies - with the exception of those described by Williamson - are not only all of a planar character, but they are even restricted to the simple point maximum orientation pattern of the clay minerals. Next to nothing is known, on the other hand, about the occurrence and the possible influence on strength of other orientation patterns of clay minerals as well as of other important soil constituents such as skeleton grains and pores. The same applies to the spatial distribution analysis of soil fabric patterns.

In other words, it seems logical to attempt to develop soil fabric analysis as an aid to soil mechanics in its widest sense along the same or similar lines as have been proved suitable and valuable in rock fabric analysis, particularly of tectonites. It has to be stressed, however, that such studies should not be restricted to artificially deformed soils. Natural soils, even before loading, invariably possess a characteristic fabric. It is certainly true that an appreciable number of the primary fabric characteristics of natural soils is due to illuviation, dispersion and flocculation processes, etc. Many natural soils, on the other hand, have been affected by processes such as seasonal swelling and shrinking, hillside creep, sliding, etc. and can, therefore, be expected to show in their fabric evidence of stress induced movements. The recognition of this evidence is almost undeveloped. The same applies to the study of the influence of such stress-induced fabric features (which, in analogy to rock tectonites, can be expected to be of a much more complex character than the simple point maximum orientation pattern) on the mechanical behaviour of soils.

To develop soil fabric analysis into a valuable and powerful tool for soil mechanics or, more generally, soil physics, these problems must be studied systematically. Such studies can only be of real value, however, if they can be expressed in a quantitative or at least semi-quantitative way.

SUMMARY

One of the principal aims of soil mechanics is to predict the mechanical behaviour of soils under specified stress conditions, i.e. primarily stress-strain relations of soils. Strain analysis is described as the branch of fabric analysis which is concerned with the recognition, description, and recording of stress-induced movements, and the kinematical analysis of their pattern. Soil fabric analysis as an ally of soil mechanics is considered to be closely associated with the fabric analysis of deformed rocks or tectonites. The importance of studying both preferred orientation and spatial distribution of fabric elements is discussed. It is stressed, at the same time, that fabric studies as an aid to soil mechanics should include natural soil fabrics prior to artificial loading in regard to the possible influence of natural, stress-induced fabric characteristics on the ultimate mechanical behaviour of soils.

REFERENCES

Altemüller, H.J., 1962. Gedanken zum Aufbau des Bodens und seiner begrifflichen Erfassung. Z. Kulturtech., 3: 323–336.

Brewer, R., 1960. Cutans. Their definition, recognition, and interpretation. J. Soil Sci., 11: 280–292.

Buessem, W.R. and Nagy, B., 1954. The mechanism of the deformation of clay. Proc. Natl. Conf. Clays Clay Minerals, 2nd - Natl. Acad. Sci. Natl. Res. Council, Publ. 327: 480–491.

Casagrande, A., 1932. The structure of clay and its importance in foundation engineering. J. Boston Soc. Civil Engrs., 19 (4): 168–208.

Hvorslev, M.J., 1937. Über die Festigkeitseigenschaften gestörter bindiger Böden. Ingeniørvidenskab. Skrifter, Ser. A. (45): 1–159.

Hvorslev, M.J., 1960. Physical components of the shear strength of saturated clays. Am. Soc. Civil. Engrs.,Res. Conf. Shear Strength Cohesion Soils, Boulder, pp.169–273.

Ingles, O.G., 1963. The shatter test as an index of strength for soil aggregates. Tewkesbury Symp. Fracture, Univ. Melbourne, pp.1–19.

Knopf, E.B. and Ingerson, E., 1938. Structural petrology. Geol. Soc. Am., Mem. 6: 270 pp.

Kubiena, L., 1938. Micropedology. Ames, Iowa, 243 pp.

Lafeber, D., 1962. Aspects of stress-induced differential movements of fabric elements in mineral soils. Proc. Australian Road Res. Board, 1st, Canberra, pp. 1059–1067.

Lafeber, D., 1963. On the spatial distribution of fabric elements in rock and soil fabrics. Proc. Austr. New Zealand Conf. Soil Mech. Found. Eng., 4th, Adelaide, pp.185–199.

Mitchell, J.K., 1956. The fabric of natural clays and its relation to engineering properties. Proc. Highway Res. Board, 35: 693–713.

Rosenqvist, I.Th., 1955. Investigations in the clay-electrolyte-water system. Publ. Norg. Geotek. Inst., 9: 125 pp.

Sander, B., 1939. Gefügekunde und ihre Anwendungen. Z. Angew. Mineral., 1: 285–312.

Sander, B., 1950. Einführung in die Gefügekunde der geologischen Körper. 2. Die Korngefüge. Springer, Wien, 409 pp.

Terzaghi, K., 1925. Erdbaumechanik auf bodenphysikalischer Grundlage. Leipzig/ Vienna, 399 pp.

Trollope, D.H. and Chan, C.K., 1960. Soil structure and step-strain phenomenon. Proc. Am. Soc. Civil. Engrs., J. Soil Mech. Found. Div., 86: 1–39.

Trollope, D.H., 1960. The fabric of clays in relation to shear strength. Proc. Austr. New Zealand Conf. Soil Mech. Found. Eng., 3rd,Sidney, pp. 197–202.

Williamson, W.O., 1954. The effects of rotational rolling on the fabric and drying shrinkage of clay. Am. J. Sci., 252: 129–143.

ZUR TECHNIK DER MIKROMORPHOLOGISCHEN UNTERSUCHUNG NICHTGETROCKNETER UNTERWASSERBÖDEN

WERNER KRAUSE

Staatliche Versuchsanstalt für Grünlandwirtschaft und Futterbau, Abteilung Höhenlandwirtschaft, Donaueschingen (Deutschland)[1]

METHODISCHE VORBEMERKUNGEN

Die Schwierigkeit, wasserdurchtränkten Boden ausreichend zu härten, bildet ein bedeutendes methodisches Problem der Mikromorphologie. Einige hydrophile Substanzen, z.B. Agar oder Gelatine, bringen eine gewisse Verfestigung zustande, doch genügt diese nicht den Anforderungen der Dünnschlifftechnik. Besondere Hindernisse bestehen für die "nasse Einbettung" von Unterwasserboden, in dem das Wasser nicht nur die feste Substanz durchtränkt, sondern darüber hinaus durch den Auftrieb einen formenden Einfluss auf das Hohlraumsystem ausübt, der im terristrischen Boden fehlt.

Ein älterer Vorschlag zur Anfertigung von Mikropräparaten aus wassergesättigtem Boden (Haarløv und Weis-Fogh, 1952) empfiehlt vorläufige Verfestigung durch Agar und Härtung durch Gefrierenlassen. In Abänderung dieses Vorschlages erscheint es auch erwägenswert, Unterwasserboden mit Hilfe eines versenkbaren Kühlaggregats an Ort und Stelle zum Gefrieren zu bringen und dann in mehr oder weniger ungestörter Lagerung zu sammeln.

Bei den Arbeiten, über die hier berichtet werden soll, mussten chemische Methoden und Vereisung wegen des hohen technischen Aufwandes ausser Betracht bleiben. Versucht wurde ausschliesslich, Proben von Unterwasserboden in ungestörter Lagerung und natürlicher Durchfeuchtung zu entnehmen sowie die Transportmöglichkeit zu klären. Als geeignet erwiesen sich beiderseits offene Glasröhren von 2–5 cm Weite, mit denen der Boden ausgestochen und ohne Strukturzerstörung herausgehoben werden kann. Auch bei dünnflüssiger Konsistenz gelingt das Herausheben, wenn die Glasröhre nach dem Einstechen am oberen Ende unter Wasser mit einem Gummistopfen verschlossen wird. Sobald das offengebliebene untere Ende aus dem Boden herausgezogen ist, wird es ebenfalls unter Wasser verschlossen. Der Boden kann in der Glasröhre mit einiger Vorsicht transportiert werden.

Das Verfahren lässt sich auch in Gewässern von mässiger Tiefe anwenden, wenn die Glasröhre an einer Stange befestigt und am oberen Ende mit einem Ventil versehen ist, das sich beim Heben des Geräts unter Wasser schliesst. Nicht selten bewirkt der dichtgelagerte mineralische Unterboden stehender Gewässer einen festen Verschluss des unteren Röhrenendes, der wenigstens für kurze Zeit ausreichende Abdichtung bietet, sogar wenn die Röhre an die Luft gebracht wird.

---

[1] Direktor: Dr. F. Brünner.

Dass bei diesem Verfahren, abgesehen von einer schwachen, in ihrem Ausmass überblickbaren Kompression bei besonders lockeren Böden, die Struktur nicht zerstört wird, lässt sich am Ausbleiben von Aufwirbelungen erkennen. Selbst leichte, locker gelagerte Böden organischer Herkunft in klaren Gewässern können bei ruhigem vorgehen entnommen werden, ohne dass die Grenzschicht zwischen Boden und Wasser, die oft durch eine auffallend gefärbte Bakterien- oder Diatomeenhaut bezeichnet ist, aufgelöst oder verformt wird. Zugunsten der Untersuchung ungehärteten Unterwasserbodens spricht ausserdem seine nicht selten zu beobachtende gallertige Konsistenz, die es gestattet, kleine festbegrenzte Stücke mit einem Löffel herauszuheben und in wassergefüllten Gefässen ohne Beschädigung zu transportieren. Sie tritt auf, wenn schleimabscheidende Organismen wie Bakterien, Diatomeen und Cyanophyceen in grosser Menge am Aufbau des Bodens beteiligt sind.

Zur mikroskopischen Untersuchung werden kleine Bodenmengen durch lange Pipetten aus dem Sammelgefäss entnommen. Liegt dünnflüssige Konsistenz vor, kann es sich empfehlen, nach Sicherstellung ausreichender Proben aus der obersten Schicht, diese mit einem Heber ganz abzusaugen und erst dann mit der Entnahme aus der folgenden Schicht zu beginnen. Bei Böden mit unverbundenen kleinen Teilchen werden Aufschlämmungen, bei kohärenter Lagerung Bruchstücke des Bodens mikroskopiert.

Die Durchmusterung des auf beschriebene Weise nass gesammelten Unterwasserbodens bietet selbstverständlich nicht alle Vorzüge der Dünnschlifftechnik. Die Präparate sind nicht dauerhaft. Die räumliche Lage der festen Teilchen verändert sich spätestens in dem Augenblick, in dem kleine Mengen auf den Objektträger gebracht werden. Dabei spielt die Sonderung nach dem spezifischen Gewicht eine grosse Rolle. Die Präparate sind dicker als Schliffe.

Ein Urteil über das Verfahren muss die Schwierigkeiten des Objekts in Betracht ziehen. Leichter, locker gelagerter Unterwasserboden organischer Herkunft bildet ein äusserst subtiles Gleichgewichtssystem, zu dessen morphologischen Kriterien das nahezu schwerelose Schweben der organischen wasserdurchtränkten Partikel gehört. Ein chemisches Härtungsmittel muss daher, wenn Strukturverfälschungen ausgeschaltet bleiben sollen, ausser anderen Eigenschaften ein spezifisches Gewicht besitzen, das dem des Wassers nahekommt. Trotzdem wären angesichts der Labilität des Gleichgewichts Verlagerungen wahrscheinlich.

In dieser Beziehung erscheint das Gefrierenlassen weniger bedenklich. Doch müsste das Ausmass der möglichen Druckdeformation geprüft werden. Allgemein gegen das Vereisen spricht die unbequeme Anwendungsweise.

Angesichts der beträchtlichen Hindernisse, die jeder stark eingreifenden Manipulation entgegenstehen, erscheint es nicht abwegig, Unterwasserboden in möglichst unverändertem Zustand zu untersuchen und als Voraussetzung nur die Sorgfalt der manuellen Behandlung und die kritische Prüfung der Ergebnisse zu fordern. Das Bemühen um brauchbare Einbettungs- und Härtungstechnik wird durch dieses Vorgehen nicht verdrängt. Jede Beschäftigung mit nassen Unterwasserböden wird den Wunsch hervorrufen, diese äusserst labilen Strukturen auch zu fixieren. Weitere Diskussion des grundsätzlichen Für und Wider erscheint erst am Platze, wenn einige Ergebnisse der eingeschlagenen Arbeitsrichtung vorgelegt sind.

UNTERWASSERBODEN IN DURCHLICHTBETRACHTUNG UNTER DEM MIKROSKOP

Vorbemerkungen: Die Photos zeigen Bodenaufschlämmungen, deren relativ grosse Tiefe die Schärfe der Aufnahmen beeinträchtigt. Bei unmittelbarer Betrachtung unter dem Mikroskop lässt sich die Tiefenschärfe auf das ganze Präparat ausdehnen.

Die Bodenproben stammen aus der Rheinaue in der Umgebung von Kappel (Kreis Lahr) in Baden–Württemberg. Die Gewässer liegen in kalkreicher Umgebung. Neben aussergewöhnlich sauberen, durch Grundwasseraustritt gespeisten Quellabflüssen und Teichen kommen hoch eutrophierte, getrübte Wasserläufe vor. Mit Ausnahme der Abb.2 und Tafel F6 zeigen alle Photos die oberste Bodenschicht aus 0–1 cm Tiefe.

Im Boden aus einem oligotrophen, kalt-stenothermen, langsam bewegten Quellabfluss ist Phanerogamen Bewuchs (Reinbestand von *Phragmites communis*) hochproduktiv, aber einseitig zusammengesetzt. Die Schattenwirkung auf das Gewässer ist stark. Die Bodensubstanz ist äusserst locker gelagert, und wirbelt flockig doch ohne diffuse Trübung auf bei geringer Störung. Sie ist schnell sedimentierend.

In Abb.1 sind die drei typischen Komponenten dieses Bodens in repräsentativer Mengenverteilung abgebildet. Man kann die folgenden Komponenten erkennen:

(1) Scharf geformte zylindrische Körper mit annähernd ebenen, selten abgerundeten Endflächen, als Tierlosung gedeutet. Sie formt etwa 60% der Substanzmenge. Die Feinstruktur kann man in Abb.2 deutlich sehen.

(2) Ungeformte feinzerteilte Substanz, die etwa 30% der Substanzmenge formt. Dem äusseren Bild zufolge ist es das gleiche Material wie in den Zylindern.

(3) Frische Phanerogamenreste, überwiegend in Splitterform, etwa 10% der Substanzmenge formend.

Alle Partikel dieses Bodens sind ohne Bildung nebeneinandergelagert. Zwischen gekreuzten Nicols erfolgt fast vollständige Auslöschung. Die zylindrischen Körper erscheinen tiefschwarz. In geringer Menge ($<1\%$ der Gesamtsubstanz) treten doppelbrechende kleinste Mineralfragmente auf.

Makroskopisch dem Oberboden ähnlich, aber dichter gelagert ist die Probe von Abb.2. Sie zeigt keine diffuse Trübung beim Aufwirbeln. Unter dem Mikroskop sieht man überwiegend gerundete Aggregate, vereinzelt Zylinder (Abb.2, links unten). Es gibt hellbraune hyaline Grundmasse mit wolkig verteilten mittelbraunen und punktförmigen dunkelbraunen Einschlüssen. Die Partikel sind überwiegend bindungslos gelagert. Zwischen gekreuzten Nicols sind sie dem Oberboden (Abb.1) ähnlich. Der Anteil des doppelbrechenden Mineralstaubs ist $>1\%$, aber im ganzen noch sehr gering.

Abb.3 zeigt einen von Algenfäden durchzogenen Oberboden, der aus brauner Bakterienhaut besteht. Die Lagerung stammt aus stehendem Wasser über grauen Tonuntergrund und ist wenig gestört. Unter dem Mikroskop erkennt man geschlossenes, bei schwacher Vergrösserung nahezu strukturlose Decke die von lichtdurchlässigen Bahnen durchzogen sind. Diese Bahnen sind bei starker Vergrösserung als abgestorbene Algenfäden zu erkennen. Zwischen gekreuzten Nicols sieht man viele Mineralstäubchen in Brown'scher Bewegung. Grössere Stücke doppelbrechender Substanz sind kaum vorhanden.

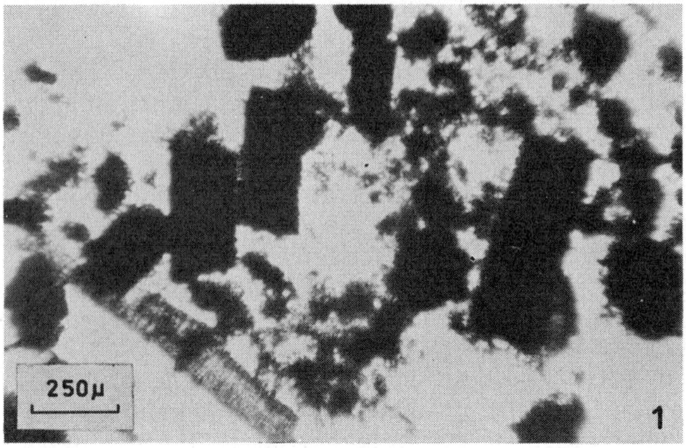
250µ
1

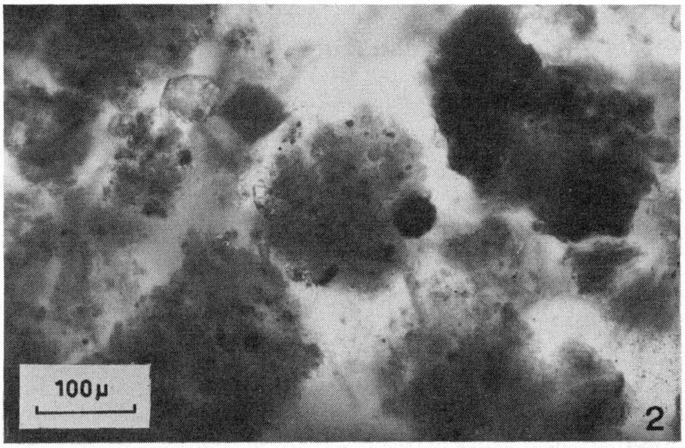
100µ
2

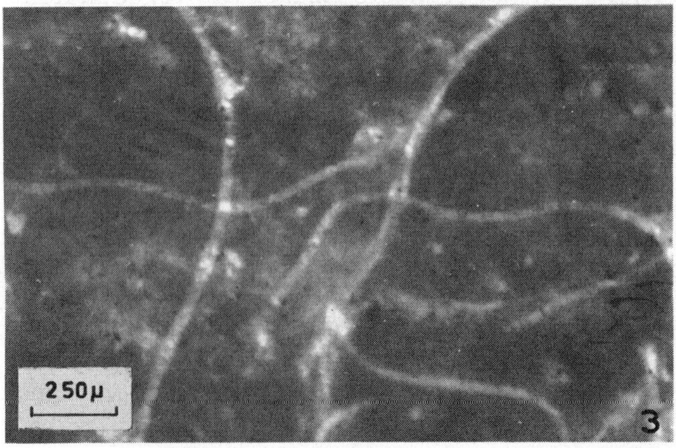
250µ
3

Ein gleiches Präparat wie in Abb.3, aber schwach gequetscht, zeigt Abb.4. Die Zerteilung der Kahmhaut in gebogenen, annähernd parallel laufenden Strängen ist eine für diesen Boden charakteristische Absonderungsform. Die Stränge sind aus distinkten, miteinander verklebten, stark lichtbrechenden, nur bei stärkster Vergrösserung erkennbaren kleinen Körpern (Bakterien) zusammengesetzt. Mehrmals findet man in der Achse der Stränge einen dünnen Blaualgenfaden.

Schwärzlich–grauer, flockiger Boden aus einem oligotrophen Teich mit relativ grosser Massenproduktion der Phanerogamen zeigt Abb.5. Sie ist schnell sedimentierend, und beim Aufwirbeln ergibt sich wenig diffuse Trübung. Unter dem Mikroskop erkennt man grosse, aus verbundenen Kleinpartikeln zusammengesetzte Aggregate sowie Diatomeen-Kolonien mit Schleimabsonderung. Die photographierte Diatomeen-Siedlung ist überdurchschnittlich gross. Zwischen gekreuzten Nicols zeigt sich, dass doppelbrechende Mineralien in Staubkörnung relativ stark vertreten sind. In den grossen Aggregaten ist Kalziumkarbonat erkennbar.

Ein Teil der Diatomeen-Kolonien, in stärkerer Vergrösserung gibt Abb.6. Am rechten Bildrand sieht man stark lichtbrechende granulierte Schleimanhäufung.

Tafel F4 zeigt gallertartigen–hyalinen Boden aus einem hoch-oligotrophen Quellteich mit sehr geringer Massenproduktion der höheren Pflanzen. Es gibt nur kleine Characeen in geringer Individuenzahl. Unter dem Mikroskop sieht man eine farblose Grundmasse des Bodens die stark durchsetzt ist mit roten und weissen Schwefelbakterien, gelbgrünen Chlorobakterien, blaugrünen Cyanophyceen und kleinen Grünalgen. Die Farben sind bereits makroskopisch zu erkennen. In Tafel F5 erkennt man auch noch grosse rötliche Kolonien von Rhodobakterien und ausserdem Diatomeen.

Ein gleicher Boden wie im Tafel F4 zeigt Tafel F5. Erkennbar sind eine Kolonie eines gelbgrünen, niedrigen Organismus (Chlorobakterium?), Diatomeen und eine Kalkkonkretion. Die Mikroflora ist hier aussergewöhnlich formen- und individuenreich. Ausser den in Tafeln F4 und F5 abgebildeten Organismen treten häufig auch fädige Algen und Bakterien auf.

Im Gegensatz zu dem anschliessend behandelten, ebenfalls dicht besiedelten Standort sind die Mikroorganismen des Bodens in Tafeln F4 und F5 sehr klein und zum grossen Teil unbeweglich. Tiere kommen in äusserst geringer Anzahl vor.

Ein stark abweichendes morphologisches Bild bietet der Boden, der sich in Buchten und sonstigen Stillwasserstrecken hocheutrophierter, getrübter Fliessgewässer ablagert. Seine unbelebte Grundmasse besteht aus grossen, mehrfach zusammengesetzten. gelappten Aggregaten, die untereinander durch kräftige, fädige Algen, vor allem *Oscillatoria limosa*,

---

Abb.1. Die drei typischen Komponenten eines lockeren gelagerten Bodens aus einem 30 cm tiefen, oligotrophen, kalt-stenothermen Quellabfluss.

Abb.2. Gerundete und zylindrische Körper mit erkennbarer Feinstruktur. Entnahmestelle wie in Abb.1. Die Probe kommt aus einer Tiefe von 15 cm.

Abb.3. Kahmhautartiger, aus Bakterien gebildeter, von Algenfäden durchzogener Oberboden in einer Baugrube. Kein Phanerogamenbewuchs.

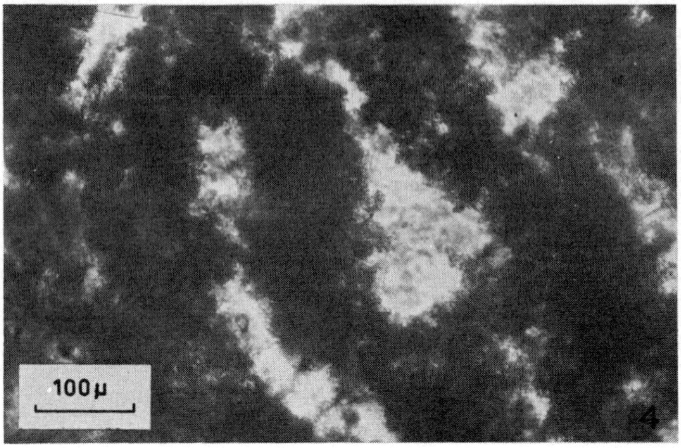
100 μ
4

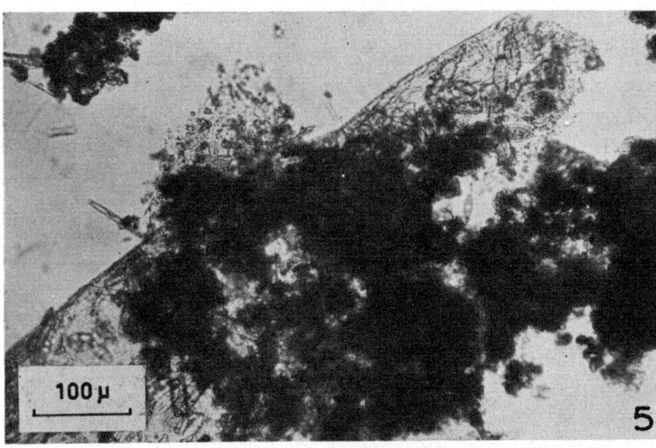
100 μ
5

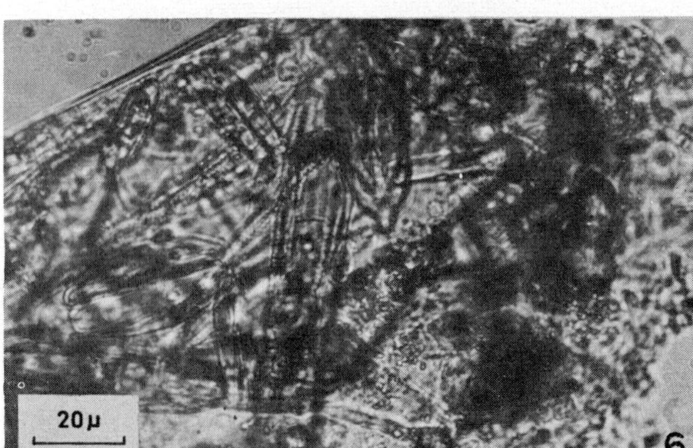
20 μ
6

Melosira varians, bandartige Diatomeen-Kolonien sowie durch mehrere aussergewöhnlich lange solitäre Diatomeen-Arten verbunden sind. Zugleich leben in diesem Boden grosse, vielfach stark bewegliche Tiere, insbesondere Thekamöben, schnellschwimmende grosse Protozoen, Tardigraden und Nauplien in hoher Individuenzahl. Auch die grossen solitären Diatomeen bewegen sich kräftig; oft schleppen sie erhebliche Mengen unbelebter Substanz mit sich herum. Beim Aufwirbeln bringt dieser Boden eine lang anhaltende diffuse Trübung zustande.

Zwischen gekreuzten Nicols (Tafel F6) zeigt sich als Grundmasse grau aufleuchtender kohlensaurer Kalk in feinflockiger Verteilung, dazu ein nicht unbeträchtlicher Anteil auslöschender organischer Substanz und eine Beimischung von $>$ 10% doppelbrechender Mineralfragmente, die hier im Gegensatz zu allen anderen behandelten Böden auch in grober Körnung auftreten.

UNTERWASSERBODEN IN AUFLICHTBETRACHTUNG UNTER DER BINOKULARLUPE

Ebenso wie in Glasröhren lässt sich Unterwasserboden in Küvetten aus plangeschliffenem Glas entnehmen. Der eingeschlossene, relativ grosse "Monolith" kann im Auflicht mit der Binokularlupe studiert und photographiert werden. Der Anwendungsbereich der direkten Durchmusterung nichtgetrockneten Bodens erweitert sich damit auf grosse Objekte und Hohlraumstrukturen. Zu prüfen bleibt allerdings, in welchem Ausmass die Hohlräume in unmittelbarer Nähe des Glases durch mechanische Beanspruchung während des Einstechens verändert werden. Vergleichsversuche mit versenkten Sedimentiergefässen dürften von Nutzen sein. Dass beim Ausstechen Veränderungen vor sich gehen, zeigt sich an der Verschleppung von Material aus höheren in tiefere Horizonte, die zwar selten grosses Ausmass annimmt, aber häufig in Gestalt gerader, senkrecht nach unten führender Striche zum Ausdruck kommt, die die Farbe des Oberbodens zeigen.

DISKUSSION DER ERGEBNISSE

Auf den Photos stehen sich mehrere morphologisch stark unterschiedene, charakteristische Aggregatformen gegenüber: isoliert gelagerte zylindrische oder gerundete Körper in Abb.1 und 2, dichte Bakterienhaut und ihre Absonderungsformen in Abb.3 und 4, zusammengesetzte Aggregate und Diatomeenkolonien in Abb.5 und 6, Ansammlungen vielfarbiger kleinster Mikroorganismen in einer farblosen organogenen Grundmasse neben vereinzelten

---

Abb.4. Gleicher Boden wie in Abb.3. Zerteilung in gewundenen, noch zusammenhängenden Strängen.

Abb.5. Grosses Bodenaggregat und Diatomeen-Kolonie im flockig gelagerten Oberboden stammend aus einem rund 60 cm tiefen kalt-stenothermen Quellteich. Bewuchs von *Hippuris vulgaris* und *Chara contraria*.

Abb.6. Teil der Diatomeen-Kolonie aus Abb.5, stark vergrössert, Schleimabscheidung deutlich erkennbar.

Kalkkonkretionen in Tafel F4 und Tafel F5, doppelbrechende Mineralfragmente diffuse Kalkausscheidungen und scharf begrenzte Partikeln organischer Herkunft in Tafel F6. Hinzu kommt der ohne Abbildung beschriebene, durch ein Girlandenwerk aus Fadenalgen und grossen Solitär-Diatomeen ausgezeichnete Boden eines verschmutzten Fliessgewässers. Die vorliegenden Ergebnisse stellen bei weiterer Anwendung der beschriebenen Methode sichere Einblicke in die Morphologie der Kleinformen des Unterwasserbodens in Aussicht. In den Fällen (vgl. Abb.3–5), in denen sich die Kleinaggregate zu grösseren, festgefügten Gebilden vereinigen, sind auch diese der Untersuchung zugänglich.

Durch Auszählen grösserer Proben von bekanntem Volumen, bei deren Entnahme auf die Möglichkeit der Schwerkraftsonderung Rücksicht genommen werden muss, dürfte sich ein ausreichend richtiges Bild des Mengenverhältnisses der verschiedenen, nebeneinander auftretenden Bodenkomponenten gewinnen lassen. Eine Vorstellung vermittelt Abb.1, die als Ausschnitt aus einem geeigneten Präparat gelten kann.

Nicht ohne weiteres erwartet werden darf die zuverlässige Aufklärung der Strukturen, die von isoliert gelagerten Teilchen gebildet werden. Das gilt vor allem dort, wo die Strömung, deren Kräfte in den benutzten Sammelgefässen unwirksam werden, bei der Ausbildung der Strukturen formend mitwirkt. Doch mag eine verfeinerte Sammeltechnik, die sich versenkter Objektträger oder flacher Sedimentierschalen bedient, noch weiterführen. Zu den wesentlichen natürlichen Eigenschaften des Unterwasserbodens, besonders in flachen Gewässern, gehört Aufwirbelung durch die Strömung oder durch Tiere (Schwarzwild, Enten) mit anschliessender neuer Sedimentation. Eine Sammelmethode, die sich diese normale Umlagerung nutzbar macht, indem sie sedimentierende Bodenteilchen in mikroskopierfähigen Behältern auffängt, hat Aussicht, den unverfälschten Strukturen nahezukommen.

Abschliessend sei auf einen prinzipiellen Vorteil eingegangen, den die Beobachtung des Unterwasserbodens im frischen Zustand bietet. Zahl und Formenreichtum der lebenden Organismen ist hier um vieles grösser als unter terrestrischen Bedingungen; niedere Pflanzen können sogar, wie es für den auf Tafel F4 und F5 dargestellten Boden zutrifft, die Hauptmasse der Substanz bilden. Die Präparation sollte daher Farben, Formen und Bewegung der Organismen möglichst wenig beeinträchtigen.

Beispiele, an denen die Vielseitigkeit im Bereich der Bodenorganismen deutlich wird, bieten die beiden oben beschriebenen stark belebten Böden (Tafel F4 und F5; Text S.365). Der erstgenannte, der einem tiefen, unbewegten, nicht verschmutzten, kalt-stenothermen Quellteich ohne oberirdischen Zufluss entstammt, produziert sehr geringe Mengen höherer Pflanzen. Auch niedere Tiere und Fische treten äusserst spärlich auf. Der Bewuchs des Gewässergrundes besteht überwiegend aus Characeen, hier und da liegen vereinzelte, nicht zum Blühen gelangende Sprosse von *Utricularia vulgaris* am Boden. Die Pflanzen ändern ihr Aussehen im gleichbleibenden Medium (Wassertemperatur im Winter +6° C – +8° C, im Sommer +12° C – +14° C) während des ganzen Jahres nicht merklich. Der Untergrund ist sauerstoffarm, wie sein starker Schwefelwasserstoffgeruch erweist. Im Boden dieses unbewegten, keiner nennenswerten Nährstoffzufuhr und keiner stimulierenden Periodizität unterliegenden, dem Gedeihen chlorophyllführender Pflanzen wenig günstigen Gewässers vegetiert eine formen-, farben- und individuenreiche Welt aus Bakterien und niederen Algen, deren Erscheinungsbild und Lebensweise stark vom Alltäglichen abweicht. Viele unter ihnen (vgl. dazu Lauterborn, 1916) gewinnen ihre Energie aus der Oxydation von Schwefelwasserstoff. Der geringen Ergiebigkeit dieser Energiequelle mag es zuzuschreiben sein, dass im Boden fast ausschliesslich kleine bis winzige, unbewegliche oder langsam bewegliche, systematisch isoliert stehende Organismen scharf geschieden von den Bedingungen der Oberwelt ihr Dasein fristen. Auch bei ihnen darf nach den bisherigen, noch lückenhaften Beobachtungen das Fehlen starker jahreszeitlicher Periodizität als wahrscheinlich angenommen werden, was ebenso wie bei den höheren Pflanzen im Einklang mit der Starrheit der physikalischen Bedingungen steht.

Einen grossen Kontrast zu diesen Zuständen bietet der Boden der hocheutrophen Fliessgewässer, in dem reichliche, vielseitig zusammengesetzte, aus einem grossen Einzugsgebiet zusammengeschwemmte Mineralsubstanz, stickstoffhaltige Sinkstoffe und im Gefolge des periodischen Wechsels der Wasserstände auch Sauerstoff zusammentreffen. In diesem Boden, in dem die normalen Energiequellen pflanzlicher und tierischer Organismen auf das stärkste fliessen und in dem der alljährliche Wechsel zwischen Austrocknung, mittlerem Wasserstand und reissendem Hochwasser keine Fixierung eines bestimmten ökologischen Zustandes zulässt, leben grosse, schnellwüchsige und kräftiger Bewegung fähige Algen und Tiere, deren Massenentfaltung, wie von *Oscillatoria limosa*, *Melosira varians* und den *Nauplius*-Larven sicher bekannt ist, im Laufe der Jahreszeiten stark schwankt.

Aus diesen Andeutungen folgt zunächst, dass die mikromorphologische Untersuchung des Unterwasserbodens der Zusammenarbeit mit der Hydrobiologie bedarf. Neues wird damit allerdings nicht ausgesprochen, denn der Bodenkunde ist die dynamische Betrachtungsweise, die sich den vielseitigen funktionellen Verflechtungen ihres Gegenstandes zuwendet, seit langem vertraut. Im Zusammenhang der vorliegenden Erörterungen soll es mehr darauf ankommen, die besonders starke Beteiligung lebender Organismen am Aufbau des Unterwasserbodens in Erinnerung zu bringen und daraus zu folgern, dass eine Arbeitsmethode, die diesen Boden im lebenden Zustand untersucht, einer wesentlichen Eigenschaft ihres Gegenstandes gerecht wird.

SUMMARY

A micromorphological investigation technique of non-dried sub-aqueous soils

The difficulty of obtaining thin sections of water saturated soil without first drying it has necessitated microscopic examinations of untreated, subaqueous soils. The main point for success in this work consists in collecting undisturbed samples of the soft, unstable material. For this purpose, the author cut out the soil by means of open glass tubes of 2–5 cm diameter. These are subsequently closed by stoppers of india rubber before lifting over the surface of the water. The soil can be collected in this way, and also transported, without disturbance. Small samples for microscopic examination can be taken out of the tubes by narrow glass pipes.

This method gives a survey of the morphology and relative quantity of small and medium-sized aggregates which are found in the preparations without damaging them. The micro-organisms of the soil are also available for investigation in their living state. However, no just idea of the soil cavity is given by this method. It seems probable that by laying out small glass plates or cups on the bottom of the water, samples of recently sedimented soil which show natural structure and are suitable for microscopic investigation can be collected.

The author emphasizes that the possibility of observing plants and animals of the soil in their living state is a significant advantage of the method. Nevertheless, it is desirable to find methods for hardening wet soil for thin section technique.

REFERENCES

Haarløv, N. and Weis-Fogh, T., 1952. A microscopical technique for studying the undisturbed texture of soils. Oikos, 4 (1952) : 44–57.

Lauterborn, R., 1916. Die sapropelische Lebewelt. Verhandl. Naturhist.–Med. Ver. Heidelberg, 13 : 395–481.

FARBENBILDER

Tafel F4: Farblos-gallertartiger Oberboden aus einem rund 50 cm tiefen kalt-stenothermen Quellteich mit Bewuchs von *Chara aspera*. Rötlichbraune Rhodobakterien und Diatomeen.

Tafel F5: Boden wie im Tafel F4. Eine Kolonie gelbgrüner Organismen (niedere Algen oder Chlorobakterien), Diatomeen und eine Kalkkonkretion sind erkennbar.

Tafel F6: Unterboden aus einem stark eutrophierten, getrübten Fliessgewässer aus einer Bucht mit schwacher Strömung. Zwischen gekreuzten Nicols sind ein hoher Anteil doppelbrechender Mineralien in grober Körnung eine Grundmasse aus feingeflockten Kalkabscheidungen und distinkt geformte Körper mit Auslöschung erkennbar.

DIE ANWENDUNG DES PHASENKONTRASTVERFAHRENS BEI DER UNTERSUCHUNG VON BODENDÜNNSCHLIFFEN

H.-J. ALTEMÜLLER

Institut für Bodenbearbeitung der Forschungsanstalt für Landwirtschaft, Braunschweig-Völkenrode (Deutschland)[1]

EINLEITUNG

Das Phasenkontrastverfahren ist heute wohl in den meisten Anwendungsgebieten der Mikroskopie im Gebrauch und hat an ihren Untersuchungsmethoden wesentlichen Anteil. Wie man aus einer grossen Zahl von Publikationen ersehen kann, wurden die Vorteile des Verfahrens vor allem in den biologischen Wissenschaften rasch erkannt und vielfältig genutzt. Auch in der Mineralogie und Petrographie wurden schon frühzeitig Untersuchungen mit dem Phasenkontrastverfahren durchgeführt, doch ist die bis heute vorliegende Literatur nicht sehr umfangreich. Es bestand hier vielleicht kein so allgemeines Bedürfnis nach neuen mikroskopischen Verfahren, denn für die Untersuchung der Gesteine und grobkörnigen Minerale bietet das Polarisationsmikroskop hervorragende Möglichkeiten und für die Untersuchung der Tone wurden andere Methoden und Geräte entwickelt, die zum Teil auch ganz vom Mikroskop abgelenkt haben.

Grundlegende Arbeiten über die Anwendung des Phasenkontrastverfahrens zur Mineraluntersuchung verdanken wir vor allem Smithson (1946), Piller (1952), Thaer (1954), Correns und Piller (1955), Schmidt (1955), Juda und Medenbach (1959). Sie zeigen uns, dass die Mineralbestimmung an Körnern von etwa 10– $< 1\ \mu$ Aequivalentdurchmesser mit Hilfe der Phasenkontrasteinrichtung wesentlich erleichtert wird. Die polarisationsoptischen Bestimmungsmethoden werden mit zunehmender Kleinheit der Objekte unsicher und sind auch nicht mehr in vollem Umfange anwendbar. Zur Identifizierung der Minerale dienen deshalb in diesem Grössenbereich vor allem die Eigenschaften der Lichtbrechung. Diese können im Phasenkontrast mit sehr grosser Genauigkeit in bequemer Weise bestimmt werden. Bei der Untersuchung von Sedimenten und in der Staubforschung hat das Phasenkontrastverfahren deshalb einen festen Platz gewonnen. In neueren Arbeiten wurden die Methoden auch zur Untersuchung der Feinschluff- und Grobtonfraktionen in Erdböden angewandt (Kalk, 1959; Meyer und Kalk, 1959; Smithson, 1961). Damit wurden Anregungen gegeben, die lichtmikroskopische Mineralbestimmung in der Bodenkunde wieder mehr zu beachten.

Um die Minerale jedoch genau und nach Grössenklassen getrennt voneinander untersuchen zu können, ist eine sorgfältige Dispergierung und Fraktionierung des Bodenmaterials unumgänglich notwendig. Verunreinigungen mit groben Körnern oder ungenügende Zerteilung würden die Untersuchung

[1] Direktor: Prof. Dr. H. Frese.

erschweren oder sogar unmöglich machen. Am gut vorbereiteten Präparat hat man sodann die Möglichkeit, durch treffende Auswahl des Immersionsmittels die für jede interessierende Mineralart günstigsten Beobachtungsbedingungen herzustellen. Auf diese Weise lassen sich quantitative Ergebnisse über die Mineralzusammensetzung erarbeiten. Um den Wechsel von Immersionsflüssigkeiten einzuschränken oder ganz zu vermeiden, kann man durch Variation der Temperatur und (oder) der Wellenlänge des beleuchtenden Lichtes die Brechungseigenschaften des Immersionsmediums in verhältnismässig weiten Grenzen verändern (Juda und Medenbach, 1959).

Über die quantitative Bestimmung und Identifizierung der Gemengteile hinaus besteht in der Bodenkunde grösstes Interesse, die Minerale in ihrer gegenseitigen Anordnung im natürlich gewachsenen Gefüge untersuchen zu können. Die mikromorphologischen Untersuchungen an Dünnschliffen haben mit aller Deutlichkeit gezeigt, in welch grossem Masse das Bodengefüge Abbild einer gesetzmässigen Entwicklung ist, die jedes Einzelteil in charakteristischer Weise mit einbezieht.

Aus der Arbeit an Bodendünnschliffen ergab sich somit der Anstoss, einmal zu überprüfen, inwieweit das Phasenkontrastverfahren auch an Dünnschliffen anwendbar ist, um die Möglichkeiten des Verfahrens bei der Untersuchung von ungestörten Gefügen kennenzulernen und dem Problem der Mineralbestimmung im Dünnschliff nachzugehen. Mit diesen Fragen soll sich die folgende Darstellung befassen.

ALLGEMEINE GRUNDLAGEN

Das Phasenkontrastverfahren bietet die Möglichkeit, farblose Objekte, die im normalen Hellfeldmikroskop oft nur unbefriedigend sichtbar werden, unter bestimmten Bedingungen kontrastreich abzubilden. Über die theoretischen Grundlagen des Verfahrens und den technischen Aufbau eines Phasenkontrastmikroskops unterrichten zahlreiche Publikationen, von denen einige schon eingangs erwähnt wurden. Hier sei nur das Wichtigste in kurzen Zügen wiedergegeben.

(1) Im gewöhnlichen Durchlichtmikroskop treffen die von der Lichtquelle ausgehenden und durch den Kondensor gerichteten Lichtwellen in das zu untersuchende Präparat und erfahren dort, je nach der Beschaffenheit der einzelnen Anteile, eine vielfältige Beeinflussung durch Absorption, Reflexion, Refraktion, Beugung und Phasenveränderung. Soweit es sich um farblose Objekte handelt, deren optische Dichte, ausgedrückt durch die Brechzahl (Brechungsindex), nur wenig von der Brechzahl des Einbettungsmediums abweicht und die ausserdem möglichst dünn sind (Grössen bis einige µ), können die Absorptions-, Reflexions- und Refraktionsphänomene ohne wesentlichen Nachteil ausser acht gelassen werden. Es bleiben also Beugung und Phasenveränderung für die weitere Darstellung zu berücksichtigen.

(2) Beugung entsteht im Präparat dort, wo Bereiche verschiedener Brechzahl aneinandergrenzen, und äussert sich in einer vom geometrischen Strahlengang der beleuchtenden Lichtwellen unabhängigen, allseitigen Ausbreitung von Lichtwellen, die man als im Präparat entstandene Elementarwellen auffassen kann. Wir bezeichnen sie im folgenden kurz als Beugungswellen und unterscheiden sie von denjenigen Lichtwellen, die ohne Beugung

durch das Präparat gelangt sind.

(3) In Abhängigkeit von den Eigenschaften der Linsensysteme (Kondensor und Objektiv), wirken die gebeugten und die nicht gebeugten Wellen hinter dem Präparat aufeinander ein und lassen in der hinteren Brennebene des Objektives und in der Bildebene des Mikroskopes durch Interferenz charakteristische Bilder entstehen. Das erste ist ein verändertes Abbild der Lichtquelle und das zweite das vergrösserte Abbild des Präparates.

(4) Das Bild in der hinteren Brennebene des Objektives sieht man beim Einblick in den Tubus nach Herausnahme des Okulars. Es wird von der scharf abgebildeten Aperturblende des Kondensors umgrenzt. Das Präparat ist nicht sichtbar, doch sind die Beugungswellen an diesem Bild voll beteiligt und wirken unter bestimmten Voraussetzungen auch erkennbar ein, so dass daraus wertvolle Hinweise auf den Bau eines Objektes gewonnen werden können. Die sogenannten Achsenbilder der Minerale sind dafür bekannte Beispiele.

(5) Zieht man die Aperturblende im Kondensor zu, so verkleinert sich das Bild der Lichtquelle in der hinteren Objektivbrennebene. Das gebeugte Licht folgt dieser Änderung jedoch nicht, sondern bleibt rings um die Blendenöffnung - bei unregelmässig gebauten Objekten, z.B. als diffuser Lichthof - über die hintere Brennebene verteilt. Gebeugtes und ungebeugtes Licht werden durch diesen Vorgang voneinander getrennt.

Dies ist eine wichtige Voraussetzung für das Phasenkontrastverfahren, die in den speziellen Kondensoren durch eine ringförmige Blendenöffnung erfüllt wird. In der hinteren Brennebene des Objektives wird dadurch ein schmaler Lichtring abgebildet.

(6) Um die Bildentstehung als Interferenzvorgang zu verstehen, ist es erforderlich, auch den Schwingungszustand der Lichtwellen zu beachten. Wir lassen hierzu die Beugungserscheinungen vorerst wieder beiseite und stellen fest, dass in einem Präparat au, Anteilen ungleicher Brechzahl und Dicke das Licht in verschiedenem Masse verzögert wird. Stellt man sich - um diesen Sachverhalt anschaulich zu machen - zwei Wellenzüge nach Art der Sinuskurven vor, von denen eine im Präparat durch ein Mineralkorn und die andere durch das Einbettungsmedium gegangen sei, dessen Brechzahl etwas geringer als die des Mineralkornes angenommen wird, so wird die erste Welle stärker verzögert als die zweite und erscheint hinter dem Objekt gegen die letztere um einen gewissen Betrag verschoben. Beide Wellen weisen dann zum jeweils gleichen Zeitpunkt verschiedene Schwingungsphasen auf.

(7) Um die Phase einer Schwingung näher zu bezeichnen, transponiert man die entsprechende Sinuskurve in eine Kreisfigur, worin ein Zeiger als Vektor eingesetzt wird, dessen Länge die maximale Schwingungsweite und dessen Richtung die Phase für eine bestimmte Zeit zum Ausdruck bringt. Aus den Vektoren zweier Schwingungen mit verschiedener Phase im gleichen Zeitpunkt ergibt sich dann sinnfällig ein Winkel, den man im Bogen- oder Winkelmass angibt. Der so gekennzeichnete Unterschied zweier Schwingungsphasen wird kurz als Phasendifferenz bezeichnet.

(8) Die aus einem Einzelobjekt (z.B. Mineralkorn) und dem umgebenden Einbettungsmedium resultierende Phasendifferenz φ kann bei bekannter Brechzahl und Dicke der Anteile als konstanter Zahlenwert berechnet werden. Dies wird später noch näher erläutert. Für die Kontrastbeurteilung eines mikroskopischen Bildes ist dieser Wert von Bedeutung. Man muss allerdings wissen, dass der Bildkontrast nicht unmittelbar aus dieser Phasendifferenz

Objekt-Medium hervorgeht. Es liegt zwar in der Absicht des Phasenkontrastverfahrens, Phasendifferenzen in Helligkeitsunterschiede umzuwandeln, doch wird dies nur insoweit erreicht, als die durch das Objekt gehenden Lichtwellen auch Beugungswellen entstehen lassen. Dies trifft aber nur bei sehr kleinen Teilchen im ganzen Grössenbereich zu. Somit ist für die Kontrastentstehung in der Bildebene unmittelbar diejenige Phasendifferenz massgebend, die zwischen Beugungswellen und nicht gebeugten Lichtwellen entsteht; sie wird mit Ψ bezeichnet. Ihre Abhängigkeit von der Phasendifferenz φ (Objekt-Medium) ergibt sich aus der Formel:

$$\Psi = \frac{\varphi}{2} \pm 90^\circ$$

Bei kleineren Brechzahl- oder Dickenunterschieden im Präparat sind die Werte für φ ebenfalls klein. Die Phase der Beugungswellen ist dann entsprechend der Formel um etwa 90° gegen das nicht gebeugte Licht verändert. Aus der Interferenz von Wellenzügen, die etwa um 90° gegeneinander verschoben sind, können sich jedoch, wie man leicht graphisch ermittelt, keine grossen Intensitätsmaxima oder -minima ergeben. Der Bildkontrast bleibt also gering, wie es im gewöhnlichen Hellfeld der Fall ist. Eine Steigerung des Kontrastes ist aber möglich, wenn die Schwingungsphasen gleichsinnig (Phasendifferenz 0°) oder entgegengesetzt (Phasendifferenz 180°) laufen.

(9) Im Phasenkontrastmikroskop werden die Beugungswellen unabhängig von den nicht gebeugten Lichtwellen durch eine in der hinteren Objektivbrennebene befindliche optische Platte zusätzlich um etwa 90° in ihrer Phase verschoben und dadurch die erwünschten Interferenzbedingungen hergestellt. Optimal wird der Kontrast jedoch erst dann, wenn auch die Intensitäten, d.h., die Amplituden, der Schwingungen aufeinander abgestimmt sind. Da die Intensität der nicht gebeugten Lichtwellen bei kleinen Phasendifferenzen Objekt-Medium stets höher ist als die Intensität der Beugungswellen, muss sie durch Absorption geschwächt werden. Man erreicht dies durch eine ringförmige Aufdampfschicht in der hinteren Brennebene, welche den dort abgebildeten Lichtring (vgl. Punkt 5) des ungebeugten Lichtes genau überdeckt.

(10) Der Phasenkontrast ist positiv, wenn Objekte mit grösserer Brechzahl als das Einbettungsmedium dunkler als das Umfeld, und negativ, wenn dieselben heller als das Umfeld erscheinen. Alle folgenden Ausführungen beziehen sich auf positiven Phasenkontrast.

PHASENDIFFERENZ UND BILDKONTRAST

Für die praktische Arbeit ist es nützlich, die Entstehung und Veränderung des Bildkontrastes in Abhängigkeit von den im Präparat gegebenen Phasendifferenzen einer weiteren Betrachtung zu unterziehen. Aus den bei der Interferenz entstehenden Helligkeitsmaxima und -minima kann der Kontrast K berechnet und zahlenmässig ausgedrückt werden. Im einzelnen geben die Darstellungen von Menzel (1957) und Michel (1957) darüber weiteren Aufschluss. Aus dieser Rechnung entsteht eine Zahlenfolge von $K = 0$, ohne jeden Kontrast, bis $K = 1$, dem Ausdruck für höchstmöglichen Kontrast. Die Abhängigkeit von den Phasendifferenzen φ (Objekt-Medium) lässt sich in Form einer Kurve darstellen und man erhält ein anschauliches, wenn auch stark vereinfachtes Bild über die Zusammenhänge zwischen den genannten Grössen.

In Anlehnung an Michel (1957) sind in Abb.1 zwei Kurven aufgezeichnet. Wie man sieht, steigt der Kontrast im Hellfeld langsam von $K = 0$ an und erreicht etwa bei einer Phasendifferenz $\varphi = 120°$ ein Maximum. Ausreichende Beobachtungsbedingungen sind aber bereits bei einem Kontrast von etwa $K = 0{,}6$, d.h., bei einer Phasendifferenz $\varphi = 50°$ gegeben. Darunter wird das Bild unbefriedigend. Im Phasenkontrast steigt die Kontrastkurve wesentlich steiler an und erreicht in Abb.1 schon bei $\varphi = 20°$ ihr Maximum. Gute Beobachtungsbedingungen sind hier schon von sehr kleinen Phasendifferenzen ($\varphi < 10°$) an gegeben und machen den grossen Vorteil des Phasenkontrastverfahrens deutlich. Der Kontrast fällt allerdings hinter dem Maximum wieder rasch ab und schneidet die Hellfeldkurve bei $\varphi = 60°$, d.h., der Kontrast ist dabei etwa gleich gross wie im Hellfeld. Er nimmt dann bis $\varphi = 180°$ weiter ab und geht in den negativen Bereich über. Dann kommt es zur Kontrastumkehr (Inversion) und die vordem dunkler abgebildeten Teilchen erscheinen nun heller als ihr Umfeld. Ausserdem kommen weitere Interferenzerscheinungen hinzu, so dass die Kontrastbeurteilung erschwert oder auch unmöglich wird.

Der Verlauf einer Phasenkontrastkurve ist allerdings nicht allgemein festgelegt, sondern ist von der Stärke der Phasenveränderung im Objektiv und, wie Gabler (1955) gezeigt hat, auch von der Stärke des Absorptionsringes abhängig. Der Hersteller hat es also in der Hand, dem Phasenkontrastmikroskop verschiedene Eigenschaften zu verleihen. Die zusätzliche Phasenänderung beträgt heute wohl allgemein $\pm 90°$. Dagegen sind die Absorptionsringe verschieden; 64%, 75% und 85% Absorption sind gebräuchliche Werte.

Die Vor- und Nachteile sind kurz folgende:

(1) Geringe Absorption: Bei kleinen Phasendifferenzen langsamer Kontrastanstieg, aber verhältnismässig spät folgende Kontrastumkehr. Wirksamer Phasenkontrast ist also in einem breiten Bereich gegeben, dafür ist der Kontrastgewinn nicht sehr gross.

(2) Hohe Absorption: Bei kleinen Phasendifferenz steiler Kontrastanstieg, also hoher Kontrastgewinn; dafür aber früher folgende Kontrastumkehr. Wirksamer Phasenkontrast bleibt auf einen schmalen Bereich beschränkt.

Nach Michel (1957) ist das Phasenkontrastverfahren nur dann von praktischem Wert, wenn die in den zu untersuchenden Präparaten auftretenden Phasendifferenzen so gering sind, dass der Kontrast im Hellfeld unbefriedigend bleibt; strenggenommen also zwischen etwa 5–60° Phasendifferenz. Dies trifft ohne weiteres zu, solange es lediglich darauf ankommt, die Objekte in bestem Kontrast abzubilden. Bei Mineraluntersuchungen im Körnerpräparat oder Dünnschliff sind aber Angaben über die relative Lichtbrechung oft wichtiger als die bestmögliche Abbildung. Es wird also die Frage gestellt, ob, z.B., ein bestimmtes Korn höhere, gleiche oder geringere Lichtbrechung aufweist als das Einbettungsmedium. Solche Feststellungen lassen sich auch ausserhalb der günstigsten Kontrastbedingungen machen. Andererseits kommt es bei der genauen Bestimmung von Brechzahlen auch darauf an, durch Auswahl geeigneter Immersionsmittel den Kontrast $K = 0$ herzustellen, um die Gewähr für völlige Brechzahlgleichheit zwischen Korn und Medium zu haben.

Die hierbei oft entstehenden Dispersionseffekte werden später noch erwähnt.

Letzten Endes ist aber in allen Fällen eine gute Anpassung zwischen den Brechzahlen der Einzelobjekte und dem Einbettungsmedium anzustreben. Dies kann jedoch in Objekten, die aus verschiedenartigen Teilen zusammengesetzt sind, immer nur für gewisse Anteile erreicht werden. Besonders in Dünnschliffen, wo man in der Wahl der Einbettungsmittel noch starken Beschränkungen unterliegt, können sich dabei Schwierigkeiten ergeben.

Wir kommen damit wieder zum engeren Fragenkreis und wollen untersuchen, inwieweit man die Kontrastbedingungen für ein Dünnschliffpräparat im voraus angeben kann. Hierzu müssen die für ein Schliffpräparat gültigen Phasendifferenzen nach folgender Formel berechnet werden:

$$\varphi = d\ (n_M - n_O)\ \frac{360}{\lambda}$$

Darin: φ = die gesuchte Phasendifferenz im Winkelmass; d = die Dicke des Objektes, ausgedrückt in derselben Einheit wie λ; n_M = die Brechzahl des Einbettungsmittels; n_O = die Brechzahl des Objektes; λ = die Wellenlänge des beleuchtenden Lichtes.

Bei Untersuchungen im gewöhnlichen Licht wird die mittlere Wellenlänge für "weisses" Licht λ = 550 mμ eingesetzt. Das Formelglied 360/λ erhält dann den Wert 0,65. Die Brechzahl des Einbettungsmittels Vestopal-H kann mit n = 1,56 ebenso als fester Wert eingesetzt werden. In Abhängigkeit von den Verarbeitungsbedingungen variiert dieser Wert zwar, doch kann ausser acht bleiben, solange es sich nur um eine allgemeine Information handelt.

Für beliebige Minerale lassen sich nun unter Berücksichtigung verschiedener Dicken die gesuchten Phasendifferenzen berechnen. Bei doppeltbrechenden Mineralen müssen dabei jeweils die Hauptbrechzahlen berücksichtigt werden und es ergeben sich bei einachsigen Mineralen zwei und bei zweiachsigen Mineralen drei entsprechende Phasendifferenzen. In Tabelle I sind einige Beispiele gegeben. Die Auswahl ist willkürlich und vor allem nach solchen Brechzahlen zusammengestellt, die sich kontinuierlich anschliessen, um eine übersichtliche Anordnung zu ermöglichen. Es sind aber auch Minerale enthalten, die häufig gemeinsam vorkommen, wie Orthoklas, Albit, Quarz, Kaolinit und Muscovit. Die Brechzahlen wurden dem Tabellenwerk von Tröger (1959) entnommen.

Man sieht nun, dass die Brechzahlen der Minerale Orthoklas, Albit und Quarz niedriger, diejenigen von Anorthit grösser sind als n = 1,56; bei Kaolinit teils niedriger, teils gleich; bei Muscovit teils niedriger, teils höher. Calcit, dem eine sehr starke Doppelbrechung eigen ist, zeigt einen niedrigen und einen hohen Wert; beide sind die Extreme auf der Tabelle. Die Brechzahldifferenzen $(n_M - n_O)$ sind teilweise positiv, teilweise negativ. Bei Kaolinit entsteht ein Wert ca. Nul.

In den folgenden Spalten sind die Phasendifferenzen im Winkelmass für Dicken von 1–20 μ aufgeführt. Die Phasenwinkel nehmen mit ansteigender Dicke um so rascher zu, je grösser die Differenz $(n_M - n_O)$ ist. Vor allem bei

TABELLE I

Beziehungen zwischen Brechzahl und Phasendifferenz für einige bekannte Mineralen

Mineralart	Brechzahl (n) (λ = 550 mμ)	Brechzahldifferenz: $n_M - n_O$ (n_M = 1,560)	Phasendifferenz φ im Winkelmass bei verschiedener Dicke							
			1 μ	2 μ	5 μ	8 μ	10 μ	12 μ	15 μ	20 μ
Calcit	$\varepsilon = \alpha$ 1,486	+ 0,074	48	97	242	387	484	581	726	968
Orthoklas	α 1,518	+ 0,042	27	55	137	220	275	329	412	549
	β 1,522	+ 0,038	25	50	124	199	248	298	373	497
	γ 1,524	+ 0,036	23	47	118	188	235	282	353	471
Albit	α 1,527	+ 0,033	22	43	108	173	216	259	324	432
	β 1,531	+ 0,029	19	38	95	152	190	227	284	379
	γ 1,538	+ 0,022	14	29	72	115	144	173	216	288
Quarz	$\omega = \alpha$ 1,544	+ 0,016	10	21	52	84	105	125	157	209
	$\varepsilon = \gamma$ 1,553	+ 0,007	5	9	23	37	46	55	69	91
Kaolinit	α 1,553	+ 0,007	5	9	23	37	46	55	69	91
	β 1,559	+ 0,001	1	1	3	5	6	8	10	13
	γ 1,560	$\pm$ 0,000	0	-	-	-	-	-	-	-
Muscovit	α 1,552	+ 0,008	5	10	26	42	52	63	78	105
	β 1,582	- 0,022	14	29	72	115	144	173	216	288
	γ 1,588	- 0,028	18	37	92	146	183	220	275	366
Anorthit	α 1,577	- 0,017	11	22	56	89	111	133	167	222
	β 1,585	- 0,025	16	33	82	131	163	196	245	327
	γ 1,590	- 0,030	20	39	98	157	196	235	294	392
Calcit	$\omega = \gamma$ 1,658	- 0,098	64	128	320	513	641	769	961	1282

Calcit werden rasch sehr hohe Werte erreicht, während sie bei Quarz unter denselben Bedingungen nur langsam ansteigen.

Nach der Kontrastkurve in Abb.1 ist maximaler Kontrast im Bereich

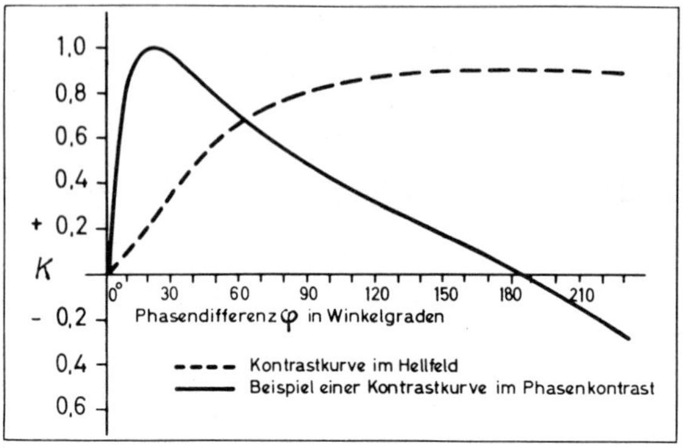

Abb.1. Beziehungen zwischen der Phasendifferenz φ und dem Bildkontrast *K*; dargestellt für Hellfeld und Phasenkontrast (nach Michel, 1957).

von $\varphi = 10°$ bis etwa $60°$ gegeben. Die Mineralkörner erscheinen dann deutlich heller oder dunkler als ihr Umfeld, je nachdem, ob die Brechzahldifferenz $(n_M - n_O)$ positive oder negative Werte ergibt. Dieser Kontrast ist bei einer Schliffdicke von 20 µ praktisch in keinem Falle gegeben. Auch wenn wir den Bereich ausdehnen und suchen, welche Minerale die Bedingungen des Phasenkontrastes wenigstens unterhalb $\varphi = 180°$, dem angenommenen Inversionspunkt, erfüllen, ist die Auswahl sehr gering. Unter bestimmten Umständen könnten allenfalls einzelne Quarzkörner und Muscovit noch einigermassen brauchbare Bilder entstehen lassen, doch wäre dies allein ohne Sinn. Kaolinitminerale von 20 µ Dicke sind nur sehr selten zu finden. Bei normaler Dünnschliffdicke sind also kaum irgendwelche Vorteile mit dem Phasenkontrast zu erwarten. Wer sich trotzdem die Mühe macht, solche Schliffe zu betrachten, findet unübersichtliche Bilder, die von geringem Wert sind.

HERSTELLUNG UND MESSUNG SEHR DÜNNER SCHLIFFPRÄPARATE

Die praktische Erfahrung hat gezeigt, dass man bei sorgfältiger Arbeit und unter Beachtung gewisser Regeln auch wesentlich dünnere Schliffe bis unter 10 µ Dicke herstellen kann. Die wichtigste Voraussetzung ist allerdings eine einwandfreie Durchtränkung und Härtung der Proben, und es empfiehlt sich, diesen Vorarbeiten besondere Aufmerksamkeit zu widmen. Die einzelnen Mineralkörnchen müssen in möglichst vollkommener Weise in der Probe fixiert sein, denn jedes aus dem Schliff ausgerissene Korn kann erheblichen Schaden im Schliffbild anrichten, sofern es gröber als das verwendete Schleifmittel ist.

Für den Feinschliff haben sich als Trägerplatten solche aus Plexiglas sehr bewährt (Altemüller, 1962). Sie sind weicher als Glas und nehmen das Schleifmittel besser an als dieses. Leider kann man auf den glatten

Oberflächen solcher Platten nicht beliebig feine Schleifkörnungen verwenden. Pulver von 10–15 μ Korngrösse erlauben noch einwandfreies Schleifen. Bei Körnungen um 7 μ ist die Adhäsion der Proben bisweilen schon zu stark und es entstehen vermehrt Kratzer. Diese und feinere Körnungen arbeiten erst wieder auf Dünnem Poliertuch einwandfrei. Da aber auch hartes Poliertuch in jedem Fall weicher ist als die glatte Plexiglasplatte, schleift sich entsprechend den Härteunterschieden in der Probe ein mehr oder weniger ausgeprägtes Relief heraus. Die besten Ergebnisse werden deshalb mit sehr harten und schneidfreudigen Schleifmitteln erzielt. Korund ist nur für die Vorarbeiten geeignet, solange auf glatter Trägerplatte gearbeitet wird. Sehr zu empfehlen sind vor allem Diamant-Schleifpasten, die bei empfindlichen Proben mit 15 μ Korngrösse auf Platten und mit 3 μ Korngrösse auf hartem Tuch angewandt werden. Mit diesen Schleifmitteln gelingen ausgezeichnete Präparate. Nur in wenigen Fällen sind auch andere Körnungen, z.B. 7 μ (2. Feinschliff) und 1 μ (2. Politur), noch anzuwenden.

Eine maximale Schliffdicke von etwa 10 μ wird angestrebt, wobei die Randzonen dann zwangsläufig noch dünner werden und die wichtigsten Objektzonen zur Untersuchung darstellen. Bei etwa 6 μ (bei gröberer Schleiftechnik auch früher) brechen die Skelettkörnchen nach und nach aus. Wenn dieses Stadium erreicht ist, hat es keinen Sinn, weiter zu schleifen, da die ausgerissenen Körnchen den raschen Zusammenbruch des Präparates herbeiführen. Der Erhaltungszustand der Objektanteile in den Randzonen bestimmt also praktisch das ganze Präparat. Daraus folgt wiederum die Forderung nach möglichst ebenen Schliffflächen.

Zur Messung der Dicke sind zwei Methoden erforderlich. Am genauesten ist die Dickenmessung auf Grund der Bestimmung von Gangunterschieden bekannter Minerale, vor allem Quarz, die mit Kompensatoren ermittelt werden. Damit aber tatsächlich der maximale Gangunterschied erfasst wird, ist es nötig, solche Körner zu suchen, deren Schnittlage genau parallel zur optischen Achse getroffen wurde. Kontrolle des Achsenbildes ist dabei nötig. Die Skelettkörner sind allerdings fast immer dicker als die feineren Anteile. Nur gelegentlich, so z.B. an regulär aufgewachsenen Kaolinit-Aggregaten, sind ebenfalls aus den Gangunterschieden Dicken zu ermitteln. Man muss aber berücksichtigen, dass die Doppelbrechung dabei nicht immer konstant ist. Auf Grund solcher Messungen wurden Unterschiede von etwa 8 μ der Skelettkörner gegenüber etwa 4 μ der tonigen Anteile gefunden. Häufig sind die letzteren auch bis 1 oder 2 μ Schliffdicke noch zu verfolgen.

Als zweite Messmethode kommt die Tiefenmessung mit starkem Immersionsobjektiv und Ablesung an der Mikrometerschraube hinzu. Dabei folgt man den einschlägigen Vorschriften der Lehrbücher.

Ein Blick auf die Tabelle I zeigt, dass bei 8 μ Dicke bereits für einen weiten Bereich von Brechzahlen bzw. Phasendifferenzen einigermassen brauchbare Kontrastbedingungen gegeben sind. Dabei kommt noch der Umstand zu Hilfe, dass bei den feinkörnigen Anteilen durchaus nicht immer die Schliffdicke für den Kontrast massgebend ist, sondern die Korngrösse. So können einzelne Partikel auch bei relativ grosser Brechzahldifferenz noch in gutem Kontrast erscheinen, wenn sie in genügend feinen Korngrössen vorliegen. Die praktischen Ergebnisse müssen zeigen, was man von Fall zu Fall erreichen kann.

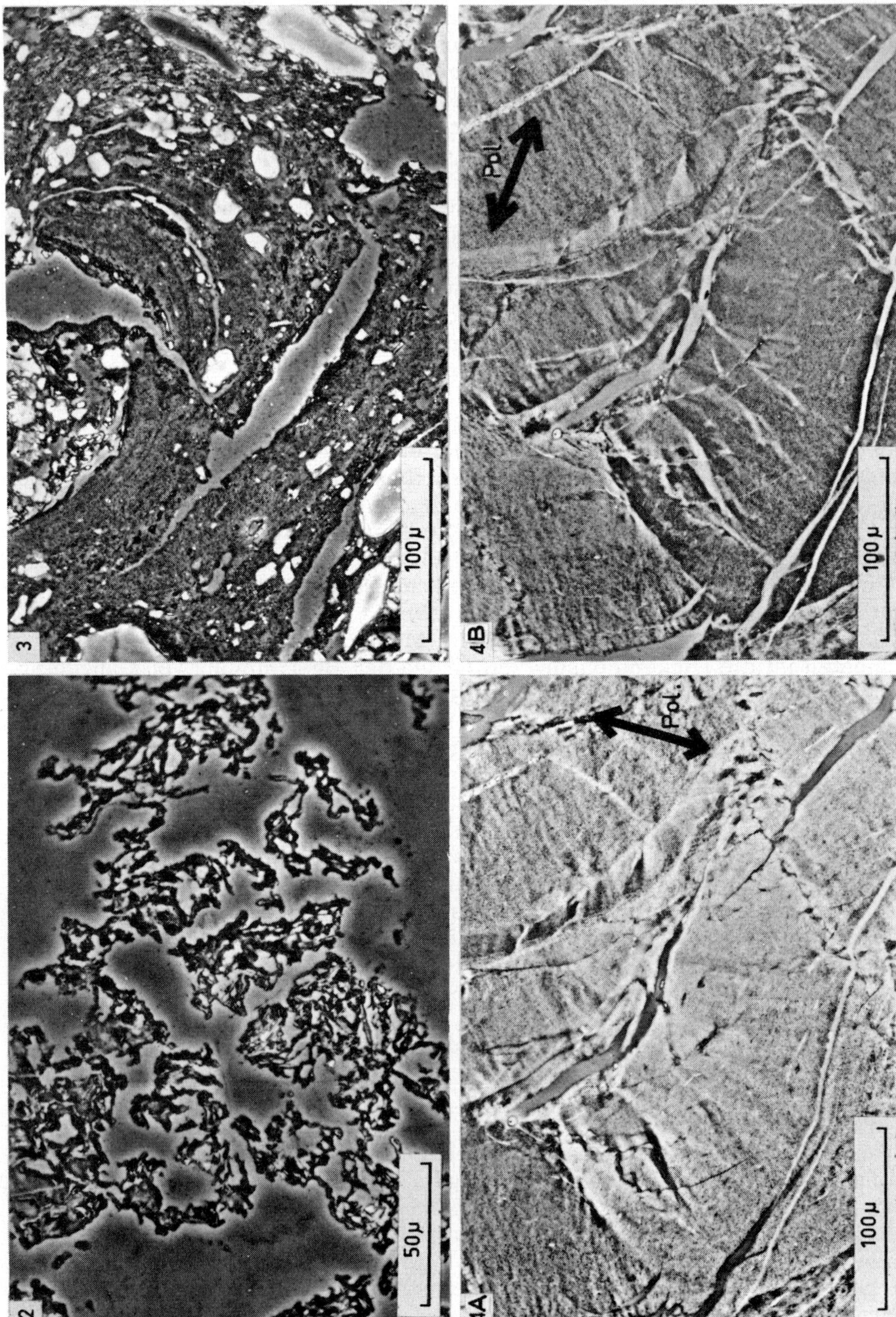
2
50μ
3
100μ
4A
Pol.
100μ
4B
Pol.
100μ

VERBESSERUNG DES BILDKONTRASTES

Da den vorausgegangenen Überlegungen nur farblose Objekte zugrunde gelegt wurden, ist jetzt die Tatsache zu berücksichtigen, dass Bodendünnschliffe, vor allem durch Eisenverbindungen und Humusstoffe, oft kräftige Farben enthalten. Trotzdem ist es nicht zutreffend, solche Schliffpräparate ausschliesslich als Absorptionsobjekte zu bezeichnen. Die Unterscheidung von Absorptions- und Phasenobjekten, die in zahlreichen Abhandlungen aus didaktischen Gründen getroffen wird, ist an Bodenschliffen nicht ohne weiteres möglich. Reine Absorptionsobjekte, die, bei Übereinstimmung der Brechzahl mit dem Einbettungsmittel, nur lichtabsorbierend wirken und dadurch farbig erscheinen, sind nur selten anzutreffen. Dagegen sind absorbierende und phasenverzögernde Eigenschaften häufig miteinander gekoppelt, und nicht selten ist es allein eine Frage der Vergrösserung, inwieweit man in scheinbar homogen gefärbten Tonen die färbenden Anteile, z.B. Hämatitkriställchen, als kleinste Partikel in einer farblosen Grundmasse unterscheiden kann. Diese ist aus verschiedenen Tonmineralien zusammengesetzt, die reine Phasenobjekte darstellen, wie im übrigen auch die gröberen Minerale, soweit sie ungefärbt sind, zu den reinen Phasenobjekten zählen. Ebenso gehören Bleichhorizonte, Bleichzonen usw. als Ganzes zu dieser Art von Objekten. Aber auch homogene Färbungen können mit dem Phasenkontrastverfahren noch nutzbringend untersucht werden, wenn die Absorption so gering ist, dass die Kontrastbildung nicht wesentlich gestört wird.

Im Laufe der Untersuchung geht man üblicherweise von schwachen zu stärkeren Vergrösserungen voran und muss dann zwangsläufig dünnere Partien der Schliffe aufsuchen, um klare Bilder zu erhalten, in denen sich die Anteile möglichst wenig gegenseitig überlagern. Dabei werden auch farbige Objekte mehr und mehr durchscheinend und die gesuchten Details verschwinden in einem unbefriedigenden flauen Kontrast. Aus diesem Grunde ist es oft leichter, z.B. den Flockungszustand einer Braunerde bei schwacher

---

Abb.2. Eisenhydroxydreiche Flockungsaggregate. Alpiner Pseudogley aus Silikatschutt (Glimmerschiefer). Hochtor, Grossglocknerstrasse, Österreich. Dünnschliff 2940, Phasenkontrast, Objektiv Leitz Apochromat Pv 40/0,70 n, Heine-Kondensor.

Abb.3. Eisenhydroxydreiche Tonanlagerung im Bt-Horizont einer Parabraunerde. Lösslehm über Goldshöfer Sand, nördl. Mutlangen Kr. Schwäb. Gmünd. Dünnschliff 2129, einfach polarisiertes Licht, Phasenkontrast, Objektiv Leitz Achromat Pv 25/0,50 n, Heine-Kondensor.

Abb.4. Eisenfreie Tonanlagerung in Bleichzonen aus einem fossilen Pseudogley. Buntsandstein-Solifluktionsschutt, Schlitz, Hessen. Dünnschliff 769, einfach polarisiertes Licht, Phasenkontrast, Objektiv Leitz Achromat Pv 25/0,50 n, Heine-Kondensor. Einbettungsmedium $n_D = 1{,}556$.

A. Wirksame Schwingungsrichtung senkrecht zur Sedimentationsebene; heller Kontrast.
B. Wirksame Schwingungsrichtung parallel zur Sedimentationsebene; dunkler Kontrast.

Vergrösserung am dickeren Schliff abzuschätzen, als mit starken Objektiven nach einzelnen "Flocken" zu suchen, die sich in mangelhaftem Kontrast verlieren. Da sich die Koagulation der Tonteilchen jedoch in feinsten mikroskopischen und submikroskopischen Grössenbereichen abspielt, muss letzten Endes die Vergrösserungsmöglichkeit des Mikroskopes ausgenützt werden können. Mit dem Phasenkontrastverfahren lassen sich Flockungsaggregate in vielen Fällen eindeutig erkennen und messen. Besonders interessante Bilder sind in den eisenreichen Braunerden, wie z.B. den Lockerbraunerden und alpinen Rasenbraunerden, und manchen Eisenpodsolen zu finden. Da diese Böden stark gefärbt sind, wird die Flockung als Ganzes auch im Hellfeld sichtbar und die statistische Isotropie solcher Gefüge wird zudem aus der Beobachtung bei gekreuzten Polarisatoren erkannt. Die kleinsten, oft nur ein bis zwei μ grossen Flockungsaggregate werden jedoch erst im Phasenkontrast deutlich abgebildet. Abb.2 gibt ein Beispiel solcher Aggregate aus einem Fe-Anreicherungshorizont eines alpinen Pseudogleys. Man sieht, wie sich die kleinsten Einheiten zu grösseren Aggregaten zusammensetzen, die wiederum zu einem lockeren Verband angeordnet sind, so dass man drei Ordnungen des Gefüges unterscheiden kann. Infolge des hohen Eisengehaltes entsteht der optimale Kontrast nur in extrem dünnen Schliffen. Er wird dann aber durch die lockere Anordnung der Aggregate in nahezu idealer Weise ausgeprägt, denn die Forderung, dass jedes Einzelobjekt allseitig vom Immersionsmittel umgrenzt sein soll, wird hier weitgehend erfüllt.

KONTRAST AN GROSSFLÄCHIGEN OBJEKTEN

Zumeist sind die Tonanteile nicht so vollkommen aggregiert, sondern schliessen sich zu dichten Massen zusammen. Man denke z.B. an die Tonanreicherungen in Leitbahnen, die vor allem aus Parabraunerden und ähnlichen Bodentypen bekannt geworden sind. Solche Tonanreicherungen bilden ziemlich einheitliche Objekte mit relativ grosser Flächenausdehnung. Wo sie an das Einbettungsmittel angrenzen, entsteht ein deutlicher Randkontrast, der von einem entgegengesetzt hell oder dunkel erscheinenden Hof (Halo) begleitet ist. In Abb.3 ist dies gut zu erkennen. Die Tonmasse erscheint aber auch im Inneren etwas dunkler als das umgebende Medium. Hier entsteht ein Summeneffekt aus mehreren Einzelwirkungen der Gemengteile. Einige der eingelagerten Minerale sind noch mit eigenem Kontrast erkennbar, so z.B. die hier weiss abgebildeten Körnchen und Splitter aus Quarz und Feldspat. Einige hochlichtbrechende Minerale erscheinen infolge Kontrastumkehr ebenfalls weiss. Man kann sie bei direkter Beobachtung im Mikroskop gut erkennen und von den anderen Mineralen unterscheiden. Die sehr zahlreichen dunklen Pünktchen, welche die gesamte Tonmasse durchsetzen, sind ebenfalls Minerale höherer Lichtbrechung, die aber infolge ihrer geringen Grösse guten Dunkelkontrast bilden. Der insgesamt dunkle Kontrast würde aber daraus allein noch nicht erklärt werden. Er wird vielmehr durch feinst verteilte Eisenhydroxyde verursacht, welche die Lichtbrechung der Tonmasse im ganzen erhöhen. Die überraschende Beobachtung, dass in der ganzen Fläche der Tonanlagerung ein ziemlich einheitlicher Kontrast entsteht, lässt sich im wesentlichen dadurch erklären, dass von jeder Stelle der Tonmasse auch Beugungswellen bestimmter Phasenverzögerung ausgehen. In wieweit hier die Durchtränkung mit dem Einbettungsmittel mitwirkt, kann vorerst noch nicht beantwortet werden.

Im Gegensatz dazu erscheinen die grösseren Mineralkörner - einige sind in Abb.3 abgebildet - nicht mehr in einheitlichem Kontrast. Im Zentrum werden die Lichtwellen zwar in der Phase verzögert, aber die Beugung ist so gering, dass diese Wellen weiterhin dem Strahlengang der ungebeugten Lichtwellen folgen. Es entsteht nur das mittlere Grau, das durch den Absorptionsring verursacht wird. Zentrum und Randpartien bilden infolgedessen verschiedene Kontraste.

Ergänzend sei noch darauf hingewiesen, dass die Haloerscheinungen der in den Tonmassen eingeschlossenen grösseren Mineralkörnchen sich weit weniger nachteilig auswirken als man hätte befürchten können. Aus der dicht gedrängten Lagerung ergeben sich noch durchaus übersichtliche Verhältnisse.

KONTRASTERSCHEINUNGEN AN DOPPELBRECHENDEN OBJEKTEN

Tonanlagerungen der beschriebenen Art zeigen häufig starke Doppelbrechung und leuchten daher unter dem Polarisationsmikroskop bei gekreuzten Polarisatoren hell auf. Die aus den Einzelgemengteilen gegebenen Brechzahlen wirken infolge der oft submikroskopischen Grösse der Teilchen zusammen, so dass die ganze Tonmasse, soweit gute Teilchenregelung vorliegt, durch eine mittlere maximale und eine mittlere minimale Brechzahl charakterisiert werden kann. Diese Brechzahlen bilden mit der Brechzahl des Einbettungsmittels verschiedene Phasendifferenzen, die man im Phasenkontrast unter Umständen deutlich unterscheiden kann. Voraussetzung ist dazu, dass ein Polarisator oder Analysator in den Strahlengang eingeschaltet wird. Es werden dann jeweils nur diejenigen Schwingungsrichtungen aus dem Objekt wirksam, die der Schwingungsrichtung des Polarisators parallel laufen, während die senkrecht dazu verlaufenden Schwingungen ohne Einfluss sind.

Sind beide mittleren Brechzahlen grösser als die Brechzahl des Einbettungsmittels, wie es bei eisenreichen Tonen oft der Fall ist (s. Abb.3), so ist der Effekt des Polarisators nicht sehr gross, denn der Dunkelkontrast bleibt für beide Schwingungsrichtungen erhalten. Man beobachtet lediglich eine Verstärkung oder Abschwächung der Kontraste, vor allem an der Grenze zum Medium. Diese Verhältnisse können sich völlig ändern, wenn das Eisen in Gestalt von mikroskopisch sichtbaren Kriställchen eingelagert ist oder - noch eindeutiger - wenn eisenfreie, gebleichte Tonanlagerungen vorliegen. Die Minimale Lichtbrechung ist dann häufig niedriger als die Lichtbrechung des Mediums und es entsteht bei geeigneter Stellung des Polarisators ein Kontrast heller als das Medium. Dreht man den Polarisator langsam in andere Richtungen, so kehrt sich, entsprechend der wirksamen Schwingungsrichtung des Objektes, der Kontrast um. In Abb.4A und 4B wird dies mit zwei Bildern, die bei verschiedener Stellung des Polarisators aufgenommen wurden, demonstriert. Da die Tonschichten etwas gekrümmt sind, werden nicht an allen Objektteilen zugleich optimale Kontrastbedingungen erreicht, die Erscheinungen sind jedoch trotzdem eindeutig.

Man muss allerdings die Frage stellen, ob sich ein Teil der Erscheinungen möglicherweise auch aus der Differenz zwischen maximaler und minimaler Brechzahl, d.h., dem Gangunterschied, erklären lassen. Zieht man diejenigen Bereiche innerhalb der Tonschichten in Betracht, wo in verschiedener Richtung geregelte Teilgefüge mit dementsprechend unterschiedlicher wirksamer Brechzahl aneinandergrenzen, so erscheint diese Möglichkeit nicht abwegig. Die Gangunterschiede, die ja ebenfalls Phasendifferenzen zweier Wellen darstellen, würden dann dieselben Kontrasteffekte hervorrufen wie die Phasendifferenzen Objekt-Medium. Wir nehmen jedoch an, dass diese Effekte zumindest geringer sind und von den Wirkungen der Phasendifferenzen Objekt-Medium überdeckt werden. Es müssten anderenfalls auch ähnliche Kontrastunterschiede im Innengefüge auftreten, wenn beide Hauptbrechzahlen höher oder beide niedriger als das Medium sind. Das ist jedoch nicht der Fall.

Liegen die Hauptbrechzahlen der Tonanlagerung niedriger als die Brechzahl des Einbettungsmediums, so erscheint das Objekt bei beliebiger Drehung des Polarisators heller als das Medium. Die verschiedenen Phasendifferenzen wirken sich dann nur noch in Helligkeitsschwankungen aus, wie dies mit umgekehrtem Kontrast bei insgesamt höheren Brechzahlen zu beobachten ist.

ZUR MINERALBESTIMMUNG IM DÜNNSCHLIFF

Trotz mancher Schwierigkeiten bei der Bestimmung von Mineralen in Bodendünnschliffen ergibt sich doch aus Gründen, die schon in der Einleitung angedeutet wurden, die Notwendigkeit, durch Anwendung mehrerer Methoden die Untersuchungsmöglichkeiten zu erweitern. Das Phasenkontrastverfahren kann hierbei wesentliche Hilfe leisten, doch muss man infolge der festen Einbettung der Schliffe gewisse Einschränkungen in Kauf nehmen. So ist es z.B. nicht möglich, in ein und demselben Präparat nacheinander verschiedene Immersionsmedien anzuwenden. Man kann aber ersatzweise Parallelproben in verschiedenen Harzen einbetten und davon Dünnschliffe anfertigen. Dabei wird jedoch der Mangel an geeigneten Harzen spürbar (vgl. Altemüller, im Druck), weshalb dieser Weg noch kaum beschritten ist.

Sehr wertvoll wäre es, wenn man aus den Kontrasterscheinungen einigermassen genaue Aussagen über die Phasendifferenzen machen könnte. Dann liessen sich bei bekannter Präparatdicke die gesuchten Brechzahlen ermitteln. Tatsächlich sind aber nur qualitative Aussagen möglich. Lediglich bei angenäherter Brechzahlgleichheit zwischen Objekt und Medium lassen sich genaue Bestimmungen durchführen. Da die Dispersionskurve des Polyesterharzes Vestopal-H steiler verläuft als die Dispersionskurve der Minerale, treten Farbeffekte auf, die meist zu einer verschiedenartigen Blaufärbung der Körner oder Aggregate führen, deren Umgebung in rötlichen und orangeroten Farbtönen erscheint. Bei doppelbrechenden Mineralen entscheidet die wirksame Brechzahl über die vorherrschenden Effekte, wenn bei eingeschaltetem Polarisator oder Analysator beobachtet wird. Günstige Beispiele liefern farblose Glimmer in Schnitten senkrecht zur Basis. Bei Drehung des Polarisators wird dann abwechselnd die kleinste oder annähernd die grösste Brechzahl wirksam und bei etwa gleicher Brechzahl mit dem Einbettungsmittel entstehen die blauen Farbtöne. Auch Kaolinitaggregate erscheinen im Dünnschliff in Vestopal-H oft deutlich blau gefärbt mit orangefarbenem Hof. Gelegentlich entstehen auch abweichende Kontraste. Das rührt entweder daher, dass die Brechzahlen verschiedener Kaolinite nicht gleich sind, oder dass Unterschiede beim Einbettungsmedium zu suchen sind. Für Kaolinit gibt Tröger (1959) Werte für n_α von 1,553–1,563 und für n_γ von 1,560–1,570 an. Beim Vestopal können die Brechzahlen vor allem in Abhängigkeit vom Styrolanteil in noch weiterem Umfange schwanken.

Mischungen von Vestopal-H (im Lieferzustand) mit gleichen Teilen Styrol ergeben ein Polymerisat mit Brechzahlen: n = 1,57–1,58; bei grösseren Anteilen Styrol steigen die Werte bis über 1,59 an. Da die mechanischen Eigenschaften solcher Produkte nicht günstig sind und im Verlauf der Präparation Quellungen eintreten (Altemüller, 1962), lassen sich diese Brechungsverhältnisse kaum nutzen. Für zahlreiche Objekte wären sie auch wenig günstig und es ist besser, niedrigere Werte anzustreben. Nicht verdünntes Vestopal-H polymerisiert unter Luftabschluss zu Blöcken von etwa n_D = 1,560. Bei langsamer Polymerisation an der Luft kann noch ein Teil des in der Fabrikation beigegebenen Styrols verdunsten, so dass sich die Werte bis etwa n_D = 1,555 oder noch weiter erniedrigen können. Glücklicherweise werden solche verhältnismässig günstigen Brechzahlen auch dann erreicht, wenn nach vorausgegangener starker Verdünnung mit Styrol (die der besseren Durchtränkung der Proben dient) bei sehr langsamer Polymerisation dieser Zusatz wieder verdunstet. Bei einheitlicher Verarbeitung liegen die Brechzahlen dann auch in Serien von Einbettungen ziemlich gleichmässig im Bereich von n_D = 1,555–1,560. Zur Bestimmung der Brechzahlen eignet sich die Immersionsmethode, wobei man entweder einen Tropfen der benutzten Flüssigkeit auf den nicht gedeckten Dünnschliff gibt oder einige Splitter des Harzes vom Block abschabt und in das Immersionsmittel einstreut und abdeckt. Bei der Auswahl von Flüssigkeiten ist darauf zu achten, dass keine Veränderung des Vestopals durch Quellung eintritt. Mischungen von Olivenöl und Zimtöl haben sich für kurzfristige Beobachtungen als brauchbar erwiesen. An grösseren, angeschliffenen und polierten Blockproben mit genügend breitem Harzrand ist auch eine direkte Messung im reflektierten Licht mit dem Abbe-Refraktometer (Zeiss) möglich.

Einige Beispiele von Mineralen, deren Brechzahlen nahe dem Einbettungsmittel liegen, sind in Abb.5, 6 und 7 zu finden. Die Kaolinitaggregate der Abb.5 kommen zahlreich im verwitterten Stubensandstein der mittleren Keuperformation vor. Es ist anzunehmen, dass sie aus einer älteren Verwitterung von Feldspäten entstanden sind, die in stark korrodierten Körnern und kleinen schluffgrossen Splitterchen zahlreich vorhanden sind. Interessanterweise sind aber nur wenige dieser Feldspatreste - es handelt sich durchweg um solche mit geringer Lichtbrechung als Quarz - in direktem Kontakt mit den Kaolinitaggregaten zu finden. Die letzteren sind in lockerem Verband angeordnet, scheinen aber ziemlich stabil zu sein. Dünne Schlieren von verlagerten Tonteilchen, die ringsum in unmittelbarer Nachbarschaft vorkommen, unterscheiden sich durch stärkere Doppelbrechung deutlich und scheinen von anderer Beschaffenheit zu sein. Die Lichtbrechung der Aggregate ist im abgebildeten Präparat insgesamt etwas höher als die Brechzahl des Vestopals. Der relativ starke Kontrast im Bild entspricht jedoch nicht ganz dem visuellen Eindruck, denn er wurde durch die Aufnahmetechnik

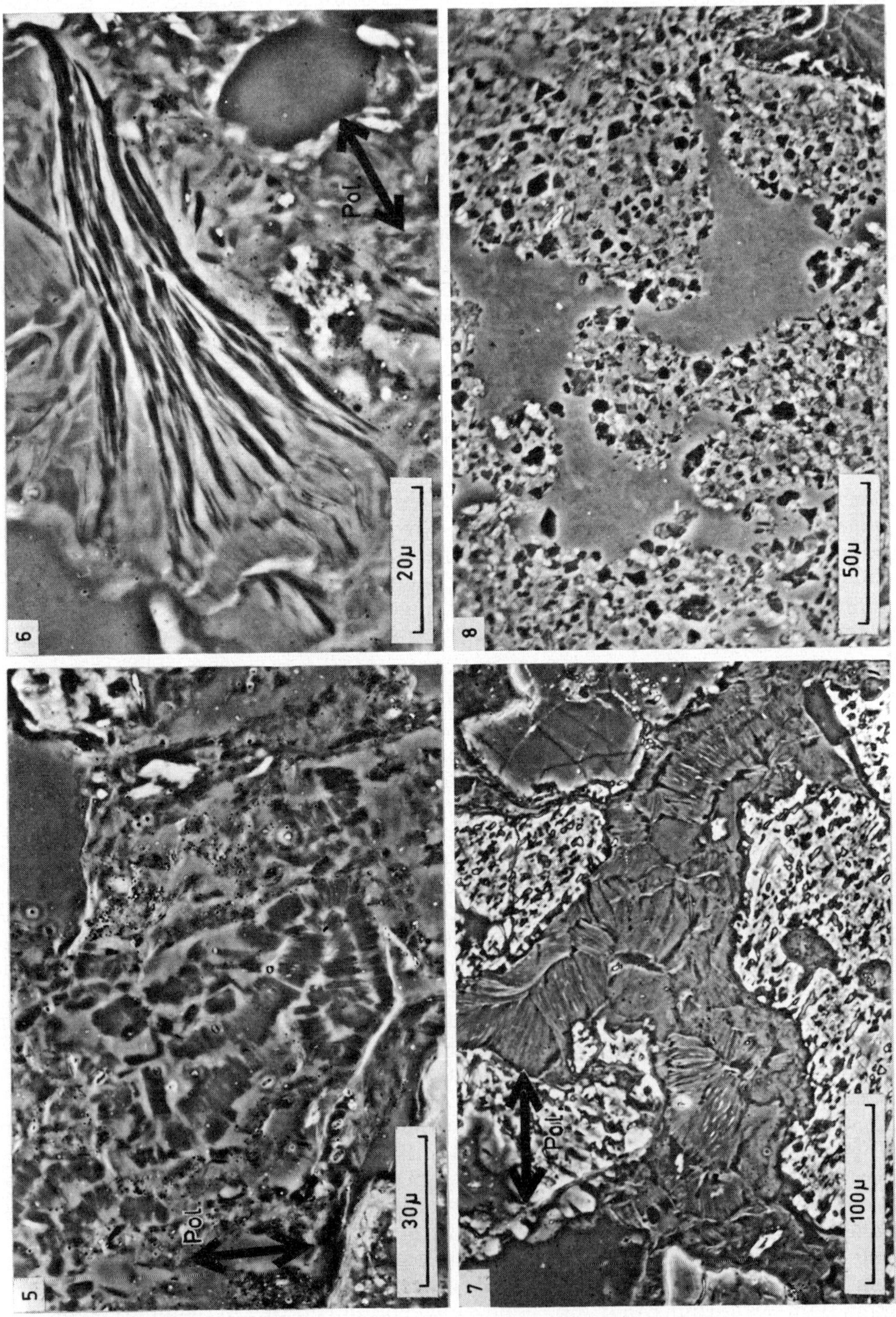
6
Pol.
20µ
8
50µ
5
Pol.
30µ
7
Pol.
100µ

bewusst gesteigert.

Besonders interessant ist die Verwitterung der Glimmer, die im gleichen Material in verschiedenen Stadien beobachtet werden kann. In Abb.6 ist zu sehen, wie sich ein Glimmerpaket fächerartig aufweitet und die Kontrasterscheinungen dabei vom Dunkelkontrast zum schwachen Hellkontrast wechseln. Die Schwingungsrichtung des Polarisators geht der Spaltrichtung parallel; dadurch wird die mittlere bis grösste Brechzahl des Objektes in ihrer Kontrastwirkung sichtbar. Die hellen Zonen in unmittelbarer Nähe der Glimmerplättchen sind allerdings Haloeffekte und überdecken etwas die anschliessenden Partien. Am linken Hand des Fächers erkennt man in zarten Streifen eine weitere Ordnung von Schichten, deren Lichtbrechung deutlich geringer ist. Auch die Doppelbrechung nimmt stark ab. Im Polarisationsmikroskop wurde mit dem Berek-Kompensator eine Doppelbrechung von etwa $\Delta = 0{,}020$ für die Stellen mit Dunkelkontrast und eine solche von $\Delta = 0{,}003$ für die fein geschichtete Partie am linken Rand gemessen. Als Unsicherheitsfaktor geht in diese Bestimmung die Dickenmessung ein, die mit einem Leitz Achromat Oel 100/1,30 durch Tiefenfokussierung unter Berücksichtigung der Brechzahl des Einbettungsmediums und des Immersionsöles, usw. mit etwa 3 μ bestimmt wurde. An sich ist diese Messung nicht schwierig und liefert bei geöffneter Aperturblende und bei Verwendung eines strengen Grünfilters ($\lambda = 540$ mμ) auch sehr gut reproduzierbare Werte. Es ist aber anzunehmen, dass bedingt durch die Rauhigkeit der Oberflächen des Schliffes die Messung eher zu grosse als zu kleine Werte liefert. Das würde bedeuten, dass die Doppelbrechung in Wirklichkeit etwas höher einzusetzen wäre (z.B. bei 2 μ Schliffdicke $\Delta = 0{,}032$ und 0,006). Dies sind nur allgemein orientierende Angaben, die hier nicht vertieft werden sollen. Ein Verwitterungsprodukt mit derart geringer Doppelbrechung kann jedoch nicht - oder zumindest

---

Abb.5. Geldrollenartige Aggregate von Kaolinit. Im Mikroskop dunkelblaue Dispersionsfarbe mit schwach orangefarbenem Halo. Verwitterung aus Stubensandstein, Weiler am Kocher, Württ. Dünnschliff 1208, einfach polarisiertes Licht, Phasenkontrast, Objektiv Leitz Fluorit-Oelimmersion Pv 70/1,15 n, Heine-Kondensor, Gelbfilter. Einbettungsmedium $n_D = 1{,}556$.

Abb.6. Stark verwitterter Glimmer und Neubildung kaolinitartiger Tonminerale. Pseudogley aus Stubensandstein, Hirschhof bei Ellwangen, Württemberg. Dünnschliff 1273A, einfach polarisiertes Licht, Phasenkontrast, Objektiv Leitz Fluorit-Oelimmersion Pv 70/1,15 n, Heine-Kondensor, Gelbgrünfilter. Einbettungsmedium $n_D = 1{,}559$.

Abb.7. Kaolinitartige Tonminerale in Geldrollenaggregaten mit glimmerartigen Zwischenschichten, angrenzend Quarz und korrodierte Feldspäte. Verwitt. Stubensandstein wie Abb.6. Dünnschliff 1273A, einfach polarisiertes Licht, Phasenkontrast, Objektiv Zeiss Achromat Ph 25/0,45, Ringkondensor, Grünfilter. Einbettungsmedium $n_D = 1{,}559$.

Abb.8. Tonmergel mit rhomboedrischen Calcit-Kriställchen in verschiedenem Kontrast. C-Horizont, Smonitza bei Čačak-Mojsinje, Jugoslavien (Probenahme Prof. Dr. Antič). Dünnschliff 1612, einfach polarisiertes Licht, Phasenkontrast, Objektiv Leitz Apochromat Pv 40/0,70 n, Heine-Kondensor.

nicht zum grössten Teil - aus Illit bestehen. Die Vermutung geht hier eher auf Kaolinit. Ein weiteres Stadium dieser Verwitterungsbildung aus demselben Schliffpräparat ist in Abb.7 zu finden. Die Übergänge sind in mehreren Stadien anzutreffen, so dass die Entwicklung aus Glimmern kaum in Zweifel zu ziehen ist. Wohlgeordnete Schichtpakete füllen hier fast die gesamten Kornzwischenräume aus. Die Kontraste sind nicht ganz einheitlich. Einige der helleren Streifen fallen durch höhere Doppelbrechung auf. Von diesen Streifen abgesehen, wurde eine maximale Doppelbrechung von etwa $\Delta = 0{,}006$ bestimmt, bei einer Schliffdicke von etwa 4 μ. Es wird angenommen, dass hier Tonminerale mit Wechsellagerung von Kaolinit mit Glimmer vorliegen, die Jasmund (1955; s. dort weitere Lit.) als letzte Verwitterungsprodukte von Biotit auffasst. Dies mag für unsere Darstellung genügen.

Dass die geringen Schliffdicken tatsächlich erreicht werden, soll noch in Abb.8 Beweis finden, die einen Dünnschliff aus einem carbonatreichen, mergelartigen C-Horizont einer Smonitza zeigt. Die rhomboedrischen Carbonatkriställchen (Calcit) sind in einer hellgrau abgebildeten Ton-Grundmasse deutlich mit dunklem und hellem Kontrast zu sehen. Aus Tabelle I ist zu entnehmen, dass erst bei Korndicken unter 2 μ optimale Kontraste, wie sie hier ausgebildet sind, erreicht werden.

Auch für grössere Körner ergeben sich manche Erleichterungen für die Abschätzung der Lichtbrechungseigenschaften. In Lössen ist es oft schwer, sich einen qualitativen Eindruck über den Anteil an Feldspaten zu machen. Im Phasenkontrast wird dies sehr erleichtert, da die am meisten vorkommenden Minerale dieser Gruppe geringere Lichtbrechung als Quarz aufweisen und deshalb Kontrastunterschiede zeigen. Diese Fragen sollen an anderer Stelle weiter ausgeführt werden.

SCHLUSSFOLGERUNGEN

Wie vorliegende Untersuchung zeigt, kann das Phasenkontrastverfahren mit gutem Erfolg an Bodendünnschliffen angewandt werden, wenn folgende Voraussetzungen erfüllt werden:

(1) Das Einbettungsmittel muss hinsichtlich seiner Lichtbrechungseigenschaften den zu untersuchenden Objekten angepasst sein. Für zahlreich Zwecke sind mittlere Brechzahlen zwischen $n = 1{,}54$ und 1,56 gut geeignet. Vestopal-H ist bei entsprechender Verarbeitung sehr gut verwendbar und ermöglicht bei Brechzahlen zwischen $n_D = 1{,}555$ und 1,560 interessante Unterscheidungen durch verschiedene Kontraste an zahlreichen Mineralen. Die Prüfung weiterer Einbettungsmittel ist zu empfehlen.

(2) Die Präparationsarbeiten erfordern besondere Sorgfalt durch einwandfreie Einbettung, saubere Schleifarbeit und ausgezeichnete Politur mit Diamantkörnungen, um möglichst ebene Oberflächen zu erzielen. Besonders wichtig ist die Herstellung dünnster Schliffe. Dicken von 2–6 μ können in den tonigen Anteilen ohne Beschädigung erreicht werden. Schluff- und Sandkörnchen bleiben dabei meist etwas dicker.

(3) Die Brechzahl des Einbettungsmittels wird vor dem Abdecken der fertigen Dünnschliffe bestimmt.

(4) Untersuchungen an doppelbrechenden Objekten im Phasenkontrast werden mit eingeschaltetem Polarisator oder Analysator durchgeführt.

Durch das Phasenkontrastverfahren werden feinste Unterschiede der Lichtbrechung oder der Dicke in verschiedenen Kontrasten sichtbar. Die Lichtbrechung der Objekte kann geschätzt und bei annähernder Brechzahlgleichheit mit dem Einbettungsmedium bestimmt werden. An doppelbrechenden Objekten entstehen eindrucksvolle Kontrastunterschiede besonders dann, wenn die kleinere Brechzahl niedriger und die grosse Brechzahl höher liegt als die Brechzahl des Einbettungsmittels. Diese Kontrastunterschiede sind besonders bei Verwitterungsstudien von Bedeutung. Darüber hinaus werden in zahlreichen Fällen Objektdetails kontrastreich abgebildet, die im normalen Hellfeld oder bei gekreuzten Polarisatoren nur ungenügend erkennbar sind. Der Vorteil der Untersuchung im Dünnschliff ist besonders darin zu sehen, dass alle Objekte in ihrem natürlichen, ungestörten Verband untersucht werden können.

SUMMARY

The application of phase contrast microscopy to the investigation of thin sections of soils

Phase contrast is an interference phenomenon which appears when using special microscopic accessories under definite conditions of illumination and preparation of the objects to be investigated. This method has been described for determination of refractive properties of silt and clay minerals in single grain preparations in several publications. Whether the phase contrast can also be used within the investigation of undisturbed soil samples in thin sections is the object of this paper.

In practical work, the basic principles of contrast formation and directions for application must be well known. Therefore, a short introduction with explanation of some of the main properties is given. The relationships between contrast phenomena and phase difference (as caused by differences of refractive properties or thickness in the sections) are discussed and illustrated by contrast curves (Fig.1) and by some numerical examples of common minerals (Table I).

For phase contrast investigations, two conditions must especially be fulfilled: (1) The refractive index of the embedding medium should be adapted to the soil constituents of main interest; (2) the sections must be ground extremely thin.

As an embedding medium, the polyester resin Vestopal-H is very suitable. The refractive indices are between n_D = 1,555–1,560, so that interesting contrast conditions are given for some important constituents. The mechanical properties allow fine grinding to less than 10 μ thickness. In marginal parts of the slides in plasmic domains, even 2–6 μ thickness is attainable without damage, whereas due to the difference in hardness the skeleton grains remain somewhat thicker. To avoid such differences as far as possible, diamond pastes for grinding and polishing are recommended.

It is shown that phase contrast microscopy is an excellent completion to other microscopic methods. Objects which in ordinary or polarized light have only faint outlines, often show good phase contrast. The phenomenon of flocculation, e.g., can be better determined (Fig.2) and small particles, which otherwise would be overlooked, will be visible. Also the contrast of larger objects, such as coarse mineral grains or deposits of migrated clays (Fig.3) is discussed.

Birefringent objects are investigated by use of a polarizer or analizer. The contrast formation then depends on the vibrating direction of the polar. The most interesting phenomena are visible when the maximum refractive index of the object is higher and the minimum index is lower than the index of the medium. Then a dark contrast becomes a bright one and vice versa, according to the azimutal position of the polar (Fig.4A, 4B).

These properties facilitate not only the recognition of various arrangements of particles or fabric units but also the determination of minerals in their natural situation. This is illustrated by pictures of kaolinite (Fig.5) and the weathering of mica (Fig.6, 7). In good sections, the very high double refractive calcite is also visible in good contrast conditions. The preparation represented in Fig.8 is estimated to be about 2 μ thick.

LITERATUR

Altemüller, H.-J., 1962. Verbesserung der Einbettungs- und Schleiftechnik bei der Herstellung von Dünnschliffen mit Vestopal. Z. Pflanzenernähr. Düng. Bodenk., 99 (144) : 164–177.

Altemüller, H.-J., im Druck. Die morphologische Untersuchung des Bodengefüges. In: Scharrer–Linser (Redakteur), Handbuch der Pflanzenernährung und Düngung. Springer, Wien.

Correns, C.W. und Piller, H., 1955. Mikroskopie der feinkörnigen Silikatminerale. In: H. Freund (Redakteur), Handbuch der Mikroskopie in der Technik. Umschau Verlag, Frankfurt/Main, 4 (1) : 697–780.

Gabler, F., 1955. Positiver oder negativer Phasenkontrast. Mikroskopie, 10 (1955) : 119–124.

Jasmund, K., 1955. Die silicatischen Tonminerale. Verlag Chemie, Weinheim, 125 S.

Juda, J. und Medenbach, K., 1959. Untersuchungen von Feinstäuben nach der λ-Variationsmethode. Z. Wiss. Mikroskop., 64 : 218–226.

Kalk, E., 1959. Die Verwendung des Polarisationsmikroskopes bei Phasenkontrast und allseitig gerader Dunkelfeldbeobachtung in der Bodenmineralogie. Zeiss-Mitt., 1 : 354–363.

Menzel, E., 1957. Phasenkontrast-Verfahren. In: H. Freund (Redakteur), Handbuch der Mikroskopie in der Technik. Umschau Verlag, Frankfurt/Main, 1 (1) : 317–356.

Meyer, B. und Kalk, E., 1959. Die Mineralbestimmung in Tonfraktionen des Bodens mit Hilfe der Phasenkontrast- und Grenzdunkelfeldmikroskopie. Landwirtsch. Forsch., 12 : 3–12.

Michel, K., 1957. Die Mikrophotographie. Springer, Wien, 740 S.

Piller, H., 1952. Die Phasenkontrastmikroskopie als Hilfsmittel zur Bestimmung feinkörniger, speziell dünner, transparenter Minerale. Heidelberger Beitr. Mineral. Petrog., 3 : 307–334.

Schmidt, K.G., 1955. Die Phasenkontrastmikroskopie in der Staubtechnik. Staub, 41 : 436–467.

Smithson, F., 1946. Phase-contrast microscopy for mineralogy. Nature, 158 : 621 pp.

Smithson, F., 1961. The microscopy of the silt fraction. J. Soil Sci., 12 : 145–157.

Thaer, A., 1954. Ein Beitrag zur lichtmikroskopischen Mineralbestimmung in Feinstäuben, insbesondere des Kohlenbergbaues. Staub, 38 : 555–570.

Tröger, W.E., 1959. Optische Bestimmung der gesteinsbildenden Minerale. 1. Bestimmungstabellen. Schweizerbart, Stuttgart, 147 S.

MAKING COLOUR PRINTS FROM MINIATURE FILM, IN PARTICULAR FOR MICROMORPHOLOGICAL PURPOSES

M.C. NATER

Soil Survey Institute, Wageningen (The Netherlands)

INTRODUCTION

Owing to its nature, soil micromorphology requires a great many visual aids. The micromorphologist often has to provide detailed illustrations to his articles and lectures by means of microphotographs of the soil (Plate G1). For his articles, he often uses paper prints and transparencies, preferably in colour. At lectures, he usually projects colour transparencies. Generally speaking, these projected pictures show more details and the colours are more natural than in printed equivalents. The one great drawback of projection as compared with prints is that the reader can inspect and compare the latter as long as he likes, whereas at a lecture the various pictures are only seen in brief succession and there is no possibility of making an immediate comparison. This drawback can be overcome by using large-size colour prints from miniature film at lectures. These prints are also very suitable for exhibition purposes.

We first produced large-size colour prints for duplicating hand-coloured maps at our Institute; the process was afterwards found suitable for micromorphology. This process will be discussed in the following pages. We will confine ourselves to the method adopted by the Soil Survey Institute regarding the apparatus, types of material and processes, although it should be clearly understood that the use of other apparatuses and materials need not give inferior results.

We chose the negative–positive process, in which the following stages should be distinguished in the production of these colour prints: (1) the photography technique; (2) negative development; (3) the production of colour prints.

THE PHOTOGRAPHY TECHNIQUE

For the production of photographs we employ: (a) a microscope, including a light-source; (b) a thin section; (c) a 24 x 36 mm negative film; (d) a photoelectric cell; (e) a camera housing.

Re a. The optics should be of a very high standard. The entire photograph should be sharp, for the magnification factor we use for colour prints is generally on the large side. If we have to increase this factor still further by using only the centre of the negative with marginal distortion, the magnification of the "granular structure" will be too high. For this reason, the photographs are taken with a Zeiss Standard WL Pol microscope provided with 8x and 12.5x Komplan eye-pieces and 4/0.16, 10/0.32 and 25/0.65 Planapochromate objectives (Fig.1).

Fig.1. Zeiss Standard Wl Pol microscope provided with a Komplan eye-piece Planapochromate objectives, Leica camera housing and photoelectric cell which is connected with exposure meter.

The light-source employed consists of a 6 V tension lamp built into the base of the microscope in the usual way. When taking colour transparent positive or transparent negative photos, care should always be taken to ensure that the voltage on the transformer does not differ from 6 V, as otherwise colour shifts are bound to occur.

Re b. The thin section is known and will not be further defined here.

Re c. We employ a Kodak Kodacolor-X, 24 x 36 mm negative film. This is a daylight film with a built-in colour separation mask, which improves the ultimate colour reproduction. In order to compensate the daylight film for artificial light, we selected a Rollei B5 filter, which is arranged on the field diaphragm.

Re d. Since only minute differences of colour are involved in colour microphotography, it is essential to choose the proper source and time of exposure if the negative is to be of optimum quality. Since these times cannot be measured with an ordinary exposure meter, it is necessary to use a photoelectric cell; we employ the Zeiss photoelectric cell in connection with an exposure meter. The cell is connected to the microscope. The meter has several amplification stages.

Re e. A camera housing is required for exposing and transporting the film. For this purpose, we use the Leica M2; the optic is removed, after which it can be arranged on the microscope by means of an extension piece. After the detail of the thin section to be photographed has been determined, the exposure time is measured with the above-mentioned photoelectric cell. The shutter time ascertained is then set on the camera and the photograph can be taken, with or without crossed nicols or compensator.

NEGATIVE DEVELOPMENT

The exposed films are developed by the Kodak C-22 development process, using nitrogen in a thermostatically heated unit. A nitrogen distributor is arranged in the separate $3\frac{1}{2}$-gallon tanks.

Since the C-22 development process is obtainable through the usual trade channels, we will omit a discussion of the various processing stages. However, some attention should be paid to the heating of the tanks and the use of nitrogen administration.

Thermostatic heating of the tanks

The various negative developing and washing baths should have a constant temperature of 22° C. We place our tanks in a vessel, which is about 70% full of water kept at a constant temperature by means of thermostatic heating elements (Fig.2).

In order to obtain uniform negatives without blotches, the processing solutions are continuously agitated by the nitrogen unit. If manual development is used without nitrogen, there is a very great risk of blotches appearing, and of course this is fatal to good colour prints.

Nitrogen unit

For this purpose, we use the Kodak Burst Valve Unit with a nitrogen cylinder accessory (Fig.3). A nitrogen distributor (Fig.4) connected to this unit gives a burst of nitrogen (Fig.5) from the bottom of the tank at intervals that can be regulated by the control unit timer (e.g., a 1-sec burst once every 10 sec), thus ensuring regular development of the films.

We use nitrogen in preference to oxygen, as the latter has the drawback of clinging to the film in the form of bubbles, at which places development is absent. Nitrogen does not have this property, and also has the advantage of

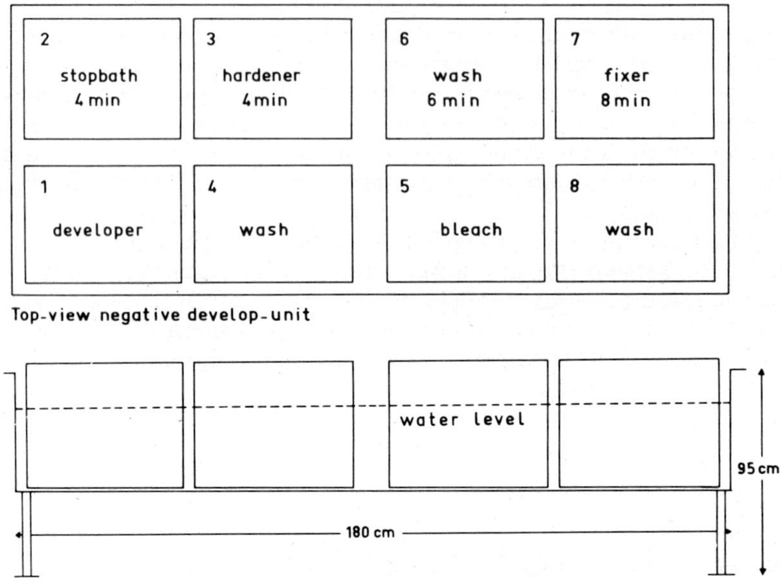

Front-view negative develop-unit FIG. 2

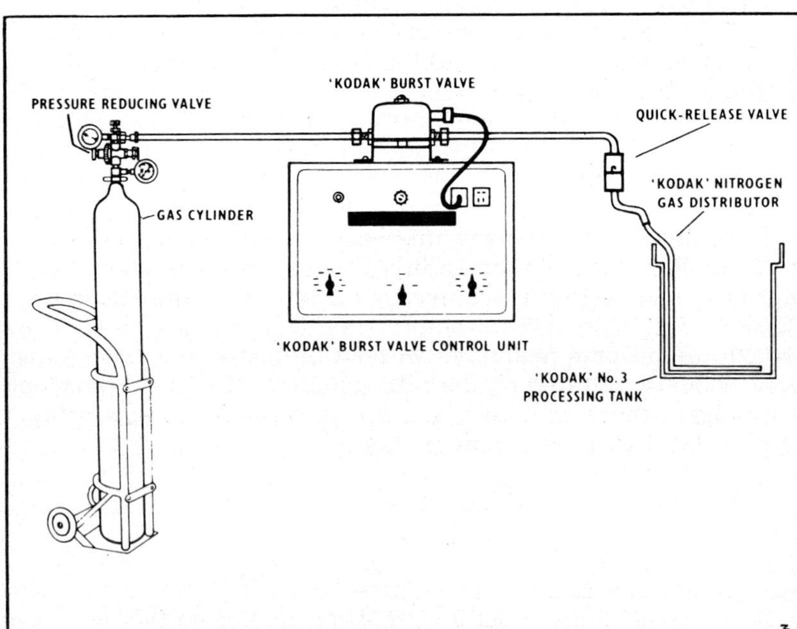

Fig.3. Kodak Burst Valve Unit with nitrogen cylinder accessory. (Taken with permission from "Kodak Burst Valve and Control Unit".)

Fig.4. Nitrogen distributor in a tank.

Fig.5. The nitrogen distributor in action.

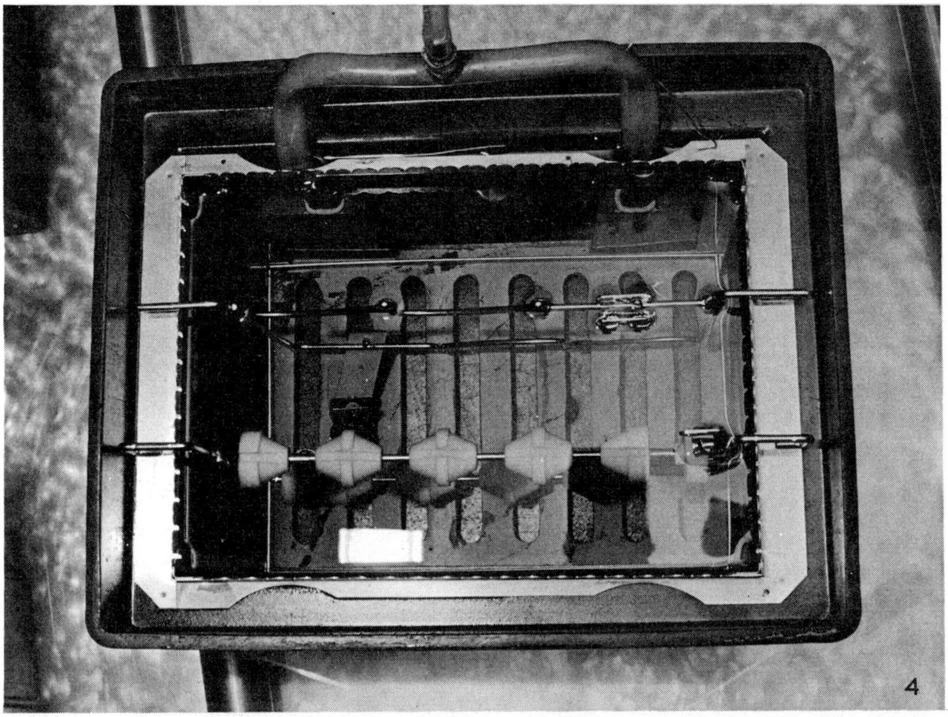
4

5

removing the oxygen introduced in the developer by the material. As a result, the developer oxidises more slowly and can be kept longer.

The entire developing process consists of eight baths, viz., five tanks of developer chemicals and three washing tanks (Fig.2).

In order to obtain completely washed negatives, we also employ nitrogen agitation during the intermediate and final washing steps. This also benefits the keeping property of the following development bath.

After this development process, which takes about 56 min, the negatives may be dried in a drying cabinet or at room temperature. In both cases, special care should be taken to prevent dust and scratches as the film is particularly sensitive to these.

THE PRODUCTION OF THE COLOUR POSITIVE

Before discussing this, it is advisable to give a very brief illustration of certain details of the negative–positive process by means of Table I.

TABLE I

The negative–positive process

Original	black	yellow	red	magenta	blue	cyanic	green	white
Colour negative	white	blue	cyanic	green	yellow	red	magenta	black
Colour positive	black	yellow	red	magenta	blue	cyanic	green	white

It can be seen from this table that the colours of the negative are in a complementary relationship to the original and that the photo is the same as the original. At any rate, this is the theory; however, practice shows that colour differences still occur if colour paper enlargements are made from the colour negative without further correction. This is due to the fact that the manufacturer is unable to exactly adjust the emulsions of the negative and positive material to each other. This drawback can be overcome by using colour correction filters during enlargement, thereby correcting the colours (Fig.6). These colour filters are obtainable in yellow, magenta and cyanic. These three colours are further subdivided into 11 stages, from light to dark, varying from 5–99%, thus providing a total of 33 colour filters for applying corrections. Each pack of enlargement paper contains instructions on filtering and processing, which means that only an enlarger with a filter slide is suitable for taking colour photographs.

For this purpose, we use the Leitz Focomat IIa Color with a 1 : 4.5/50 m objective, with which we can make negative enlargements from microfilm up to a size of 50 x 60 cm; the enlargement factors vary from 8–20x greater than the negative.

In addition to this enlarger, we must also use a constant voltage transformer for equalizing electricity fluctuations during exposure. This apparatus is of the 250 W constant voltage transformer type connected to the exposure timer of the enlarger. Without this constant voltage transformer, uncontrollable colour differences can occur as a result of fluctuations in the voltage.

The colour prints are then produced as follows: (a) exposure of the

Fig.6. Set of colour correction filters.

sensitive material, viz., Kodak Professional Ektacolor paper; (b) development; (c) drying and glossing.

Exposure of the sensitive paper

In the first place, one or more test strips are always made in order to determine the exposure time and the filtering. For this purpose, the filters are inserted in the enlarger between the light source and condensor and the film is negatively adjusted, after which it can be exposed. It is not easy to find the right filtering as this is determined by the photographer's wide experience and skill.

Development

This is done by the Kodak P122 paper developing process, a nine-bath system. When only small numbers and sizes are involved, the developing can be done in flat vessels on the bain-marie system. The problems then are to keep the baths at the right temperature, and to agitate the photographs during development; the latter should be done by hand in order to obtain uniform photographs.

As this is impossible to do when large numbers and sizes are involved, we purchased a unit in which we develop the photographs in separate vertical tanks on the bain-marie system (Fig.7). These tanks each have a capacity of

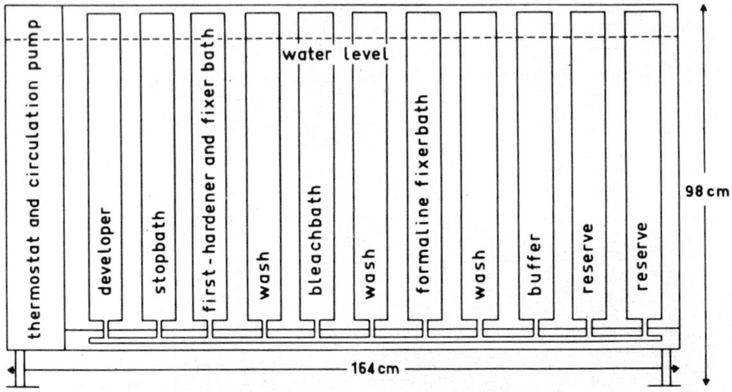

Front-view paper develop-unit FIG. 7

10 U.S. gallons. Heating is entirely thermostatically set at 29° C and nitrogen is also supplied to the separate tanks from the bottom, in the same way and with the same unit as described on p.393.

To prevent photographs from settling to the bottom of the tanks, they are clamped in hangers lined with petticoat gauze, so that the developer solutions have free access to the photographs.

If several photographs must be developed simultaneously, they can be placed in the hanger in pairs (with their backs to each other). By using several hangers at the same time, in our case nine, a maximum of eighteen 60 x 60 cm photographs can be developed simultaneously. This is a great time-saving factor as each development step takes 22 min. We develop our colour photographs in the above-mentioned vertical unit at a temperature of 29° C. This fairly high temperature can only be used in vertical units with nitrogen agitation; when colour photographs are developed in flat vessels at this temperature, it is impossible to prevent colour spots. The temperature prescribed for the flat-vessel method is 24° C. The total developing time is then 44 min, this time being halved for a temperature of 29° C.

In order to make the fullest possible use of the P122 paper developer, we employ the following method.

At each development step, part of the developer is "consumed". This consumption is made up, for example, by removing two parts of the developer solution or solutions from the tank or tanks and adding three parts of regenerator instead. This keeps the developing chemicals constant, and we are able to process 300–400 50 x 60 cm sheets in our 10 gallon solutions without any noticeable colour shifts occurring. This regeneration method differs for each make and one should consult the directions enclosed with the chemicals

For washing the photographs, we prefer water of 29° C. Since large amounts are required, we purchased a 7.2 kW continuous-flow boiler which supplies the water via thermo-mixing taps.

We believe that all these provisions enable the prevention of any unexpected occurrence. They are expensive, but on the other hand, when the temperature fluctuation in the developing bath is more than 1/4° C and in the other baths more than 2° C, the process cannot be properly controlled and this considerably impairs the quality of the final product.

Drying and glossing

Every photograph, whether black-and-white or coloured, has to be dried after processing and if necessary also glossed, depending on the purpose for which it is made.

The difference between the two processes is that the wet black-and-white photograph has all the grey tints in the proper gradation; drying and especially glossing increase the effect of contrast and sharpness. Wet Ektacolor paper has a bluish-grey tint and the colours do not become visible until the photograph is entirely dry. This drying can be done without equipment, but in this case the photographs are not entirely flat and the drying time is fairly long.

There are several reasons why it is advisable to gloss the photographs. We do it on the rotating Kodak Rapid glossing machine, which has a drum width of 58 cm. This has the advantages that the photographs can be dried and glossed in a short time, and that they have distinctly better sharpness and are entirely flat.

Should, however, the gloss be a drawback for certain purposes, there are sprays and foils on the market which can be used to give the photographs a mat surface.

Finally, we must make the following observation on the keeping qualities of the colour photographs. Prolonged exposure of the colour photograph to bright daylight, especially to direct sunlight, should be avoided as it causes bleaching.

SUMMARY

The equipment used for making colour prints from miniature film should be of the highest technical and optical standards.

These prints are made at the Soil Survey Institute, Wageningen, by means of the negative–positive process.

In the first place, photographs of the thin section are made on daylight 24 x 36 mm negative film via the microscope. These negatives are developed by an eight-bath system at a temperature of 22° C with nitrogen administration, the total time being 56 min.

After being dried, the negatives are enlarged on colour-sensitive paper by a microfilm enlarger provided with a filter slide. The correct filters to be used for the colour correction can be ascertained by means of a number of test strips, after which the actual exposure can be made, care being taken to ensure that it is kept as constant as possible by means of a regulator.

The exposed photograph is then developed by a nine-bath system at a temperature of 29° C, also with nitrogen administration, the total time being 22 min.

After the final wash, the photograph is glossed on a rotating glossing machine.

To obtain good results, an operator with wide experience and skill is required.

COLOURED ILLUSTRATION

Plate G1. Example of a colour print from miniature film: different coloured Algae.

A QUANTITATIVE MICROSCOPICAL METHOD OF PYRITE DETERMINATION IN SOILS

L.J. PONS

Soil Survey Institute, Wageningen (The Netherlands)

INTRODUCTION

Sulphides producing strong acids during oxidation play a very important part in initial soil formation (Pons, 1960) or in ripening of muds to soils (Pons and Zonneveld, in press). When soft muds contain sulphides (pyrite) in excess of total bases ($CaCO_3$ and bases adsorbed to the clay complex), they are transformed into cat clays upon oxidation after drainage. The conditions of this cat clay formation are discussed by Van Beers (1962).

Ripening under brackish conditions, however, according to Pons (1963), results in "pseudo cat clays" with the same favourable physical properties as cat clays, but without poisonous chemical conditions. Pseudo cat clays also develop from fresh muds with about equal quantities of pyrite and bases.

Even small quantities of pyrites in reduced muds are important, however. The acids formed upon oxidation mobilize and leach part of the bases adsorbed to the soil complex (Pons, 1960).

It is very important to know the pyrite content together with the other characteristics of the soil, especially for reclamation of new land and improvement of drainage of old empoldered land. It is not difficult to estimate the clay content of a soil by "feel" and the carbonate content by effervescence with hydrochloric acid. For the pyrite content or the total content of sulphides, however, up to now only chemical laboratory methods have been available.

In this paper we will describe a rapid and accurate microscopical method, which also gives some information about the genesis of the soils.

THE GENESIS OF PYRITE

Harmsen (1954) found that marine deposits which have never been oxidized always contain a certain percentage of sulphides, mainly in the form of pyrite. Probably two processes are responsible for this formation. The pyrite found in marine sediments is formed by concentrations of S-containing proteins from dead marine organic matter.

As soon as reducing conditions prevail and recently dead organic matter is available, sulphate reducing bacteria can reduce the sulphates from the sea water. Pyrite is then formed with iron compounds from the mud.

Based on these two processes, Pons (1960) distinguished the following kinds of pyrite occurring in soils and sediments:

(1) *Primary pyrite*. This kind of pyrite occurs in floating silt and in recently deposited and not yet overgrown marine sediments. It is also present in ripened subsoils which have never been rooted in coastal plains. The photographs of Fig.1 show pyrite bodies of different shapes and dimensions, mixed with clay and silt particles of various kinds of reduced soil materials.

(2) *Secondary pyrite*. This part of the pyrite is formed in the deposited sediments after the vegetation has settled, but before the soils are ripened or when they are only in the first stages of ripening. The pyrite bodies of secondary pyrite are always observed in the cells of tissue remnants of the organic matter, as shown in Fig.2.

(3) *Tertiary pyrite*. When ripened or nearly ripened soils with humus or peaty topsoils are inundated with brackish water containing sulphates, and at the same time reducing conditions prevail in the soil, much pyrite can be accumulated, especially in the upper horizons. Microscopically this kind of pyrite cannot be distinguished from secondary pyrite.

Pyrite is easily identified microscopically because of its shape and size and by the fact that pyrite is really black. Fig.1 and 2 show samples of soils with the most common forms of pyrite.

According to Harmsen (1954) and our own results, only a very small amount of sulphides is monovalent sulphides. The greater part is bivalent and even polyvalent and, according to Crommelin (personal communication), has mainly the crystal shapes of pyrite and markasite. Both are totally opaque. Confusion with other opaque bodies is not likely because of their

---

Fig.1. Typical forms and occurrence of primary pyrite bodies.

A. Location: Arakonie, along Nickerie river, Surinam, about 70 km south of the coast. Peaty, silty marine clay from a brackish marine deposit below a peat layer at a depth of 2 m. A totally reduced, soft, potential cat clay. Between the clay and silt particles, a number of pyrite spheres, bunch-of-grapes and framboidal bodies of different dimensions (about 2–25 μ) are visible; x 100.

B. Location same as in A. A very large pyrite body with a bunch-of-grapes or framboidal form with a diameter of about 30 μ is visible; x 400.

C. Location: Acaribo along Surinam river, about 30 km south of the coast. Peaty, heavy marine clay from a brackish estuarine deposit in the subsoil of a riverine soil. A totally reduced, fairly soft, potential cat clay layer, containing 189 m.e. S per 100 g dry soil. A very clear pyrite cube is visible; x 200.

D. Location: Arakonie along the Nickerie river. Peat with some thin clay layers. An estuarine clay with a high content of organic matter and a medium content of pyrite. A pyrite sphere with a diameter of 18 μ is visible; x 200.

E. Location: De Goede Hoop along Coppename river, Surinam, about 10 km south of the coast. Totally reduced silty clay with a fairly high pyrite content. Lumps of clay with the pyrite bodies are mixed with silt and clay. Diameters of the pyrite bodies are up to about 10 μ; x 200.

F. Location same as in A. Silty clay with a high content of pyrite bodies. Part of the pyrite is still arranged, probably in its original arrangement in plant tissues. On the left side, many small pyrite grains (1–2 μ) are visible, probably held together by nearly disappeared cells; x 200.

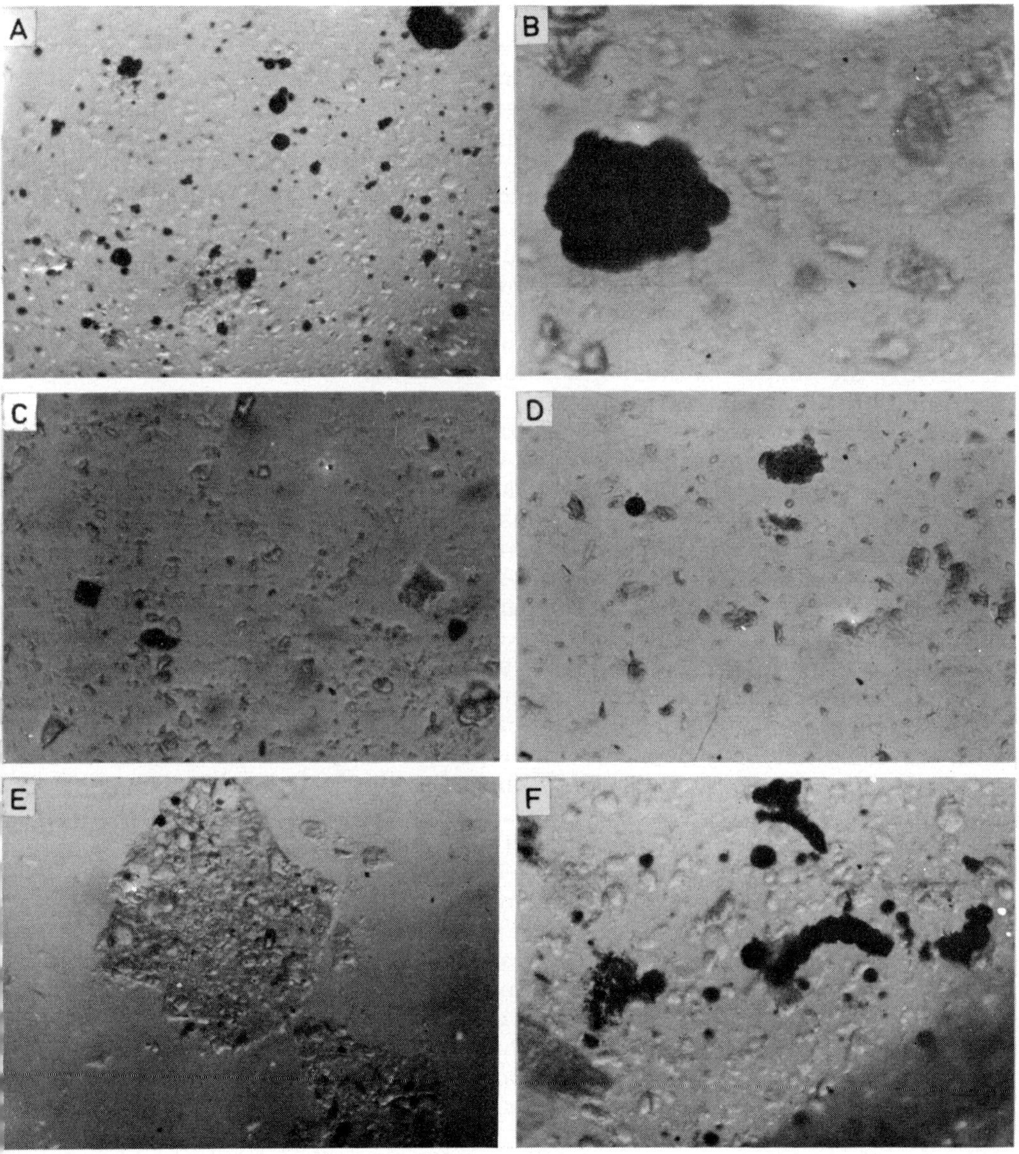
A
B
C
D
E
F

typical shape and size. They are mostly rounded spheres (Fig.1A, D, F; 2A, D, F), but sometimes clear cubes are observed (Fig.1C; 2B). It is very probable that a large part of the apparently rounded spheres are also more cubic, which could be checked with a microscope of very high quality. Sometimes the bodies are somewhat weathered and broken and blunt splinters are present (Fig.1C; 2A). Very typical are the rounded to oval framboidal (like a bunch-of-grapes) spheres (Fig.1A, F; 2A, F). Especially in sediments with much pyrite, large framboïdal bodies may be present with diameters of about 40 μ as shown in Fig.1B.

Some particles are very small and have sizes from 1–2 μ. Most of the pyrite bodies have dimensions between 2–20 μ.

Sometimes opaque organic matter is present, especially in Smonitza, such as black profiles with relatively low contents of organic matter. Microscopically the shapes of these particles of organic matter are more irregular and totally different from the pyrites.

Secondary pyrites are most easily seen in dead plant tissues. The cells may be totally filled with small black spheres (1–2 μ), which in most places have grown together (Fig.1F; 2A, C, D, F) to shapes resembling a bunch-of-grapes. After some time, a process of recrystallization seems to take place in these framboïdal spheres, which then appear more concentrated and more regular in outline (Fig.1A, F; 2A, F).

Destruction of a part of the organic material liberates at least a part of the pyrite, which is afterwards mixed with the clay, and which from this moment on is no longer recognizable as secondary pyrite. In Fig.2F, for example, this process has partly taken place.

The pyrite spheres may remain in this particular shape for a very long time, for they are very abundant in the permanently reduced deep subsoils of Pleistocene Coropina soils. This kind of pyrite body is also described by

---

Fig.2. Organic matter and occurrence of secondary pyrite bodies.

A. Location: Arakonie along the Nickerie river. Same layer as in Fig. 1A. A lump of organic matter of unknown origin with many pyrite spheres in the cells. A great many seem to be more or less cubic. On the right-hand size is a large pyrite sphere of about 50 μ diameter; x 200.

B. Location: same as in 2A, same layer. A lump of dead plant tissue. The pyrite bodies in the cells are all clearly cubic; x 200.

C. Location: Paramaribo, mud banks along the river, overgrown with young parwa (*Avicennia nitida*). A dead root of *Avicennia nitida* with secondary pyrite formation in the surface cells. In and between the oblong cells, strings of pyrite spheres are visible; x 100.

D. Location: Coppename river, mud banks with mangrove vegetation (*Rhizophora*). The small dead *Rhizophora* roots contain several kinds of pyrite bodies. The short oblong cells are partly filled up with strings of spheres and framboidal spheres; x 400.

E. Location: Coppename river, mouth Wayambo river. Mud with roots of moko moko. The rounded to oval cells contain no pyrites. The visible pyrite spheres occur in the cutans of clay lumps which are around the roots

F. Location: Arakonie along the Nickerie river. Old peaty marine clay (potential cat clay) below 2 m of peat. Tissues of unknown dead organic matter with cells filled up with pyrite spheres; x 200.

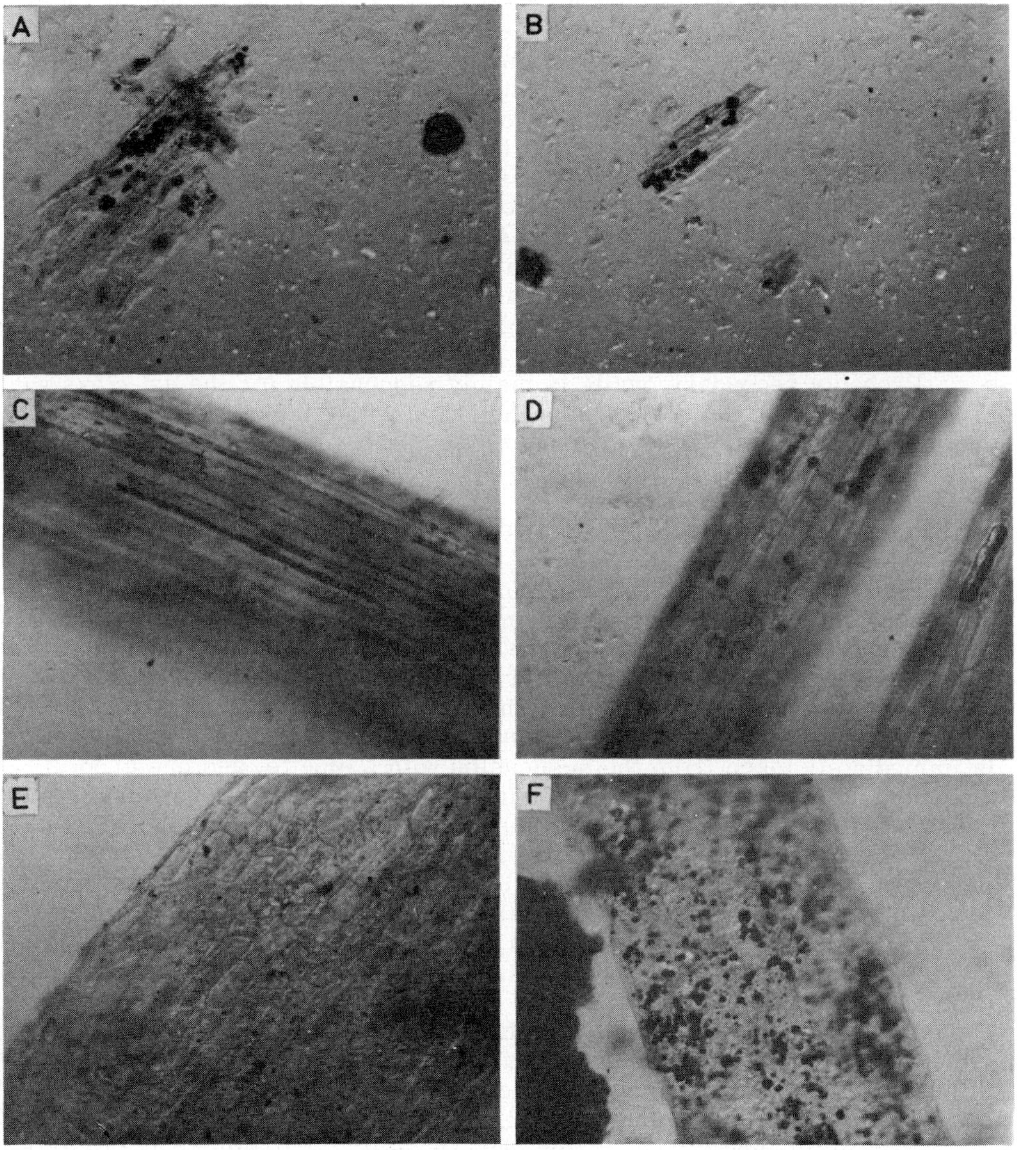
A
B
C
D
E
F

Love (1962) as framboïdal spheres of 3–20 μ diameter. He considers them to be microfossils from *Pyritosphaera barbaria* and found them in recent muds, as well as in Marl slates and Kupferschiefers.

After erosion of secondary pyrite containing muds and soils and after destruction of the organic matter, secondary pyrite is no longer distinguishable from primary pyrite and in fact has really become primary pyrite.

The primary pyrite consists of small pyrite particles mixed with clay and silt and probably forming a part of the humus-clay floc. These pyrites are mostly cubic or oval bodies, and sometimes also framboïdal.

Tertiary pyrite has about the same appearance as secondary pyrite. The only difference is that it occurs in remnants of all kinds of organic matter and not only in that originating from mangroves or other tidal pioneer vegetations.

Probably secondary pyrite is mainly responsible for the cat clays and pseudo cat clays in the tropics. In temperate areas, high pyrite contents are due mainly to primary pyrite (Pons, 1963).

A QUANTITATIVE MICROSCOPICAL METHOD

The greatest difficulty in the rapid quantitative microscopical method is to get a constant amount of soil in a homogeneous layer under a microscope cover glass. The following procedure has proved to be the best.

A quantity of 100 mg dry soil or a comparable quantity of moist soil is weighed and mixed in a beaker with a quantity of 10 ml water. A homogeneous suspension is made by stirring with a policeman. A small quantity is immediately taken from the suspension with a pipette, and one drop is placed on an object glass. It is very important that the soil suspension in the drop be the same as in the beaker, so quick work is essential.

The drop should contain about 0.5 mg soil, as this proves to be the most suitable quantity for microscopical examination. Nearly all the water is removed by evaporation and a small drop of glycerine or another non-evaporating liquid is added. With this new liquid, a very good suspension is made; this is covered with a cover glass of a known surface area, in our case 300 mm^2.

In the microcopic slide, which has been made as homogeneous as possible, the pyrite bodies are measured with an ocular micrometer at a magnification of 200 x. The dimensions generally vary between 1–20 μ. They are counted in four size classes: <2, 2–7, 7–15 and >15 μ.

Five strips are counted, each 360 μ wide (the diameter of the field of vision of the microscope) and 2,000 μ long. The strips are chosen to obtain a representative estimate of the pyrite content in the total microscopic slide

By multiplying the numbers of pyrite bodies in four size classes with the average weights of the pyrite bodies within these size classes, pyrite weights can be calculated (s.g. = ca. 5) per 5 times 360 x 2,000 sq.μ and per 300 mm^2 (i.e., per 0.5 mg dry soil). From this, the weight % of the pyrite is computed.

The size class < 2 μ is not important from a quantitative point of view, but it is an important fraction for consideration of genesis. It characterizes sediments with marine influence.

In some sediments with great quantities of pyrites, sometimes bodies of 20–30 μ also occur (Fig.1A, B). These bodies have much influence on the weight % and in most cases are characteristic for cat clays.

Counting in microscopic slides with much pyrite in the cells of organic matter gives difficulties because of the heterogeneity of the soil suspension.

Grinding samples with coarse organic matter to liberate the pyrite may solve this problem.

COMPARISON OF MICROSCOPICALLY ESTIMATED PYRITE WEIGHTS AND CHEMICAL PYRITE DETERMINATIONS

To test the described microscopical method, we investigated a number of samples for which chemical analyses of pyrite content were available. We used two groups of samples from Guianan coastal clays. The chemical data in each group were determined in a different way:

Group 1 consisted of 27 soil samples originating from British Guiana and sampled during the U.N.-B.G. soil survey. The chemical analysis was made by Cate Jr., F.A.O. soil chemist. He used the method of oxidizing the soil with H_2O_2 and measuring the SO_4-contents before and after oxidation.

Group 2 consisted of 13 soil samples from Surinam, the same as were used for the investigations of cat clay formation under fresh and saline conditions (Pons, 1963). The Soil Laboratory at Oosterbeek, The Netherlands, made the pyrite analysis by oxidizing the soil with strong acids and also measuring the SO_4 contents before and after oxidation.

In Fig.3A, the microscopical and the chemical data for the samples of B.G. are compared. The pyrite contents vary from nearly zero to about 90 m.e. S per 100 g dry soil. The microscopically determined quantities roughly fit in with the chemical data, although wide variations occur.

Fig.3B gives the results of the Surinam samples. In this case, there is a very good correlation between the two methods. The regression lines (a) in both graphs have the same gradient. We assume that the H_2O_2 oxidized the pyrite to a varying degree in the different samples. This would be an explanation for the irregularities of the correlation in Fig.3A.

In both graphs the lines (b) are also drawn, which show the theoretical relationship between the pyrite measured chemically and estimated visually. In both cases, at low pyrite content more pyrite is seen under the microscope than is measured chemically. At high pyrite content, the theoretical and determined microscopical quantities fit very well, especially in Fig.3B.

After some practice, it is very easy to estimate in a short time whether or not a dangerous quantity of pyrite is present and we believe that on the basis of this method a simple field test will also become possible.

SUMMARY

Marine deposits always contain sulphides as pyrite bodies in varying quantities. They are formed partly by the concentration of S-containing proteins from marine organic matter and are mixed with the clay and silt particles (primary pyrites). After sedimentation, they can also be formed in dead

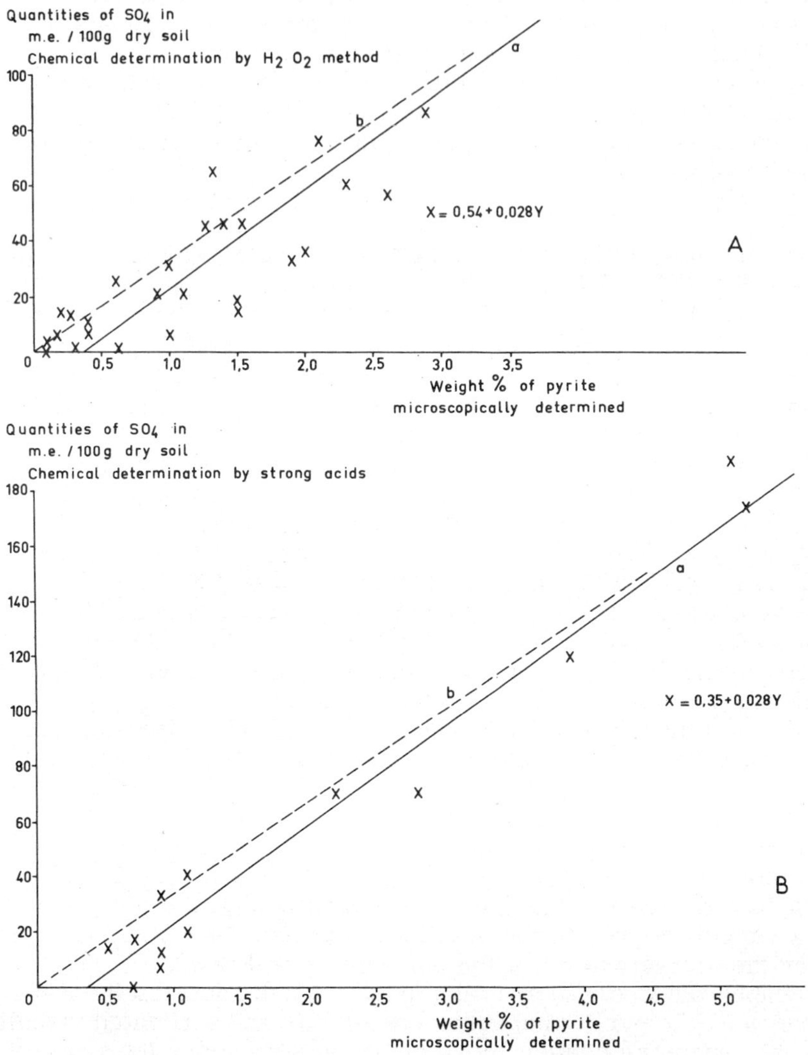

Fig.3. The correlation of chemically and microscopically determined pyrite contents. a = line indicating the microscopical estimated quantity of pyrite; b = quantity of pyrite calculated from the chemical data. A. Samples from British Guiana. B. Samples from Surinam.

organic matter of mangrove vegetations by bacterial reduction processes of sulphates from the seawater (secondary pyrites).

In initial soil formation, the oxydation of sulphides to strong acids is very important, not only when much pyrite is present (formation of cat clays and pseudo cat clays) but also when only small quantities occur (they mobilize and leach part of the bases of the soil complex).

The pyrite bodies are small (1–25 μ) and occur, as spheres, framboïdal or bunch-of-grapes-like spheres, and cubes and splinters of cubes; they are easily to recognize because of their typical shapes and because they are totally opaque.

A method has been developed to microscopically count the numbers of pyrite bodies in four size classes from which the weight % of the pyrite can be multiplied.

REFERENCES

Harmsen, G.W., 1954. Observations on the formation and oxidation of pyrites in the soil. Plant Soil, 5 : 324–348.

Love, L.G., 1962. Biogenic primary sulphide of the Permian Kupferschiefer and Marl slate. Econ. Geol., 57 : 350–366.

Pons, L.J., 1960. Het voorkomen en de vorming van pyriet in diverse mariene afzettingen in Nederland en een verklaring van het ontstaan van het knipverschijnsel. Soil Surv. Inst., Wageningen, unpublished rept.

Pons, L.J., 1963. Pyrites as a factor controlling chemical ripening and formation of cat clays, with special reference to the coastal plain of Surinam. Trans. Congr. Agr. Res. Guyanas, Paramaribo, 1963, in press.

Pons, L.J. and Zonneveld, I.S., in press. Initial soil formation (ripening) and soil classification in alluvial (marine and fluviatile) and peat deposits. Intern. Inst. Land Reclamation Improvement, Bull.

Van Beers, W.F.J., 1962. Acid sulphate soils. Intern. Inst. Land Reclamation Improvement, Bull., 3 : 31 pp.

A METHOD TO STUDY THE DISTRIBUTION OF BIOPORES IN SOILS

L. VAN DER PLAS and S. SLAGER

Institute of Regional Pedology, Geology and Mineralogy, Agricultural State University, Wageningen (The Netherlands)

INTRODUCTION

At present the earth sciences have advanced to the point where quantitative information is not only wanted by a few eccentrics, but also by a large number of investigators, including those who deal with the practical aspects of these sciences. As soon as practical problems are involved much value is attached to efficient procedures. It is possible to envisage measuring procedures which give reliable information on the size frequency distribution of pores in a soil sample. However, these methods are cumbersome, and the gathering of a sufficient amount of observations is a time-consuming affair. The alternative is to measure the porosity of the sample as a whole. Measuring porosity by current methods can be rapidly done, but in a number of cases it does not give the information needed. The same porosity can be produced by a large number of rather small pores or by a small number of rather large pores. Therefore the investigator has to acquire a well-balanced combination of adequate and reliable information and an efficient technique. Finally, the reproducibility and the reliability of the counting result must be given due consideration, for the days no longer exist when we can fool ourselves and others with such concepts as "mean" or "probable" error.

In this paper a counting and measuring technique will be described that enables one to gain an insight into the distribution of pores larger than 200 μ over the area of a soil profile. The pore distribution is of the greatest interest for problems of root development. It has been established that roots of higher order will only enter a pore with a rigid wall if the diameter is more than 200 μ (Wiersum, 1957). In sandy subsoils such pores are generally formed by biological activity only, and they are therefore named biopores (Slager, 1964).

EXPERIMENTAL

The observations on the structural properties of a soil profile are carried out with a stereomicroscope and with "peels"[1] of a soil profile. The peels are made according to the description of Jager (1964). The stereomicroscope must meet the following minimum requirements:

[1] In American literature this term is often used for the description of monoliths. As the term is of no importance here it will not be discussed any further.

(a) The stereomicroscope must be fitted to a stand allowing for a linear and continuous movement in the horizontal plane over a minimum distance of about 50 cm (Fig.1).

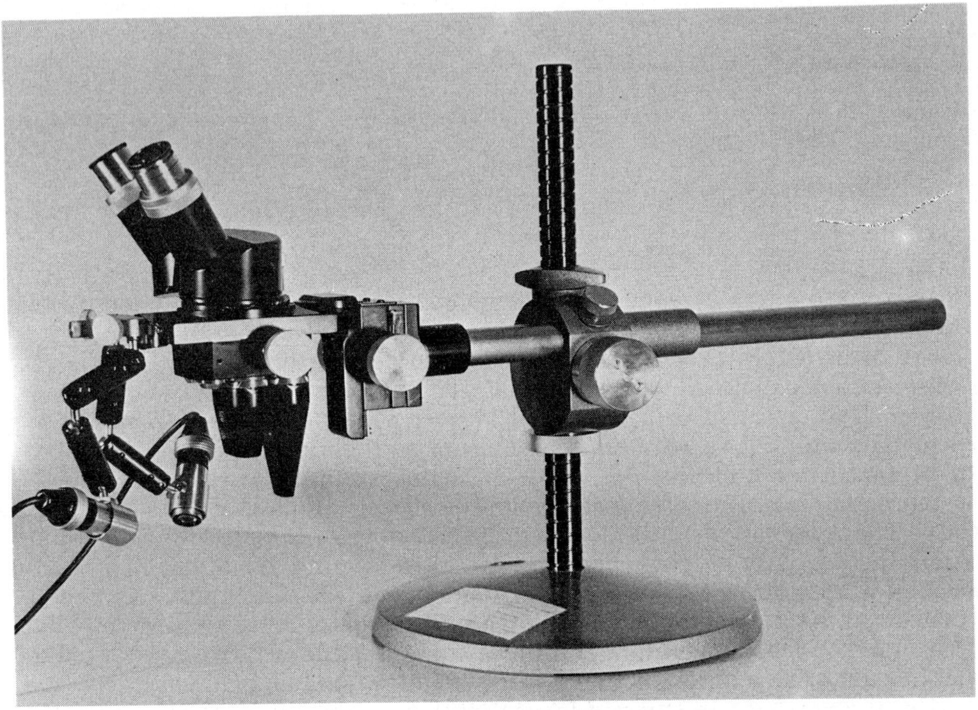

Fig.1. Leitz stereomicroscope of Greenough and Leitz stand type US II.

(b) As the object, i.e., the soil profile, is observed with incident light, an illuminating device is needed that will move with the travelling microscope and not produce an annoying cast shadow.

(c) One of the oculars must be equiped with a micrometer.

(d) The optical system must allow for an easy interchange of objectives without moving the apparatus from its present position, so that the measuring scheme is not disturbed.

(e) The optical system must be so constructed that the distance of the apparatus from the object is about the same for all objectives.

As far as we now know, the combination of the Leitz stereomicroscope of Greenough and the Leitz stand type US II, which we used, is the only device meeting the requirements mentioned above (cf. Fig.1, 2).

The object under study, such as the peel or any other undisturbed sample of a soil profile, rests on a horizontal table; the microscope moves over the surface to be sampled. A grid is laid over the sample area in order to provide the operator with a reference system while measuring and counting. In our case a wooden frame, fixed with a centimeter grid of thin threads, was used (Fig.2).

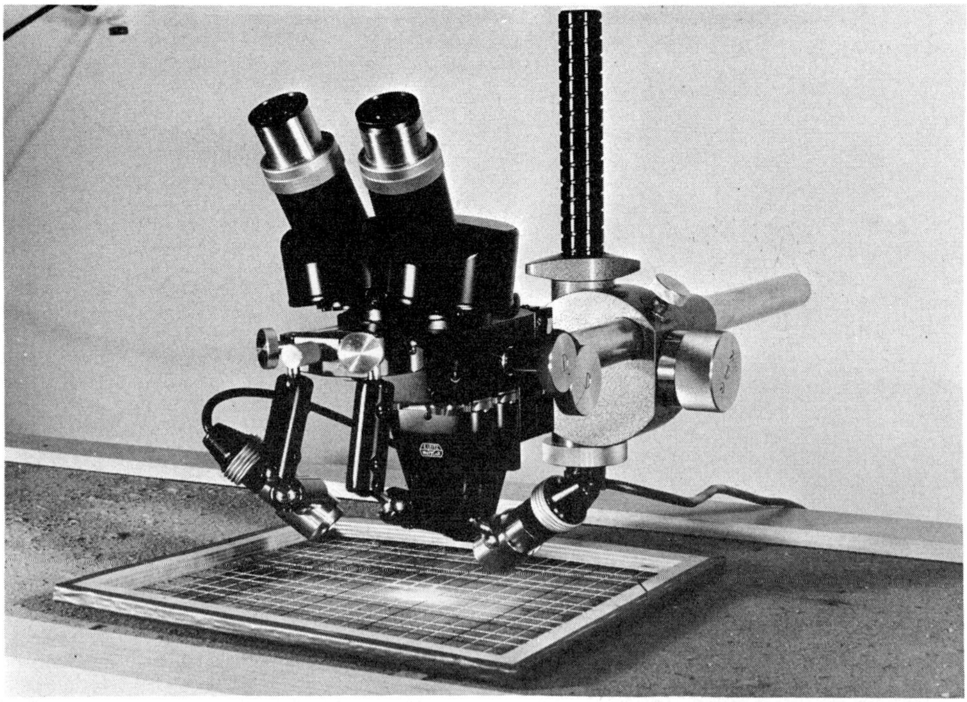

Fig.2. Microscopic device over soil peel with grid frame.

MEASURING TECHNIQUES

When measuring pores with various diameters and a variety of shapes, the measuring technique must have such properties that the result will not show any correlation between the object properties and the properties of the measuring technique. These aspects of measuring should be given due consideration, because otherwise the results may have no practical meaning unless one is willing to go through the tedious procedure of finding adequate correction factors (Van der Plas, 1962). For the purpose at hand the ribbon counting technique should be used. This technique applies a band systematically laid over the sample area. The band's area determines whether a pore is counted or not. Only those pores are counted whose mid-points are situated within the band. In order to meet the requirements of the ribbon counting procedure, the microscope is moved over the sample area in a direction normal to the depth of the profile. The sides of the ribbon are determined by two marks of the ocular-micrometer. In order to reach this situation the micrometer scale should be set normal to the direction of measuring.

Because the distance between two marks on the micrometer as projected on the object is different for different combinations of ocular and objective, measuring should always be done with the same combination of optical parts. In our case we used a magnification of 50 x, the scale unit being 25 μ. In order to avoid correlation effects, the ribbon should be several times wider

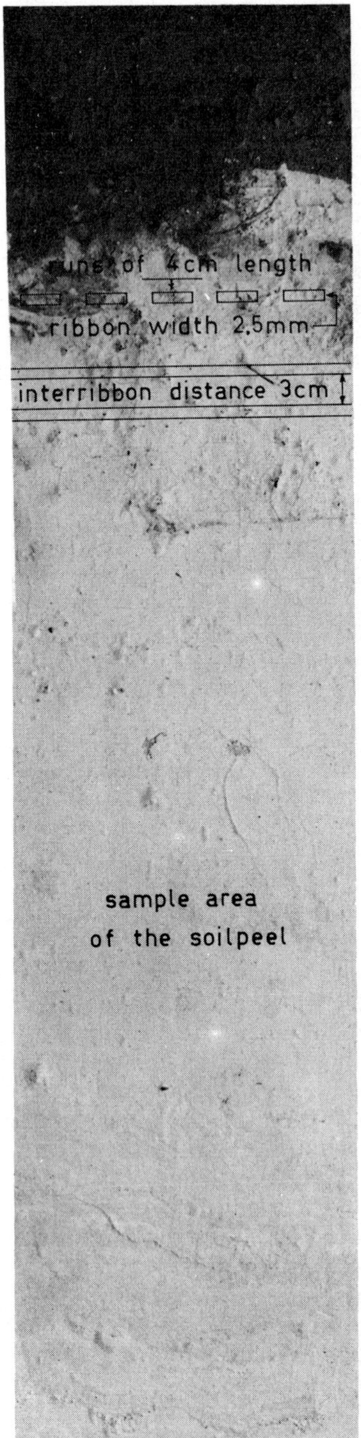

Fig.3. A schematic picture of a soil peel covered with the systematic grid of ribbons, with an inter-ribbon distance of 3 cm. The ribbons are 2.5 mm wide. At least 5 runs of 4 cm length are situated in every ribbon. The counting result is expressed in the number of pores larger than 200 μ/cm^2.

than the diameter of the larger pores present; therefore the width of the ribbon is taken as 2.5 mm.

Counting is done in runs. This method has advantages because in this way heterogeneities in the horizontal plane do not go undetected. For that reason the counting results are written down every time a ribbon of a length of 4 cm is finished; at least five such runs are measured in each horizon (Fig.3). The counting result is written down as the number of biopores larger than 200 μ/cm². Pores smaller than 200 μ are not counted. The runs are taken 3 cm apart.

If the need is felt to have some more information on the region between two specific runs in the graphical plot of the counting results, it is possible to count runs between the two previous ones. Such additional information may be wanted, e.g., if a specific run shows a large increase of pores/cm² in comparison to the previous run.

PRACTICAL ASPECTS OF THE COUNTING PROCEDURE

It is commonly felt that one of the most important aspects of a quantification method is the establishment of mutually exclusive classes of objects. In plain language this means that in this case it is necessary to first describe exactly what is meant by, and counted as, a pore. This must be described in such a way, that the reader is able to apply the concept to his own problem or modify it to fit his case. The transference of criteria in the case of counting pores in a soil profile has been found to be rather difficult, and until now we have followed the procedure of having skilled operators training unskilled operators. They work together on the same profile and check each other's observations, and from time to time they count standard areas in order to correct "bad habits" developed during a longer period of measuring.

Although it is felt that the criteria are difficult to describe, some aspects are given in the following enumeration. The following pores are counted in the procedure:

(1) Pores that in all probability are the end of a burrow cut normal to its axis; (a) if they have circular outlines and a diameter larger than 200 μ; (b) if they have irregular outlines, solid walls, and fit a circle with a diameter of 200 μ; (c) if they have irregular outlines and a solid wall plastered with loose material (e.g., clay particles or silt particles loosely attached to the wall of such pores), and a 200 μ circle fits the outline of the rigid wall.

(2) Burrows, cut parallel to their length axis and with a visible width of at least 200 μ.

(3) Burrows cut obliquely, but only if it can be ascertained that the pore at least has a normal diameter fitting a circle of 200 μ.

These traces of burrows may show the relics of plant roots, which constitutes no reason for excluding them from the counting result. Profiles often show cracks and vugs which are not counted. Partly-filled burrows are counted only if the remaining vug fits a 200 μ circle. Funnel-shaped holes are counted if the minimum diameter fits the 200 μ circle.

TABLE I

Count of the "biopore trend" and "biopore spread" of a soil peel by two operators, showing trend, spread and reproducibility of profile Blerick 451

Depth (cm)	First operator, A								Second operator, B							
	A	B	C	D	E	$\bar{x}^1$	min.	max.	A	B	C	D	E	$\bar{x}^1$	min.	max.
3	12	11	3	2	13	8,2	2	13	9	12	4	3	12	8,0	3	12
6	7	8	6	10	8	7,8	6	10	8	7	7	9	8	7,8	7	9
9	5	2	7	2	4	4,0	2	7	4	2	7	2	4	3,8	2	7
12	11	4	4	4	6	5,8	4	11	11	6	4	5	7	6,6	4	11
15	7	3	2	2	3	3,4	2	7	6	4	2	1	4	3,4	1	6
18	8	8	10	8	7	8,2	7	10	9	8	11	8	5	8,2	5	11
21	3	5	4	1	1	2,8	1	5	4	3	6	2	1	3,2	1	6
24	1	2	2	1	2	1,6	1	2	1	2	2	1	1	1,4	1	2
27	1	6	1	3	4	3,0	1	6	1	7	2	2	4	3,2	1	7
30	5	0	4	8	7	4,8	0	8	6	0	3	8	9	5,2	0	9
33	7	3	7	9	11	7,4	3	11	7	3	9	9	12	8,0	3	12
36	25	19	6	9	13	14,4	6	25	20	15	7	11	11	12,8	7	20
39	17	8	14	19	16	14,8	8	19	16	6	12	18	14	13,2	6	18
42	17	13	15	24	21	18,0	13	24	15	11	15	26	16	16,6	11	26
45	11	12	21	24	21	17,8	11	24	13	15	20	23	22	18,6	13	23
48	19	22	23	28	31	24,6	19	31	14	25	27	25	32	24,6	14	32
51	21	25	19	22	29	23,2	19	29	20	21	14	18	25	19,6	14	25
54	14	19	24	25	14	19,2	14	25	12	21	27	29	18	21,4	12	29
57	13	14	21	30	21	19,8	13	30	12	20	21	32	21	21,2	12	32
60	12	10	16	23	13	14,8	10	23	12	11	16	21	12	14,4	11	21
63	9	12	13	17	15	13,2	9	17	8	9	12	15	14	11,6	8	15
66	9	10	19	17	9	12,8	9	19	10	11	21	17	10	13,6	10	21
69	17	15	21	16	8	15,4	8	21	15	13	24	15	11	15,6	11	24
72	17	18	16	17	10	15,6	10	18	14	21	18	15	10	15,6	10	21
75	19	10	10	25	32	19,2	10	32	18	12	10	26	32	19,6	10	32
78	11	4	11	12	16	10,8	4	16	13	6	10	12	19	12,0	6	19
81	8	14	11	11	13	11,6	8	14	10	6	11	8	15	10,0	6	15
84	7	13	16	15	15	13,2	7	16	7	11	16	14	11	11,8	7	16
87	7	7	10	16	12	10,4	7	16	6	8	11	15	12	10,2	6	15
90	5	6	10	17	17	11,0	5	17	7	6	10	16	18	11,4	6	18
93	11	16	10	12	13	12,4	11	16	12	18	7	9	9	11,0	7	18
96	5	8	11	7	10	8,2	5	11	4	7	14	7	7	9,8	4	14
99	9	10	9	6	8	8,4	6	10	10	10	7	7	10	8,8	7	10
102	5	6	9	6	14	8,0	5	14	7	8	11	3	13	8,4	3	13

[1] Column $\bar{x}$ lists the mean of the 5 runs A, B, C, D, E, covering a band with an area 1 cm^2 and a length of 4 cm each in these bands of 4/0.25 cm. Only pores with a diameter larger than 200 μ were counted.

PRESENTATION OF THE MEASURING RESULTS

In problems involving both presentation and reliability of measurements, one should first consider the properties of the results desired. It is useless to spend a great deal of time in carrying out laborious measurements and afterwards use the results in a qualitative way only. This remark is not thought superfluous. A number of papers, for example, give lists of accurate granulometric analyses or chemical analyses of soil fractions. Comparison of the lists of analyses with the conclusions reveals that the lists are hardly used. Such conclusions could just as well be based on a few field estimates between finger and thumb.

The first aspect of the measurements is the number of pores per unit of surface area, since the mechanical resistance for the plant roots decreases as the number of pores larger than 200 μ increases. However, the average number, though interesting for a comparison of a number of soil profiles and their agricultural value, is not always sufficient for answering questions concerning possible root development. For instance, the detection of a zone in the soil profile that is less permeable for plant roots is not enough. The properties of such zones, such as whether they are continuous or sometimes broken or even frequently broken over short distances, are also important. Therefore the presentation of the counting results should emphasize the spread of the measurements. This, among others, is one of the reasons why counting is done in runs and why a minimum of 5 runs per ribbon is advised. If the measurements are compiled, the information on spread is just lost.

Furthermore, the trend of the increase or decrease of the number of pores per unit of surface area is considered to be as important as the absolute figure. In order to meet these requirements, the counting method has to be sensitive, for the increase and decrease of objects and the operator's error of the trend needs to be rather low. Such a counting technique must have a high reproducibility, i.e., the criteria for the mutually exclusive classes must be unambiguous.

The right side of Table I (results of operator B) shows the counts on a real peel of a soil from Blerick (451); the counts are graphically represented in Fig.4. The solid line connects the arithmetic means of each set of five runs. The dotted lines on both sides of the mean connect the extreme values observed in a set of five runs. In this way the "average biopore trend", of importance for classification and comparison, is represented in combination with the "biopore spread" in each depth level.

Counting results not only reflect the distribution of pores over the sampling area, but are also influenced by fluctuations due to chance. In order to suppress such fluctuations one may calculate smoothed values from a set of measurements. Such a set of smoothed values will better reflect the general direction taken by the number of pores in the profile that is subject to change from top to bottom of the profile.

In statistical literature a number of sophisticated methods are reported for the calculation of smoothed values, cf. Whittaker and Robinson (1924, p.285) and Mellor (1955, p.321). In our case, however, a rather simple procedure was thought to suffice. Such a method is the calculation of the arithmetic means of any combination of two successive observations at a time, and the plotting of these values. It will be seen that this is similar to determining the mid-points of the lines connecting two successive observations. Connecting these points gives a kind of trend line. One should, however, be very careful with the interpretation of such smoothed values.

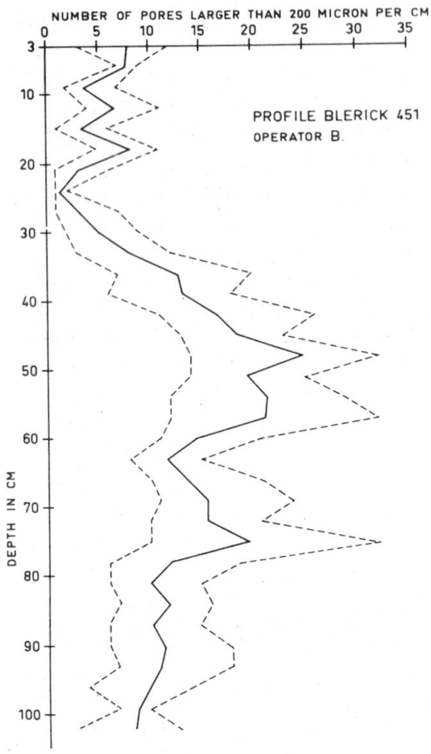

Fig.4. "Biopore trend" and "biopore spread" of profile Blerick 451.

THE REPRODUCIBILITY OF THE METHOD

The possibility of finding the same results in counting the same area is influenced by a number of factors. Some of these factors can be isolated from the others and their contribution to the deviation can be measured. Other influences are correlated with each other and it is only possible to measure their total result. We measured two aspects of the reproducibility, viz.:

(a) To what extent is one operator able to reproduce his own observations after a few days, in both cases measuring exactly the same area? It was found that a trained operator can reproduce his results within a margin of error of 10%.

(b) How large is the deviation of the results of two trained operators measuring exactly the same area? This difference generally proved to be within 10% of the number of pores measured in the area (cf. Table I).

In carrying out these experiments we found that trained operators showed a tendency to develop certain habits during routine counts. It became evident that after a certain time their criteria for the classes of holes to be involved in the result changed slightly. These "bad habits" became evident while counting a "standard area". In order to cure such "bad habits" and to train the operators to produce the same kind of results, we developed a scheme which may be given in three commandments:

(a) Do not let an untrained operator work alone without adequate and frequent control.

(b) Let two operators change place while counting, and let them compare their results frequently.

(c) Provide a number of standard areas that are measured frequently.

ACKNOWLEDGEMENTS

The investigations, the results of which are presented here, are part of a larger research program. Several students have participated in this program. The authors are indebted to Messrs. C.O. van der Veen, J.D. Kalma, G.J. Hofstede and V.J.G. Houba, students of the Wageningen Agricultural University. The authors gratefully acknowledge the stimulating discussions carried on with Prof. Dr. C.H. Edelman.

SUMMARY

A method is described which enables one to gather quantitative information on the distribution of pores larger than 200 μ in soils by means of soil peels and a stereoscopic microscope. The apparatus and the measuring technique are discussed in detail as well as the presentation of the results and the reproducibility of the method. Special attention has been paid to the definition of the pores to be counted.

REFERENCES

Jager, A., 1964. Preparation of soil monoliths. Soil Surv. Papers, Neth. Soil Surv. Inst., 2, in press.

Mellor, J.W., 1955. Higher Mathematics for Students of Chemistry and Physics, 4 ed. Dover, U.S.A., 641 pp.

Slager, S., 1964. A study of the distribution of biopores in some sandy soils in The Netherlands. In: A. Jongerius (Editor), Soil Micromorphology. Elsevier, Amsterdam, pp.421–426.

Van der Plas, L., 1962. Preliminary note on the granulometric analysis of sedimentary rocks. Sedimentology, 1 (2) : 145–157.

Wiersum, L.K., 1957. The relationship of the size and structural rigidity of pores to the penetration of roots. Plant Soil, 9 (1) : 75–85.

Whittaker, E.T. and Robinson, G., 1924. The Calculus of Observations. Blackie, London, 395 pp.

A STUDY OF THE DISTRIBUTION OF BIOPORES IN SOME SANDY SOILS IN THE NETHERLANDS

S. SLAGER

Agricultural State University, Wageningen, Institute of Regional Pedology, Geology and Mineralogy (The Netherlands)

INTRODUCTION

When considering the properties of plant roots in relation to soil porosity according to Wiersum (1957) the following statements are of fundamental importance:

(1) Most plant roots which penetrate more or less deeply into the subsoil have a diameter exceeding 200 μ.

(2) Pores larger than 200 μ only occur in sand with a single grain structure if the sand has a grain size of at least 800 μ.

(3) The penetration of plant roots in sandy subsoils with a single grain structure is inhibited by the absence of larger pores.

According to Wiersum (1957) this barrier to root development is absent if the sand is mixed with organic matter and/or inorganic colloidal particles, for structural elements are formed in those cases.

One must also note the following considerations:

(a) According to our experience a possible development of structural elements is restricted to the topsoil in sandy soils.

(b) Sandy soils in The Netherlands generally have a grain size smaller than 800 μ.

(c) Root development is observed in a number of sandy subsoils. From them we conclude that, at least in a number of cases, pores larger than 200 μ occur in sandy subsoils of another type than the ones mentioned above.

These pores larger than 200 μ have been proved to exist by studying these sandy subsoils systematically by means of soil peels and a stereoscopic microscope (Van der Plas and Slager, 1964). Judging from a number of properties such as shape, we assume that these pores were formed by biological activity, and therefore we call them biopores.

The number of biopores may vary considerable with depth in the same soil, and at the same depth in different soils. In this paper the distribution of biopores in some typical soil profiles will be discussed.

OBSERVATIONS

Profile Rolde 335
Land use: arable land. Parent material: cover sand. Hydrology: well-drained. Classification: old arable land cover on acid brown forest soil.[1]

---

[1] According to the U.S.D.A. Soil Classification (1960): Intergrade Typorthod-Plaggudent (Ed.)

Soil horizons

0–17 cm: Ap1, dark gray, topsoil, cover, clear and smooth, on
17–37 cm: Ap2, dark gray, cover, clear and irregular, on
37–59 cm: B2b, brown diffuse and smooth, on
59–80/94 cm: B3b, pale brown, clear and wavy, on
> 80/94 cm: Cb, whitish gray, with an undisturbed sedimentary structure.

Bioporosity: cf. Fig.1.

The topsoil has a high amount of biopores per cm^2, the bioporosity slowly increasing towards the ploughsole. The ploughsole is thin, shallow and rather porous. Below this ploughsole the bioporosity increases to a maximum at about 35 cm. Below this level it gradually decreases with depth, without reaching a comparitively low value near the bottom of the profile.

Profile Haaren 378

Land use: arable land. Parent material: cover sand. Hydrology: well to somewhat excessively drained. Classification: old arable land cover on acid brown forest soil.[1]

Soil horizons

0–18 cm: Ap1, brownish gray, topsoil, cover, gradual and smooth, on
18–42 cm: Ap2, brownish gray, cover,gradual and smooth, on
42–57/65 cm: Ap3b, brownish gray, zone of tillage in the original A-horizon and top of the original B-horizon, clear and irregular, on
57/65–72/77 cm: B2b, light brown, clear and irregular, on
>72/77 cm: B3b, pale brown, lacking patches of the undisturbed sedimentation structure until at least 1.25 m below the surface.

Bioporosity: cf. Fig.2.

The biopore trend shows no systematic increase or decrease over the depth of the cover, i.e., from the surface until a depth of 42 cm. In this zone, however, there are several relative minima and maxima, with a small and large spread respectively. Below this level the bioporosity increases to a maximum at 60–65 cm, followed by a decrease. At a depth of one meter below the surface the same number of biopores has been observed as near the surface. Both the high bioporosity and the large spread in the bioporosity in the B-horizon are striking features in this profile.

Profile Venray 260

Land use: arable land. Parent material: cover sand. Hydrology: excessively drained. Classification: old arable land cover on truncated acid brown forest soil.[1]

Soil horizons

0–26 cm: Ap1, dark brownish gray, topsoil, cover, abrupt and smooth, on
26–59/65 cm: Ap2, brownish gray, cover, clear and irregular, on
59/65–88/95 cm: B2b, light grayish brown, the top of which is the zone of tillage in the truncated original B-horizon, clear and wavy, on
>88/95 cm: B3b, light yellowish brown, containing both patches of the disturbed and the disturbed sedimentary structure.

Bioporosity: cf. Fig.3.

The graphical presentation of the bioporosity shows a minimum only for the recent ploughsole. Otherwise no systematic differences in bioporosity are noticed from the top to the bottom in this profile.

Profile Rolde 337

Vegetation: *Callunetum* Parent material: cover sand. Hydrology: excessively drained. Classification: humus podzol as a secondary development in an acid brown forest soil.[2]

---

[1] According to the U.S.D.A. Soil Classification (1960): Plaggudent (Ed.)
[2] According to the U.S.D.A. Soil Classification (1960): Humod (Ed.)

Soil horizons
0–12 cm: A1, black, gradual and smooth, on
12–21 cm: A2, dark gray, clear and smooth, on
21–27 cm: B1, black, clear and smooth, on
27–36 cm: B2h, dark brown, gradual and smooth, on
36–60 cm: B2, light brown with dark brown humus fibers, smooth and gradual, on
> 60 cm: B3, pale brown, containing both patches of undisturbed and disturbed sedimentation structure.

Bioporosity: cf. Fig.4.

The amount of biopores decreases more or less gradually with depth. Biopores are even present at greater depth though few in number. This profile is characterized by a maximum of bioporosity at the surface, and a low level of bioporosity in the rest of the profile.

Profile Someren 19
Land use: arable land. Parent material: cover sand. Hydrology: imperfectly drained. Classification: wet A–C profile.[1]

Soil horizons
0–15/23 cm: Ap, dark brown, topsoil, irregular and abrupt, on
15/23–32/38 cm: ApCg, light brown patches in a whitish gray matrix, irregular and abrupt, on
> 32/38 cm: Cg, whitish gray, the first patches of the undisturbed sedimentary structure are noticed in the top of this layer. Lower than about 85 cm no traces of disturbance of this structure have been noticed.

Bioporosity: cf. Fig.5.

The number of biopores decreases more or less gradually with depth. Striking features of this profile are the maximum number of biopores near the surface, the low level of bioporosity in the rest of the profile, and the shallow depth at which the bioporosity tends to zero.

INTERPRETATION AND DISCUSSION

For the interpretation of the graphs we will first discuss the properties of the acid brown forest soil under natural vegetation, i.e., under forest. We did not succeed in finding an example of such a profile since most of these soils have been reclaimed by man. This large scale action strongly suggests that these soils are considered to be the best sandy soils, from an agricultural point of view, in The Netherlands. Such a forest profile would show a high bioporosity near the surface. The number of biopores is assumed to decrease very fast with depth near the surface (A-horizon); lower on it will remain constant or decrease gradually (B-horizon), and become zero only at a very great depth (C-horizon). Rode (1962, p.187, table 38; p.371, table 88) showed that the same trend exists for the total porosity in several profiles. In the profile under consideration there is a high biological activity of animals and plant roots up to a great depth. The number of soil horizons is small, and the horizon-boundaries are diffuse.

After reclaiming this profile the biological activity will soon decrease. This implies that biopores which are obliterated will not be substituted by new ones, resulting in a decrease of bioporosity. Decrease of bioporosity is expected to be rather strong in the topsoil. Here, after reclamation, the high level biological regeneration is substituted by a lower level mechanical regeneration. The annual ploughing loosens the soil mass, enabling the roots

[1] According to the U.S.D.A. Soil Classification (1960): Umbraquept (Ed.)

to penetrate. The settling of the soil mass around the growing roots will lead to the formation of new biopores in the course of the year. At the ploughsole level this regeneration of the structure does not occur. Below the ploughsole the number of biopores increases gradually 5^0 as to reach a maximum value at some depth. Below this maximum value the bioporosity will again gradually decrease with depth.

While studying soils of this type we found several variations within the standard pattern just described. The ploughsole, for example, may vary in thickness, depth, and in the amount of biopores per cm^2. The maximum value in the graph may occur at different depth levels, while the amount of biopores in the B-horizon, as well as the amount of biopores in the subsoil, may vary widely. These variations may, on the one hand, be correlated with depth and the nature of soil tillage, and with hydrological properties of the original profile on the other.

In The Netherlands a group of man-made soils has been formed on these acid brown forest soils. They are known as "old arable land soils", cf. Edelman (1950). They have been gradually heightened over centuries by fertilizing with earth manure.

In profile Rolde 335 (Fig.1), for instance, we found the bottom of the black zone of tillage at a depth of about 35 cm; at the same depth the maximum amount of biopores occurs. A similar phenomenon was observed in most of the soils studied in this group. The tillage is supposed to have taken place until this depth; the original surface was at a depth of 15–20 cm. The initial rather simple mode of soil tillage has caused a strong reduction in the bioporosity in the topsoil of the original profile, but has not lead to the formation of a ploughsole. The original surface has been heightened by a relatively thin cover, characterized by a bioporosity of equal mean and large spread. At the bottom of the recent topsoil the ploughsole is found to be thin, shallow and rather porous.

In profile Haaren 378 (Fig.2) the bottom of the black zone of tillage occurs at about 60 cm below the surface. The original profile has been heightened by 40–45 cm. The cover is characterized by the absence of a systematical trend of bioporosity with depth. However, there is a distinct spread in bioporosity, varying with depth. Several ploughsoles can be distinguished in this zone, a phenomenon we labelled "ploughsole succession".

In the group of soils under consideration the basic pattern of bioporosity is always the same, notwithstanding a distinct variation.

An exception has to be made for a number of old arable land soils which are now situated on the boundary of the reclamation area. These soils provided the source of earth manure for the arable regions which are now more in the centre of the reclamation area; afterwards they were also used as arable land. This explains why the top part of the original profile is now absent. Profile Venray 260 (Fig.3), for example, lacks both the original A-horizon and part of the B-horizon. During the initial tillage the rather

---

Fig.1–5. Graphical presentation of the distribution of biopores in some soil profiles described in the text. The solid line connects the arithmetic average of a set of five observations at one depth level. The dotted lines on both sides of the averages connect the extreme values observed in such a set of five observations. In this way "average biopore trend" and "biopore spread" are represented at each depth level, cf. Van der Plas and Slager (1964).

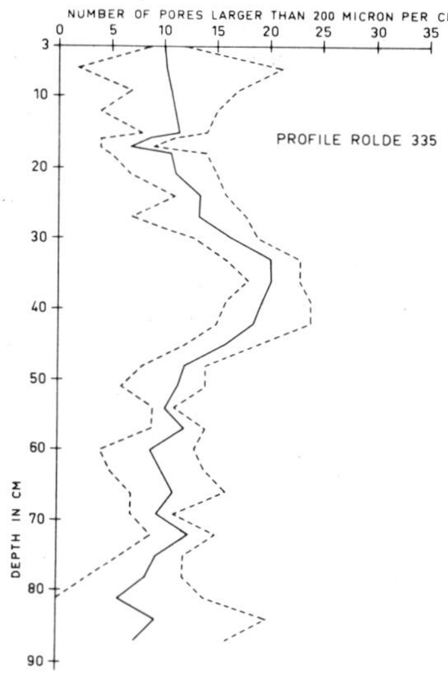
NUMBER OF PORES LARGER THAN 200 MICRON PER CM^2
0 5 10 15 20 25 30 35
3 10 20 30 40 50 60 70 80 90
DEPTH IN CM
PROFILE ROLDE 335

1.

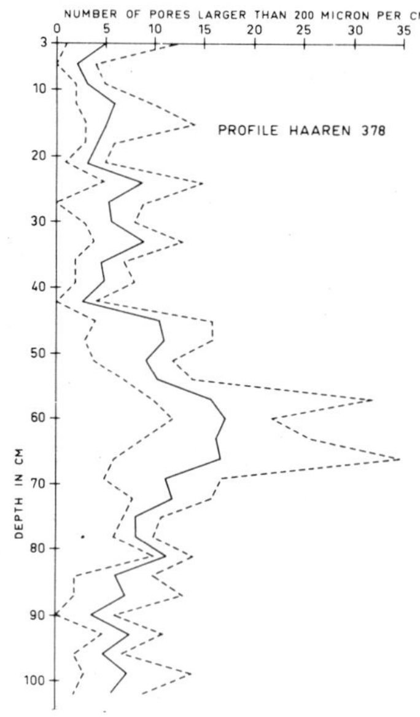
NUMBER OF PORES LARGER THAN 200 MICRON PER CM^2
0 5 10 15 20 25 30 35
3 10 20 30 40 50 60 70 80 90 100
DEPTH IN CM
PROFILE HAAREN 378

2.

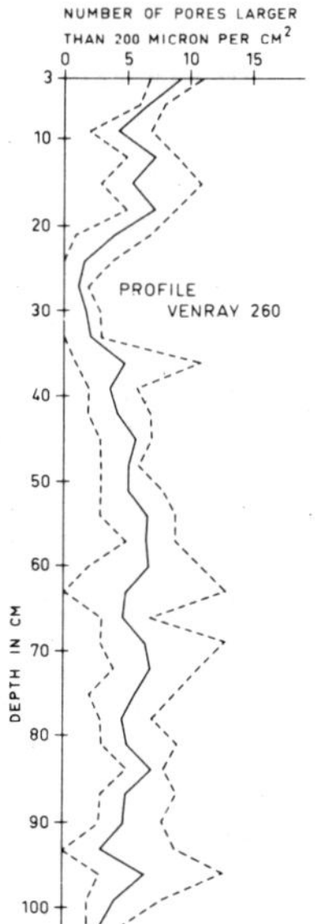
NUMBER OF PORES LARGER THAN 200 MICRON PER CM^2
0 5 10 15
3 10 20 30 40 50 60 70 80 90 100
DEPTH IN CM
PROFILE VENRAY 260

3.

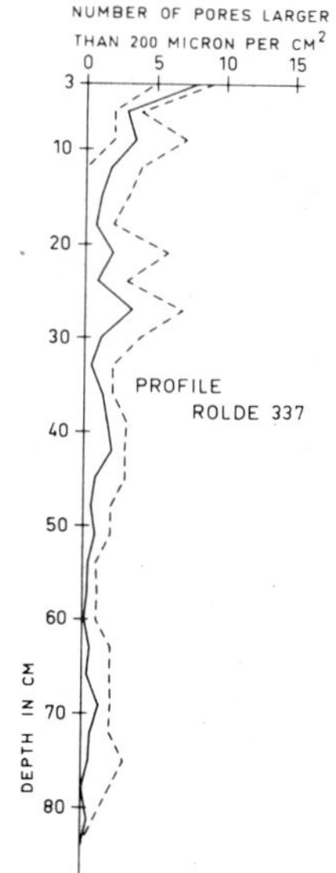
NUMBER OF PORES LARGER THAN 200 MICRON PER CM^2
0 5 10 15
3 10 20 30 40 50 60 70 80 90
DEPTH IN CM
PROFILE ROLDE 337

4.

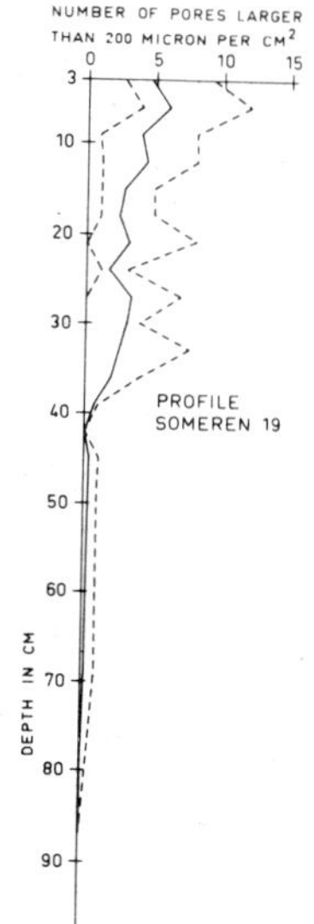
NUMBER OF PORES LARGER THAN 200 MICRON PER CM^2
0 5 10 15
3 10 20 30 40 50 60 70 80 90 100
DEPTH IN CM
PROFILE SOMEREN 19

5.

high porosity of the B-horizon was reduced, followed by the heightning of the profile. The resulting bioporosity does not show distinct systematic variations from the top to the bottom of the profile.

Profile Rolde 337 (Fig.4), a degenerated acid brown forest soil, now a humus podzol, shows a distinct disturbance of the sedimentary structure below a compact B2h-horizon. This phenomenon cannot be correlated with recent soil development. This soil is supposed to have been situated under forest and it must have had a high bioporosity originally. The biological activity was later minimized by devastation, deforestation and the growth of heather. Together all these influences resulted in the formation of a humus podzol.

Finally, the graph of profile Someren 19 (Fig.5) characterizes a soil with a low initial bioporosity due to the fact that it may be considered rather wet for strong and deep biological activity. Consequently the formation of biopores has been severely hampered.

FINAL REMARKS

Mitscherlich (1905, p.32) stressed that from the practical point of view soil texture and soil colour are the most important characteristics of a soil. Up till now much value has been attached to the separation of the profile into pedogenetic horizons; according to colour, texture, etc. Soil structure seems difficult to distinguish in sandy soils. As soon as a soil horizon is homogeneous according to colour and texture it is generally supposed to be homogeneous throughout. Biopore spread shows that apparently homogeneous soil horizons are not always so in respect to root penetration. Comparison of biopore trend, soil colour and soil texture reveals that in many cases a change in colour or texture does not correspond to a change in bioporosity; on the other hand many changes in bioporosity are not correlated with changes in colour and texture. During the investigations presented here, the impression was strong that both for the purpose of soil classification and for land use classification more attention has to be paid to soil porosity, in addition to the traditional items now included in the profile description.

Until quite recently the definition of the A-horizon usually included a remark concerning biological activity, stating that biological activity reaches a maximum value in the top of the A-horizon. The highest biological activity generally leads to the highest bioporosity. Consequently, this statement may give rise to misunderstandings. Observations prove this statement to be true for sandy soils now under natural vegetation; in The Netherlands; in profiles under tillage, however, it does not hold.

The question as to whether a profile with a high bioporosity is better from an agricultural point of view than one with a low bioporosity cannot be answered easily. The mere fact that a high bioporosity is present in a profile indicates that the profile has or has had a high degree of biological activity, i.e., that there are or have been circumstances favouring root development in the profile. It goes without saying that the mechanical resistance towards root development decreases simultaneously with the increase of the number of biopores.

ACKNCWLEDGEMENTS

The investigations whose results are presented here are part of a larger research program. Several students participated in these investigations. The author is greatly indebted to Messrs. C.O. van der Veen, J.D. Kalma, G.J. Hofstede, V.J.G. Houba, B.J. van den Berg and J. Bouma, students at the Wageningen Agricultural University.

The author gratefully acknowledges the stimulating discussions with Prof.Dr. C.H. Edelman and Ir. K.J. Hoeksema.

Finally, the author wishes to express his sincere gratitude to Dr. L. van der Plas, who greatly improved this manuscript.

SUMMARY

In this paper the penetration of roots in sandy subsoils and the concept of bioporosity is first discussed. The biopore trend and biopore spread of five sandy soils are presented together with some of the traditional macromorphological properties. The origin of the properties of these profiles is discussed in detail. Finally, the importance of soil porosity investigations in relation to soil description, soil classification and land use classification is stressed.

REFERENCES

Edelman, C.H., 1950. Soils of The Netherlands. North-Holland, Amsterdam, 178 pp.
Mitscherlich, E.A., 1905. Bodenkunde für Land- und Forstwirte. Parey, Berlin, 364 pp
Rode, A.A., 1962. Soil Science. Israel Program for Scientific Translations, Jerusalem, 517 pp.
Van der Plas, L. and Slager, S., 1964. A method to study the distribution of biopores in soils. In: A. Jongerius (Editor), Soil Micromorphology, Elsevier, Amsterdam, pp. 411–419.
Wiersum, L.K., 1957. The relationship of the size and structural rigidity of pores to the penetration by roots. Plant Soil, 9 : 75–85.

ZUR ERMITTLUNG DES DREIDIMENSIONALEN AUFBAUES DER BODENSTRUKTUR MIT HILFE MIKROMORPHOMETRISCHER METHODEN

WALTER BECKMANN

Abteilung Bodenkunde der Bundesforschungsanstalt für Forst- und Holzwirtschaft, Reinbek bei Hamburg (Deutschland)

EINLEITUNG

Bei Untersuchungen der natürlichen Bodenstruktur hat sich die Methodik der mikromorphometrischen Strukturanalyse bereits bewährt (Beckmann, 1962; Geyger, 1962; Kubiena et al., 1961, 1963). Die Auswertungen wurden vorwiegend an Bodendünnschliffen verschiedener Formate durchgeführt. Darüber hinaus sind aber auch sog. fluoreszierende Bodenanschliffe nach Werner (1962) für quantitative Strukturuntersuchungen hervorragend geeignet. Nun hat sich ausserdem gezeigt, dass eine kombinierte Auswertung von Bodendünnschliffen und von Bodenanschliffen die Möglichkeit bietet, auch sichere quantitative Angaben über den dreidimensionalen Aufbau der Bodenstruktur zu machen (Beckmann, 1964). Diese Möglichkeiten werden im folgenden benutzt, um den räumlichen Bau der Bodenstruktur im A-Horizont von drei Wiesenböden zu beschreiben. Die Untersuchungen beziehen sich in erster Linie auf die räumliche Verteilung der Bodenhohlräume, ihre bevorzugte Richtung und ihre Breitenfrequenz sowie die Formen und die räumliche Verteilung der Bodenaggregate.

Die untersuchten Böden befinden sich im Südschwarzwald in der Gemarkung Hottingen (Top. Karte Nr. 8313 und 8314). Sie liegen auf dem Schwemmfächer, den die Hauensteiner Murg beim Austritt aus einer Engstrecke ihres Laufes aufgeschüttet hat. Die Lage der Profile im Gelände gibt die Abb.1. Ausgangsmaterial der Böden sind junge Sedimente aus Verwitterungsmaterial von Gneisen und Graniten. Die übrigen allgemeinen Daten zu den Standorten gibt die Tabelle I.

UNTERSUCHUNGSMETHODEN

Zunächst wurden vertikal entnommene Rahmenproben 40 x 60 x 80 mm im ganzen in der üblichen Weise mit Vestopal getränkt (Altemüller, 1962); jedoch war hier dem Tränkharz ein Fluoreszenzfarbstoff beigemischt (Beckmann, 1964; Werner, 1962). Nach Aushärtung wurden die Proben so zerschnitten, dass sich vier bis fünf Scheiben von 5 x 50 x 70 mm Grösse je Probe ergaben. Daraus konnten dann leicht durch Planschleifen und Polieren etwa zehn Anschliffe hergestellt werden; sie ergeben eine auswertbare Fläche von jeweils 50 x 70 mm. Diese Anschliffe haben Abstände von 5 mm (= Dicke der Scheiben) oder etwa 1 mm (= Schnittverluste beim Sägen) (Beckmann, 1964). Für viele Untersuchungen genügten schon diese parallelen

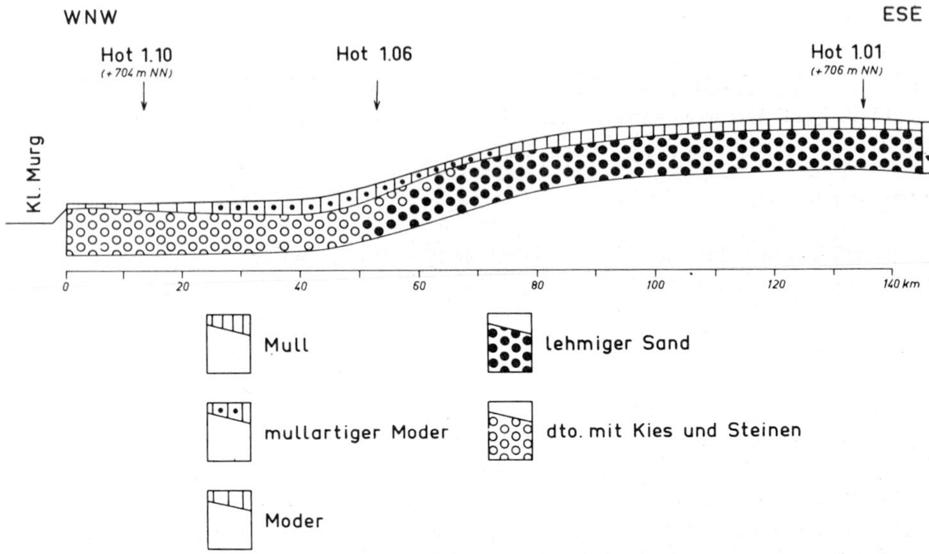

Abb.1. Lage der untersuchten Böden im Gelände. Schnitt durch den westlichen Teil eines Schwemmfächers. Breite der Kl. Murg etwa 2 m.

Anschliffe; mehrfach waren jedoch ihre Abstände noch zu gross, so dass auch das sog. "Abschleifverfahren" benutzt wurde (Beckmann, 1964). Dieses besteht darin, dass von einem fluoreszierenden Anschliff mehrfach eine genau bestimmbare dünne Schicht mit der "Blohm"-Flächenschleifmaschine (Geyger, 1962) abgeschliffen wird. Jeder so neu entstehende Anschliff wird photographiert. Aus einer Serie solcher Bilder lässt sich der räumliche Bau von Hohlräumen und Aggregaten im Boden wie auch die Form von festen Teilen, die in die übrige feste Bodensubstanz eingeschlossen sind, exakt konstruieren. Es wurden auf diese Weise mehrere Anschliffe hergestellt, die einen Abstand von 100 μ hatten.

Alle Anschliffe wurden zunächst unter dem Stereomikroskop auf den räumlichen Bau der Bodenstruktur hin qualitativ untersucht.

Für die mikromorphometrische Untersuchung wurden zunächst von den Anschliffen Strukturphotogramme im Massstab $\times$ 5 und $\times$ 10 angefertigt. Alle Photogramme wurden mit dem Detailphotometer vermessen (Kubiena et al., 1963). Ein Photogramm hat etwa DIN A 4 - Format, die Messfläche des Photometers hat einen Durchmesser von 6 cm, so dass sich pro Bild zwölf Messungen ergaben. Die Messstellen sind auf allen Photogrammen genau orientiert. Anschliessend wurden die Hohlraumbreitenfrequenzen mit dem Teilchengrössenanalysator (Zeiss) bestimmt (Kubiena et al., 1963; Schluge, 1959).

Die Ergebnisse dieser Anschliffauswertungen geben zunächst einen Uberblick über die Makro- und Grobstruktur des Bodens (vgl. Geyger, 1964). Sie ergeben aber gleichzeitig Anhaltspunkte dafür, wieviele Grossdünnschliffe für die anschliessende Untersuchung der Feinstruktur des Bodens

TABELLE I

Allgemeine Standortsdaten[1]

Lufttemperatur (1881–1930)

Januar	- 1°	Frosttage:	100
Juli	+16°	Eistage:	30
Jahr	+ 7°	Tage >10° C:	140

Niederschläge (1881–1930)

Jahr: 1.400 mm:		Schneeanteil:	20%
> 1,0 mm:	130 Tage	Schneefall:	40 Tage
>10,0 mm:	50 Tage	Schnee = 0 cm:	70 Tage

Grundwasser (1955–1959; nur für Hot 1.01)

Januar	-1,93 m	Maximum:	-0,72 m (April)
Juli	-1,59 m	Minimum:	-2,19 m (Oktober–November)
Jahr	-1,63 m		

Chemisch-physikalische Bodenuntersuchungen (nur Hot 1.01)

Tiefe (cm)	pH (KCl)	Steine (%)	Grobsand (%)	Feinsand (%)	Schluff (%)	Ton (%)
5	4,8	0	10,4	50,3	31,3	8,0
20	4,8	9,8	18,0	41,0	31,2	0
30	5,0	48,4	20,3	21,1	8,4	1,8
50	5,1	24,0	24,8	38,0	9,0	4,2
70	5,1	46,9	17,4	26,9	6,4	2,4

[1] Nach Klima-atlas von Baden-Württemberg; Grundwasserdaten nach unveröffentlichten Untersuchungen der Schluchseewerk-AG in Freiburg (Breisgau). Vgl. auch Abb.1.

notwendig sind und welche Anschliffe für die Herstellung dieser Grossschliffe verwendet werden sollen.

Hier wurde für jeden A-Horizont ein vertikal orientierter Grossschliff 5 x 7 cm angefertigt (Geyger, 1962). Von jedem Grossschliff wurden in je zehn horizontalen Reihen jeweils zehn Strukturphotogramme im Massstab x 50 hergestellt, so dass nahezu die ganze Schlifffläche bedeckt wurde. Alle Strukturphotogramme wurden mit dem Gesamtphotometer vermessen (Kubiena et al., 1961). Die Anordnung der Messstellen ist in der Legende der Abb.6 miteingezeichnet.

Alle Orte innerhalb der Bodenstruktur, an denen Messungen durchgeführt wurden, sind somit in vertikaler und horizontaler Richtung eindeutig bestimmt. Der räumliche Bau der Bodenstruktur lässt sich daher durch die vergleichende Betrachtung aller Messergebnisse charakterisieren: denjenigen aus den Photogrammen eines (vertikal orientierten) Anschliffes und denjenigen an "gegenüberliegenden" Orten in parallelen Anschliffen oder Grossschliffen.

UNTERSUCHUNGSERGEBNISSE

Die Humusform der Braunerde auf dem Rücken des Schwemmfächers ist Mull (Kubiena, 1953); der A-Horizont ist gut gekrümelt, locker und hohlraumreich. Die stereoskopische Anschliffuntersuchung hat gezeigt, dass die Strukturelemente des Bodens weitgehend sehr gut miteinander zusammenhängen; die feste Bodensubstanz wird von einem mehr oder weniger engmaschigen Netz von untereinander verbundenen Hohlraumsystemen durchzogen. Die Form dieser Hohlräume ist sehr verschieden: Im unteren Teil des A-Horizontes finden sich spaltartige Hohlräume, die tief in die feste Bodensubstanz eingreifen, die aber sehr unregelmässigen Verlauf haben (Tafel F8); im oberen Teil überwiegen kompliziert geformte Höhlungen. Gemeinsam ist allen Strukturpartien dieses Horizontes, dass die Hohlräume keine bevorzugte Richtung erkennen lassen; vielmehr durchziehen sie den Boden kreuz und quer in allen Richtungen.

Die Wände der Hohlräume sind durch viele Nischen und Vorsprünge sehr stark gegliedert, so dass die eigentliche Breite der Hohlräume auf kleinstem Raum sehr stark wechselt (Abb.2); auch grössere Hohlräume verengen sich oft bis auf ganz kleine Durchlässe. Hohlräume, die auf längere Strecken gleichmässig weit geöffnet sind, finden sich nur als frische Gänge von Bodentieren. Viele solcher Gänge sind allerdings oft mit Kleintierlosung fast vollständig angefüllt (Tafel F7). Die direkte stereoskopische Anschliffuntersuchung lässt zudem erkennen, dass diese Losungsformen im allgemeinen überhaupt die einzigen kleineren wirklich losen Bodenaggregate sind; Absonderungsformen etwa, die allseits durch Spalten von der übrigen festen Bodensubstanz abgetrennt sind, fehlen weitgehend. Das Innere von grösseren Komplexen fester Bodensubstanz ist dicht; hier finden sich nur wenige kleine, meist rundliche Poren (Abb.2, Tafel F7, F8).

Die mikromorphometrische Strukturanalyse der Anschliffe ergab zunächst das Verhältnis Hohlraum : feste Bodensubstanz für die Makro- und Grobstruktur. Die Mittelwerte dieser Messungen sind für alle Anschliffe gemeinsam in Tabelle II zusammengestellt.

Es zeigt sich, dass über die ganze Probe hin etwa 40% Hohlraumgehalt vorhanden ist. Die Unterschiede, die die Tabelle für verschiedene Tiefen angibt, sind nicht signifikant. Für eine genaue Charakterisierung des räumlichen Aufbaues der Bodenstruktur sind daher die gemessenen Einzelwerte von grösserer Wichtigkeit. Die Abb.4 gibt aus diesem Grunde drei Wertefolgen der detailphotometrischen Auswertung in graphischer Darstellung an. Ausgewählt wurde je eine typische Messreihe für drei verschiedene Tiefen im Horizont.

Im Gegensatz zu den Mittelwerten ergibt sich jetzt, dass der Gehalt an Hohlräumen innerhalb kleinerer Ausschnitte aus der Gesamtstruktur sehr stark streut. Es finden sich Werte zwischen 10% und 78%. Die Abb.4 zeigt aber auch, dass diese Streuungen, die sich in der Steigung jeder Kurve niederschlagen (Kubiena et al., 1961), in den verschiedenen Tiefen sehr ähnlich sind. Gleiche Charakteristika weisen auch die übrigen Wertefolgen aus dieser Probe auf. Da sie ja verschiedene parallele Anschliffe umfassen, ist zu schliessen, dass in diesem Horizont sowohl vertikal als auch horizontal gesehen eine ähnliche Heterogenität der Bodenstruktur vorhanden ist; es finden sich also keine grösseren extrem lockeren oder besonders dichten Strukturpartien.

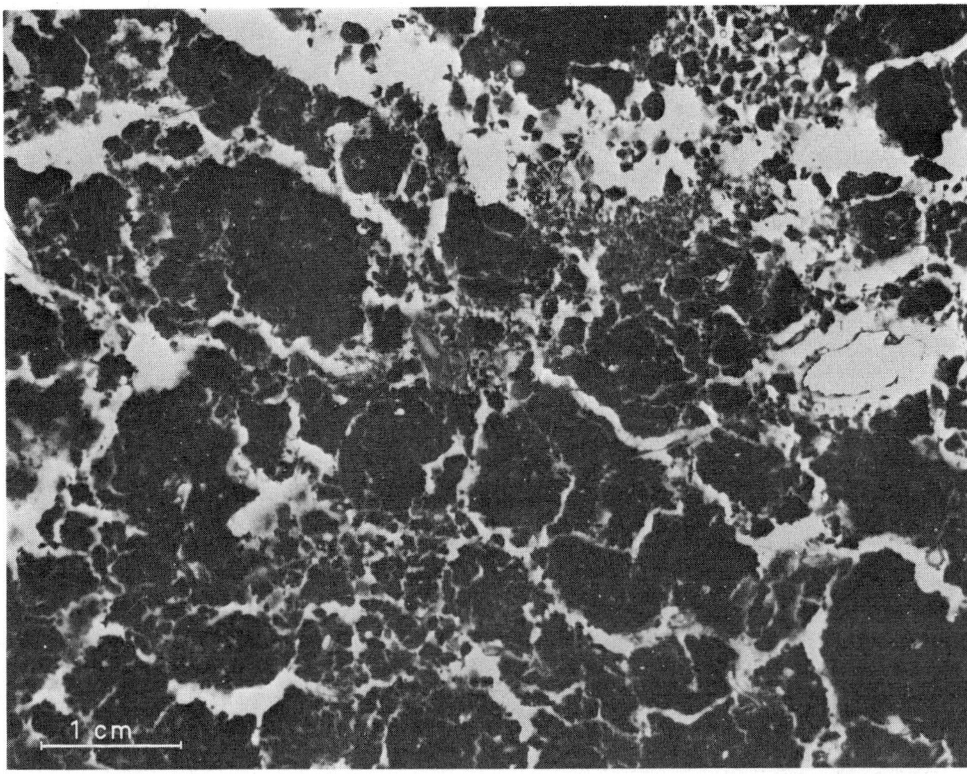

Abb.2. Strukturphotogramm von einem vertikalen fluoreszierenden Anschliff. Probe Hot 1.01 (vgl. Abb.1). Hohlräume von sehr verschiedener Breite durchziehen die ganze Struktur und greifen in verschiedenen Hohlraumsystemen ineinander. Neben grösseren rauhwandigen Aggregaten finden sich viele Anhäufungen von Losungen.

TABELLE II

Mittlerer Hohlraumgehalt[1]

Probe	Hohlraumgehalt (%)			
	1–3 cm	3–5 cm	5–7 cm	Ø
Hot 1.01	40,6	37,8	41,8	40,0
Hot 1.06	54,3	36,7	35,8	42,1
Hot 1.10	64,7	41,0	20,3	42,0

[1] Höhlenräume >200 μ.

In ähnlicher Weise ist die räumliche Verteilung bestimmter Hohlraumbreiten zu untersuchen. Die Tabelle III gibt zunächst einmal die Messergebnisse aus mehreren Anschliffen verschiedener Abstände (5 mm und 1 mm).

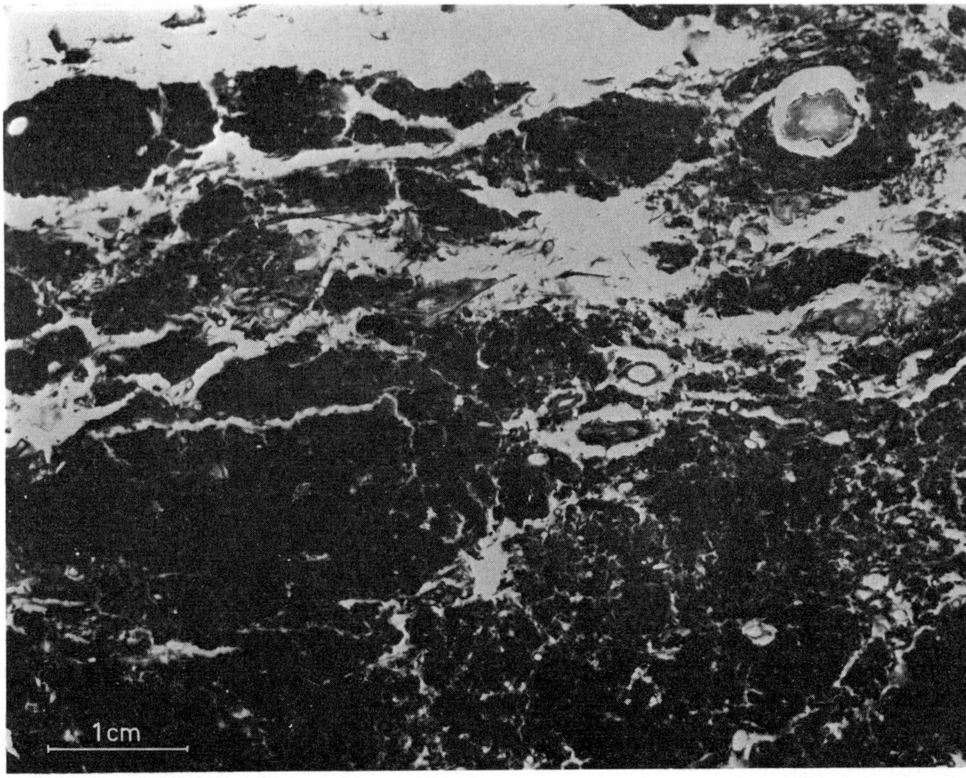

Abb.3. Strukturphotogramm eines vertikalen fluoreszierenden Anschliffes. Probe Hot 1.10 (vgl. Abb.1). Das Bild lässt die deutliche Zonierung der Bodenstruktur erkennen. Im oberen Teil überwiegen breite spaltartige Hohlräume mit vorwiegend horizontaler Richtung. Darunter ist eine dichtere Zone mit sehr feinen in verschiedenen Richtungen verlaufenden Hohlräumen zu erkennen. Diese Zonen wiederholen sich mehrfach.

Die Messungen wurden nach Horizonttiefen unterteilt und für alle Anschliffe zusammengefasst. Die Werte lassen erkennen, dass in 1–5 cm Tiefe dieser Probe 1.01 die Hohlraumbreitenfrequenz eine ähnliche ist. Erst in 5–7 cm Tiefe nehmen grössere Hohlraumbreiten ($>$ 1.000 μ) stark zu. In diesem Boden ist also eine wesentliche Änderung in der Verteilung der Hohlraumbreiten in vertikaler Richtung erst ab 5 cm Tiefe zu beobachten.

In Tabelle IV sind nun die Ergebnisse aus fünf Anschliffen im Abstand von jeweils 100 μ getrennt angegeben. Die Messergebnisse zeigen, dass in allen fünf Anschliffen gleiche Hohlraumbreitenfrequenzen vorhanden sind. Beim Fortschreiten im Boden in horizontaler Richtung ändert sich also die Strukturbildung nicht grundsätzlich.

Aufgrund dieser für jeden parallelen Anschliff getrennt vorliegenden Messreihen wurde nun jener Anschliff ausgewählt, der für die ganze Probe als repräsentativ angesehen werden kann; von ihm wurde ein Grossschliff angefertigt. Die Mittelwerte der gesamtphotometrischen Auswertung von

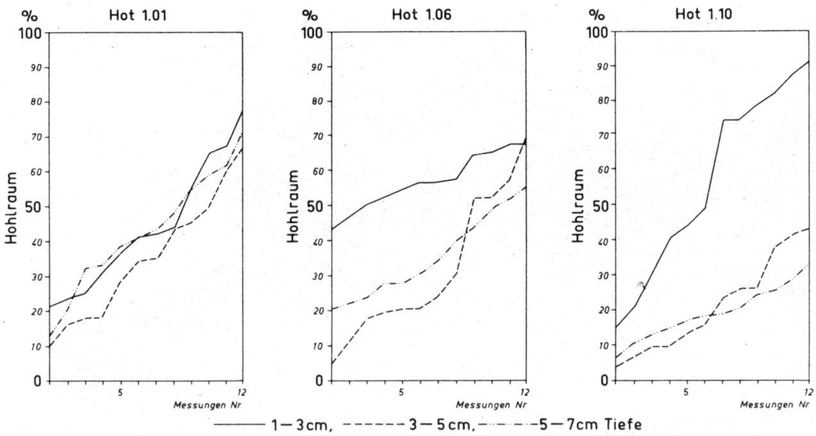

Abb.4. Ausgewählte Wertefolgen der photometrischen Anschliffauswertung. Die Werte wurden nach zunehmenden Hohlraumgehalt geordnet und in dieser Anordnung von 1–12 numeriert.

TABELLE III

Hohlraumbreitenfrequenzen[1]

Probe	Tiefe	200–500 μ (%)	500–1.000 μ (%)	1.000-2.000 μ (%)	2.000 μ ∅
Hot 1.01	1–3 cm	66,8	24,7	7,6	0,9
	3–5 cm	68,7	23,2	6,7	1,4
	5–7 cm	51,6	31,1	14,5	2,8
	1–7 cm	62,4	26,3	9,6	1,7
Hot 1.06	1–3 cm	48,7	31,5	16,3	3,5
	3–5 cm	57,5	20,4	15,6	6,5
	5–7 cm	61,1	31,5	7,1	0,3
	1–7 cm	55,7	27,8	13,0	3,4
Hot 1.10	1–3 cm	35,1	28,4	20,0	16,5
	3–5 cm	80,6	16,2	3,2	0,0
	5–7 cm	78,9	12,8	2,4	5,2
	1–7 cm	64,9	19,1	8,5	7,2

[1] Hohlräume > 200 μ.

Strukturphotogrammen im Massstab 50 ×, die von diesem Grossschliff stammen, sind in Tabelle V zusammengestellt. Sie erfassen jetzt Hohlräume grösser als 20 μ. Die Messwerte sind in vertikaler und horizontaler Richtung

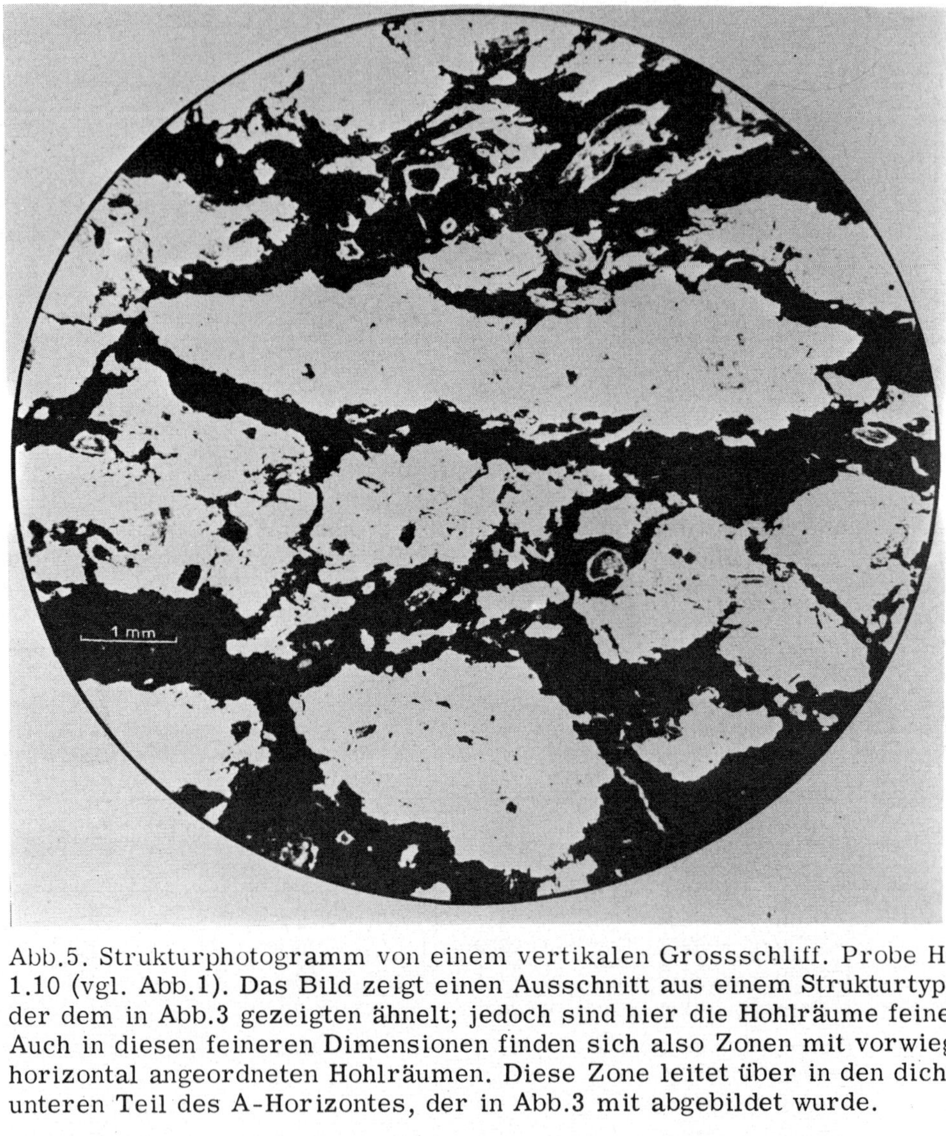

Abb.5. Strukturphotogramm von einem vertikalen Grossschliff. Probe Hot 1.10 (vgl. Abb.1). Das Bild zeigt einen Ausschnitt aus einem Strukturtyp, der dem in Abb.3 gezeigten ähnelt; jedoch sind hier die Hohlräume feiner. Auch in diesen feineren Dimensionen finden sich also Zonen mit vorwiegend horizontal angeordneten Hohlräumen. Diese Zone leitet über in den dichten unteren Teil des A-Horizontes, der in Abb.3 mit abgebildet wurde.

geordnet worden (vgl. die Legende der Abb.6).

Die Detailuntersuchung in höherer Vergrösserung gestattet natürlich eine feinere Unterteilung der Schichten des Horizontes. Zum Vergleich mit den Anschliffuntersuchungen sind die Mittelwerte für die dortigen Horizontunterteilungen mitangegeben. (Die Differenzen ergeben sich dadurch, dass jetzt auch Hohlräume von 20–200 μ gemessen werden können und dass natürlich sehr grosse Hohlräume hier unberücksichtigt bleiben). Es ist zu erkennen, dass generell auch hier eine relativ gleichmässige Verteilung des mittleren Hohlraumgehaltes gegeben ist. Die Unterschiede sind wiederum über die ganze Probe 1.01 hin nicht signifikant. Der Vergleich zwischen

TABELLE IV

Hohlraumbreitenfrequenzen[1]; vertikale Anschliffe mit Abstand 100 µ

Anschliff Nr. 1–7 cm Hot 1.01	200–500 µ (%)	500–1.000 µ (%)	1.000–2.000 µ (%)	2.000 µ Ø
1	59,8	20,1	11,5	8,6
2	73,8	14,0	6,1	6,1
3	64,9	21,1	7,0	7,0
4	67,2	14,8	12,6	5,4
5	67,5	16,4	11,2	4,9

[1] Hohlräume 200 µ.

den in horizontaler Richtung geordneten Mittelwerten und den in vertikaler Richtung geordneten ergibt auch keine wesentlichen Unterschiede.

Auch hier führt eine Analyse der Einzelwerte weiter, besonders für den späteren Vergleich mit den anderen Proben. In der Abb.6 sind die Mittelwerte der Tabelle V zusammen mit den maximalen und minimalen Einzelwerten des Gesamthohlraumgehaltes graphisch dargestellt, wiederum getrennt für vertikale und horizontale Richtung. Die Abb.6 zeigt, dass beide Mittelwert-Kurven sowohl ähnliche Richtung als auch ähnliche Lage in ihrem Koordinatensystem haben; d.h. dass in beiden Richtungen ähnlicher Gesamtgehalt und ähnliche Verteilung von Hohlräumen und fester Bodensubstanz zu beobachten ist. Entsprechende Ergebnisse zeigen auch die Maxima und Minima: Der Streifen um die Mittelwert-Kurve hat aufgrund des überall stark variierenden Hohlraumgehaltes wechselnde Breite, aber die Form dieser Streifen ist in vertikaler Richtung eine ähnliche wie in horizontaler. Auch bei zusätzlicher Berücksichtigung der geringeren Hohlraumbreiten sind also keine extrem lockeren oder besonders dichten Strukturpartien auszuscheiden.

Aus der gemeinsamen Betrachtung aller Messergebnisse sowie des Bildmaterials, d.h. der Strukturphotogramme, ergibt sich daher: Der volumenmässige Anteil der Bodenhohlräume streut innerhalb der Struktur kleinräumig sehr stark; diese Streuung ist aber bis etwa 5 cm Tiefe weniger durch das Vorkommen verschiedener Hohlraumbreitenfrequenzen bedingt, als vielmehr durch eine unregelmässige Verteilung und Anordnung der Hohlräume. Erst ab 5 cm Tiefe ändern sich diese Verhältnisse derart, dass eine deutliche Zunahme grösserer Hohlraumbreiten festzustellen ist. Eine Zonierung bestimmter Hohlraumvolumina und auch eine von den Hohlräumen bevorzugte Richtung sind nicht festzustellen. Der Vergleich der Strukturphotogramme ergibt ferner, dass auch die vorkommenden Aggregattypen und -grössen eine ähnliche unzonierte Anordnung haben.

Die Humusform des Bodens, der sich am Fusse des Schwemmfächers befindet (Hot 1.06), ist als ein feuchter mullartiger Moder anzusprechen (Kubiena, 1953). Infolge austretenden Hangwassers sind hier bedeutend feuchtere Standortsverhältnisse vorhanden. Daher ist die Zersetzung des anfallenden pflanzlichen Materials schlechter als bei Hot 1.01. Dies zeigt sich bei der Anschliffbeobachtung mit dem Stereomikroskop an einem hohen

TABELLE V

Mittlerer Hohlraumgehalt[1]

Probe	Zeilen (hor.)	Hohlraum (%)	Spalten (vert.)	Hohlraum (%)
Hot 1.01	a	44,2	1	33,6
	b	44,2	2	36,9
	c	40,2	3	40,2
	d	47,1	4	39,6
	e	38,2	5	35,1
	f	42,2	6	37,1
	g	34,8	7	40,2
	h	33,5	8	44,6
	i	38,5	9	39,8
	k	45,8	10	47,1
			11	41,0
Mittelwert für	1–3 cm:	57,1		
	3–5 cm:	40,6		
	5–7 cm:	39,3		
Hot 1.06	a	63,5	1	49,2
	b	54,6	2	43,5
	c	54,0	3	46,1
	d	52,0	4	50,1
	e	46,1	5	50,5
	f	30,5	6	54,1
	g	57,2	7	48,6
	h	43,7	8	50,4
	i	70,1	9	51,4
			10	49,5
Mittelwert für	1–3 cm:	54,4		
	3–5 cm:	42,9		
	5–7 cm:	38,8		
Hot 1.10	a	66,7	1	38,0
	b	46,2	2	48,5
	c	56,1	3	48,0
	d	46,8	4	42,5
	e	53,1	5	46,8
	f	46,4	6	39,4
	g	33,6	7	39,7
	h	28,0	8	33,9
	i	24,8	9	43,8
	k	32,3	10	41,4
	l	30,3		
Mittelwert für	1–3 cm:	56,3		
	3–5 cm:	45,0		
	5–7 cm:	28,8		

[1] Hohlräume $>20\ \mu$.

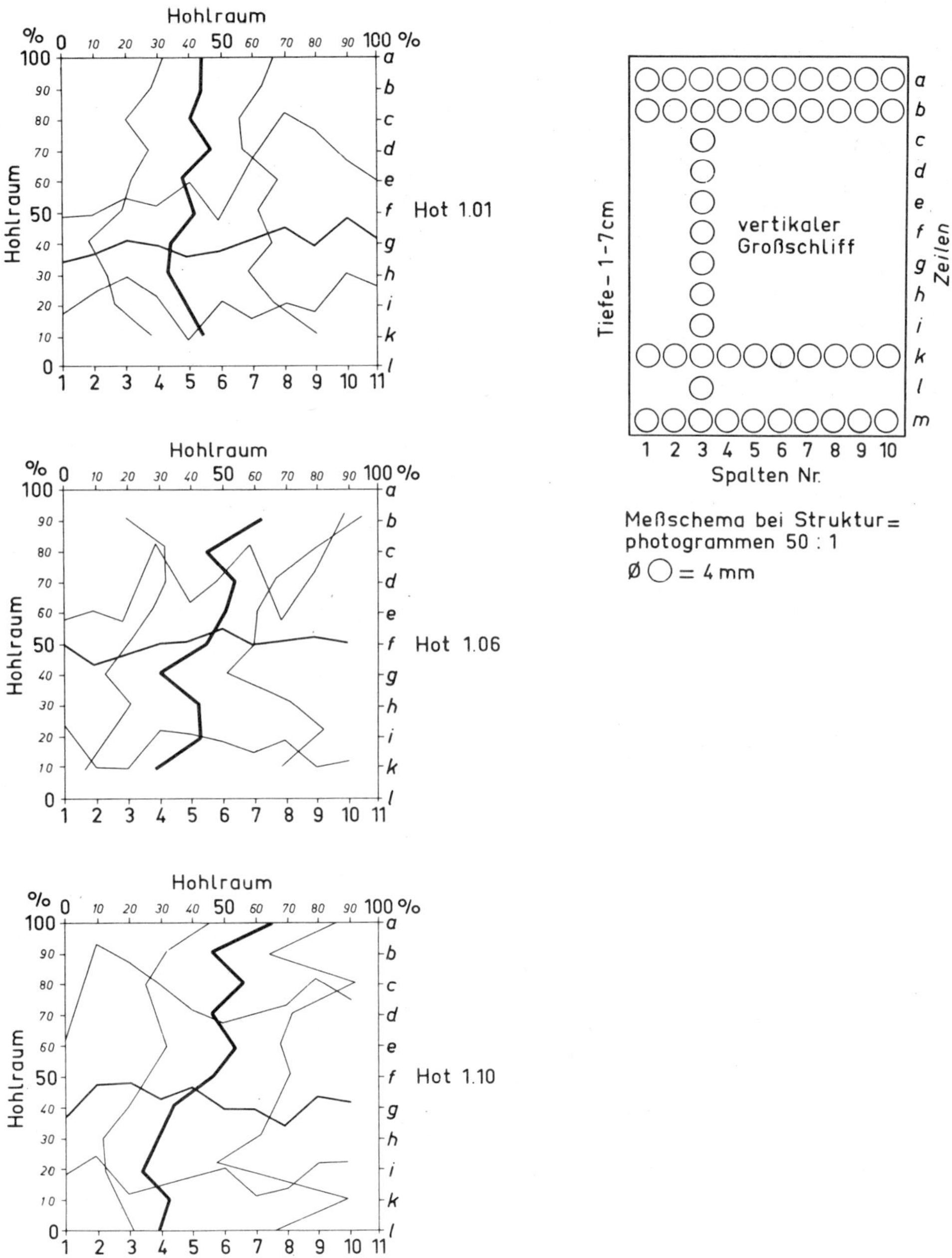

Abb.6. Graphische Darstellung der Mittelwerte und der maximalen und minimalen Einzelwerte der photometrischen Messungen an Grossschliffen. Die Werte wurden in vertikale Spalten und horizontale Zeilen geordnet (vgl. Schema).

Anteil halb- oder unzersetzter Pflanzenreste innerhalb der Struktur sehr deutlich (Abb.3). Die feste Bodensubstanz ist in diesem Horizont im allgemeinen sehr dicht, jedoch zeigen sich Zonen mit grossen und meist spaltartig ausgebildeten Hohlräumen. Sie führen meist oberflächenparallel und mit weithin gleichbleibender Breite durch den Boden. Sie sind vorwiegend durch Schrumpfung beim Austrocknen entstanden. Sie werden aber von vielen (lebenden) Pflanzenwurzeln durchzogen, so dass der Zusammenhang der Gesamtstruktur erhalten bleibt (Tafel H1). Die Hohlraumsysteme, die hier gebildet werden, sind deutlich abgetrennt von jenen, die innerhalb anderer grösserer Komplexe der festen Bodensubstanz vorhanden sind; diese letzteren Hohlraumsysteme sind bedeutend feiner, die Breite der Einzelhohlräume ist geringer und mehr wechselnd als die jener in den Spalträumen, auch ist die Richtung der Hohlraumführung hier sehr unterschiedlich. Die Wandausbildung der festen Bodensubstanz ist ähnlich wie bei Hot 1.01, desgleichen die Aggregatbildung.

Die Mittelwerte der photometrischen Auswertung der Anschliffphotogramme sind in der Tabelle II zusammengestellt. Es zeigt sich, dass der mittlere Hohlraumgehalt des gesamten Horizontes mit 42,1% ähnlich dem in der Probe Hot 1.01 ist; es zeigt sich aber auch, dass hier nun grössere Unterschiede in verschiedenen Tiefen vorhanden sind; eine deutliche Abnahme des Hohlraumgehaltes ist in 3–7 cm Tiefe zu beobachten.

Das ist auch in den Wertefolgen in Abb.4 zu erkennen. Darüber hinaus wird hier nun deutlich, dass auch die Verteilung der Hohlräume in kleineren Ausschnitten der Gesamtstruktur sehr unterschiedlich ist: In 1–3 cm haben wir bei hohem Hohlraumgehalt eine relativ gleichmässige Verteilung der Hohlräume (schwach geneigte Kurve), in 3–5 cm streuen die Einzelwerte sehr stark (steile Kurve), in 5–7 cm Tiefe haben wir wieder ähnliche Verteilung wie in 1–3 cm, aber allgemein geringeren Hohlraumgehalt. Da die Wertefolgen aller parallelen Anschliffe gleiche Ergebnisse bringen, ist daraus zu schliessen, dass eine deutliche - mehr oder weniger oberflächenparallele - Zonierung der Heterogenität der Bodenstruktur vorhanden ist.

In gleicher Weise wie vorher sind jetzt die Hohlraumbreiten bestimmt worden. Die Werte sind in Tabelle III zusammengestellt. Es ergibt sich, dass die schmalen Hohlräume am stärksten vertreten sind, dass aber auch viele breitere Hohlräume vorhanden sind. Innerhalb der einzelnen Bodenschichten fällt in 3–5 cm Tiefe der hohe Anteil von Hohlräumen mit grösserer Breite als 2.000 μ auf. Es ist dies die Zone, in der die breiten spaltartigen Schwundrisse häufig sind.

Die auffälligen vertikalen Unterschiede der Hohlraumbreitenfrequenz innerhalb der Struktur finden sich stets auch beim Fortschreiten in horizontaler Richtung; daher kann geschlossen werden, dass eine deutliche horizontale Zonierung breiterer Hohlräume in 3–5 cm Tiefe vorhanden ist. Diese Hohlraumbreiten sind - wie der Vergleich der Ergebnisse aus parallelen Anschliffen ergibt - zudem oberflächenparallel angeordnet.

Diese Beobachtungen sollen für die Feinstruktur geprüft werden. Die Tabelle V gibt zunächst die Mittelwerte des Gesamthohlraumgehaltes aus Strukturphotogrammen × 50. Es zeigt sich Übereinstimmung darin, dass der Hohlraumgehalt mit zunehmender Horizonttiefe deutlich abnimmt. Horizontal gesehen sind innerhalb der Schicht von 3–5 cm die stärksten Streuungen vorhanden. Die Zusammenstellung der Messergebnisse in den vertikalen Spalten ergibt aber auch hier geringere Unterschiede der Mittelwerte.

In der Abb.6 sind dann wieder die verschiedenen Mittelwerte zusammen mit den Maxima und Minima graphisch dargestellt. Während die vertikalen Messspalten eine ausgeglichene Kurve liefern, zeigen die horizontalen Messzeilen starke Streuungen; die Werte der Zeile b (2 cm Tiefe), d (3,5 cm) sowie h–i (6,5 cm) weisen auf horizontale Zonen von grösserem Hohlraumgehalt hin. Die schon vorher erkannte Zonierung des Hohlraumgehaltes findet sich also in mehreren feineren Streifen mehrfach innerhalb der Struktur. Die Zonen haben offenbar, wie die Maxima in den Messspalten 4 und 7 zeigen, wechselnde vertikale Ausdehnung.

Zusammenfassend ist aus diesen Ergebnissen zu schliessen: Der mittlere Gesamthohlraumgehalt lässt für einzelne Schichten des Horizontes eine gewisse Zonierung erkennen. Diese Zonen sind mehr oder weniger oberflächenparallel. Sie unterscheiden sich ausserdem durch den verschiedenen Anteil breiter Hohlräume. Die grossen Hohlraumfraktionen ergeben sich aus Hohlräumen mit vorwiegend oberflächenparallelem Verlauf. Verbindungen in verschiedene Tiefenregionen des Horizontes sind im allgemeinen nur durch sehr schmale Hohlräume gegeben (Abb.3). Die Detailuntersuchungen der Feinstruktur zeigen zudem, dass sich diese Zonierungen auch in feineren Bereichen wiederholen (Abb.5). Die vorkommenden Aggregattypen zeigen ähnliche Zonierung: Innerhalb der groben Strukturpartien finden sich vorwiegend Bröckel, die durch eine gewisse Schrumpfung des Materiales bei Austrocknung entstanden sind; in den übrigen Strukturpartien überwiegen rauhwandige und kompliziert geformte grosse Komplexe fester Bodensubstanz.

Im Boden Hot 1.10 sind - vor allem durch die gröbere Bodenart bedingt - weniger feuchte Standortsverhältnisse zu beobachten. Im Stereomikroskop lässt sich daher am Anschliff schon erkennen, dass hier eine Struktur vorliegt, die mehr der von Hot 1.01 als der von 1.06 ähnelt. Der A-Horizont ist relativ gut gekrümelt; nur in den obersten Schichten finden sich zahlreiche Absonderungen innerhalb von grösseren spaltartigen Hohlräumen. Darunter sind unregelmässig geformte grosse Komplexe fester Bodensubstanz vorhanden. Die hier vorliegenden Hohlraumsysteme sind kompliziert geformte Höhlungen mit unregelmässig verlaufenden Einzelhohlräumen. Viele frische und ältere Gänge, meist mit Losung angefüllt, sind zu erkennen. Mit zunehmender Tiefe wird die Struktur bald sehr dicht, die Hohlräume sind nur noch klein und bilden keine durchgehenden Systeme mehr.

Die Mittelwerte der photometrischen Anschliffauswertung sind wieder in der Tabelle II zusammengestellt. Es ist zu erkennen, dass sich über die ganze Probe hin ein ähnlicher Mittelwert wie in den beiden vorherigen Proben ergibt. Die Unterteilung in dünnere Schichten ergibt aber deutliche Unterschiede: In 5–7 cm Tiefe ist nur noch ein Drittel der Hohlräume der obersten Schichten vorhanden. Die drei Wertefolgen in Abb.4 zeigen das auch, lassen darüber hinaus aber erkennen, dass die Schicht von 1–3 cm bedeutend heterogener aufgebaut ist als die beiden tieferen, die sich weitgehend ähneln.

Die in Tabelle III zusammengestellten Hohlraumbreitenfrequenzen zeigen dann deutlich, dass in dieser obersten Bodenschicht auch ein sehr hoher Anteil an besonders breiten Hohlräumen vorhanden ist. Nach unten schliesst sich dann eine Zone an, die überwiegend nur sehr schmale Hohlräume aufweist. Dass diese Hohlraumbreitenverhältnisse nicht zufällig sind, zeigt die Zusammenstellung von Ergebnissen aus 6 parallelen Anschliffen mit jeweils nur 100 μ Abstand (Tabelle VI). Hier sind die Mittelwerte für

TABELLE VI

Hohlraumbreitenfrequenzen[1]; vertikale Anschliffe mit Abstand 100 μ

Anschliff Nr. 1–7 cm Hot 1.10	200–500 μ (%)	500–1.000 μ (%)	1.000–2.000 μ (%)	2.000 μ Ø
1	87,6	9,0	2,1	1,3
2	88,7	9,2	2,1	0
3	87,5	10,1	2,4	0
4	84,9	11,1	4,0	0
5	84,5	10,5	4,1	0,9

[1] Hohlräume >200 μ.

den ganzen Horizonte angegeben; sie stimmen weitgehend mit denen in Tabelle III überein.

Um diese Ergebnisse wiederum für feinere Schichten des Horizontes zu prüfen, wurde ein Grossschliff in der vorher angegebenen Weise photometrisch ausgewertet. Die Mittelwerte sind in der Tabelle V mitgeteilt. Die Abnahme des Gesamthohlraumgehaltes nach der Tiefe ist auch hier deutlich. Die einzelnen Mittelwerte zeigen aber, dass diese Abnahme sehr deutlich abgegrenzt ist: Oberhalb lassen sich auch in feineren Schichten keine Strukturpartien mit entscheidend geringerem Hohlraumgehalt erkennen. Der vorher genannte sehr geringe Anteil an sehr breiten Hohlräumen in 3–5 cm Tiefe bedeutet also keine absolute Verdichtung in dieser Schicht.

Dass auch in diesem Boden die Streuung der Mittelwerte in den Messspalten weitaus geringer ist als in den Messzeilen, wird ebenfalls aus Tabelle V deutlich. Dies ist auch aus der Zusammenstellung aller Mittelwerte mit den Maxima und Minima zu erkennen (Abb.6).

Zusammenfassend werden aus allen Messergebnissen und der vergleichenden Betrachtung der Strukturphotogramme folgende Schlüsse möglich: Im Gesamthohlraumgehalt finden sich deutliche Abgrenzungen etwa ab 5 cm Tiefe; hier nimmt der Hohlraumgehalt um rund 50% ab. Oberhalb 5 cm sind auch in feineren Strukturpartien keine absolut dichten Zonen vorhanden, jedoch ist in 2–3 cm Tiefe eine Zone mit auffällig wenigen breiten Hohlräumen festzustellen. Wie der Vergleich der Ergebnisse aus allen parallelen Anschliffen erkennen lässt, sind diese Zonen mehr oder weniger oberflächenparallel.

Die räumliche Verteilung verschiedener Hohlraum- und Aggregattypen hält sich vorwiegend auch an diese Zonierung; in der sehr lockeren obersten Schicht finden sich vor allem spaltartige und gleichmässig breit durch die ganze Struktur ziehende Hohlräume; die dichten Partien zeigen unregelmässige und weitaus feinere Hohlraumsysteme. In den lockeren Partien überwiegen Absonderungsaggregate; in den dichten sind fast nur sehr grosse und kompliziert geformte Komplexe fester Bodensubstanz vorhanden; hier finden sich auch viele mehr oder weniger verklebte Losungen.

SUMMARY

In this paper, previously developed methods of micromorphometric analysis of soil structure have been applied to the study of spatial structure of soil. Studies were carried out by making use of both thin sections and fluorescent polished soil sections. Special consideration was given to spatial distribution, chief direction, and variation in width of the pore spaces in the soil, as well as to the different types of soil aggregates and their distribution. The method applied enables us not only to study the soil sample as a whole, but also to separately study narrow (horizontal) layers ranging from 5–0.5 cm in depth. Thus, small portions of the soil structure of special interest can be individually examined. These detailed investigations play a special part in studies concerning soil water system and penetrability of soil by plant roots.

The A-horizons of three meadow soils in the southern part of the Black Forest, situated at the top, on the slope and at the bottom of an alluvial fan, were studied. When examining the soil samples (the volume of which was about 200 cm^3), we observed that the proportion of pore space in the soil sample taken as a whole was equal in all three A-horizons under investigation. Moreover, no significant variations in the width of pore spaces were observed. Essential (quantitative) differences in the soil structure became obvious, however, when studying different smaller layers of the horizons concerned. Results of this investigation are as follows:

In the soil at the top of the alluvial fan, pore spaces of different widths are found widely distributed in all layers of the A-horizon, and pore space systems extend through the soil in all directions. Therefore, the form of structure is highly heterogeneous everywhere in the soil, and it lacks large zones of extremely dense or loose structure.

At the bottom of the alluvial fan, the soil structure is of a different nature in as much as there are many wide pore spaces, formed like splits, which are principally arranged in a horizontal direction, thus giving rise to a laminiform structure held together by a dense root system. The layers with a high proportion of pore space are interrupted by soil zones of greater density; in these zones, very fine pore space systems prevail. When subdividing the horizon into 10 layers of 0.5 cm depth each, we observed three zones especially rich in pore space separated by layers (about 1 cm thick) with a smaller number of pore spaces. All the layers under investigation were found to run more or less parallel to the soil surface.

The soil beyond the alluvial fan resembles that just described in respect to the structure of the A-horizon; but here all zones of an extremely high or a very small proportion of pore space per volume of soil are nearer together. There are also three very thin layers of soil with many wide pore spaces formed like splits. From a depth of 5 cm downward, the horizon is very dense in structure. Here, only rather fine, narrow pore spaces were observed. All zones run parallel to the soil surface also in this form of structure.

LITERATUR

Altemüller, H.J., 1962. Verbesserungen der Einbettungs- und Schleiftechnik bei der Herstellung von Bodendünnschliffen mit Vestopal. Z. Pflanzenernähr., Düng., Bodenk., 99 : 164–177.

Beckmann, W., 1962. Zur Mikromorphometrie von Hohlräumen und Aggregaten im Boden. Z. Pflanzenernähr., Düng., Bodenk., 99 : 129–139.

Beckmann, W., 1964. Möglichkeiten der stereoskopischen Untersuchung im Innern der natürlichen Bodenstruktur. Zeiss-Mitt., im Druck.

Geyger, E., 1962. Zur Methodik der mikromorphometrischen Bodenuntersuchung. Z. Pflanzenernähr., Düng., Bodenk., 99 : 118–129.

Geyger, E., 1964. Mikromorphometrische Untersuchungen über den Einfluss bestimmter Pflanzengemeinschaften auf die Strukturbildung im Boden. In: A. Jongerius (Editor), Soil Micromorphology. Elsevier, Amsterdam, pp.445–457.

Klima-atlas von Baden-Württemberg, 1953. Deutscher Wetterdienst, Bad Kissingen.

Krause, W., 1959. Über die natürlichen Bedingungen der Grünlandberieselung in verschiedenen Landschaften Südbadens mit Ausblick auf den Wirtschaftserfolg. Z. Acker Pflanzenbau, 107 : 245–274.

Kubiena, W.L., 1953. Bestimmungsbuch und Systematik der Böden Europas. Enke, Stuttgart, 392 S.

Kubiena, W., Beckmann, W. und Geyger, E., 1961. Zur Methodik der photogrammetrischen Strukturanalyse des Bodens. Z. Pflanzenernähr., Düng., Bodenk., 92 : 116–126.

Kubiena, W., Beckmann, W. und Geyger, E., 1963. Mikromorphometrische Untersuchungen an Hohlräumen im Boden. Anales Edaf. Agrobiol. Madrid, 22 (11, 12) : 551–688.

Manig, M. und Schirmer, H., 1961. Das Klima des südlichen Schwarzwaldes. Einflüsse der Änderungen des natürlichen Abflusses auf das Klima. Ber. Deut. Wetterdienstes, 11 (77).

Odén, S., 1957. Förslag till klassifikation av markporer. Kgl. Skogs Lantbruksakad. Tidskr., 96 : 297–313.

Schluge, H., 1959. Der Teilchengrössenanalysator nach Endter. Zeiss-Werkzeitschrift, 33 : 68–73.

Werner, J., 1962. Über die Herstellung fluoreszierender Anschliffe. Z. Pflanzenernähr., Düng., Bodenk., 99 : 144–150.

FARBENBILDER

Tafel F7. Aufnahme eines Anschliffausschnittes; Probe Hot 1.01 (vgl. Abb.1); Massstab etwa x 55. Ausschnitt aus einer Strukturpartie, die aus sehr vielen verschiedenen grossen, meist rundlich geformten Losungsaggregaten aufgebaut ist.

Tafel F8. Aufnahme eines Anschliffausschnittes; Probe Hot 1.01 (vgl. Abb.1); Massstab etwa x 110. Das Bild zeigt, dass innerhalb dieser Struktur spaltartige Hohlräume (Mitte) dicht neben kompliziert geformten Höhlungen vorkommen. Bei den Spalten sind die Wandungen der festen Bodensubstanz relativ glatt und wenig gegliedert, an anderer Stelle sind sie knollignierig geformt und durch viele Vorsprünge und Nischen gegliedert (links).

Tafel H1. Aufnahme eines Anschliffausschnittes; Probe Hot 1.10 (vgl. Abb.1); Massstab etwa x 55. Ausschnitt aus einer Strukturpartie mit vorwiegend oberflächenparallel verlaufenden Spalten. Das Bild zeigt, dass hier aber durch eine recht dichte Durchwurzelung doch ein sehr guter Zusammenhalt der festen Bodensubstanz gewährleistet ist.

MIKROMORPHOMETRISCHE UNTERSUCHUNGEN ÜBER DEN EINFLUSS BESTIMMTER PFLANZENGEMEINSCHAFTEN AUF DIE STRUKTURBILDUNG IM BODEN, 1

Wiesenstandorte im Überschwemmungsbereich der Oberelbe bei Hamburg

ERIKA GEYGER

Abt. Bodenkunde der Bundesforschungsanstalt für Forst- und Holzwirtschaft, Reinbek bei Hamburg (Deutschland)

EINLEITUNG

Wenn man unter natürlichen Bedingungen den Einfluss verschiedener Pflanzengemeinschaften auf die Bodenstruktur möglichst rein erfassen will, eignen sich dazu am besten solche Wuchsorte, die in allen wesentlichen anorganischen Standortsfaktoren übereinstimmen und nur durch unterschiedlichen Pflanzenbewuchs differenziert sind. Diese Voraussetzungen sind an bestimmten vergleichbaren Feuchtwiesenstandorten im Aussendeichsland der Oberelbe weitgehend erfüllt. Die folgenden Untersuchungen beziehen sich auf zwei Pflanzengemeinschaften dieses Gebietes.

Das Untersuchungsgelände befindet sich im Naturschutzgebiet Heuckenlock (Deutsche Grundkarte 1 : 5.000, Blatt Moorwerder, Neuland) auf dem sog. "Grossen Sand". Dieser liegt etwa +2 m über N.N., d.h., ungefähr 25 cm über dem M.H.W. der Süderelbe. Daher wird die Bodenoberfläche häufig bei höherem Elbwasserstand überflutet und erhält so eine stets erneute natürliche Nährstoffzufuhr (Meyer, 1957). Das Grundwasser schwankt in Abhängigkeit vom Elbwasserstand erheblich; jedoch bewirken weder das Oberflächen- noch das Grundwasser eine länger andauernde Vernässung des Bodens. Der Grund hierfür liegt in der relativ groben Bodenart dieser jungen Flusssedimente.

Die Vegetation besteht vorwiegend aus Glatthafer- und Kohldistelwiesen (*Molinio-Arrhenatheretea*, *Arrhenatheretum* und *Cirsio-Polygonetum*) (Ellenberg, 1963), die hier aber infolge der geschilderten Standortsverhältnisse mit Flutrasenpflanzen (*Agropyro-Rumicion*) durchsetzt sind. Sie gedeihen üppig; die Grösse der assimilierenden Oberflächen, die als Mass für ihre Leistungsfähigkeit quantitativ ermittelt wurde (Geyger, 1964), betrug schon bei einmaligem Schnitt das 10- bis 15-fache der Bodenfläche. Diese Wiesen werden - wie an derartigen Aussendeichsstandorten üblich - nur unregelmässig genutzt, so dass, z.B., die Blätter und Stengel des Sandreitgrases *Calamagrostis epigeios*, das 20–30% des Grasanteils ausmacht, oftmals als schwerzersetzlicher Bestandesabfall dem Boden aufgelagert sind. An den Bodenprofilen dieser Wiesen fallen vor allem die schwache Durchwurzelung und die geringe Mächtigkeit des Humushorizontes auf; schon nach wenigen cm geht der graubraune sandige A-Horizont in gelben Grobsand über.

Ganz anders ist die Humusbildung unter einer Pestwurzherde (*Petasites officinalis* mit fast 100% Deckungsgrad), die inmitten dieser Wiesen eine Fläche von 300–400 m^2 einnimmt. Sie besiedelt die gleiche Stelle nach eigenen Beobachtungen seit mindestens zehn Jahren, vermutlich aber schon weit länger. Da hier nie gemäht wird, bleibt der üppige, stickstoffreiche Bestandesabfall am Ort. Je m^2 Bodenfläche werden jährlich 15–25 Stengel von 5–10 cm Dicke und 120 cm Länge nebst Blättern von 50–70 cm Durchmesser gebildet. Daraus konnte sich eine besondere Humusform bilden: Der A-Horizont ist 10–20 cm tief, zeigt dunkelbraune Farbe und sehr hohlraumreiche Struktur. Erst in tieferen Schichten gleicht das Bodenprofil dem der übrigen Wiesen.

Diese auffallenden Strukturunterschiede im Humushorizont, die bei der Übereinstimmung der übrigen Standortsfaktoren auf den Einfluss des verschiedenartigen Pflanzenbewuchses zurückzuführen sein dürften, sollen im folgenden mikromorphometrisch untersucht werden.

UNTERSUCHUNGSMETHODEN

Die mikromorphometrische Strukturuntersuchung wurde nach der in den letzten Jahren von uns entwickelten Methodik durchgeführt (Beckmann, 1962; Geyger, 1962; Kubiena et al., 1961, 1963). Über die bisher geübte Analyse an Grossschliffen hinausgehend wurden auch die Möglichkeiten der Auswertung von fluoreszierenden Anschliffen ausgenutzt (Werner, 1962; Beckmann, 1964a, 1964b). Zur Herstellung der Anschliffserien wurden horizontal und vertikal entnommene Rahmenproben aus dem A-Horizont in ungestörter Lagerung verwendet; sie wurden in üblicher Weise mit Vestopal getränkt (Altemüller, 1962) und in Scheiben von 5 mm Dicke zerschnitten (Beckmann, 1964b). Die Anschliffe wurden zunächst unter dem Stereomikroskop mikromorphologisch untersucht, um einen Einblick in den räumlichen Aufbau der Bodenstruktur zu gewinnen.

Für die quantitative Untersuchung wurden an den Anschliffserien Strukturphotogramme mit der Vergrösserung 5 : 1 hergestellt (Beleuchtung mit U.V.-Licht und Aufnahme durch ein gelbes Sperrfilter) (Beckmann, 1964b). Auf den Anschliff-Photogrammen erscheint die feste Bodensubstanz einschliesslich aller Minerale schwarz, die Hohlräume weiss (Abb.1, 2). Die Auswertung erfolgte in entsprechender Weise wie bei Strukturphotogrammen von Dünnschliffen. Die Messergebnisse gaben Kenntnis von den natürlichen Strukturverhältnissen im makroskopischen Grössenbereich bis hinab zu 0,5 mm Teilchengrösse oder Hohlraumbreite. Da für die Untersuchungen von jeder Bodenprobe 10–12 Anschliffe mit genau bekanntem Abstand in horizontaler oder vertikaler Richtung im Boden vorlagen, konnte an insgesamt etwa 1.000 cm^3 Bodenvolumen die Heterogenität der Strukturbildung in oberflächenparalleler Richtung und nach der Tiefe zu geklärt werden.

Daraus ergab sich die Zahl der notwendigen Grossschliffe für die quantitative Untersuchung der feineren Strukturelemente. Von ihnen wurden - jetzt im Durchlicht - Serien von Strukturphotogrammen im Massstab 20 : 1 und 50 : 1 hergestellt. Da die untersuchten Böden fast reine Quarzsande sind, wurden die zur Messung bestimmten Photogrammserien im zirkularpolarisierten Licht bei halb-gekreuzten Nikols aufgenommen;

dadurch ist die Trennung der Quarzkörner (schwarz) von den Hohlräumen (grau) leicht möglich. Photometrische Bestimmungen können an solchen Photogrammen nicht durchgeführt werden; daher wurde der Gesamthohlraumgehalt aus den Ergebnissen der Hohlraumbreitenmessungen berechnet.

Die Hohlraumbreiten wurden mit dem Teilchengrössenanalysator (Zeiss) ermittelt, wobei nach dem in Reinbek benutzten Verfahren (Kubiena et al., 1963) in der Längserstreckung der Hohlräume in regelmässigen Abständen die Hohlraumbreiten gemessen wurden. Durch die kombinierte Verwendung der Photogrammserien von Anschliffen und Dünnschliffen konnten alle Hohlräume zwischen 5.000 µ–20 µ Breite erfasst werden. Sie wurden in Breitenfraktionen gemäss Odén (1957) unterteilt:

5.000–2.000 µ : Makroporen
2.000– 200 µ : Grobporen
200– 20 µ : Mittelporen

Die Hohlraumbreitenfrequenzen wurden hier auf Volumanteile der verschiedenen Fraktionen am gesamten Bodenvolumen umgerechnet; so ergab sich ausserdem der Gesamtanteil an Hohlräumen bzw. fester Bodensubstanz.

Die "Innere Oberfläche" der Struktur, d.h., die Summe der Grenzflächen zwischen den Hohlräumen von 5.000–20 µ Breite und der festen Bodensubstanz, wurde ermittelt, indem die Länge der Umrisslinien aller Hohlräume auf den Strukturphotogrammen durch Nachfahren mit einem Distanzrädchen gemessen und auf cm^2 Fläche je cm^3 Bodenvolumen umgerechnet wurde (Kubiena et al., 1963).

UNTERSUCHUNGSERGEBNISSE

Vergleich der qualitativen mikromorphologischen Befunde

Unter der Kohldistelwiese erreicht der Humushorizont nur geringe Mächtigkeit. Er erscheint makroskopisch betrachtet ziemlich dicht (Abb.1) und nur von wenigen grösseren Hohlräumen durchsetzt; sie befinden sich meist in der Umgebung sperriger Pflanzenreste oder lebender Wurzeln. Die stereoskopische Anschliffuntersuchung zeigt, dass die Wände dieser Hohlräume rauh und unregelmässig sind, oft ragen die scharfen Kanten von Quarzkörnern in die Hohlräume hinein. Viele Quarzkörner sowie kleine rundliche Losungsstücke liegen auch lose am Grunde der Hohlräume. Im übrigen hängt die feste Bodensubstanz scheinbar überall eng zusammen.

Bei der Betrachtung mit stärkerer Vergrösserung zeigt sich jedoch ein unerwartet lockerer Aufbau der Feinstruktur. Die Wiedergabe eines Strukturphotogrammes vom Dünnschliff, Originalmassstab 20 : 1, in Abb.3 lässt viele rauhwandig umgrenzte und krummlinig in den verschiedensten Richtungen verlaufende Hohlräume erkennen. Bei noch stärkerer Vergrösserung (Abb.5) wird die unregelmässige Zusammenfügung verschiedener Strukturelemente sichtbar: Splittrige, schlecht zersetzte Pflanzenreste sind mit verklebten Losungsstücken von komplizierten Umrissformen und mit vielen kantigen Mineralkörnern zu einem relativ lockeren Feingefüge verbunden. An anderen Stellen ist die feste Bodensubstanz zu grösseren Komplexen vereinigt, die ebenfalls viele Quarzkörner enthalten und nur von

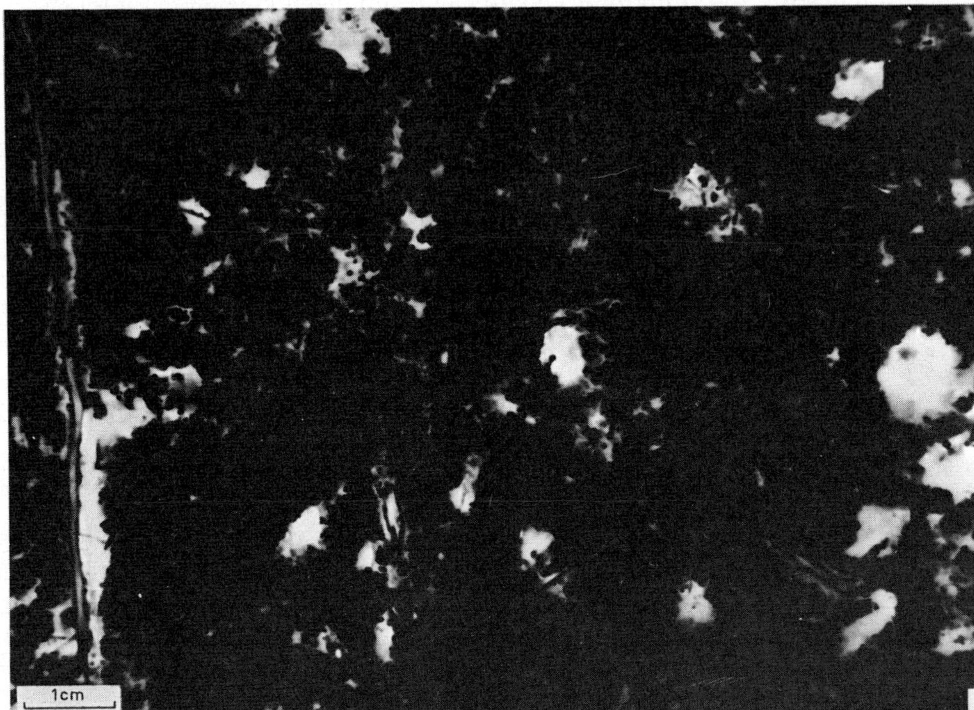
1cm
1

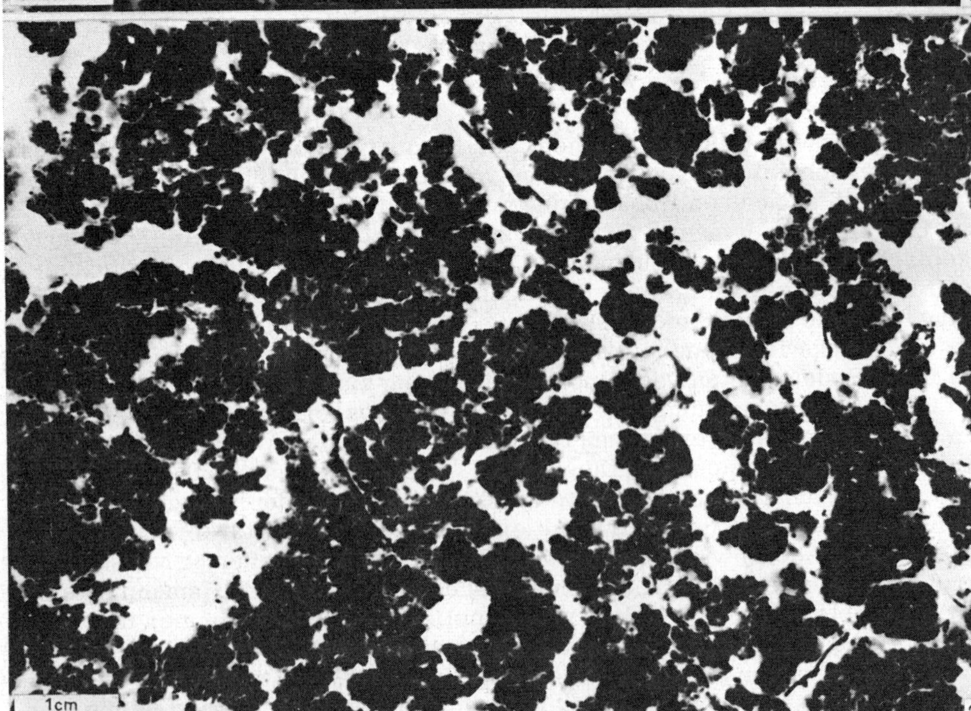
1cm
2

kürzeren krummlinigen Spalten durchsetzt sind.

Unter der Pestwurzherde zeigt der bedeutend mächtigere Humushorizont vor allem in der Grobstruktur einen wesentlich anderen Strukturaufbau. Schon die makroskopische Betrachtung lässt erkennen, dass viel mehr und weit grössere Hohlräume vorhanden sind (Abb.2). Die stereoskopische Untersuchung des räumlichen Aufbaues ergibt die für einen solchen Sandboden überraschende Tatsache, dass hier ein Schwammgefüge vorliegt, dass also die festen Bestandteile über viele breite und schmale Brücken miteinander verbunden sind und die Hohlräume bei stark wechselnder Breite ebenfalls ein zusammenhängendes Hohlraumgefüge bilden. Der Hauptunterschied gegenüber typischen Schwammgefügen, wie sie in lehmigen Böden oft zu beobachten sind, liegt in der Rauhwandigkeit der Grenzflächen zwischen Hohlräumen und fester Bodensubstanz, die durch die vielen kantig und spitz hervorragenden Quarzkörner erzeugt wird.

Durch die stereoskopische Untersuchung der Anschliffe wird die richtige Deutung der auf den Dünnschliff-Phtogrammen abgebildeten Strukturelemente erleichtert (Abb.4). Viele der allseits von Hohlräumen getrennten Teile fester Bodensubstanz können in anderen Schnittebenen durchaus wieder durch Brücken mit anderen Teilen zusammenhängen. Nur bei den abgerundeten, z.T. durch Spalten gleichmässiger Breite abgetrennten Losungsstücken ist anzunehmen, dass sie noch mehr oder weniger lose im Gefüge liegen.

Die Betrachtung der Feinstruktur (Abb.6) ergibt, dass die Porosität im Innern der grösseren festen Teile relativ gering ist; meist sind nur schmale krummlinige Spalten vorhanden, die in vielen Fällen unvollständige Verklebungsstellen von Losungsbrocken sind, denn sie verlaufen oft parallel zu deren gewölbten Aussenflächen. Die Losungsstücke sind um ein vielfaches grösser als die unter der Kohldistelwiese.

Die Unterschiede der beiden betrachteten A-Horizonte lassen sich ausserdem an den Farbaufnahmen von Dünnschliff-Ausschnitten ablesen. Unter der Kohldistelwiese (Tafel G2) sind die Pflanzenreste sperrig und wenig zersetzt; z.T. sind noch die Zellstrukturen völlig erhalten. Der koprogene Anteil liegt im Form kleiner rundlicher oder verklebter Losungsreste mit Durchmessern unter 100 μ vor. Die Quarzkörner liegen grösstenteils lose oder nur angeklebt zwischen diesen Losungsresten, sie haben dünne bräunliche Hüllen. Unter der Pestwurzherde (Tafel G3) dagegen ist die gesamte organische Substanz durch grössere Bodentiere zu stabilen

---

Abb.1. Strukturphotogramm von einem horizontalen fluoreszierenden Anschliff. A-Horizont in 1 cm Tiefe unter Kohldistelwiese. Die Makrostruktur erscheint dicht; nur in der Umgebung lebender Wurzeln oder unzersetzter Pflanzenreste finden sich grössere Hohlräume; sie sind meist im Querschnitt zu sehen und haben unregelmässige rauhe Wandungen.

Abb.2. Strukturphotogramm von einem horizontalen fluoreszierenden Anschliff. A-Horizont in 1 cm Tiefe unter Pestwurzherde. Das Bild zeigt die Makrostruktur; sie ist netzartig von einem System wechselnd breiter Hohlräume durchzogen; die feste Bodensubstanz besteht vorwiegend aus grösseren Losungsbrocken, die mehr oder weniger miteinander verkittet sind und ein grobmaschiges Schwammgefüge bilden.

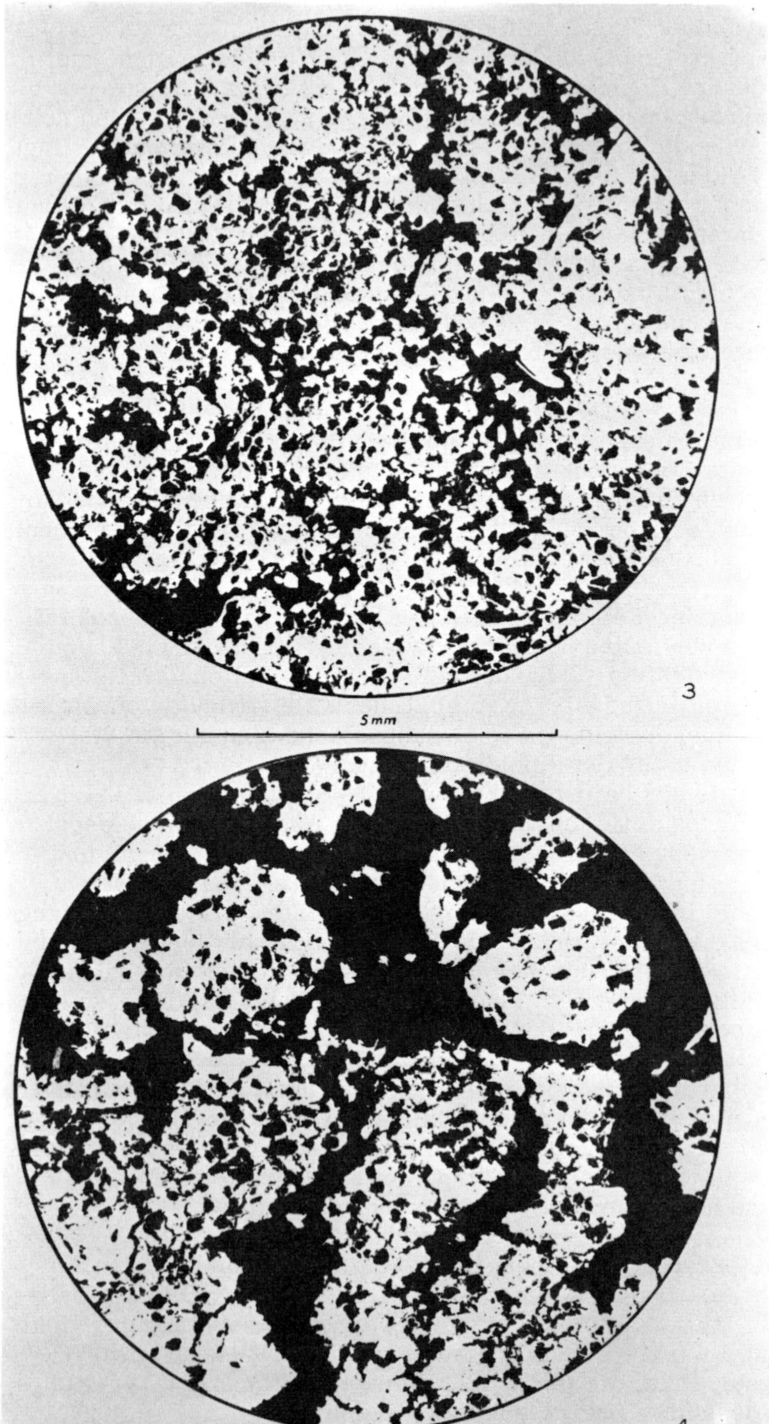
3
5 mm
4

tiefbraunen Losungsbrocken verarbeitet, in die die Quarzkörner meist fest eingebaut sind. Die Art der Zusammenfügung dieser Teile zu einem rauhwandigen Schwammgefüge zeigt das Stereo-Bildpaar (Tafel G4, 5).

Vergleich der quantitativen Ergebnisse

Bei beiden untersuchten A-Horizonten ist die Durchwurzelung ähnlich schwach: Unter der Kohldistelwiese sind lebende Wurzeln mit 1,6% am Bodenvolumen beteiligt, unter der Pestwurzherde mit 1,74%. Der Anteil fester Bodensubstanz liegt bei der Kohldistelwiese über 80%, bei der Pestwurzherde unter 50%. Die Anzahl der Quarzkörner ist bei Bezug auf das Gesamtvolumen dieser Bodenschicht unter der Kohldistelwiese etwa doppelt so gross wie unter der Pestwurzherde, dabei reicht unter der letzteren diese relativ sandarme Schicht ungefähr 15 cm tiefer, während unter der Kohldistelwiese der Quarzanteil mit der Tiefe noch erheblich zunimmt.

Die Hohlraumverhältnisse sind durch die Ergebnisse der Hohlraumbreitenmessungen charakterisiert. Der Anteil der verschiedenen Porenweiten ist in der Tabelle I zusammengestellt.

TABELLE I

Volumanteile verschiedener Hohlraumbreitenfraktionen am gesamten Bodenvolumen

Hohlraumbreitenfraktionen	A-Horizont	
	unter Kohldistelwiese	unter Pestwurzherde
Makroporen 5.000–2.000 μ	0,6%	10%
Grobporen 2.000– 200 μ	6,5%	25%
Mittelporen 200– 20 μ	12,5%	16%
Gesamthohlraum	19,6%	51%

Abb.3. Strukturphotogramm eines horizontalen Grossschliffes. A-Horizont in 1 cm Tiefe unter Kohldistelwiese. Die abgebildete "Grobstruktur" ist bereits weniger kompakt, sie wird von krummlinigen Hohlräumen in allen Richtungen durchzogen. In den grösseren sind z.T. lose Aggregate zu erkennen. Die feste Bodensubstanz enthält sehr viele Quarzkörner.

Abb.4. Strukturphotogramm eines horizontalen Grossschliffes. A-Horizont in 1 cm Tiefe unter Pestwurzherde. Die abgebildete "Grobstruktur" zeigt einige abgerundete Losungsbrocken mit dichter Innenstruktur und wenigen Quarzkörnern. Die übrigen festen Teile sind vielgestaltig und rauhwandig, sie hängen über breite und schmale Brücken mehrfach miteinander zusammen. Die Hohlräume sind über grössere Strecken recht breit.

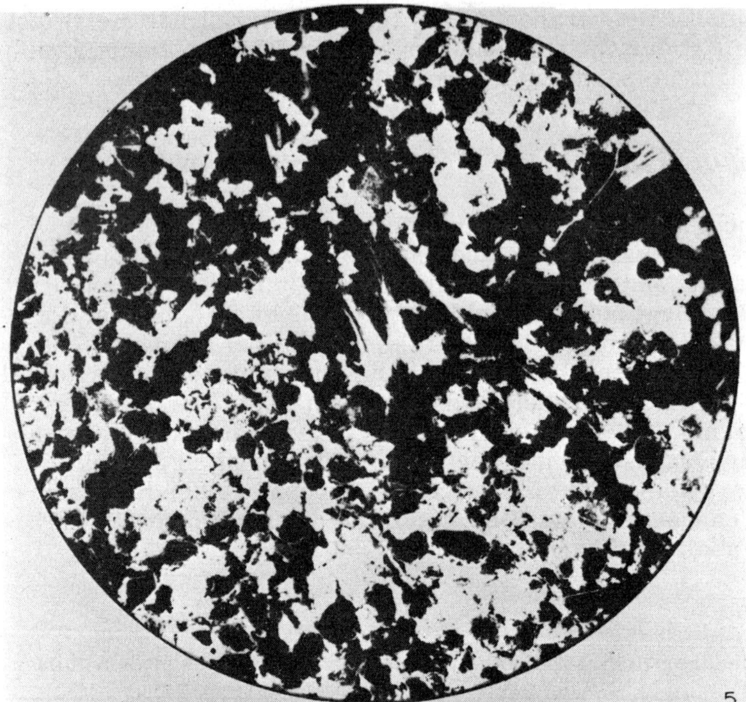

5

1 mm

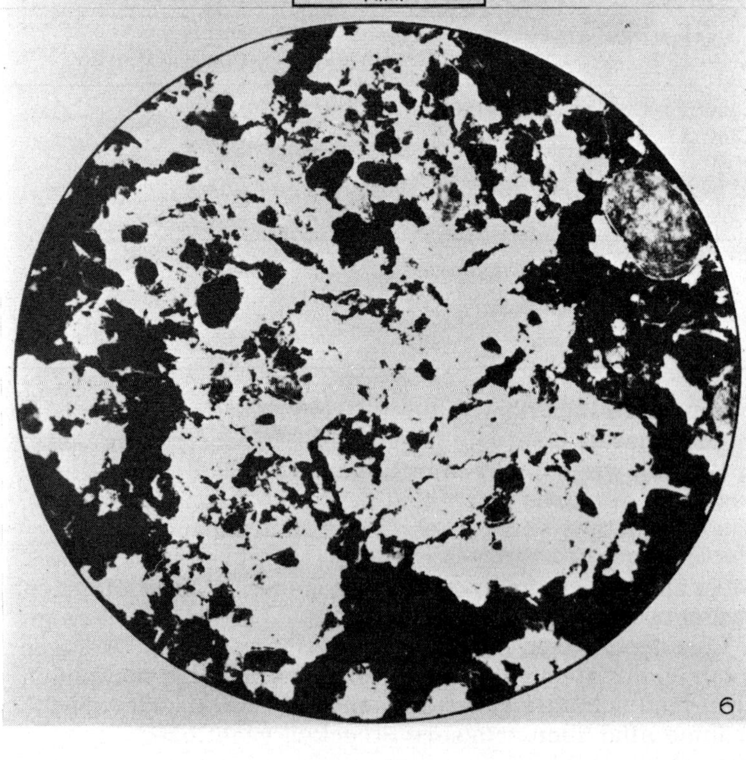

6

Wie aus der Tabelle I deutlich hervorgeht, wird der grosse Unterschied im Gesamthohlraumvolumen hauptsächlich durch weitlumige Poren über 200 μ bewirkt: Unter der Pestwurzherde ist der Anteil an Makroporen siebzehnmal so gross, der Anteil an Grobporen viermal so gross wie unter der Kohldistelwiese. Demgegenüber ist der Unterschied bei den Mittelporen nur gering.

Entsprechendes kann auch an der Grösse der inneren Oberfläche der Bodenstruktur in den verschiedenen Messbereichen abgelesen werden, denn es handelt sich ja um die Grenzflächen zwischen den Hohlräumen verschiedener Breite und der von ihnen durchzogenen festen Bodensubstanz. Die jeweiligen Messergebnisse zeigt die Tabelle II.

TABELLE II

"Innere Oberfläche" der Bodenstruktur in cm^2/cm^3 Bodenvolumen

Erfasste Hohlraumbreiten	A-Horizont	
	unter Kohldistelwiese	unter Pestwurzherde
5.000–500 μ ("Makrostruktur)	1,80	11,3
2.000–100 μ ("Grobstruktur")	3,87	57,3
500– 20 μ ("Feinstruktur")	49,3	29,1

Abb.5. Strukturphotogramm eines horizontalen Grossschliffes. A-Horizont in 1 cm Tiefe unter Kohldistelwiese. Die hier gezeigte "Feinstruktur" ist aus vielen, z.T. lose liegenden Quarzkörnern, splittrigen Pflanzenresten mit gut erhaltener Struktur und vielgliedrigen verklebten Losungsaggregaten aufgebaut. Durch die komplizierten Umrissformen der durch viele Brücken verbundenen Losungsreste entsteht ein lockeres Feingefüge mit vielgestaltigen Hohlräumen wechselnder Breite.

Abb.6. Strukturphotogramm eines horizontalen Grossschliffes. A-Horizont in 1 cm Tiefe unter Pestwurzherde. Dieser Ausschnitt aus der Feinstruktur zeigt die rauhwandigen Umgrenzungen der festen Bodensubstanz sowie ihre ziemlich dichte Innenstruktur. Die kurzen und schmalen Hohlräume im Innern verlaufen häufig parallel zu den gewölbten Aussenflächen; es handelt sich vermutlich um unvollständige Verklebungsstellen der Losungsstücke. In sie sind die Mineralkörner grösstenteils fest eingebaut.

Bei Erfassung der Makrostruktur ist die innere Oberfläche unter der Pestwurzherde ungefähr sechsmal so gross, bei Erfassung der Grobstruktur sogar fünfzehnmal so gross wie unter der Kohldistelwiese. In den feineren Dimensionen kehrt sich das Verhältnis jedoch um. Hier ist die innere Oberfläche unter der Kohldistelwiese fast doppelt so gross. Nur ein Teil davon wird aus den Oberflächen loser oder herausregender kantiger Mineralkörner gebildet; der grössere Anteil ist auf die Oberflächen der vielgliedrigen verklebten Losungsaggregate zurückzuführen. Diese sind viel kleiner als die Losungsformen unter der Pestwurzherde, daher haben sie trotz ihres geringeren Anteils am Bodenvolumen eine relativ viel grössere Oberfläche (vgl. Abb.5). Die weit grösseren Losungsbrocken unter der Pestwurzherde haben zudem eine ziemlich dichte Innenstruktur; ausserdem sind hier die Quarzkörner grösstenteils fest eingebaut (Abb.4, 6).

Ursachen für die ermittelten Strukturunterschiede

Um den Ursachen für die auffälligen Strukturunterschiede im Humushorizont der beiden Pflanzengemeinschaften näherzukommen, müssen alle die Einwirkungen auf die Bodenstruktur, die bei beiden in ähnlicher Weise vorhanden sind, ausgeschlossen werden. Wie eingangs erwähnt, stimmen die anorganischen Standortsbedingungen weitgehend überein. Vor allem sind die Wasser- und Nährstoffversorgung für beide Standorte gleich günstig; demzufolge ist das Wurzelwerk unter beiden Beständen nur gering ausgebildet und nimmt trotz erheblicher morphologischer Verschiedenheit wenig unmittelbaren Einfluss auf die Bodenstrukturbildung. (Ob allerdings die chemische Beschaffenheit der organischen Wurzelsubstanz als Nahrung für die Bodentiere unterschiedlich zu bewerten ist, bedürfte näherer bodenzoologischer Untersuchungen.)

Der Hauptunterschied besteht offenbar in der Art und Menge des für die Humusbildung zur Verfügung stehenden Bestandesabfalles: Bei der Kohldistelwiese wird er entweder durch Mahd teilweise entfernt oder besteht vorwiegend aus den schwerzersetzlichen Blättern und Stengeln von *Calamagrostis epigeios*. Die Losungsreste der kleinen Bodentiere, die die hieraus entstehende Feinmoderschicht besiedeln, sind wenig stabil gegen Verschlämmung. Demzufolge ist die beobachtete relativ lockere Feinstruktur auf die obersten cm des A-Horizontes beschränkt.

Die Pestwurz dagegen bildet einen eiweissreichen, leicht zersetzlichen Bestandesabfall, der im Herbst auf einmal dem Boden aufgelagert wird, jedoch bereits im Laufe des Winters (vermutlich mitbewirkt durch das Zusammenspiel von Durchfeuchtung und Frost) in eine stark zerkleinerte braune Masse umgewandelt wird. Durch die Qualität und Quantität dieser anfallenden organischen Substanz konnte sich ein besonders reges Bodenleben entwickeln. An der Bodenoberfläche herrscht während des Sommers durch die nicht durch Mahd unterbrochene starke Beschattung (rel. Helligkeit nur 1–2%) ein ausgeglichenes Mikroklima. Die intensive Verarbeitung des organischen Materiales durch Lumbriciden, Insektenlarven und nässeertragende Arthropoden führt zu dem beobachteten stabilen Grobschwammgefüge.

Für die Leistungsfähigkeit der beiden Pflanzengemeinschaften hinsichtlich ihrer oberirdischen Stoffproduktion spielen allerdings diese Strukturunterschiede bei den derzeitigen Standortsverhältnissen keine entscheidende Rolle; die Kohldistelwiese hat sogar aufgrund der gestaffelten Anordnung ihrer Assimilationsorgane eine weit grössere assimilierende Gesamtoberfläche als die Pestwurzherde mit ihrem nahezu einschichtigen Blätterdach. Offenbar kann der Luftbedarf der wenigen Pflanzenwurzeln auch aus dem geringeren Hohlraumvolumen des Bodens unter der Kohldistelwiese gedeckt werden, da die grobe Bodenart der tieferen Schichten längere Vernässungsperioden ausschliesst.

SUMMARY

Micromorphometrical investigations on the influence of certain plant associations on the formation of soil structure

This study deals with the structural differences in the A-horizon under two plant communities growing on sites of equal ecological conditions. For examining spatial structure of the soil, polished sections have been observed through the stereoscopic microscope; for studying micromorphology of soil structure investigations have been carried out by means of thin-section microscopy. The quantitative analysis of soil structure, being based on both photograms of polished soil sections and of thin sections, thus, covers pore widths ranging from 5.000–20 μ. The following results have been obtained from the macro- and micromorphometric investigations of the soil structure:

The A-horizon is only of small depth under a *Cirsio-Polygonetum* cover with a high proportion of *Calamagrostis epigeios*; the macrostructure is dense except for a few larger pore spaces within the reach of the scarcely developed root system; the microstructure, however, shows less compactness, the texture being marked by bulky plant residues, loose quartz grains and small dropping-aggregates of great diversity. Pore spaces are mostly narrow, diameters being smaller than 200 μ. The inner surface within the range of macro- and mesostructure is very small, that of microstructure is relatively large (about 50 cm^2/cm^3 soil volume).

On equal site conditions, under a *Petasites officinalis* cover, providing for a high amount of albuminous plant residues always remaining on the site, a stabile spacious "spongy fabric" is being formed under the influence of soil fauna. The quartz grains being fixed within the compact soil material often form extremely rough surfaces, thus giving this kind of "spongy fabric" quite a typical appearance. More than 50% of the soil volume consist of pore spaces, most of which are wider than 200 μ. Under this type of vegetation cover the inner surface of the meso-structure is largest, with 57 cm^2/cm^3 soil volume, whereas that of the microstructure is, by far, inferior to that observed under *Cirsio-Polygonetum* cover.

LITERATUR

Altemüller, H.-J., 1962. Verbesserung der Einbettungs- und Schleiftechnik bei der Herstellung von Bodendünnschliffen mit Vestopal. Z. Pflanzenernähr., Düng., Bodenk., 99 : 164–177.

Beckmann, W., 1962. Zur Mikromorphometrie von Hohlräumen und Aggregaten im Boden. Z. Pflanzenernähr., Düng., Bodenk., 99 (2–3) : 129–139.

Beckmann, W., 1964a. Zur Ermittlung des dreidimensionalen Aufbaues der Bodenstruktur mit Hilfe mikromorphometrischer Methoden. In: A. Jongerius (Editor), Soil Micromorphology. Elsevier, Amsterdam, pp.429–444.

Beckmann, W., 1964b. Möglichkeiten der stereoskopischen Untersuchung im Innern der natürlichen Bodenstruktur. Zeiss Mitt., im Druck.

Ellenberg, H., 1963. Vegetation Europas mit den Alpen. In: H. Walter (Redakteur), Einführung in die Phytologie. Ulmer, Stuttgart, 4 (2) : 943 S.

Geyger, E., 1962. Zur Methodik der mikromorphometrischen Bodenuntersuchung. Z. Pflanzenernähr., Düng., Bodenk., 99 (2–3) : 118–129.

Geyger, E., 1964. Methodische Untersuchungen zur Erfassung der assimilierenden Gesamtoberflächen von Wiesen. Thesis, Hamburg, im Druck.

Kubiena, W.L., 1943. L'investigation mikroscopique de l'humus. Z. Weltforstwirtsch., 10 (7–10) : 387–410.

Kubiena, W.L., 1953. Bestimmungsbuch und Systematik der Böden Europas. Enke, Stuttgart, 392 S.

Kubiena, W.L., Beckmann, W. und Geyger, E., 1961. Zur Methodik der photogrammetrischen Strukturanalyse des Bodens. Z. Pflanzenernähr., Düng., Bodenk., 92 (2) : 116–126.

Kubiena, W.L., Beckmann, W. und Geyger, E., 1963. Mikromorphometrische Untersuchungen an Hohlräumen im Boden. Anales Edafol. Agrobiol. (Madrid), 22 (11–12) : 551–567.

Meyer, F.H., 1957. Über Wasser- und Stickstoffhaushalt der Röhrichte und Wiesen im Elballuvium bei Hamburg. Mitt. Staatsinst. Allgem. Botan. Hamburg, 11 : 137–203.

Odén, S., 1957. Förslag till klassifikation av markporer. Kgl. Skogs-hantbruksakad. Tidsskr., 96 : 297–313.

Werner, J., 1962. Über die Herstellung fluoreszierender Bodenanschliffe. Z. Pflanzenernähr., Düng., Bodenk., 99 : 144–150.

FARBENBILDER

Tafel G2. Dünnschliffaufnahme aus dem Humushorizont unter der Kohldistelwiese. Grosser Sand, N.S.G. Heuckenlock an der Süderelbe oberhalb Hamburgs. Bildausschnitt: 1,2 x 1,2 mm. Das Bild zeigt eine lose Anhäufung von Quarzkörnern mit dünnen braunen Hüllen. Typisch ist, dass die Mineralkörner und die teilweise fast unzersetzten Pflanzenreste an vielen Stellen mit Material verklebt sind, das aus mehr oder weniger veränderter Kleintierlosung besteht.

Tafel G3. Dünnschliffaufnahme aus dem Humushorizont unter der Pestwurzherde. Grosser Sand, N.S.G. Heuckenlock an der Süderelbe oberhalb Hamburgs. Bildausschnitt: 1,2 x 1,2 mm. Das Bild zeigt Teile des sehr hohlraumreichen Schwammgefüges. Die feste Bodensubstanz besteht vorwiegend aus ziemlich gut erhaltenen grossen dunkelbraunen Losungsstücken, in denen die Pflanzenreste teilweise nur grob zerkleinert sind.

Tafel G4,5. Stereobildpaar aus einem horizontalen Anschliff von Humushorizont unter der Pestwurzherde. Grosser Sand, N.S.G. Heuckenlock an der Süderelbe oberhalb Hamburgs. Die grüne Färbung im Bild beruht auf dem Fluoressenzfarbstoff, der dem Einbettungsmittel beigemischt wurde. Hier ist der räumliche Bau des Schwammgefüges zu erkennen. Mineralkörner und übrige feste Bodensubstanz sind fest miteinander verbunden; an den Wandungen der hohlraumreichen Gefügekomplexe ragen oft die scharfen Ecken der Quarzkörner heraus; dadurch entsteht eine deutliche Rauhwandigkeit, die für dieses Schwammgefüge charakteristisch ist. (Um den Stereo-Eindruck der Bilder zu erreichen, müssen sie durch ein einfaches Bildstereoskop betrachtet werden.)

MIKROMORPHOLOGISCHE BEOBACHTUNGEN AN MELIORIERTEN BÖDEN

H. BORCHERT

Institut für Landeskultur der Justus Liebig Universität, Giessen (Deutschland)[1]

EINLEITUNG

Im Zuge der Flurbereinigung werden im Land Hessen verschiedentlich Flächen vor der Zuteilung an die Bauern melioriert. Zu den üblichen Meliorationsmassnahmen der Vorflutregelung und Dränung werden - vorerst versuchsweise - auch Maulwurfdränung, Tiefenlockerung mit gleichzeitiger Kalkung und Tiefpflugkultur hinzugezogen. Das Institut für Landeskultur der Universität Giessen ist zum Teil in diese Meliorationsversuche eingeschaltet. Ihm obliegt die Beobachtung der veränderten Bodenverhältnisse im Feld und die physikalische wie chemische Untersuchung dieser Böden im Labor über die Dauer von mehreren Jahren. Seit Frühjahr 1963 ist die Methode der mikroskopischen Bodengefügeuntersuchung in den Arbeitsplan des Instituts aufgenommen worden. Die Bodenschliffe werden im Institut nach dem Plexigum-Verfahren hergestellt (Borchert, 1962, 1964). Die Beobachtungszeit mit Hilfe dieser Untersuchungsmethode ist also relativ kurz. Trotzdem soll im vorliegenden über mikroskopische Beobachtungsergebnisse von Maulwurfdränversuchsfeldern berichtet werden, da diese Teilergebnisse doch schon richtungweisend sein können.

Die Versuchsflächen mit Maulwurfdränung befinden sich im Vogelsberg. Das Bodenmaterial besteht aus degradiertem Löss auf Basalt. Bodentypologisch unterscheiden sich die Versuchsfelder voneinander durch Lage und menschlichen Eingriff.

Die Maulwurfdräne wurden im Herbst 1962 gezogen und liegen etwa 50 cm tief.

ERGEBNISSE UND DISKUSSION

Das Versuchsfeld Laubach

Von den hier erwähnten Versuchsfeldern ist das Feld Laubach als Pflanzenstandort am günstigsten zu werten. Trotzdem leidet der in einer Mulde liegende Boden, der als Parabraunerde anzusprechen ist, unter zeitweiliger Staunässe. Obwohl der Lösslehm tiefgründig ist und in Dräntiefe noch kein Basalteinfluss vorliegt, durchsickert das Wasser nur langsam den Boden. Im Schliffbild sind im Unterboden ausser den bekannten "wabenförmigen"

[1] Direktor: Prof. Dr. H. Weber.

Poren des Lösses kaum andere Hohlräume zu beobachten. Stattdessen sieht man häufig horizontale 1–2 mm dicke und mehrere mm breite Absätze von blankem ausgewachsenen Schluff-Feinsand-Material mit solchen aus tonigem Material wechseln. Wie an anderer Stelle bereits beschrieben (Borchert, 1961) lässt diese Erscheinung auf starke Durchfeuchtung schliessen, die eine Kornsortierung mit sich bringt. Die Eisenkonkretionsbildung ist aber gering und beschränkt sich auf ein bis einige mm grosse, runde braune Konkretionen.

Ein Jahr nach dem Ziehen des Dräns war der Gang verfüllt, aber der Lockerungswinkel an der Profilwand an seiner etwas dunkleren Farbe noch erkennbar. Im Schliffbild lassen sich dunklere, erhalten gebliebene Krumenaggregate von dem sie umgebenden hellen Bodenmaterial aus Schluff–Feinsandkörnern zerriebener Aggregate unterscheiden. Bodentiere durchwühlten bevorzugt diesen lockeren Bereich und hinterliessen rundlich–ovale Hohlräume. Hier kann man mit Blümel (1956) von einem entstandenen "Schwammgefüge" des Bodens sprechen (Abb.1). Eisenausscheidungen sind nicht vorhanden. Die Schliffprobe 50 cm neben dem durch die Dränung entstandenen Lockerungsbereich lässt keine Gefügeveränderungen erkennen.

Das Versuchsfeld Ruppertenrod

Handelt es sich bei dem Boden von Laubach um einen "staubsandigen Lehm" (nach Schönhals, 1954), so stellt das zweite Versuchsfeld auf einer Flachkuppe bei Ruppertenrod, einen "sandig-grusigen bis tonigen Lehm" dar. Der Einfluss des Basalts reicht bis in Dräntiefe. Der Tongehalt beträgt in der Krume 17%, in 40 cm Tiefe 27%, in 60 cm Tiefe 33%, der Schluffgehalt entsprechend 73%, 63% und 67%[1]. Die in Laubach festgestellten hellen Feinsand-Absätze fehlen. Die Aggregate in der Krume sind durch hohen Schluffgehalt dem Zerfall eher ausgesetzt, erhalten aber andererseits durch organische Substanz und schwache Eisenverklebung etwas Festigkeit. Zum Unterboden hin widerstehen gebildete Aggregate durch höheren Tongehalt und häufigere Eisenausscheidungen besser der Mikroerosion, die Wandungen von Wurzel- und Tierhohlräumen sind gefestigt und haben keine durch lose gebundene Körnung zwischen gekreuzten Nicols aufleuchtende Säume. Die Tonsubstanz zeigt nur geringe Verlagerungstendenz. Der Boden ist in der Krume bröckelig

[1] Die hier mitgeteilten Ergebnisse physikalischer und chemischer Bodenuntersuchungen sind einer noch nicht veröffentlichten Arbeit des Diplomlandwirts W. Mertin aus dem Institut für Landeskultur des Justus Liebig-Universität Giessen entnommen.

Abb.1. Gut ausgebildetes Schwammgefüge.

Abb.2. Gangfüllung eines Erddräns aus eckigen Aggregaten in loser Lagerung.

Abb.3. Gangfüllung eines Erddräns aus zusammenrutschenden Aggregaten und in die Hohlräume fliessendem Schluff-Feinsandmaterial.

Abb.4. Dichtfliessender Schluff-Feinsandboden mit Bodentierhohlräumen, die mit dunklerem Boden ausgefüllt sind.

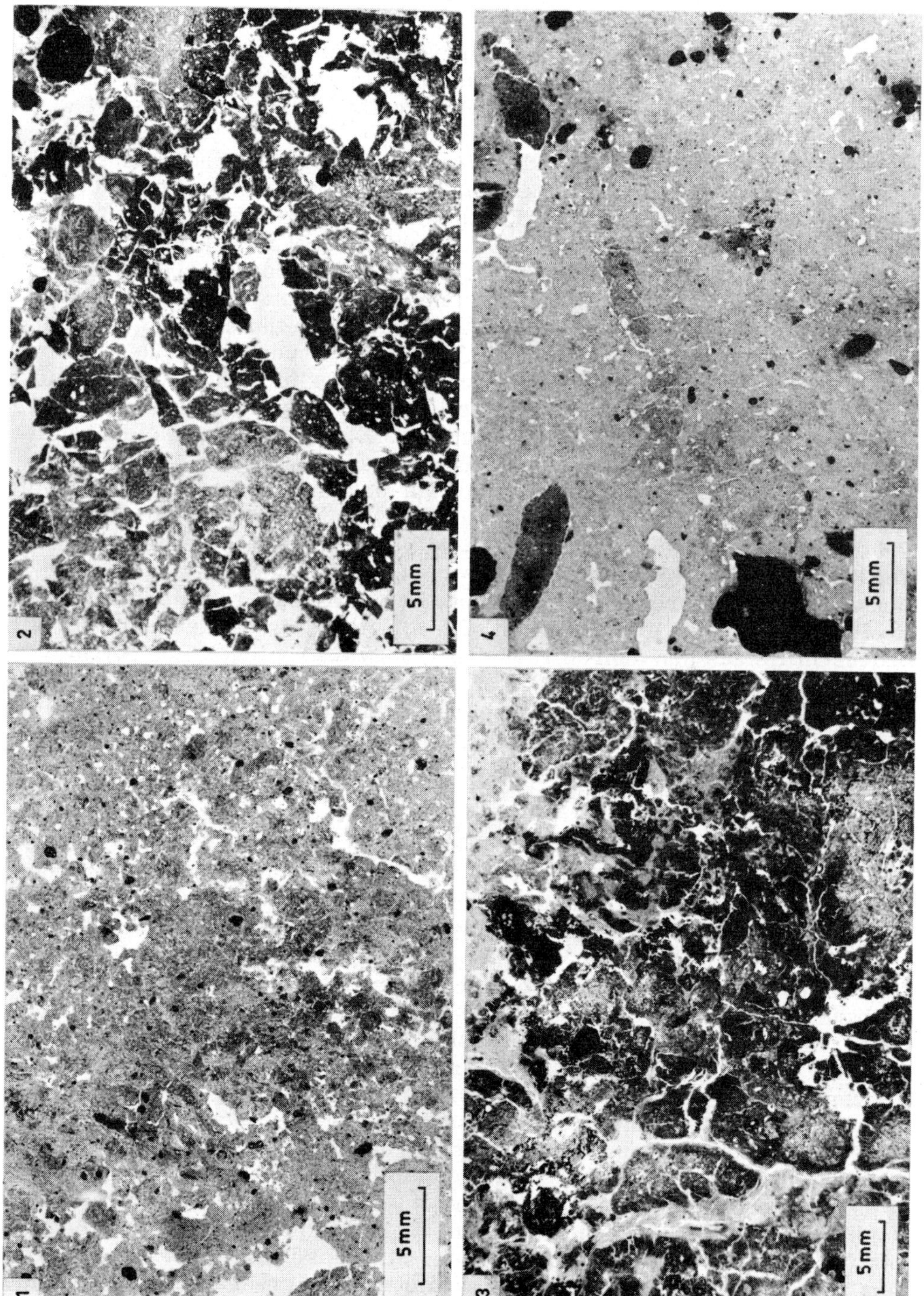
2
5mm
4
5mm
1
5mm
3
5mm

die Bröckel sind klein und relativ dicht. Im tonigeren Unterboden entstehen durch die Trockenrisse grössere, polyedrische bis plattig-polyedrische Gefügeelemente, die im Innern ausgeprägte Feinporosität aufweisen. Plattiges Gefüge und die dem Löss eigene Feinporosität führen zur ungenügenden vertikalen Wasserbewegung. Die Pseudogley Merkmale treten bereits so deutlich hervor, dass der Boden als Pseudogley–Braunerde typisiert werden darf.

Die Eisenkonkretionen sind in der Krume gerundet und bräunlich bis opak. Im Staunässeleiter tritt die tüpfelige oder saumartige Form an Porenwänden in den Vordergrund, die Färbung ist schwärzlich. Zur Staunässesohle hin treten wieder bräunliche, kompakte Eisenkonkretionen hinzu. Diese Beobachtung deckt sich mit der von Zakosek (1960) beschriebenen. Um jedoch zu einem endgültigen Urteil zu kommen, müssen noch weitere Untersuchungen angestellt werden. Mit dieser schwarzen tüpfelförmigen Eisenkonkretionsbildung, scheint eine grünlich-graue Grundfärbung des Bodens und eine relative Anreicherung von tonigem Material einherzugehen. Diese Grünfärbung und die beobachtete tüpfelige Konkretionsform beschreibt auch Blümel (1962) als "zahlreiche, kleinflockige Eisen(III)-Ausscheidungen in einem schweren Go-Horizont eines Hanggleyes".

Im Oktober 1962 wurden die Maulwurfdräne in einer Tiefe von etwa 50 cm gezogen. Auf die Bodenfeuchtigkeit beim Dränziehen, die Plastizitätszahl und die übrigen Bodenuntersuchungen soll hier nicht näher eingegangen werden, weil diese Messergebnisse von Diplomlandwirt Mertin in einer gesonderten Arbeit veröffentlicht werden. Nur soviel sei bereits gesagt, dass die Porenanteile der groben Poren sich in dem Gefügebild der zugehörigen Bodenschliffe aus den Jahren 1962 und 1963 wiederspiegeln. So betrug in 1962 der Anteil der Poren $> 50\ \mu$ in der Krume: 8 Vol.%, und in 1963: 10 Vol.% In der gelockerten Zone in 30–35 cm Tiefe waren in 1962: 7 Vol.% und in 1963: 8 Vol.% Poren $> 50\ \mu$ vorhanden. In dem nicht gelockerten Boden, 50 cm vom Maulwurfdrän entfernt, erhöhte sich dieser Porenanteil zwischen 1962 und 1963 durch die Melioration von 1,9 Vol.% auf 6,6 Vol.%. Der Anteil der groben Poren von 1,6 Vol.% im Untergrund, in 60 cm Tiefe, blieb dagegen konstant. Das Gefügebild der gelockerten Zone zeigt neben mehr oder weniger locker gelagerten Eisenkonkretionen und Basaltmineralresten abgerundete und eckige Querschnittsformen von ebenfalls locker gelagerten Aggregaten. Die Gangfüllung des nach einem halben Jahr zugefallenen Maulwurfdräns besteht aus vielen Aggregaten, die im Schliff als Querschnitte wie Trümmerstücke wahllos und nur teilweise durch ineinander verschmolzene Haftbrücken verbunden nebeneinander liegen (Abb.2). Ihre durch das Herausbrechen aus dem früheren Bodenverband zerfransten Ränder leuchten zwischen gekreuzten Nicols als Leuchtsaum von reinen Schluff–Feinsand-Körnern auf. Tonverlagerungen sind nicht zu erkennen. Ein Schwammgefüge liegt nicht vor.

Durch Verstopfung eines Abflussrohres stand in einem Drän so viel Wasser, dass der Boden bis in die Krume durchnässte. Die Gangfüllung an dieser Stelle ergab folgendes Bild: Im Schliff lassen sich deutliche Tonverlagerungen und -ansammlungen verfolgen, auch weissliche Streifen von in Hohlräumen sich ergiessendem Schluff–Feinsandmaterial, das von Aggregaten herrührt, die durch Mikroerosion in der Staunässe zerfielen. Die im Frühjahr noch trümmerartig nebeneinanderliegenden Aggregate sind zusammengerutscht und miteinander so verbacken, dass sie nur noch kleine Hohlräume zwischen sich frei lassen (Abb.3).

Das Versuchsfeld Oberseibertenrod

Ein weiteres Versuchsfeld mit Maulwurfdränen liegt bei Oberseibertenrod auf einem Nordhang. Nach Schönhals (1954) ist das Bodenmaterial ein "staubsandiger Lehm". Der Boden hat einen hohen Schluffgehalt mit 77% in der Krume, 80% in Dräntiefe und 54% im Unterboden. Der Tongehalt beträgt entsprechend 16%, 14% und 32%. Bodentypologisch stellt das Material einen schwach podsoligen Pseudogley dar. In der Krume sind überall Losungsstückchen von Milben zu finden, dagegen tritt Regenwurmtätigkeit zurück, und auch die organische Aggregatverklebung ist gering. Aus diesen Gründen sind die Krümel weniger haltbar. Runde dunkelbraune Eisenkonkretionen sind häufig, sie werden von grossen kompakten rotbraunen Eisenkonkretionen im schluffangereicherten und tonarmen A_2-Horizont abgelöst. Die Konkretionen haben zum grössten Teil Eindellungen. Diese dürften durch Abscheuern im Boden entstehen, indem die zusammengewachsenen härteren Konzentrationskerne der Eisenausscheidungen erhalten bleiben und die weicheren Zwischenfelder bei Bodenbewegungen Eindellungen erfuhren. Da im Profil beobachtet wurde, wie stark der A_2-Horizont hangwasserführend war, darf mit Blümel (1962) angenommen werden, dass hier nicht nur oxydative Vorgänge Eisenkonkretionen entstehen lassen, sondern auch Bewegungsvorgänge von Eisenverbindungen zu Anhäufungen und Eisenausscheidungen führen. Die Konkretionsbildung setzt sich bis in den Untergrund fort. In Dräntiefe heben sich gegen die helle Grundmasse aus Schluff–Feinsand wieder ausgefüllte dunklere Tiergänge oder erhalten gebliebene Aggregate der Krume scharf ab (Abb.4). Basaltstückchen und höherer Tongehalt bestimmen das Schliffbild im Unterboden. Tonwanderung ist unter dem Mikroskop kaum zu sehen.

Die Gänge waren nach einem halben Jahr zusammengefallen. Dabei stellte es sich heraus, dass der schluffreiche A_2-Horizont über dem Gang völlig dicht geschlämmt war und die typische weiss-graue "molkenartige" Färbung annahm. Der Boden in der Lockerungszone fiel einige Tage nach dem Ziehen bereits zusammen, wie der Bodenschliff erkennen lässt. Die Gangfüllung bestand allerdings im Mai 1963 aus lockerem, zusammengewürfeltem Material wie Konkretionen und Bodenkrümel, ohne dass ein schwammähnliches Gefüge entstanden war. Die im Polarisationslicht aufleuchtenden lockeren Kornsäume der Aggregate waren im Gegenteil ein Erkennungsmerkmal für das labile Gefüge der Gangfüllung.

Ein ganz anderes Bild bot sich im Oktober 1963: Der von Tieren durchwühlte Boden bestand aus abgerundeten Aggregaten und gefestigten Hohlräumen in rundlicher Form. Es war also ein ausgesprochenes "Schwammgefüge" entstanden. Es bleibt nunmehr die Probeentnahme im Frühjahr und Herbst 1964 abzuwarten, um die Veränderung oder Festigkeit des gewonnenen Gefüges zu beurteilen. Vielleicht ist das Frühjahr ein günstigerer Zeitpunkt zum Herstellen der Erddräne, weil die Tätigkeit der Bodenfauna über Sommer und Herbst die Wirkung der Meliorationsmassnahme erhöht.

Neben der Gangfüllung bzw. der Lockerungszone behielt der Boden sein plattiges, nur feinporiges, im grossen und ganzen leicht zusammenrutschendes, dichtschliessendes Gefüge. Die häufigen Konkretionen bewirken nur wie stützende Skeletteile eine vorübergehende Erhaltung von biogenen Hohlräumen, bis Schluff–Feinsandkörner wieder nachrollen. So kommt es auch zu einer gewissen Anreicherung von Schluff und ebenfalls Ton in dem zuvor offenen Gang. Im Drän befanden sich nach einer Kornanalyse 6 Gew.% mehr Ton und 2,5 Gew.% mehr Schluff.

Das Versuchsfeld Sellnrod

Bei Sellnrod liegt ein viertes Dränversuchsfeld, und zwar am Westhang unter starkem Einfluss von Hangdruckwasser. Ab 40 cm tritt zu den rotbraunen bis opaken Eisenkonkretionen die schwarze tüpfelige Eisenausscheidung hinzu. Als "sandig–grusiger bis toniger Lehm" nach Schönhals (1954) ausgewiese ist der Boden bis in die Krume basalthaltig. Die vorliegende Pseudogley–Braunerde hat nur eine geringmächtige Lössauflage. Der Schluffgehalt beträgt in der Krume 79%, in Dräntiefe 63% und in 60 cm Tiefe 48%, der Tongehalt in den entsprechenden Tiefen 13%, 29% und 45%. Einem flachgründigen Oberboden folgt in 40 cm Tiefe ein verdichteter, tonreicher Stauhorizont mit plattigem Gefüge. Trotz relativ hohen Tongehalts sind im Schliff des Unterbodens zwischen gekreuzten Nicols kaum Tonwanderungen zu finden. Die Aggregate und Hohlräume haben auch nach Aufbrechen des Bodens durch die Melioration einen festen Bau. Die starke Trockenrissbildung im tonigen Unterboden täuscht einen durchlässigen Boden vor. Es laufen Versuche im Institut, die Trockenrissbildung bei der Schliffauswertung mit zu erfassen. Bisherige Untersuchungen mit Bodenproben aus dem Vogelsberg ergaben, dass bei einer schnellen Trocknung von 105° C geringer ausgeprägte Trocken risse in den Tränkungsproben entstehen als bei einer langsamen Trocknung bei Raumtemperatur.

Das Versuchsfeld Rauisch–Holzhausen

Aufschlussreich ist der Vergleich der Vogelsbergfelder mit dem Maulwurfdränversuchsfeld Rauisch–Holzhausen, Kreis Marburg, am Rande des Vogelsberges. Der vorliegende Braunerde–Pseudogley aus umgelagertem Löss ist toniger. Der Tongehalt beträgt im Oberboden etwa 30%, in 40 cm Tiefe 42% und zwischen 50–80 cm rund 50%. Der Schluffgehalt liegt im Oberboden bei 57%, in 40 cm Tiefe bei 47% und zwischen 50–80 cm bei 37%. Der Schluffgehalt dieses Bodens ist also weitaus geringer als der von den Vergleichsböden im Vogelsberg. Im Juli 1963 wurden die Dräne gezogen. Nach einem Vierteljahr war der Maulwurfdrän in 50 cm Tiefe noch völlig unzerstört. Im Schliffbild sind wohl einzelne Schlierenbildungen zu beobachten, jedoch fehlen Tonanlagerungen oder Ansammlungen von Tonmaterial in Hohlräumen. Die Umgrenzungen von Aggregatquerschnitten leuchten auch nicht wie bei schluffreichem Boden auf. Das Gefüge erscheint im Aufbau gefestigt. Ob allerdings der Einwand zu Recht besteht, dass der vergrösserte Gangquerschn eines zusammengefallenen Maulwurfdräns in tonhaltigen Böden eine höhere Abflussspende des Feldes ergibt als ein gleichgebliebener Maulwurfquerschn (Calaminus, 1954), kann auf dem Feld Rauisch–Holzhausen noch nicht beantwortet werden.

KONKLUSION

Die bisherigen Ergebnisse lassen erkennen, dass die Methode der mikroskopischen Bodengefügebeobachtung eine wertvolle Ergänzung der üblichen bodenphysikalischen und -chemischen Untersuchungen sein kann. Sie scheint sich besonders dafür zu eignen, die Wirkung einer kulturtechnischen Bodenverbesserung laufend zu verfolgen. Deshalb soll sie in Zukunft auf verschiedenen meliorations-bedürftigen Standorten vor und einige Jahre nach durchgeführter Melioration angewendet werden. Nach langjährigen Erfahrungen wird es u.U. möglich sein, die Meliorationswürdigkeit eines verbesserungsbedürftigen Bodens und die geeigneten Massnahmen von vornherein abzuschätzen.

SUMMARY

Micromorphological observations in meliorated soils

In connection with re-allotment in Hessen, melioration projects are being carried out. On some of these meliorated areas, the Institut für Landeskultur of the University of Giessen has started field experiments. Besides soil-physical and -chemical investigations since autumn 1962, microscopic investigations of the soil structure have also been made. Results of these investigations are in good agreement with data from the conventional analytical methods. Further, these microscopic observations have given indications and explanations of variations in the soil conditions. On earthmole-drained experimental fields with degraded loess soils in the Vogelsberg, a fast filling could be observed during the time of observation. The soil increased in looseness and porosity as the clay content and the soil-fauna activity increased. It is intended to start micro-morphological investigations on different soils before and after melioration.

LITERATUR

Blümel, F., 1956. Bodengefügeveränderungen durch Maulwurfdränung. Österr. Wasserwirtsch. 8 : 313–319.

Blümel, F., 1962. Formen der Eisenoxydhydrat-Ausscheidungen in Gleyen und Pseudogleyen. Z. Pflanzenernähr., Düng. Bodenk., 98 : 258–264.

Borchert, H., 1961. Einfluss der Bodenerosion auf die Bodenstruktur und Methoden zu ihrer Kennzeichnung. Geol. Jahrb., 78 : 439–502.

Borchert, H., 1962. Die Herstellung von Bodendünnschliffen mit dem Festlegungsmittel Plexigum M 7466. Z. Pflanzenernähr., Düng., Bodenk., 99 : 159–164.

Borchert, H., 1964. Eine Methode zur Untersuchung des Bodengefüges meliorationsbedürftiger Standorte. Fragen des Gebiets- und Bodenwasserhaushaltes, Tagung Komm. 6. Deut. Botan. Ges., Kiel, 1964, im Druck.

Calaminus, U., 1954. Die Eignung der physikalischen Bodenanalyse für die Vorausbestimmung der Lebensdauer von unverrohrten Maulwurfdränen. Dissertation, Universität Bonn, Bonn.

Schönhals, E., 1954. Die Böden Hessens und ihre Nutzung. Abhandl. Hess. Landesamtes Bodenforsch., Wiesbaden, 2 : 288 S.

Zakosek, H., 1960. Durchlässigkeitsuntersuchungen an Böden unter besonderer Berücksichtigung der Pseudogleye. Abhandl. Hess. Landesamtes Bodenforsch., Wiesbaden, 32 : 63 S.

DIE BEDEUTUNG DER MIKROMORPHOLOGIE HINSICHTLICH DER ORGANISCHEN DÜNGUNG

H.-J. ALTEMÜLLER und H.-J. BANSE

Instituten für Bodenbearbeitung[1] und Humuswirtschaft[2] der Forschungsanstalt für Landwirtschaft, Braunschweig-Völkenrode (Deutschland)

EINLEITUNG

Im Vergleich zur Mikromorphologie der Humusformen unter natürlichen Vegetationsdecken erscheint die der organischen Stoffe in Ackerböden zunächst weniger aussagekräftig. Die Humusdecke eines Waldbodens zeigt meist schon in einer einzigen Probe den weiten Bereich vom unzersetzten bis zum völlig humifizierten Material und erlaubt daraus Schlüsse über die oekologischen Gegebenheiten des Standortes zu ziehen. Im Ackerboden dagegen herrschen völlig andere Verhältnisse und man kann aus einer einzelnen Aufnahme vergleichsweise nur wenig Aussagen machen.

Die Aufgabe liegt auch nicht in erster Linie in der Charakterisierung von Standortseigenschaften als vielmehr im Studium von Abbau- und Umsetzungsvorgängen der im landwirtschaftlichen Betrieb anfallenden oder nutzbaren organischen Stoffe im Boden. Diese lassen sich mit Hilfe des Mikroskopes in vielfacher Weise untersuchen, wobei die Stadien der Verrottung durch zeitlich gestaffelte Probeentnahmen erfasst werden.

Die organischen Dünger, ihre Umsetzungen und ihre Wirkungen auf den Boden wurden bislang im wesentlichen chemisch, mikrobiologisch und physikalisch untersucht. Für die mikromorphologische Untersuchung ist hier ein weites Feld offen, das neue Erkenntnismöglichkeiten bietet und Verbindungen zwischen den einzelnen Untersuchungszweigen finden lässt.

Die hier mitgeteilten Ergebnisse gründen sich auf Untersuchungen von Probematerial aus Feld- und Laborversuchen des Instituts für Humuswirtschaft in Völkenrode. Herr Professor Frese hat in grosszügiger Weise die Einrichtungen des Instituts für Bodenbearbeitung für die Durchführung dieser mikromorphologischen Untersuchungen zur Verfügung gestellt.

Biologische Objekte können heute in vielseitiger Weise mikroskopisch erforscht werden und es sind hoch entwickelte Methoden dafür ausgearbeitet, deren man sich von Fall zu Fall bedienen wird. Überall dort aber, wo organisches Material mit Boden vermischt ist, bestehen gewisse Schwierigkeiten, z.B. Färbungen an Schnitten, Imbibitionen usw. durchzuführen. Hier kommt deshalb der Anfertigung von Dünnschliffen zentrale Bedeutung zu. Wir wollen uns im folgenden darauf beschränken, allein die Untersuchungsmöglichkeiten an Dünnschliffen aufzuzeigen, die sich an verschiedenen Objekten mit verschiedenen optischen Untersuchungsverfahren bieten.

---

[1] Direktor: Prof. Dr. Frese.
[2] Direktor: Prof. Dr. Sauerlandt.

STROH IM ACKERBODEN

Das bei der Getreideernte auf dem Acker verbleibende Stroh bietet nicht nur Schutz gegen Wasserverlust oder gegen Verschlämmung, sondern fördert ein reges Bodenleben.

Im Dünnschliff erscheint eine Strohdecke immer als weiträumiges, lose und sperrig angelegtes Gerüst aus Halmen und anderen Teilen, die in verschiedenen Querschnitten angetroffen werden. Der Kontakt mit dem Boden ist zunächst gering und wird erst nach und nach durch die Bodenfauna erhöht. Durch Regen werden wasserlösliche Stoffe aus dem Stroh in den Boden gewaschen und wirken anregend auf die Organismen- und Kleintierwelt.

Der Dünnschliff brachte den Beweis, dass die Kotablage durch Kleintiere eine Vermischung von Stroh und Boden herbeiführt. Die Kotkrümel werden nicht nur in der Strohdecke, sondern auch in Hohlräumen des Bodens abgelegt. Die Produktionsleistung der Kleintiere ist demnach grösser als in dem Dünnschliff der Strohdecke sichtbar wird. Bei eingehender Beobachtung lassen sich jedoch manche Unterscheidungen treffen. Verschlämmtes Materia wird durch die charakteristischen, meniskenartig ausgerundeten Oberfläche oder feine Schichtungen erkennbar. Lose angelagerte Aggregate bis Feinsandgrösse und einzelne Körnchen legen auch die Annahme nahe, dass sie als Staubteilchen durch die Luft verfrachtet wurden.

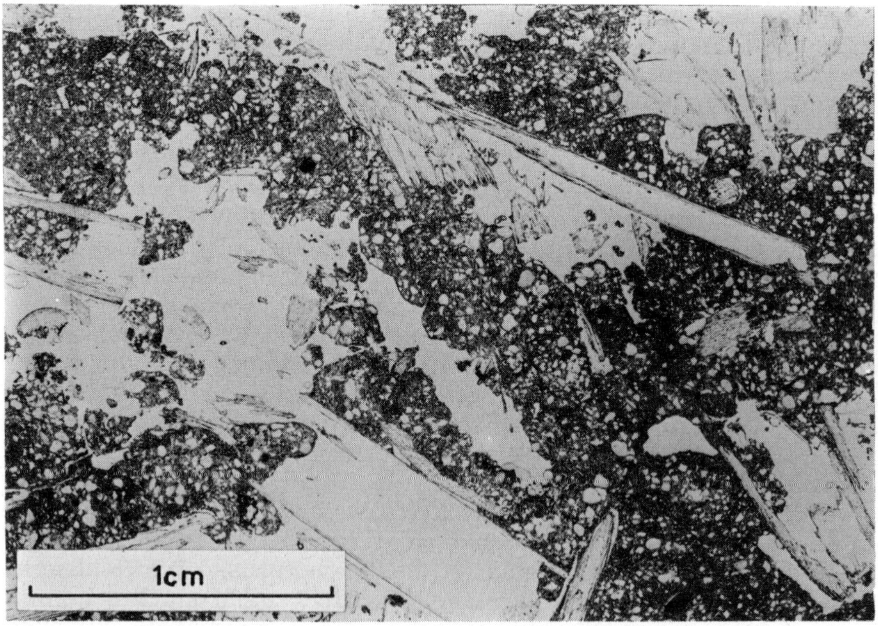

Abb.1. Strohdecke auf dem Ackerboden. Horizontaler Schnitt an der Grenze Stroh–Boden. Dünnschliff.

Abb.1 zeigt einen horizontalen Querschnitt an der Bodenoberfläche. Zwischen der kohärent-krümeligen Bodenmasse liegen die durch Kleintiere verlagerten Strohteile ungeordnet. Hohlräume sind reichlich vorhanden und bleiben auch nach Regenfällen ziemlich ungestört erhalten, da die Tropfen durch die Decke gebremst werden und bei offenem Boden kein Wasserstau eintritt. Die Bodenanteile auf der Abbildung erweisen sich grösstenteils als ineinandergedrückter Regenwurmkot. Vor allem in der Bildmitte erkennt man ovale Formen mit oft etwas dunklerer Randzone.

Die Kotformen und ihre Zusammensetzung aus mineralischen und organischen Teilen sind für das Studium des Bodenaufbaus wesentliche Hinweise. Wenn auch die Zuordnung zu bestimmten Tierarten noch immer nicht möglich ist, lassen sich doch bereits gewisse Gruppierungen vornehmen. Die Inhalte wechseln auch bei einem einzigen Individuum in Abhängigkeit von der Nahrungsaufnahme stark. So kann man rein organische, gemischte und rein mineralische Kotbällchen hintereinander aufgereiht finden. (s. Abb. bei Sauerlandt und Graff, 1959). Oft finden sich zerkleinerte Gewebestückchen, die kaum irgendwelche stoffliche Veränderung zeigen.

Werden zur Strohdecke auch anorganische Düngemittel gegeben, lassen sich deren Umsetzungen und Wirkungen ebenfalls verfolgen. In den erwähnten Versuchen waren dies Kalkstickstoff, Calciumcarbonat und Superphosphat. Fünf Monate nach Versuchsbeginn zeigte der Dünnschliff von Stroh mit Kalkstickstoff feinverteilten Kohlenstoff und Calciumcarbonat, letzteres ausschliesslich in Gestalt von radialstrahlig gebauten Sphärokriställchen. In den Parzellen mit Stroh und Calciumcarbonatzusatz waren dagegen nur rhomboedrische Carbonatkriställchen vorhanden.

Unter dem Einfluss von basisch wirksamen Düngern (z.B. Kalkstickstoff und Calziumkarbonat) wurde die Entwicklung von Pilzen, die bei der Rotte von Stroh sonst immer gegenwärtig sind, fast völlig unterbunden. Im Gegensatz dazu fördert Superphosphat durch saure Reaktion das Pilzwachstum. Hier waren alle Strohhalme stark von Pilzhyphen durchsetzt (Abb.2).

Im Dünnschliff erkennt man vorwiegend braune, aber auch farblose Hyphen mit Scheidewänden, Seitensprossen und Sklerotien. Da die Schlifflage meist nicht mit der Wachstumsebene parallel geht, sind einzelne Hyphen oft nur über kurze Strecken zu verfolgen. Fruchtkörper wurden in keinem Falle angetroffen. Sie sind nur kurzlebig und kollabieren bei der Trocknung, die der Präparation vorausgeht. Bestimmungen der Art sind deshalb im Schliff allein nicht möglich.

Vom Superphosphat selbst blieben nur uncharakteristische, optisch isotrope Aggregate von etwa Feinsandgrösse zurück, über deren Zusammensetzung nichts ausgesagt werden kann. Der Begleitstoff Gips wurde offenbar völlig gelöst.

KOMPOST

Über grosse Erfolge eines gärtnerischen Betriebes in England, in dem mit zehn-wöchiger Kompostierung von Stroh unter Zusatz von Gartenabfall, Schweinemist, Blutmehl und etwas Stickstoff gewirtschaftet wird, hat Sauerlandt (1960) berichtet. In Dünnschliffen konnte die intensive Wirkung dieses Kompostes auf den Boden an dem einheitlich lockeren, feinkrümeligen Gefüge und der gleichmässigen Durchmischung mit vererdeteten organischen Massen anschaulich dargestellt werden.

Ein weiteres Anwendungsgebiet für Dünnschliffuntersuchungen ergibt

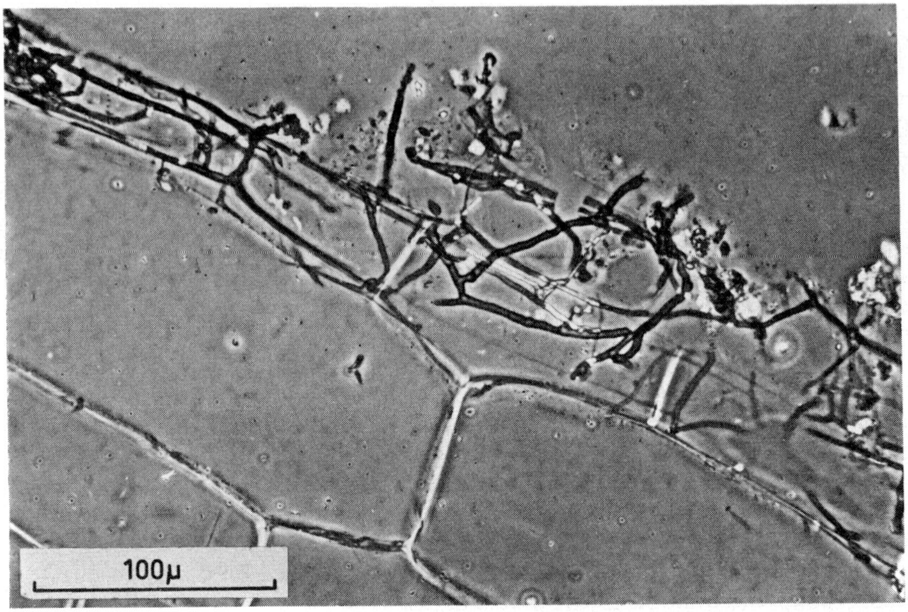

Abb.2. Wachstum von Pilzhyphen auf Parenchymgewebe von Stroh. Dünnschlif

sich bei der Verfolgung der Rottevorgänge in Kompostmieten. Von den zahlreichen Faktoren, die hier erfasst werden können, sind zu nennen: Der Grad der Vermischung der Ausgangsstoffe, Stroh oder Mist und dergleichen mit Boden und Dünger; ferner die Art der Lagerung, die stoffliche Veränderung der organischen Anteile, Zunahme der Mineralisierung, Tätigkeit von Tieren, Wirkung der Mikroflora usw. Die tiefgreifenden Veränderungen, die sich hierbei vollziehen, zeigt folgendes Bildbeispiel (Abb.3), das Kompostierungsversuchen mit Stallmist entstammt.

Ein vielseitiges und mikromorphologisch interessantes Arbeitsgebiet ergibt sich weiterhin in Verbindung mit der Müllkompostierung. Mit Hilfe von Dünnschliffen und den nachfolgend zu beschreibenden mikroskopischen Untersuchungsverfahren lassen sich über die bisher meist geübte Charakterisierung solcher Produkte hinausgehende Massstäbe zur Beurteilung dieser Bodenverbesserungsmittel finden. Die verschiedenen Aufbereitungsverfahren führen zu weitgehender Homogenisierung, Zerkleinerung und mechanisch-biologisch gebildeten Aggregaten. Die ursprünglich äusserst heterogen zusammengesetzten Ausgangsstoffe sowie das kompostierte Material zeigen in Dünnschliffpräparaten erfassbare Unterschiede. Die je nach gewählter Behandlungsmethode unterschiedliche weitere mikrobielle Umwandlung der organischen Stoffe in Kompostmieten oder im Boden kann mikromorphologis verfolgt werden. Am mit Müllkompost behandelten Boden selbst schliesslich lässt sich im Dünnschliff die Beeinflussung der Strukturverhältnisse erkenne und studieren. Der Müllkompost erwies seine Brauchbarkeit unter dem Gesichtspunkt des Erosionsschutzes in Hanglagen im Weinbaugebiet.

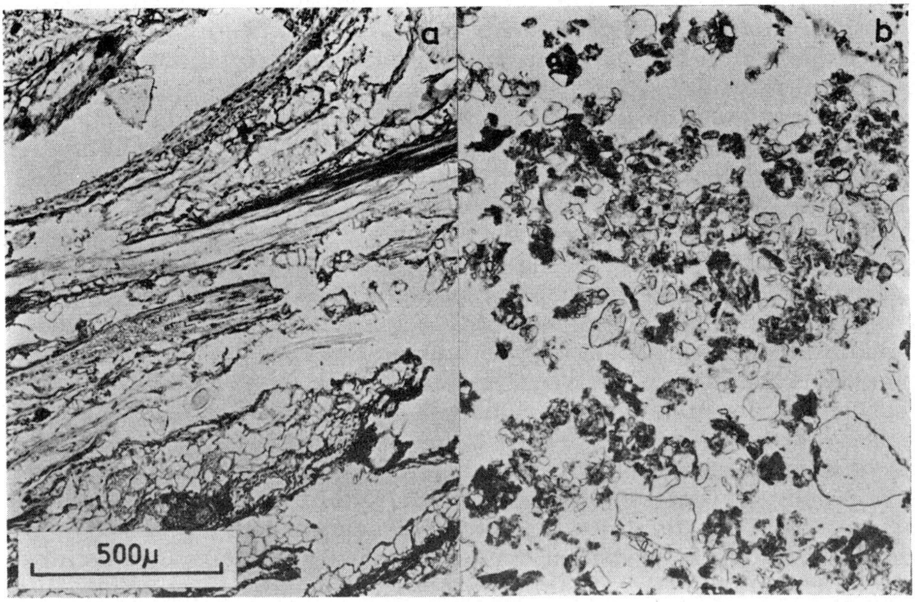

Abb.3. Mistkompost aus strohreichem Rinderfrischmist und Boden; (a) mit 5% Superphosphatzusatz; (b) mit 1% Superphosphatzusatz. Abbildung zeigt Unterschiede im Mineralisierungsgrad nach über 4jähriger Rotte. Dünnschliff.

MIKROSKOPISCHE UNTERSUCHUNGEN

Wie eingangs schon erwähnt, ist bei den mit Boden gemischten organischen Stoffen die Herstellung von Dünnschliffen besonders vorteilhaft, da auf diese Weise der natürliche Zusammenhang der Einzelobjekte am besten studiert werden kann. Als Einbettungsmittel hat sich das Polyesterharz Vestopal-H auch bei organischen Stoffen in verschiedenen Zersetzungsstadien sehr bewährt (Altemüller, 1962); gelegentlich auftretende Verzögerungen der Polymerisation und Extraktionswirkungen, die sich in gelblicher Verfärbung des Harzes äussern, brachten keine wesentlichen Nachteile. Die Verlängerung der Polymerisationszeiten, wie sie Jongerius und Heintzberger (1963) vorgeschlagen haben, erwies sich als nützlich.

Die Nachteile, die sich für manche Untersuchungen durch die starre Einbettung ergeben, können durch ergänzende Untersuchungen an Schnitten ausgeglichen werden. Sehr vorteilhaft können jedoch an den Schliffen, bei genügend sauberer und dünner Präparation, verschiedene mikroskopische Untersuchungsverfahren angewandt werden. Zur Betrachtung im Hellfeld kommen die Polarisations-, Phasenkontrast- und Fluoreszenzmikroskopie hinzu, die sich gegenseitig ergänzen. Dass jedes Ding soviel Gesichter hat, als es Betrachtungsweisen gibt (Westphal, 1963) kann hierbei ständig neu erlebt werden. Einige Beispiele aus dem Strohabbau mögen dies erläutern.

Frisches Stroh zeigt im Dünnschliff seine Teile, wie Halme, Knoten, Blätter usw. in beliebigen Querschnitten. Auffallende Elemente sind die langgestreckten grossen Parenchymzellen, und die charakteristisch verzahnten Zellen der Epidermis, ferner Tracheen mit verschieden gestalteten Verdickungen und sonstige Stützgewebe. Alle Elemente des frischen Strohes erscheinen im Hellfeld-Durchlicht praktisch farblos; zwischen gekreuzten Polarisatoren leuchten sie jedoch infolge starker Doppelbrechung hell auf. Die Doppelbrechung entsteht im wesentlichen aus dem geordneten Bau der Zellwand-Cellulose. Im Gegensatz zu den mineralischen Kristallen sind die optischen Merkmale in der Pflanzenzelle nicht konstant, sondern wechseln durch Einlagerungen in der Zellwand. Diese stellt einen sogenannten Mischkörper dar, in dem Stoff- und baubedingte Eigenschaften zusammenwirken, die sich mit jeder Änderung eines Bestandteiles auch in ihrer Gesamtheit ändern. Dabei spielen neben den natürlichen Einlagerungen in der Zellwand grundsätzlich auch die optischen Eigenschaften der Einbettungsmittel eine Rolle, doch sind infolge der dichten Packung der Micelle in der trockenen Zellwand diese letzteren Einflüsse gering (Frey, 1926), besonders bei Brechzahlen, die in der Nähe derjenigen des Objektes selbst liegen, wie dies in unseren Präparaten der Fall ist.

So weisen die stark verdickten Sklerenchym-Zellen z.T. eine sehr hohe, andere eine sehr geringe Doppelbrechung auf. Die Baupläne sind ebenfalls verschieden, doch soll dies hier beiseite gelassen werden.

Im Phasenkontrast erscheint der grösste Teil der Zellwände heller als das Einbettungsmedium (Brechzahl etwa n = 1,56), (Abb.4), das heisst, die

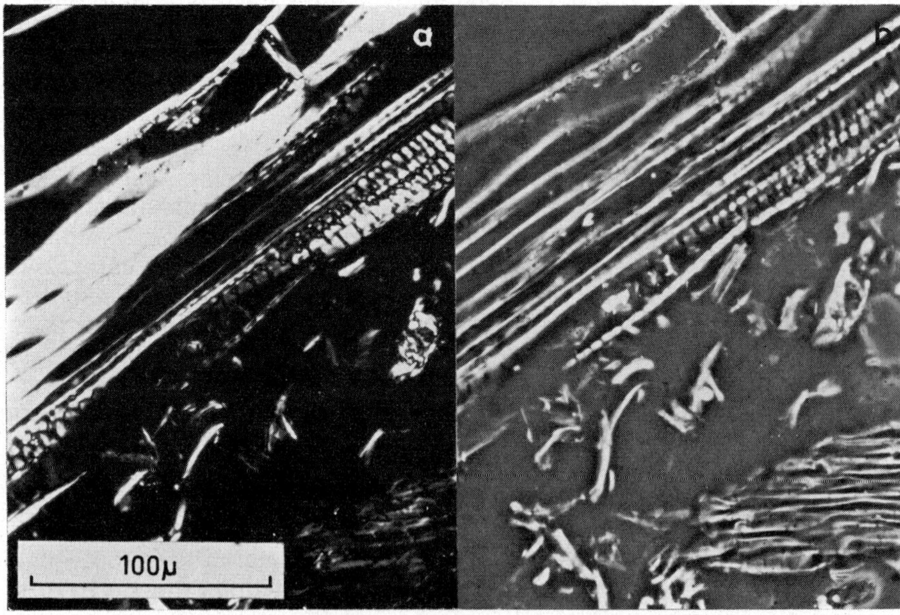

Abb.4. Dünnschliff von frischem Stroh; (a) zwischen gekreuzten Polarisatoren (starke Doppelbrechung); (b) im Phasenkontrast; Stroh erscheint heller als das Umfeld (mittlere Brechzahl niedriger als die Brechzahl des Mediums). (Objektive: Leitz Achromat 25 Pol und Leitz Pv 25 mit Heine-Kondensor).

mittlere Lichtbrechung der Objekte ist niedriger als die des Einbettungsmittels. Bei Einschaltung eines Polarisators oder Analysators zeigt sich, dass die maximale Lichtbrechung jedoch höher ist als die des Einbettungsmittels, da parallel der Schwingungsrichtung des Polarisators die Zellwände dunkler als das Umfeld erscheinen.

Die Sklerenchymzellen verhalten sich zum Teil anders. Bei ihnen sind vielfach alle Lichtbrechungswerte niedriger.

Diese Beobachtungen sind wesentlich und müssen weiter vertieft werden, da sich die Eigenschaften beim Abbau möglicherweise verändern. Kubiena (1943, 1953) hat schon mehrfach darauf hingewiesen, dass die Doppelbrechung des pflanzlichen Gewebe mit der Verrottung abnimmt. Die von Mohtadi (1962) veröffentlichten Bilder von Strohschliffen in verschiedenen Rottestadien zeigen dies ebenfalls. Die braunen, mehr und mehr von Rotteprodukten überlagerten Gewebe sind immer undeutlicher gegliedert. Wie Abb.5 zeigt, sind nur noch wenige Partien in einigermassen geordnetem Verband.

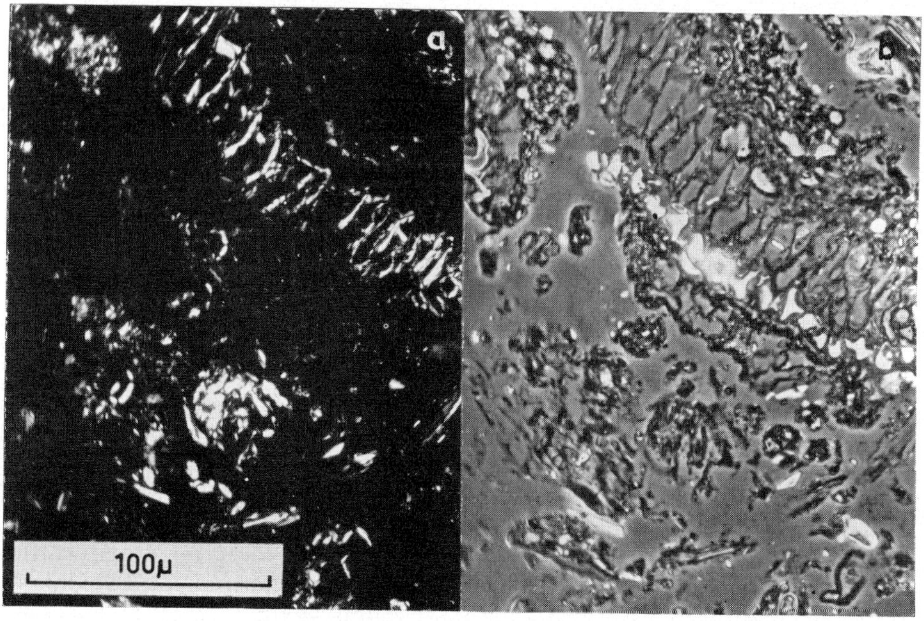

Abb.5. Dünnschliff von verrottetem Stroh; (a) zwischen gekreuzten Polarisatoren (weniger doppelbrechende Anteile); (b) im Phasenkontrast, Stroh erscheint dunkler als das Umfeld (mittlere Brechzahl scheinbar höher als die Brechzahl des Mediums). (Objektive wie Abb.4).

Im Phasenkontrast ist zu beobachten, dass fast alle Teile nun dunkler als das Umfeld erscheinen. Nach der Theorie des Phasenkontrastes (s. Altemüller, 1964) bedeutet dies eine allgemeine Zunahme der Lichtbrechung. Durch die absorbierenden Eigenschaften der braunen Rotteprodukte wird der Kontrast zwar beeinflusst, doch ist die Störung in dem sehr dünnen Schliff gering. Starke Absorption bewirkt sogar Aufhellung im Phasenkontrastbild.

Diese immerhin überraschende Beobachtung erfährt eine wesentliche Ergänzung durch Untersuchungen bei hoher Vergrösserung. Abb.6 zeigt stark verrottetes Zellgewebe im Querschnitt im Phasenkontrast. Das vor dem

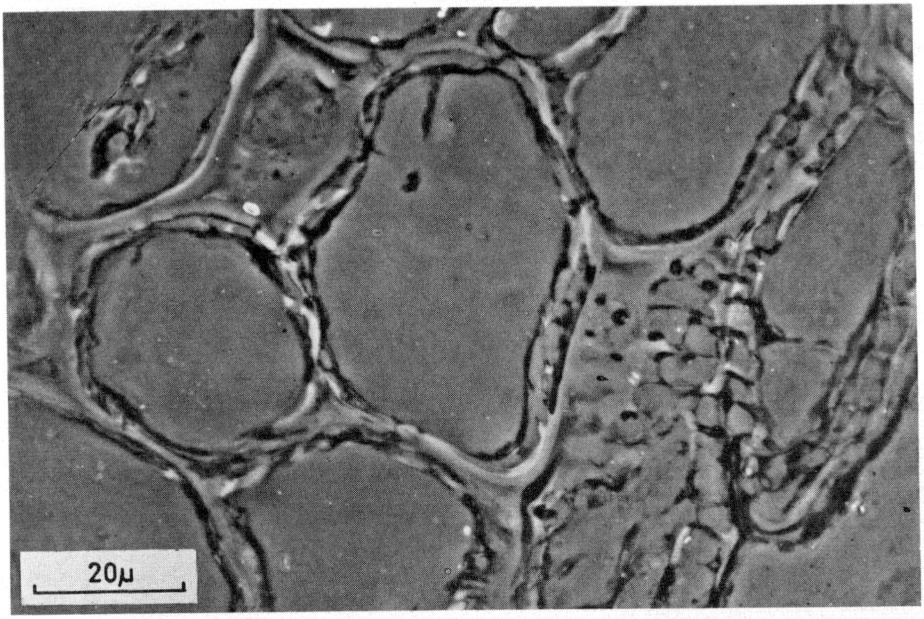

Abb.6. Dünnschliff von verrottetem Stroh im Phasenkontrast. Die Zellwände zeigen Partien von unzersetzter Cellulose mit hellen und angelagerten Huminstoffen mit dunklem Kontrast. (Objektiv: Leitz Pv Fluorit Oel 70 mit Heine-Kondensor).

einheitlich dunkel erscheinende Gewebe ist nun deutlich gegliedert in Partien mit hellem und dunklem Kontrast. Nehmen wir an, es handle sich bei den hellen Partien um ungestörte Zellwandteile und diese Annahme ist nach dem oben geschilderten berechtigt, so sieht man deutlich die netzartig ausgehöhlte Zellwände im verrotteten Stroh vor sich. Die dunklen (im Hellfeld schwach bräunlichen) Huminstoffe sind ringsum filmartig oder fädig angelagert. Sie überdecken bei schwächerer Vergrösserung die unzersetzten Zellwandreste so vollständig, dass diese Anteile ganz übersehen werden. Die Lichtbrechung und Doppelbrechung bleibt aber in den kleinsten noch unzersetzten Zellwandpartien - soweit dies bisher feststellbar ist - in wesentlichen Merkmalen erhalten.

Wir kommen damit einem Resultat entgegen, welches Pauli (1960) aus fluoreszenzmikroskopischen Untersuchungen gewonnen hat. Nach seinen Feststellungen liess sich in stark verrottetem Maiskolbenmehl und Eukalyptus-Sägemehl nach Extraktion der Rotteprodukte stets noch unzersetzte Zellulose nachweisen, woraus der Schluss gezogen wurde, dass es zwischen unzersetzten Pflanzenresten und den neugebildeten Huminverbindungen "kein Übergangsstadium gibt, das als wohlcharakterisierte Zwischenform gelten könnte".

Mit dem Fluoreszenzmikroskop lassen sich diese Beobachtungen auch am Dünnschliff machen. Frisches Stroh zeigt eine relativ starke, vorherrschend bläuliche Primärfluoreszenz. Im Laufe der Verrottung nehmen die fluoreszierenden Anteile ebenso wie die doppelbrechenden Anteile mengenmässig ab. Die Fluoreszenzfarbe ändert sich dabei nach graublau, bisweilen auch nach gelb oder weisslichgelb, während die dunklen Rotteprodukte eine sehr schwache rötlich-braune Fluoreszenz aufweisen. In sehr dünnen Schliffen lassen sich ähnlich wie mit dem Phasenkontrastmikroskop zersetzte und frische Zellwandteile nebeneinander unterscheiden. Mit zunehmender Zersetzung bilden sich fluoreszierende Abbauprodukte, die im Einbettungsmittel in Lösung gehen, wodurch dieses selbst aufleuchtet. Inwieweit sich aus diesen Beobachtungen Gesetzmässigkeiten ergeben, müssen weitere Untersuchungen zeigen. Die Anwendung der verschiedenen mikroskopischen Untersuchungsverfahren dürfte sich jedoch als wertvoll erweisen, um neue Vorstellungen über die Abbauvorgänge im Boden zu gewinnen.

SUMMARY

The significance of micromorphology for the investigation of organic manure

The process of soil organic matter transformation, as well as its influence on soil fabric, can be demonstrated and investigated by thin-sections and microscopical methods. The decomposition of straw left on the soil surface proceeds with the participation of the soil fauna mixing the straw with mineral soil particles, mainly earthworm casts. Under acid conditions caused by fertilizers, i.e., superphosphate, the network of fungi mycelium in straw debris can be made visible in thin-sections. Samples of stable manure mixed with soil and different quantities of superphosphate showed different degrees of decomposition in thin-sections after more than four years of rotting time. By combinations of different microscopical methods, i.e., bright field, dark field, polarized light, fluorescent illumination or phase contrast, the process of decomposition can be characterized by distinguishing various stages.

LITERATUR

Altemüller, H.–J., 1962. Verbesserung der Einbettungs- und Schleiftechnik bei der Herstellung von Dünnschliffen mit Vestopal. Pflanzenernähr., Düng., Bodenk., 99 : 164–177.

Altemüller, H.–J., 1964. Die Anwendung der Phasenkontrastmikroskopie bei der Untersuchung von Boden-Dünnschliffen. In: A. Jongerius (Ed.), Soil Micromorphology, Elsevier, Amsterdam, pp.371–390.

Frey, A., 1926. Die mikroskopische Struktur der Zellmembranen. Jahrb. Wiss. Botan., 65 : 195–233.

Jongerius, A. and Heintzberger, G., 1963. The preparation of mammoth-sized thin sections. Soil Surv. Papers, Neth. Soil Surv. Inst., 1 : 37 pp.

Kubiena, W.L., 1943. Die mikroskopische Humusuntersuchung. Z. Weltforstwirtsch. 10 : 387–410.

Kubiena, W.L., 1953. Bestimmungsbuch und Systematik der Böden Europas. Stuttgart, 392 S.

Mohtadi, S., 1962. Beitrag zur Isolierung von pflanzenphysiologisch wirksamen Stoffen während der Rotte von Stroh. Diss. Giessen, 117 S.

Pauli, F., 1960. The fluorescence microscop in humus research. Mikroskopie, 15 : 139–142.
Sauerlandt, W., 1960. Zum Aufbau eines gärtnerischen Kulturbodens mittels Strohkompost. Org. Landbau, 3 : 49–52.
Sauerlandt, W. und Graff, O., 1959. Die Strohdecke auf dem Ackerboden. Landbauforsch. Völkenrode, 9, (3, 4) : 57–60.
Westphal, A., 1963. Einführung in die Reflexmikroskopie. Thieme, Stuttgart, 112 S.

EINFLUSS VERSCHIEDENER FRUCHTFOLGEN UND NUTZUNGSWEISEN DES LEICHTEN BODENS AUF DIE MIKROSTRUKTUR UND EINIGE PARAMETER[1]

B. SWIETOCHOWSKI und B. JABLONSKI

Lehrstuhl für Allgemeinen Acker- und Pflanzenbau der Landwirtschaftlichen Hochschule[2], Wroclaw (Polen)

EINFÜHRUNG

Die in neuerer Zeit über die Mikromorphologie des Bodens angestellten Untersuchungen (Altemüller und Frese, 1962) haben eine bessere Kenntnis sowohl von der Zusammensetzung des Bodens und seiner chemischen Eigenschaften wie von dem Bodenleben oder von der Bodenphysiologie zum Erfolg gehabt. Der grösste Teil der durchgeführten Untersuchungen betraf jedoch die fruchtbaren mittleren oder schweren Böden. Nur wenige der Arbeiten behandeln die sandigen Böden oder Sande. Der Hauptgrund dafür ist nicht das grössere Interesse der Bodenforscher für bessere Böden, sondern die angetroffene Schwierigkeit bei der Untersuchung des losen Materials, welches Sand oder schwach lehmiger Sand ist. Leichte Böden, besonders Sande, unterscheiden sich von anderen Böden durch ihre nicht deutliche Struktur, so dass viele Forscher Sandböden als strukturlose Böden bezeichnen (Wiliams, 1959).

Es hat jedoch den Anschein, dass eine solche Bezeichnung nicht gerecht ist, denn sogar in losem Sande mit 6–8% abschwemmbaren Teilen kann man eine deutliche Struktur unterscheiden, wenn sich nur Mikroflora und Humussubstanzen vorfinden. Gibt es nahezu keine Mikroflora und ist der Humusgehalt sehr niedrig, so kann man natürlich schwerlich von Struktur reden, andererseits wäre es jedoch zuviel gesagt, solch ein Gebilde als Boden zu bezeichnen. Man kann aber darin einig werden, dass in solch einem wenig aggregierten Bodenmaterial keine dauernde Makrostruktur auftritt. Wenn die Oberschicht durch das Pflügen des Bodens einen neuen Bau bekommt, hat man es eigentlich zu tun mit einer neuen Textur und nicht mit einer neuen Makrostruktur.

Auf Sandböden verdichtet sich diese Textur sehr schnell unter dem Einfluss der Schwerkraft und Wetterfaktoren, wodurch der Boden wiederum seine frühere, einheitliche Mikrostruktur bekommt. Aus diesem Grunde kamen die Schwierigkeiten der Strukturuntersuchungen solcher Böden mit der üblich angewendeten Methode (Nassiebverfahren) wie nach Tiulin (zitiert nach Di Gleria et al., 1962), Sawwinow (zitiert nach Di Gleria et al., 1962) und anderen. Wenn auch die Absonderung der Bodenkrümel von 2–4 mm Durchmesser - der Aggregate, die landwirtschaftlich den grössten Wert haben (Birecki und Gastol, 1961) - aus diesem Boden gelingt, so sind sie

[1] Diese Untersuchungen wurden teilweise finanziert durch "United States Department of Agriculture, Agricultural Research Service".
[2] Leiter: Prof. Dr. Dr. h.c. Boleslaw Swietochowski.

doch nicht dauerhaft und können Ephemeriden genannt werden. Auch die Porenvolumenbestimmung nach Kopecky gibt keine befriedigende Resultate, weil im Falle der Austrocknung des Bodens eine Entnahme der Bodenproben ohne Strukturbeschädigung unmöglich ist. Es scheint also, dass nur mikromorphologische Untersuchungen nach der Dünnschliffmethode, tauglich sein können.

Polen besteht für 42,7% aus besseren oder schlechteren Sandböden, wovon die Hälfte angebaut wird und 15% mit Wald bedeckt sind. Auf Grund dieser Ziffern wird es deutlich, dass für diese Böden grosses Interesse besteht. Die in dem Lehrstuhl für Allgemeinen Acker- und Pflanzenbau der Landwirtschaftlichen Hochschule in Wroclaw und Lehranstalt für Ackerbau und Fruchtfolge des Instituts für Ackerbau, Düngung und Bodenkunde durchgeführten Untersuchungen, beziehen sich dann auch zu einem grossen Teil auf diese Sandböden.

Als Hauptziel dieser Arbeit wurden folgende Probleme näher untersucht:

(1) Ob in Dünnschliffen die Parameter, die die Mikrostruktur der Sandböden charakterisieren, ausgezählt werden können.

(2) Ob es möglich sei, mit diesbez. ausgezählten Parametern die Abhängigkeit der Mikrostruktur von der Nutzungsweise des leichten Bodens zu unterscheiden, auf dieselbe Weise wie mit anderen Parametern, also, ob man die Veränderungen, die durch verschiedene agrotechnische Massnahmen hervorgerufen wurden, erfassen kann.

UNTERSUCHUNGSMETHODIK

Da die Mikrostrukturuntersuchungen mit leicht bestimmbaren Parametern verbunden werden sollten, umfassten unsere Untersuchungen neben der Mikrostrukturbestimmung auch den Wasser- und Humushaushalt und die Dynamik der freien Aminosäuren in einzelnen Objekten der Feldversuche.

Dünnschliffe wurden nach der Kubienamethode (Kubiena, 1937) in der Altemüller-modifikation (Altemüller, 1962) hergestellt. Zu diesem Zweck wurde der polnische Polyesterharz "Polimal 100" der mit seinen Eigenschaften dem deutschen Harze "Vestopal-H" ähnelt, angewendet.

Der Total-Kohlstoffgehalt in den einzelnen Humusfraktionen wurde nach der kolorimetrischen Westerhoffmethode bestimmt, wobei die Aufteilung der Humusfraktionen nach der Sven Odénmethode in der Modifikation nach Miklaszewski (1959) durchgeführt wurde. Die Bestimmung des freien Aminosäuren-Gehalts im Boden wurde nach der chromatographischen Methode, nach Miklaszewski, durchgeführt (Swietochowski et al., 1964).

Durch Austrocknung im Trockenschrank wurde die Bodenfeuchtigkeit in Prozenten bestimmt.

UNTERSUCHUNGSERGEBNISSE

Im Jahre 1959, wurde in Swojec bei Wroclaw auf leicht lehmigen Sandboden ein Fruchtfolgeversuch angelegt, in dem verschiedene intensive Fruchtfolgen, schwarzer Brache und Ödland mit natürlichen Pflanzenbestand verglichen wurden. Die verschiedenen Fruchtfolgeintensitäten betrafen die Einführung verschiedener Arten und Zahl von Leguminosen und mineralischer Düngung in der Rotation. Zu dieser Untersuchung wurde nur die sogenannte "extensive" und "intensive" Fruchtfolge genommen. Auf der "intensiven" Fruchtfolge wurden 1959 Kartoffeln auf Stallmist, 1960 - Hafer, 1961 - Lupine zur Samengewinnung, 1962 - Roggen mit Winternachfrucht als Gründüngung für Kartoffeln 1963, mit Stallmistgabe angebaut. Die Mineraldüngung wurde in doppelter Gabe NPK in allen Jahren gegeben. In der "extensiven" Fruchtfolge wurden folgende Gewächse eingesetzt: Kartoffeln mit Stallmistdüngung, im zweiten Jahr - Hafer, im dritten und vierten Jahr Roggen und dann wieder Kartoffeln auf Stallmist. Die Mineraldüngung bestand jedes Jahr aus einer normalen Gabe NPK. Die schwarze Brache wurde so bearbeitet, dass kein Unkraut wachsen konnte.

Auf Ödland blieb nach der letzten Ernte im Jahre 1958 der Pflanzenbestand unangerührt. Es trat eine natürliche Veränderung der Unkrautpflanzengemeinschaft ein, sowohl qualitativ (Artenzusammenstellung) wie quantitativ. Die Zahl der Arten verringerte sich vom Jahr zu Jahr, besonders die der monokarpischen Arten. In polikarpischen Arten kam es zu einer Artenveränderung, aber nicht zu einer Mengenveränderung. Die Individuenzahl der monokarpischen Arten vergrösserte sich bis einschliesslich 1961 und fiel dann stark ab. Die Zahl der polikarpischen Arten vergrösserte sich bis einschliesslich 1960 und ging dann langsam zurück. Diese nur langsame Senkung in dem Bestand der polikarpischen Arten nach 1960, war die Ursache, dass im Jahre 1963 ihre Zahl um das 3-fache grösser als die der monokarpischen Arten war. Die Pflanzenzahl im 1961 war um das 5-fache höher als 1959, fiel jedoch 1963 zum Ausgangspunkt zurück. Sicher haben die Veränderungen in der Pflanzengemeinschaft einen Einfluss auf das Bodenmilieu gehabt.

Das Wasser im Boden

Die angegebenen Resultate stammen aus dem Jahr 1963, welches das fünfte Jahr der Versuche war und das erste der zweiten Rotation. Diesem Jahr ging ein ziemlich trockener Herbst voran und ein Winter mit wenig Schneefall, was zu einer niedrigen Bodenfeuchtigkeit führte. Die starken Regenfälle im März und im Mai vergrösserten den Wasservorrat. Danach folgte die normale Verminderung der Bodenfeuchtigkeit im Sommer. Da es in den Monaten Juli und August nur wenig Regen gab, dauerte die Bodendürre fort bis an die Regenfälle Ende August, Anfang September. Dieser Verlauf wurde in Abb.1 vorgetragen.

Wie aus dieser Abbildung hervorgeht, war der Wasservorrat für Ödland und Brache grösser als für die Fruchtfolgen. Die Unterschiede sind besonders gross im Mai und im Juni. Dagegen waren die Unterschiede zwischen den beiden Fruchtfolgen und zwischen Ödland und Brache nicht deutlich.

In Abb.2 wurde die Dynamik des Total-Kohlstoffgehaltes und die Summe des Kohlstoffes der Fulvosäuren-, Hymatomelansäuren- und Huminsäurenfraktionen, also der leichtlöslichen Fraktionen, und der freien Aminosäuren im Boden, vorgestellt.

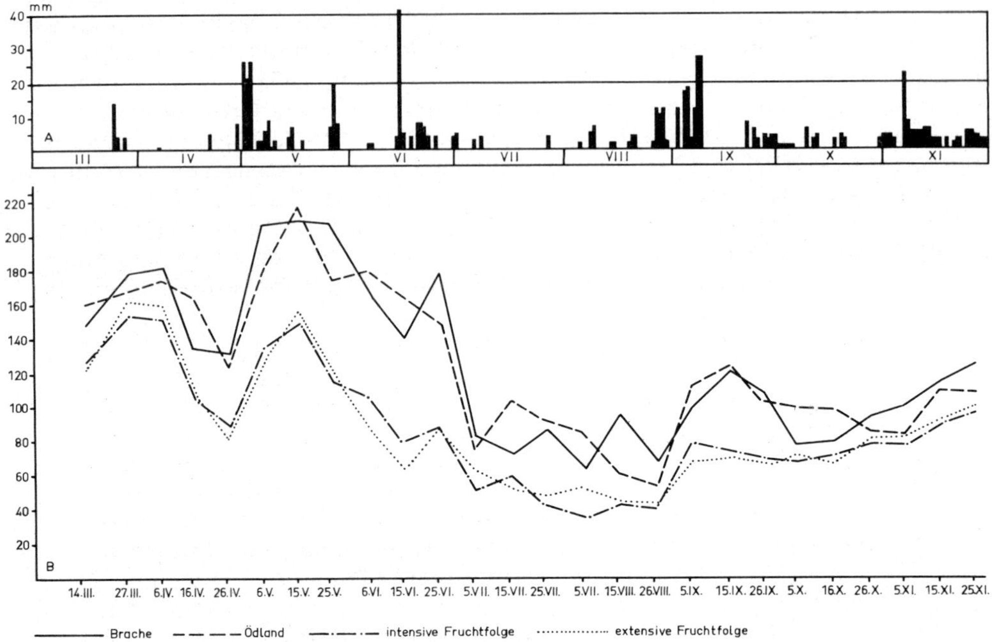

Abb.1.A. Regenfallsummen. B. Wasserdynamik in mm in einer Bodenschicht von 0–100 cm eines auf verschiedene Weise ausgenutzten leichten Bodens.

Die Humuswirtschaft

Wie man aus Abb.2A sehen kann, war die Dynamik des Humuskohlstoffes im leichtlehmigen Sandboden sehr gross und betrug von 800–1400 mg/100 g Boden, nicht nur auf bewachsenem Boden, sondern auch auf Brache. Die Produktion der Humussubstanz war im sandigen Boden, in Relativzahlen während des Jahres, sehr gross. Der Zuwachs des Humuskohlstoffes war am grössten im August–September, als die Bodenfeuchtigkeit nach der zweimonatlichen Dürre stieg; auf dem Ödland war der Kohlstoffgehalt am niedrigsten und am höchsten unter Kartoffeln auf der intensiven Fruchtfolge. Auf der extensiven Fruchtfolge wurde der höchste Kohlstoffgehalt Endc September beobachtet. Die Löslichkeitskurve der Humussäuren stellt sich anders vor. Der grösste Gehalt wurde nach dem Winter bis einschliesslich Mai festgestellt, dann verminderte er in der Trockenperiode (Juli, August). Die Regenfälle im frühen Herbst erhöhten deren Zahl, aber nur in den Frucht folgen mit Kartoffeln und nicht auf Brache und Ödland. Die vorgeführten Daten weisen darauf hin, dass Höhe und Dynamik des totalen Kohlstoffgehaltes und der leichtlöslichen Humusfraktionen verschieden war, abhängig der Bodenbenutzung. Ähnlich grosse Veränderlichkeit des Humusgehaltes in der Abhängigkeit von Wetterverlauf auf dem selben Boden wurde auch in anderen Untersuchungen festgestellt (Jablonski, 1963).

Der summarische Gehalt freier Aminosäuren in der Untersuchungszeit (Abb.2B) zeigte, dass die Summenkurven der 13 freien Aminosäuren ähnlich

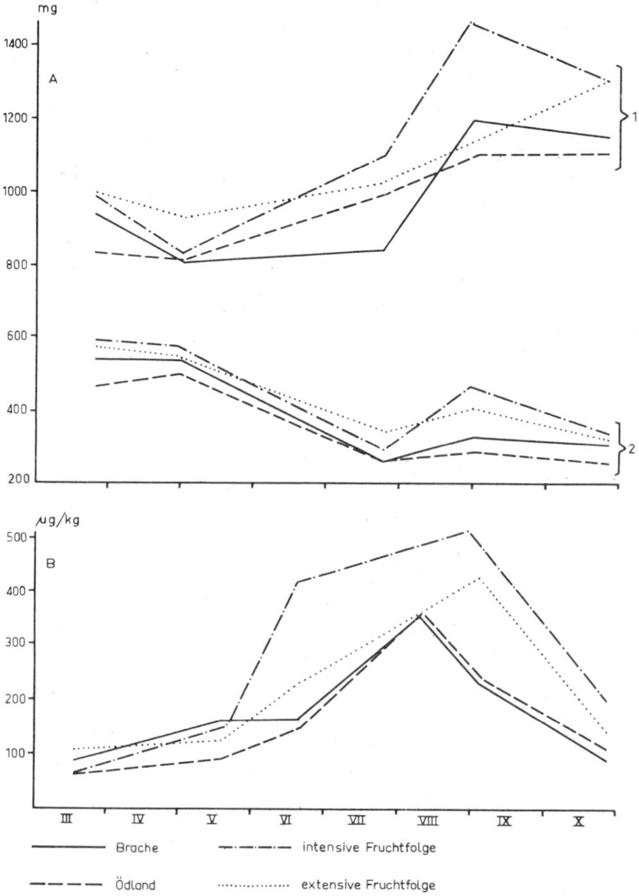

Abb.2. Die Humus (A) und "freie Aminosäuren"-Dynamik (B) eines auf verschiedene Weise ausgenutzten leichten Bodens. 1 = Total Kohlstoffgehalt des Humus (mg/100 g Boden). 2 = Summen des Kohlstoffes der Fulvo-Hymatomelan- und Huminsäuren (mg/100 g Boden).

denen des totalen Humuskohlstoffes sind. Nach dem Winter ist die Zahl (ausgedrückt in µg/kg Boden), ziemlich klein. Sie steigt im Mai und Juni, aber erst Ende August, Anfang September erreicht sie ihren Höhepunkt um dann bis zum Winter zu fallen. Der niedrigste Gehalt ist für alle Objekte auf gleichem Niveau, bei höheren Gehalten sind die Unterschiede gross. Das Maximum wird für Objekte mit Kartoffeln im August erreicht. Für Brache und Ödland etwas früher. Den höchsten Gehalt der freien Aminosäuren treffen wir in der intensiven Fruchtfolge an, einen kleineren in der extensiven, ausgenommen im Frühjahr. Den niedrigsten Gehalt zeigte dagegen die Brache. Man kann daraus folgern, dass die Dynamik der freien Aminosäuren hauptsächlich durch die Mikroflora hervorgerufen wurde. Unterschiede zwischen Brache und bewachsenen Böden wurden von der Lebenstätigkeit der Gefässpflanzen verursacht. Die spärliche Pflanzenwelt auf Ödland, welches fünf Jahre hindurch nicht gedüngt worden war, konnte den Verlauf der Kurven, im Vergleich zu Brache, nicht ändern.

Mehr interessant ist der Kurvenverlauf für einzelne Aminosäuren. Auf Grund der Ergebnisse von Miklaszewski der dreizehn Aminosäuren untersuchte, wurden drei am meisten interessante Aminosäuren zur Untersuchung genommen, nämlich: Lysin, Asparaginsäure und die Summe von Serin und Glykokol. Die Ergebnisse wurden in Abb.3 zusammengestellt.

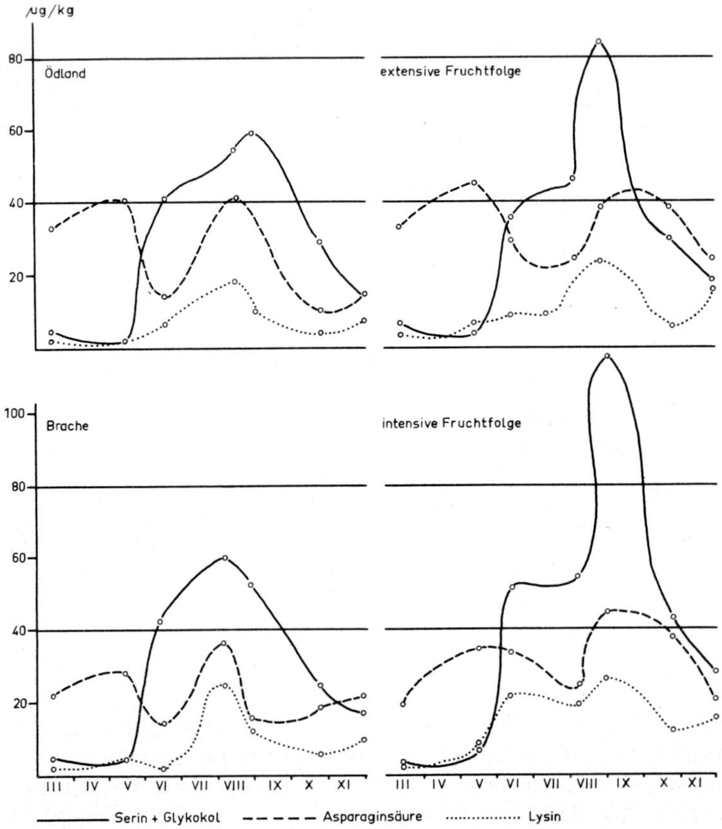

Abb.3. Die Dynamik einiger freien Aminosäuren in einem auf verschiedene Weise ausgenutzten Boden.

Bei der Betrachtung der Aminosäurenkurven der einzelnen Objekte sehen wir, dass der Verlauf für die einzelnen Objekte ganz verschieden ist. Das Maximum für Lysin fällt bei Brache und Ödland auf die Hälfte von August, bei der "extensiven" Fruchtfolge auf Anfang September und in der "intensiven" auf die Hälfte von Juni und das zweite, wie in der "extensiven" Fruchtfolge, auf Anfang September. Die Asparaginsäurekurven haben auf allen Objekten zwei Maxima, in verschiedenen Zeiten. Am interessantesten ist der Kurvenverlauf für Serin und Glykokol. Auf Brache ohne Pflanzenbedeckung ist der Kurvenverlauf kuppelförmig, ausgedehnt in der Trockenzeit und fällt beiderseits bei der Feuchtigkeitserhöhung desto stärker, je grösser der Wasservorrat im Boden ist (Mai). Bei dem Pflanzenbestand auf anderen

Objekten erfolgt die Ausdehnung der Kurven Ende August–Anfang September. Diese Ausdehnung war desto grösser, je üppiger der Pflanzenbestand war: sie war also schwächer ausgedrückt auf Ödland als auf den Kartoffelfeldern. Auf der intensiven Fruchtfolge war die Ausdehnung zweimal grösser als auf der extensiven. Es sei unterstrichen, dass die Entwicklung der Kartoffeln im letzten Falle üppiger war und der Mehrertrag von Kartoffeln auf der intensiven Fruchtfolge hinsichtlich der extensiven 120.000/ha betrug.

Da jedoch die Bestimmungen in der Zeit nur siebenmal wiederholt wurden, ist unserer Meinung nach die Zahl der Wiederholungen zu klein und müssten diese wenigstens je zehn Tage oder öfters durchgeführt werden. Die Zahl der Wiederholungen bei einer Probenahme könnte dann begrenzt werden. Es ist selbstverständlich, dass die Proben immer wieder aus denselben Stellen vom Feld entnommen werden sollen. Erst dann, bei einer so erhaltenen langen Reihe von Ergebnispunkten, kann man die richtige Kurvenlinie konstruieren.

Die angeführten Ergebnisse zeigten, dass der Wasservorrat, Humus- und freier Aminosäurengehalt, verschieden ist in einem auf verschiedene Weise gebrauchten leichten Boden. Ähnliche Unterschiede zeigten auch andere Parameter (Swietochowski, 1963).

Die mikromorphometrische Untersuchung

Untersuchungen über die Mikrostruktur des Bodens wurden an Dünnschliffen unternommen. Zahlreiche Beobachtungen der Dünnschliffe (einige Hunderte) erhalten aus verschiedenen und auf verschiedene Weise gebrauchten Böden, zeigten, dass in leichten Böden die Humusverteilung anders verläuft als auf mittleren und schweren Böden. In leichten Böden tritt Humus in Form abgesonderter Ansammlungen, mehr oder weniger gleichmässig zwischen den mineralen Teilen der konstanten Bodenphase verteilt, auf, dagegen werden auf schweren Böden die mineralischen Bestandteile des Bodens plasmenartig umgeben und kleinere Poren ausgefüllt. Eine Erklärung dafür konnte der niedrige Gehalt an abschwemmbaren und kolloidalen, mineralischen Fraktionen von Montmorylonittyp sein, mit welchen der Humus organo-minerale Verbindungen bilden könnte. Auf Grund dieses Auftretens von Humus auf sandigen Böden in leicht identifizierbaren Anhäufungen, wurde versucht, eine zahlenmässige Ergreifung der durch Humus, konstante Phase und Bodenporen eingenommenen Oberflächen durchzuführen.

(1) Auf Grund der von Jablonski ausgearbeiteten Punktzahlmethode zur Bestimmung des Porenvolumens leichter Böden (Jablonski, 1962) und ähnlichen Methoden (Geyger, 1962; Kubiena et al., 1961), wurden Messungen an Dünnschliffen durchgeführt mit einem Mikroskop, welches mit einem mikrometrischen Okular, Produktion PZO, ausgestattet war. Das Präparat wurde unter dem Mikroskopobjektiv einer Linie entlang, je einen Strich der Okularskala seitlich verschoben. Bei jeder Verschiebung wurden die unter dem Schnittpunkt der zwei Kreuzfäden angetroffenen Humus- oder Mineralsubstanzen oder Poren, notiert. Nachdem die Zählung, die ganze Linie entlang, fertig war, wurde das Bodenpräparat um 1,5 mm höher verschoben und die Zählung erneut. Im Ganzen wurden so zehn Linien bearbeitet, d.h., dass bei einer 30-maliger Vergrösserung 800–1.300 Punkte, abhängig der

Preparatgrösse, gezählt sind. Stellt man diese Zahl auf 100%, so kann man aus den Ergebnissen den prozentuellen Anteil der mineralen Phase, Humussubstanz und Poren berechnen.

Aus den erhaltenen Daten hat sich hervorgezeigt, dass im Boden aus den untersuchten Objekten des Fruchtfolgeversuches die mineralische Fraktion 43–51% des Volumes der Mikrostruktur einnimmt (das Mittlere = 46%), die Humusfraktion 16–20% (das Mittlere = 18%) und die Bodenporen 31–49% (das Mittlere = 36%). Die Ergebnisse zeigten weiter, dass diese Elemente in der Zeit einer grossen Veränderung unterliegen. Von Mai bis Juni wurde eine Senkung des Humusgehaltes auf allen Objekten festgestellt. Diese Senkung war sehr deutlich auf Ödland und Brache, schwächer auf der extensiven und gering auf der intensiven Fruchtfolge.

Weitere Untersuchungen zeigten, dass im Monat Juli eine Erhöhung des Humusgehaltes auf drei Objekten erfolgte (Ödland, Brache und extensive Fruchtfolge), wobei auf der intensiven Fruchtfolge eine Senkung des Humusgehaltes weiterhin anhielt.

Deutlich kommt hier die Abhängigkeit der Mikrostruktur des leichten Bodens von der Nutzungsweise und dem Bestimmungstermin zum Vorschein, jedoch Schlussfolgerungen dürfen noch nicht daraus gezogen werden, denn die Zahl der bearbeiteten Schliffproben war noch zu gering.

Das angewendete Punktzahlverfahren ermöglicht, wie man ersehen kann, die Ergreifung der gegenseitigen Mengenverhältnisse der untersuchten Bodenelemente, kennzeichnet aber noch nicht genügend die Mikrostruktur.

(2) Da bei der direkten Beobachtung der Mikrostruktur deutlich die Unterschiede zwischen den Objekten hervortreten, besonders in der Grösse und Zahl der Humusanhäufungen und Bodenporen (Abb.4), wurde für die zahlenmässige Ergreifung dieser Unterschiede eine besondere Methode angewendet.

Die Humusanhäufung- und Bodenporengrössebestimmung wurde mit 100x Vergrösserung vorgenommen. Das Präparat wurde länglich der Messungslinien verschoben und die Zahl der Humusanhäufungen und Bodenporen notiert, mit gleichzeitiger Messung der Elementenlänge. Als Klassengrösse wurde je 0,08 mm angenommen ($<$ 0,08 mm; 0,08–0,16 mm; 0,16–0,24 mm, etc., bis $>$ 0,96 mm). Auf diese Weise erhielt man dreizehn Klassen.

Die betreffenden Ergebnisse der Humusanhäufungen (oder Bodenporen) für jede Klasse multipliziert mit der Klassengrösse, wurden in Prozenten ausgedrückt, wobei als 100% die Gesamtlänge der zehn Messlinien angenommen wurde (Tabelle I).

Die Daten in Tabelle I, worin drei Aufnahmen aller Objekte verglichen werden, weisen darauf hin, dass im untersuchten Boden zwei Maxima auftreten. Das erste (deutliche) fällt den kleinen Phasengrössen zu, - das zweite (weniger deutlich) den grösseren Klassen. Dies betrifft sowohl die Humusanhäufungen als auch die Bodenporen. Man kann also zwei Arten von Humusanhäufungen und von Poren unterscheiden. Abhängig des Untersuchungstermins kommen quantitative Verschiebungen vor, d.h., grössere Humusanhäufungen oder Poren schwinden oder erscheinen gerade in grösserer Zahl. Aus den Daten der Tabelle I kann man vermuten, dass im Monat Juli neue Strukturen mit grösseren Humusanhäufungen und sehr zusammengesetzten Mikroaggregaten entstanden.

Für bessere Charakterisierung der Unterschiede zwischen den

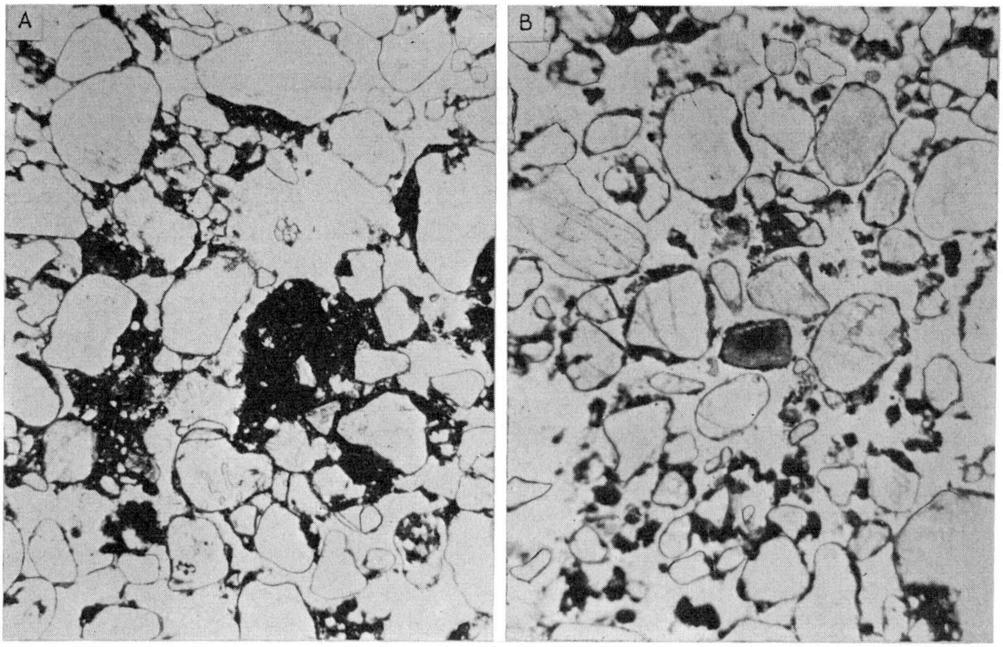

Abb.4. Dünnschliffbilder (23. Juni 1963). A. Ödland; B. Kartoffelfeld der intensiven Fruchtfolge.

Terminen und Objekten wurde der prozentuale Anteil der Humusanhäufungen und Bodenporenflächen ohne Klassenaufteilung in Tabelle II zusammengestellt. Wie man aus Tabelle II ersehen kann, war der Mittelgehalt an Bodenporen von Mai bis Juni gestiegen (nach grösseren Regenfällen, Abb.1), um dann im trockenen Juli zu fallen. Für Humusanhäufungen verlaufen die Veränderungen anders. Auf Ödland stieg die Summe der Humusanhäufungen während der ganzen Zeit (Pflanzenwachstum ohne Menscheneingriff). Auf Brache blieb die Summe unverändert, auf der extensiven Fruchtfolge dagegen stieg sie im Juni (also umgekehrt wie die Porengrösse). Auf der intensiven Fruchtfolge verringerte sich die Zahl im Monat Juli stark, was zu einer Mittelwertverringerung für diesen Termin führte.

Bei der Betrachtung der Werte für Poren (Tabelle II) könnten wir eine grosse Veränderlichkeit der Objekte in den verschiedenen Terminen beobachten. Mit Ausnahme der intensiven Fruchtfolge stieg die summarische Grösse der Poren im Juni, fiel im Juli aber wieder herab, besonders auf Brache, weniger auf extensiver Fruchtfolge.

Bei der Betrachtung der beiden Strukturkomponenten in einzelnen Objekten sehen wir, dass das Ödland sich mit der höchsten summarischen Grösse der Humusansammlung und Porengrösse auszeichnete, Brache mit grosser Humusansammlung und mittlerer Porengrösse, extensive Fruchtfolge mit mittlerer Humusansammlung und Porengrösse und zuletzt die intensive Fruchtfolge mit der kleinsten Humusansammlung und mittleren Porengrösse.

TABELLE I

Ausführliche Zahlencharakteristik der Mikrostrukturelemente in einem auf verschiedene Weise ausgenutzten leichten Boden[1]

Objekte	Grössenklassen (mm)					
	0,08 0,16	0,16 0,24	0,24 0,32	0,32 0,40	0,40 0,48	0,48 0,56
	Humus					
15. Mai 1963						
Ödland	2,9		4,1	3,7	1,5	2,2
Brache	3,3		3,8	4,0	2,9	1,5
Fruchtfolge extensive	3,6		4,8	3,2	1,1	0,8
Fruchtfolge intensive	4,1		4,0	2,3	1,5	0,6
18. Juni 1963						
Ödland	3,7	5,3	4,2	0,5	0,9	1,8
Brache	2,8	5,0	4,0	3,3	1,6	0,8
Fruchtfolge extensive	2,7	6,0	2,6	0,9	1,4	-
Fruchtfolge intensive	3,8	6,5	3,6	2,3	2,0	0,4
23. Juli 1963						
Ödland	3,7	6,5	2,3	2,3	2,3	0,9
Brache	4,6	6,4	4,1	2,6	1,1	-
Fruchtfolge extensive	3,5	5,6	4,8	2,4	0,9	0,4
Fruchtfolge intensive	3,0	5,0	1,0	0,6	0,7	-
	Poren					
15. Mai 1963						
Ödland	4,1		5,2	4,1	3,3	1,8
Brache	2,8		4,9	4,6	2,6	2,3
Fruchtfolge extensive	3,4		3,3	5,1	3,2	2,5
Fruchtfolge intensive	5,3		5,8	6,6	1,8	2,5
18. Juni 1963						
Ödland	1,4	4,5	6,0	3,3	2,1	3,9
Brache	1,8	4,8	4,0	6,1	2,1	3,6
Fruchtfolge extensive	1,5	5,1	3,8	2,5	2,0	1,4
Fruchtfolge intensive	2,4	4,3	2,1	2,7	1,8	1,7
23. Juli 1963						
Ödland	2,8	5,1	2,5	1,8	0,8	0,5
Brache	2,3	2,6	1,4	0,6	0,2	1,3
Fruchtfolge extensive	2,6	4,9	2,6	1,5	0,4	0,4
Fruchtfolge intensive	2,4	4,5	3,2	2,4	1,4	1,7

[1] Die Zahlen bedeuten die totale Länge der Humusansammlungen und totale Länge der Bodenporen in einzelnen Grössenklassen, umgezählt zur totalen Länge der Messungslinien.

0,56 0,64	0,64 0,72	0,72 0,80	0,80 0,88	0,88 0,96	>1,04	Gesamt	Objekte
							15. Mai 1963
1,3	1,0	-	-	-	-	16,7	Ödland
1,5	2,5	-	-	-	-	19,4	Brache
0,9	1,6	-	-	-	-	16,0	Fruchtfolge extensive
0,7	1,2	-	-	-	-	14,4	Fruchtfolge intensive
							18. Juni 1963
-	-	-	-	-	0,8	17,2	Ödland
0,6	0,4	0,8	-	-	0,6	19,9	Brache
0,3	-	0,4	-	-	1,0	15,3	Fruchtfolge extensive
-	-	-	-	-	-	18,6	Fruchtfolge intensive
							23. Juli 1963
0,8	1,2	-	0,4	-	-	20,4	Ödland
0,5	-	0,3	-	-	-	19,6	Brache
-	0,3	0,3	-	0,8	0,4	19,4	Fruchtfolge extensive
0,3	-	-	-	-	0,4	11,0	Fruchtfolge intensive
							15. Mai 1963
1,0	3,1	-	-	-	-	22,6	Ödland
2,4	2,8	-	-	-	-	22,4	Brache
2,3	4,8	-	-	-	-	24,6	Fruchtfolge extensive
1,8	1,2	0,5	-	0,6	-	26,1	Fruchtfolge intensive
							18. Juni 1963
1,2	2,8	0,5	-	-	9,5	35,2	Ödland
2,6	2,6	0,4	0,9	-	3,2	32,1	Brache
1,1	4,7	0,7	0,8	0,9	2,9	27,4	Fruchtfolge extensive
0,8	0,6	-	0,4	-	-	16,8	Fruchtfolge intensive
							23. Juli 1963
0,8	-	1,0	0,8	-	0,5	16,6	Ödland
-	-	-	-	-	-	8,4	Brache
0,3	0,6	-	-	-	-	13,3	Fruchtfolge extensive
0,8	1,2	0,7	-	-	0,4	18,7	Fruchtfolge intensive

TABELLE II

Die Zahlencharakteristik der Veränderungen in der Mikrostruktur in einem auf verschiedene Weise ausgenutzten leichten Boden in Prozenten der gesamten Messungslinie

Datum der Probeentnahme	Humusansammlungen				
	Ödland	Brache	Fruchtfolgen		Mittelwert
			extensive	intensive	
15. Mai 1963	16,6	19,3	15,9	14,5	16,6
18. Juni 1963	17,6	19,9	15,3	18,5	17,8
23. Juli 1963	20,5	19,5	19,4	11,0	17,6
Mittelwert	18,3	19,6	16,9	14,7	

Schlussbemerkung

Die dargestellten Untersuchungen auf vier Objekten des Fruchtfolgeversuches zeigten, dass auf leichtem Boden die Prozesse, verbunden mit langdauernder Nutzung des Bodens, sich ausprägen in Parametern wie Wasserkapazität, Humusgehalt, freier Aminosäurengehalt und in der Bodenmikrostruktur; die in Dünnschliffen erkennbaren Unterschiede konnten zahlenmässig ausgedrückt werden. Es muss jedoch unterstrichen werden, dass wir bei den mikromorphologischen Untersuchungen einige Schwierigkeiten gehabt haben. Erstens war es unmöglich bei der Herstellung der Dünnschliffe eine Sammelprobe anzuwenden. Jede Probe muss gesondert genommen und bearbeitet werden, was die Häufigkeit der Probeentnahme begrenzt. Weitere zahlenmässige Ergreifung ähnlicher Mikrostrukturen würde eine zweite Schwierigkeit sein. An einem einzigen Dünnschliff müssten ungefähr 1.000 Messungen durchgeführt werden, um die drei Zahlen für mineralische Bestandteile, Humus und Poren, zu erhalten. Zum Vergleich einiger Objekte während einer gewissenen Zeit und in einigen Wiederholungen, würde die Zahl der Messungen hunderttausend betragen.

Aus den zitierten Ergebnissen geht jedoch hervor, dass die angewendete Messmethode einen Vergleich der Mikrostruktur auf leichten Böden ermöglicht. Die Methode ist noch nicht zu Ende bearbeitet worden. Es muss noch festgestellt werden, in wieviel Wiederholungen die Bodenproben entnommen werden sollen um representativen Werte zu bekommen. Es scheint, dass bei günstigen Bedingungen für das Bodenleben (Feuchtigkeit und Temperatur), die Bodenproben massgebend sind.

SUMMARY

The influence of different crop rotations and different utilization of light soil on micrc structure and some soil parameters

The 1963 investigations on a crop rotation experiment on sandy loam (begun in 1959) were carried out at the Experimental Institute Swojec of the

Poren					Datum der Probeentnahme
Ödland	Brache	Fruchtfolgen		Mittelwert	
		extensive	intensive		
22,6	22,2	24,5	26,0	24,8	15. Mai 1963
35,1	31,9	27,2	16,6	27,6	18. Juni 1963
16,6	8,3	13,3	18,6	14,2	23. Juli 1963
24,8	20,8	21,7	20,4		Mittelwert

College of Agriculture, Wroclaw. The following objects were selected for investigation: fallow with natural plant succession, black fallow, a potato field with extensive crop rotation, and a potato field with intensive crop rotation.

The investigations comprised the following determinations: water supply to 100 cm depth, contents and qualities of humus, contents of free amino acids, and micrometric measurements on thin sections for quantitative characterization of the differences in microstructure.

These investigations allowed us to draw the following conclusions:

(1) Thin section studies give a good insight into the structure of sandy soils. When combined with the determination of other parameters, they enable us to understand some processes occurring in light soils under the influence of agrotechnical operations.

(2) It appears that it is possible to compare and characterize the different structures by means of micromorphometrical technique.

(3) When examining the micromorphology of sandy soils, it is necessary to apply replications over a long period to learn the dynamics of their microstructures.

(4) Because characterization of microstructure by means of thin sections is very time consuming, determinations of other parameters, such as total carbon content of humus and the content of free aminoacids in soil, should be made at the same time, so that micromorphological examinations can be limited to a few periods determined by the results of other investigations.

(5) The differences in humidity, humus, free aminoacids, nitrates and nitrites in the examined soil are more strongly influenced by the course of weather (precipitations and temperature) than by technological processes (manuring, crops).

Thus, as soil parameters are variable, they should be examined dynamically.

(6) It is necessary to base a determination of the influence of man on soil properties on several parameters.

(7) The variability of any parameters resulting from different activities of man on light soils appears to be more pronounced in the periods of high

values of these parameters, and less as their value decreases; thus, the intensity of investigations should be greater during the period of higher values of the parameters.

LITERATUR

Altemüller, H.–J., 1962. Verbesserung der Einbettungs- und Schleiftechnik bei der Herstellung von Bodendünnschliffen mit Vestopal. Z. Pflanzenernähr. Düng. Bodenk., 99 : 164–177.

Altemüller, H.–J. und Frese, H., 1962. Arbeiten aus dem Gebiet der Mikromorphologie des Bodens. Chemie Verlag, Weinheim, 11 : 252 S.

Birecki, M. und Gastol, J., 1961. Die Charakteristik einiger Bestandteile der Bodenkrümel unter verschiedenen Anbaupflanzen. Roczniki Nauk Rolniczych, 84 : 367–408 (polnisch).

Di Gleria, J., Klimes–Szmik, A. und Dvoracsek, M., 1962. Bodenphysik und Bodenkolloidik. Fisher, Jena, 795 S.

Geyger, E., 1962. Zur Methodik der mikromorphometrischen Bodenuntersuchung. Z. Pflanzenernähr. Düng. Bodenk., 99 : 118–129.

Jablonski, B., 1962. Die Anwendung von Bodendünnschliffen zur Porenvolumenbestimmung. Roczniki Gleboznawcze, 11 : 109–116 (polnisch).

Jablonski, B., 1963. Untersuchungen über den Humusgehalt und die Humusdynamik in leichten Böden. Zeszyty Probl. Postepow Nauk Rolniczych, 40a : 125–140. (polnisch).

Kubiena, W.L., 1937. Herstellung von Dünnschliffen von Böden in ungestörter Lagerung. Zeiss Nachrichten, 2 (3) : 81–91.

Kubiena, W.L., Beckmann, W. und Geyger, E., 1961. Zur Methodik der photogrammetrischen Strukturanalyse des Bodens. Z. Pflanzenernähr. Düng. Bodenk., 92 : 116–126

Miklaszewski, S., 1959. Versuche zur Bestimmung der Fruchtbarkeit von Sandböden als Teste und Indikators der physikalischen und chemischen Eigenschaften der Humusverbindungen, Zeszyty Nauk Wyzszej Szkoly Rolniczej Wroclawiu, Rolnictwo, 25 : 3–20 (polnisch).

Swietochowski, B., 1963. Eine Methode zur direkten Bestimmung des Zuwachses an Nitrat- und Ammoniakstickstoff im Boden. Zentr. Bakteriol., Parasitenk., Abt. II : 259–267.

Swietochowski, B. und Miklaszewski, S., 1964. Freie Aminosäuren als Indikator der biologischen Prozesse im Boden. Zeszyty Nauk Wyzszej Szkoly Rolniczej Wroclawiu, im Druck (polnisch).

Wiliams, W., 1950. Grundlagen der Landwirtschaft. Panstwowe Wydawnictwo Rolnicze i Lesne, Warszawa, 648 S. (polnisch).

Worobjew, S.A., Jegorow, W.E. und Kisjelew, A.N., 1951. Übungen für Acker- und Pflanzenbau. Panstwowe Wydawnictwo Rolnicze i Lesne, Warszawa, 302 S. (polnisch).

THE MORPHOLOGY OF HUMIC GLEY SOILS (ORTHIC HAPLAQUOLLS) UNDER DIFFERENT LAND USE

A. JONGERIUS and A. JAGER

Soil Survey Institute, Wageningen (The Netherlands)

INTRODUCTION

The aim of nearly every micromorphological investigation carried out in The Netherlands is to support soil survey. Because The Netherlands is a very intensively agricultural as well as an economically extremely important horticultural country, it is quite understandable that many large-scale mappings are performed. Such maps, of course, ought to provide as much agricultural information as possible. For this reason in our country micromorphological investigations are not directed at pedogenesis alone, but also are increasingly devoted to the study of many agricultural problems concerning hydrology, soil management, rooting, etc.

The following pages deal with an example of this applied micromorphology. A current problem is involved; namely, the extent to which certain map units are suitable for tulip culture. This culture was formerly confined to a few "classic", light-textured areas. However, in the post-war years, a considerable infiltration and expansion of the culture took place in clayey regions, particularly in the northern part of the province North Holland. It appears that certain soils in this area are well-suited to the purpose of tulip culture; nevertheless, there are several other soils about which we have only a rough knowledge. Of course, this applies in particular to those soil types on which tulip culture has hardly begun or is done on a very small scale. It is of the greatest importance to learn the extent to which these soils are suitable for this culture. Because the tulip is very susceptible to water excess - or, probably more correctly, to oxygen shortage - it is obvious that this problem, among others, must be tackled, from the morphological point of view.

We emphasize that we do not attempt to give a complete report of our investigations in this paper, but only an idea about a method of research that we are developing: the "morphological sample-area research". Therefore, only some typical aspects are presented out of the total complex of results.

THE AREA

The area in which the investigations took place is situated east of the village of Hauwert, about five miles north-northeast of Hoorn. It was - and to a certain extent still is - a typical grassland region, in which there was little outdoor horticulture. Tulip culture has recently increased very rapidly, partly in already existing horticultural grounds, but for the greater part on ploughed grassland plots.

The soils, which according to the American soil classification (Soil Survey Staff, U.S.D.A., 1960) are predominantly orthic haplaquolls[1], developed from marine deposits, viz., the Westfrisian sediments I (20–25% clay[2]) and II (25–50% clay) with a clayey layer of varying heaviness on top. The latter contains about 25% organic matter in grassland and 14–17% in horticultural soil. The top-sides of both Westfrisian sediments are marked by very dark A_1-horizons which contain about 20–25% clay and 5% organic matter and about 30% clay and 25% organic matter respectively.

The oldest Westfrisian sedimentation period, in particular, brought about a pronounced creek and basin pattern with differences in height up to 1 m. This subsoil is locally either very clayey (up to 46% clay) or sandy (1–7% clay).

The yearly precipitation amounts to about 675 mm; spring and early summer are relatively dry, and late summer and autumn are wet. The mean temperature in January is 0° C, and that in July is approximately 20° C.

METHODS

To obtain a profound insight into morphological dynamics of a soil under different land use, one must make a comparative profile study. This study must include a complete sequence of standard profiles; in our case, this sequence ranges from the old grassland soil to the old horticultural soil used for tulip culture. The profiles may be spread over the entire relevant region if its soil conditions are fairly well known; otherwise, an exact choice of standard profiles is impossible. Another disadvantage of a widely diffused location of the study objects is that there may be rather large local differences, e.g., in manuring levels, etc.

For these reasons, we applied a different procedure. The investigations were concentrated in a comparatively small sample area (about 10 ha) which was representative for the whole region.

The field activities in the sample area consisted of making a detailed soil map, a relief map and a structure map (spade mapping), and of making observations on grassland vegetation, periodical measurements of groundwater levels, studies of some experimental yields of tulip fields and a minute study of land use and soil management. From all this information standard profiles were selected. Their morphological dynamics were followed by means of both field and laboratory investigations. The latter included the making of mammoth-sized thin sections (Jongerius and Heintzberger, 1963), qualitative investigations of these, micromorphometrical studies (Jongerius, 1963), analyses of earthworm populations and several complementary physical methods, such as the determination of soil consistency, water stability of soil structure and phases (soil–water–air) ratios.

---

[1] The grassland soils are orthic haplaquolls, intergrading to histic haplaquolls; the histiculture soils orthic haplaquolls.

[2] Clay: the mineral fraction $< 2\ \mu$, as determined by means of granular analysis.

Because it is impossible to discuss the total complex of results within the scope of this paper, this will be published elsewhere (*Boor en Spade*). We confine ourselves here mainly to the most striking morphological facts.

THE MACROSTRUCTURE

It appeared that structure mapping had already supplied a great deal of information about the morphologic dynamics of the orthic haplaquolls and their suitability for tulip culture.

The map units are structure profiles. A structure profile is a vertical succession of structure types, which is determined by agricultural factors, e.g., hydrology, soil cultivation, etc.

The old grassland

In these, four structure profiles were found, viz., two profiles in a topographically, relatively high position and two in a low one. One would expect the high types to represent fairly dry conditions and the low types wet conditions. However, this is only partly the case; in the lower reaches of the terrain, a structure profile with features of relative drought is met with.

It is apparent from previous investigations (Jongerius, 1957) that the shape and size of structure elements and their eventual compound nature and grade of structure are important hydrological criteria. Broadly speaking, in clay soils, blocklike simple elements with a high grade of structure point to at least periodical drought, whereas blocklike simple elements with a low grade of structure indicate wet soil conditions. Hydrological intermediary conditions are characterized by compound prisms.

In fact, the macromorphological features of the different profiles are:

(1) Low and fairly wet; 0–20 cm: small subangular blocky elements, low grade of structure; 20–30 cm: strongly developed prisms composed of small subangular blocky elements which have a low grade of structure.

(2) Low and fairly dry; 0–20 cm: strongly developed prisms composed of subangular blocky elements with a medium grade of structure; 20–30 cm: small subangular blocky elements with a high grade of structure.

(3) High and fairly dry; 0–20 cm: same as top-layer of type 2; 20–40 cm: small strongly developed prisms composed of blocklike elements with a medium grade of structure; the elements of lower order are of the subangular blocky type in the top part of the layer and of the blocky type in the bottom part.

(4) High and dry; 0–35 cm: small subangular blocky elements with a high grade of structure.

These four morphotypes generally have the same subsoil structures, viz., a fine-porous homogenized layer with small vertical root channels and some earthworm voids, underlain with the almost unaltered stratified West-frisian sediment I.

The justness of the hydrological evaluation of the profiles is affirmed by the results of grassland vegetation investigations made by Mr. Th. van Egmond (Soil Survey Institute, Wageningen).

The remarkable phenomenon that periodically too-dry soils are situated even in very low sites is due to the fossil A_1 -horizon of the West-frisian sediment II. This layer, which mainly occurs between 20–30 cm below the surface, appears to be strongly developed in structure profiles 2 and 4, but only weakly in profiles 1 and 3. Under wet conditions, this layer is very compact, due to its granular composition and to its content and form of organic matter. When it dries, it crumbles into small fragments, having a high degree of structure, which means that if the layer is more than about 5 cm thick the capillary water movement in the profile is interrupted at this stage.

The horticulture land

During the first year after the grassland has been ploughed, the cultivatea layer usually maintains a very porous structure; however, a more or less pronounced structure deterioriation then sets in (crust formation, increasing compactness, etc.).

Because of this phenomenon, one would expect the older horticultural soils to have a very bad structure, yet we found a large variety in the structure of the cultivated topsoil, such as two adjacent fields, one of which showed a nearly crumbly and the other a totally solid structure. These two cases are extremes from long series of mapped structure profiles. Such structure mapping gives only a snapshot, because all the differentiated morphotypes can pass into each other. Nevertheless, mapping is of great value in obtaining better insight into the factors determining soil structure; it has revealed that:

(1) The cultivated orthic haplaquoll of this area had a strong tendency towards structure deterioration. A wet autumn and winter and a wet thaw cause rapid puddling of the top soil and shifting of the components. However, even under less extreme circumstances, the top soil inclines towards gradually increasing compactness.

(2) Even under only humid soil conditions, cultivation causes a distinct plough-sole.

(3) Soil structure deterioration is considerably promoted by rotary cultivation; however, it is slowed down if the soil is left very "rough" in autumn.

(4) Some of the crops grown in the region may have a favorable influence on soil structure. A good example is potato culture. The necessary culture measures, as well as the fairly deep cultivation during harvest, eliminate the advancing compactness, at least if the work is done under dry soil conditions. At the same time, the crop protects the soil surface for a longer period against the deteriorating action of rain showers. On the contrary, beet culture, for instance, is very dangerous to these soils, particularly because the crop is usually harvested while the soil is very wet.

(5) The changes in texture within the profile have, as a rule, less influence on the formation of the structure profile, as was the case under grassland. This is due to the fact that the buried A_1 -horizon of the West-frisian sediment II is either entirely, or for the most part, broken up and mixed with the top layer, depending on soil tillage.

(6) The compacted soil, when cultivated under dry conditions, may break into small pieces which may even have a crumbly appearance.

THE MICROMORPHOLOGY OF THE SOIL

Microscopical investigations of thin sections of these soils disclosed immediately what is in our opinion the main reason for their susceptible structures: the top soil appears to have an "anmooric" character. In thin section, the clay–humus complex shows a typical greyish shade and is interspersed with very small, mainly dark plant remains (Plate H2). This anmooric character is in good accordance with the sedimentological origin and the low position of the area. We may assume that the region was for the greater part a swamp, up to at least the Middle Ages but probably much longer. Because the drainage, and therefore also the grass vegetation, of these soils was gradually improved, a mull-formation started: the top soil lost its extreme anmooric character, but did not develop beyond the anmooric phase. After the grassland was ploughed, the top soil developed a more pronounced anmooric character. This is probably largely a result of the mixing of the A_1-horizon of the Westfrisian sediment II with the top soil since this vegetation layer is a real anmoor (Plate H3), as is that of the Westfrisian sediment I.

As we have stated several times (Van Heuveln et al., 1960; Jongerius and Pons, 1962a, 1962b), anmoor soils have a relatively low water stability. Under a grass vegetation, this is not conspicuous. If the soil is used for agricultural or horticultural purposes, however, the material at the surface is shifted into its components, which may eluviate to a reasonable depth; the material not exposed directly to disintegration by rain gradually puddles when wet. The features of these processes are clearly visible in thin section: the scattered appearance of humus cutans on pore walls and along cracks (Plate H5), humic fibres in voids and cracks filled up with sand which, in its turn, is also illuviated from the surface, a pronounced micro-heterogeneity (mosaic-like structure of clayey and sandy domains) in the top soil, etc.

Thin sections show that the puddling of the soil material starts at the bottom of the cultivated layer. This is easy to comprehend, because practically every cultivation causes a thicker or thinner compacted horizon in these clayey anmooric soils. In this way, the excess water stagnates and puddling starts. Water stagnation may, in some cases, also be caused by the AC-horizon of Westfrisian sediment II when this horizon has a high clay content since it is then very compact as a result of self-mulching, a phenomenon which is clearly visible in thin section.

The micromorphological results make it easy to understand the above-mentioned macromorphological phenomena and even the fact that the compact soil may break into small pieces under favourable conditions, since this is due to the micro-heterogeneity of the soil material.

A detailed discussion of the micromorphological differences between grassland and horticultural soils would take too much room; but it is clear that the grassland soils are for instance much more porous than the compact morphotypes of horticultural soils, at least in the top layers. As has already been mentioned, the soils under horticulture may also temporarily possess very favourable soil structures. This is demonstrated by Fig.1 and 2, in which is given the results of micromorphometrical investigations of a grassland soil and a good tulip soil. The measurements were performed on mammoth-sized thin sections with the aid of a Zeiss integration eyepiece I; however, the opticvolumetric determinations of large cracks and pores were done on negative photographs (enlarged ten times), with a Zeiss TGZ 3 particle size analyser (Jongerius, 1963). Of special interest in Fig.2 are the data of the plough-sole; these data excellently demonstrate the effect of soil tillage under humid conditions in this soil. The plough-sole is formed by shallow-ploughing only once!

The figures contain many other interesting data; for instance, the percentage of recent free organic matter in the top layers of the tulip soil is considerably higher than in those of the grassland soil. This is partly a result of organic manure and crop remains which have been ploughed in. It is also due to the periodical anaerobic conditions which occur during a period of structure deterioration. Another phenomenon clearly visible from the micromorphometrical data is that the material of the A_1-horizon of the Westfrisian sediment II in the tulip soil is indeed mixed with the material of the top layers, but is not incorporated in it.

Many other very interesting conclusions, for instance, about the influence of soil fauna on the profiles and the distribution of iron oxides (Plate H4), $CaCO_3$, etc., can be drawn from Fig.1 and 2, but this must be left to the interested reader.

CONCLUSIONS

The haplaquoll soils of Hauwert are very susceptible to structure deterioration. However, in climatically normal years, this can be prevented wholly or for the greater part by excellent soil management. This means that the soils have restricted suitability for tulip culture. In a normal year, the yield can be very good, both quantitatively and qualitatively, but in years with a wet winter season, considerable yield depressions may be expected.

For this reason, it is better to concentrate tulip culture in this area on freshly ploughed grassland soils; however, this land must be reseeded with grass immediately after the tulip bulbs are harvested in order to prevent soil structure deterioration. After at least 5 years, but preferably not until after about a decade, one may again utilize such grassland soils during one year for tulip culture, etc.

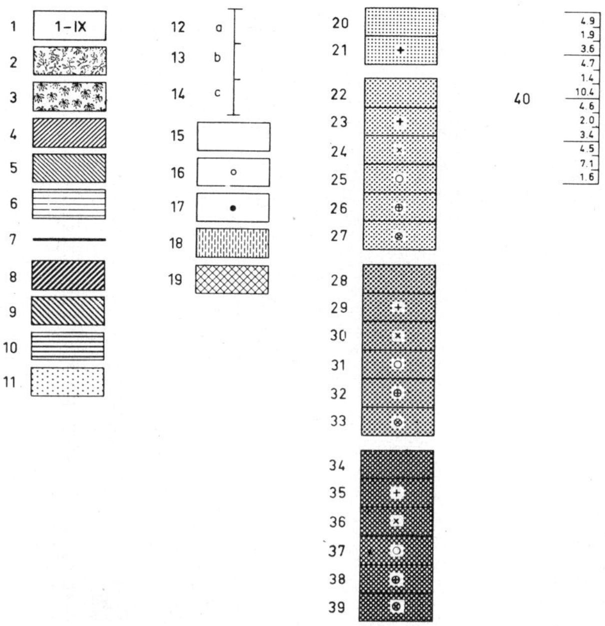

1 = Profile horizons. 2 = Organic matter, brown (recent). 3 = Organic matter, black (remnants of Westfrisian "anmooric" II). 4 = Mixture of recent organic matter and fine mineral fractions. 5 = Mixture of fine mineral fractions and a little recent organic matter. 6 = Fine mineral fractions (C material of Westfrisian sediment II). 7 = Dividing line between materials of Westfrisian sediments I and II respectively. 8 = Mixture of fossil organic matter (Westfrisian sediment I) and fine mineral fractions. 9 = Mixture of fine mineral fractions and a little fossil organic matter (Westfrisian sediment I). 10 = Fine mineral fractions (C material of Westfrisian sediment I). 11 = Sand. 12 = Formed by soil animals (mainly earthworms). 13 = Mixed as a result of stress or desintegration. 14 = Biological or mechanical unchanged. 15 = Iron oxides, diffuse. 16 = Iron concretions, non-opaque. 17 = Iron concretions, opaque. 18 = Glauconite. 19 = Calcium carbonate (shells and shell fragments). Primary pores: 20 = 10–30 μ, 21 = 30–100 μ. Biopores: 22 = 10–30 μ, 23 = 30–100 μ, 24 = 100–300 μ, 25 = 300–500 μ, 26 = 500–1,000 μ, 27 = > 1,000 μ. Cracks: 28 = 10–30 μ, 29 = 30–100 μ, 30 = 100–300 μ, 31 = 300–500 μ, 32 = 500–1,000 μ, 33 = > 1,000 μ. Cavities caused by soil cultivation: 34 = 10–30 μ, 35 = 30–100 μ, 36 = 100–300 μ, 37 = 300–500 μ, 38 = 500–1,000 μ, 39 = > 1,000 μ. 40 = Multiplication-factors to calculate the relative volume percentage of the corresponding components in the a, b and c part of the horizon.

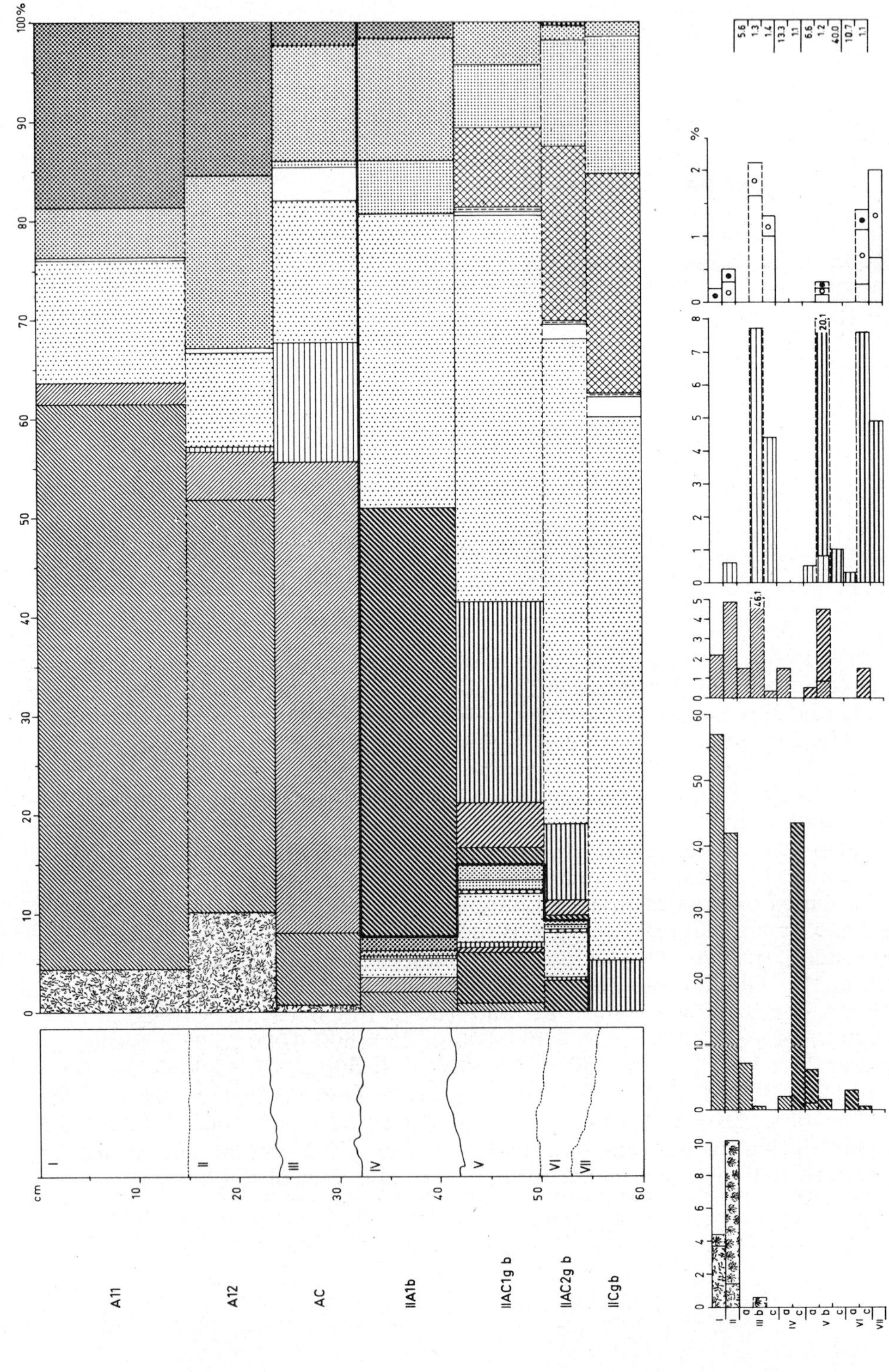
100%
A11
A12
AC
IIA1b
IIAC1g b
IIAC2g b
IICgb
cm
I
II
III
IV
V
VI
VII
5.6
1.3
1.4
13.3
1.1
6.6
1.2
40.0
10.7
1.1
46.1
20.1
%

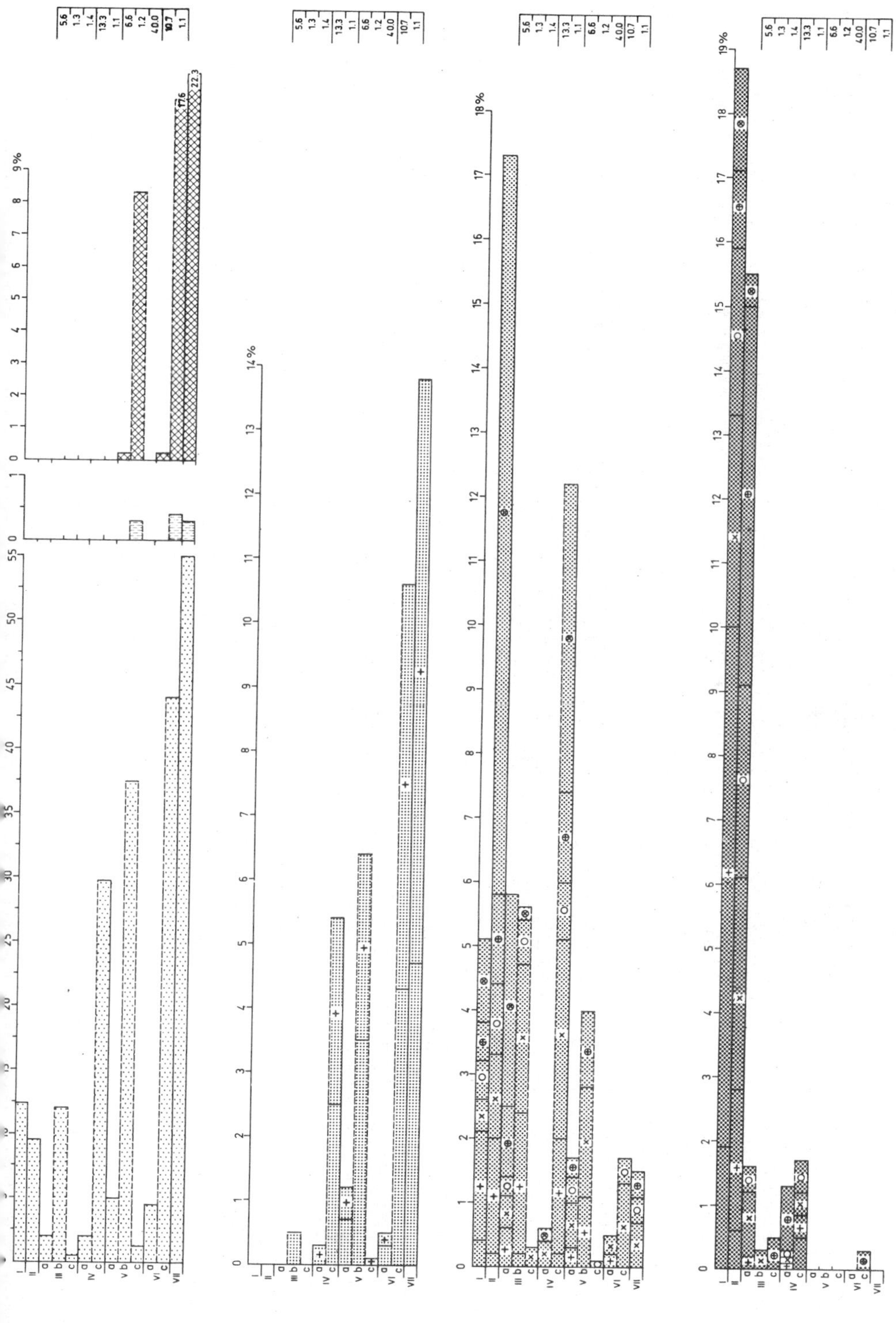

Fig.1. Results of optic-volumetric measurements on mammoth-sized thin sections of a grassland profile (orthic haplaquoll, intergrading to histic haplaquoll; structure profile 2). Legend see p.497.

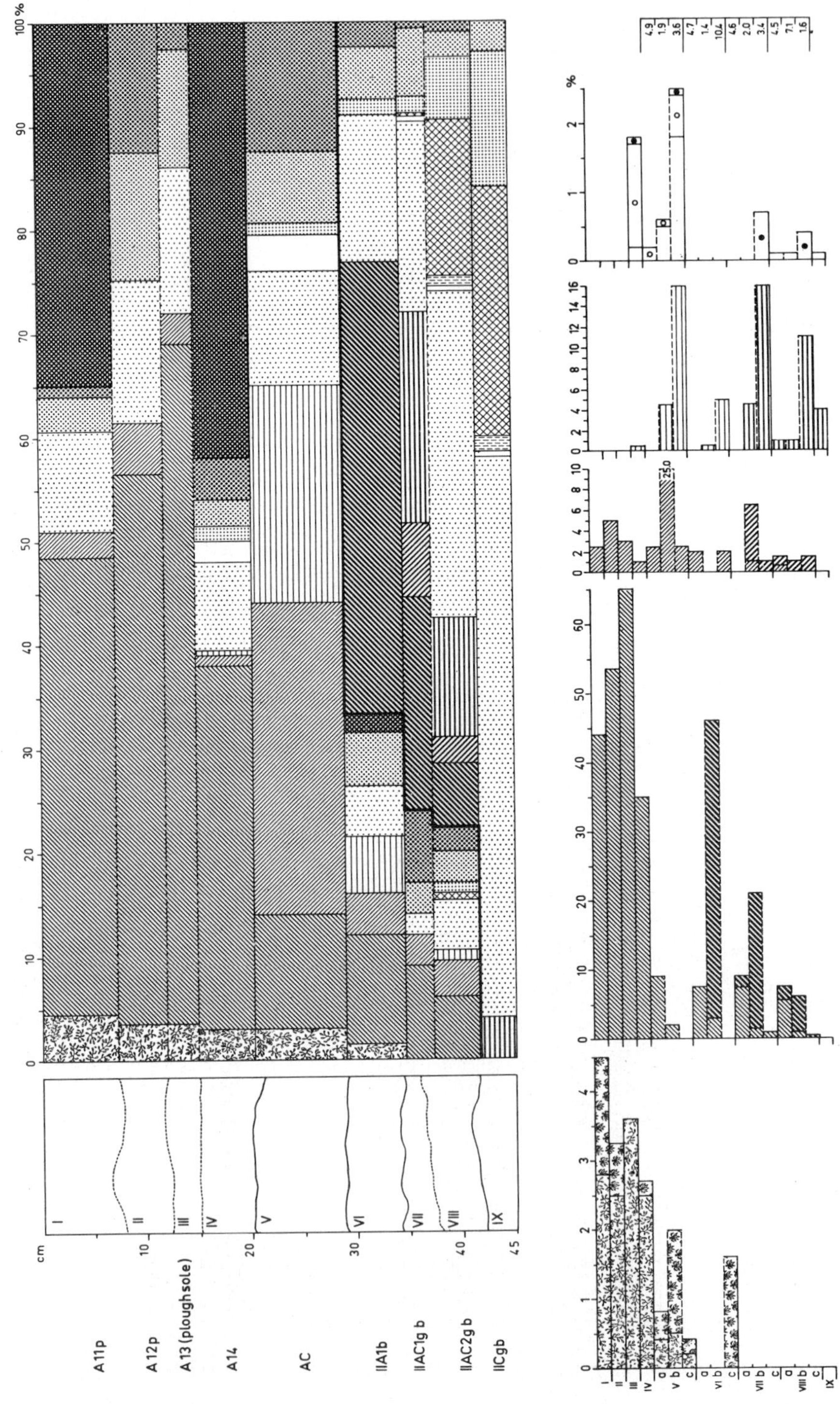
A 11p
A 12p
A 13 (ploughsole)
A 14
AC
IIA1b
IIAC1g b
IIAC2g b
IICgb
cm
I
II
III
IV
V
VI
VII
VIII
IX
100 %
%
25.0
4.9
1.9
3.6
4.7
1.4
10.4
4.6
2.0
3.4
4.5
7.1
1.6

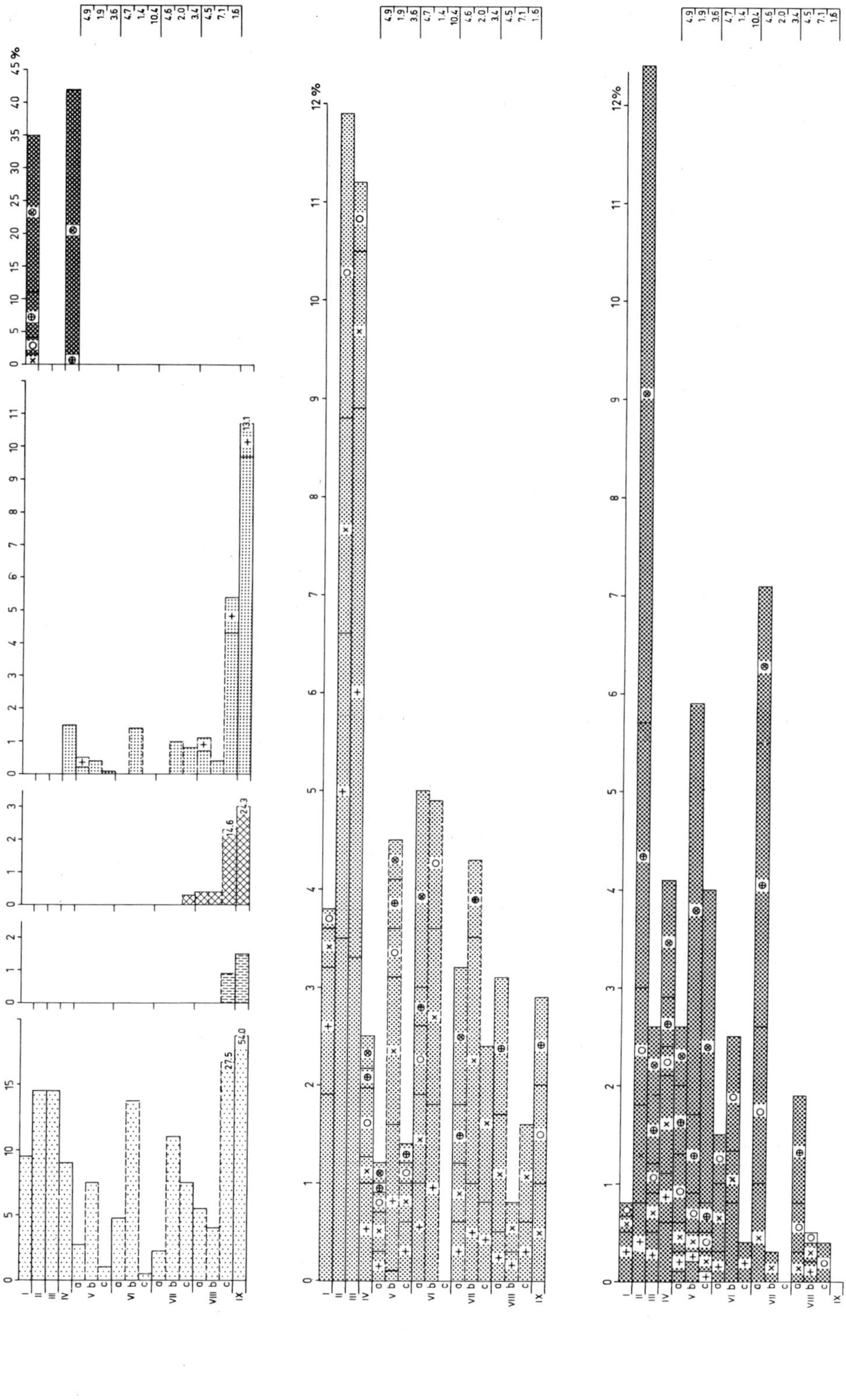

Fig.2. Results of optic-volumetric measurements on mammoth-sized thin sections of an old horticultural soil (orthic haplaquoll). Legend see p.497.

SUMMARY

This paper deals with a morphological investigation of the suitability of orthic haplaquolls near Hauwert (North Holland) for tulip culture. The area is a typical old grassland region, in which the said culture recently extended considerably into the already existing horticultural grounds, as well as to ploughed grassland soils.

The investigations were performed in a sample area of about 10 ha which was representative for the entire region.

Among other things, a structure map of the sample area was made. The map units are structure profiles. A structure profile is a vertical succession of structure types, which is determined by agricultural factors. It appeared that the soil under horticulture shows a great variation in soil structure, from very porous to very compact. This was determined by factors such as the nature of cultivation, hydrological conditions during cultivation, the preceding crop, weather conditions etc.

This behaviour is mainly due to the "anmooric" character of the organic matter, which was shown by micromorphological investigations.

Finally, an example is given of the characterization of a grassland soil and a soil under horticulture by means of micromorphometrical methods.

The conclusion of the investigations was that the soil is suited to tulip culture if soil management is very good and weather conditions, especially during winter and spring, are reasonable. It is, however, safer to concentrate the tulip culture in this area on freshly ploughed grassland soils, which must be reseeded with grass after the tulip bulbs are harvested.

REFERENCES

Jongerius, A., 1957. Morfologische onderzoekingen over de bodemstructuur. Thesis, Agricultural State University, Wageningen - Neth. Soil Surv. Inst., Bodemk. Studies, 2 : 93 pp. Also: Verslag. Landbouwk. Onderzoek., 63.12: 93 pp.

Jongerius, A., 1963. Optic-volumetric measurements on some humus forms. In: J.J. Doeksen and J. van der Drift (Editors), Soil Organisms. North-Holland, Amsterdam, 137–148.

Jongerius, A. and Heintzberger, G., 1963. The preparation of mammoth-sized thin sections. Soil Surv. Papers, Neth. Soil Surv. Inst., 1 : 37 pp.

Jongerius, A. and Pons, L.J., 1962a. Soil genesis in organic soils. Boor Spade, 12 : 156–168.

Jongerius, A. and Pons, L.J., 1962b. Einige mikromorphologische Bemerkungen über den Vererdungsvorgang im niederländischen Moor. Z. Pflanzenernähr., Düng., Bodenk., 97 : 243–255.

Soil Survey Staff, U.S.D.A., 1960. Soil classification, a comprehensive system, 7th Approximation. Soil Conservation Service, U.S. Dept. Agr., Washington, D.C., 265 pp.

Van Heuveln, B., Jongerius, A. and Pons, L.J., 1960. Soil formation in organic soils. Trans. Intern. Congr. Soil Sci., 7th, Madison, 1960, 4 : 195–204.

COLOURED ILLUSTRATIONS

Plate H2. Microphotograph of a fairly puddled structure. The matrix consists of a clay-humus complex with an anmooric tendency. Many rounded iron-concretions. A_{1p}-horizon of orthic haplaquoll in use for horticulture. Plain light; x 34.

Plate H3. Microphotograph of a buried A_1-horizon of profile developed in the Westfrisian sediment II. The humus form is a typical anmoor. Plain light; x 90.

Plate H4. Microphotograph of small round iron-concretions (formed by iron-bacteria?) C-horizon of orthic haplaquoll intergrading to histic haplaquoll under old grassland. Incident light; x 90.

Plate H5. Microphotograph of a humic illuviation cutan on a ped. The brown amorphous organic matter is interspersed with very tiny opaque organic particles. A_{12p}-horizon of orthic haplaquoll use for horticulture. Plain light; x 70.

ÜBER DIE STELLUNG DER MIKROMORPHOLOGIE IM RAHMEN DER BODENKUNDE

K. ZIMMERMANN

Institut für Bodenkunde[1] der Rheinischen Friedrich-Wilhelms-Universität, Bonn (Deutschland)

EINLEITUNG

Diese Arbeit soll die Beziehungen und Abhängigkeiten zeigen, die die Bodenmikromorphologie mit dem Gesamtrahmen der Bodenkunde verbinden. Hierdurch werden ihre Grundprinzipien klarer erkennbar sowie die Hauptlinien ihrer Untersuchungsmethodik. Um zunächst den bodenkundlichen Gesamtrahmen abzustecken, ist es erforderlich, etwas näher auf den Bodenbegriff und die Einteilung der Bodenkunde einzugehen. Es werden dabei bewusst moderne Anschauungen wiedergegeben, da nur auf ihrer Grundlage eine wirklich umfassende Mikromorphologie möglich ist. Im Anschluss daran wird die Stellung der Mikromorphologie innerhalb dieses bodenkundlichen Gesamtrahmens herausgearbeitet. Ferner werden die Grundzüge ihrer Methodik dargestellt, die ganz entscheidend von dieser Stellung der Mikromorphologie im Rahmen der Bodenkunde bestimmt werden.

DIE HEUTIGE AUFFASSUNG VOM BODEN

In dem Wort "Mikromorphologie" steckt das aus dem Griechischen abgeleitete Kunstwort "Morphologie". Im Griechischen bedeutet ἡ μορφή die Gestalt, die Form. Morphologie ist also im ursprünglichen Sinne die Lehre von den Gestalten und Formen. Wird die Gestalt bzw. die Form ganz konkret-anschaulich verstanden, so ist eine "Morphologie" in diesem Sinne nur bei konkreten, d.h., körperlichen Gegenständen möglich. So ist es verständlich, dass in den Wissenschaften, die einen körperlichen Gegenstand als Forschungsobjekt haben, die Befassung mit dessen Aufbau "Morphologie" genannt wird (Beispiel: die Botanik). Allerdings wird dabei unter dem Begriff "Aufbau" nicht nur - wie es der ursprünglichen Wortbedeutung ("Form", "Gestalt") entsprechen würde - die *geometrische* Seite, sondern auch die *stoffliche* Komponente verstanden.

In Analogie hierzu ist es auch in der Bodenkunde üblich geworden, von "Morphologie" zu sprechen (Bodenmorphologie). Auch der Boden stellt - wenigstens nach der heute allgemein anerkannten Auffassung - ein körperliches, unzufällig aufgebautes Gebilde dar. Früher hat man ihn allerdings lange Zeit durchweg als formlose Masse angesehen. So definiert z.B. Mitscherlich (1913, S.6) den Boden als blosses "*Gemenge* von pulverförmigen,

[1]Direktor: Prof. Dr. Dr. E. Mückenhausen.

festen Teilchen, Wasser und Luft". Diese Auffassung und Betrachtungsweise ist zwar auch heute noch überall dort durchaus berechtigt und zweckmässig, wo das Bodenmaterial aus dem ursprünglichen Verbande herausgenommen und im groben Sinne als blosser "Stoff" (z.B. Schüttungsmaterial für den Tiefbau usw.) behandelt wird. Eine "Morphologie" wäre hier - abgesehen von Spezialproblemen - im allgemeinen überflüssig.

Im Gegensatz zu diesen speziellen Teilgebieten der angewandten Bodenkunde beschäftigt sich die "reine" oder selbstständige Bodenkunde jedoch gerade mit dem Boden an seinem ursprünglichen Standort. Hier handelt es sich tatsächlich um ein *körperliches*, in seinem jeweiligen Aufbau *unzufällig* zustandegekommenes Gebilde. Ausserdem besitzt der Boden an diesem Standort eine Besonderheit, auf die immer wieder nachdrücklich hingewiesen werden muss: Der Boden stellt in seinem jeweiligen Aufbau kein starres Gerüst dar. Vielmehr laufen in ihm unter dem Einfluss der Standortbedingungen ständig charakteristische Bewegungen und Veränderungen ab. Es ist deswegen bereits wiederholt der Versuch gemacht worden, den Boden als ein Analogon zu den *Organismen* zu betrachten, bei denen bekanntlich neben dem körperlichen Aufbau (Morphologie) typische Lebensvorgänge (Physiologie) unterschieden werden können. Gewissermassen das entgegengesetzte Extrem zu dieser letzten Auffassung vom Boden ist lange Zeit von den Geologen vertreten worden. Sie sahen den Boden lediglich als eine Art *Gestein* an. Zwar billigten sie ihm damit (im Gegensatz zu Mitscherlich; siehe oben) Gestaltcharakter zu. Die dauernden Bewegungen und Veränderungen des Bodens, seien sie biologisch oder abiologisch bedingt, blieben jedoch bei dieser Auffassung unberücksichtigt.

Bei näherer Betrachtung steht der Boden in seinem Wesen nun tatsächlich *zwischen* den Organismen und den Gesteinen. Er stellt dabei ein Phänomen ganz eigener Art dar, das man sich in seinem vollen Umfang zutiefst bewusst machen muss, wenn man den Boden als Gegenstand einer selbstständigen Wissenschaft ("reine" Bodenkunde) richtig verstehen will. Um die hauptsächlichsten Charakterzüge dieses Phänomens herauszuarbeiten, sollen hier die Ähnlichkeiten bzw. die Unterschiede des Bodens gegenüber den Organismen und Gesteinen kurz und summarisch zusammengefasst werden:

(1) Der Boden zeigt wie die Organismen dauernde und dabei - wenigstens bis zu einem gewissen Grade - typisierbare Bewegungen und Veränderungen. Da sie jedoch nicht von einem "Organisations"-Zentrum aus gesteuert werden, kann man sie nicht als "*Lebens*vorgänge" bezeichnen.

(2) Im Gegensatz zu den Organismen haben die verschiedenen Böden als systematische Einheiten - wenigstens nach den Seiten hin - keine typisierbare Aussengestalt. Vielmehr sind sie in der Horizontalen theoretisch unbegrenzt. Diese Eigentümlichkeit haben sie mit den Gesteinsarten in der Petrographie oder - deutlicher noch - mit den artspezifischen Kristallgittern in der Mineralogie gemeinsam, die im Prinzip, und zwar nach allen Seiten hin, unendliche Ausdehnung haben (vgl. Kleber, 1959, S.22 ff.). Da der Boden ausserdem als Ganzes wie die Gesteinsarten bzw. Kristallgitter, zumindest in der Horizontalen, Gefügecharakter hat, ist er als anschauliches körperliches Gebilde - in Analogie zu der artspezifischen Elementarzelle in der Mineralogie bzw. zum artspezifischen Elementarkörper in der Petrographie (vgl. Kleber, 1959, S.23; Hoenes, 1955, S.383) -

ebenfalls nur durch eine Art "Elementarkörper" darstellbar.

(3) Gemeinsam mit den Gesteinen wie mit den Organismen hat der Boden einen unzufälligen, anschaulich erfassbaren, "inneren" Aufbau. Im Gegensatz zum Gestein zeigt der Boden jedoch, wie bereits erwähnt, typisierbare und für ihn ganz und gar zu seinem augenblicklichen Wesen gehörende aktuelle Bewegungen und Veränderungen dieses Aufbaues.

Nach allem stellt also der Boden eine ganz besondere Art "Gefügekörper" dar, der nicht nur - wie jeder andere typische Gefügekörper auch - durch eine charakteristische, unzufällig zustandegekommene, geometrisch-stoffliche Ausbildung als solche, sondern gerade auch durch deren eigentümliche aktuelle Beweglichkeit und Veränderlichkeit gekennzeichnet ist.

DIE EINTEILUNG DER HEUTIGEN BODENKUNDE

Die "reine" oder selbstständige Bodenkunde beschäftigt sich - im Gegensatz zur angewandten Bodenkunde - mit dem soeben in seinem Wesen charakterisierten Boden *allein um seiner selbst Willen*. Sie hat ihn als unabhängigen wissenschaftlichen Gegenstand und sucht ihn in allen seinen Teilaspekten - dabei immer von ihm selbst ausgehend - zu erfassen. Das Thema erfordert es, etwas näher auf die Einteilung dieser Bodenkunde einzugehen.

Genauso wie beim Bodenbegriff ist auch im Hinblick auf die Bodenkunde und ihre Einteilung eine Entwicklung festzustellen. Abgesehen von der Abhängigkeit von den technologischen Fächern (Land- und Forstwirtschaft, Ingenieurswesen) ist der Boden auch sonst lange Zeit nicht als selbstständige Bildung behandelt worden (z.B., die Bodenkunde als Teilgebiet der Geologie). In allen diesen Fällen wurde die Einteilung der Bodenkunde jeweils entsprechend dieser Abhängigkeit vorgenommen.

Selbst seit sich die Berechtigung der Eigenstellung des Bodens durchgesetzt hat, lebte und lebt z.T. auch heute noch eine gewisse Abhängigkeit von anderen Wissensgebieten fort. Das offenbart sich in einer Einteilung der Bodenkunde, die als Teildisziplinen Fächer wie Bodenphysik, Bodenchemie, Bodenmineralogie und Bodenbiologie als Elemente enthält. Diese Einteilung zeigt, dass - überspitzt ausgedrückt - im Grunde nicht der Boden selber, sondern immer nur die Forschungsgegenstände dieser *speziellen* naturwissenschaftlichen Teildisziplinen betrachtet werden: Es werden lediglich die einzelnen physikalischen Erscheinungen, die chemischen Stoffe, die Minerale und die Organismen behandelt. Zwar kommen sie im Boden vor, jedoch werden sie stets für sich, d.h., isoliert betrachtet. Die eigentlichen, meist in komplexer Weise daraus zusammengesetzten, spezifisch bodeneigenen Phänomene bleiben dagegen unberücksichtigt.

Eine solche Einteilung scheint auf den ersten Blick allerdings den Vorteil einer streng logischen Gliederung zu haben: Das Einteilungsprinzip deckt sich mit dem allgemein anerkannten Einteilungsprinzip der Naturwissenschaften. Dennoch ist diese Gliederung nicht zweckmässig. Abgesehen davon, dass sie nicht auf bodeneigenen Kriterien beruht, ist sie (und zwar gerade deswegen!) nicht in der Lage, das Gesamtgebiet der Bodenkunde lückenlos zu überstreichen. Die Teilgebiete der Bodengenetik, der Bodensystematik und der Bodengeographie werden nicht erfasst. Versucht man nun, diese Gebiete mit den zuerst genannten "naturwissenschaftlichen"

Einzeldisziplinen zu einem einheitlichen Einteilungsschema zu verschmelzen, so stellt man fest, dass sie nicht recht zu einander passen wollen. Es ergeben sich Überschneidungen und sonstige Unstimmigkeiten (vgl. Scheffer, 1962, S.20). Das liegt daran, dass die Abgrenzung solcher Teilgebiete wie Bodengenetik, Bodensystematik und Bodengeographie auf einem ganz anderen Einteilungsprinzip beruht. Dieses Einteilungsprinzip - so sieht man bei näherer Betrachtung - wird nun allerdings in der Tat den eingangs gestellten Anforderungen gerecht: Es geht von den Gegebenheiten des Bodens selber aus und überstreicht ausserdem zwanglos, ohne Bruch und Überschneidungen, das Gesamtgebiet der Bodenkunde.

Beim Boden handelt es sich, wie erwähnt, um ein unzufällig entstandenes, in charakteristischer Weise aufgebautes, körperliches Gebilde, das bestimmte typisierbare Bewegungen und Veränderungen zeigt. Bei einem jeden solchen körperlichen und dabei veränderlichen Objekt lässt sich eine ganz bestimmte Anzahl von allgemeinsten Aspekten erkennen, durch die eine vollständige Beschreibung und Charakterisierung möglich ist. Diese Aspekte sind durch die folgenden fünf Fragen gekennzeichnet:

(1) Wie ist das Objekt aufgebaut?

(2) Welche gegenwärtigen Veränderungen ("Bewegungen") zeigt es?

(3) Wie ist es entstanden?

(4) Welche typischen Spielarten dieses Objektes ("Formen" im übertragenen Sinne) sind zu erkennen und wie kann man sie zum Zwecke der Übersicht zu einem sinnvollen System ordnen?

(5) Wo finde ich diese verschiedenen Formen?

Es ist leicht zu erkennen, dass die drei zuletzt genannten Fragen in den Teildisziplinen "Bodengenetik" (Frage 3), "Bodensystematik" (Frage 4) und "Bodengeographie" (Frage 5) ihre Entsprechung finden. Übrig bleiben zunächst die beiden ersten Fragen. Es sind jedoch, wie man leicht sieht, bereits seit langem Ansätze auch zur Ausscheidung dieser beiden noch fehlenden Teildisziplinen vorhanden. Nur ist ihre klare Zusammenfassung und z.T. auch ihre Abgrenzung gegenüber den drei zuerst genannten Teilgebieten noch nicht so deutlich. Da es gerade auch auf diese beiden Teildisziplinen im Hinblick auf das hier gestellte Thema ankommt, muss dieser Punkt noch etwas eingehender behandelt werden. Am besten dient dazu wiederum die geschichtliche Betrachtung.

Bereits seit langem ist versucht worden, in Analogie zu der Ausscheidung einer "Morphologie" und einer "Physiologie" in den biologischen Wissenschaften auch in der Bodenkunde entsprechende Teilgebiete auszuscheiden (vgl. Stebutt, 1930, S.2 ff.; Laatsch, 1938, S.V ff.; 1957, S.2 ff.). Laatsch nennt das erste Gebiet "Anatomie", während Stebutt keinen besonderen Namen dafür einführt, sondern nur allgemein von der "Lehre vom Boden als organisiertem Körper" spricht. Das zweite Teilgebiet wird von beiden Autoren übereinstimmend als "Bodendynamik" bezeichnet.

Es ist die Frage, ob man das erste Teilgebiet nicht ebenfalls wie in den biologischen Wissenschaften "Morphologie" nennen sollte. Die Bezeichnung "Morphologie" bezieht sich auf den tatsächlichen Gegenstand der Betrachtung (den Bodenaufbau), während "Anatomie" nur die Methode der Untersuchung (das "Aufschneiden") angibt. Allerdings ergeben sich gewisse terminologische Schwierigkeiten bei der Verwendung der Bezeichnung "Morphologie", auf die jedoch bewusst erst später eingegangen werden soll (vgl. S.513). Hier wird jedenfalls die Teildisziplin, die sich mit dem Aufbau des Bodens beschäftigt, aus den zuerst genannten Gründen - und unbeschadet der später nötig werdenden Differenzierung - "Bodenmorphologie" genannt.

Infolge der inzwischen eingetretenen Entwicklung ergibt sich bei der genauen Umgrenzung des zweiten Teilgebietes, das der Physiologie in den biologischen Wissenschaften entspricht und das - wie erwähnt - mit "Bodendynamik" bezeichnet worden ist, ein Problem. Es liegt in folgendem begründet: Unter dem Begriff "Bodendynamik" werden - soweit er geneuer präzisiert wird - heute noch nach wie vor sowohl die aktuellen (den Boden gewissermassen als systematische Einheit im gegenwärtigen Zeitraum *nicht* umwandelnden, sondern im Gegenteil gerade zu seinem Wesen gehörenden) Vorgänge als auch die bodengenetischen Prozesse verstanden bzw. sogar als gleichbedeutend angesehen (vgl. Laatsch, 1957, S.VI; Scheffer und Schachtschabel, 1960, S.250, 264). Auf der anderen Seite wird ebenfalls heute die Bodengenetik allgemein als eine *selbstständige* Teildisziplin der Bodenkunde betrachtet (vgl. Mückenhausen, 1962, S.27; Scheffer, 1962, S.22 ff.). Es offenbart sich hier also eine gewisse Unlogik: Derselbe Gegenstand (die bodengenetischen Prozesse), der in einer gesonderten Teildisziplin (der Bodengenetik) bereits behandelt wird, ist gleichzeitig in derselben Form auch noch einmal Gegenstand einer anderen Teildisziplin (der Bodendynamik) hier allerdings in Gemeinschaft mit den aktuellen Vorgängen des Bodens. Es erhebt sich daher die Frage, ob man die aktuellen Prozesse des Bodens jetzt nicht ebenfalls in einer für sie allein reservierten Teildisziplin zusammenfassen sollte, wie es bereits von Kubiena im Ansatz gemacht worden ist ("Phänologie" des Bodens; Kubiena, 1948, S.87; 1953, S.369; 1962a, S.2).

Bei näherer Betrachtung erweist sich eine solche Zusammenfassung der aktuellen Vorgänge des Bodens zu einer eigenen Disziplin sogar als dringend notwendig: Die in der Bodensystematik aufgestellten systematischen Einheiten werden in Anbetracht der langsamen Bodenentwicklung mit Recht - wenigstens für einen gewissen Zeitraum - als etwas Bleibendes angesehen (vgl. Kubiena, 1948, S.168). Sie lassen sich deswegen auch durchaus als Kartierungseinheiten für bestimmte grosse Kartenwerke, die für eine gewisse Dauer bestimmt sind, verwenden. Diese in der Gegenwart als konkrete Naturkörper in der Landschaft auffindbaren Bodeneinheiten sind nun nicht nur durch ihren charakteristischen *Aufbau*, sondern auch - wie bereits angedeutet - durch ihr charakteristisches augenblickliches *Verhalten*, d.h., eben durch die *aktuellen Vorgänge*, gekennzeichnet. Gerade für die Anwendungsseite ist dieser letzte Aspekt oft von ausschlaggebender Bedeutung. Hier interessiert nicht so sehr, dass sich der fragliche Boden etwa im Laufe von Jahrhunderten von einem mässig ausgeprägten in einen stark ausgeprägten Pseudogley umwandeln wird. Vielmehr ist sein aktuelles Verhalten, z.B., die Eigenart des Wechsels zwischen Nässe-, Feucht- und Trockenphase, zwischen Verbreiung und Verhärtung usw. von grösstem Interesse (vgl. Mückenhausen, 1958). Es müsste deswegen eigentlich geradezu selbstverständlich sein, diese aktuellen Prozesse nun auch in einer gesonderten Teildisziplin der Bodenkunde für sich allein zu behandeln.

Mit diesen Ausführungen soll selbstverständlich nichts über den Wert der Bodengenetik gesagt werden. Dieser ist unbestritten. Er liegt nur auf einem anderen Gebiete als der der "Bodendynamik" (im Sinne dieser aktuellen Prozesse): Es ist zunächst allein die so verstandene Bodendynamik, die

- in Verbindung mit der Bodenmorphologie - zu der *unmittelbaren* und vollständigen Erfassung des aktuellen Bodens als *konkretem korperlichem Phänomen* führt. Zu einem vollen *Verständnis* allerdings, *warum* der Boden den betreffenden Aufbau und z.T. auch das zu beobachtende aktuelle Verhalten zeigt, verhilft dagegen selbstverständlich erst die Bodengenetik. Erst mit diesem Verständnis ist auch die vollkommene Einsicht in den Boden als solchen sowie in die natürlichen bodensystematischen und bodengeographischen Zusammenhänge möglich. Umgekehrt muss aber der Boden zunächst immer auch ganz klar und deutlich als konkretes, unmittelbar-anschauliches und aktuelles Phänomen, d.h. also - abgesehen von seinem Aufbau - auch in seinem *aktuellen* Verhalten, erfasst werden, damit eben konkrete Böden und nicht immer nur abstrakte "Entwicklungszustände" gesehen werden.

Nach allem erscheint es tatsächlich notwendig und zweckmässig, eine eigene Teildisziplin für die aktuellen Prozesse des Bodens auszugrenzen. Es erhebt sich die Frage, ob man für diese neue, wesentlich enger gefasste Teildisziplin weiterhin die Bezeichnung "Bodendynamik" beibehalten soll. Von Scheffer (1962, S.21) ist neuerdings die Bezeichnung "Metabolik" vorgeschlagen worden, allerdings ausdrücklich in einer *noch umfassenderen* Bedeutung, als sie der ursprüngliche Begriff "Bodendynamik" gehabt hat. Ausserdem ist der Ausdruck "Metabolik" durch den Begriff "Metabolismus" in der Botanik einseitig auf die Energie- und Stoffwechsel-Prozesse festgelegt. Andererseits hat die Bezeichnung "Metabolik" den Vorzug, dass sie neu ist. Es wird also auch die terminologische Schwierigkeit umgangen, die darin liegt, dass Kubiena (1953, S.363) zwischen der "Biologie des Bodens" und der "Dynamik des Bodens" unterscheidet. Dennoch soll hier für die Teildisziplin, die sich mit den aktuellen Vorgängen des Bodens beschäftigt, zunächst weiterhin - und zugegebenermassen aus Gründen der "Gewöhnung" - die Bezeichnung "Bodendynamik" verwendet werden. Darunter werden sowohl die biologisch als auch die nicht-biologisch bedingten Prozesse verstanden.

Abschliessend lässt sich also eine Einteilung der "reinen" Bodenkunde in fünf sachlich bedingte Teildisziplinen erkennen, die den oben auf Seite 508 genannten fünf Teilaspekten vollkommen entsprechen (diese Einteilung steht weitgehend in Einklang mit der in 1962 von Scheffer veröffentlichten Gliederung der Bodenkunde): (1) Bodenmorphologie; (2) Bodendynamik; (3) Bodengenetik; (4) Bodensystematik; (5) Bodengeographie.

Die *Bodenmorphologie* beschäftigt sich mit dem Aufbau des Bodens. Ausgehend von den Bauverhältnissen, wie sie dem Betrachter jeweils als Augenblicksbild entgegentreten, werden durch Vergleich und Typisierung die charakteristischen, gewissermassen "bleibenden" Bauformen herausgeschält. Der Begriff "Aufbau" umfasst dabei nicht nur die geometrische Ausbildung der Baueigentümlichkeiten (= Gefüge; vgl. Sander, 1948, S.2), sondern auch den "Stoff", der diese geometrische Ausbildung besitzt.(vgl. Kubiena, 1953, S.368). Das Fachgebiet der Bodenmorphologie ist also umfassender als das der bodenkundlichen Gefügekunde. Zu ihr gehören Teilgebiete wie Profilmorphologie, Humusformen, Bodengefüge, mineralogische und chemische Zusammensetzung des Bodens (einschliesslich der chemischen Zusammensetzung und Konstitution der Humusstoffe) usw., also Teilgebiete, die heute lehrbuchmässig noch an ganz verschiedenen Stellen behandelt werden.

Die *Bodendynamik* beschäftigt sich mit den aktuellen Einzelvorgängen im Boden sowie den Gesetzmässigkeiten und Bedingungen, die ihnen zugrundeliegen. Nach dem vorangegangenen erübrigt sich eine nähere Umschreibung dieser Teildisziplin. Bodenmorphologie und Bodendynamik zusammen ergeben ein vollständiges Bild vom Boden, so wie er im gegenwärtigen Zeitraum dem Betrachter draussen im Gelände unmittelbar (als körperliches und dabei "bewegliches" Phänomen) entgegentritt.

Die *Bodengenetik* beschäftigt sich mit den Entstehungs- und Entwicklungs-Prozessen des Bodens sowie den Gesetzmässigkeiten und Bedingungen, die ihnen zugrundeliegen. Die Kenntnis dieser Tatbestände führt zum vollständigen Verständnis der Böden. Sowohl die Eigenart ihres Aufbaues als z.T. auch die Eigenart ihrer Dynamik wird verständlich, sobald die Entstehung der Böden sowie ihre Stellung im Rahmen der grösseren Entwicklungszusammenhänge bekannt ist (vgl. Kubiena, 1948; Mückenhausen, 1962, S.2 ff.).

Die *Bodensystematik* versucht aufgrund dieser jeweils vollständigen Erkenntnis der Böden die charakteristischen Bodenformen (systematische Einheiten) zu erkennen und sie zum Zwecke der Übersicht in einem sinnvollen System zu ordnen (vgl. Kubiena, 1948, p.157 ff.; Mückenhausen, 1962, p.28 ff.).

Die *Bodengeographie* beschäftigt sich schliesslich mit der Verbreitung dieser charakteristischen Bodenformen auf der Erde sowie mit den Gesetzmässigkeiten und Bedingungen, die ihr zugrundeliegen (vgl. Ganssen, 1957).

Selbstverständlich stehen die genannten Teildisziplinen nicht ganz so isoliert da, wie es hier in Anbetracht der knappen Schilderung vielleicht den Anschein hat. In Wirklichkeit bestehen zwischen ihnen vielerlei Querverbindungen und Wechselbeziehungen. Im ganzen fügen sie sich jedoch harmonisch - d.h., ohne Überschneidungen und ohne Lücken - zu dem Gesamtbild des Bodens als eines unmittelbar erfassbaren Naturkörpers, als eines Gliedes in einem grösseren Entwicklungs- und Gliederungszusammenhang sowie als eines über die ganze Erde hinweg vorkommenden geographischen Phänomens zusammen.

Der hier zunächst allein betrachteten *reinen* Bodenkunde kann schliesslich noch die *angewandte* Bodenkunde gegenübergestellt werden (vgl. Scheffer, 1962, S.24), die den Boden als Faktor der Pflanzen- und Tierökologie (= 6, "ökologische" Bodenkunde) und als Gegenstand der Technologie (= 7, "technologische" Bodenkunde) behandelt. Ihr muss entsprechend den jeweiligen Anwendungsgebieten eine eigene spezielle Unterteilung gegeben werden.

BODENMORPHOLOGIE UND BODENMORPHOLOGISCHE UNTERSUCHUNG

Es mag nach diesen Ausführungen den Anschein haben, als wenn die Stellung der Mikromorphologie im Rahmen der Bodenkunde jetzt ohne weiteres klar ist: Die Mikromorphologie würde sich als ein Teilgebiet der soeben enger umschriebenen Disziplin "Bodenmorphologie" darstellen. Leider ist der Begriff "Morphologie" nicht ganz so eindeutig, wie es bisher bewusst - der Klarheit wegen - geschildert wurde. "Morphologie" kann nämlich - besonders in der Formulierung "morphologische Untersuchung" oder auch "morphologisches Prinzip" (vgl. Kubiena, 1962a, S.3) - durchaus einen wesentlich *allgemeineren* Inhalt haben. Da diese Frage für die Beantwortung des Themas entscheidende Bedeutung hat und um den Begriff "Morphologie" in dieser erweiterten Bedeutung von allem ihm von mancher Seite zugeschriebenen "Mystizismus" frei zu halten, muss hier etwas näher auf diesen Punkt eingegangen werden.

In jeder Wissenschaft, die es mit einem körperlichen und dabei veränderlichen Gegenstand zu tun hat (z.B., die Biologie; die Petrologie, und zwar in petrogenetischer Behandlung; die Tektonik, usw.) kann man grundsätzlich zwei allgemeinste Arten der wissenschaftlichen Betrachtung unterscheiden:

(1) Die eine befasst sich - und zwar auf grösstmögliche Vollständigkeit bedacht - mit der anschaulich erkennbaren oder vorstellbaren Körperlichkeit des Gegenstandes. Ihre Hilfsmittel sind die möglichst detaillierte Beschreibung und die genaue bildliche Darstellung.

(2) Die andere abstrahiert bewusst von dieser - zumindest von der vollständigen - Anschauung. Stattdessen sucht sie an Hand von ausgewählten, wohldefinierten und vielfach bewusst vereinfachten Einzelgegebenheiten, die verschiedenen "physikalisch"-kausalen Zusammenhänge zu erkennen, die den anschaulich fassbaren Phänomenen zugrundeliegen. Bei strengster Abstraktion und Verallgemeinerung führt sie im Endergebnis zu einer streng mathematisch-funktionalen Formulierung der erkannten Zusammenhänge und Gesetzmässigkeiten.

Sander (1948, S.VI) bezeichnet diese beiden Grundformen wissenschaftlicher Betrachtung mit *morphologischer* bzw. *physikalischer* Befassungsart. Beide befruchten sich gegenseitig und können sogar für sich allein letztlich nicht existieren. Nur durch dauerndes wechselseitiges Geben und Nehmen zwischen ihnen ist es möglich, zu der vollständigen wissenschaftlichen Erkenntnis zu gelangen.

Auch in der Bodenkunde können diese beiden allgemeinsten wissenschaftlichen Befassungsarten unterschieden werden. Verfolgt man die hier im Zusammenhang mit dem Thema in erster Linie interessierende "morphologische" Befassung, d.h., die Beschäftigung mit der anschaulich erkennbaren bzw. anschaulich vorstellbaren Körperlichkeit des Bodens, so macht man folgende Feststellung: Es lässt sich tatsächlich nicht nur der gewissermassen "bleibende" Aufbau des Bodens anschaulich erfassen ("Bodenmorphologie" im bisher gebrauchten Sinne; vgl. S.510). Vielmehr sind auch die aktuellen Prozesse im Boden ("Bodendynamik") und sogar die bodengenetischen Prozesse ("Bodengenetik") als unmittelbar anschauliche oder doch in dieser Weise vorstellbare "Bewegungs"-Phänomene "morphologisch" erfassbar.

Die morphologische Erfassung dieser Prozesse kann in dreierlei Form vor sich gehen:

(1) Einmal sind die betreffenden Vorgänge - wenn auch nur zu einem kleinen Teile - der *unmittelbaren* Beobachtung zugänglich.

(2) Sodann können sie durch die Erfassung der *Spuren* erkannt werden, die diese Prozesse im Bodenaufbau hinterlassen.

(3) Schliesslich ist es durch Erfassung der einzelnen *Teilphasen* (z.B., in Form von Entwicklungsreihen; vgl. Mückenhausen, 1962, S.9) möglich, den Gesamtprozess gewissermassen wie beim kinematographischen Film

aus den Einzelbildern zusammenzusetzen. Es sei vermerkt, dass auch chemische und physikalisch-chemische Vorgänge sich in Form von Modellvorstellungen gleichfalls als körperliche Bewegungsvorgänge anschaulich darstellen lassen (vgl. z.B. für die Pufferungsreaktionen: Laatsch, 1957, S.82).

Durch Zusammenschau dieser drei genannten morphologisch erfassbaren Teilaspekte des Bodens (Bodenaufbau, bodendynamische und bodengenetische Prozesse) kann man sich nun in der Bodensystematik sogar bis zu einem gewissen Grade die Eigenart der verschiedenen charakteristischen Bodenformen (= bodensystematische Einheiten) in einem anschaulichen Gesamtbild vor Augen stellen (vgl. S.511). Ferner ist es in der Bodengeographie - besonders durch die Vermittlung der Bodenkartierung - möglich, die einzelnen Bodenindividuen in der Landschaft und ihre Vergesellschaftung ganz konkret-körperlich zu erfassen (vgl. Laatsch, 1957, S.3 ff.). Schliesslich ist auch in den angewandten Fächern der Bodenkunde ("ökologische" bzw. "technologische" Bodenkunde) eine solche "morphologische" Betrachtungsweise durchaus anwendbar. Es werden dabei speziell die Mechanismen, die die in Frage stehenden, die "Anwendungsseite" interessierenden, Zusammenhänge regieren, unmittelbar-anschaulich erfasst. Der Wert dieser durch keine andere Untersuchungsmethode zu ersetzenden Betrachtungsweise für die Lösung bestimmter "Anwendungs"-Probleme ist ohne weiteres deutlich.

Wie man sieht, geht der Begriff "Bodenmorphologie" in dieser erweiterten Bedeutung also quer durch alle Teildisziplinen hindurch, die weiter oben als Gliederungselemente der heutigen Bodenkunde dargestellt wurden (vgl. S.510). Für die "physikalische" Befassung würde man bei entsprechender Prüfung zu dem gleichen Ergebnis kommen. Nach allem hat also der Begriff "Bodenmorphologie" eine doppelte Bedeutung:

(1) Einmal bezeichnet er die spezielle Teildisziplin der Bodenkunde, die sich mit dem (gewissermassen "bleibenden") Bodenaufbau beschäftigt. Dies Gebiet verdankt, genauso wie die anderen erwähnten Teildisziplinen der Bodenkunde auch, ihre Ausgliederung ganz bestimmten, *objektbedingten* Kriterien.

(2) Zum anderen bezeichnet der Begriff "Bodenmorphologie" jene ganz allgemeine "morphologische" Betrachtungsweise, die alles zu erfassen sucht, was am Boden irgendwie anschaulich-körperlich in Erscheinung tritt ("Körperlichkeit" des Bodens). Auch sie betrachtet dabei sowohl die geometrische wie auch die "stoffliche" Seite der Phänomene (vgl. S.510). Sie hat - im Gegensatz zu der vorerwähnten Teildisziplin "Bodenmorphologie" überwiegend *methodischen* Charakter.

Da es sich bei dieser zuletzt genannten Bodenmorphologie eindeutig um den umfassenderen Begriff handelt, könnte man diese Befassungsart mit "Bodenmorphologie im weiteren Sinne" (i.w.S.) bezeichnen. Die zuerst genannte Teildisziplin der Bodenkunde wäre dann die "Bodenmorphologie im engeren Sinne" (i.e.S.).

Als die ganz allgemeine (universelle) morphologische Befassung mit dem Boden hat die Bodenmorphologie i.w.S. stets mit den folgenden drei Grundgegebenheiten des Bodens zu rechnen, die damit zugleich die Grundzüge ihrer Methodik entscheidend bestimmen:

(1) Mit der *Langsamkeit* der Mehrzahl der sich im Boden abspielenden *Bewegungen und Veränderungen*. Die meisten dieser Prozesse und sogar

der "bleibende" Bodenaufbau (vgl. S.510) können also nicht unmittelbar beobachtet werden. Vielmehr müssen sie, wie bereits erwähnt, aus den Spuren, die sie im Bodenaufbau hinterlassen, oder aus den beobachtbaren Teilphasen (in Form von Augenblicksbildern oder von Entwicklungsreihen) nach und nach zu einem Gesamtbild zusammengefügt werden.

(2) Mit der *Undurchsichtigkeit des Bodens*. Um die morphologischen Phänomene im Innern des Bodens erfassen zu können, muss dieser in geeigneter Weise aufgeschlossen werden ("Aufschliessungsmethodik"). Dabei kann auf zwei verschiedenen Wegen vorgegangen werden: (a) Einmal kann man den Boden im Querschnitt betrachten (z.B. die geglättete Wand einer Profilaufgrabung; "Querschnittsmethode"). (b) Zum anderen kann man den Boden vorsichtig zerlegen (z.B. das behutsame Freilegen der natürlichen Gefügekörper des Bodens bei der Gefügeansprache am Bodenprofil; "Zerlegungsmethode").

(3) Mit dem *weiten Spektrum der Grössenbereiche*, in denen die morphologischen Phänomene des Bodens ausgebildet sind. Z.B. können die Steinringe bestimmter arktischer Rohböden im Dekameter-Bereich liegen (10^3 cm). Die Moleküle, Atome und Ionen, die in Form von Modellvorstellungen ebenfalls durchaus Gegenstand einer morphologischen Betrachtung sein können, reichen in den Ångström-Bereich hinein (10^{-8} cm). Es ergibt sich somit als Dimensions-Spektrum für die Bodenmorphologie der ungeheuer ausgedehnte Bereich von zwölf Grössenordnungen. Selbstverständlich kann dieser Grössenbereich nicht auf einmal und mit dem blossen Auge allein überblickt werden. Vielmehr ist das nur in Teilabschnitten und nur unter Heranziehung von verschiedensten Hilfsmitteln möglich. So können, z.B., die Steinringnetze mancher arktischer Strukturböden erst vom Flugzeug aus oder im Luftbild in ihrer ganzen Eigenart erkannt werden. Die individuelle Umgrenzung und die Vergesellschaftung der Böden in der Landschaft wird meist erst auf der Bodenkarte unmittelbar anschaulich. Beides sind *verkleinernde*, den Überblick ermöglichende Hilfsmittel. Reichen die morphologischen Phänomene auf der anderen Seite in Dimensionen hinein, die vom unbewaffneten Auge wegen ihrer Kleinheit nicht mehr erkannt werden können, so müssen *vergrössernde* Hilfsmittel herangezogen werden (Lupe, Lichtmikroskop, Elektronenmikroskop, usw.; vgl. Michel, 1957). Von einer bestimmten Dimension ab oder auch bei bestimmten Objekten versagen schliesslich auch diese Hilfsmittel. Jetzt können nur noch *indirekte* physikalische oder physikalisch-chemische Methoden weiterhelfen. Sie führen aber immerhin noch zu modellmässigen Vorstellungen, die der morphologischen Betrachtung zugänglich sind.

DIE MIKROMORPHOLOGIE ALS TEIL DER BODENMORPHOLOGISCHEN UNTERSUCHUNG

Nach dem vorangegangenen wird nun die Stellung der Mikromorphologie im Rahmen der Bodenkunde ohne weiteres erkennbar. Sie ist ein Teil jener - methodisch zu verstehenden - universellen "morphologischen" Befassung mit dem Boden (Bodenmorphologie i.w.S.), von der im vorigen Abschnitt ausführlich die Rede war. Und zwar ist sie aus folgendem Grunde eben nur ein Teil davon: Es handelt sich bei ihr lediglich um die Betrachtung eines

Ausschnittes aus dem weitgefächerten Gesamtspektrum der Dimensionen, in denen die körperlich-anschaulich erfassbaren Phänomene des Bodens ausgebildet sind (vgl. S.516). Es ist der Ausschnitt, der nur mit optischen Hilfsmitteln, in erster Linie mit Hilfe des Lichtmikroskopes, erkannt werden kann.

Abgesehen von dieser dimensionsmässigen Einschränkung hat die Mikromorphologie sonst prinzipiell alle Besonderheiten mit der Gesamt-Bodenmorphologie (i.w.S.) gemeinsam. Sie lässt sich also auch genauso wie diese in allen jenen Teildisziplinen der Bodenkunde anwenden, die weiter oben eingehend beschrieben wurden (vgl. S.510, 511). In diesem Sinne kann man also auch folgerichtig eine Mikromorphologie des Bodens "i.e.S." unterscheiden, die sich mit dem gewissermassen als *bleibend* ("starr") angesehenen, mikroskopischen Bodenaufbau beschäftigt. Auch hier tritt also das Wort "Morphologie" in seiner doppelten Bedeutung auf.

Allgemein muss jedoch die Bezeichnung "Mikromorphologie" in der umfassenderen, d.h., in der sich auf die universelle "morphologische" Betrachtungsmethode beziehenden, Bedeutung verwendet werden. Sonst bleiben wesentliche Anwendungsgebiete der Mikromorphologie unberücksichtigt. In diesem Sinne hat das Gebiet der Mikromorphologie also tatsächlich nur methodischen Charakter. Es erhebt sich deswegen die Frage, ob im Rahmen der Bodenkunde ein reines Methodengebiet überhaupt die Berechtigung zu einem gewissen Eigendasein hat.

Die Frage lässt sich nur in Anbetracht der in allen Wissenschaften zu beobachtenden und leider nicht zu vermeidenden Spezialisierung beantworten. Der Spezialist auf dem Gebiete der Mikromorphologie benötigt für seine Arbeit ein so grosses Mass an methodischen und fachlichen Spezialkenntnissen (das letztere in Bezug auf die Hilfswissenschaften, z.B., die mikroskopische Petrographie, Mineralogie, Zoologie, Botanik, die mikroskopische Gefügekunde, usw.), dass eine gewisse Sonderstellung seines Fachgebietes durchaus gerechtfertigt ist. Es ist dieselbe Sonderstellung, die auf anderem Gebiete dazu führt, dass z.B. eigene Symposia über die Anwendung der Radioisotopen in der Bodenkunde usw. abgehalten werden.

Diese Sonderstellung ist jedoch nur sehr bedingt und erstreckt sich lediglich auf die Methodik und die durch sie bestimmte Auswahl der erforderlichen Spezialkenntnisse. Genau wie das rein methodisch bedingte Sondergebiet der Isotopenanwendung kann auch die Mikromorphologie nicht ohne den - gewissermassen unmittelbar vom Forschungsobjekt "Boden" und seiner Eigenart diktierten, oben ausführlich geschilderten - Gesamtrahmen der Bodenkunde auskommen. Vielmehr erfährt sie erst durch ihn die Richtung, in die die Untersuchung jeweils zielen muss. Die sachlich bedingten Teildisziplinen der Bodenkunde werden also gewissermassen zu den Hauptanwendungsgebieten der Mikromorphologie. Der ständige Kontakt mit dem Gesamtrahmen der Bodenkunde bewahrt sie ausserdem vor einem sterilen, abgekapselten Sonderdasein. Er weist ihr den Weg zu fruchtbarer Mitarbeit an der grossen Gemeinschaftsaufgabe der Bodenkunde, nämlich der umfassenden, wissenschaftlichen Erforschung des Bodens in allen seinen Teilaspekten.

Die Mikromorphologie als Teil der bodenmorphologischen Gesamtuntersuchung setzt nun dort ein, wo das Auflösungsvermögen des unbewaffneten Auges seine Grenzen hat. Von einem normalsichtigen Auge können

Bauelemente in ihrer vollständigen *äusseren Formausbildung* etwa noch bis zu einem Durchmesser von 1 mm (10^{-1} cm) hinab erfasst werden. Innenstrukturen in diesen Bauelementen werden jedoch nur noch als undifferenzierte Punkte und Linien, und zwar bis zu einer Grösse von etwa 0,1 mm Durchschnitt (10^{-2} cm) hinab wahrgenommen. Was in der Grösse noch weiter darunter liegt, erscheint als mehr oder weniger homogenes *Mischbild*, an dem nur noch bestimmte "Qualitäten des Aussehens" (z.B., trübe, klar; matt, glänzend; "flockig", homogen-durchscheinend; die verschiedenen Farben usw.) erkannt werden können.

Die untere Grenze des mikromorphologischen Forschungsbereiches hängt davon ab, welche optischen Hilfsmittel herangezogen werden. Nimmt man, z.B., noch das Elektronenmikroskop zu Hilfe, das durchaus zum Instrumentarium der Mikromorphologie gerechnet werden könnte, so ergibt sich als unterste Grenze der Auflösung etwa 1 mμ (10^{-7} cm; vgl. Reimer, 1959, S. 296). Strukturen dieser Dimension werden jedoch ebenfalls nur noch - und zwar jetzt aus physikalisch-optischen, in den optischen Geräten liegenden Gründen - als undifferenzierte Punkte oder Linien wahrgenommen. In ihrer genauen äusseren Formausbildung sind die Baueigentümlichkeiten erst erkennbar, wenn sie mindestens eine Grössenordnung darüber liegen (10 mμ = 10^{-6} cm; vgl. Von Borries und Kausche, 1940). Bei Verwendung des Lichtmikroskopes, das gemeinhin als das eigentliche Untersuchungsinstrument in der Mikromorphologie angesehen wird, liegen die entsprechenden Grenzen in der Grössenordnung von 0,1 μ (10^{-5} cm; unterste Grenze der absoluten Auflösung, und zwar bei kurzwelligem Licht) bzw. 1 μ (10^{-4} cm; unterste Grenze der genauen Formauflösung).

Es ergibt sich somit, z.B., für die "lichtmikroskopische" Mikromorphologie ein Dimensionsspektrum von 4 bis 5 Grössenordnungen (10^{-1}–10^{-5} cm Im Rahmen der bodenmorphologischen Gesamtuntersuchung (Bodenmorphologie i.w.S.) ist diese *Dimensions*begrenzung tatsächlich das einzige Definitionsmerkmal der Mikromorphologie. Welche objektiv gegebenen Baueigentümlichkeiten des Bodens (z.B., etwa bestimmte Mineralkörner, bestimmte spezifisch bodeneigene Bauelemente, bestimmte Organismen oder Organismenreste, usw.) jeweils in diesen Beobachtungsausschnitt hineinreichen, ist nicht von vornherein bestimmt. Vielmehr muss man sich immer wieder ganz klar machen, dass der Boden als körperliches Gebilde eine von den gröbsten bis zu den feinsten Dimensionen hin in sich völlig zusammenhängende ("ineinandergeschachtelte") und in dieser Weise unauflösbare morphologische Einheit darstellt. Es gibt in ihr keinen irgendwie gearteten und stets auf der gleichen Grössenstufe vorkommenden Gliederungshiatus, auf dem am Ende doch die obere sowie auch die untere Grenze der Mikromorphologie beruhen könnte. Das von der Mikromorphologie betrachtete Spektrum der Baueigentümlichkeiten ist also ein rein *zufälliger*, allein durch die oben angegebenen Dimensionsgrenzen der Beobachtungshilfsmittel diktierter Ausschnitt aus der Gesamtheit der morphologischen Phänomene des Bodens.

Diese Eigentümlichkeit der Mikromorphologie hat - wie man leicht sieht - eine ganz bedeutsame Konsequenz: Mikromorphologische Untersuchungen können niemals isoliert betrieben und ihre Ergebnisse niemals isoliert gewertet werden. Sie sind nicht nur, sondern sie *müssen* sogar immer Teil einer gesamtmorphologischen Erfassung des Bodens sein. Es hat ihnen deswegen stets eine Erforschung des grösseren morphologischen

Zusammenhanges - sei es durch den Mikromorphologen selber, sei es in Zusammenarbeit mit einem Feldbodenkundler - voranzugehen. Auf der anderen Seite führt die Mikromorphologie bei der morphologischen Erfassung des Bodens in vielem noch nicht zu der abschliessenden Erkenntnis (z.B., bei der "stofflichen" Identifizierung mancher Bauelemente). Hier müssen, soweit nun auch die elektronenmikroskopischen Hilfsmittel versagen, die indirekten Methoden der Physik, der Chemie und der physikalischen Chemie (z.B., Röntgenanalyse, Elektronenbeugung, chemische und physikalisch-chemische Erforschung der Konstitution usw.) ansetzen.

Umgekehrt stellt die Mikromorphologie in ihrer dimensionsmässigen Mittelstellung geradezu das *Bindeglied* zwischen den genannten "gröberen" und den genannten "feineren" morphologischen Erfassungsmethoden dar. In dieser Rolle kann sie durch nichts ersetzt werden. Tatsächlich ist die vollständige morphologische Erfassung des Bodens nur durch die Zusammenarbeit *aller* Teilgebiete der Bodenmorphologie (i.w.S.), und zwar dabei immer durch ein Voranschreiten von den gröberen zu den feineren Dimensionen, zu erreichen.

DIE METHODIK DER MIKROMORPHOLOGIE

Abschliessend soll ein kurzer Abriss der Methodik der Mikromorphologie gegeben werden. Hierdurch werden die Beziehungen und Abhängigkeiten noch deutlicher, die zwischen der Mikromorphologie und dem Gesamtrahmen der Bodenkunde bestehen. Ausserdem geschieht es, um zu zeigen, dass es sich bei der Mikromorphologie - genau wie bei der Gesamt-Bodenmorphologie (i.w.S.) - tatsächlich um eine ganz und gar wissenschaftlich-exakte Forschungsmethode der Bodenkunde handelt. Allerdings ist ihr Wesen, wie oben bereits näher ausgeführt wurde, von grundsätzlich anderer Art als das der kausalanalytischen "physikalischen" Befassung (im Sinne von Sander; vgl. S. 512). Deswegen ist es prinzipiell unmöglich, die Morphologie mit den Massstäben *dieser* ("physikalischen") Befassungsart zu messen und das Zurücktreten von "Mass und Zahl" bzw. das Fehlen einer mathematisch-funktionalen Behandlung der kausalen Zusammenhänge zu monieren. Vielmehr muss auch hier die "physikalische" Befassung soweit wie möglich *neben* und in *Wechselbeziehung* zur "morphologischen" Betrachtungsweise treten. Erst dann kann es zu der umfassenden wissenschaftlichen Erkenntnis der Phänomene des Bodens kommen.

Wie bei jeder morphologischen Untersuchung müssen auch bei der Mikromorphologie zwei grundsätzlich verschiedene Untersuchungsschritte klar auseinandergehalten werden:

(1) Der erste ist die vollständige und exakte *Erfassung der objektiv gegebenen morphologischen Daten.*

(2) Der zweite besteht darin, diese Daten erschöpfend und widerspruchsfrei auszudeuten (*Interpretation der Daten*).

Während sich der zweite Untersuchungsschritt nicht wesentlich von jeder anderen naturwissenschaftlichen Schlussfolgerung auch unterscheidet, ist das bei dem ersten Untersuchungsschritt anders. Die objektiv gegebenen Daten sind in der Mikromorphologie - wie überall in der Bodenmorphologie - in der Regel keine durch einfache Zahlen ausdrückbare "Analysenwerte".

Die morphologischen Phänomene des Bodens sollen gerade in ihrer *umfassenden anschaulich-körperlichen Eigenart und Komplexheit* erfasst werden. Deswegen verbietet sich ihre Reduzierung auf einige wenige ausgewählte und dabei bewusst idealisierte Grössen und deren höchstmöglich-exakte messtechnische Erfassung, wie sie bei den meisten physikalischen und chemischen Analysen üblich und vom Prinzip her sogar unbedingt erforderlich ist. Vielmehr sind es die Methoden der *Typisierung* sowie der *Schätzung* bzw. *statistisch-quantitativen Erfassung*, die angewandt werden müssen. Bei richtiger Handhabung dieser Methoden kommt man jedoch dabei prinzipiell zu genauso objektiv feststehenden "Analysendaten", wie sie bei den oben genannten anderen naturwissenschaftlichen Untersuchungsmethoden gewonnen werden. Lediglich der Grad der messtechnischen Festlegung ("Messgenauigkeit") ist nicht so hoch wie dort. In Anbetracht der meist sehr komplexen und mehr oder weniger stark streuenden Natur der zu erfassenden Phänomene wäre sie jedoch sogar nicht einmal immer sinnvoll.

Die Methodik der Mikromorphologie soll nun kurz dargestellt werden, indem die soeben geschilderten beiden Haupt-Untersuchungsschritte jeweils in ihren einzelnen Teilabschnitten betrachtet werden:

(1) Erfassung der Daten: (a) Aufschliessungsmethodik; (b) Beobachtungsmethodik; (c) Darstellungsmethodik.

(2) Interpretation der Daten: (a) Bestimmung der Ausdeutungsrichtung; (b) Bestandsaufnahme ("Katalog") der Deutungsmöglichkeiten; (c) Abwägung der Deutungsmöglichkeiten und Schlussfolgerung. Wie man leicht sieht, gilt diese Gliederung nicht nur speziell für die Mikromorphologie sondern - in den Grundzügen - auch für die bodenmorphologische Gesamt-Untersuchung. Der im folgenden gegebene kurze Abriss der Methodik der Mikromorphologie ist somit zugleich im Prinzip eine Schilderung der allgemeinen Methodik der Bodenmorphologie (i.w.S.).

Erfassung der Daten

Aufschliessungsmethodik

Die Undurchsichtigkeit des Bodens erfordert auch im mikromorphologischen Bereich eine Aufschliessung des Bodenkörpers für die Untersuchung. Wiederum wie bei jeder bodenmorphologischen Untersuchung kommt dafür entweder die "Querschnittsmethode" oder die "Zerlegungsmethode" in Frage.

Die erste wird in der Mikromorphologie durch den Anschliff bzw. durch den Dünnschliff repräsentiert. Infolge der Fragilität des Bodens muss in der Regel eine Imprägnierungshärtung dieser "Querschnitts"-Präparation vorausgehen. Wie bei jeder Querschnittsbetrachtung wird ein ausgezeichneter und übersichtlicher Einblick in das "*Innere*" des untersuchten Körpers ermöglicht. Die genaue geometrische Anordnung und gegenseitige Zuordnung der Bauverhältnisse ist auf einen Blick erkennbar und dadurch der exakten Vermessung verhältnismässig leicht zugänglich. Die Erfassung ist jedoch stets nur *zwei*dimensional.

Diesen Nachteil vermeidet die "Zerlegungsmethode". Sie ist in der Mikromorphologie durch die mikroskopische Betrachtung von vorsichtig herauspräparierten Flächen oder Bodenstücken bzw. von sogenannten

"Trümmer"- oder "Einzelkornpräparaten" vertreten (vgl. Kubiena, 1938, S.69–77). Mit Ausnahme bei gewissen durchsichtigen Anteilen kann mit dieser Methode der Bau jedoch immer nur "von aussen" erfasst werden. Ausserdem wird bei dem erforderlichen, im besten Falle sukzessiven Zerlegen des Baukörpers der ursprüngliche morphologische Verband (zunehmend) zerstört. Die genaue räumliche Anordnung und gegenseitige Zuordnung der Baueigentümlichkeiten lässt sich also niemals auf einen Blick erfassen. Vielmehr muss sie erst aus dem Gedächtnis (oder an Hand von sukzessive gemachten Einzelaufnahmen) nach und nach rekonstruiert werden. Die Schwierigkeiten für die exakte Vermessung der Bauverhältnisse liegen bei diesem Verfahren also auf der Hand. Nach allem ergibt sich die Notwendigkeit, für eine vollständige, dreidimensionale und nach Möglichkeit auch messtechnische Erfassung der morphologischen Phänomene des Bodens die beiden genannten Haupt-Aufschliessungsmethoden, nämlich die "Querschnittsmethode" und die "Zerlegungsmethode", stets miteinander zu kombinieren.

Beobachtungsmethodik

Hinsichtlich der Beobachtungsmethodik ist die Mikromorphologie, wie geschildert wurde, gerade durch die Verwendung optischer Hilfsmittel gekennzeichnet, die also für sie spezifisch sind. Die Methodik ergibt sich zwanglos aus der Anwendung von Lupe und Mikroskop sowie aus den verschiedenen Arten der Mikroskopiertechnik. Es seien hier nur die Methoden der Stereomikroskopie und der normalen "zweidimensionalen" Mikroskopie, der Mikroskopie mit polarisiertem und mit gewöhnlichem Licht, der Auflicht- und Durchlichtmikroskopie, der Hellfeld-, Dunkelfeld-, Phasenkontrast-, Interferenzkontrast- und Fluoreszenzmikroskopie sowie die verschiedenen optischen bzw. kristalloptischen Messmethoden genannt (vgl. Michel, 1957; Kleber, 1959). Auch hier erweist es sich als zweckmässig, verschiedene Methoden je nach Aufgabenstellung nebeneinander zu verwenden.

Darstellungsmethodik

Für die Erfassung der objektiv gegebenen Daten genügt es nicht, die durch geeignete Aufschliessungs- und Beobachtungsmethodik sichtbar gemachten morphologischen Phänomene des Bodens nur subjektiv zu registrieren. Vielmehr müssen sie auch - vollständig und exakt - in einer *objektiven* Form *dargestellt* werden. Erst diese Form gestattet eine interne Dokumentation bzw. eine Veröffentlichung der Beobachtungsergebnisse und damit die Selbstkontrolle bzw. den Nachvollzug durch andere Beobachter.

Wie bereits erwähnt, ist es meist nicht möglich, die beobachteten morphologischen Daten in einfacher Weise durch "Mass und Zahl" festzuhalten. In der Regel ist nur die Möglichkeit einer mehr oder weniger ausführlichen Beschreibung mit Worten sowie die bildliche Darstellung (Zeichnung bzw. Mikrophoto) gegeben. Beides sollte der besseren Anschauung wegen möglichst weitgehend miteinander kombiniert werden. Für eine bessere Objektivierung der Beschreibung wäre es - in ähnlicher Weise, wie es für die makroskopische Profilbeschreibung bereits in ziemlich perfekter Form verwirklicht ist - sehr zweckmässig, wenn allgemein nach einem einheitlichen Schema verfahren würde. Die Erarbeitung eines solchen

Standard-Beschreibungsschemas könnte also sehr zur weiteren Präzisierung der mikromorphologischen Methodik beitragen.

Alles, was bei der Beschreibung nach Grösse, Eigenart und "Stoff" exakt identifiziert werden kann (z.B. Minerale, Organismen und Organismenreste, Hohlräume, usw.), sollte auch in dieser bestimmten Form niedergelegt werden. Leider wird mit diesen genau bestimmbaren Elementen in der Regel nur ein beschränkter Teil der Baueigentümlichkeiten des Bodens eindeutig charakterisierbar. Gerade die spezifisch bodeneigenen morphologischen Gegebenheiten (z.B., die Gestalt und die Eigenart vieler typischer Bauelemente des Bodens, die Eigenart der Bodengefüge, usw.) sind damit in der Regel nicht zu erfassen.

Hier offenbart sich - wie jeder Mikromorphologe aus eigener Erfahrung weiss - immer wieder das systematisch-terminologische Dilemma auf dem Gebiete der Bodenmorphologie: Die typischen Gefüge, Bauelemente und "Qualitäten des Aussehens" (vgl. hierzu S.516), die dem Beobachter bei der morphologischen Untersuchung entgegentreten, müssen in der Beschreibung mit vollkommen eindeutigen und allgemeingültigen Bezeichnungen versehen werden, wenn die Darstellung der Bauverhältnisse "objektiv" sein soll. Nun ist zwar in den zahlreichen, bis jetzt vorliegenden Arbeiten (vgl. insbesondere die wegweisenden Veröffentlichungen von Kubiena, 1935, 1938, 1948, 1953, 1956, 1962b, usw.) bereits ein beträchtlicher Teil des mikromorphologischen Formenschatzes beschrieben und mit inzwischen auch allgemein anerkannten Namen belegt worden. Dennoch geschieht es immer wieder, dass man sich vor neue Formen gestellt sieht, für die eine eingeführte Bezeichnung noch nicht zur Verfügung steht. Es fehlt also - trotz aller Ansätze - noch immer eine umfassende und einheitliche Systematik bzw. Nomenklatur der mikroskopischen Bauverhältnisse des Bodens. Die gemeinschaftliche Erarbeitung eines solchen Begriffssystems ist vielleicht die entscheidenste Aufgabe für die Mikromorphologie im Hinblick auf die weitere Präzisierung ihrer Methodik.

Interpretation der Daten

Sind die soeben beschriebenen drei Teilschritte (Aufschliessung des Bodens, visuelle Beobachtung und objektive Darstellung der erkannten Baueigentümlichkeiten) in der richtigen Weise durchgeführt worden, so stellen die damit erfassten morphologischen Daten ein für allemal feststehende Beobachtungstatsachen dar (objektiv gegebene Daten). Sie sind nun einer gedanklich-logischen Interpretation zugänglich, bei der u.U. noch andere feststehende Daten oder begründete Hilfsannahmen hinzugezogen werden können. Die im folgenden kurz geschilderten drei Teilschritte dieser Interpretation werden sich in der Praxis der mikromorphologischen Untersuchung wohl stets mehr oder weniger durchdringen. Auch werden sie z.T. bereits bei der eigentlichen Erfassung der Daten - zumindest im Ansatz (Notizzettel) - vollzogen werden. Darin liegt auch durchaus ein Vorteil: Man wird bereits unmittelbar bei der visuellen mikroskopischen Untersuchung zu mancher neuen Beobachtung angeregt, die sonst möglicherweise nicht gemacht worden wäre. Trotzdem sollte man im Interesse einer exakten und vorurteilsfreien Untersuchung die Teilschritte, wenigstens gedanklich, sauber auseinanderhalten.

Bestimmung der Ausdeutungsrichtung

Wenn die Festlegung der geplanten Untersuchungsrichtung z.T. auch bereits Art und Umfang der zu erfassenden morphologischen Daten bestimmt, so beeinflusst sie doch in erster Linie die Interpretation. Es sind zwei Haupt-Untersuchungsarten zu unterscheiden:

(1) Die *allgemeine* (oder vollständige) mikromorphologische Untersuchung, bei der - ohne eine von vornherein feststehende spezielle Fragestellung - möglichst alle irgendwie nur erkennbaren morphologischen Daten erfasst und im Anschluss daran, so vollständig wie es nur irgend geht, ausgedeutet werden.

(2) Die *spezielle* mikromorphologische Untersuchung, bei der es nur um die Klärung eines bestimmten Problems geht (z.B., das Vorhandensein oder Fehlen von Durchschlämmungserscheinungen) und von vornherein nur die in dieser Hinsicht in Frage kommenden morphologischen Daten erfasst werden.

Die speziellen Fragen lassen sich dabei stets den sieben Sachgebieten unterordnen, die als Teildisziplinen der Bodenkunde ausgeschieden wurden (vgl. S.510, 511): (1) Bodenmorphologie (i.e.S.); (2) Bodendynamik; (3) Bodengenetik; (4) Bodensystematik; (5) Bodengeographie; (6) "ökologische" Bodenkunde; (7) "technologische" Bodenkunde. Diese sieben Teilgebiete stellen ausserdem insgesamt das natürliche und von vornherein vollständige Gliederungsschema für die Interpretation im Falle der allgemeinen mikromorphologischen Untersuchung dar.

Bestandsaufnahme ("Katalog") der Deutungsmöglichkeiten

Innerhalb der jeweiligen Untersuchungsrichtung, sei es bei der allgemeinen, sei es bei der speziellen mikromorphologischen Untersuchung, ergeben sich bei der Interpretation im konkreten Einzelfall fast immer mehrere Deutungsmöglichkeiten. Die aus den morphologischen Daten ablesbaren Tatbestände (insbesondere die zu beobachtenden kinematischen oder relativ-zeitlichen sowie die an Entwicklungsreihen ablesbaren, genetischen Gegebenheiten) sind meist mehr oder weniger komplex. Vielfach stehen zunächst bestimmte Kriterien für die eindeutige Entscheidung nicht ohne weiteres zur Verfügung, und so existieren dann für eine gewisse Zeit mehrere, sich mehr oder weniger widersprechende Deutungen nebeneinander. Das liegt in der Natur des Untersuchungsobjektes begründet und ist kein Schade. Vielfach regt ein solcher Widerspruch sogar zu äusserst fruchtbaren neuen Fragestellungen an, die u.U. zu vertieften Einsichten führen. Aus diesem Grunde und aus dem Bestreben, nun wirklich *alle* Gesichtspunkte bei der Interpretation zu berücksichtigen, ist es zweckmässig, alle theoretisch überhaupt denkbaren Deutungen möglichst vollständig zusammenzustellen ("Katalog" der Deutungsmöglichkeiten). Dadurch wird es erreicht, dass der Blick nicht von vornherein durch eine vorgefasste Meinung eingeengt wird und das endgültige Ergebnis der morphologischen Untersuchung durch eine voreilige, nicht genügend fundierte Interpretation leidet.

Abwägung der Deutungsmöglichkeiten und Schlussfolgerung

Diese endgültige Abwägung ist im Grunde ein ganz gewöhnlicher gedanklich-logischer Akt. Selbstverständlich muss er in Anbetracht der komplexen Tatbestände, mit denen man es bei der morphologischen Untersuchung zu tun hat, mit ganz besonderer Sorgfalt und grösstmöglicher Kritik vollzogen werden. Sind die Kriterien für die endgültige Entscheidung nicht oder nicht allein durch die unmittelbar erfassten morphologischen Daten eindeutig gegeben, so kann man sonstige feststehende Tatsachen oder sogar gewisse begründete Hilfsannahmen heranziehen. In diesen - übrigens gar nicht so seltenen Fällen - sind die Resultate dann zwar nicht mehr morphologisch exakt bewiesene Tatsachen. Trotzdem stellen sie, als die nach allem bisher Bekannten *wahrscheinlichsten* Deutungen, schlechthin die wissenschaftliche Wahrheit dar. Voraussetzung dafür ist selbstverständlich, dass tatsächlich alle in Frage kommenden Gesichtspunkte bei der endgültigen Entscheidung berücksichtigt wurden und dass sich aus keiner der bekannten feststehenden Tatsachen irgendwelche Widersprüche ergeben haben. Betrachtet man in diesem Zusammenhang übrigens kritisch den Bestand an wissenschaftlich gesicherten Erkenntnissen bei den anderen naturwissenschaftlichen Fächern, so sieht man, dass die Mikromorphologie mit dieser Eigenart ihrer Resultate durchaus nicht allein dasteht. Auf Schritt und Tritt - sogar in der exaktesten Naturwissenschaft, der Physik - begegnet man dieser gewissen "Vorläufigkeit" der Ergebnisse. Wird in jedem Falle der Gedankengang, der zu den endgültigen Deutungsergebnissen geführt hat, genau und vollständig wiedergegeben, so handelt es sich bei der Mikromorphologie, da auch die jederzeitige Nachprüfbarkeit ihrer Resultate gegeben ist - um ein ganz und gar wissenschaftlich-exaktes Forschungsgebiet der Bodenkunde.

SUMMARY

Concerning the position of micromorphology within the context of soil science

This theoretical paper is intended as a contribution to the clarification of the basic principles of soil micromorphology. The ideas put forward here may also possibly be useful in assisting in the further development of micromorphological methodology. The following five problems are treated:

(1) The term *soil* in modern soil science. Only the concept of soil as an "organized" body composed in a characteristic way and characteristically changeable permits a morphological treatment.

(2) Sub-division of modern soil science. Based on general inherent soil criteria a natural sub-division of soil science results in seven component disciplines, as follows: (a) soil morphology; (b) soil dynamics; (c) soil genetics; (d) soil systematics; (e) soil geography; (f) "ecological" soil science; (g) "technological" soil science.

(3) Soil morphology and soil morphological investigation. Besides the special discipline "soil morphology" (soil morphology in the narrower sense) as conditioned by the object, there is the more comprehensive methodologically conditioned field of soil morphological investigation (soil morphology in the wider sense). This must take into consideration three basic facts:

(a) that many processes in the soil are taking place at a very slow rate (necessity of its indirect deduction); (b) that soil is not transparent (necessity of "opening" the soil); (c) that the morphological phenomena of soil show a wide range of spatial dimensions (necessity of using, for example, optical means).

(4) Micromorphology as a part of soil morphological investigation. Due to its limitation to the microscopic range, micromorphology in the widest sense is a special field of the universal soil morphological investigation (soil morphology in the wider sense). Fields of application are here as well as there the seven component disciplines of soil science which were mentioned above.

(5) The methodology of micromorphology. The methodology of micromorphology basically coincides with that one of general soil morphology (in the wider sense). As has been shown, it is an entirely and scientifically exact research field of soil science.

LITERATUR

Ganssen, R., 1957. Bodengeographie. Koehler, Stuttgart, 219 pp.

Hoenes, D., 1955. Mikroskopische Grundlagen der technischen Gesteinskunde. In: H. Freund (Herausgeber), Handbuch der Mikroskopie in der Technik. Umschau-Verlag, Frankfurt am Main, 4 (1) : 321–695.

Kleber, W.L., 1959. Einführung in die Kristallographie. Verlag Technik, Berlin, 332 S.

Kubiena, W.L., 1935. Über das Elementargefüge des Bodens. Bodenkundliche Forsch., 4 : 380–412.

Kubiena, W.L., 1938. Micropedology. Collegiate Press, Ames, Iowa, 243 pp.

Kubiena, W.L., 1948. Entwicklungslehre des Bodens. Springer, Wien, 215 S.

Kubiena, W.L., 1953. Bestimmungsbuch und Systematik der Böden Europas. Enke, Stuttgart, 392 S.

Kubiena, W.L., 1956. Zur Mikromorphologie, Systematik und Entwicklung der rezenten und fossilen Lössböden. Eiszeitalter Gegenwart, 7 (1956) : 102–112.

Kubiena, W.L., 1962a. Wesen, Ziele und Anwendungsgebiete der mikromorphologischen Bodenforschung. In: H.-J. Altemüller und H. Frese (Herausgeber), Arbeiten aus dem Gebiete der Mikromorphologie des Bodens. Verlag Chemie, Weinheim/ Bergstrasse, S.1–13.

Kubiena, W.L., 1962b. Die taxonomische Bedeutung der Art und Ausbildung von Eisenoxydhydratmineralien in Tropenböden. In: H.-J. Altemüller und H. Frese (Herausgeber), Arbeiten aus dem Gebiete der Mikromorphologie des Bodens. Verlag Chemie, Weinheim/Bergstrasse, S.95–103.

Laatsch, W., 1938. Dynamik der deutschen Acker- und Waldböden. Steinkopff, Dresden, 270 S.

Laatsch, W., 1957. Dynamik der mitteleuropäischen Mineralböden. Steinkopff, Dresden, 280 S.

Michel, K., 1957. Die Mikrophotographie. Springer, Wien, 740 pp.

Mitscherlich, E.H., 1913. Bodenkunde für Land- und Forstwirte. Parey, Berlin, 317 S.

Mückenhausen, E., 1958. Der Wasserhaushalt der Pseudogleye und dessen Bedeutung für die Pflanze. Verhandl. II, IV Komm. Intern. Bodenkundlichen Ges., 2 : 105–111.

Mückenhausen, E., 1962. Entstehung, Eigenschaften und Systematik der Böden der Bundesrepublik Deutschland. DLG-Verlag, Frankfurt am Main, 148 S.

Reimer, L., 1959. Elektronenmikroskopische Untersuchungs- und Präparationsmethoden. Springer, Berlin, 300 S.

Sander, B., 1948. Einführung in die Gefügekunde der geologischen Körper. 1. Allgemeine Gefügekunde und Arbeiten im Bereich Handstück bis Profil. Springer, Wien, 215 S.

Scheffer, F., 1962. Die bodenkundliche Forschung und ihre Aufgaben im Rahmen der Schwesterdisziplinen. Z. Pflanzenernähr., Düng., Bodenk., 98 : 18–29.
Scheffer, F. und Schachtschabel, P., 1960. Lehrbuch der Agrikulturchemie und Bodenkunde. 1. Bodenkunde. Enke, Stuttgart, 332 S.
Stebutt, A., 1930. Lehrbuch der allgemeinen Bodenkunde. Borntraeger, Berlin, 518 S.
Von Borries, B. und Kausche, G.A., 1940. Über mikroskopische Bestimmung der Form und Grössenverteilung von Goldkolloiden. Kolloid-Z., 90 : 132–141.

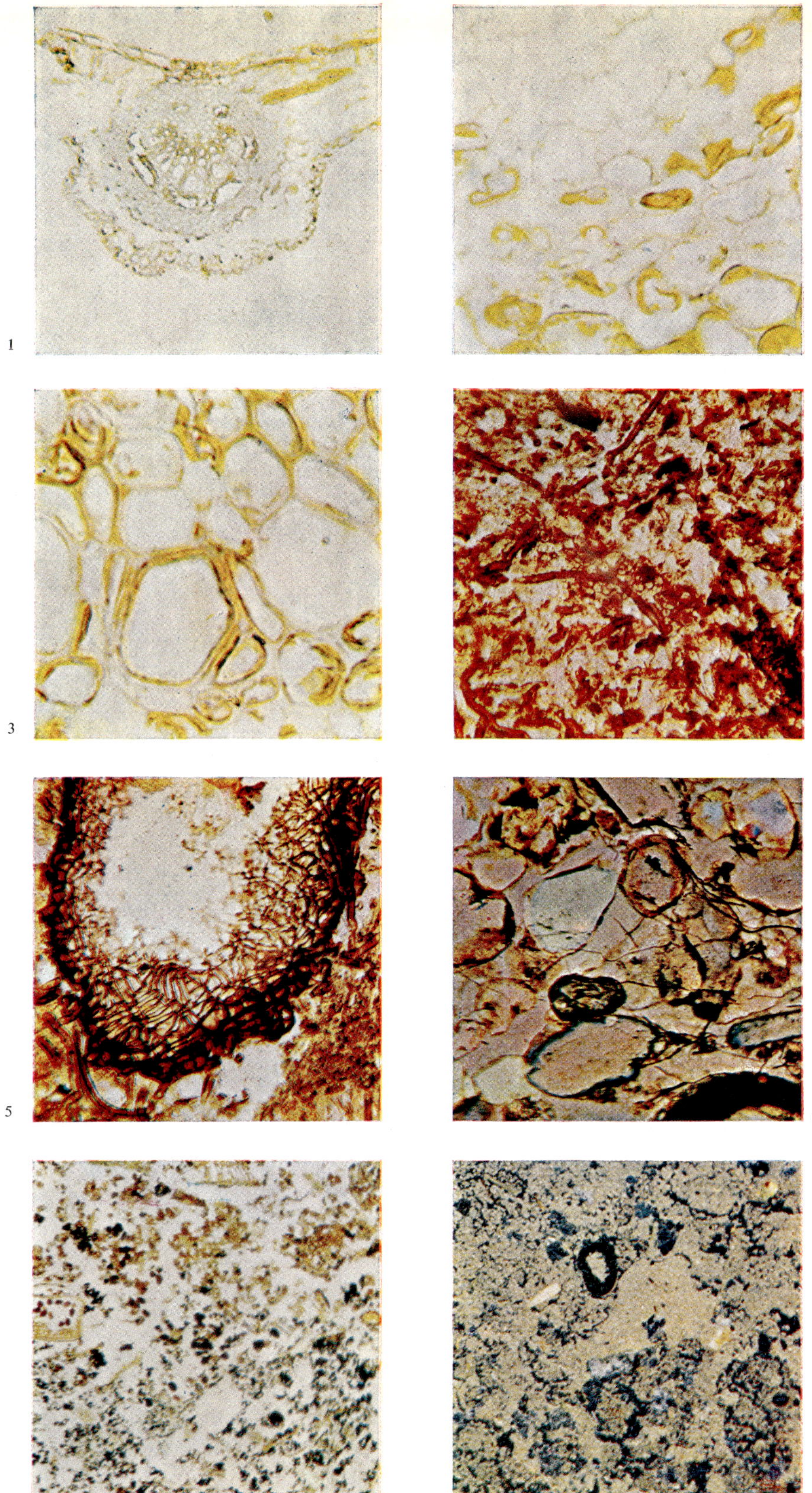

PLATE A

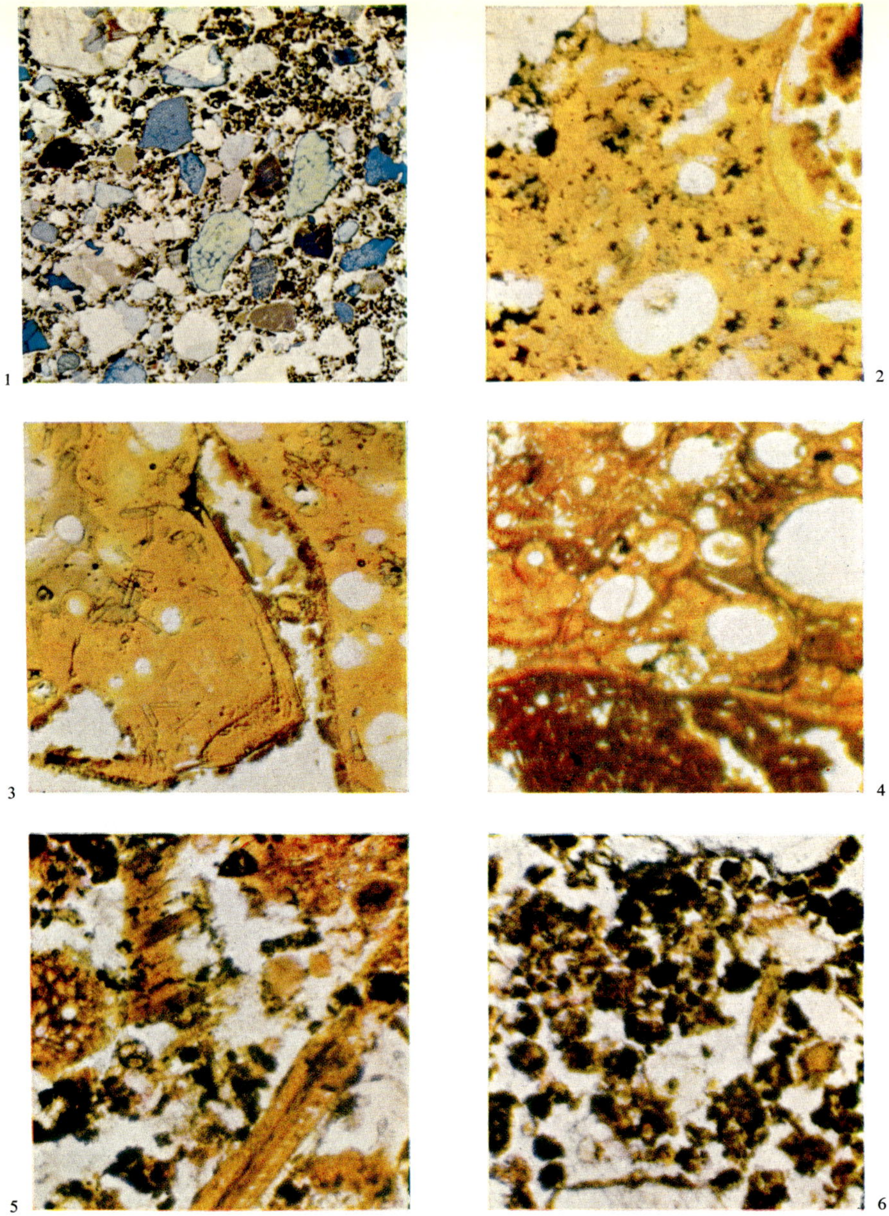

PLATE B

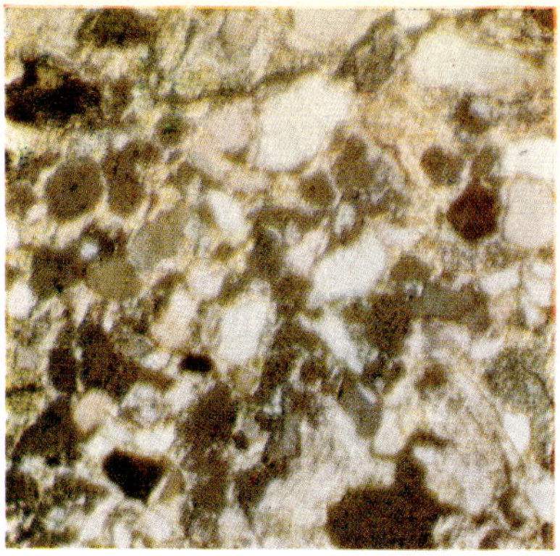
1

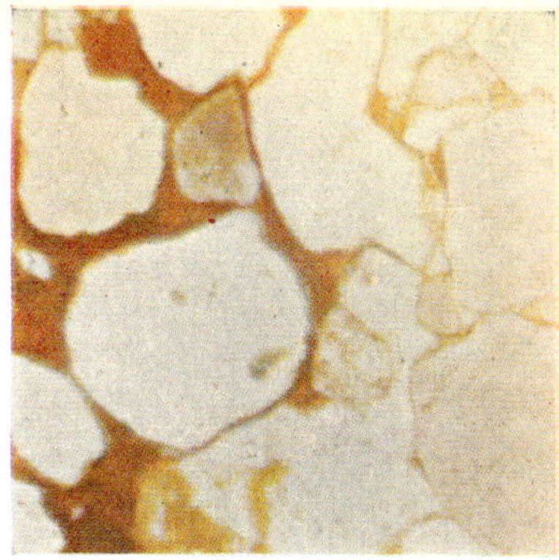
2

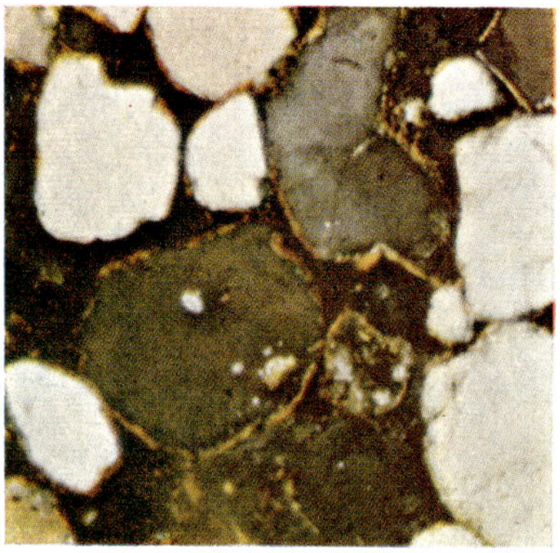
3

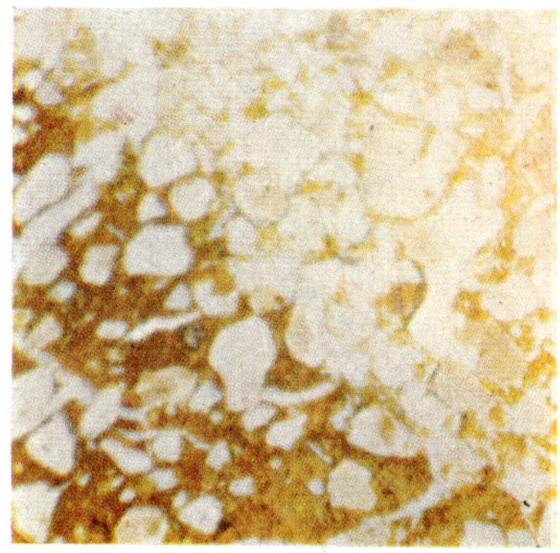
4

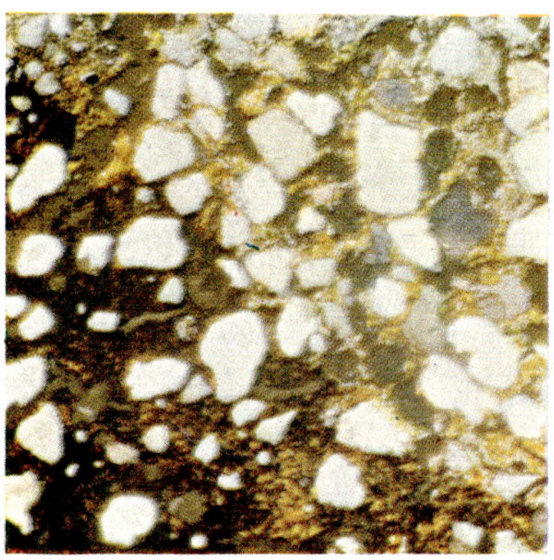
5

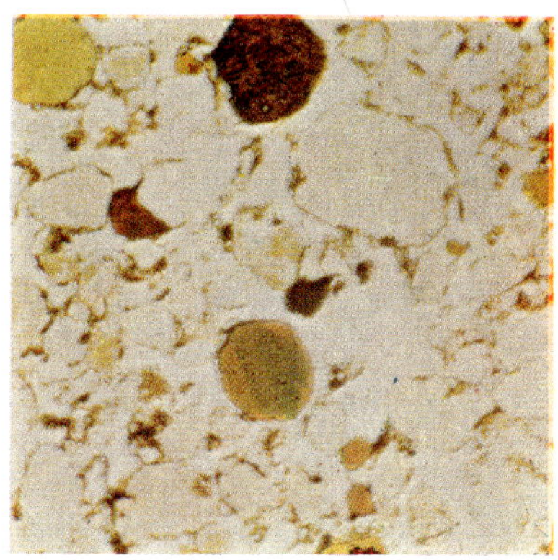
6

7

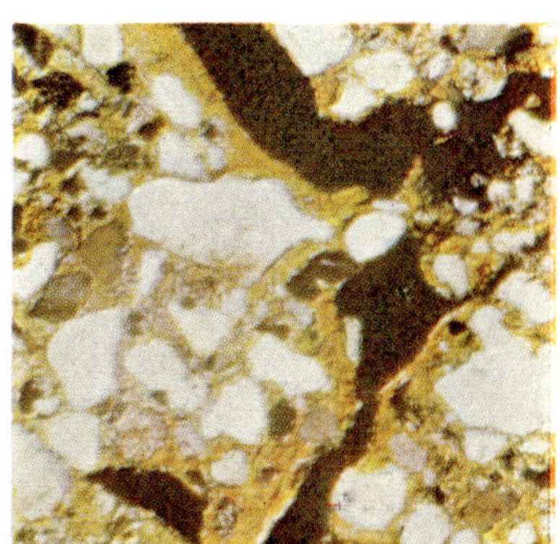
8

PLATE C

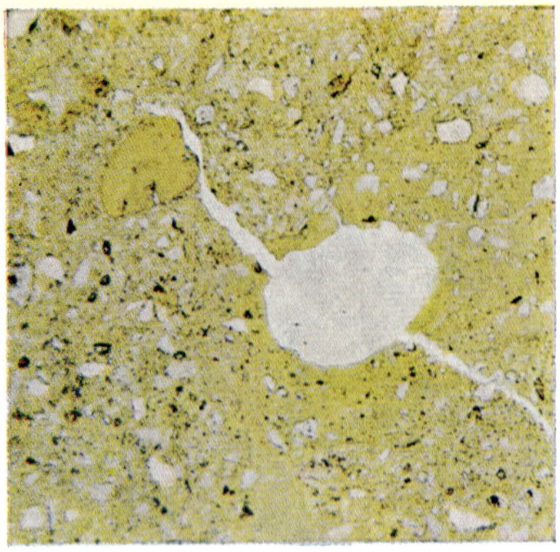
1

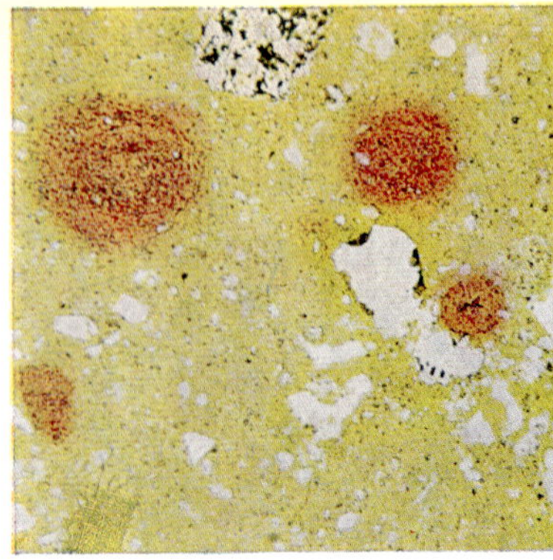
2

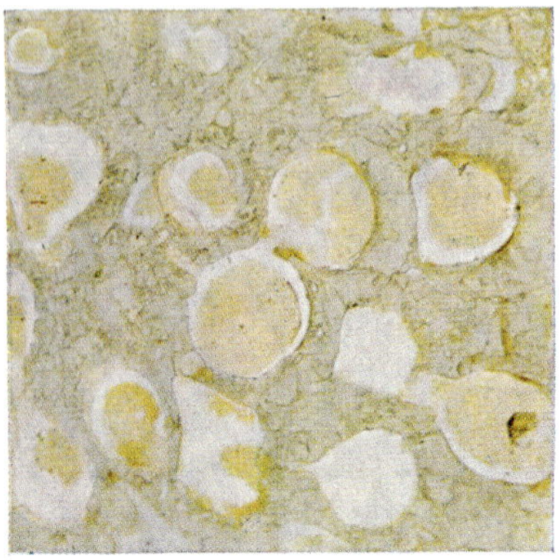
3

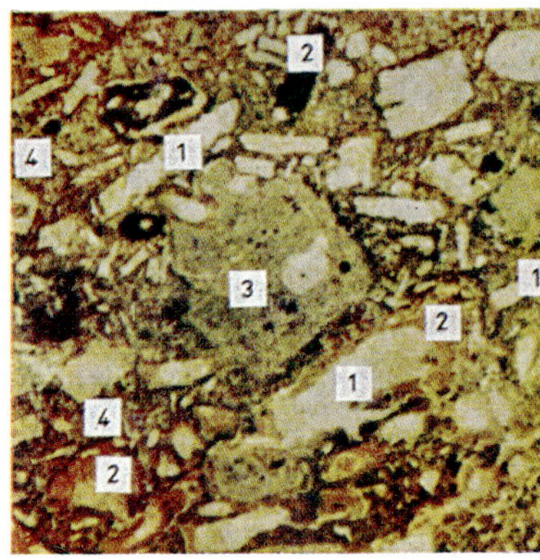

4

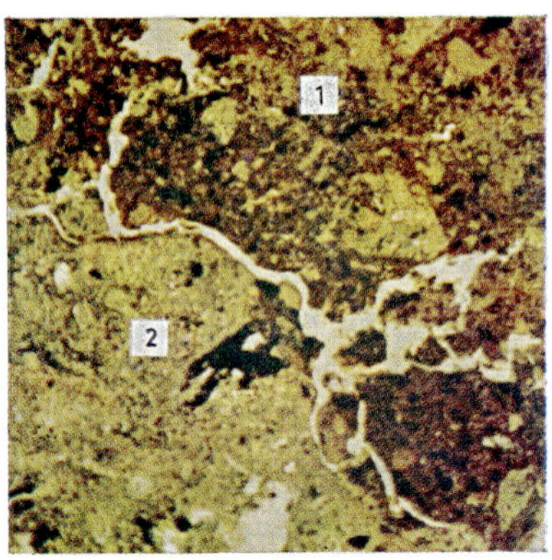

5

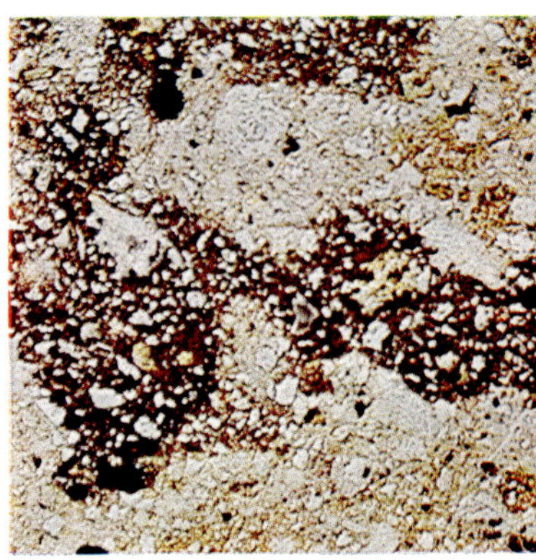
6

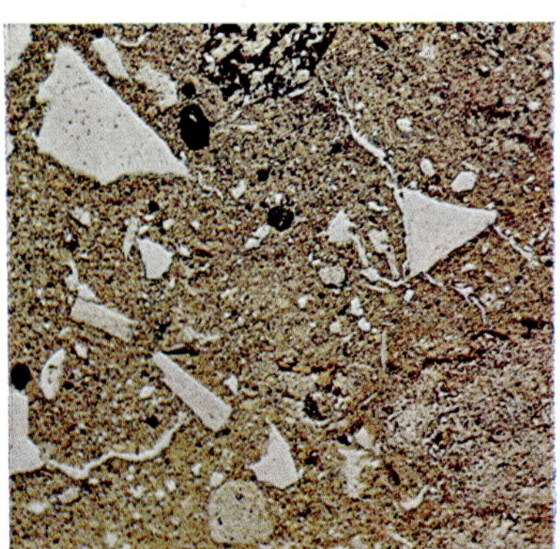
7

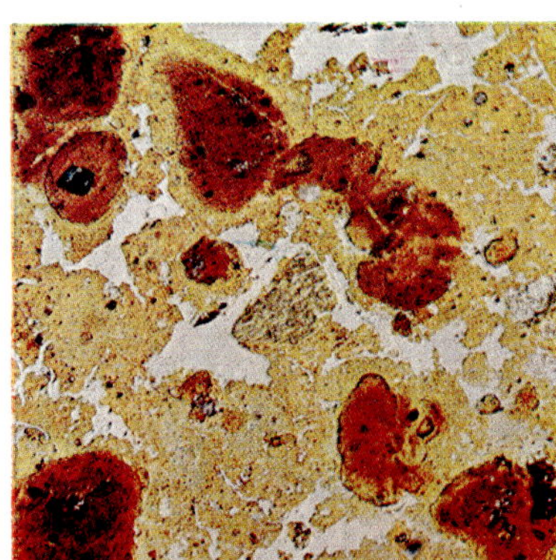
8

PLATE D

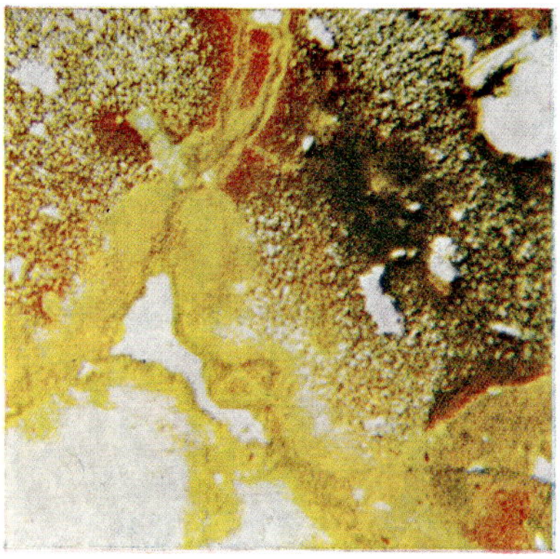
2

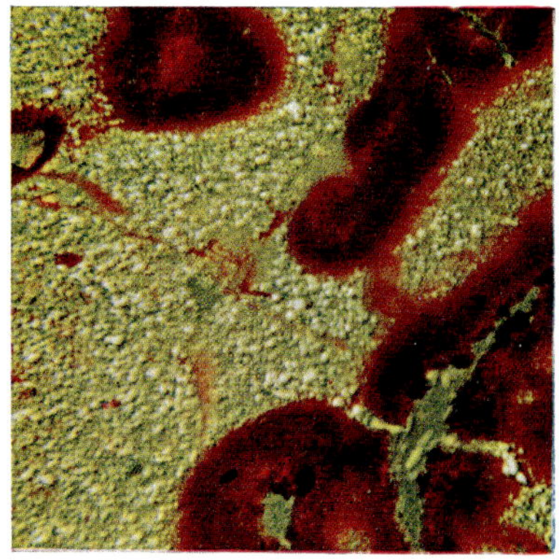
3

1

PLATE E

PLATE F

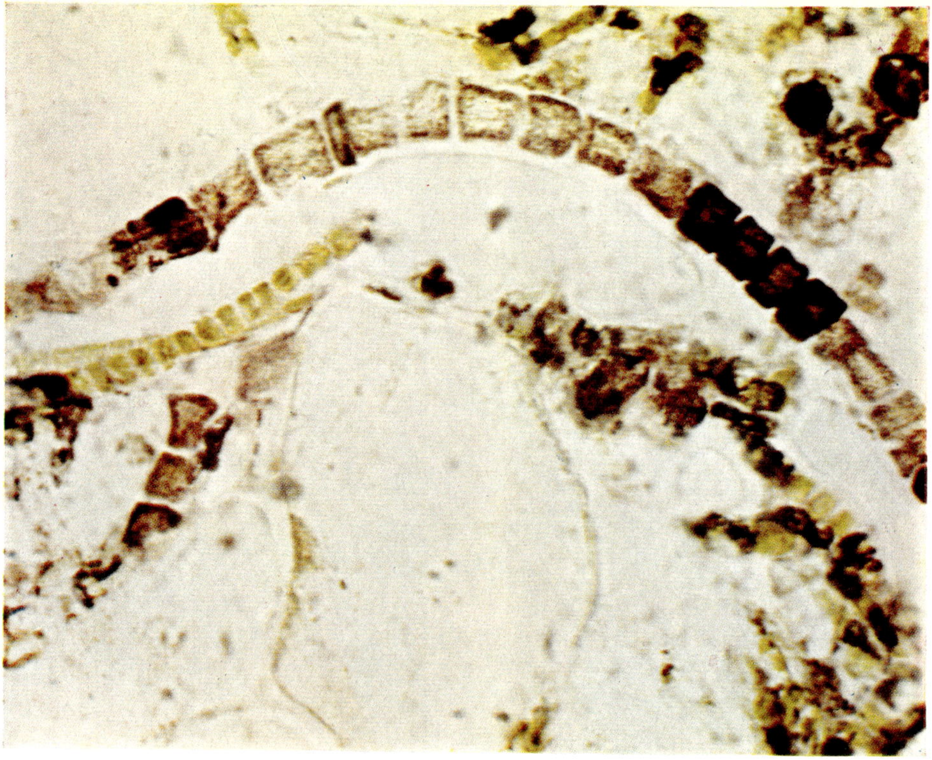

1

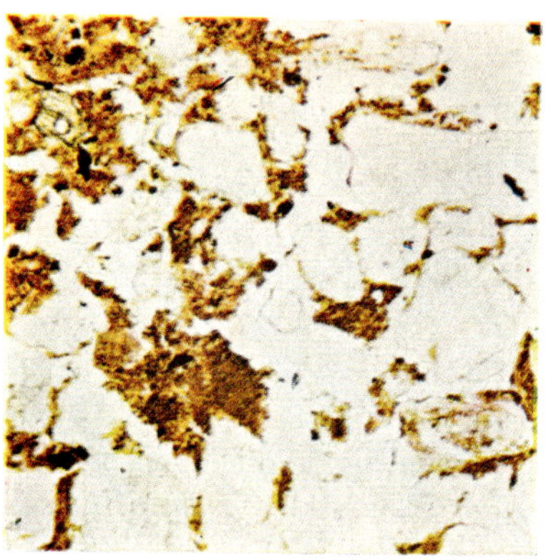

2

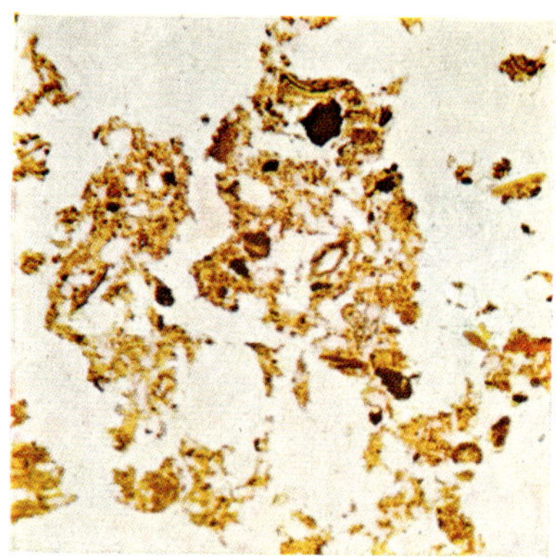

3

4

5

PLATE G

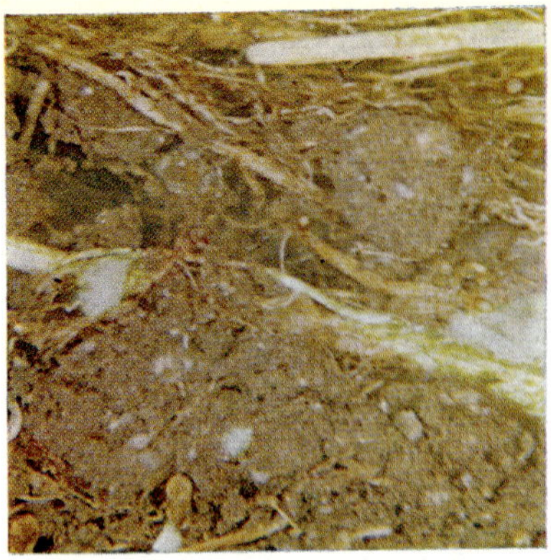

1

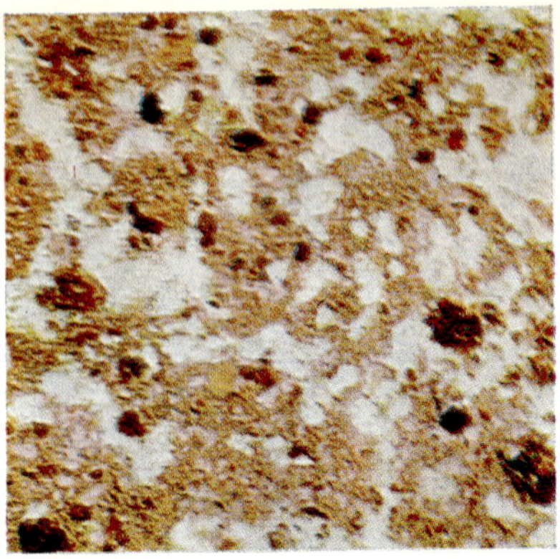

2

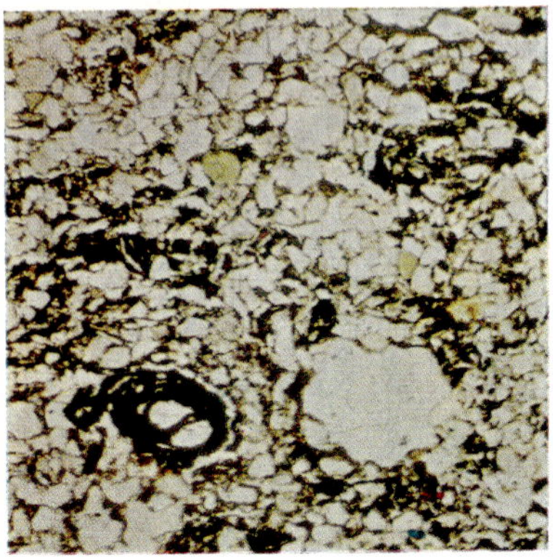

3

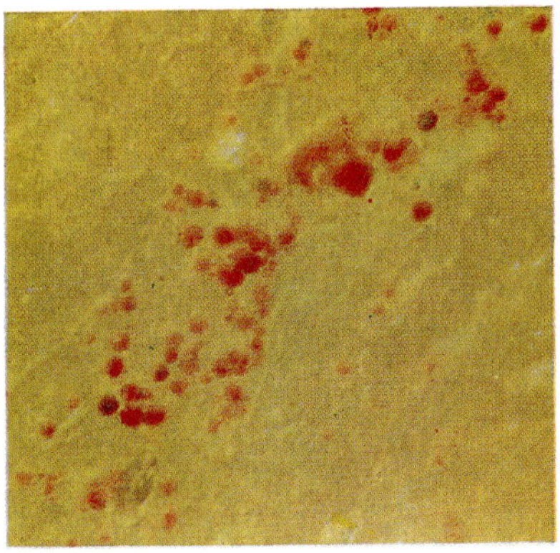

4

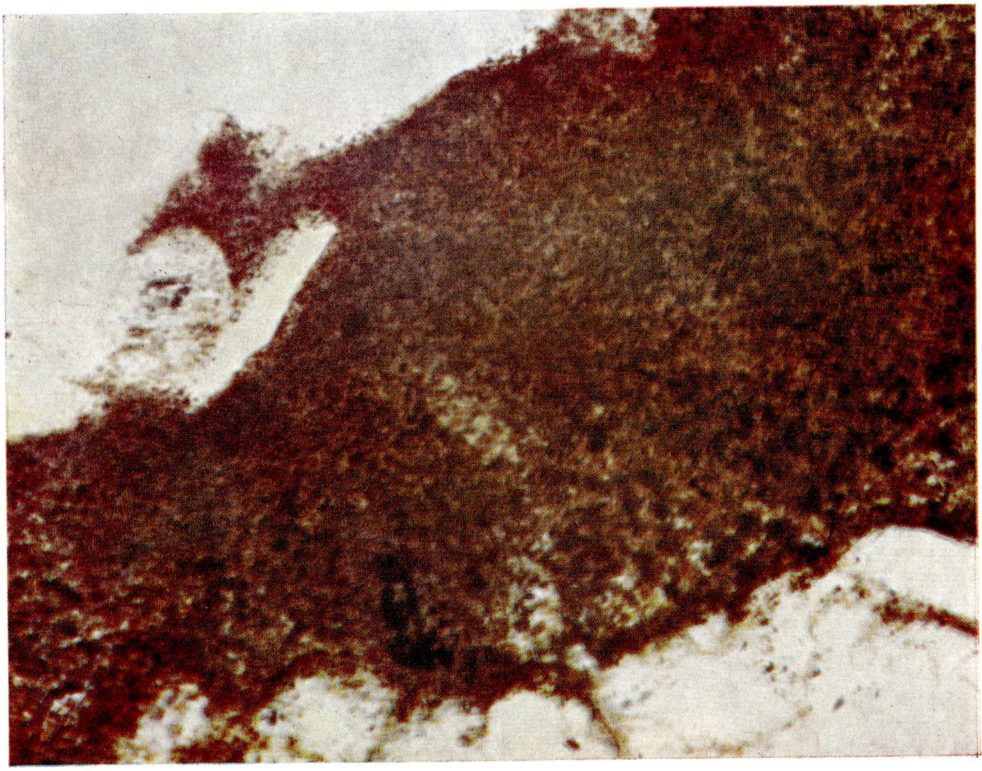

5

PLATE H

AUTHOR INDEX

INDEX I

According to main topics of papers

INDEX II

According to main areas concerned

PRINTED IN THE NETHERLANDS